Malware Analysis Using Artificial Intelligence and Deep Learning

Mark Stamp · Mamoun Alazab ·
Andrii Shalaginov
Editors

Malware Analysis Using Artificial Intelligence and Deep Learning

Editors
Mark Stamp
Department of Computer Science
San Jose State University
San Jose, CA, USA

Mamoun Alazab
College of Engineering, IT & Environment
Charles Darwin University
Darwin, NT, Australia

Andrii Shalaginov
Faculty of Information Technology
and Electrical Engineering
Norwegian University of Science
and Technology
Gjøvik, Norway

ISBN 978-3-030-62584-9 ISBN 978-3-030-62582-5 (eBook)
https://doi.org/10.1007/978-3-030-62582-5

This Springer imprint is published by the registered company Springer Nature Switzerland AG
The registered company address is: Gewerbestrasse 11, 6330 Cham, Switzerland

Preface

Artificial intelligence (AI) is changing the world as we know it. From its humble beginnings in the late 1940s as little more than an academic curiosity, AI has gone through multiple boom and bust cycles. With recent advances in machine learning (ML) and deep learning (DL), AI has finally taken root as a fundamental transformative technology. The changes wrought by AI already affect virtually every aspect of daily life, yet we are clearly only in the early stages of an AI-based revolution.

In the field of information security, there is no topic that is more significant than malware. The sheer volume of malware and the cost of dealing with its consequences are truly staggering. It is therefore timely to consider ML, DL, and AI in the context of malware analysis.

The chapters in this book apply numerous cutting-edge AI techniques to a wide variety of challenging problems in the malware domain. The book includes no less than 8 survey articles, which can serve to bring a reader quickly up to speed with the current state of the art. The heart of the book consists of 11 chapters that are tightly focused on AI-based techniques for malware analysis. We have also included 6 chapters where AI is applied to information security topics that are not strictly malware, but are closely related.

We are confident that this book will prove equally valuable to practitioners working in the trenches and to researchers at all levels. New and novel techniques as well as clever applications abound, yet we have strived to make the material accessible to the widest possible audience. It is our fervent hope—and firm belief—that the tools and techniques presented in the chapters of this book will play a major role in taming the malware threat.

San Jose, USA
Darwin, Australia
Gjøvik, Norway
December 2020

Mark Stamp
Mamoun Alazab
Andrii Shalaginov

Contents

Surveys

A Selective Survey of Deep Learning Techniques and Their Application to Malware Analysis

Mark Stamp

Abstract In this chapter, we consider neural networks and deep learning, within the context of malware research. A variety of architectures are introduced, including multilayer perceptrons (MLP), convolutional neural networks (CNN), recurrent neural networks (RNN), long short-term memory (LSTM), residual networks (ResNet), generative adversarial networks (GAN), and Word2Vec. We provide a selective survey of applications of each of these architectures to malware-related problems.

1 Introduction

In this chapter, we discuss a variety of topics related to deep learning, with the primary focus on popular neural networking-based architectures. We survey various malware-related applications of each architecture considered. Each topic is discussed in some detail, with additional references for further reading provided in all cases.

This chapter can be viewed as a companion to the survey [78], which covers classic machine learning techniques and their applications in cybersecurity research. Our focus here is on neural networks and deep learning, and with respect to applications, we focus most of our attention on malware-related topics, but we do mention other applications within the broader information security domain.

For the sake of completeness, we begin with an introduction to artificial neural networks (ANNs), which includes a brief history of neural networks. We then introduce a wide variety of architectures and techniques, including convolutional neural networks (CNN), recurrent neural networks (RNN), long short-term memory (LSTM), residual networks (ResNet), and generative adversarial networks (GAN). We also discuss related techniques, such as word embeddings—including Word2Vec. We also briefly mention ensemble techniques and transfer learning in passing.

M. Stamp (✉)
San Jose State University, San Jose, CA, USA
e-mail: mark.stamp@sjsu.edu

M. Stamp et al. (eds.), *Malware Analysis Using Artificial Intelligence and Deep Learning*, https://doi.org/10.1007/978-3-030-62582-5_1

2 A Brief History of ANNs

The concept of an artificial neuron [26, 82] is not new, as the idea was proposed by McCulloch and Pitts in the 1940s [52]. However, modern computational neural networking really begins with the perceptron, which was first proposed by Rosenblatt in the late 1950s [68].

An artificial neuron with three inputs is illustrated in Fig. 1. In the original McCulloch-Pitts formulation, $X_i \in \{0, 1\}$, $w_i \in \{+1, -1\}$, and the output $Y \in \{0, 1\}$. The threshold T determines whether the output Y is 0 (inactive) or 1 (active), based on $\sum w_i X_i$. The thinking was that a neuron either fires or it does not (thus, $Y \in \{0, 1\}$), and the inputs would come from other neurons (thus, $X_i \in \{0, 1\}$), while the weights w_i specify whether an input is excitatory (increasing the chance of the neuron firing) or inhibitory (decreasing the chance of the neuron firing). Whenever $\sum w_i X_i > T$, the excitatory response wins, and the neuron fires; otherwise, the inhibitory response wins and the neuron does not fire.

A *perceptron* is considerably less restrictive than a McCulloch–Pitts artificial neuron, as the X_i and w_i can be real-valued. Since we want to use a perceptron as a binary classifier, the output is generally taken to be binary. McCulloch and Pitts chose such a restrictive formulation because they were trying to model logic functions. At the time, it was felt that encoding elementary logic into artificial neurons would be the key step to constructing systems with artificial intelligence. However, that point of view has certainly not panned out, while the additional generality offered by the perceptron formulation has proven extremely useful.

Given a real-valued input vector $X = (X_0, X_1, \ldots, X_{n-1})$, a perceptron can be viewed as a function of the form

$$f(X) = \sum_{i=0}^{n-1} w_i X_i + b,$$

Fig. 1 Artificial neuron

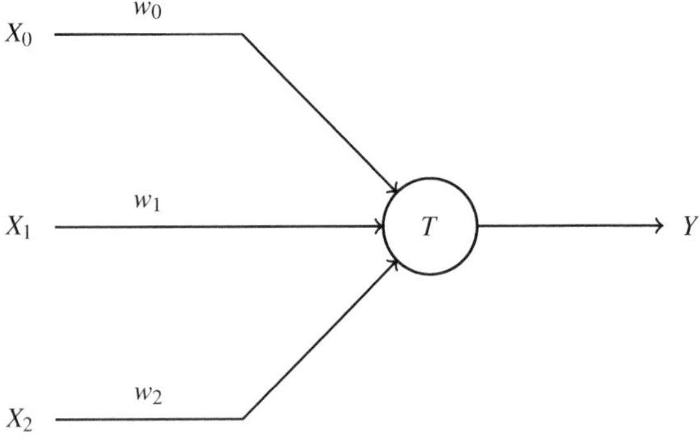

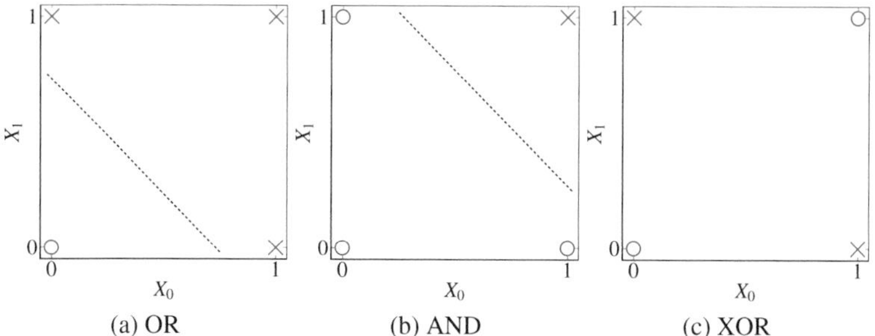

Fig. 2 OR and AND are linearly separable but XOR is not

that is, a perceptron computes a weighted sum of the components. Based on a threshold, a perceptron can be used to define a binary classifier. For example, we could classify a sample X as "type 1" provided that $f(X) > T$, for some specified threshold T, and otherwise classify X as "type 0."

In the case of two-dimensional input, the decision boundary of a perceptron defines a line

$$f(x, y) = w_0 x + w_1 y + b. \tag{1}$$

It follows that a perceptron cannot provide ideal separation in cases where the data itself is not linearly separable.

There was considerable research into ANNs in the 1950s and 1960s, and that era is often described as the first "golden age" of AI and neural networks. But the gold turned to lead in 1969 when an influential work by Minsky and Papert [55] emphasized the limitations of perceptrons. Specifically, they observed that the XOR function is not linearly separable, which implies that a single perceptron cannot model something as elementary as XOR. The OR, AND, and XOR functions are illustrated in Fig. 2, where we see that OR and AND are linearly separable, while XOR is not.

As the name suggests, a multilayer perceptron (MLP) is an ANN that includes multiple (hidden) layers in the form of perceptrons. An example of an MLP with two hidden layers is given in Fig. 3, where each edge represents a weight that is to be determined. Unlike a single-layer perceptron, MLPs are not restricted to linear decision boundaries, and hence an MLP can accurately model the XOR function. However, the perceptron training method proposed by Rosenblatt [68] cannot be used to effectively train an MLP [44]. To train a single perceptron, simple heuristics will suffice, assuming that the data is linearly separable. From a high-level perspective, training a single perceptron is somewhat analogous to training a linear SVM, except that for a perceptron, we do not require that the margin (i.e., minimum separation) be maximized. However, training an MLP would appear to be challenging since we have hidden layers between the input and output, and it is not clear how changes to the weights in these hidden layers will affect each other, let alone the output.

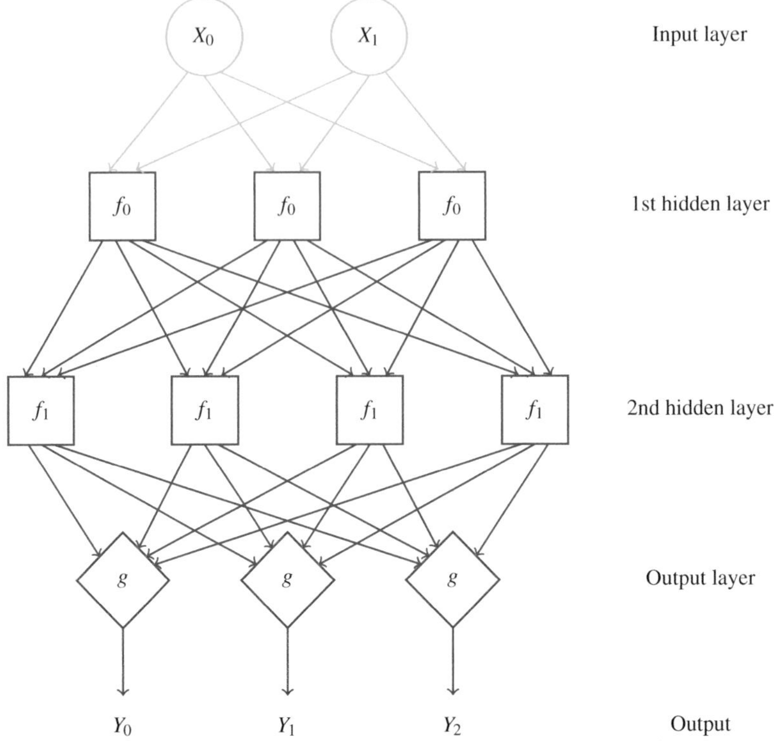

Fig. 3 MLP with two hidden layers

As an aside, it is interesting to note that for SVMs, we deal with data that is not linearly separable by employing a soft margin (i.e., we allow for training errors) and by the use of the so-called "kernel trick," where we map the input data to a higher dimensional feature space using a (nonlinear) kernel function. In contrast, perceptrons (in the form of MLPs) overcome the limitation of linear separability by the use of multiple layers. For an MLP, it is almost as if the nonlinear kernel function has been embedded directly into the model itself through the use of hidden layers, as opposed to a user-specified explicit kernel function, as is the case for an SVM.

One possible advantage of the MLP approach over an SVM is that for an MLP, the equivalent of the kernel function is, in effect, derived from the data and refined through the training process. In contrast, for an SVM, the kernel function is selected by a human, and once selected it does not change. In machine learning, removing those pesky humans from the learning process is a good thing. However, a possible tradeoff is that significantly more training data will likely be needed for an MLP, as compared to an SVM, due to the greater data requirement involved in learning the equivalent of a kernel function.

As another aside, we note that from a high-level perspective, it is possible to view MLPs as combining some aspects of SVMs (i.e., specifically, nonlinear decision boundaries) and HMMs (i.e., hidden layers). Also, we will see that the backpropagation algorithm that is used to train MLPs includes a forward pass and backward pass, which is eerily reminiscent of the training process that is used for HMMs.

As yet another aside, we note that an MLP is a *feedforward neural network*, which means that there are no loops—the input data and intermediate results feed directly through the network. In contrast, a recurrent neural network (RNN) can have loops, which gives an RNN a concept of memory but can also add significant complexity.

In the book *Perceptrons: An Introduction to Computational Geometry*, published in 1969, Minsky and Papert [55] made much of the perceived shortcoming of perceptrons—in particular, the aforementioned inability to model XOR. This was widely viewed as a devastating criticism at the time, as it was believed that successful AI would need to capture basic principles of logic. Although it was known that perceptrons with multiple layers (i.e., MLPs) can model XOR, at the time, nobody knew how to efficiently train MLPs. Minsky and Papert's work was highly influential and is frequently blamed for the relative lack of interest in the field—a so-called "AI winter"—that persisted throughout the 1970s and into the early 1980s.

By 1986, there was renewed interest in ANNs, thanks in large part to the work of Rumelhart, Hinton, and Williams [70], who developed a practical means of training MLPs—the method of backpropagation. For details on backpropagation, see [80], for example.

It is worth noting that there was another "AI winter" that lasted from the late 1980s through the early 1990s (at least). The proximate cause of this most recent AI winter was that the hype far outran the limited successes that had been achieved. Although deep learning has now brought ANNs back into vogue, your author (a doubting Thomas, and proud of it) is not convinced that the current artificial intelligence mania will prove any less artificial than previous AI "summers" which, on the whole, yielded mostly disappointment. Some of the ridiculous statements being made today [28] lead your eminently sensible author to believe that the hype is already hopelessly out of control.[1]

Next, we discuss deep learning, which builds on the foundation of ANNs. We can view the relationship between ANNs and deep learning as being somewhat akin to that of Markov chains and HMMs, for example. That is, ANNs serve as a basic technology that can be used to build a powerful machine learning technique, analogous to the way that an HMM is built on the foundation of an elementary Markov chain. But, before we get into the details of deep learning, we consider the topic from a high-level perspective.

---

[1] In stark contrast to the nonsensical hype that envelopes far too much of the discussion of deep learning and (especially) AI, there does exist some clear-headed thinking that points to the great transformative potential of learning technology in the real world, rather than the world of science fiction. For a fine example of this latter genre, see the intriguingly titled article, "Models will run the world" [14]. (Spoiler alert: "Models will run the world" is *not* about world domination by skinny women in swimsuits).

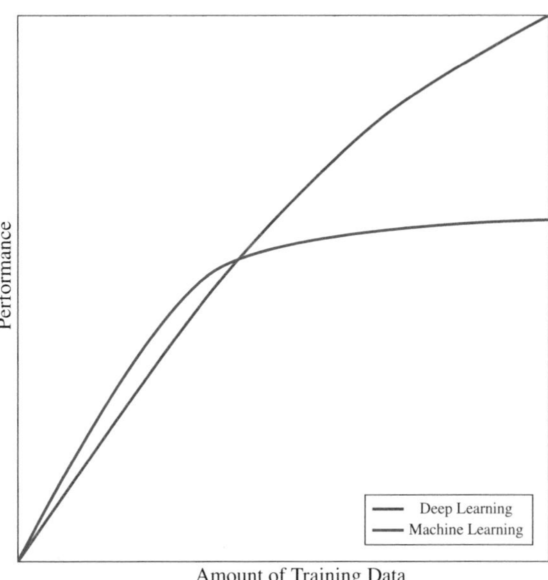

Fig. 4 Model performance as a function of the amount of training data

3 Why Deep Learning?

It is sometimes claimed that the major advantage of deep learning arises when the amount of training data is large. For example, the tutorial [35] gives a graph similar to that in Fig. 4, which purports to show that deep learning will continue to achieve improved results as the size of the dataset grows, whereas other machine learning techniques will plateau at some relatively early point. That is, models generated by non-deep learning techniques will "saturate" relatively quickly, and once this saturation point is reached, more data will not yield improved models.[2] In contrast, deep learning is supposed to continue learning, essentially without limit as the volume of training data increases, or at least it will plateau at a much higher level. Of course, even if this is entirely true, there are practical computational constraints since more data requires more computing power for training.

[2]If any learning model truly saturates, then adding more data will be counterproductive beyond some point, as the work factor for training on larger datasets increases, while there is no added benefit from the resulting trained model. It would therefore be useful to be able to predetermine a "score" of some sort that would tell us approximately how much data is optimal when training a particular learning model for a given type of data.

4 Decisions, Decisions

The essence of machine learning is that when training a model, we minimize the need for input from those fallible humans. That is, we want our machine learning models to be *data-driven*, in the sense that the models learn as much as possible directly from the data itself, with minimal human intervention. However, any machine learning technique will require some human decisions—for HMMs, we specify the number of hidden states; for SVMs, we specify the kernel function; and so on.

For ANNs in general, and deep learning in particular, the following design decisions are relevant [22].

- The *depth* of an ANN refers to the number of hidden layers. The "deep" in deep learning indicates that we employ ANNs with lots of hidden layers, where "lots" seems to generally mean as many as possible, based on available computing power.
- The *width* of an ANN is the number of neurons per layer, which need not be the same in each layer.
- In an MLP, for example, nonlinearity is necessary, and this is achieved through the *activation functions* (also known as transfer functions). Most activation functions used in deep learning are designed to mimic a step function—examples include the sigmoid (or logistic) function

$$f(x) = \frac{1}{1+e^{-x}},$$

the hyperbolic tangent

$$f(x) = \tanh(x) = \frac{e^x - e^{-x}}{e^x + e^{-x}},$$

the inverse tangent (also known as arctangent)

$$f(x) = \tan^{-1}(x),$$

and the rectified linear unit (ReLU)

$$f(x) = \max\{0, x\} = \begin{cases} x & \text{if } x > 0 \\ 0 & \text{otherwise.} \end{cases}$$

Note that the softmax function is a generalization of the sigmoid function to multiclass problems.

The graph of each of the activation functions given above is illustrated in Fig. 5. As of this writing, ReLU is the most popular activation function. Numerous variants of the ReLU function are also used, including the leaky ReLU and exponential linear unit (ELU).

- In addition to activation functions, we also specify an *objective function*. The objective function is the function that we are trying to optimize and typically represents the training error.
- A *bias node* may be included (or not) in any hidden layer. Each bias node generates a constant value and hence is not connected to any previous layer. When present, a bias node allows the activation function to be shifted. In the perceptron example given in (1), the bias corresponds to the y-intercept b.

For the sake of comparison with our favorite non-deep learning technique, the depth of an HMM can be viewed as the order of the underlying Markov model. Typically, for HMMs, we only consider models of order one (in which case, the current state depends only on the previous state), but it is possible to consider higher order models. The width of an HMM might be viewed as being determined by N, the number of hidden states. But, regardless of the order of the model or the choice of N, there is really only one hidden layer in any HMM. The fact that an HMM is based on linear operations implies that adding multiple hidden layers would have no effect, as the multiple layers would be equivalent to a single layer. Furthermore, the A and B matrices of an HMM can be viewed as its activation functions (with the B matrix corresponding to the output layer), and $P(\mathcal{O} \mid \lambda)$ corresponds to the objective function in an ANN. Note that these functions are all linear in an HMM, while at least some of the activation functions must be nonlinear in any true multilayer ANN, such as an MLP.

Neural networks are trained using the backpropagation algorithm, which is a special case of a more general technique known as reverse mode automatic differentiation. For additional details on the topic of backpropagation, see, for example, [80].

The remainder of this paper is focused on various neural network based architectures and related topics. For each topic covered, we discuss research in the field of malware analysis.

5 Multilayer Perceptrons

We have already discussed multilayer perceptrons (MLP) in some detail. MLPs are in some sense one of the most generic neural networking architectures—when someone speaks of a neural network in general, there is a good chance that they have an MLP in mind.

5.1 Overview of MLPs

Recall that Fig. 3 is an example of an MLP with two hidden layers. Each edge in the figure represents a weight that is to be determined via training, and backpropagation

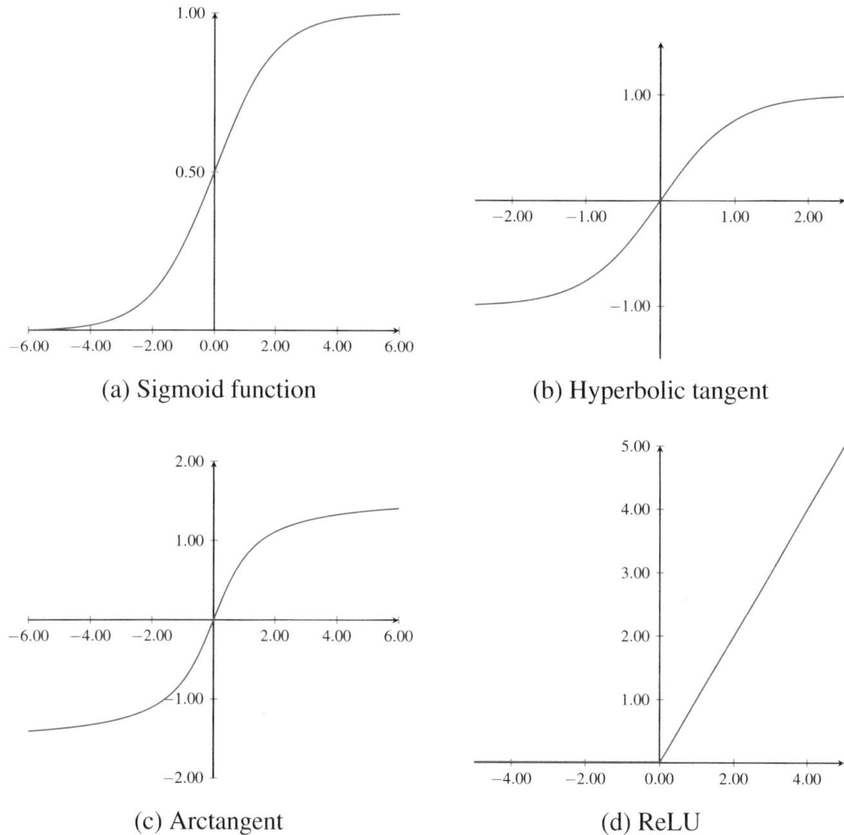

Fig. 5 Activation functions

is an efficient and effective way to train such a network. The advantage of an MLP is that it is not restricted to linear decision boundary.

5.2 *MLPs in Malware Analysis*

MLPs are extremely popular, and in most fields, they are one of the first learning techniques considered. Information security is no exception, as MLPs have been applied to nearly every security problem where deep learning techniques are applicable. Not surprisingly, large numbers of malware research papers employ MLPs. For example, in [5] MLPs are trained on progressively more generic malware families, yielding quantifiable results on the inherent tradeoff between the generality of the training data and accuracy. The research in [74] shows that a straightforward ensemble of various learning algorithms—including MLPs—can generate significantly stronger results than any of the component techniques. The paper [85] uses MLPs as part of an Android malware detection technique.

Another field in information security where MLPs have played a very prominent role is in intrusion detection systems (IDS). For example, the paper [57] uses MLPs in a novel multiclass IDS approach.

6 Convolutional Neural Networks

In this section, we provide an introduction to one of the most important and widely used learning techniques—CNN. After a brief overview, we introduce discrete convolutions with the focus on their specific application to CNNs. We then consider a simplified example that serves to illustrate various aspects of CNNs.

6.1 Overview of CNNs

Generically, ANNs use fully connected layers. A fully connected layer can deal effectively with correlations between any points within the training vectors, regardless of whether those points are close together, far apart, or somewhere in between. In contrast, a CNN, is designed to deal with local structure—a convolutional layer cannot be expected to perform well when crucial information is not local. A key benefit of CNNs is that convolutional layers can be trained much more efficiently than fully connected layers.

For images, most of the important structure (edges and gradients, for example) is local. Hence, CNNs would seem to be an ideal tool for image analysis and, in fact, CNNs were developed for precisely this problem. However, CNNs have performed well in a variety of other problem domains. In general, any problem for which there exists a data representation where local structure predominates is a candidate for a CNN. In addition to images, local structure is crucial in fields such as text analysis and speech analysis, for example.

6.2 Convolutions and CNNs

A discrete convolution is a sequence that is itself a composition of two sequences and is computed as a sum of pointwise products. Let $c = x * y$ denote the convolution of sequences $x = (x_0, x_1.x_2, \ldots)$ and $y = (y_0, y_1.y_2, \ldots)$. Then the kth element of the convolution is given by

$$c_k = \sum_{k=i+j} x_i y_j = \sum_i x_i y_{k-i}.$$

We can view this process as x being a "filter" (or kernel) that is applied to the sequence y over a sliding window.

For example, if $x = (x_0, x_1)$ and $y = (y_0, y_1, y_2, y_3, y_4)$, we find

$$c = x * y = \big(x_0 y_1 + x_1 y_0, x_0 y_2 + x_1 y_1, x_0 y_3 + x_1 y_2, x_0 y_4 + x_1 y_3\big).$$

If we reverse the order of the elements of x, then we have

$$c = \big(x_0 y_0 + x_1 y_1, x_0 y_1 + x_1 y_2, x_0 y_2 + x_1 y_3, x_0 y_3 + x_1 y_4\big)$$

which is, perhaps, a slightly more natural and intuitive way to view the convolution operation.

Again, we can view x as a filter that is applied to the sequence y. Henceforth, we define this filtering operation as convolution with the order of the elements of the filter reversed. For example, suppose that we apply the filter $x = (1, -2)$ to the sequence $y = (0, 1, 2, 3, 4)$. In this case, the convolution gives us

$$\begin{aligned} c = x * y &= \big(x_0 y_0 + x_1 y_1, x_0 y_1 + x_1 y_2, x_0 y_2 + x_1 y_3, x_0 y_3 + x_1 y_4\big) \\ &= \big(1 \cdot 0 - 2 \cdot 1, 1 \cdot 1 - 2 \cdot 2, 1 \cdot 2 - 2 \cdot 3, 1 \cdot 3 - 2 \cdot 4\big) \\ &= \big(-2, -3, -4, -5\big) \end{aligned}$$

We can define an analogous filtering (or discrete convolution) operation in two or three dimensions. For the two-dimensional case, suppose that $A = \{a_{ij}\}$ is an $N \times M$ matrix representing an image and $F = \{f_{ij}\}$ is an $n \times m$ filter. Let $C = \{c_{ij}\}$ be the convolution of F with A. As in the one-dimensional case, we denote this convolution as $C = F * A$. In this two-dimensional case, we have

$$c_{ij} = \sum_{k=0}^{n-1} \sum_{\ell=0}^{m-1} f_{k,\ell} a_{i+k, j+\ell},$$

where $i = 0, 1, \ldots, N - n$ and $j = 0, 1, \ldots, M - m$. That is, we simply apply the filter F at each offset of A to create the new—and slightly smaller—matrix that we denote as C. The three-dimensional case is completely analogous to the two-dimensional case.

We could simply define filters as we see fit, with each filter designed to correspond to a specific feature.[3] But since we are machine learning aficionados, for CNNs, we let the data itself determine the filters. Therefore, training a CNN can be viewed as determining filters, based on the training data. As with any respectable neural network, we can train CNNs via backpropagation.

Suppose that A represents an image and we train a CNN on the image A. Then the first convolutional layer is trained directly on the image. The filters determined at this first layer will correspond to fairly intuitive features, such as edges, basic

[3]We see examples of filters applied to simple images in Sect. 6.3.

shapes, and so on. We can then apply a second convolutional layer, that is, we apply a similar convolutional process, but the output of the first convolutional layer is the input to this second layer. At the second layer, filters are trained based on features of features. Perhaps not surprisingly, these second layer filters correspond to more abstract features of the original image A, such as the "texture." We can repeat this convolution of convolutions step again and again, at each layer obtaining filters that correspond to features representing a higher degree of abstraction, as compared to the previous layer. The final layer of a CNN is not a convolution layer but is instead a typical fully-connected layer that can be used to classify based on complex image characteristics (e.g., "cat" versus "dog"). In addition, so-called pooling layers can be used between some of the convolutional layers. Pooling layers are simple—no training is involved—and serve primarily to reduce the dimensionality of the problem. Below, we give a simple example that includes a pooling layer.

In addition to having multiple convolutional layers, at each layer, we can (and generally will) stack several convolutions on top of each other. These filters are all initialized randomly, so they can all learn different features. In fact, for a typical color image, the image itself can be viewed as consisting of three layers, corresponding to the R, G, and B components in the RGB color scheme. Hence, for color images, the filters for the first convolutional layer will be three-dimensional, while subsequent convolutional layers can—and, typically, will—be three-dimensional as well, due to the stacking of multiple convolutions/filters at each layer. For simplicity, in our example, we only consider a black-and-white two-dimensional image, and we only apply one convolution at each layer.

Before considering a simple example, we note that there are advantages of CNNs that are particularly relevant in the case of image analysis. For a generic neural network, each pixel would typically be treated as a separate neuron, and for any reasonable size of image, this would result in a huge number of parameters, making training impractical. In contrast, at the first layer of a CNN, each filter is applied over the entire image, and at subsequent layers, we apply filters over the entire output of the previous layer. One effect of this approach is that it greatly reduces the number of parameters that need to be learned. Furthermore, by sliding the filter across the image as a convolution, we obtain a degree of translation invariance, i.e., we can detect image features that appears at different offsets. This can be viewed as reducing the overfitting that would otherwise likely occur.

The bottom line is that CNNs represent an efficient and effective technique that was developed specifically for image analysis. However, CNNs are not restricted to image data, and can be useful in any problem domain where local structure is dominant.

6.3 Example CNN

Now we turn our attention to a simple example that serves to illustrate some of the points discussed above. Suppose that we are dealing with black-and-white images,

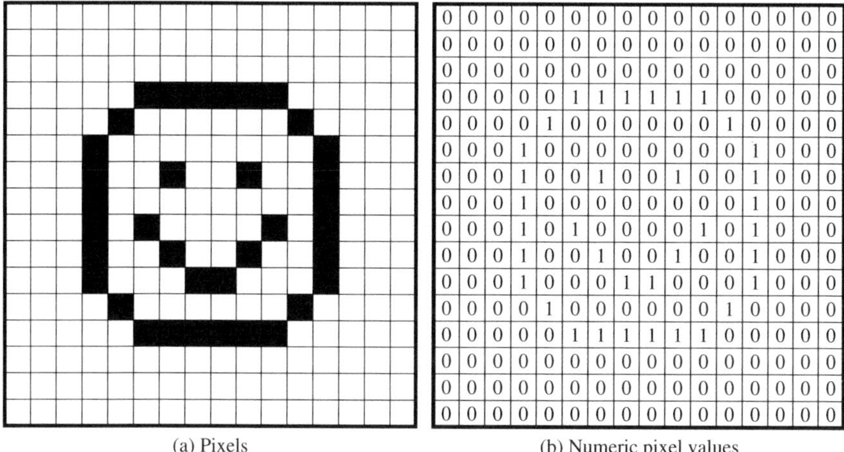

0	0	0	0	0	0	0	0	0	0	0	0	0	0	0	0
0	0	0	0	0	0	0	0	0	0	0	0	0	0	0	0
0	0	0	0	0	0	0	0	0	0	0	0	0	0	0	0
0	0	0	0	0	1	1	1	1	1	1	0	0	0	0	0
0	0	0	0	1	0	0	0	0	0	0	1	0	0	0	0
0	0	0	1	0	0	0	0	0	0	0	0	1	0	0	0
0	0	0	1	0	0	1	0	0	1	0	0	1	0	0	0
0	0	0	1	0	0	0	0	0	0	0	0	1	0	0	0
0	0	0	1	0	1	0	0	0	0	1	0	1	0	0	0
0	0	0	1	0	0	1	0	0	1	0	0	1	0	0	0
0	0	0	1	0	0	0	1	1	0	0	0	1	0	0	0
0	0	0	0	1	0	0	0	0	0	0	1	0	0	0	0
0	0	0	0	0	1	1	1	1	1	1	0	0	0	0	0
0	0	0	0	0	0	0	0	0	0	0	0	0	0	0	0
0	0	0	0	0	0	0	0	0	0	0	0	0	0	0	0
0	0	0	0	0	0	0	0	0	0	0	0	0	0	0	0

(a) Pixels (b) Numeric pixel values

Fig. 6 A 16 × 16 black-and-white image

2	−1	−1
−1	2	−1
−1	−1	2

(a) Diagonal

−1	2	−1
−1	2	−1
−1	2	−1

(b) Vertical

−1	−1	−1
2	2	2
−1	−1	−1

(c) Horizontal

−1	−1	2
−1	2	−1
2	−1	−1

(d) Anti-diagonal

Fig. 7 Examples of filters

where each pixel is either 0 or 1, with 0 representing white and 1 representing black.[4] Further, suppose that the black-and-white images under consideration are 16 × 16 pixels in size. An example of such an image appears in Fig. 6.

In Fig. 7, we give some 3 × 3 filters. For example, the output of the filter in Fig. 7a is maximized when it aligns with a diagonal segment. Figure 8 shows the result of applying the convolution represented by the filter in Fig. 7a to the smiley face image in Fig. 6.

We note that, for the convolution in Fig. 8, the maximum value of 6 does indeed occur only at the three offsets where the (main) diagonal segments are all black and the off-diagonal elements are all white. These maximum values correspond to convolutions over the red boxes in Fig. 9.

In a CNN, so-called pooling layers are often intermixed with convolutional layers. As with a convolutional layer, in a pooling layer, we slide a window of a fixed size over the image. But whereas the filter in a convolutional layer is learned, in a pooling

[4]Color and grayscale images are more complex. For grayscale, a nonlinear encoding (i.e., gamma encoding) is employed, so as to make better use of the range of values available. For color images, the RGB (red, green, and blue, respectively) color scheme implies that each pixel is represented by 24 bits (in an uncompressed format), in which case convolutional filters can be viewed as operating over a three-dimensional box that is 3 bytes deep.

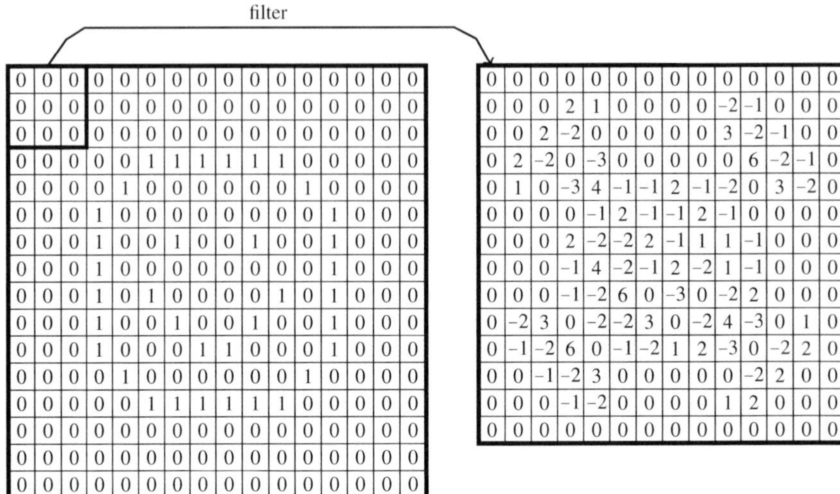

Fig. 8 First convolutional layer (3 × 3 filter from Fig. 7a)

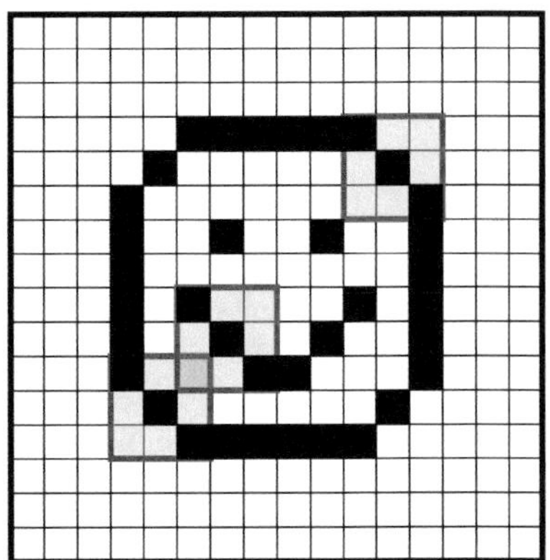

Fig. 9 Maximum convolution values in Fig. 8

layer an extremely simple filter is specified and remains unchanged throughout the training. As the name implies, in *max pooling*, we simply take the the maximum value within the filter window. An illustration of max pooling is given in Fig. 10.

Instead of a max pooling scheme, sometimes *average pooling* is used. In any case, pooling can be viewed as a downsampling operation, which has the effect of reducing the dimensionality, and thus easing the computational burden of training subsequent convolutional layers.[5] To increase the downsampling effect, pooling usually uses non-overlapping windows. Note that the dimensionality reduction of pooling could

[5]It is also sometimes claimed that pooling improves certain desirable characteristics of CNNs, such as translation invariance and deformation stability. However, this is disputed, and the current trend seems to clearly be in the direction of fully convolutional architectures, i.e., CNNs with no pooling layers [69].

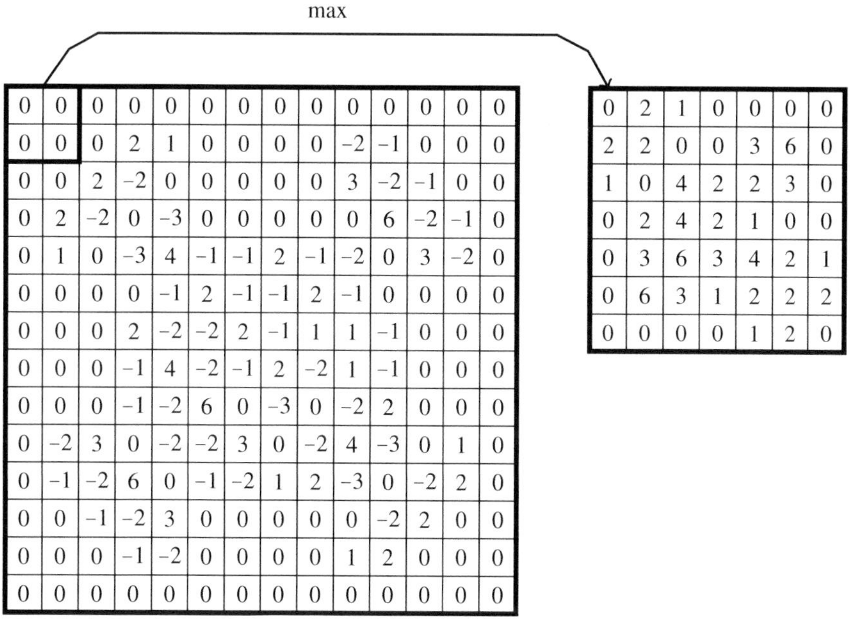

max

0	0	0	0	0	0	0	0	0	0	0	0	0	0
0	0	0	2	1	0	0	0	0	−2	−1	0	0	0
0	0	2	−2	0	0	0	0	0	3	−2	−1	0	0
0	2	−2	0	−3	0	0	0	0	0	6	−2	−1	0
0	1	0	−3	4	−1	−1	2	−1	−2	0	3	−2	0
0	0	0	0	−1	2	−1	−1	2	−1	0	0	0	0
0	0	0	2	−2	−2	2	−1	1	1	−1	0	0	0
0	0	0	−1	4	−2	−1	2	−2	1	−1	0	0	0
0	0	0	−1	−2	6	0	−3	0	−2	2	0	0	0
0	−2	3	0	−2	−2	3	0	−2	4	−3	0	1	0
0	−1	−2	6	0	−1	−2	1	2	−3	0	−2	2	0
0	0	−1	−2	3	0	0	0	0	0	−2	2	0	0
0	0	0	−1	−2	0	0	0	0	1	2	0	0	0
0	0	0	0	0	0	0	0	0	0	0	0	0	0

0	2	1	0	0	0	0
2	2	0	0	3	6	0
1	0	4	2	2	3	0
0	2	4	2	1	0	0
0	3	6	3	4	2	1
0	6	3	1	2	2	2
0	0	0	0	1	2	0

Fig. 10 Max pooling layer (2 × 2, non-overlapping)

also be achieved by a convolutional layer that uses a larger stride through the data, and in [75], for example, it is claimed that such an approach results in no loss in accuracy for the resulting CNN.

An illustration of the first convolutional layer for a color image is given in Fig. 11. In this case, a three-dimensional filter is applied over the R, G, and B components in the RGB color scheme. The example in Fig. 11 is meant to indicate that five different filters are being trained. Since each filter is initialized randomly, they can all learn different features. At the second convolutional layer, we can again train three-dimensional filters, based on the output of the first convolutional layer. This process is repeated for any additional convolutional layers.

There are several possible ways to visualize the filters in convolutional layers. For example, in [89], a de-convolution technique is used to obtain the results in Fig. 12. Here, each row is a randomly selected filter and the columns, from left to right, correspond to training epochs 1, 2, 5, 10, 20, 30, 40, and 64. From layer 4, we see that the training images must be faces. In general, it is apparent that the filters are learning progressively more abstract features as the layer increases.

A fairly detailed discussion of CNNs can be found at [38], while the paper [15] provides some interesting insights. For a more intuitive discussion, see [37], and if you want to see lots of nice pictures, take a look at [16]. More details on convolutions can be found in [61].

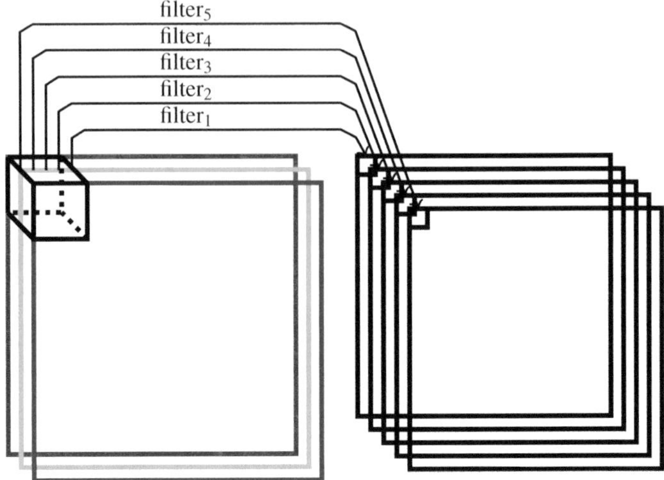

Fig. 11 First convolutional layer with stack of five filters (RGB image)

6.4 CNNs in Malware Analysis

CNNs have proven their worth in a wide variety of security-related applications. Some of these applications, such as image spam detection [1, 10, 72], are obvious and relatively straightforward applications of CNNs. However, other security domains that do not have any apparent image-based component have also had success with CNNs.

By treating executable files as images, researchers have been able to leverage the strengths of CNNs for malware detection, classification, and analysis. For example, the papers [6] and [88] treat executable files as images, and obtain the state-of-the-art result for the malware detection problem. In particular, the research in [88] makes extensive use of transfer learning, whereby the output layer of previously trained CNNs are retrained for the malware detection problem. This results in fast training times and very high malware classification accuracies.

The research in [34] compares CNNs to so-called extreme learning machines (ELM), a topic that we discuss below, in Sect. 10. The best CNN results in [34] are obtained using a one-dimensional CNN trained on the raw bytes of executable files. In [86], CNNs are successfully applied to a combination of static and dynamic features.

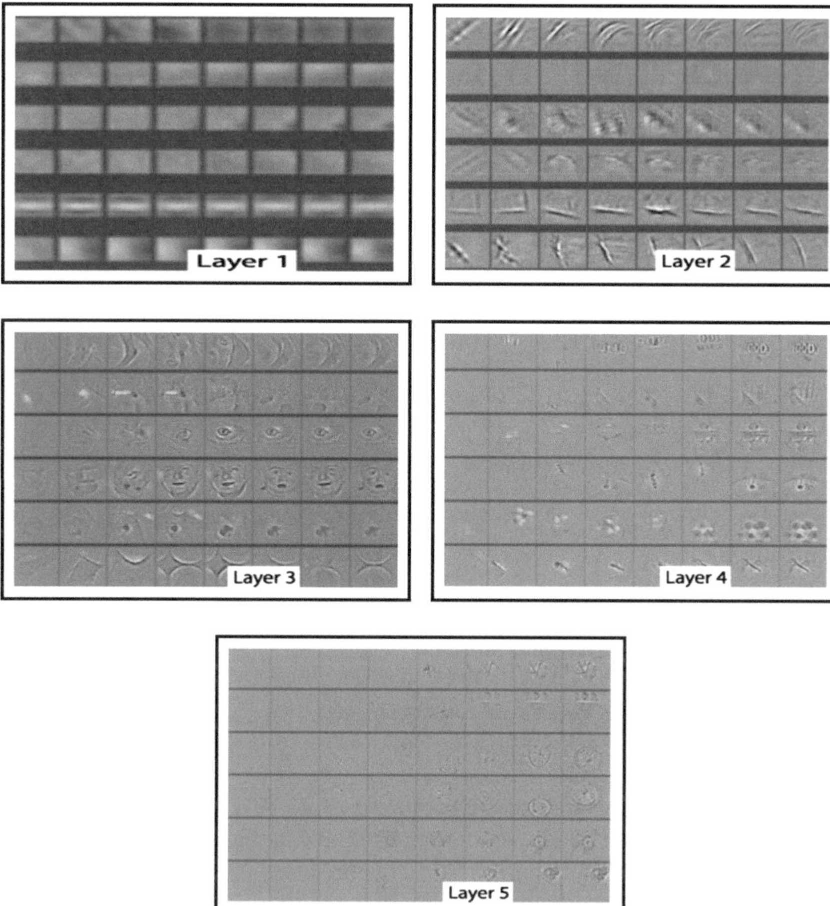

Fig. 12 Visualizing convolutions [89]

7 Recurrent Neural Networks

An example of a feedforward neural network with two hidden layers is given in Fig. 13. This type of neural network has no "memory" in the sense that each input vector is treated independently of other input vectors. Hence, such a feedforward network is not well suited to deal with sequential data.

In some cases, it is necessary for a classifier to have memory. For example, if we want to tag parts of speech in English text (i.e., noun–verb, and so on), this is not feasible if we only look at words in isolation. For example, the word "all" can be an adjective, adverb, noun, or even a pronoun, and the only way to determine which is the case is to consider the context. A recurrent neural network (RNN) provides a way to add memory (or context) to a feedforward neural network.

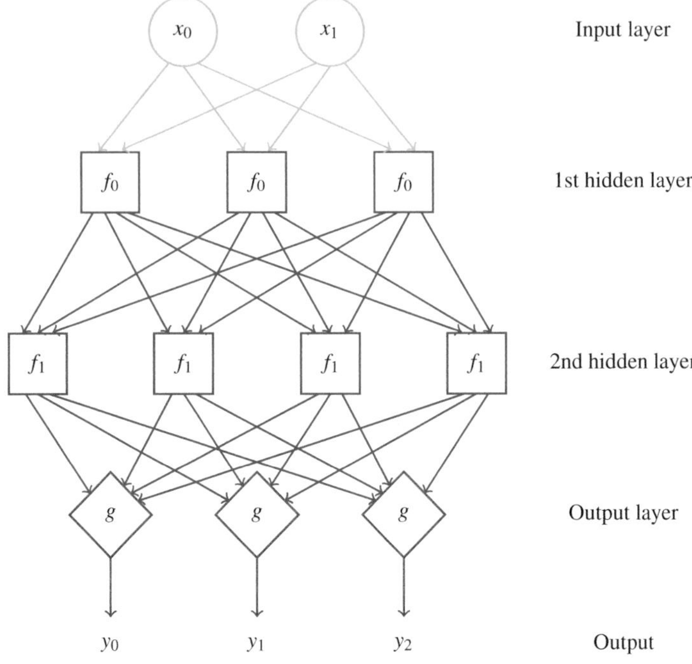

Fig. 13 Feedforward neural network with two hidden layers

To convert a feedforward neural network into an RNN, we treat the output of the hidden states as another input. For the neural network in Fig. 13, the corresponding generic RNN is illustrated in Fig. 14. The structure in Fig. 14 implies that there is a time step involved, that is, we train (and score) based on a sequence of input vectors. Of course, we cannot consider infinite sequences, and even if we could, the influence of feature vectors that occurred far back in time is likely to be minimal.

The RNN in Fig. 14 can be "unrolled," as illustrated in Fig. 15. Note that in this case, we use f to represent the hidden layer or layers, while the notation X_t is used to represent (x_0, x_1) at time step t from un-unrolled RNN in Fig. 14 and, similarly, Y_t corresponds to (y_0, y_1, y_2) at time t. From the unrolled form, it is clear that any RNN can be treated as a special case of a feedforward neural network, where the intermediate hidden layers (f in our notation) all have identical structures and weights. We can take advantage of this special structure to efficiently train an RNN using a (slight) variant of backpropagation, known as backpropagation through time (BPTT).

Before briefly turning our attention to BPTT, we illustrate some variants of a generic RNN. An RNN such as that illustrated in Fig. 15 is known as a sequence-to-sequence model, since each input sequence $(X_0, X_1, \ldots, X_{n-1})$ corresponds to an output sequence $(Y_0, Y_1, \ldots, Y_{n-1})$. In Fig. 16a, we have illustrated a many-to-one example of an RNN, that is, the case where an input sequence of the

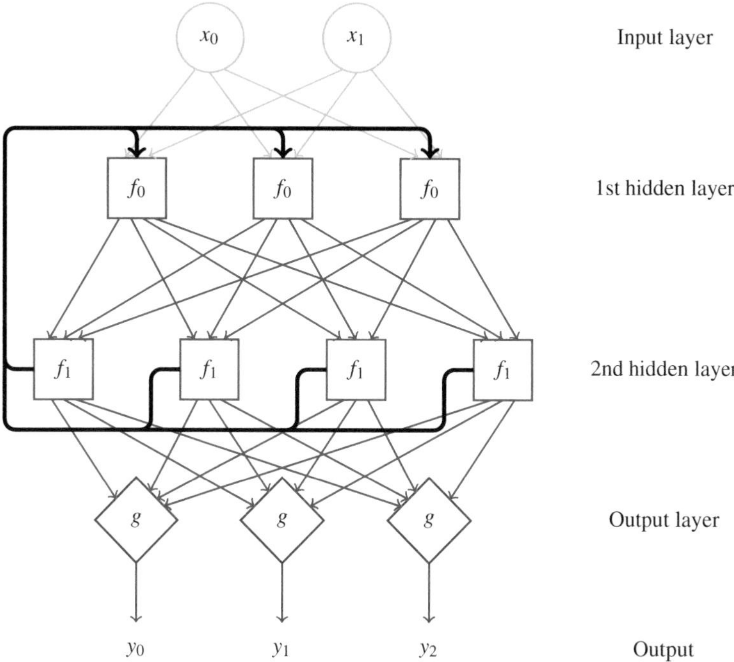

Fig. 14 Network in Fig. 13 as an RNN

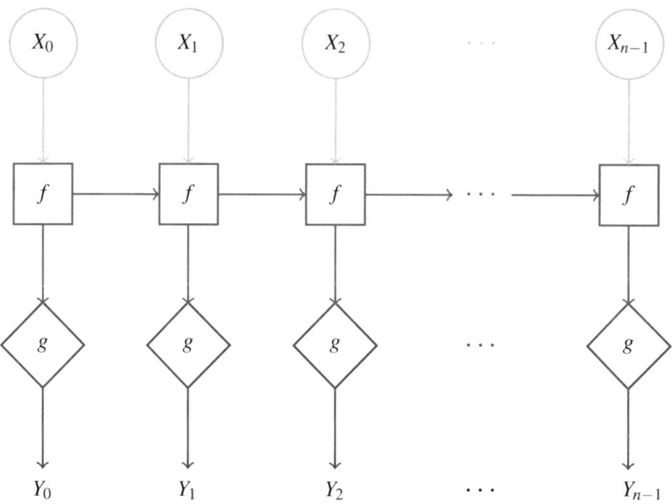

Fig. 15 Unrolled RNN (sequence-to-sequence model)

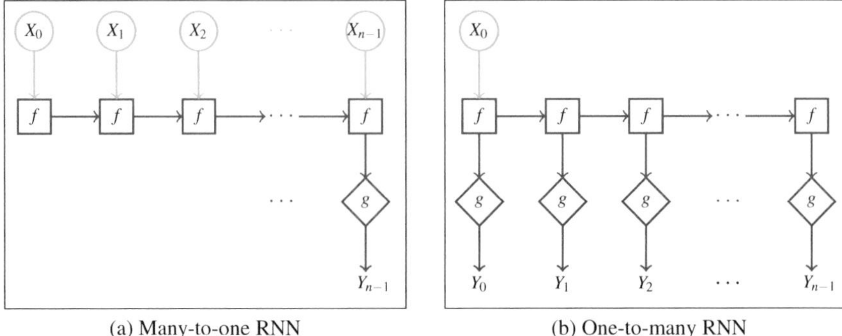

(a) Many-to-one RNN (b) One-to-many RNN

Fig. 16 Variants of the generic RNN in Fig. 15

form $(X_0, X_1, \ldots, X_{n-1})$ corresponds to the single output Y_{n-1}. At the other extreme, Fig. 16b illustrates a one-to-many RNN, where the single input X_0 corresponds to the output sequence $(Y_0, Y_1, \ldots, Y_{n-1})$.

A many-to-one model might be appropriate for part-of-speech tagging, for example, while a one-to-many RNN could be used for music generation. An example of an application where a sequence-to-sequence (or many-to-many) RNN would be appropriate is a machine translation. There are numerous possible variants of the sequence-to-sequence RNN. Also, note that a feedforward neural network, such as that in Fig. 13, can be viewed as a one-to-one RNN.

Multilayer RNNs can also be considered. This can be viewed as training multiple RNNs simultaneously, with the first RNN trained on the input data, the second RNN trained on the hidden states of the first RNN, and so on. A two-layer (sequence-to-sequence) RNN is illustrated in Fig. 17. Of course, more layers are possible, but the training complexity will increase, and hence only "shallow" RNN architectures (in terms of the number of layers) are generally considered.

7.1 Backpropagation Through Time

RNNs can be viewed as neural networks that are designed specifically for time series or other sequential data. With an RNN, the number of parameters is reduced so as to ease the training burden. This situation is somewhat analogous to CNNs, which are designed to efficiently deal with local structure (e.g., in images). That is, both CNNs and RNNs serve to make training more efficient—as compared to generic feedforward neural networks—for specific classes of problems. Backpropagation through time (BPTT) is simply an ever-so-slight variation on backpropagation that is optimized for training RNNs.

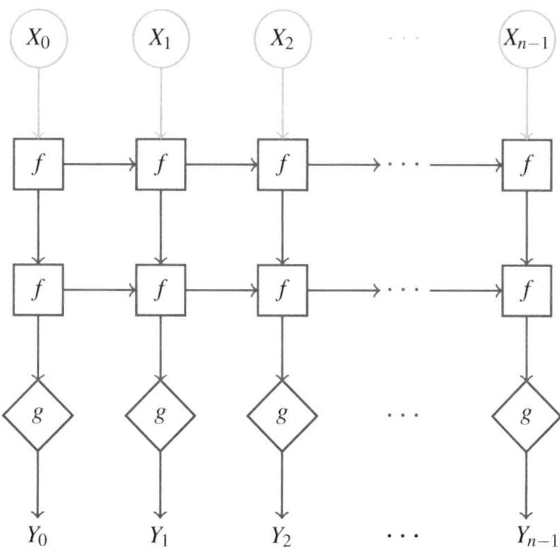

Fig. 17 Two-layer RNN

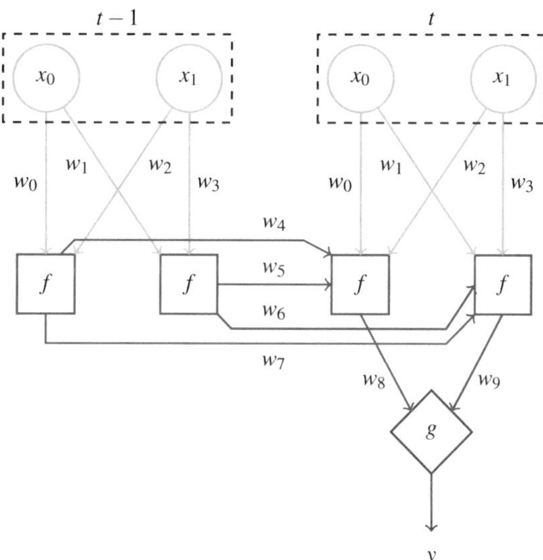

Fig. 18 Simple RNN example

In Fig. 18, we give a detailed view of a many-to-one (actually, two-to-one) RNN. In this case, we see that the 10 weights, $(w_0, w_1, \ldots, w_9)$ must be determined via training.

In Fig. 19, we give a neural network that is essentially the fully connected version of the RNN in Fig. 18. Note that in this fully connected version, there are 20 parameters to be determined. In an RNN, we assume that the data represents sequential input

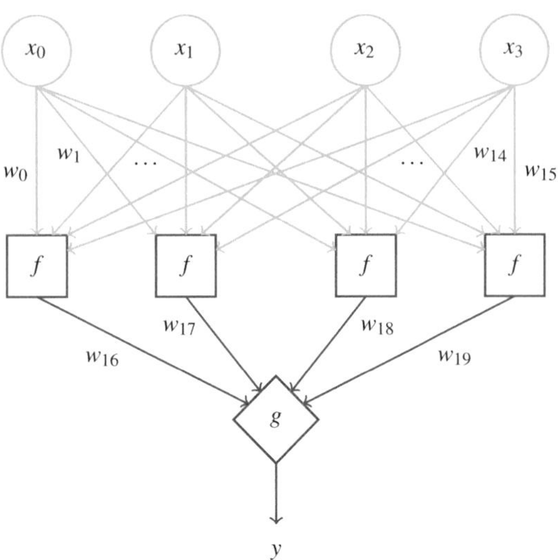

Fig. 19 Fully connected analog of Fig. 18

and hence the reduction in the number of weights is justified, since we are simply eliminating from consideration cases where the past is influenced by the future.[6]

It is well known that gradient issues are a concern when training neural networks in general, and are a particularly acute issue with generic RNNs. In an RNN, the further that we attempt to backpropagate through time, the more likely that the gradient will "explode" or "vanish" or oscillate between extremes. The details of the exploding gradient and vanishing gradient are beyond the scope of this survey; for more information on these topics, see [79], for example.

Next, we turn our attention to specialized RNN architectures that are designed to mitigate the gradient issues that plague generic RNNs. Specifically, we consider LSTM networks in some detail and we then briefly discuss a variant of LSTM known as gated recurrent units (GRU). In fact, a vast number of variants of the LSTM architecture have been developed. However, according to the extensive empirical study in [23], "none of the variants can improve upon the standard LSTM architecture significantly."

7.2 Long Short-Term Memory

In addition to being a tongue twister, LSTM networks are a class of RNN architectures that are designed to deal with long-range dependencies. That is, LSTM can deal with "gaps" between the appearance of a feature and the point at which it is needed by the model [23]. The claim to fame of LSTM is that it can reduce the effect of a

[6]Obviously, the inventors of RNNs were not familiar with *Back to the Future* or *Star Trek*, both of which conclusively demonstrate that the future can have a large influence on the past.

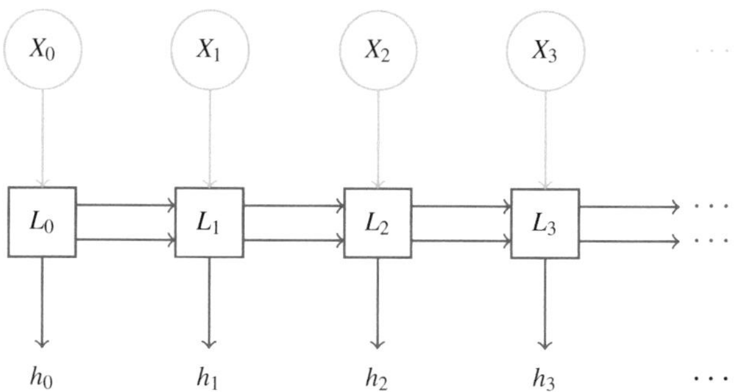

Fig. 20 LSTM

vanishing gradient, which is what enables such models to account for longer range dependencies [30].

Before outlining the main ideas behind LSTM, we note that the LSTM architecture has been one of the most commercially successful learning techniques ever developed. Among many other applications, LSTMs have been used in Google Allo [39], Google Translate [84], Apple's Siri [46], and Amazon Alexa [25]. However, recently, the dominance of LSTM may have begun to wane. ResNet has been shown to have theoretical advantages over LSTM, and it outperforms LSTM in a wide range of applications [63].

Figure 20 illustrates an LSTM. The obvious difference from a generic vanilla RNN is that an LSTM has two lines entering and exiting each state. As in a standard RNN, one of these lines represents the hidden state, while the second line is designed to serve as a gradient "highway" during backpropagation. In this way, the gradient can "flow" much further back with less chance that it will vanish along the way.

In Fig. 21, we expand one of the LSTM cells L_t that appear in Fig. 20. Here, σ is the sigmoid function, τ is the hyperbolic tangent (i.e., tanh) function, the operators "$\times$" and "$+$" are pointwise multiplication and addition, respectively, while "$\|$" indicates concatenation of vectors. The vector i_t is the "input" gate, f_t is the "forget" gate, and o_t is the "output" gate. The vector g_t is an intermediate gate and does not have a cool name, but is sometimes referred to as the "gate" gate [47], which, come to think of it, is especially cool. We have much more to say about these gates below.

The gate vectors that appear in Fig. 21 are computed as

$$f_t = \sigma\left(W_f\begin{pmatrix}h_{t-1}\\X_t\end{pmatrix} + b_f\right) \qquad g_t = \tau\left(W_g\begin{pmatrix}h_{t-1}\\X_t\end{pmatrix} + b_g\right)$$

$$i_t = \sigma\left(W_i\begin{pmatrix}h_{t-1}\\X_t\end{pmatrix} + b_i\right) \qquad o_t = \sigma\left(W_o\begin{pmatrix}h_{t-1}\\X_t\end{pmatrix} + b_o\right),$$

while the outputs are

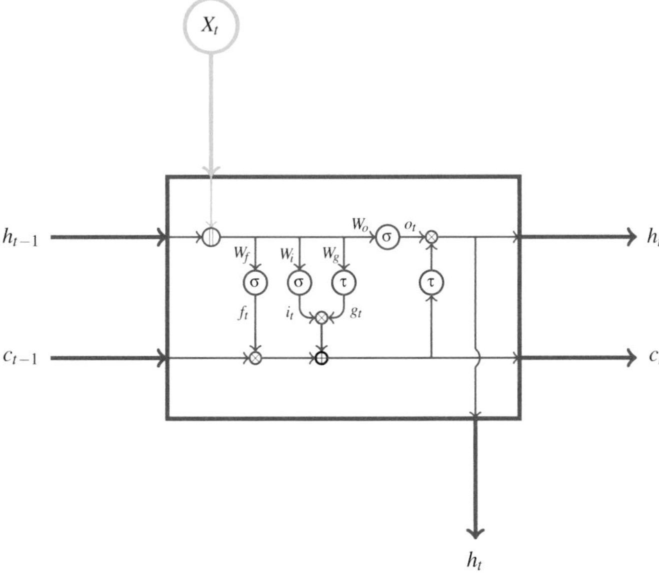

Fig. 21 One timestep of an LSTM

$$
\begin{aligned}
c_t &= f_t \otimes c_{t-1} \oplus i_t \otimes g_t \\
h_t &= o_t \otimes \tau(c_t),
\end{aligned}
$$

where "$\otimes$" is pointwise multiplication and "$\oplus$" is the usual pointwise addition. Note that each of the weight matrices is $n \times 2n$.

In matrix form, ignoring the bias terms b, we have

$$
\begin{pmatrix} i_t \\ f_t \\ o_t \\ g_t \end{pmatrix} = \begin{pmatrix} \sigma \\ \sigma \\ \sigma \\ \tau \end{pmatrix} W \begin{pmatrix} h_{t-1} \\ X_t \end{pmatrix},
$$

where X_t and h_{t-1} are column vectors of length n, and W is the $4n \times 2n$ weight matrix

$$
W = \begin{pmatrix} W_i \\ W_f \\ W_o \\ W_g \end{pmatrix}
$$

Further, each of the gates i_t, f_t, o_t, and g_t is a column vectors of length n. Recall that the sigmoid σ squashes its input to be within the range of 0 to 1, whereas the tanh function τ gives output within the range of -1 to $+1$.

To highlight the intuition behind LSTM, we follow a similar approach as that given in the excellent presentation [47]. Specifically, we focus on the extreme cases, that is, we assume that the output of each sigmoid σ is either 0 or 1, and each hyperbolic tangent τ is either -1 or $+1$. Then the forget gate f_t is a vector of 0s and 1s, where the 0s tell us the elements of c_{t-1} that we forget and the 1s indicate the elements to remember. In the middle section of the diagram, the input gate i_t and gate gate g_t together determine which elements of c_{t-1} to increment or decrement. Specifically, when element j of i_t is 1 and element j of g_t is $+1$, we increment element j of c_{t-1}. And if element j of i_t is 1 and element j of g_t is -1, then we decrement element j of c_{t-1}. This serves to emphasize or de-emphasize particular elements in the new-and-improved cell state c_t. Finally, the output gate o_t determines which elements of the cell state will become part of the hidden state h_t. Note that the hidden states h_t is fed into the output layer of the LSTM. Also note that before the cell states are operated on by the output gate, the values are first squeezed down to be within the range of -1– $+1$ by the τ function.

Of course, in general, the LSTM gates are not simply countered that increment or decrement by 1. But, the intuition is the same, that is, the gates keep track of incremental changes thus allowing relevant information to flow over long distances via the cell state. In this way, LSTM negates some of the limitations caused by vanishing gradients.

7.3 Gated Recurrent Units

As mentioned above, there are a large number of variants of the basic LSTM architecture. Most such variants are slight variants, with only minor changes from a standard LSTM. A gated recurrent unit (GRU), on the other hand, is a fairly radical departure from an LSTM. Although the internal state of a GRU is somewhat complex and, perhaps, less intuitive than that of an LSTM, there are fewer parameters in a GRU, and hence it is easier to train a GRU, and less training data is required. The wiring diagram for a GRU is given in Fig. 22.

The gate vectors that appear in Fig. 21 are computed as

$$z_t = \sigma\left(W_z\begin{pmatrix}h_{t-1}\\ X_t\end{pmatrix} + b_z\right)$$

$$r_t = \sigma\left(W_r\begin{pmatrix}h_{t-1}\\ X_t\end{pmatrix} + b_r\right)$$

$$g_t = \tau\left(W_g\begin{pmatrix}r_t \otimes h_{t-1}\\ X_t\end{pmatrix} + b_g\right),$$

while the output is

$$h_t = (1 - z_t) \otimes h_{t-1} \oplus z_t \otimes g_t,$$

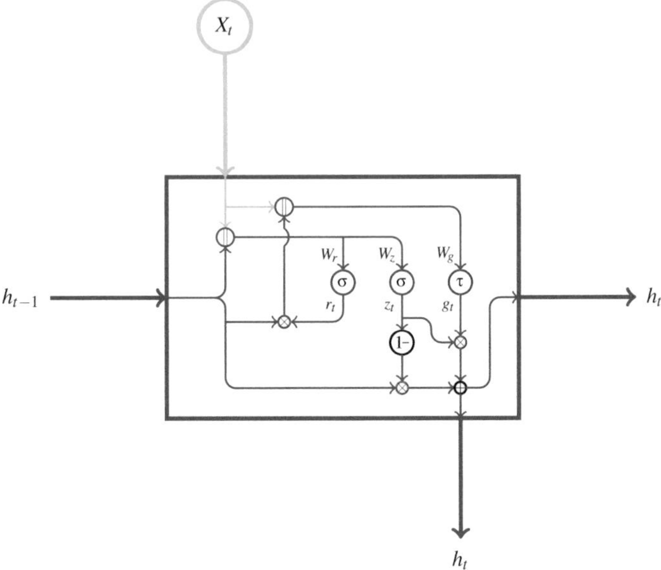

Fig. 22 One timestep of a GRU

where "$\otimes$" is pointwise multiplication and "$\oplus$" is the usual pointwise addition. Note that each of the weight matrices is $n \times 2n$.

In matrix form, ignoring the bias terms b, we have

$$\begin{pmatrix} z_t \\ r_t \\ g_t \end{pmatrix} = \begin{pmatrix} \sigma \\ \sigma \\ 0 \end{pmatrix} W \begin{pmatrix} h_{t-1} \\ X_t \end{pmatrix} + \begin{pmatrix} 0 \\ 0 \\ \tau \end{pmatrix} W \begin{pmatrix} r_t \otimes h_{t-1} \\ X_t \end{pmatrix},$$

where X_t and h_{t-1} are column vectors of length n, and W is the $3n \times 2n$ weight matrix

$$W = \begin{pmatrix} W_z \\ W_r \\ W_g \end{pmatrix}$$

Each of the gates z_t, r_t, and g_t is a column vectors of length n.

The intuition behind a GRU is that it replaces the input, forget, and output gates of an LSTM with just two gates—an "update" gate z_t and a "reset" gate r_t. The GRU update gate serves a similar purpose as the combined output and forget gates of an LSTM. Specifically, the update serves to determine what to output (or write) and what to forget. The function $1 - z_t$ in the GRU implies that anything that is not output must be forgotten. Thus, the GRU is less flexible as compared to an LSTM since an LSTM allows us to independently select elements for output and elements that are forgotten. The GRU reset gate and the LSTM input gate each serve to combine new input with previous memory.

The gating in a GRU is more complex and somewhat less intuitive as compared to that found in an LSTM. In any case, the most radical departure of the GRU from the LSTM architecture is that there is no cell state in a GRU. This implies that any memory must be stored in the hidden state h_t. This simplification (as compared to an LSTM) relies on the fact that in a GRU, the write and forget operations have been combined.

7.4 Recursive Neural Network

We mention in passing that recursive neural networks can be viewed as generalizing recurrent neural networks.[7] In a recursive neural network, we can recurse over any hierarchical structure, with trees being the archetypal example. Then training can be accomplished via backpropagation through structure (BPTS), often using stochastic gradient descent for simplicity. In contrast, a recurrent neural network is restricted to one particular structure—that of a linear chain.

7.5 Last Word on RNNs

RNNs are useful in cases where the input data is sequential. Generic RNN architectures are subject to vanishing and exploding gradients, which limit the length of the history (or gaps) that can effectively be incorporated into such models. Relatively complex RNN-based architectures—such as LSTM and its variants—have been developed that can better handle such gradient issues. These architectures have proven to be commercially successful across a wide range of products.

A good general discussion of RNNs can be found in [59], and an overview of various RNN-specific topics—with links to many relevant articles—is available at [58]. A more detailed (mathematical) description can be found in Chap. 10 of [20]. The slides at [47] provide a good general introduction to RNNs, with nice examples and a brief, but excellent, discussion of LSTM.

---

[7]Unfortunately, "recursive neural network" is typically also abbreviated as RNN. Here, we reserve RNN for recurrent neural networks and we do not use any abbreviation when referring to recursive neural networks.

7.6 RNNs in Malware Analysis

In a commercial sense, LSTMs are surely the most successful deep learning technique yet developed, so it is not surprising that LSTMs have been successfully applied to the malware detection problem [50]. Both LSTMs and GRUs—along with CNNs—are considered in [2], with the authors claiming a major improvement over relevant previous work. The paper [31] considers an adversarial attack, where the attacker can defeat a system that uses RNNs based on API calls.

There are many applications of RNNs in areas of information security outside of the malware domain. In [87], CNN and LSTM architectures are used to detect cybersecurity events, based on social networking messages. Other infosec applications of LSTMs include generating security ontologies [19], network security [49], breaking CAPTCHAs [11], host-based intrusion detection [41], and network anomaly detection [13], among others.

8 Residual Networks

At the time of this writing, residual network (ResNet) is considered the state of the art in deep learning for many image analysis problems. A residual network is one in which instead of approximating a function $F(x)$, we approximate the "residual," which is defined as $H(x) = F(x) - x$. Then the desired solution is given by $F(x) = H(x) + x$.

The original motivation for considering residuals was based on the observation that deeper networks sometimes produce worse results, even when vanishing gradients are not the cause [27]. This is somewhat counterintuitive, as the network should simply learn identity mappings when a model is deeper than necessary. To overcome this "degradation" problem, the authors of [27] experiment with residual mappings and provide extensive empirical evidence that the resulting ResNet architecture yields improved results as compared to standard feedforward networks for a variety of problems. The authors of [27] conjecture that the success of ResNet follows from the fact that the identity map corresponds to a residual of zero, and "if an identity mapping were optimal, it would be easier to push the residual to zero than to fit an identity mapping by a stack of nonlinear layers."

Whereas LSTM uses a complex gating structure to ease gradient flow, ResNet defines additional connections that correspond to identity layers. This enables ResNet to deal with vanishing gradients, as well as the aforementioned degradation problem. These identity layers allow a ResNet model to skip over layers during training, which serves to effectively reduce the minimum depth when training. Intuitively, ResNet is able to train deeper networks by, in effect, training over a considerably shallower network in the initial stages, with later stages of training serving to flesh out the intermediate connections. This approach was inspired by pyramidal cells in

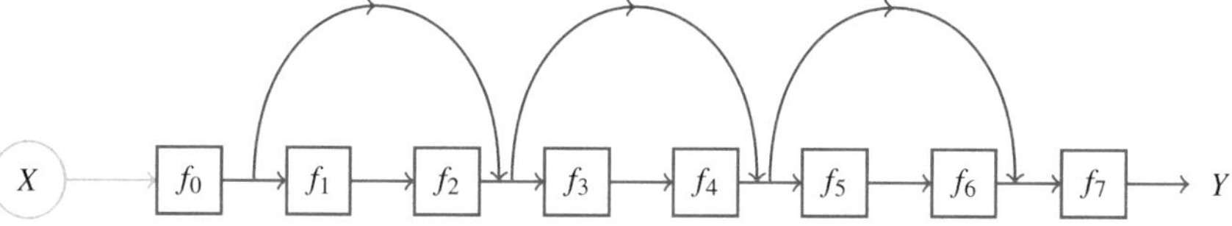

Fig. 23 Example of a ResNet architecture

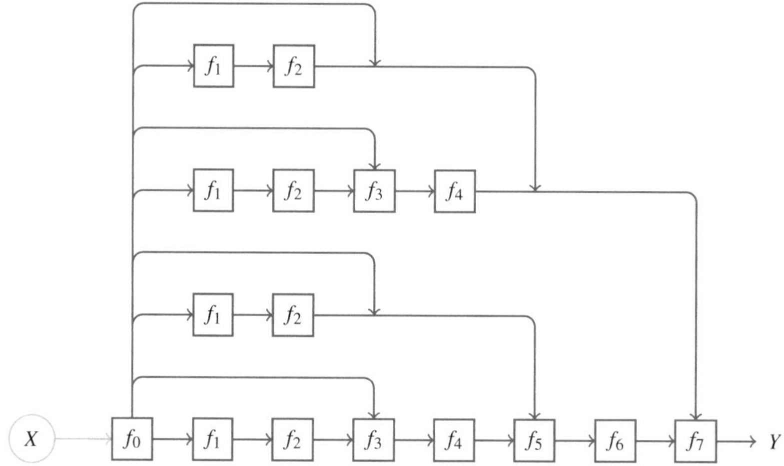

Fig. 24 Another view of the ResNet architecture in Fig. 23

the brain, which have a similar characteristic in the sense that they bridge "layers" of neurons [76].

A very high-level illustrative example of a ResNet architecture is given in Fig. 23, where each curved edge represents an identity transformation. Note that in this case, the identity transformations enable the model to skip over two layers. In principle, ResNet would seem to be applicable to any flavor of deep neural network, but in practice, it seems to applied to CNNs.

If a ResNet has N identity paths, then the network contains 2^N distinct feedforward networks. For example, the ResNet in Fig. 23 can be expanded into the graph in Fig. 24. Note that most of the paths in a ResNet are relatively short.

Surprisingly, the paper [81] provides evidence that in spite of being trained simultaneously, the multiple paths in a ResNet "show ensemble-like behavior in the sense that they do not strongly depend on each other." And perhaps an even more surprising result in [81] shows that "only the short paths are needed during training, as longer paths do not contribute any gradient." In other words, a deep ResNet architecture is more properly viewed as a collection of multiple, relatively shallow networks.

8.1 ResNet in Malware Analysis

At the time of this writing, ResNet is a relative newcomer and the level of research in the security domain is somewhat limited. Nevertheless, ResNet architectures have shown promise for dealing with the usual suspects, namely malware analysis [40, 66] and intrusion detection [43, 83].

9 Generative Adversarial Network

Let $\{X_i\}$ be a collection of samples and $\{Y_i\}$ a corresponding set of class labels. In statistics, a *discriminative* model is one that models the conditional probability distribution $P(Y \mid X)$. Such a discriminative model can be used to classify samples—given an input X of the same type as the training samples $\{X_i\}$, the model enables us to easily determine the most likely class of X by simply computing $P(Y \mid X)$ for each class label Y.

In contrast, a model is said to be *generative* if it models the joint probability distribution of X and Y, which we denote as $P(X, Y)$. Such a model is called "generative" because, by sampling from this distribution, we can generate new pairs (X_i, Y_i) that fit the probability distribution. Note that we can produce a discriminative model from a generative model, since

$$P(Y \mid X) = \frac{P(X, Y)}{P(X)}.$$

Therefore, in some sense, a generative model is inherently more general than a discriminative model.

Consider, for example, hidden Markov models (HMM) [77], which are a popular class of classic machine learning techniques. An HMM is defined by the three matrices in $\lambda = (A, B, \pi)$, where π is the initial state distribution, A contains the transition probability distributions for the hidden states, and B consists of the observation probability distributions corresponding to the hidden states. If we train an HMM on a given dataset, then we can easily generate samples that match the probability distributions of the HMM. To generate such samples, we first randomly select an initial state based on the probabilities in π. Then we repeat the following steps until the desired observation sequence length is reached: Randomly select an observation based on the current state, using the probabilities in B, and randomly select the next state, based on the probabilities in A. The resulting observation sequence will be indistinguishable (in the HMM sense) from the data that was used to train the HMM.

From the discussion in the previous paragraph, it is clear that a trained HMM is a generative model. However, it is more typical to use an HMM as a discriminative model. In discriminative mode, we determine a threshold, then we classify a given observation sequence as matching the model if its HMM score is above the specified threshold. This example shows that in practice, it is easy to use a generative model as a discriminative model.

On the other hand, while a trained SVM serves to classify samples, we could not use such a model to generate samples that match the training set. Thus, an SVM is an example of a discriminative model.

In the realm of deep learning, a discriminative network is designed to classify samples, while a generative network is designed to generate samples that "fit" the training data. From the discussion above, it is clear that we can always obtain a discriminative model from a generative model. Intuitively, it would seem that training a (more general) generative model in order to obtain a (more specific) discriminative model would be undesirable since we do not need the full generality of the model. However, reality appears to be somewhat more subtle. In [60], it is shown that for one generative–discriminative pair (naïve Bayes and logistic regression) the discriminative models do indeed have a lower asymptotic error; however, the generative models consistently converge faster. This suggests that with limited training data, a generative model might produce a superior discriminative model, as compared to directly training the corresponding discriminative model. In any case, in the realm of deep learning, discriminative models dominate, with an example of a typical application being image classification. In contrast, generative models have only recently come into vogue, with an example application being the creation of fake images.

Now, suppose that when training a discriminative neural network, in addition to the real training data, we generate "fake" training samples that follow a similar probability distribution as the real samples. Further, suppose that these fake training samples are designed to trick the discriminative network into making classification mistakes. Such samples would tend to improve the training of the network, thus making it stronger and more effective than if we had restricted the training to only the real data.

Although intuitively appealing, several problems arise when trying to implement a training technique based on fake samples. For one thing, we generally do not know the distribution of the training set, which often lives in an extremely high dimensional space of great complexity. Another issue is that during training, the discriminative network is constantly evolving, so determining samples that are likely to trick the network is a moving target. Another concern is that if the fake training samples are too difficult—or too easy—to distinguish at any point in the training process, we are unlikely to see any improvement over simply using the real training data

Several techniques have been proposed to try to take advantage of fake data so as to improve the training process. In the case of a generative adversarial network (GAN), we use a neural network to generate the fake data—a generative network is trained to defeat a discriminative network. Furthermore, the discriminative and generative networks are trained simultaneously in a minimax game. This approach sidesteps the complications involved in trying to model the probability distribution of the training samples. In fact, the generative network in a GAN simply uses random noise as its underlying probability distribution.

To summarize, a GAN consists of two competing neural networks—a generative network and a discriminative network—with the generative network creating fake data that is designed to defeat the discriminative network. The two networks are trained simultaneously following a game-theoretic approach. In this way, both networks improve, with the ultimate objective being a discriminative model (and/or a generative model) that is stronger than it would have been if it was trained only on the real training data.

We define two neural networks, namely a discriminator $D(x;\theta_d)$, and a generator $G(z;\theta_g)$, where θ_d consists of the parameters of the discriminator network, and θ_g consists of the parameters of the generator network. Here, we describe the training process in terms of images, but other types of data could be used. Also, to simplify the notation, we suppress the dependence on θ_d and θ_g in the remainder of this discussion, except where it is essential for understanding and may not be clear from context.

The generator $G(z)$ produces a fake image (based on the random seed value z) with the goal of tricking the discriminator into believing it is a real training image. In contrast, the discriminator $D(x)$ returns a value in the range of 0–1 that can be viewed as its estimate of the probability that the image x is real. For example, $D(x) = 1$ means that the discriminator is completely certain that the image is real, while $D(x) = 0$ tells us that the discriminator is sure that the image is fake, and $D(x) = 1/2$ implies that the discriminator is clueless. Note that the discriminator must deal with both real and fake images, while the generator is only concerned with generating fake images that trick the discriminator.

The generator G wins if D thinks its fake images are real. Thus, we can train G by making $1 - D(G(z))$ as close to zero as possible or, equivalently, by minimizing $\log(1 - D(G(z)))$. On the other hand, D wins if it can distinguish the fake images from real images so, ideally, when training D, we want $D(x) = 1$, when x is a real image, and $D(G(z)) = 0$ for fake images $G(z)$. Therefore, we can train D by maximizing $D(x)(1 - D(G(z))$ or, equivalently, by maximizing $\log(D(x)) + \log(1 - D(G(z)))$. We want the D and G models to be in competition, so they can strengthen each other. This can be accomplished by formulating the training in terms of the minimax game

$$\min_G \max_D \Big(E\big(\log(D(x))\big) + E\big(\log(1 - D(G(z)))\big) \Big), \tag{2}$$

where E is the expected value, relative to the implied probability distribution. Specifically, for the max over D, the expectation is with respect to the real sample distribution which has parameters θ_d, while for the min over G, the expectation is with respect to the fake sample distribution, which is specified by the parameters θ_g.

In the case of stochastic gradient descent (or ascent), at each iteration, we consider one real sample x and one fake sample $G(z)$. Then, due to the max in equation (2), we first perform gradient ascent to update the discriminator network D. This is followed by gradient descent to update generator network G. Of course, both of these steps rely on backpropagation.

Note that for the discriminator network D, the backpropagation error term involves

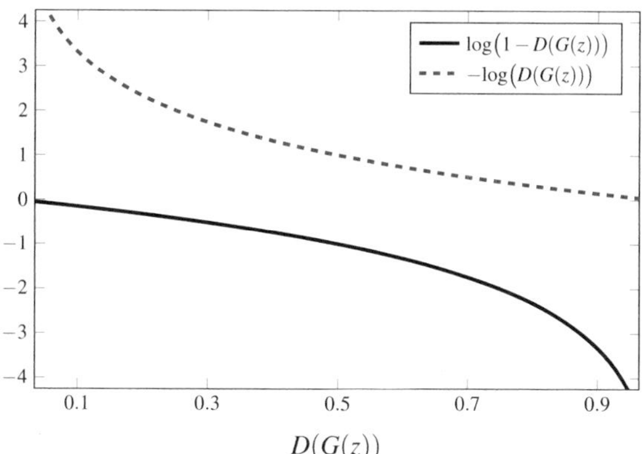

Fig. 25 Gradient of generator network G

$$\log\bigl(D(x)\bigr) + \log\bigl(1 - D(G(z))\bigr),$$

while for the generator network G, the error term involves only

$$\log\bigl(1 - D(G(z))\bigr). \tag{3}$$

Of course, in practice, we would typically use a minibatch of, say, m real samples and m fake samples at each update of D and G, rather than a strict stochastic gradient descent/ascent.

There is one technical issue that arises when attempting to train the generator network G as outlined above. As illustrated in Fig. 25, the gradient of the expression in (3) is nearly flat for values of $D(G(z))$ near zero. This implies that, early in training, when the generator network is sure to be extremely weak—and hence the discriminator can easily identify most $G(z)$ images as fake—it will be difficult for the G network to learn. From, Fig. 25, we also see that

$$\log\bigl(D(G(z))\bigr) \tag{4}$$

is relatively steep near zero. Hence, instead training G based on a gradient ascent involving equation (3), we perform gradient *descent* based on (4). Note that we have simply replaced the problem of maximizing $1 - D(G(z))$ with the equivalent problem of minimizing the probability $D(G(z))$.

The algorithm for training a GAN is summarized in Fig. 26. In some applications, letting $\texttt{iters} = 1$ works best, while in others, $\texttt{iters} > 1$ yields better results. In the latter case, we update the discriminator network D multiple times for each update of the generator network G. This implies that in such cases, the generator might otherwise overwhelm the discriminator, that is, the generator is in some sense easier to train. Finally, while a GAN certainly is an advanced architecture, it is important to realize that training reduces to a fairly straightforward application of gradient ascent.

initialize parameters θ_d and θ_g and `iters` ≥ 1
repeat
    for $k = 1$ to `iters`
        randomly select n noise samples $Z = (z_0, z_1, \dots, z_{n-1})$
        randomly select n real samples $X = (x_0, x_1, \dots, x_{n-1})$
        update θ_d by gradient ascent on
$$\sum_{i=0}^{n-1} \Big(\log\big(D(x_i)\big) + \log\big(1 - D(G(z_i))\big) \Big)$$
    next k
    randomly select n noise samples $Z = (z_0, z_1, \dots, z_{n-1})$
    update θ_g by gradient ascent on
$$\sum_{i=0}^{n-1} \log\big(D(G(z_i))\big)$$
until stopping criteria is met
return(θ_d, θ_g)

Fig. 26 GAN training algorithm

As with LSTM, there are a vast number of variations on the basic GAN approach outlined here; see [48] for a list of nearly 50 such variants. Additional sources of information on GANs include the original paper on the subject [21] and the excellent slides at [48].

9.1 GANs in Malware Analysis

GANs seem to show promise for dealing with some of the most challenging problems in information security. For example, GANs have been applied with some success to zero-day malware detection [32, 42]. In addition, the generative aspect of a GAN can be used to create challenging security problems in the "lab," thus enabling researchers to consider defenses against potential threats before those threats arise in a real-world setting [67].

10 Extreme Learning Machines

As with most aspects of ELMs, the origin of the technique is somewhat controversial. The unfortunate terminology of "Extreme Learning Machine" was apparently first used in [24]. Regardless of the origin of the technique, ELMs are essentially randomized feedforward neural networks that effectively minimize the cost of training.

An ELM consists of a single layer of hidden nodes, where the weights between inputs and hidden nodes are randomly initialized and remain unchanged throughout training. The weights that connect the hidden nodes to the output are trained, but due to the simple structure of an ELM, these weights can be determined by solving

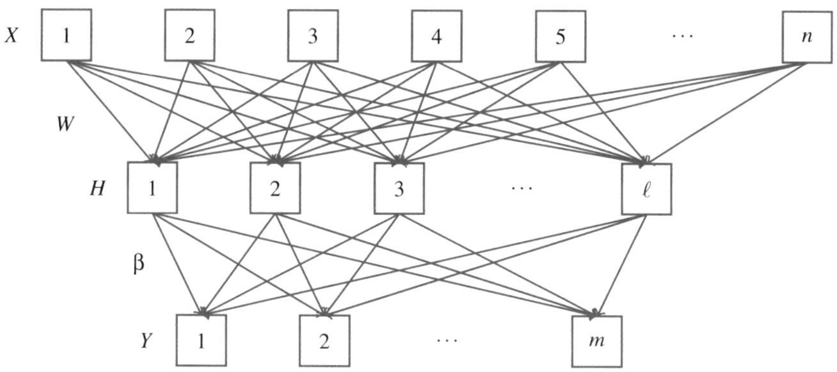

Fig. 27 Architecture of an ELM model

linear equations—more precisely, by solving a linear regression problem. Since no backpropagation is required, ELMs are far more efficient to train, as compared to other neural network architectures. However, since the weights in the hidden layer are not optimized, we will typically require more weights in an ELM, which implies that the testing phase may be somewhat more costly, as compared to a network trained by backpropagation. Nevertheless, in applications where models must be trained frequently, ELMs can be competitive.

Consider the ELM architecture shown in Fig. 27, where X denotes the input layer, H is the hidden layer, and Y is the output layer. In this example, there are N samples of the form (x_i, y_i) for $i = 1, 2, \ldots, N$, where $x_i = \left(x_{i_1}\ x_{i_2}\ \ldots\ x_{i_n}\right)^T$ is the feature vector for sample i and $y_i = \left(y_{i_1}\ y_{i_2}\ \ldots\ y_{i_m}\right)^T$ are the output labels, where T indicates the transposition operation. Then the input and output for the ELM are as $X = \left(x_1\ x_2\ \ldots\ x_n\right)^T$ and $Y = \left(y_1\ y_2\ \ldots\ y_m\right)^T$, respectively. In this example, the hidden layer H has ℓ neurons. We denote the activation function of the hidden layer as $g(x)$.

To train an ELM, we randomly select the weight matrix that connects the input layer X to the hidden layer H. We denote this randomly assigned weight matrix as $W = \left(w_1\ w_2\ \ldots\ w_\ell\right)$, where each w_i is a column vector. We also randomly select the bias matrix $B = \left(b_1\ b_2\ \ldots\ b_\ell\right)$ for this same layer. During the training phase, both W and B remain unchanged.

After W and B have been initialized, the output of the hidden layer H is given by

$$H = g(WX + B).$$

The output of the ELM is denoted as Y and is calculated as

$$Y = H\beta,$$

where β is the weight matrix for the output layer.

The values of the weights β at the hidden layer are learned via linear least squares, and can be computed using $H^\dagger$, the Moore–Penrose generalized inverse of H, as discussed below. It is worth emphasizing that the only parameters that are learned in the ELM are the elements of β.

Given that Y is the desired output, a unique solution of the system based on least squared error can be found as follows. We denote the Moore–Penrose generalization inverse of H as $H^\dagger$, which is defined as

$$H^\dagger = \begin{cases} (H^T H)^{-1} H^T \text{ if } H^T H \text{ is nonsingular} \\ H^T (H H^T)^{-1} \text{ if } H H^T \text{ is nonsingular} \end{cases}$$

Then the desired solution β is given by

$$\beta = H^\dagger Y.$$

After calculating β, the training phase ends. For each test sample x, the output Y can be calculated as

$$Y = g\big(C(x)\big)\beta,$$

where $C(x)$ is defined below. The entire training process is extremely efficient, particularly in comparison to the backpropagation technique that is typically used to train neural networks [80].

For the research reported in this paper, we use the Python implementation of ELMs given in [17]. This implementation uses input activations that are a weighted combination of two functions referred to as an "MLP" kernel and an "RBF" kernel—we employ the same terminology here. The MLP kernel is simply the linear operation

$$M(x) = Wx + B,$$

where the weights W and biases B are randomly selected from a normal distribution. This is the kernel function that is typically associated with a standard ELM.

The RBF kernel is considerably more complex and is based on generalized radial basis functions as defined in [18]. The details of this RBF kernel go beyond the scope of this paper; see [18] for additional information and, in particular, examples where this kernel is applied to train ELMs. We use the notation $R(x)$ to represent the RBF kernel. Also, it is worth noting that the RBF kernel is much more costly to compute, and hence its use does somewhat negate one of the major advantages of an ELM.

The input activations are given by

$$C(x) = \alpha M(x) + (1 - \alpha) R(x), \qquad (5)$$

where $0 \leq \alpha \leq 1$ is a user-specified mixing parameter. Note that for $\alpha = 0$ we use only the MLP kernel $M(x)$ and for $\alpha = 1$, only the RBF kernel $R(x)$ is used.

10.1 ELMs in Malware Analysis

In [34], ELMs are compared to CNNs for malware classification, and it is shown that ELMs can outperform CNNs in some cases. This is impressive since ELMs have training times that are only a small fraction of those required for comparable CNNs. ELMs have also been applied to malware detection on the Android platform in [90], where the training is based on static features, with reasonably strong results. In [71], the authors consider the effectiveness of a technique that they refer to as high-performance extreme learning machines (HP-ELM). By varying the features and activation functions of their HP-ELM architecture, they achieve high accuracy on a challenging dataset. A two-layer ELM is applied to the malware detection problem in [33]. A partially connected network is used between the input and the first hidden layer, and this layer is aggregated with a fully connected network in the second layer. The authors utilize an ensemble to improve the accuracy and robustness of the resulting ELM-based system.

11 Word Embedding Techniques

Word2Vec is a technique for embedding terms in a high-dimensional space, where the term embeddings are obtained by training a shallow neural network. After the training process, words that are more similar in context will tend to be closer together in the Word2Vec space.

Perhaps surprisingly, meaningful algebraic properties also hold for Word2Vec embeddings. For example, according to [53], if we let

$$w_0 = \text{“king”},\ w_1 = \text{“man”},\ w_2 = \text{“woman”},\ w_3 = \text{“queen”}$$

and $V(w_i)$ is the Word2Vec embedding of word w_i, then $V(w_3)$ is the vector that is closest—in terms of cosine similarity—to

$$V(w_0) - V(w_1) + V(w_2).$$

Results such as this indicate that Word2Vec embeddings capture significant aspects of the semantics of the language.

The focus of this section is Word2Vec, but before discussing this popular and effective word embedding technique, we consider a couple of alternatives. First, we discuss simple embedding strategies based on hidden Markov models. Then we briefly consider a word embedding technique the uses PCA. Finally, we discuss the main ideas behind Word2Vec.

11.1 HMM2Vec

To begin, we consider individual letter embeddings, as opposed to word embeddings. We call the letter embedding technique considered here Letter2Vec.

Recall that an HMM is defined by the three matrices A, B, and π, and is denoted as $\lambda = (A, B, \pi)$. The π matrix contains the initial state probabilities, A contains the hidden state transition probabilities, and B consists of the observation probability distributions corresponding to the hidden states. Each of these matrices is row stochastic, that is, each row satisfies the requirements of a discrete probability distribution. Notation-wise, we let N be the number of hidden states, M is the number of distinct observation symbols, and T is the length of the observation (i.e., training) sequence. Note that M and T are determined by the training data, while N is a user-defined parameter. For more details in HMMs, see [77] or Rabiner's fine paper [65].

Suppose that we train an HMM on a sequence of letters extracted from English text, where we convert all uppercase letters to lowercase and discard any character that is not an alphabetic letter or word-space. Then $M = 27$, and we select $N = 2$ hidden states, and we use $T = 50{,}000$ observations for training. Note that each observation is one of the $M = 27$ symbols (letters plus word-space). For the example discussed below, the sequence of $T = 50{,}000$ observations was obtained from the Brown corpus of English [7]. Of course, any source of English text could be used.

For one specific case, an HMM trained with the parameters listed in the previous paragraph yields the B matrix in Table 1. Observe that this B matrix gives us two probability distributions over the observation symbols—one for each of the hidden states. We observe that one hidden state essentially corresponds to vowels, while the other corresponds to consonants. This simple example nicely illustrates the concept of machine learning, as no a priori assumption was made concerning consonants and vowels, and the only parameter we selected was the number of hidden states N. Through the training process, the model learned a crucial aspect of English directly from the data. This illustrative example is discussed in more detail in [77] and originally appeared in Cave and Neuwirth's classic paper [8].

Suppose that for a given letter ℓ, we define its Letter2Vec representation $V(\ell)$ to be the corresponding row of the matrix B^T in Table 1. Then, for example,

$$\begin{array}{ll} V(\text{a}) = \big(0.13537\ 0.00364\big) & V(\text{e}) = \big(0.21176\ 0.00223\big) \\ V(\text{s}) = \big(0.00032\ 0.11069\big) & V(\text{t}) = \big(0.00158\ 0.15238\big) \end{array} \tag{6}$$

Next, we consider the distance between these Letter2Vec representations. Instead of using Euclidean distance, we measure the cosine similarity.[8]

The cosine similarity of vectors X and Y is the cosine of the angle between the two vectors. Let $S(X, Y)$ denote the cosine similarity between vectors X and Y. Then for $X = (X_0, X_1, \ldots, X_{n-1})$ and $Y = (Y_0, Y_1, \ldots, Y_{n-1})$,

---

[8]Cosine similarity is not a true metric, since it does not, in general, satisfy the triangle inequality.

Table 1 Final B^T for HMM

Letter	State 0	State 1	Letter	State 0	State 1
a	0.13537	0.00364	n	0.00035	0.11429
b	0.00023	0.02307	o	0.13081	0.00143
c	0.00039	0.05605	p	0.00073	0.03637
d	0.00025	0.06873	q	0.00019	0.00134
e	0.21176	0.00223	r	0.00041	0.10128
f	0.00018	0.03556	s	0.00032	0.11069
g	0.00041	0.02751	t	0.00158	0.15238
h	0.00526	0.06808	u	0.04352	0.00098
i	0.12193	0.00077	v	0.00019	0.01608
j	0.00014	0.00326	w	0.00017	0.02301
k	0.00112	0.00759	x	0.00030	0.00426
l	0.00143	0.07227	y	0.00028	0.02542
m	0.00027	0.03897	z	0.00017	0.00100
Space	0.34226	0.00375	–	–	–

$$S(X, Y) = \frac{\displaystyle\sum_{i=0}^{n-1} X_i Y_i}{\sqrt{\displaystyle\sum_{i=0}^{n-1} X_i^2}\sqrt{\displaystyle\sum_{i=0}^{n-1} Y_i^2}}$$

In general, we have $-1 \leq S(X, Y) \leq 1$, but since our Letter2Vec encoding vectors consist of probabilities—and hence are non-negative values—we always have $0 \leq S(X, Y) \leq 1$.

When considering cosine similarity, the length of the vectors is irrelevant, as we are only considering the angle between vectors. Consequently, we might want to consider vectors of length one, $\widetilde{X} = X/\|X\|$ and $\widetilde{Y} = Y/\|Y\|$, in which case the cosine similarity simplifies to the dot product

$$S(\widetilde{X}, \widetilde{Y}) = \sum_{i=0}^{n-1} \widetilde{X}_i \widetilde{Y}_i$$

Henceforth, we use the notation $\widetilde{X}$ to indicate a vector X that has been normalized to be of length one.

For the vector encodings in (6), we find that for the vowels "a" and "e", the cosine similarity is $S(V(\text{a}), V(\text{e})) = 0.9999$. In contrast, the cosine similarity of the vowel "a" and the consonant "t" is $S(V(\text{a}), V(\text{t})) = 0.0372$. The normalized vectors $V(\text{a})$

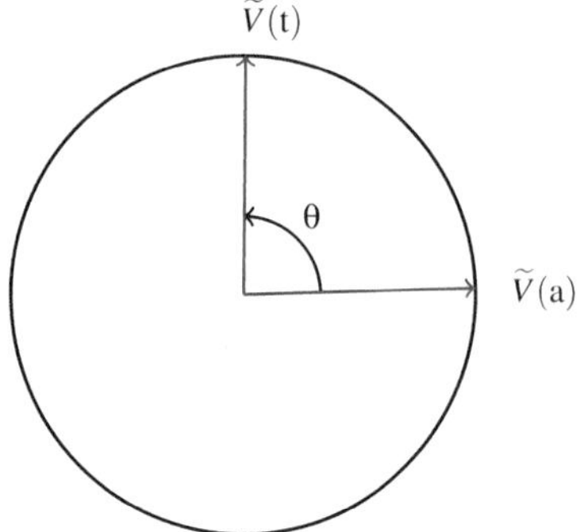

Fig. 28 Normalized vectors $\widetilde{V}$(a) and $\widetilde{V}$(t)

and V(t) are illustrated in Fig. 28. Using the notation in this figure, cosine similarity is $S(V(\text{a}), V(\text{t})) = \cos(\theta)$.

These results indicate that these Letter2Vec encodings—which are derived from a trained HMM—provide useful information on the similarity (or not) of pairs of letters. Note that we could obtain a vector encoding of any dimension by simply training an HMM with the number of hidden states N equal to the desired dimension.

Our HMM-based approach to Letter2Vec encoding is interesting, but we want to encode words, not letters. Analogous to the Letter2Vec embeddings discussed above, we could train an HMM on words and then use the columns of the resulting B matrix (equivalently, the rows of B^T) to define word vectors. The state of the art for Word2Vec uses a dataset corresponding to $M = 10{,}000$, $N = 300$ and $T = 10^9$. Training an HMM with similar parameters would be decidedly non-trivial, as the work factor is on the order of N^2T.

While the word embedding technique discussed in the previous paragraph—we call it HMM2Vec—is plausible, it has some potential limitations. Perhaps the biggest issue with HMM2Vec is that we typically train an HMM based on a Markov model of order one. This means that the current state only depends on the immediately preceding state. By basing our word embeddings on such a model, the resulting vectors would likely provide only a very limited sense of context. While we can train HMMs using models of a higher order, the work factor would be prohibitive.

11.2 PCA2Vec

Another option for generating embedding vectors is to apply PCA to a matrix of pointwise mutual information (PMI). To construct a PMI matrix, based on a specified window size W, we compute the probabilities $P(w_i, w_j)$ for all pairs of words (w_i, w_j) that occur within a window W of each other within dataset, and we also compute $P(w_i)$ for each individual word w_i. Then we define the PMI matrix as $X = \{x_{ij}\}$ as

$$x_{ij} = \log\left(\frac{P(w_j, w_i)}{P(w_i)P(w_j)}\right) = \log P(w_j, w_i) - \log P(w_i) - \log P(w_j).$$

Let X_i be column i of X. We use X_i as the feature vector for word w_i and perform PCA (using SVD) based on these X_i feature vectors. As usual, we project the feature vectors X_i onto the resulting eigenspace. Finally, by choosing the N dominant eigenvalues for this projection, we obtain word embedding vectors of length N.

It is shown in [56] that these embedding vectors have many similar properties as Word2Vec embeddings, with the author providing examples analogous to those we give in the next section. Interestingly, it may be beneficial in some applications to omit a few of the dominant eigenvectors when determining the PCA2Vec embedding vectors [45].

For more details on using PCA to generate word embeddings, see [45]. The aforecited blog [56] gives an intuitive introduction to the topic.

11.3 Word2Vec

Word2Vec uses a similar approach as the HMM2Vec concept outlined above. But, instead of using an HMM, Word2Vec is based on a shallow (one hidden layer) neural network. Analogous to HMM2Vec, in Word2Vec, we are not interested in the resulting model itself, but instead we make use the learning that is represented by the trained model to define word embeddings. Next, we consider the basic ideas behind Word2Vec. Our presentation is fairly similar to that found in the excellent tutorial [51].

Suppose that we have a vocabulary of size M. We encode each word as a "one-hot" vector of length M. For example, suppose that our vocabulary consists of the set of $M = 8$ words

$$\begin{aligned} W &= (w_0, w_1, w_2, w_3, w_4, w_5, w_6, w_7) \\ &= (\text{"for"}, \text{"giant"}, \text{"leap"}, \text{"man"}, \text{"mankind"}, \text{"one"}, \text{"small"}, \text{"step"}) \end{aligned}$$

Then we encode "for" and "man" as

$$E(w_0) = E(\text{"for"}) = 10000000 \text{ and } E(w_3) = E(\text{"man"}) = 00010000,$$

respectively.

Now, suppose that our training data consists of the phrase

$$\text{"one small step for man one giant leap for mankind".} \tag{7}$$

Table 2 Training data

Offset	Training pairs
"[one] small step …"	(one, small), (one, step)
"one [small] step for …"	(small, one), (small, step), (small, for)
"one small [step] for man …"	(step, one), (step, small), (step, for), (step, man)
"… small step [for] man one …"	(for, small), (for, step), (for, man), (for, one)
"… step for [man] one giant …"	(man, step), (man, for), (man, one), (man, giant)
"… for man [one] giant leap …"	(one, for), (one, man), (one, giant), (one, leap)
"… man one [giant] leap for …"	(giant, man), (giant, one), (giant, leap), (giant, for)
"… one giant [leap] for mankind"	(leap, one), (leap, giant), (leap, for), (leap, mankind)
"… giant leap [for] mankind"	(for, giant), (for, leap), (for, mankind)
"… leap for [mankind]"	(mankind, leap), (mankind, for)

To obtain training samples, we specify the window size, and for each offset, we use all pairs of words within the specified window. For example, if we select a window size of two, then from (7), we obtain the training pairs in Table 2.

Consider the pair "(for,man)" from the fourth row in Table 2. As one-hot vectors, this training pair corresponds to input 10000000 and output 00010000.

A neural network similar to that in Fig. 29 is used to generate Word2Vec embeddings. The input is a one-hot vector of length M representing the first element of a training pair, such as those in Table 2, and the network is trained to output the second element of the ordered pair. The hidden layer consists of N linear neurons and the output layer uses a softmax function to generate M probabilities, where p_i is the probability of the output vector corresponding to w_i for the given input.

Observe that the Word2Vec network in Fig. 29 has NM weights that are to be determined, as represented by the blue lines from the hidden layer to the output layer. For each output node ω_i, there are N edges (i.e., weights) from the hidden layer. The N weights that connect to output node ω_i form the Word2Vec embedding $V(w_i)$ of the word w_i.

As mentioned above, the state of the art in Word2Vec for English text is based on a vocabulary of $M = 10{,}000$ words, and embedding vectors of length $N = 300$. These embeddings are obtained by training on a set of about 10^9 samples. Clearly, training a model of this magnitude is an extremely challenging computational task, as there are 3×10^6 weights to be determined, not to mention a huge number of training samples to deal with. Most of the complexity of Word2Vec comes from tricks that are used to make it feasible to train such a large network with a massive amount of data.

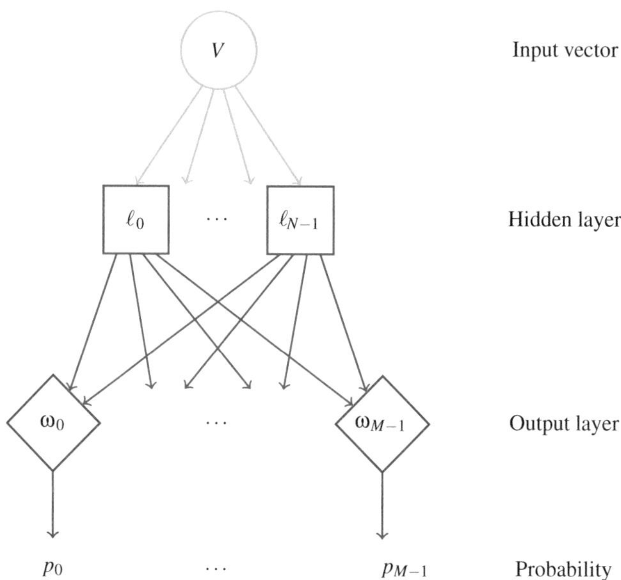

Fig. 29 Neural network for Word2Vec embeddings

One trick that is used to speed training in Word2Vec is the subsampling of frequent words. Common words such as "a" and "the" contribute little to the model, so these words can appear in training pairs at a much lower rate than they are present in the training text.

The most significant work-saving trick that is used in Word2Vec is so-called "negative sampling." When training a neural network, each training sample potentially affects all of the weights of the model. Instead of adjusting all of the weights, in Word2Vec, only a small number of "negative" samples have their weights modified per training sample. For example, suppose that the output vector of a training pair corresponds to word w_0. Then the "positive" weights are those of output node ω_0, and all of the corresponding weights are modified. In addition, a small subset of the $M - 1$ "negative" words (i.e., every word in the dataset except w_0) are selected and only the weights of the corresponding output nodes are modfied. The distribution used to select the negative subset is biased toward more frequent words.

A high-level discussion of Word2Vec can be found in [3], while a very nice and intuitive—yet reasonably detailed—introduction is given in [51]. The original paper describing Word2Vec is [53] and an immediate follow-up paper discusses a variety of improvements that mostly serve to make training practical for large datasets [54].

11.4 Word Embeddings in Malware Analysis

Word2Vec is fairly popular in the malware detection literature. For example, in [64] Word2Vec models based on machine code form the basis for a malware detection technique, while in [12], an Android malware detection scheme dubbed Droid-VecDeep uses Word2Vec results as features in deep belief networks [29]. The recent malware research in [9] considers multiple word embedding techniques (Word2Vec, HMM2Vec, and PCA2Vec) based on opcode sequences. Better results are obtained in most cases, as compared to using raw opcode sequences, which indicates that word embeddings are a useful form of feature engineering. The paper [36] considers Word2Vec and HMM2Vec embeddings for malware classification, with strong results obtained in many cases. In [62], word embeddings are used as part of a scheme that can successfully distinguish points in time where significant evolution has occurred within a malware family.

Word2Vec has proven surprisingly useful in a variety of security applications beyond the malware domain. Such applications range from network-based anomaly detection [4] to analyzing the evolution of cyberattacks [73].

12 Conclusion

In this chapter, we have provided details on a wide array of deep learning techniques that have proven useful in the field of malware analysis. We began with an introduction to the historical development of neural network-based techniques and related topics. This was followed by a discussion of several popular modern architectures. Specifically, we covered the following architectures: Multilayer perceptrons (MLP), convolutional neural networks (CNN), recurrent neural networks (RNN), long short-term memory (LSTM), gated recurrent units (GRU), residual networks (ResNet), generative adversarial networks (GAN), extreme learning machines (ELM), and Word2Vec. For each of these architectures, we cited representative examples of relevant malware-related research, and in most cases, we also mentioned other applications related to information security.

References

1. Annapurna, Annadatha, and Mark Stamp. 2018. Image spam analysis and detection. *Journal of Computer Virology and Hacking Techniques* 14 (1): 39–52.
2. Ben Athiwaratkun and Jack W. Stokes. 2017. Malware classification with LSTM and GRU language models and a character-level CNN. https://www.microsoft.com/en-us/research/wp-content/uploads/2017/07/LstmGruCnnMalwareClassifier.pdf.
3. Banerjee, Suvro. 2018. Word2vec — A baby step in deep learning but a giant leap towards natural language processing. https://medium.com/explore-artificial-intelligence/word2vec-

a-baby-step-in-deep-learning-but-a-giant-leap-towards-natural-language-processing-40fe4e8602ba.

4. Barot, Ketul, Jialing Zhang, and Seung Woo Son. 2016. Using natural language processing models for understanding network anomalies. http://ieee-hpec.org/2016/techprog2016/index_htm_files/R-w2vec-final.pdf.
5. Basole, Samanvitha, Fabio Di Troia, and Mark Stamp. 2020. Multifamily malware models. *Journal of Computer Virology and Hacking Techniques* 16 (1): 79–92.
6. Bhodia, Niket, Pratikkumar Prajapati, Fabio Di Troia, and Mark Stamp. 2019. Transfer learning for image-based malware classification. In *Proceedings of the 5th International Conference on Information Systems Security and Privacy*, ICISSP 2019, eds. Paolo Mori, Steven Furnell, and Olivier Camp, 719–726.
7. The Brown corpus of standard American English. http://www.cs.toronto.edu/~gpenn/csc401/a1res.html.
8. Cave, Robert L., and Lee P. Neuwirth. 1980. Hidden Markov models for English. In *Hidden Markov models for speech*, 16–56, IDA-CRD. New Jersey: Princeton. https://www.cs.sjsu.edu/~stamp/RUA/CaveNeuwirth/index.html.
9. Chandak, Aniket, Fabio Di Troia, and Mark Stamp. 2020. A comparison of word embedding techniques for malware classification. In *Malware analysis using artificial intelligence and deep learning*, eds. Stamp, Mark, Mamoun Alazab, and Andrii Shalaginov. Berlin: Springer.
10. Chavda, Aneri, Katerina Potika, Fabio Di Troia, and Mark Stamp. 2018. Support vector machines for image spam analysis. In *Proceedings of the 15th international joint conference on e-business and telecommunications*, ICETE 2018, eds. Callegari, Christian, Marten van Sinderen, Paulo Novais, Panagiotis G. Sarigiannidis, Sebastiano Battiato, Ángel Serrano Sánchez de León, Pascal Lorenz, and Mohammad S. Obaidat, 597–607.
11. Chen, Rui, Jing Yang, Rong-gui Hu, and Shu-guang Huang. 2013. A novel lstm-rnn decoding algorithm in CAPTCHA recognition. https://ieeexplore.ieee.org/document/6840561.
12. Chen, T., Q. Mao, M. Lv, H. Cheng, and Y. Li. 2019. Droidvecdeep: Android malware detection based on Word2Vec and deep belief network. *KSII Transactions on Internet and Information Systems* 13 (4): 2180–2197.
13. Cheng, Min, Qian Xu, Jianming Lv, Wenyin Liu, Qing Li, and Jianping Wang. 2016. MS-LSTM: A multi-scale LSTM model for BGP anomaly detection. In *2016 IEEE 24th International Conference on Network Protocols (ICNP)*, 1–6.
14. Cohen, Steven A., and Matthew W. Granade. 2018. Models will run the world. *Wall Street Journal*. https://www.wsj.com/articles/models-will-run-the-world-1534716720.
15. Cornelisse, Daphne. 2018. An intuitive guide to convolutional neural networks. https://medium.freecodecamp.org/an-intuitive-guide-to-convolutional-neural-networks-260c2de0a050.
16. Deshpande, Adit. 2018. A beginner's guide to understanding convolutional neural networks. https://adeshpande3.github.io/A-Beginner%27s-Guide-To-Understanding-Convolutional-Neural-Networks/.
17. Extreme learning machine implementation in Python. https://github.com/dclambert/Python-ELM.
18. Fernández-Navarro, Francisco, César Hervás-Martinez, Javier Sanchez-Monedero, and Pedro Antonio Gutiérrez. 2011. MELM-GRBF: A modified version of the extreme learning machine for generalized radial basis function neural networks. *Neurocomputing* 74(16): 2502–2510.
19. Gasmi, Houssem, Jannik Laval, and Abdelaziz Bouras. 2019. Cold-start cybersecurity ontology population using information extraction with LSTM. In *2019 international conference on cyber security for emerging technologies*, CSET, 1–6.
20. Goodfellow, Ian, Yoshua Bengio, and Aaron Courville. 2016. *Deep learning*. Cambridge: MIT Press. http://www.deeplearningbook.org.
21. Goodfellow, Ian J, Jean Pouget-Abadie, Mehdi Mirza, Bing Xu, David Warde-Farley, Sherjil Ozair, Aaron Courville, and Yoshua Bengio. 2014. Generative adversarial nets. In *Proceedings of the 27th international conference on neural information processing systems, NIPS'14*, vol. 2, 2672–2680.

22. Gormley, Matthew R. 2017. Neural networks and backpropagation. https://www.cs.cmu.edu/~mgormley/courses/10601-s17/slides/lecture20-backprop.pdf.
23. Greff, Klaus, Rupesh Kumar Srivastava, Jan Koutník, Bas R. Steunebrink, and Jürgen Schmidhuber. 2017. LSTM: A search space odyssey. *IEEE Transactions on Neural Networks and Learning Systems* 28 (10): 2222–2232. https://arxiv.org/pdf/1503.04069.pdf.
24. Huang, Guang-Bin, Qin-Yu Zhu, and Chee-Kheong Siew. 2004. Extreme learning machine: A new learning scheme of feedforward neural networks. In *2004 IEEE international joint conference on neural networks*, vol. 2, 985–990.
25. Gupta, Arpit. 2018. Alexa blogs: How Alexa is learning to converse more naturally. https://developer.amazon.com/blogs/alexa/post/15bf7d2a-5e5c-4d43-90ae-c2596c9cc3a6/how-alexa-is-learning-to-converse-more-naturally.
26. Hardesty, Larry. 2017. Explained: Neural networks. http://news.mit.edu/2017/explained-neural-networks-deep-learning-0414.
27. He, Kaiming, Xiangyu Zhang, Shaoqing Ren, and Jian Sun. Deep residual learning for image recognition. https://arxiv.org/pdf/1512.03385.pdf.
28. Hern, Alex. 2017. *The guardian*. Elon Musk says AI could lead to third world war. https://www.theguardian.com/technology/2017/sep/04/elon-musk-ai-third-world-war-vladimir-putin.
29. Hinton, Geoffrey. 2007. Deep belief nets. https://www.cs.toronto.edu/~hinton/nipstutorial/nipstut3.pdf.
30. Hochreite, Sepp and Jürgen Schmidhuber. 1997. Long short-term memory. *Neural Computation* 9(8): 1735–1780. http://www.bioinf.jku.at/publications/older/2604.pdf.
31. Hu, Weiwei and Ying Tan. 2017. Black-box attacks against RNN based malware detection algorithms. https://arxiv.org/abs/1705.08131.
32. Hu, Weiwei and Ying Tan. 2017. Generating adversarial malware examples for black-box attacks based on gan. https://arxiv.org/pdf/1702.05983.pdf.
33. Jahromi, Amir Namavar, Sattar Hashemi, Ali Dehghantanha, Kim-Kwang Raymond Choo, Hadis Karimipour, David Ellis Newton, and Reza M. Parizi. 2019. An improved two-hidden-layer extreme learning machine for malware hunting. *Computers and Security* 89.
34. Jain, Mugdha, William Andreopoulos, and Mark Stamp. Convolutional neural networks and extreme learning machines for malware classification. *Journal of Computer Virology and Hacking Techniques*.
35. Kaan, Can. 2018. Deep learning tutorial for beginners. https://www.kaggle.com/kanncaa1/deep-learning-tutorial-for-beginners.
36. Kale, Aparna Sunil, Fabio Di Troia, and Mark Stamp. 2020. Malware classification with hmm2vec and word2vec features. submitted for publication.
37. Kalfas, Ioannis. 2018. Modeling visual neurons with convolutional neural networks. https://towardsdatascience.com/modeling-visual-neurons-with-convolutional-neural-networks-e9c01ddfdfa7.
38. Karpathy, Andrej. 2018. Convolutional neural networks for visual recognition. http://cs231n.github.io/convolutional-networks/.
39. Khaitan, Pranav. 2016. Google AI blog: Chat smarter with Allo. https://ai.googleblog.com/2016/05/chat-smarter-with-allo.html.
40. Khan, Riaz Ullah, Xiaosong Zhang, and Rajesh Kumar. 2019. Analysis of resnet and googlenet models for malware detection. *Journal of Computer Virology and Hacking Techniques* 15 (1): 29–57.
41. Kim, Gyuwan, Hayoon Yi, Jangho Lee, Yunheung Paek, and Sungroh Yoon. 2016. LSTM-based system-call language modeling and robust ensemble method for designing host-based intrusion detection systems. https://arxiv.org/abs/1611.01726.
42. Kim, Jin-Young, Bu Seok-Jun, and Sung-Bae Cho. 2018. Zero-day malware detection using transferred generative adversarial networks based on deep autoencoders. *Information Sciences* 460–461: 83–102.
43. Kravchik, Moshe, and Asaf Shabtai. 2018. Detecting cyberattacks in industrial control systems using convolutional neural networks. https://arxiv.org/pdf/1806.08110.pdf.

44. Kurenkov, Andrey. 2015. A 'brief' history of neural nets and deep learning. http://www.andreykurenkov.com/writing/ai/a-brief-history-of-neural-nets-and-deep-learning/.
45. Levy, Omer, Yoav Goldberg, and Ido Dagan. 2015. Improving distributional similarity with lessons learned from word embeddings. *Transactions of the Association for Computational Linguistics* 3: 211–225. https://levyomer.files.wordpress.com/2015/03/improving-distributional-similarity-tacl-2015.pdf.
46. Levy, Steven. 2016. The iBrain is here—and it's already inside your phone. *Wired*. https://www.wired.com/2016/08/an-exclusive-look-at-how-ai-and-machine-learning-work-at-apple/.
47. Li, Fei-Fei, Justin Johnson, and Serena Yeung. 2017. Lecture 10: Recurrent neural networks. http://cs231n.stanford.edu/slides/2017/cs231n_2017_lecture10.pdf.
48. Li, Fei-Fei, Justin Johnson, and Serena Yeung. 2017. Lecture 13: Generative models. http://cs231n.stanford.edu/slides/2017/cs231n_2017_lecture13.pdf.
49. Li, Shixuan, and Dongmei Zhao. 2019. A LSTM-based method for comprehension and evaluation of network security situation. In *2019 18th IEEE international conference on trust, security and privacy in computing and communications*, 723–728.
50. Lu, Renjie. 2019. Malware detection with lstm using opcode language. https://arxiv.org/abs/1906.04593.
51. McCormick, Chris. 2016. Word2vec tutorial — The skip-gram model. http://mccormickml.com/2016/04/19/word2vec-tutorial-the-skip-gram-model/.
52. McCulloch , Warren S, and Walter Pitts. 1943. A logical calculus of the ideas immanent in nervous activity. *Bulletin of Mathematical Biophysics* 5. https://pdfs.semanticscholar.org/5272/8a99829792c3272043842455f3a110e841b1.pdf.
53. Mikolov, Tomas, Kai Chen, Greg Corrado, and Jeffrey Dean. 2013. Efficient estimation of word representations in vector space. https://arxiv.org/abs/1301.3781.
54. Mikolov, Tomas, Ilya Sutskever, Kai Chen, Greg Corrado, and Jeffrey Dean. 2013. Distributed representations of words and phrases and their compositionality. https://papers.nips.cc/paper/5021-distributed-representations-of-words-and-phrases-and-their-compositionality.pdf.
55. Minsky, Marvin, and Seymour Papert. 1969. *Perceptrons: An introduction to computational geometry*. Cambridge: MIT Press.
56. Moody, Chris. Stop using word2vec. https://multithreaded.stitchfix.com/blog/2017/10/18/stop-using-word2vec/.
57. Moradi, Mehdi, and Mohammad Zulkernine. A neural network based system for intrusion detection and classification of attacks. https://pdfs.semanticscholar.org/cbf2/57a638aff38eae99bf88d8e22f150d9d8c47.pdf.
58. Narwekar, Abhishek, and Anusri Pampari. 2016. Recurrent neural network architectures. http://slazebni.cs.illinois.edu/spring17/lec20_rnn.pdf.
59. Neubig, Graham. 2018. NLP programming tutorial 8 — Recurrent neural nets. http://www.phontron.com/slides/nlp-programming-en-08-rnn.pdf.
60. Ng , Andrew Y, and Michael I. Jordan. 2001. On discriminative vs. generative classifiers: A comparison of logistic regression and naïve Bayes. In *Proceedings of the 14th international conference on neural information processing systems: natural and synthetic*, NIPS'01, 841–848.
61. Olah, Christopher. 2014. Understanding convolutions. http://colah.github.io/posts/2014-07-Understanding-Convolutions/.
62. Paul, Sunhera, Fabio Di, and Troia Mark Stamp. Word embedding techniques for malware evolution detection. submitted for publication.
63. Philipp, George, Dawn Song, and Jaime G. Carbonell. 2018. The exploding gradient problem demystified — Definition, prevalence, impact, origin, tradeoffs, and solutions. https://arxiv.org/pdf/1712.05577.pdf.
64. Popov, I. 2017. Malware detection using machine learning based on Word2Vec embeddings of machine code instructions. In *2017 Siberian symposium on data science and engineering*, SSDSE, 1–4.
65. Rabiner, Lawrence R. 1989. A tutorial on hidden Markov models and selected applications in speech recognition. *Proceedings of the IEEE* 77(2): 257–286. https://www.cs.sjsu.edu/~stamp/RUA/Rabiner.pdf.

66. Rezende, E, G. Ruppert, T. Carvalho, F. Ramos, and P. de Geus. 2017. Malicious software classification using transfer learning of resnet-50 deep neural network. In *16th IEEE international conference on machine learning and applications*, ICMLA 2017, 1011–1014.
67. Rigaki, Maria, and Sebastian Garcia. 2018. Bringing a GAN to a knife-fight: Adapting malware communication to avoid detection. https://mariarigaki.github.io/publication/gan-knife-fight/.
68. Rosenblatt, Frank. 1961. Principles of neurodynamics: Perceptrons and the theory of brain mechanisms. http://www.dtic.mil/dtic/tr/fulltext/u2/256582.pdf.
69. Ruderman, Avraham, Neil C. Rabinowitz, Ari S. Morcos, and Daniel Zoran. 2018. Pooling is neither necessary nor sufficient for appropriate deformation stability in CNNs. https://arxiv.org/abs/1804.04438.
70. Rumelhart, David, Geoffrey Hinton, and Ronald Williams. 1986. Learning representations by back-propagating errors. *Nature* 323 (9)
71. Shamshirband, Shahab, and Anthony T. Chronopoulos. 2019. A new malware detection system using a high performance-elm method. In *Proceedings of the 23rd international database applications & engineering symposium*, IDEAS'19, 33:1–33:10.
72. Sharmin, Tazmina, Fabio Di Troia, Katerina Potika, and Mark Stamp. 2020. Convolutional neural networks for image spam detection. *Information Security Journal: A Global Perspective* 29 (3): 103–117.
73. Shen, Yun, and Gianluca Stringhini. 2019. Attack2vec: Leveraging temporal word embeddings to understand the evolution of cyberattacks. https://seclab.bu.edu/people/gianluca/papers/attack2vec-usenix2019.pdf.
74. Singh, Tanuvir, Fabio Di Troia, Corrado Aaron Visaggio, Thomas H. Austin, and Mark Stamp. 2016. Support vector machines and malware detection. *Journal of Computer Virology and Hacking Techniques* 12 (4): 203–212.
75. Springenber, Jost Tobias, Alexey Dosovitskiy, Thomas Brox, and Martin Riedmiller. 2014. Striving for simplicity: The all convolutional net. https://arxiv.org/abs/1412.6806.
76. Spruston, Nelson. 2019. Pyramidal neurons: Dendritic structure and synaptic integration. *Nature Reviews Neuroscience* 9: 206–221. https://www.nature.com/articles/nrn2286.
77. Stamp, Mark. 2004. A revealing introduction to hidden Markov models. https://www.cs.sjsu.edu/~stamp/RUA/HMM.pdf.
78. Stamp, Mark. 2018. A survey of machine learning algorithms and their application in information security. In *Guide to vulnerability analysis for computer networks and systems: an artificial intelligence approach*, eds. Parkinson, Simon, Andrew Crampton, and Richard Hill, chapter 2, 33–55. Berlin: Springer.
79. Stamp, Mark. 2019. Alphabet soup of deep learning topics. https://www.cs.sjsu.edu/~stamp/RUA/alpha.pdf.
80. Stamp, Mark. 2019. Deep thoughts on deep learning. https://www.cs.sjsu.edu/~stamp/RUA/ann.pdf.
81. Veit, Andreas, Michael Wilber, and Serge Belongie. Residual networks behave like ensembles of relatively shallow networks. https://arxiv.org/pdf/1605.06431.pdf.
82. Wallis, Charles. 2017. History of the perceptron. https://web.csulb.edu/~cwallis/artificialn/History.htm.
83. Wu, Peilun, Hui Guo, and Nour Moustafa. 2020. Pelican: A deep residual network for network intrusion detection. https://arxiv.org/pdf/2001.08523.pdf.
84. Wu, Yonghui, Mike Schuster, Zhifeng Chen, Quoc V. Le, Mohammad Norouzi, Wolfgang Macherey, Maxim Krikun, Yuan Cao, Qin Gao, Klaus Macherey, Jeff Klingner, Apurva Shah, Melvin Johnson, Xiaobing Liu, Łukasz Kaiser, Stephan Gouws, Yoshikiyo Kato, Taku Kudo, Hideto Kazawa, Keith Stevens, George Kurian, Nishant Patil, Wei Wang, Cliff Young, Jason Smith, Jason Riesa, Alex Rudnick, Oriol Vinyals, Greg Corrado, Macduff Hughes, and Jeffrey Dean. 2016. Google's neural machine translation system: Bridging the gap between human and machine translation. https://arxiv.org/abs/1609.08144.
85. Xu, Ke, Yingjiu Li, Robert H. Deng, and Kai Chen. 2018. Deeprefiner: Multi-layer android malware detection system applying deep neural networks. In *2018 IEEE European symposium on security and privacy*, Euro SP, 473–487.

86. Xue, Di, Jingmei Li, Tu Lv, Weifei Wu, and JiaXiang Wang. 2019. Malware classification using probability scoring and machine learning. *IEEE Access*, 91641–91656.
87. Yagcioglu, Semih, Mehmet Saygin Seyfioglu, Begum Citamak, Batuhan Bardak, Seren Guldamlasioglu, Azmi Yuksel, and Emin Islam Tatli. 2019. Detecting cybersecurity events from noisy short text. https://arxiv.org/abs/1904.05054.
88. Yajamanam, Sravani, Vikash Raja Samuel Selvin, Fabio Di Troia, and Mark Stamp. 2018. Deep learning versus gist descriptors for image-based malware classification. In *Proceedings of the 4th international conference on information systems security and privacy*, ICISSP 2018, eds. Mori, Paolo, Steven Furnell, and Olivier Camp, 553–561.
89. Zeiler, Matthew D, and Rob Fergus. 2014. Visualizing and understanding convolutional networks. https://cs.nyu.edu/~fergus/papers/zeilerECCV2014.pdf.
90. Zhang, Wei, Huan Ren, Qingshan Jiang, and Kai Zhang. 2015. Exploring feature extraction and ELM in malware detection for Android devices. *Advances in Neural Networks, ISNN*, eds. Hu, Xiaolin, Yousheng Xia, Yunong Zhang, and Dongbin Zhao, 489–498.

Malware Detection with Sequence-Based Machine Learning and Deep Learning

William B. Andreopoulos

Abstract In this chapter, we review sequence-based machine learning methods that are used for malware detection and classification. We start by reviewing the datatypes extracted from code: static features and dynamic traces of program execution. We review recent research that applies machine learning on opcode and API call sequences, call graphs, system calls, registry changes, information flow traces, as well as hybrid and raw data, to detect and classify malware. With a focus on metamorphic malware, we discuss Hidden Markov Models (HMMs) and Long Short-Term Memory (LSTM) networks. We describe their input formats, such as one-hot encoding and vector embeddings, the architecture of the machine learning models, the training process, and the output formats. Finally, we discuss commercial and open-source tools that are used for data extraction from software.

1 Introduction

Malware is software that is designed to disrupt or damage computer systems. According to Symantec, more than 669 million new malware variants were detected in 2018, which was an increase of more than 80% from 2017, with such trends continuing into 2019 and 2020 [41, 42]. Every day, there are at least 350,000 instances of new malware being created and detected. Additionally, 81% of all ransomware infections target businesses and organizations, making malware infections very costly. Malware and web-based attacks are the two most costly attack types—companies spent an average of US $2.4 million in defense in 2018–2019 [30, 48]. Clearly, malware detection is a critical task in computer security.

Malware detection can be based on static or dynamic software features, or a combination of those, or raw data. Static features are extracted from static files, while dynamic features are extracted during code execution or emulation. Static approaches often use features such as calls to external libraries, strings, and byte

W. B. Andreopoulos (✉)
Department of Computer Science, San Jose State University, San Jose, USA
e-mail: william.andreopoulos@sjsu.edu

M. Stamp et al. (eds.), *Malware Analysis Using Artificial Intelligence and Deep Learning*, https://doi.org/10.1007/978-3-030-62582-5_2

sequences for classification. Other static approaches extract higher level information from binaries, such as sequences of API calls or opcode information.

Signature-based detection that uses static data is widely used within commercial antivirus software. While this method is used widely in commercial antivirus tools and is capable of detecting specific malware families efficiently, it fails to detect new malware. Therefore, modern antivirus tools go beyond static signature-based detection and can detect unknown malwares more accurately using dynamic data. Using dynamic data, it is able to detect unknown malwares according to their behavior [10, 16].

Sequential features extracted from malware source code analysis have been used for the classification of malware with deep learning approaches. Sequences used in malware analysis have been used for LSTMs, as well as HMMs. Both features and sequences can be extracted by performing either static or dynamic analysis. There are many tools that can extract either or both data types. We provide an overview of these tools later in this chapter.

This chapter starts with describing the distinction between static and dynamic data for malware detection. Then we give an overview of the data type representation, recent malware detection methods, and tools for extraction of static and dynamic data. Finally, we describe how sequence-based deep learning algorithms can be used for malware detection.

2 Data Extraction

2.1 Static Data

Machine learning models for malware detection and classification can be trained on static features or attributes that are extracted from executable files [13]. Static analysis involves analyzing the malware software or code without actually executing the program. Static analysis involves disassembling software and representing some of its attributes as features for input to a machine learning tool [16]. Approaches to perform static analysis usually employ the executable binary file, while others use the source code file. Examples of static features include opcodes, API calls, control flow graphs, and many others. Specific features for training the machine learning model include extracting opcode sequences after disassembling the binary file, or extracting the control flow graph from the assembly file, extracting API calls from the binary, as well as extracting byte code sequences from the binary executable file [2, 9, 21, 28, 38, 44] (Figs. 1 and 2).

```
mov     ax,0000h
add     [0CB12h],cl
push    ds
add     [si+0D09h],dh
and     [bx+si+4C01h],di
push    sp
push    7219h
and     [bx+si+71h],dh
```

Fig. 1 Static opcode sequence example. Consider an example based on the assembly code snippet is shown above; the following sequences of length 2, also named bi-grams or 2 grams, can be generated: s1 = (mov, add), s2 = (add, push), s3 = (push, add), s4 = (add, and), s5 = (and, push), s6 = (push, push), and s7 = (push, and). Because most of the common operations that can be used for malicious purposes require more than one machine code operation, this example uses sequences of two opcodes, instead of individual opcodes [35]

2.2 Dynamic Data

Dynamic Analysis is the analysis of a software's behavior that is performed while executing the program. Some of the data that can be obtained through dynamic analysis are API calls, system calls, instruction traces, taint analysis, registry changes, memory writes, and information flow tracking. Dynamic analysis uses the tasks

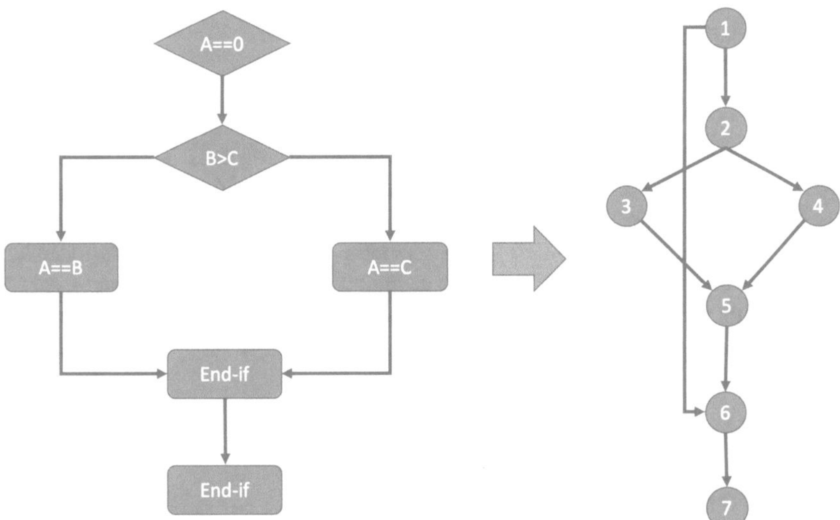

Fig. 2 Static control flow graph (static): A Control Flow Graph (CFG) is the graphical representation of control flow or computation during the execution of programs or applications. Control flow graphs are mostly used in the static analysis, as they can accurately represent the flow inside of a program unit

performed by a program while it is being executed in a virtualized environment [16]. Dynamic analysis is known to provide more accurate malware behavior detection results, but the data can be time-consuming to extract. There are several techniques and approaches followed to perform dynamic analysis.

As an example, this is an API system call extracted dynamically [16]:

```
fork(); getpid(); ioctl(); read(); write(); wait(); exit();
```

In order to input this to a sequence-based neural network we can represent it in one-hot encoded format. One-hot encoding is created by replacing the ith system call with an n-vector of zeros and a '1' in the ith position.

2.3 Hybrid Analysis

Besides static and dynamic analysis techniques, another analysis technique is the hybrid analysis of malware, which combines the advantages of both static and dynamic analyses [34]. Hybrid analysis technique is the combination of static and dynamic analysis techniques. In hybrid analysis, static analysis is done before the execution of a program and then dynamic analysis is done selectively based on the information obtained from static analysis. Dynamic analysis can be tedious due to multi-path execution. Static analysis can be used to selectively choose the path of execution for dynamic analysis. Often hybrid analysis results in increased accuracy and efficiency [34].

2.4 Alternative Approaches That Use Raw Data

In many cases the extraction of these sequences and features for LSTMs and HMMs can be costly, so approaches using raw bytes are preferred, if comparable accuracy can be obtained. For example, byte n-grams have been successfully used as features [43]. Also, it is possible to treat executable files as images and apply image analysis techniques. Images of malware executables have previously been used for classification using convolutional neural networks [25].

2.5 Evaluation of Malware Detection Accuracy

For API call sequences and opcode sequences, three test case scenarios have been implemented in previous work:

- The first case uses dynamic analysis data alone for both training and testing.
- The second case uses dynamic analysis data for training and static analysis data for testing.

- The third case uses static analysis data alone for both training and testing.

It is widely known that dynamic analysis data is more accurate and is therefore popular for training purposes. On the other hand, using dynamic data for training and testing is less efficient due to the complexity of extracting data in a dynamic fashion from software.

3 Recent Research Examples

In this section, we give an overview of recent research on malware detection and classification.

In [51], a system is proposed that extracts the API sequences from a Portable Executable (PE) file format. Then Objective-Oriented Association (OOA) based mining is done for malware classification. The system parses PE files and generates OOA rules efficiently for classification using FP Growth and a frequent-pattern tree. The system was tested on a large collection of PE files obtained from the anti-virus laboratory of KingSoft Corporation to compare various malware detection approaches. The accuracy and efficiency of the OOA system outperformed anti-virus software, such as Norton AntiVirus and McAfee VirusScan, as well as previous data mining-based detection systems that employed Naive Bayes, Support Vector Machine (SVM), and Decision Tree techniques.

In [31], malware is analyzed by abstracting the frequent itemsets in API call sequences. The authors focused on the usage of frequent messages in API call sequences, They hypothesized that frequent itemsets consisting of API names and/or API arguments could be valuable for identifying the behavior of malware. The authors clustered a dataset of malware binaries, demonstrating that using the frequent itemsets of API call sequences can achieve high precision for malware clustering while reducing the computation time.

In [30], a kernel object behavioral graph is created and graph isomorphism techniques and weighted common graph technique are used to calculate the hotpath for each malware family. And the unknown malware is then classified into whichever malware family has similar hotpaths.

In [12], dynamic instruction sequences are logged and are converted to abstract assembly blocks. Data mining algorithms are used to build a classification model using feature vectors extracted from the above data. For malware detection, the same method is used and scored against the classification model.

In [3], the authors propose a malware detection technique that uses instruction trace logs of executables collected dynamically. These traces are then constructed as graphs. The instructions are considered as nodes and the data from the instruction trace is used to calculate the transition probabilities. Then a similarity matrix is generated between the constructed graphs using different graph kernels. Finally, the constructed similarity matrix is input to an SVM for classification.

In [47], the authors presented a malware detection technique using dynamic analysis where fine-grained models are built to capture the behavior of malware using system calls information. Then they use a scanner to match the activity of any unknown program against these models to classify them as either benign or malware. The behavior models are represented in the form of graphs. The vertices denote the system calls and the edges denote the dependency between the calls where the input of one system call (vertex) depends on the output of another system call (vertex).

In [1], the authors propose a run-time monitoring based malware detection tool that extracts statistical features from malware using spatio-temporal information from API call logs. The spatial information is the arguments and return values of the API calls and temporal information is the sequence of the API calls. This information is used to build formal models that are fed to standard machine learning algorithms for malware detection.

From all the research given above, it is evident that dynamic analysis is a good source of information for malware behavior. Even though producing dynamic data incurs an execution overhead, a more accurate model can be obtained from dynamic analysis than using static analysis alone.

3.1 Hybrid Analysis

Hybrid analysis tools are developed for using the accuracy benefit that dynamic analysis offers and the static analysis' advantage of time complexity.

HDM Analyzer uses both static analysis and dynamic analysis techniques in the training phase and performs only static analysis in the testing phase. By combining static and dynamic analyses, HDM Analyzer achieved a better accuracy and time complexity than static and dynamic analysis methods alone [18]. The authors extracted a sequence of API calls for dynamic analysis, which is one of the most effective features for describing the behavior of a program.

In [9], the authors propose a framework for classification of malware using both static and dynamic analysis. They define the features or characteristics of malware as Mal-DNA (Malware DNA). Mal-DNA combines static, dynamic, and hybrid characteristics. Besides extracting the static features of malware, they extract dynamic data with debugging based behavior monitoring. Then they classify malware using machine learning algorithms.

In a slightly modified version, [19] apply n-grams method to the API calls extracted and use the above as a feature set for malware classification. Application Programming Interface (API) call sequences are commonly used features in intelligent malware detection systems. An API call sequence captures the activities of a program and, hence, it is useful data for mining of malicious behavior. Different order of each API call in a sequence may mean a different behavior model. Therefore, the order of API calls is an important issue to analyze malware behavior. The paper proposes a feature extraction approach for modeling malware behavior that extracts

API call sequences by dynamic analysis of executing programs. The novelty of the approach is utilizing n-grams to preserve the order of API calls.

In [17], a set of program API calls is extracted and combined with the control flow graphs (CFGs) to obtain a new representation model called API-CFG, where API calls form the edges in the control flow graph. This API-CFG is trained by a learning model and used as a classifier during the testing stage for malware detection. The behavior of a program is represented by a set of API calls. Therefore, a classifier can be employed to construct a learning model with a set of programs' API calls. Finally, an intelligent malware detection system is developed to detect unknown malwares automatically. This approach is capable of classifying benign and malicious code with high accuracy. The results show a statistically significant improvement over n-grams-based detection method.

In [23], dynamic malware detection is done using registers values set analysis. In this paper, a novel method is proposed based on similarities of binaries behaviors. At first, run-time behavior of the binary files is found and logged in a controlled environment tool. The approach assumes that the behavior of each binary can be represented by the values of memory contents in its run-time. That is, values stored in different registers while the malware is running in the controlled environment can be a distinguishing factor to discriminate it from those of benign programs. Then, the register values for each Application Programming Interface (API) call are extracted before and after API is invoked. After that, the changes of registers values throughout the executable file are traced to create a vector for each of the values of EAX, EBX, EDX, EDI, ESI, and EBP registers.

In [32], a new runtime kernel memory mapping method called allocation-driven mapping is introduced, which identifies dynamic kernel objects, including their types and lifetimes. The method works by capturing kernel object allocation and deallocation events. A benefit of kernel-based malware analysis includes providing a temporal view of kernel objects by performing a temporal analysis of kernel execution. Their system includes a temporal malware behavior monitor that tracks malware behavior by the manipulation of dynamic kernel objects. Allocation-driven mapping is shown to reliably analyze malware behavior by guiding the analysis only to the events relevant to a malware attack.

In [36], the API call sequences and assembly code are combined and a similarity-based matrix is produced that determines whether a portion of code has traces of a particular malware. This research showed good results by using API call sequences and Opcode sequences to give a good description of the behavior of a malware.

An orthogonal approach to the monitoring of function calls during the execution of a program, is the analysis of how the program processes data. The goal of information flow tracking is to propagate and track "taint-labeled" data throughout the system, while a program manipulating this data is executed. The data that should be tracked is specifically marked (tainted) with a corresponding label. Assignment statements, for example, usually propagate the taint-label of the source operand to the target [16].

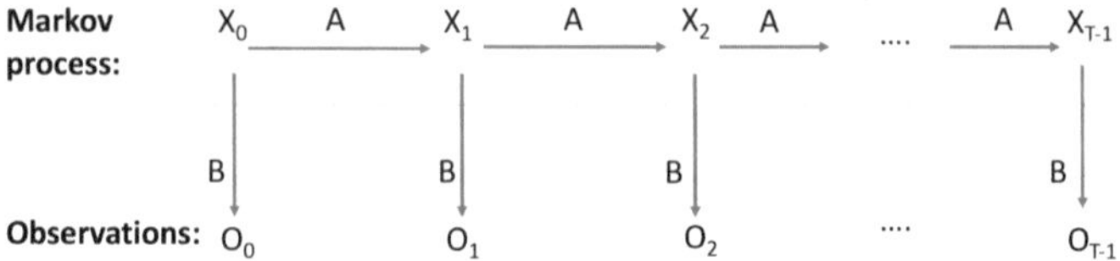

Fig. 3 Illustration of a generic Hidden Markov Model [40]

4 HMM Architecture

The HMM is based on augmenting the Markov chain. A Markov chain is a model that represents the probabilities of sequences of random variables, called states, each of which can take on values from some set. These sets can be words, or tags, or symbols representing anything, like the alphabet. In the case of malware, the sets of values may be opcodes or API calls.

A Markov chain makes an assumption that for predicting the future in a sequence, the current state is all that matters. The states before the current state only impact the future via the current state. For instance, to predict the next word you could consider the current word, but you should not examine previously seen words. Similarly, to predict if the states in a sequence produced by a piece of software constitute malware, you could use the states immediately preceding a state, but not states seen in the distant past.

HMMs have been used for malware detection. A Hidden Markov Model (HMM) is a machine learning model to represent probability distributions over a sequence of observations [22]. The HMM satisfies the markov property, i.e., the current state t is dependent only on $t - 1$ and is independent of all states prior to $t - 1$. A HMM is motivated by the idea of considering both observed events (such as words that we see in the input) and hidden events (such as part-of-speech tags) that we think of as causal factors in our probabilistic model. An HMM is specified by the following components:

- $Q = q_1 q_2 \ldots q_N$ a set of N states
- $O = o_1 o_2 \ldots o_T$ a sequence of T observations, each one drawn from a vocabulary $V = v_1, v_2, \ldots, v_V$
- $A = a_{11} \ldots a_{ij} \ldots a_{NN}$ a transition probability matrix A, each a_{ij} representing the probability of moving from state i to state j, s.t. $P_{j=1..N} a_{ij} = 1 \forall i$
- $B = b_i(o_j)$ a sequence of observation likelihoods, also called emission probabilities, each expressing the probability of an observation o_j being generated from a state i
- $\pi = \pi_1, \pi_2, \ldots, \pi_N$ an initial probability distribution over states. π_i is the probability that the Markov chain will start in state i. Some states j may have $\pi_j = 0$, meaning that they cannot be initial states.

4.1 Training for Malware Detection

The basic steps followed for performing malware detection using HMM are as follows. First, we select the observation data that the model should be trained for. In this case, the observed sequences represent software states that may originate from malware data. The observed sequences can be API calls sequence or opcode sequences. A model is trained with the above-observed sequences. After convergence, we get an accurate model that best fits the observed sequences. Next, we score a set of malware and benign files against the trained model. If the scores are higher than a predetermined threshold, the scores above the threshold can be classified as files from the malware family and ones that are below the threshold can be classified as files from the benign family [5] (Fig. 3).

4.2 Metamorphic Malware Detection

The use of static data is insufficient when dealing with the advanced malware obfuscation techniques such as code relocation, mutation, and polymorphism [11]. In [37], an opcode-based software similarity measure was developed, showing excellent results for metamorphic malware detection and classification. In [14], the metamorphic malware detection is done based on function call graph analysis. In [29], multiple sequence alignment algorithms from bioinformatics lead to viral code signatures that generalize successfully to previously known polymorphic variants of viruses.

The detection of metamorphic malware became more effective due to the application of Markov models and HMMs to malware detection. Profile Hidden Markov Models (PHMMs) are known for their success at detecting relations between DNA and protein sequences. When applied for malware detection it has been found that PHMMs can effectively detect metamorphic malware [40] and HMMs have also been successful in this regard. In [5], HMMs were used for malware classification. The HMM clustering results classify the malware samples into their appropriate families with good accuracy, providing a useful tool in malware analysis and classification.

5 LSTM Architecture

Long Short-Term Memory networks—usually just called "LSTMs"—are a special kind of Recurrent Neural Network (RNN), capable of learning long-term dependencies in sequences of events. They work well on a large variety of problems. Besides malware detection, they have also been widely used for sequence classification in other fields such as text mining and biology.

In this section, we analyze how LSTM sequence-based deep learning methods may be used for malware classification [24]. While static signature-based malware detection methods are quick, static code analysis can be vulnerable to code obfuscation techniques. LSTMs offer the benefit that they don't rely on static analysis and can analyze a short snapshot of the runtime behavior. Behavioral data collected during file execution is more difficult to obfuscate but takes a long time to capture (typically up to 5 min). This often means that the malicious payload has likely already been delivered by the time it is detected.

In [39], LSTMs were applied to microarchitectural event traces captured through on-chip hardware performance counter (HPC) registers. The proposed LSTM approach achieved up to 11% higher detection accuracy compared to other sequence-based classification, such as HMM-based approaches in detecting obfuscated malware.

References [27, 46, 49] used sequences of API system calls for training an LSTM. In [27], they trained the LSTM model to learn from the most informative of sequences from the API-dataset based on their relative ranking as determined by Term Frequency–Inverse Document Frequency (TF–IDF) recommended features. They were able to achieve accuracy as high as 92% in detecting malware and benign code from an unknown test API-call sequence.

Reference [8] extracted 3-grams of opcode sequences and API call sequences. They then used attention mechanisms to identify API system calls that are more important than others for determining whether a file is malicious. They report this approach gave an accuracy that was 12% and 5% higher than conventional malware detection models using convolutional neural networks and skip-connected LSTM-based detection models, respectively.

Reference [39] used dynamic data in the form of microarchitectural event traces captured through on-chip hardware performance counter (HPC) registers. They combined this with localized feature extraction from image binaries corresponding to the application binaries. Using this advanced approach, an accuracy of 94% and nearly 90% is achieved in detecting normal and metamorphic malware created through code relocation obfuscation technique.

An LSTM variation is to use coupled forget and input gates. This variation on the LSTM is called the Gated Recurrent Unit, or GRU [7]. It combines the forget and input gates into a single "update gate." It also merges the cell state and hidden state. Instead of separately deciding what to forget and what to add new information to, GRUs make those decisions together. GRUs only forget when they are going to input something in its place. GRUs only input new values to the state when they forget something older. GRUs have also been used for malware detection [6].

5.1 LSTM Training

Input format: A single training data element consists of the label and an input word: xseq—a subsequence of a fixed size sampled randomly from the full original sequence. Reference [45] used one-hot encoding of logged API call sequences reflecting process behavior. One-hot encoding is a method for transforming categorical data to numerical data by representing the ith categorical value from the universe as a numerical vector of zeros and a '1' in the ith position.

Reference [50] compared opcode embedding against one-hot encoding methods. One-hot encoding is simple to use and with the rather small number of Android opcodes, the sparseness of one-hot encoding does not cause a negative impact on efficiency. Another method is learning opcode embedding from data samples. Using opcode embedding achieves better malware detection results than one-hot encoding since opcode embedding captures the opcode semantic information better compared to one-hot encoding. The embedding idea comes from word vector learning in NLP, such as word2vec. Opcode embedding helps to learn the semantic information of opcode sequences and mine for malicious behaviors [50].

Reference [26] also input word embeddings derived from opcode sequences to LSTMs for malware detection and malware classification. Their evaluation results showed their proposed method can achieve an average AUC of 0.99 and an average AUC of 0.987, respectively.

Generally, an LSTM or RNN takes an input sequence of a fixed size for training or classification. An input sequence that is of a shorter size than the input layer of the LSTM can be padded with special characters [6]. Otherwise, it can either be trimmed or subsampled to derive samples of the desired size.

Reference [33] used trimming of sequences to detect whether or not an executable is malicious based on a short snapshot of behavioral data. They collected ten numerical machine activity data metrics (e.g., CPU and memory usage) as feature inputs, which are continuous numeric values, allowing for a large number of different machine states to be represented in a small vector of size 10. They used an ensemble of RNNs to build an RNN model able to predict whether an executable is malicious or benign within the first 5 s of execution with 94% accuracy. This was one of the first works to predict malicious code during execution. Previous dynamic analysis research collected data for around 5 min per sample [33].

In some works training the LSTM involved a subsampling of sequences from the original sequence. Specifically, training on a set of labeled sequences occurs by subsampling a number (say, 50–100) of short fixed-length sequence samples. Then training happens on each sample. For instance, [4] used samples extracted from Windows files. They trained a multiclass classification RNN, more specifically a LSTM on the dataset. This model for analyzing unstructured data was tested on unseen programs and the accuracy reached 67.60%, including six classes with five different types of malware.

The subsampling is chosen to ensure a fair representation of smaller and larger sequences. The number (m) of fixed-length subsequences sampled from each executable code sample should be proportional to the logarithm or square root of the sequence length, such that longer sequences will contribute more samples, but do not overwhelm the training. Each sample is a different subsequence xseq associated with the originating sequence. The sequence xseq can be one-hot encoded if the number of possible values is small.

Output format: The output from the final neuron is in a range from 0 to 1. This value is used to discriminate between a positive or negative classification value; in

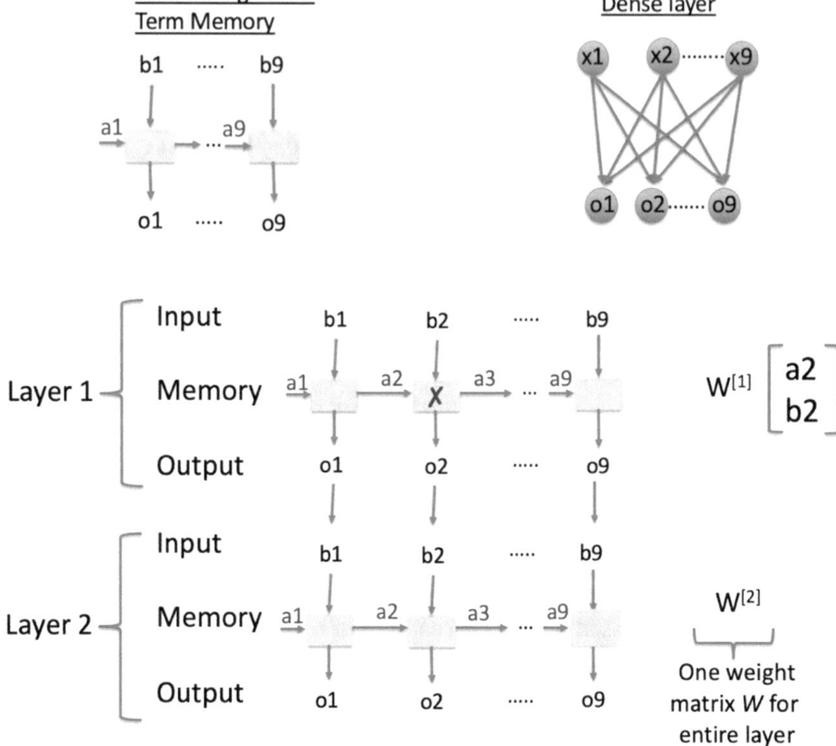

Fig. 4 An LSTM and a layered LSTM architecture

other words, if it is predicted to be malware or not malware. A neural network is a function

$$F(x_{seq}, x_f|\theta) = y. \tag{1}$$

that accepts sequences of points x_{seq} and the feature vector x_f. The function F also depends implicitly on the DL model parameters θ, which are determined during the training process. The output of the neural network, score y, is computed using the softmax function, which ensures that y satisfies $y \in [0, 1]$. By convention, the higher the score is for the sequence, the more likely the sequence is to be a true malware (Fig. 4).

Gradient Descent and Backpropagation

The training cycles are typically repeated for a specified number of epochs (such as 30 training cycles): The loss (error) is computed as target—actual output. E is the loss overall, computed by averaging loss over all instances of the training set.

The gradient of the loss is computed at the position we end up at. For neuron o_j at layer j:

$$\frac{\partial E}{\partial w_{ij}} = \frac{\partial E}{\partial o_j} \frac{\partial o_j}{\partial \text{net}_j} \frac{\partial \text{net}_j}{\partial w_{ij}} \qquad o_j = \varphi(\text{net}_j) = \varphi\left(\sum_{k=1}^{n} w_{kj} o_k\right).$$

During gradient descent the weights are adjusted by the learning rate η:
∇w_{ij} = the product of $-\eta$ and the gradient of loss
∇w_{ij} changes w_{ij} such that E decreases in next epoch
Backpropagation then adjusts all weights from outer to the inner layers.

As hyperparameters, the model can be trained with a binary cross-entropy loss function and Adam optimizer. Usually, there are dropout layers to reduce overfitting.

Another possible architecture involves a hybrid of LSTM and a dense neural network. As input in this case, a single training data element consists of the label and two input words: x_{seq}—a subsequence of a fixed size sampled randomly from the full original sequence and x_f—a vector containing features extracted from the full sequence. The values of the feature vector x_f can be normalized to be bound within $[-1, 1]$. This architecture takes the sequences as input on one branch; and the features as matrices or vectors in another branch [24, 25] (Fig. 5).

6 Tools

6.1 IDA Pro

IDA Pro is a popular disassembler for generating assembly language source code from executables. It can also be used as a debugger. IDA Pro can be used to generate .asm files from which opcodes and windows API calls can be extracted. Also, IDA Pro is useful for collecting the instruction trace logs of executables.

6.2 OllyDbg

OllyDbg is a 32-bit disassembler and debugger, which is second best to IDA Pro. OllyDbg has limited features compared to IDA Pro.

6.3 Ether

Ether is an open-source tool for malware analysis that has been developed via hardware virtualization extensions and resides completely outside of the target OS. This disables the detection of guest software components. Many recent viruses can detect a debugger or a virtual environment during execution. Ether malware analysis tool overcomes this problem and hence provides a benefit as a tool for malware detection [15].

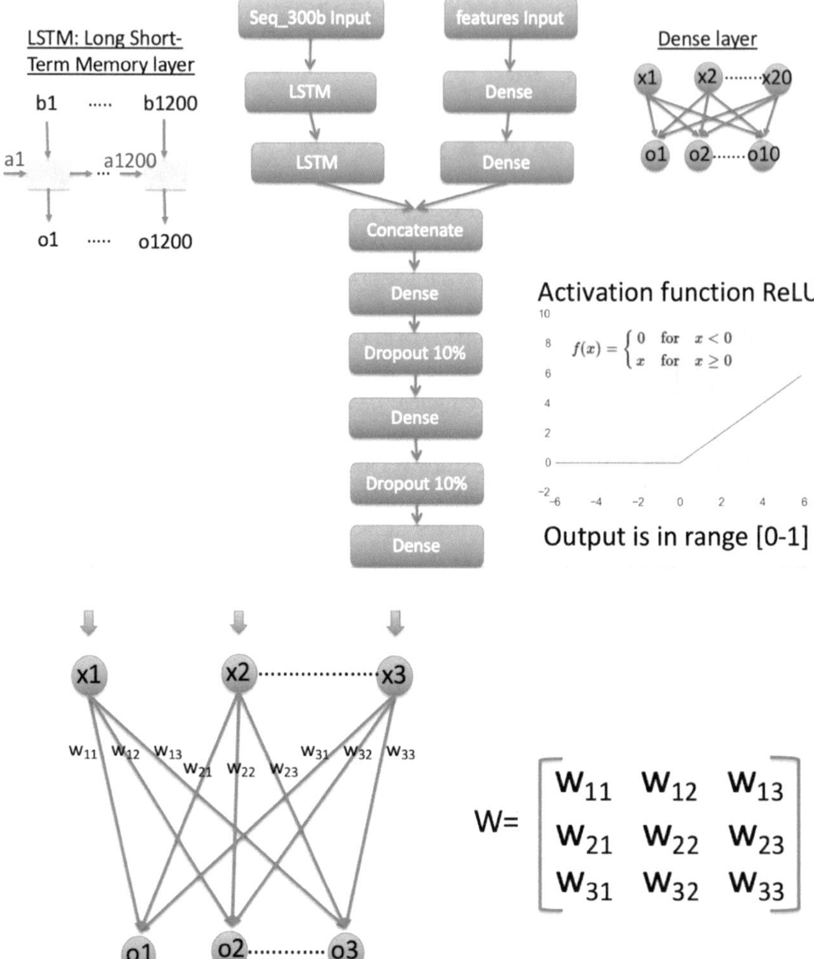

Fig. 5 A hybrid LSTM and dense architecture. A neuron in the dense layer has all-to-all connectivity and connections are described as a two-dimensional numerical matrix. Dropout layers are used to reduce overfitting

6.4 API Logger

API Logger is a tool that logs all API calls that meet the restrictions of the inclusion and exclusion lists. The inclusion list specifies which libraries need to be included and exclusion list specifies which libraries or functions can be ignored [20].

6.5 WinAPIOverride

WinAPIOverride is an advanced API monitoring software. Its main distinction is the ability to manually extract API calls by deciding the flow of the program during execution. In that sense, it fills the gap between classic API monitoring software and debuggers.

6.6 API Monitor

API Monitor is a software tool that also helps in monitoring and controlling API calls made by applications of processes. This tool needs to be run inside a virtual machine to analyze a malware and cannot be run in a sandboxed environment (www.rohitab.com/apimonitor. Accessed 07/14/2020).

6.7 BSA

Buster Sandbox Analyzer (BSA) is a tool that can decide if processes exhibit malicious activities based on dynamic analysis. In addition to analyzing the behavior of running processes, BSA keeps track of the changes made to the system, such as registry changes. The tool runs inside a Sandbox that protects the system from getting infected while executing the malware. BSA can generate API trace calls for win32 executables. Other tools similar to BSA that are useful for tracing in sandboxed environments are CWSandbox and Norman Sandbox (http://bsa.isoftware.nl. Accessed 07/14/2020).

7 Conclusion

In this chapter, we have compared data representations for static and dynamic datasets that can be used to classify and detect malware. The extracted data is input to a sequence-based machine learning or deep learning tool. Sequence-based machine learning provides a non-signature-based malware detection method that can effectively classify new and unknown types of malware, as well as metamorphic malware. Both static and dynamic datasets contain data types that are sequence-based. Using dynamic data offers a benefit over static data for detecting obfuscated code. The data can be trained upon and classified by sequence-based machine learning tools, such as HMMs and LSTMs. The sequence-based approaches are in contrast to training upon and classifying using feature-based machine learning tools, such as dense or convolutional neural networks, which often employ images or raw data from malware.

References

1. Ahmed, Faraz, Haider Hameed, M. Zubair Shafiq, and Muddassar Farooq. 2009. *Using spatio-temporal information in API calls with machine learning algorithms for malware detection*, 55. New York City: ACM Press.
2. Alqurashi, Saja, and Omar Batarfi. 2016. A comparison of malware detection techniques based on hidden Markov model. *Journal of Information Security* 07 (03): 215–223.
3. Anderson, Blake, Daniel Quist, Joshua Neil, Curtis Storlie, and Terran Lane. 2011. Graph-based malware detection using dynamic analysis. *Journal in Computer Virology* 7 (4): 247–258.
4. Andrade, Eduardo de O, José Viterbo, Cristina N. Vasconcelos, Joris Guérin, and Flavia Cristina Bernardini. 2019. A model based on lstm neural networks to identify five different types of malware. *Procedia Computer Science* 159: 182–191.
5. Annachhatre, Chinmayee, Thomas H. Austin, and Mark Stamp. 2015. Hidden Markov models for malware classification. *Journal of Computer Virology and Hacking Techniques* 11 (2): 59–73.
6. Athiwaratkun, B, and J. W. Stokes. 2017. Malware classification with lstm and gru language models and a character-level cnn. In *2017 IEEE international conference on acoustics, speech and signal processing (ICASSP)*, 2482–2486.
7. Cho, Kyunghyun, Bart van Merriënboer, Caglar Gulcehre, Dzmitry Bahdanau, Fethi Bougares, Holger Schwenk, and Yoshua Bengio. 2014. Learning phrase representations using RNN encoder–decoder for statistical machine translation. In *Proceedings of the 2014 conference on empirical methods in natural language processing (EMNLP)*, 1724–1734, Doha, Qatar. Association for Computational Linguistics.
8. Choi, Sunoh, Jangseong Bae, Changki Lee, Youngsoo Kim, and Jonghyun Kim. 2020. Attention-based automated feature extraction for malware analysis. *Sensors* 20 (10): 2893.
9. Choi, Y.H, B.J. Han, B.C. Bae, H.G. Oh, and K.W. Sohn. 2012. Toward extracting malware features for classification using static and dynamic analysis. In *IEEE conference publication.*
10. Christodorescu, M, S Jha, S A Seshia, D Song, and R E Bryant. 2005. *Semantics-aware malware detection*, 32–46, IEEE.
11. Christodorescu , Mihai, and Somesh Jha. 2003. Static analysis of executables to detect malicious patterns. In *Proceedings of the 12th conference on USENIX security symposium - volume 12*, SSYM'03, 12. USA: USENIX Association.
12. Dai, Jianyong, Ratan Guha, and Joohan Lee. 2009. Efficient virus detection using dynamic instruction sequences. *Güncel Pediatri* 4 (5).
13. Damodaran, Anusha, Fabio Di Troia, Corrado Aaron Visaggio, Thomas H. 2017. Austin, and Mark Stamp. A comparison of static, dynamic, and hybrid analysis for malware detection. *Journal of Computer Virology and Hacking Techniques* 13(1): 1–12.
14. Deshpande, Prasad. 2013. Metamorphic detection using function call graph analysis.
15. Dinaburg, Artem, Paul Royal, Monirul Sharif, and Wenke Lee. 2008. *Ether: Malware analysis via hardware virtualization extensions*, 51. New York City: ACM Press.
16. Egele, Manuel, Theodoor Scholte, Engin Kirda, and Christopher Kruegel. 2012. A survey on automated dynamic malware-analysis techniques and tools. *ACM Computing Surveys* 44 (2): 1–42.
17. Eskandari, Mojtaba, and Sattar Hashemi. 2012. A graph mining approach for detecting unknown malwares. *Journal of Visual Languages and Computing* 23 (3): 154–162.
18. Eskandari, Mojtaba, Zeinab Khorshidpour, and Sattar Hashemi. 2013. Hdm-analyser: A hybrid analysis approach based on data mining techniques for malware detection. *Journal of Computer Virology and Hacking Techniques* 9 (2): 77–93.
19. Eskandari, Mojtaba, Zeinab Khorshidpur, and Sattar Hashemi. 2012. *To incorporate sequential dynamic features in malware detection engines*, 46–52, IEEE.
20. Fasikhov, R. The api logger tool. http://blackninja2000.narod.ru/rus/api_logger.html. Accessed 14 July 2020.
21. Gandotra, Ekta, Divya Bansal, and Sanjeev Sofat. 2014. Malware analysis and classification: A survey. *Journal of Information Security* 05 (02): 56–64.

22. Ghahramani, Zoubin. 2001. An introduction to hidden Markov models and bayesian networks. *International Journal of Pattern Recognition and Artificial Intelligence* 15 (01): 9–42.
23. Ghiasi, Mahboobe, Ashkan Sami, and Zahra Salehi. 2012. *Dynamic malware detection using registers values set analysis*, 54–59, IEEE.
24. Hr, Sandeep. 2019. *Static analysis of android malware detection using deep learning*, 841–845, IEEE.
25. Jain, Mugdha, William Andreopoulos, and Mark Stamp. 2020. Convolutional neural networks and extreme learning machines for malware classification. *Journal of Computer Virology and Hacking Techniques*.
26. Lu, Renjie. 2019. Malware detection with lstm using opcode language. ArXiv:abs/1906.04593.
27. Mathew, J, and M A Ajay Kumara. 2020. API call based malware detection approach using recurrent neural network – LSTM. In *Intelligent systems design and applications*, *Advances in intelligent systems and computing*, eds. Abraham, Ajith, Aswani Kumar Cherukuri, Patricia Melin, and NiketaEditors Gandhi, vol. 940, 87–99. Springer International Publishing.
28. Moser, Andreas, Christopher Kruegel, and Engin Kirda. 2007. *Limits of static analysis for malware detection*, 421–430, IEEE.
29. Naidu, Vijay, Jacqueline Whalley, and Ajit Narayanan. 2017. Exploring the effects of gap-penalties in sequence-alignment approach to polymorphic virus detection. *Journal of Information Security* 08: 296–327.
30. Park, Younghee, Douglas S. Reeves, and Mark Stamp. 2013. Deriving common malware behavior through graph clustering. *Computers and Security* 39: 419–430.
31. Qiao, Yong, Yuexiang Yang, Lin Ji, and Jie He. 2013. *Analyzing malware by abstracting the frequent itemsets in API call sequences*, 265–270, IEEE.
32. Rhee, Junghwan, Ryan Riley, Xu Dongyan, and Xuxian Jiang. 2010. Kernel malware analysis with un-tampered and temporal views of dynamic kernel memory. In *Recent advances in intrusion detection*, *Lecture notes in computer science*, eds. Somesh Jha, Robin Sommer, and Christian Kreibich, vol. 6307, 178–197. Berlin: Springer.
33. Rhode, Matilda, Pete Burnap, and Kevin Jones. 2018. Early-stage malware prediction using recurrent neural networks. *Computers and Security* 77: 578–594.
34. Roundy, Kevin, A., and Barton P. Miller. 2010. Hybrid analysis and control of malware. In *Recent advances in intrusion detection*, *Lecture notes in computer science*, eds. Somesh Jha, Robin Sommer, Christian Kreibich, vol. 6307, 317–338. Berlin: Springer.
35. Runwal, Neha, Richard M. Low, and Mark Stamp. 2012. Opcode graph similarity and metamorphic detection. *Journal in Computer Virology* 8 (1–2): 37–52.
36. Shankarapani, Madhu K., Subbu Ramamoorthy, Ram S. Movva, and Srinivas Mukkamala. 2011. Malware detection using assembly and api call sequences. *Journal in Computer Virology* 7 (2): 107–119.
37. Shanmugam, Gayathri, Richard M. Low, and Mark Stamp. 2013. Simple substitution distance and metamorphic detection. *Journal of Computer Virology and Hacking Techniques* 9 (3): 159–170.
38. Shijo, P.V., and A. Salim. 2015. Integrated static and dynamic analysis for malware detection. *Procedia Computer Science* 46: 804–811.
39. Shukla, Sanket, Gaurav Kolhe, Sai Manoj P D, and Setareh Rafatirad. 2019. Stealthy malware detection using rnn-based automated localized feature extraction and classifier. In *2019 IEEE 31st international conference on tools with artificial intelligence (ICTAI)*, 590–597, IEEE.
40. Stamp, M. A revealing introduction to hidden Markov models. tutorial. www.cs.sjsu.edu/~stamp/RUA/HMM.pdf. Accessed 14 July 2020.
41. Symantec. Symantec Internet security threat report (ISTR) Volume 23. Technical report, Symantec, 03 2018.
42. Symantec. Symantec Internet security threat report (ISTR) Volume 24. Technical report, Symantec, 02 2019.
43. Tabish, S. Momina, M. Zubair Shafiq, and Muddassar Farooq. 2009. Malware detection using statistical analysis of byte-level file content. In *Proceedings of the ACM SIGKDD workshop on cybersecurity and intelligence informatics - CSI-KDD '09*, eds. Chen, Hsinchun, Marc Dacier,

Marie-Francine Moens, Gerhard Paass, and Christopher C. Yang, 23. New York City: ACM Press.

44. Le Thanh, Hieu. 2013. Analysis of malware families on android mobiles: detection characteristics recognizable by ordinary phone users and how to fix it. *Journal of Information Security* 04 (04): 213–224.
45. Tobiyama, S, Y. Yamaguchi, H. Shimada, T. Ikuse, and T. Yagi. 2016. Malware detection with deep neural network using process behavior. In *2016 IEEE 40th annual computer software and applications conference (COMPSAC)*, vol. 2, 577–582.
46. Vinayakumar, R, K P Soman, Prabaharan Poornachandran, and S Sachin Kumar. 2018. Detecting android malware using long short-term memory (lstm). *Journal of Intelligent and Fuzzy Systems* 34 (3): 1277–1288.
47. Wang, Xiaofeng. 2009. Effective and efficient malware detection at the end host. In *USENIX security symposium*, 351–366.
48. Wong, A. Symantec internet security threat report highlights. www.techarp.com/cybersecurity/2019-symantec-istr-highlights/. Accessed 14 July 2020.
49. Xiao, Xi, Shaofeng Zhang, Francesco Mercaldo, Guangwu Hu, and Arun Kumar Sangaiah. 2017. Android malware detection based on system call sequences and lstm. *Multimedia Tools and Applications* 78 (4): 1–21.
50. Yan, Jinpei, Yong Qi, and Qifan Rao. 2018. Lstm-based hierarchical denoising network for android malware detection. *Security and Communication Networks* 1–18: 2018.
51. Ye, Yanfang, Dingding Wang, Tao Li, Dongyi Ye, and Qingshan Jiang. 2008. An intelligent pe-malware detection system based on association mining. *Journal in Computer Virology* 4 (4): 323–334.

Review of the Malware Categorization in the Era of Changing Cybethreats Landscape: Common Approaches, Challenges and Future Needs

Andrii Shalaginov, Geir Olav Dyrkolbotn, and Mamoun Alazab

Abstract Malicious software threats have been known to Information Security professionals for over several decades since the dawn of computers. Developers of such software have been keeping up with technologies addressing known and unknown vulnerabilities for successful infection. With the growing amount of devices connected to the Internet, it has become apparent that the categorization of millions of malware samples is an emerging challenge. Malware labelling has become a significant challenge in the light of a large number of malware samples appearing daily. Many researchers and anti-virus vendors developed their unique naming methods that do not contribute to efficient incident response and remediation of the malware infections on a global scale. In this paper, first, we provide a view on the modern approaches to malware categorization concerning the needs of malware detection and analysis, specifically focusing on general *modus operandi* and automated analysis. Then, we review the State of the Art technical reports from the antivirus on the existing labelling initiatives and their usage by vendors. Finally, we give practical insight into future needs and current challenges of the naming schemes using ground truth knowledge. This review aims at bridging a knowledge gap between the existing labelling approaches, threats and malware functionality and problems related to large-scale malware classification.

A. Shalaginov (✉) · G. O. Dyrkolbotn
Norwegian University of Science and Technology, Trondheim, Norway
e-mail: andrii.shalaginov@ntnu.no

G. O. Dyrkolbotn
e-mail: geir.dyrkolbotn@ntnu.no

M. Alazab
Charles Darwin University, Palmerston City, Australia
e-mail: alazab.m@ieee.org

M. Stamp et al. (eds.), *Malware Analysis Using Artificial Intelligence and Deep Learning*, https://doi.org/10.1007/978-3-030-62582-5_3

1 Introduction

"Malware", *"malicious software"* or broadly speaking *"computer viruses"* is an umbrella term that is assigned to software designed to harm end-user as opposite to a *"goodware"* or *"benign software"* such as office and entertainment programs, Internet browsers and games. Further, there are several ways how actual malware can be called from the perspective of functionality, threats to end-user and distribution model, as for example, *Trojan*, *Worm* or *Ransomware* [66]. In general, *malware labelling* is a non-trivial problem and might vary depending on the analysis approach, discovered artefacts and similarity to other previously discovered samples [8]. To name a malware, one can use either (i) cryptographic hash sum such as MD5[1] equal to *0c5e15ea8c92f33396fe3fb85d7a7fbf* or (ii) naming convention *Trojan:Win32/Detplock* to describe a specific malware species [76]. One can see that the hash sum approach is a robust machine-readable standard, however, not appropriate for a human expert. On the other hand, the second approach is far more appropriate for a malware analyst to make a decision on what kind of threats malware poses and what kind of vulnerabilities it uses based on the malware type and family specification. However, sometimes it is not clear based on what the malware categories are created since the names often are machine-generated and do not reflect peculiarities of actual malicious behaviour.

Malware analysis and defences recently became an emerging topic over the last few decades—covering many areas and attracting multiple companies and researchers to work on more secure platforms and solutions. However, the history of malware spans back to the twentieth century. Commonly, 1949 is considered to be the year when the theoretical foundations of self-replicating automata have been set up [34]. The contemporary malware era started with the well-known Morris worm in 1980 [18]. The real "boom" of malware infection reflected widespread development of desktop and further mobile OS in early 2000 [36, 51, 60].

VirusTotal was established in 2004 and now can be considered as a *de-facto* standard in the Information Security community [77]. It provides reporting of the malware detection from 70 anti-virus vendors, which in addition to extensive Threats Intelligence and community reporting, giving the most extensive publicly-available malware awareness. As per 6th of January 2020, there have been reported 1,304,817 distinct files submitted to VirusTotal, while 803,259 files were labelled as malicious by one or more anti-virus vendors [78]. At the same time, as per January 2020, VirusShare collection offers access to 34,339,374 different malware samples available through the website [75]. This initiative started back in 2011, and now has collected over 400 versatile archives with different examples of malware.

Most of the researchers in the Information Security community work on the techniques used to identify and detect malware samples among others, answering the question *How to identify malicious software using known labels?*. This includes the construction of specific features and finding attributes, yet mostly focuses on automated malware detection to facilitate human experts. However, we can see a

[1]Message-Digest algorithm for 128-bit hash sum.

considerable need for research initiatives to answer the question *How to best label malware?*. CARO (Computer Antivirus Research Organization) created a naming scheme back in 1990 [11], which was supposed to be a stepping stone for malware naming standardization. However, only Microsoft mostly uses CARO approach in their products [41] with Trend Micro moving in that direction since 2018. In the literature, there can be found several sources mentioning these challenges, yet offering no comprehensive overview or even solution.

The scope of this research is (i) to reveal existing challenges that complicate malware identification and cyber threat intelligence services. To our knowledge, the topic has been approached by many researchers, while in most cases the emphasis is *only* on a few most common malware classes rather than on a multi-faceted approach. Moreover, we aim at (ii) providing a high-level overview of existing methods to malware naming by anti-virus vendors. There can be seen a clear gap in the literature regarding challenges in malware naming and categorization, which this paper is designed to address. Finally, we (iii) try to project the current needs of anti-virus domain concerning existing approaches based on the real-world examples from VirusTotal and large-scale malware naming problem.

This paper is organized as follows. The Sect. 2 introduces the history of malware categorization, distribution of works and their quantitative overview. Further, Sect. 3 gives an overview of the most common approaches to malware categorization, used characteristics and supplementary software that can be found in the literature. Section 4 presents an in-depth review of the commonly used malware taxonomy by different Anti-Virus vendors. Survey of industrial malware naming standards and initiatives is given in Sect. 5. The comparative analysis and outlined challenges of the malware categorization methods are presented in Sect. 6. Section 7 offers insights into current practical implications, especially in the era of Big Data. Finally, Sect. 8 give final remarks and recommendations towards the future development of the malware naming and taxonomy.

2 Background: From Malware Developers to Malware Analysts

This section presents a high-level insight into the problem of cyberattacks boosted by the development of new malware in addition to the existing general methods of malware analysis and cybersecurity awareness. Development of Information and Communications Technologies (ICT) boosted the number of adversarial activities targeting users, as well as large organizations over the last few years. The main reason for that is the growing number and complexity of both hardware and software products available to end-users. Multiple security vulnerabilities, lack of cybersecurity awareness and weak data protection mechanism created a concrete stepping stone for paradigms such as *Attacks-as-a-Service* and *Malware-as-a-Service*. Having 29 billions of connected devices by 2022 [24], one will expect malware infections and cyberattacks in general to be major threats. Therefore, it is important to provide a quick incidence response as soon as possible based on the available cyber threats intelligence, including malware naming information.

2.1 Severity of Malware Infection and Modus Operandi

By shortening development and delivery cycles, sacrificing cybersecurity standards number of vulnerabilities may appear in the end-products as shown by OWASP Top 10 [80]. This can be considered as one of the root-cause problems related to cyber threats in connection with malicious software that tends to exploit well-known, yet not fixed, or zero-day vulnerabilities. Over the decades, there have been found infection of hundreds of thousands of malware samples, some of those becoming notorious for their elaborate attacking mechanism and caused damages. Since malware is considered as one of the cyberattacks, the same reasons for committing those apply *financial gain and political motivation* [49].

Adversarial actors use common mechanisms of exploiting the systems when preparing and developing malware. Having strong cybersecurity knowledge, they can learn about the target's valuable assets, possible exploitation scenarios and weak links in the system. The target's cybersecurity watch mechanism with such knowledge cannot even detect malicious intentions if they are not explicitly active. By now, the importance of awareness of cyberattacks have been recognized by governments and companies, such as Cisco [15], Symantec [68], Akamai [2] Check Point [12], UK National Cyber Security Center [73], etc. Symantec, as many other companies defined, beside huge attacks landscape, the generally-recognized routine of performing cyberattacks:

1. *Reconnaissance*—attackers retrieve as much information as possible about the target from public sources.
2. *Incursion*—active phase of delivering malware through various technical means and social engineering.
3. *Discovery*—mapping of the internal infrastructure and security measures, staying "low and slow".
4. *Capture*—active malware infection or sensitive information access.
5. *Exfiltration*—getting the information and other incentives out of the target's network back to the attacker via covert anonymized channels.

It can be seen that adversarial actors often need a lot of background information about hardware and software in the target's network to be able to find and analyze corresponding possible vulnerabilities and protection weaknesses. This process can take days or months depending on the complexity, zero-day vulnerabilities and human factor. On multiple occasions over the last decade, it was shown that some of the malware requires thorough development by a large group of people to be efficient on the incursion. One of the notorious is *ZeuS* botnet [53], which had the version 1.0 detected in 2006 and later mainstream versions up to 1.4 running through 2013 and later. Moreover, there have been developed multiple forks from the mainstream of Zeus, such as Citadel, LEAK and Skynet. This is an example of how powerful the adversarial malware developers community target MS Windows OS. The reason for such popularity is multi-fold: ranging from general simplicity for everyday utilization, range of users' privileges and up to elaborate security features and control of software being installed [1].

2.2 Detection and Approach Strategy

Similarly to real life, the fight against cybercrime, malware, in this case, begins with reports and indications of malicious activities. There have been suggested several methods on how to approach the cyberattacks in general.

Lockheed Martin's Cyber Kill Chain [38] was originally a military term, later formulated in a framework from 2011, having the following phases: *Reconnaissance, Weaponization, Delivery, Exploitation, Installation, Command and Control, Actions and Objective*. Corresponding defending action should be taken during each identified stage to disrupt the attacker's efforts.

NIST Cybersecurity Framework [20] is a policy framework released in 2014, for public sector consisting of five main functions: *Identify, Protect, Detect, Respond, Recover*.

Beside specific guidelines, there exists a general approach to malware discovery that includes routine starting from malicious activities being discovered to the point when malware signature is being published to end-point security solutions, for example, suggested by Securosis [57]:

1. Malicious activities are recorded
2. Likely relevant artefacts have been unidentified
3. Discovery of binary files and possible cause of infection
4. Search for hash sums in known-to-be-benign and known-to-be-malicious
5. Reverse engineering and internal logic understanding of found files
6. Assignment of most likely malware category, malware family and variant
7. Creation of Anti-Virus signatures and awareness campaigns

To address the Analysis part, there have been created several guidelines and thorough routines with similar tools as elaborately studied by Sikorski et al. [64]. Further, we can see an enormous growth of anti-malware awareness campaigns and diversity of protective software products to combat such infections on the market of digital services. Furthermore, malware is not just a software anymore, yet also more low-level such as Hardware Trojans as studied by Tehranipoor et al. [71]. While most anti-virus detection mechanisms successfully use behavioural-based and signature-based detection, this is out of scope in this study. We will only focus on the aspects related to categorization.

One of the core issues in cybersecurity that makes the possible development of malware is so-called vulnerabilities. Either software development bugs or unforeseen applications, make a user to execute restricted commands or to gain access to sensitive information [39]. To address issues of malware naming and information sharing, there has been developed *Common Vulnerabilities and Exposures (CVE)* standard that creates a public entry for known vulnerabilities [19], where the details are taken from the National Vulnerability Database (NVD). The CVE was developed and maintained by MITRE Corporation [44].

2.3 Preliminary Analysis and Dissection

One of the biggest challenges that cybersecurity community faced a few decades ago is how to efficiently and in a fast manner perform the analysis: (i) to identify malicious components and functionality, (ii) to assign the corresponding cluster among existing malware groups and (iii) to develop a recommendation on how to prevent malware infection based on the analysis results. There have been suggested a few community-accepted approaches according to the SANS report by Distler et al. [22], by Damshenas et al. [21], Kendall et al. [31] and Zeltser [84]:

- *static properties analysis* aimed to study the characteristics of malware files without executing them. Different aspects of files can be investigated such as headers, possible encrypted parts, present strings, bytes, opcodes and API n-grams, Portable Executable header features, strings and others [26, 55, 74, 83].
- *dynamic behavioural analysis* considers different parts of the executed malware sample influence of different factors present in the target system, as shown by Kendall et al. [31]. Multiple activities such as network traffic, registry keys and disk usage patterns, API-calls and instruction tracing, and memory layout investigation are explored to find out what differentiates malware from non-malware according to Egele et al. [23]. To collect such information, one can use either specialized sandboxes like Cuckoo [27] or utilize any Virtual Machines such as VirtualBox accompanied by monitoring software.

2.4 Malware Categorization and Cybersecurity Awareness

Malware categorization becomes more and more important, in particular, in response to critical incidents. A growing amount of malware threats and malware variants [6] require maintainable of large-scale datasets and knowledge databases. Without having a systematic approach to labelling and taxonomizing new and existing pieces of malicious software, one may do the same job over. Resolution of such issues was first attempted by the Computer Antivirus Research Organization (CARO) with their naming scheme as defined by Skulason in 1991 [65]. Therefore, it is essential to have a high-level overview of the contemporary malware categories, methods and standards for more efficient similarity-based malware detection and further analysis. The most extensive available collection of anti-virus reports is available through VirusTotal, which can demonstrate how complex, diverse malware categorization used by different companies. One of the large-scale malware studies showed that out of 60+ AV vendors on VirusTotal, Microsoft follows easy-to-use CARO naming for their reports [62]. We will only look into how malware is categorized by different vendors and what are the current and future challenges.

There can be seen several works that attempted to summarize and provide an overview of standard approaches to malware categorization. We will introduce each of the works with their findings below. One of those is a report written by Hardikars [28] published by SANS Institute in 2008. The paper aimed at giving a detailed overview of viruses, a self-replicated type of malware. The author briefly acknowledges different malware categories as a worm, logic bomb, backdoor, trojan, spyware and botnet [5] in addition to virus. Further, all existing at that time viruses were classified into the following categories: memory-based, target-based, obfuscation technique-based and payload-based with multiple specific subcategories.

3 Malware Classification: State of the Art

This section gives an overview of the malware classifications used in scientific research. To understand how the classification has been developing over the years, the history of the malware analysis research, corresponding used characteristics, type of classification and mentioned naming schemes would be referenced to. To the authors' knowledge, there have not been any major scientific reviews that address aspects of malware naming and categorization. Besides, only Anti-Virus vendors and cybersecurity companies focus on these aspects.

Malware developers have been employing advanced techniques in their software to remain unnoticed as long as possible and to cause as much harm as possible. They can use fake Windows certificates, zero-day vulnerabilities and default software settings, etc. as described by Wu et al. in 2016 [82], making it challenging to notice abnormalities. Further, a set of obfuscation methods is applied, such as encryption, polymorphism, metamorphism, dead code insertions, or instruction substitution [55] to conceal the real functionality logic of the software. In addition to this, MS Windows is a known target of many attacks crafted by famous viruses such as Stuxnet, Duqu and Flame [7]. Multiple market share surveys suggest that more than 50% of desktop computers and laptops users utilize MS Windows as an Operating System (OS) [67]. At the same time, nearly 2% of the users still have Windows XP installed, which is no longer a supported OS version [48].

3.1 Characteristics-Based Detection for Multinomial Classification

There has been a tremendous interest in the academic community towards intelligent malware classification to the ability of Machine Learning methods to perform fast and efficient classification without human interaction. However, previous works mostly focus on the intelligent differentiation of a file between *benign* or *malicious* type [58]. This is a *binary classification*, where a heap of malware samples are classified against a collection of goodware. Cohen [17] suggested in 1987, that no algorithm will be able to detect all computer viruses confidently. This assertion was strengthened by Chess et al. [14]. As a result, we can assume that no methods can achieve 100% classification accuracy on large-scale sets. Bragen [10] applied Machine Learning on opcode sequences and achieved 95% accuracy with the tree-based RandomForest method. Kolter et al. [33] used 1,971 malicious files and 1,651 benign, while Bragen used only 992 malicious and 771 benign. Markel et al. [37] used PE32 header data in malware and benign files detection on Decision Tree, Naive Bayes, and Logistic Regression. The authors achieved a 0.97 F-score in binary classification. Furthermore, Shankarapani et al. [63] applied PE32 file parser to extract static features for similarity analysis. Overall, 1,593 samples were acquired for binary classification in that study.

In contrast to binary, *multinomial classification* can be described as detection of whether a malware belongs to a particular family or type. There exist several *malware categories*, (trojan, backdoor, etc.) and *malware families*, (Poison, Ramdo, etc.), which are commonly defined by the Information Security community. A *malware category* is a general type of malware that uses a certain kind of approach to exploit a system and gain illegal access, such as a *worm*, which is a self-replicating code that can spread over email, or *ransomware* that encrypts files and requires a financial ransom to be paid [13]. On the other hand, a *malware family* is a specific sub-category that uses a particular vulnerability or targets specific software versions. For example, considering the *worm* category, we can distinguish the *p2p worm* family such as *Spybot* from *removable drive worm* like *Autorun!inf* [81]. Cohen [17] suggested in 1987, that there are no algorithms that will be able to confidently detect all possible computer viruses. This statement was strengthened by Chess et al. [14]. This type of classification can be seen appearing in scientific literature, mostly starting from 2008. Rieck et al. [52] studied 14 different malware families extracted from 10,072 unique binaries. The authors achieved, on average 88% accuracy in family detection using individual SVM for each one. Further, Zhang et al. [87] explored binary classification using binary sub-sets of 450 viruses and goodware based on the 2-gram analysis.

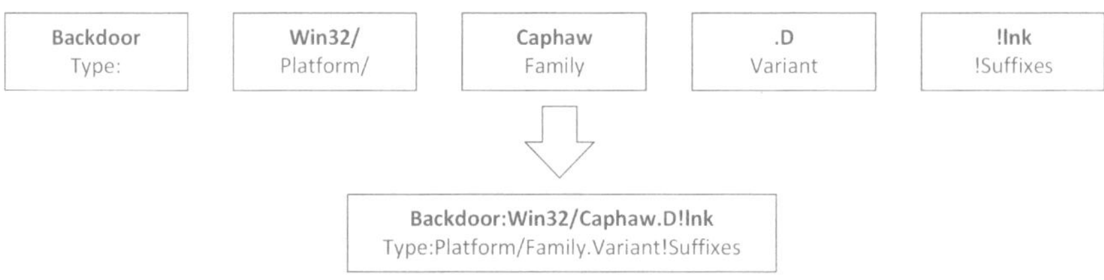

Fig. 1 CARO malware naming scheme [41]

3.2 Commonly Used Malware Naming

There exists several malware types (like *trojan, backdoor*, etc.) and families (like *Poison, Ramdo*, etc.), which are commonly defined by the Information Security community. In 1991, the Computer Antivirus Research Organization (CARO) proposed a standardized naming scheme for malware [11]. Although CARO states that this naming scheme is "widely accepted", we found that from all the vendors on *VirusTotal*, Microsoft is the only one that complies with this. It is, therefore, challenging to establish a typical pattern in scanner results across anti-virus databases. An example of a widely used approach is CARO naming, shown in Fig. 1.

To the authors' knowledge, there has been no adequately performed a scientific study that provides a complete taxonomy of malware or relevant naming schemes. However, there can be found non-academic blog entries or Anti-Virus vendors reports with different malware species descriptions and dissections. Mushtaq [47] gave an overview of the fraction malware samples considering only 20 species, mostly families. Another list of families is published by Microsoft as a part of the description of the Windows Malicious Removal Tool [43] starting from 2005 up until March 2020. Finally, The Malware Database website offers an extensive collection of different information pages, structured adequately for each particular malware family and category [81]. However, what we can see is that in the scientific community, authors usually mix up both families and categories, and typically consider fewer samples than exist in the wild [52, 87].

3.3 Auxiliary Software Tools and Research Datasets

The de-facto standard online resource in the malware analysis community is VirusTotal that was launched in 2004, and now offers access as a single entry point to 70 anti-virus vendors' databases. Over the years it started offering Public and Private APIs, also used by major security labs and companies. However, as it was mentioned before, only Microsoft, in most cases, uses CARO labelling scheme, which can be parsed into malware family and type accordingly. Other vendors use their approaches to labelling, often confusing and misleading. To overcome this limitation, there has developed a tool called "AVCLASS: A Tool for Massive Malware Labeling" and released in 2016 [3, 56]. The main goal of this tool is to process VirusTotal reports to be able to extract the most likely malware family considering the variety of the formats of the virus analysis reports.

Another important aspect that should be mentioned is community-maintained datasets, which are used to foster malware analysts collaboration, results from dissemination and malware detection mechanism testing. *VirusShare* [75] gives access to 34,503,473 malware samples as of 05.02.2020. Those are not categorized and are available in archives containing up to 131,072 files in 374 archives. *VxHeaven* [79] also offers an overview of computer viruses in a very categorized and dissected manner. Besides, there has been created a new solution called *VirusTotal Intelligence* that can be used to extract binary samples of a specific malware that the user defines.

4 Analysis of Community—and Commercially—Accepted Malware Taxonomies

As we can see, the diversity of malware threats became a real Big Data problem with nearly 0.5M malware samples being detected by VirusTotal everyday [78]. Over the last decades, there has been a growth in hardware and software technologies, subsequently making appropriate categorization of malware samples a very cumbersome and non-trivial task. It led to cases where one malware has been categorized completely differently by various AV vendors. This also might indicate that malware can belong to multiple categories, as mentioned by Kaspersky [30].

This section provides a high-level overview of the most commonly used approaches in reviewed literature to categorize malware in various aspects of used technologies and internal functionality. We first start from the general categories and then will go more into more specific functionality-oriented taxonomies. It was also provided with an overview of relevant Anti-Virus vendors' approaches to malware categorization

4.1 Overall Software Category

The most general malware categorization approach is according to the type of software, as stated by F-secure [25]:

1. *Clean/"Goodware"*—the software that does not pose any risk to the user and perform predefined benign functions like office programs or browsers.
2. *Potentially Unwanted Applications (PUP) & Unwanted Applications/"Grayware"* —the software that can be considered as unwanted, depending on the user and environment they are being used, typically performs actions without consent such as advertising or programs parameters changes.

3. *Harmful/"Malware"*—the software that has a negative impact either on user's data or devices functionality and is designed specifically by adversarial actors.

Further, speaking of PUP, *ClamAV* [16] offers following thorough definition of the sub-types (10): Packed, PwTool, NetTool, P2P, IRC, RAT, Tool, Spy, Server, Script.

4.2 Risk Level/Threat Level

This is another categorization that is used by Anti-Virus vendors to indicate the danger of using a particular piece of software. For example, Microsoft Security Essentials [42] differentiates the following security levels, which are also included in the reports on discovered malware.

1. *Low*—potentially unwanted programs, which, however, might have some benign functionality with some malicious intentions.
2. *Medium*—software that might harm the user's privacy.
3. *High*—potentially harmful programs that may misuse personal information or make unauthorized system changes.
4. *Severe*—well-known malware species.

4.3 Malware Targets/Platforms/Operating Systems

Another important categorization of the malware is the target, for which such software was developed, including types, formats and platforms. Each Operating System and Platform dictates the way how it can be attacked, which depends on multiple factors, security controls, user privileges, file system access mechanisms, frequency of updates and general quality of manufacturer maintenance. Depending on this, adversaries may exploit particular know or zero-day vulnerabilities when the attack reconnaissance phase is finished, and some information is known about the end-user system.

F-Secure categorization includes 49 general programming languages and platform types [25]: AM, Android, ACAD, BAT, Boot, ChromeOS, CM, CS, DOS, HLP, HTML, IDA, INF, INI, iPhoneOS, MSIL, Java, JS, Linux, MacOS, MMS, OM, OS/2, OSX, PM, PalmOS, Perl, PHP, PPM, PUM, REG, SH, SMS, Solaris, SymbOS, SVL, SWF, Unix, VBS, W16, W32, W64, W128, WM, WinCE, WinHEX, WMA, WMV, XM.

Microsoft uses more elaborate approaches and actually differentiates three following items [41]:

1. *Operating Systems* (22): AndroidOS, DOS, EPOC, FreeBSD, iPhoneOS, Linux, MacOS, MacOS_X, OS2, Palm, Solaris, SunOS, SymbOS, Unix, Win16, Win2K,

Table 1 Top 10 desktop and mobile OS with corresponding version distributions

Desktop OS (%)		Desktop OS version (%)		Mobile OS (%)		Mobile OS version (%)	
Windows	88.07	Windows 10	48.96	Android	69.99	Android 8.1	13.91
MacOS	9.44	Windows 7	32.37	iOS	29.21	Android 8.0	10.28
Linux	1.87	Mac OS X 10.14	4.80	Unknown	0.65	Android 9.0	10.12
ChromeOS	0.41	Windows 8.1	3.95	Series 40	0.06	Android 7.0	7.31
Unknown	0.19	Windows XP	1.96	Windows phone OS	0.04	Android 6.0	7.08
BSD	0.02	Mac OS X 10.13	1.78	Linux	0.03	iOS 12.1	5.69
		Linux	1.36	RIM OS	0.02	Android 7.1	5.35
		Mac OS X 10.15	0.95	Symbian	0.01	Android	4.97
		Mac OS X 10.12	0.81	Bada	0.00	iOS 12.3	4.64
		Windows 8	0.67	Windows	0.00	iOS 12.4	4.55

Win32, Win64, Win95, Win98, WinCE, WinNT. This list will change upon the advancement and appearance of the new OS.

2. *Scripting Languages* (30): ABAP, ALisp, AmiPro, ANSI, AppleScript, ASP, AutoIt, BAS, BAT, CorelScript, HTA, HTML, INF, IRC, Java, JS, LOGO, MPB, MSH, MSIL, Perl, PHP, Python, SAP, SH, VBA, VBS, WinBAT, WinHlp, WinREG. Moreover, Windows binaries can be categorized as *EXE* (executable) and *DLL* (Dynamically-Linked Library).
3. *Macros* for Microsft Office (12): A97M, HE, O97M, PP97M, V5M, W1M, W2M, W97M, WM, X97M, XF, XM.
4. *Other file types* (9): ASX, HC, MIME, Netware, QT, SB, SWF, TSQL, XML.

So, it can be seen that there is a significant separation in malware taxonomies related to what kind of OS and platforms that can run. To give an overview of the end-user OS market share as of February 2020, we looked into the market share of different platforms provided by the company Net Marketshare [48]. The overall statistics are represented in the Table 1 for mobile and desktop devices. It can be seen that Windows takes 88.07% of all installed desktop OS, while Android occupies 69.99% of devices. Surprisingly, most of the vulnerabilities and malware infections are affiliated with these OS. We do not consider the enterprise cloud server solutions at this moment due to generally lower cybersecurity awareness among private end-users.

4.4 Malware Type/General Categories

This category defines the general type of malware. Hardikar [28] suggested the following categorization based on their functionality aspects as defined in the SANS report:

1. *Memory-based* (resident, temporary, swapping, non-resident, user and kernel).
2. *Payload-based* (no, non-destructive, destructive, dropper).
3. *Obfuscation techniques-based* (no, ecryption, oligo-, meta-, poly-morphism, stealth, armouring, tunnelling, retrovirus).
4. *Target-based* (compiled (file, boot sector), interpreted (macro, script), multipartite).

These are also affiliated with the functional classification, dissemination methods and behavioural aspects1. This is the most commonly used approach by anti-virus vendors, researchers and malware labs. Further, it can be divided into sub-families and sub-types. However, this is the complete list of 31 names given by *Microsoft* [41]: Adware, Backdoor, Behavior, BrowserModifier, Constructor, DDoS, Exploit, Hacktool, Joke, Misleading, MonitoringTool, Program, PWS, Ransom, RemoteAccess, Rogue, SettingsModifier, SoftwareBundler, Spammer, Spoofer, Spyware, Tool, Trojan, TrojanClicker, TrojanDownloader, TrojanNotifier, TrojanProxy, TrojanSpy, VirTool, Virus, Worm. It is worth mentioning, that over the last few years there were removed three obsolete malware types such as Dialer, DoS, TrojanDropper.

Classification provided by *F-secure* [25] is more generic and consists of 12 malware types: Application, Adware, Trackware, Hack-Tool, Monitoring-Tool, Spyware, Backdoor, Exploit, Trojan, Rootkit, Worm, Virus.

Symantec offers a different approach to categorization, based on the reporting and policies definition:

- *Virus/Threat types* [70] (7): Crack, Damaged, False positive, Joke, Malicious, Speculative, Phish.
- *Policies-oriented configurations* [69] (46): Ad-supported Program, Adware, Adware Bundler, Adware Installer, Attack, Backdoor, Botnet, Browser hijacker, Browser plug-in, Critical Spyware Web site, Custom Restricted Lists 1, 2, and 3, Destroyer, Dialer, Downloader, Exploit, Hack tool, Joke program, Keylogger, Major Spyware Website, Malicious behaviour, Minor Spyware Website, Misleading Application, P2P, Parental control, Password Hijacker, Phishing, Potentially Unwanted Software, RAT, Remote access, Rogue Security Program, Rootkit, Security Assessment Tool, Security risk, Spammer, Spyware, Spyware Marketing and Tools, Stealth Notifier, Surveillance, System Monitor, Tracking Software, Trackware, Trojan, Trojan FTP, Unclassified Critical Spyware, Unclassified Spyware, Worm.

Avira uses a predefined set of malware type prefixes (69) in addition to malware family name [4]: ABAP, ACAD, AM, A97M, APM, ASM, Bash, BAT, BDC, BDS, Boo, CSC, Csh, DIAL, DOS, DR, EML, Game, HLLx, HLP, HTML, INF, INI, IRC:,

JAVA, Joke, JS, JSc, Kit, Linux, MIRC, O2000M, 097M, OS2, P2000M, P97M, Palm, Perl, PDF, PHP, PIF, PP, Sh, SPR, SWF, Sys, TR, Unix, UWS, VBS, Vgen, VXD, W16, W2000, W2000M, W32, W95, W97M, WB, WIN, WIN2k, WM, Worm, Wscr, X2000M, X97F, X97M, XF, XM.

TrendMicro used the following set of malware type prefixes (34) prior to making CARO compliant naming [72]: ADW, ALS, ATVX, BAT, BHO, BKDR, CHM, COOKIE, DIAL, [DOS, DDOS], ELF, EXPL, GENERIC, HKTL, HTML, IRC, JAVA, JOKE, JS, NE, PALM, PE, PERL, RAP, REG, RTKT, SPYW, SYMBOS, TSPY, TROJ, UNIX, VBS, WORM, [W2KM, W97M, X97M, P97M, A97M, O97M, WM, XF, XM, V5M, X2KM, X97M].

4.5 Malware Family/Functionality-Specific Categories

Depending on the specific functionality, exploitation and code similarity methods, malware types are divided into more specific sub-types, also called families. For example, the Ransomware malware type contains the following families: Cryptolocker, WannaCry, Cryptowall, etc. Some of the family names have a clear meaning and denotes the malicious software they are assigned to. However, the majority are more like a machine-constructed derivative from some other names. However, the total amount of malware families is not well-defined. From the 2016 study, it was discovered 10,362 among nearly 328,000 Windows malware samples found in the first ten archives from VirusShare [26, 62]. Further, Malware Wiki2 has one of the most comprehensive collections of malware types that are further divided into malware families. There are nearly 3,000 malware families' descriptions, including history, exploited vulnerabilities, artefacts and caused harm [81]. Some of the families are no longer relevant since the software has been patched or just replaced with new versions.

4.6 System and Digital Forensic-Related Artefacts

Another way of categorizing malware is by the traces that they leave in the system [35, 64]. This also defines the ability of Anti-Virus to detect such malware based on available signatures.

1. *Disk-related*—file operations, registry modification. However, there is a new trend, file-less malware, which makes tracking of such artefacts less efficient.
2. *Network-related*—interaction with C&C centers, port scanning, etc.
3. *Memory-related*—any kind of memory activities such that read/write, new processes, etc.

4.7 Malware Variants

Malware developers are using sophisticated techniques to avoid detection such as obfuscation, encryption and packers changes. However, the internal functionality of malware stays the same. Therefore, it is important to denote such changes such as variant ".BC" that will naturally follow variants ".BB" and ".BA" according to Microsoft [41].

4.8 Malware Name Suffix

Final categorization is according to different supplementary optional material, like threats campaign, dropper, library identifier, etc. as defined by Microsoft [41] (19): .dam, .dll, .dr, .gen, .kit, .ldr, .pak, .plugin, .remnants, .worm, !bit, !cl, !dha, !pfn, !plock, !rfn, !rootkit, @m, @mm.

4.9 Binary Compilation Timestamps/Timeline

One of the challenges is that there is no clear and trustworthy information on when the malware was exactly created [60], making timeline analysis irrelevant. However, one can consider using time-related fields from VirusTotal reports as an indicator of when the malware first draw attention and was uploaded to the website (Fig. 2).

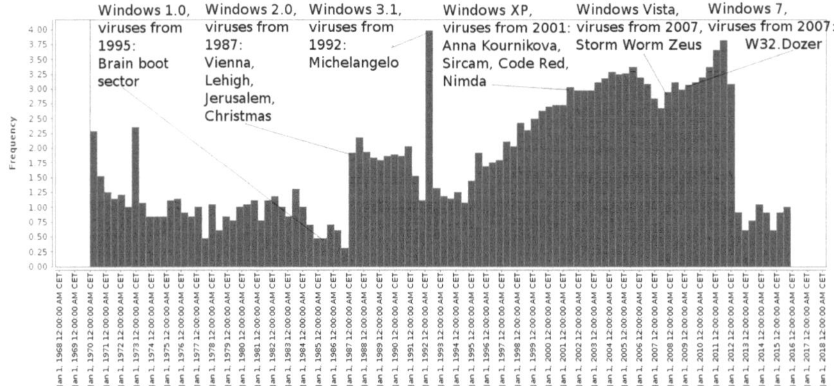

Fig. 2 Distribution of malware creation timestamps from VirusShare_00000 archive [60]

4.10 Country/Adversarial Groups Origins

Malware developers use advanced methods to hide their origins (country or region-specific), sometimes pretending to be from one or another geographic location based on user language in the source code or comments [54]. Because any information containing in malware may be fake, such origins definitions are rather speculations than "ground truth".

5 Review of the Existing Anti-virus Naming Schemes

The section presents an overview of the Anti-Virus vendors standards and malware naming initiatives. To our knowledge, there has not been done relevant academic adaptation for the research, even though, those approaches have been mentioned by cybersecurity researchers. To our knowledge, there are no commonly used standards. However, CARO seems to have a long-standing initiative that is accepted by at least Microsoft and Trend Micro.

Malware Naming in principle is a cumbersome task because one malware can be classified differently from the perspective of the functional taxonomy. The challenge with malware categorization and malware naming, in general, had been seen already for many decades and several researchers raised this topic. Impre [29] indicated difficulties of proper classification as Big Data problem with 700 million malware samples in Q1 2017. Besides existing malware analysis techniques, the author analyzed a few existing approaches to malware naming accepted by Anti-Virus vendors. There are different needs for *incident response, tracking malware relations and preliminary analysis* that will dictate with a scheme to use. Zeltser [85, 86] identified that there were two major efforts made to address the changing nature of the malware landscape since mid-2000. As a response to multiple malware infections, analysts started assigning eye-catching names to malware campaigns like CryptoLocker or WannaCry. The author shows an example of how *Duqu* was named after "DQ" prefix that the malware used. Mo [46] performed a study of 30 most common malware samples using Metascan engine. It was shown that there is a considerable lack of consistency and agreement between different Anti-Virus engines. Some of the vendors call, for example, trojan as a worm and vice-versa. Mitre tried to address those issues by suggesting new solutions in addition to the CARO naming scheme.

5.1 Computer Antivirus Research Organization (CARO)

Computer Antivirus Research Organization (CARO) had been established in 1990, to facilitate the study and research in malware analysis [11]. They published the so-called CARO naming scheme or Virus Naming Convention of computer viruses.

The original idea was to reduce confusion in malware names. Many anti-virus vendors subsequently used this naming scheme to establish their standards. Since 2007, CARO workshop happens every year, where security experts meet to discuss challenges and opportunities.

The main idea of this naming is to have a clear separation into *Type, Platform, Family, Variant and Additional Information* that is used to mention additional useful information. The Type part indicates a general malware category such that Trojan Horse, Rootkit, Worm, etc. According to Microsoft [41], there are 31 malware types and 22 Operating Systems. The Family part should indicate what a specific malware group based on the similarity in code or functionality. Such similarity helps researchers to get more details regarding virus from previously analyzed members of this family. Despite the long existence of such a scheme and the involvement of multiple Anti-Virus vendors, it was not widely used in the industry. For now, only Microsoft applies this method to labelling their malware signatures. The general structure of the naming looks as follows:

Listing 1 CARO naming convention
```
[<type>://][<platform>/]<family>[.<group>][.<length
    ↪ >].<variant>[<modifiers>][!<comment>]
```

5.2 *Common Malware Enumeration (CME)*

Common Malware Enumeration (CME) is another approach to establish a standard in malware naming with a particular focus on merging the indexing approaches utilized by different anti-virus vendors [45]. It was announced in 2005, as a result of inconsistencies and lack of communication between anti-virus vendors and very similar to Common Vulnerabilities and Exposures (CVE), also developed by MITRE. In contrary to the per-file naming of viruses, CME offers a unique treat identification that is independent of variants or number of files that are assigned to a particular family or attack. By identifying a major outbreak, there is a unique code that is being assigned in the form of CME-N, where N is an integer. Meanwhile, anti-virus vendors can assign such CME identification to their naming of malware. However, the threats identifiers on the website show the last update was in 2007, and the report by Bontchev [9] stated a large number of weaknesses and problems with CME in comparison to CARO naming. Despite the efforts and promising results, we can conclude that the CME scheme is not in use anymore. CME uses unique identifiers that are further linked to specific malware infections detected by different Anti-Virus Vendors.

Listing 2 CME identifier
```
CME-540: Symantec \rightarrow "w32.zotob.e", Sophos \
    ↪ rightarrow "W32/Tpbot-A", \dots
```

5.3 *Malware Attribute Enumeration and Characterization (MAEC)*

Malware Attribute Enumeration and Characterization (MAEC) is a community-developed malware naming models based on the behavioural information, system artefacts and specific relationships between malware samples [32]. It is maintained by MITRE and can be reinforced by Structured Threat Information Expression (STIX) cyber observable high-level objects.[2] An example of Zeus botnet samples is shown below

Listing 3 MAEC malware description


```
{
    "type":"package",
    "id":"package--f53adac8-c416-42c6-6fbc-7
       ↪ b6ef8876fc5",
    "schema_version":"5.0",
    "maec_objects":  [
          {
                " type ":"malware-family",
                " id ":"malware-family--df91014d-0c2e
                   ↪ -4e01-b8a5-d8c32bb038e6",
                " name ":  {
                      "value":"Zeus",
                      "confidence":90
                }
          }
    ]
}
```


5.4 *Malware Information Sharing Platform (MISP)*

Malware Information Sharing Platform (MISP) [50] is developed with a primary goal of facilitation of incident response, where malware classification schemes with machine-readable tags human-readable descriptions are essential components. An example of botnet classification taxonomy is shown below (Fig. 3).

6 Analysis of Existing Approaches to Malware Categorization

As we have shown below, over a few decades, since 1990s, there have been developed multiple schemes to approach malware naming. However, the main challenge

[2] https://maecproject.github.io/documentation/overview/.

Fig. 3 MISP taxonomy [50]

in malware naming is diversity in malware functionality, attack vectors and several malicious software samples. The optimal way is to use multi-faceted taxonomy that can provide an accurate description of malware categories, as well as facilitate human-understandable labelling.

From the literature, we can see that the problem of malware naming attracted the attention of a few companies, while still there is no unified format that may facilitate faster and better malware detection based on previous knowledge. One of the studies [29] provided a recommendation of choosing CARO naming for general classification due to its solid foundation, while for incident response, there might be a need to employ additional mechanisms. OPSWAT [46] stated that CARO consortium did not convince other Anvi-Virus vendors to use this approach. While others tried their solutions like CME, they did not succeed over CARO. BitDefender [8] in 2006, state that CME is a very perspective and intelligent approach, however, has major limitations concerning tremendous speed that Anvi-Virus industry is moving forward. Finally, in 2005, one of the founders of CARO—Vesselin Bontchev [9], highlighted a few major challenges in Anti-Virus naming conventions. Generally, Bontchev advocates for the usage of CARO due to well-through naming standards and restrictions to avoid confusion and mixed naming. Further, he criticizes CME initiative due to many limitations that might easily lead to malware being named without CME identifier and vice-versa, one malware being assigned to several CME identifiers such as Zotob.E has two different CME numbers. Finally, 73% of the vulnerabilities in the SANS RISK bulletin has no CVE numbers making CME less applicable.

It is notable how the AV industry has grown: from 5,000 new malware programs per month according to Bontchev on 13 October 2005 to 595,010 distinct malware files per day identified by at least one vendor in VirusTotal as of 10th of February, 2020 [78]. We can guesstimate around 3,563% increase in daily malware samples appearance over 15 years. To the authors' knowledge, not every AV vendor has adopted CARO naming: only Microsoft consistently use it over many years and TrendMicro have recently adopted it in July 2018.

7 Practical Implications of Malware Naming in the Light of Big Data

The general challenge is that a single malware samples can be attributed to multiple malware categories and families. To demonstrate the example of malware labelling, we refer to the VirusTotal scanning result of the file named "VirusShare_0c5e15ea8c92f33396fe3fb85d7a7fbf" and MD5 hash sum 0c5e15ea8c92f33396fe3fb85d7a7fbf [76]. A fraction of 26 Anti-Virus reports is presented in the Fig. 4.

However, the diversity of the used naming approaches draw significant attention. There were few examples of consistency among Anti-Virus vendors such as *Trojan.Downloader.Zlob.ABKL* is used by the 7 following companies: Ad-Aware, Emsisoft, eScan, Arcabit, BitDefender, F-Secure, GData. Furthermore, among all reports, 49 out of 70 AV classified these files as malicious software with the following keywords (types) used in labels: "Zlob"—28 vendors, "Trojan"—27, "Win32" or "W32"—21, "Downloader"—19 and "Generic"—5. This drastically changes the efforts of malware analyst to find the most likely classification result. Finally, **only** Microsoft specified malware family name "Detplock" has used the CARO naming scheme approach: *Trojan:Win32/Detplock*. Trend Micro adopted CARO naming in 2018 [40]; however, this company named the aforementioned malware samples as Undetected/Non-malicious. As a result of these considerations, we have to put forward the following needs (i) fostering cooperation between AV vendors through public dialogue involving domain experts, (ii) unification of the malware naming format, possible towards CARO scheme, (iii) utilization of advancing processing mechanisms like *AVCLASS* tool to extract as close label as possible.

The main problem with malware naming is the consistent growth in the known malware samples pool. VirusTotal receives more than a million files scanning requests

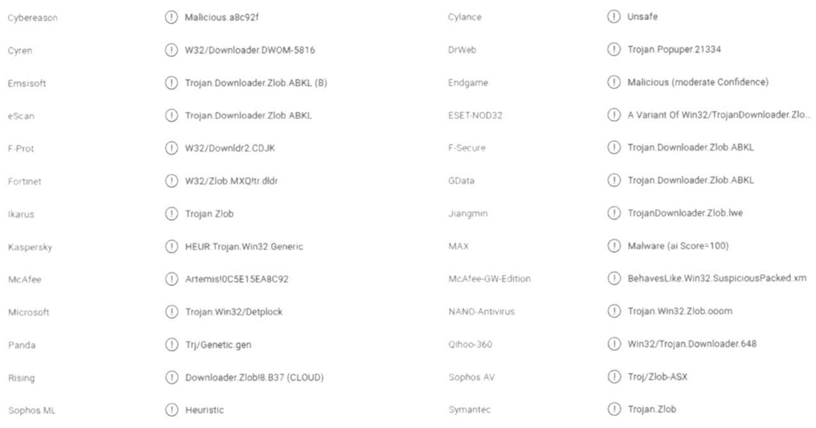

Cybereason	Malicious.a8c92f	Cylance	Unsafe
Cyren	W32/Downloader.DWOM-5816	DrWeb	Trojan.Popuper.21334
Emsisoft	Trojan.Downloader.Zlob.ABKL (B)	Endgame	Malicious (moderate Confidence)
eScan	Trojan.Downloader.Zlob.ABKL	ESET-NOD32	A Variant Of Win32/TrojanDownloader.Zlo...
F-Prot	W32/Downldr2.CDJK	F-Secure	Trojan.Downloader.Zlob.ABKL
Fortinet	W32/Zlob.MXQ!tr.dldr	GData	Trojan.Downloader.Zlob.ABKL
Ikarus	Trojan.Zlob	Jiangmin	TrojanDownloader.Zlob.lwe
Kaspersky	HEUR:Trojan.Win32.Generic	MAX	Malware (ai Score=100)
McAfee	Artemis!0C5E15EA8C92	McAfee-GW-Edition	BehavesLike.Win32.SuspiciousPacked.xm
Microsoft	Trojan:Win32/Detplock	NANO-Antivirus	Trojan.Win32.Zlob.ooom
Panda	Trj/Genetic.gen	Qihoo-360	Win32/Trojan.Downloader.648
Rising	Downloader.Zlob!8.B37 (CLOUD)	Sophos AV	Troj/Zlob-ASX
Sophos ML	Heuristic	Symantec	Trojan.Zlob

Fig. 4 Example of 26 Anti-Virus naming reports for the executable file with MD5 has sum 0c5e15ea8c92f33396fe3fb85d7a7fbf [76]

Fig. 5 Statistics of the malware types and families 328,000 Windows PE32 malware samples [59, 61]

Top Malware families		Top Malware categories	
Label	Number	Label	Number
onlinegames	11,129	trojan	76,932
small	9,284	trojandownloader	52,479
hupigon	7,743	backdoor	45,499
renos	7,423	pws	38,598
agent	7,022	virtool	25,710
zlob	6,933	worm	23,809
vb	6,805	trojanspy	15,239
vbinject	6,502	trojandropper	12,046
obfuscator	5,939	virus	11,947
vundo	5,875	rogue	8,096
frethog	5,593	adware	5,294
delf	4,833	browsermodifier	3,051
winwebsec	4,273	trojanclicker	1,962
farfli	4,137	trojanproxy	1,502
bifrose	4,068	spammer	886
zbot	3,941	dialer	806
c2lop	3,892	monitoringtool	806
alureon	3,702	hacktool	730
delfinject	3,631	ransom	671
startpage	3,553	exploit	566
bancos	3,406	ddos	491
banload	3,399	program	362
vobfus	3,319	constructor	262
lolyda	2,841	dos	177
injector	2,646	spyware	169
gamania	2,636	joke	98
tibs	2,378	settingsmodifier	70
taterf	2,341	softwarebundler	52
allaple	2,283	trojannotifier	7
lmir	2,254	tool	4
ircbot	2,160	spoofer	4
banker	2,137	flooder	3
hotbar	2,079	remoteaccess	3
bho	2,065	nuker	3
poison	2,039	misleading	3

per day, with more than half of this amount classified as malicious samples by at least one of the anti-virus vendors. One of the extensive studies in 2016, investigated naming of 400,000 of Windows Portable Executable 32bit (PE32) files [26, 62]. Based on the analysis of the JSON reports from VirusTotal, it became clear that there is no unified approach for malware labelling bringing the task of proper malware type and family labelling to an almost impossible one. However, 328,000 malware samples were labelled as malicious by Microsoft, which also meant that the CARO naming scheme was used. Therefore, it is considered as one of the most extensive multinomial malware studies [59]. The statistics of the malware types and families are shown in the Table 5.

Further, Fig. 6 presents how the 35 general malware types and 10,362 more specific malware families are distributed based on the count of each category. The most

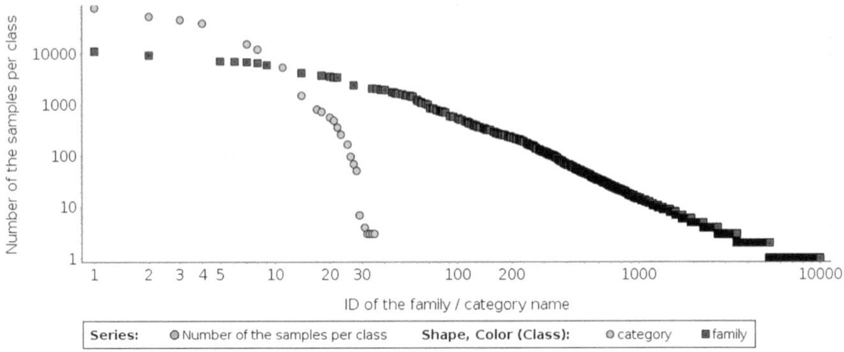

Fig. 6 Distribution of malware types and families over 328,000 Windows PE32 malware samples [59]

frequent malware type is "trojan" detected in 76,932 malware files, while examples like "flooder" and "remoteaccess" were only found in 3 cases. While most of the malware types of names can be explained, the process is not that trivial with malware families. For example, "bho" corresponds to Browser Hijack Object (BHO); however, "Lolyda" family does not represent anything except that we know that it is a gaming trojan.

So, from the most extensive collection of malware reports by different anti-virus in VirusTotal, there can be seen a high level of irregularity and diversity. For example, *Trojan Downloader* called *Zlob* [76] shows that each Anti-Virus vendor assigned a unique name, which is not only differed by malware name, but also by the name components format and order.

8 Discussions and Conclusions

The amount of malware being discovered everyday grows exponentially. The number of platform variants, programming languages, exploitation ways, vulnerabilities and delivery techniques is huge considering the heavy development of new hardware and software in recent decades. We discovered that there is a growing interest in multi-class malware detection since 2009. The evolution of malware is affected by the current usage of technologies. Despite multiple attempts since 1990s to unify the approach of malware categorization and labelling, only a few comply with the community-accepted CARO naming scheme. Another discovery of this study is, despite the existence of CARO, there is still no unified standard for malware taxonomy since every malware lab and anti-virus vendors use their practices while keeping the general names in the reports, so it can be correlated globally with others. This can be seen from VirusTotal's 70 anti-virus reports, as a de-facto standard tool for malware analysis. Then, we practically confirmed that the challenge with categorization is that manual analysis of a large number of samples may not be feasible in a short timeframe to issue adequate protective measures and proper unique identification within the range of existing families. Finally, some of the malware is no longer usable and outdated, while others use zero-day vulnerabilities and have not been discovered yet. As a future direction, we see a need for more work towards the unification of malware labelling, better cyber threats intelligence exchange and general public-oriented cybersecurity awareness campaigns to reduce the risk of malware infection.

References

1. Abraham, Shawn. 2018. Why windows get more virus attacks than mac or linux. https://www.malwarefox.com/windows-virus-attacks/. Accessed 25 Mar 2020.
2. Akamai. Cyberattacks. https://www.akamai.com/uk/en/resources/cyber-attacks.jsp. Accessed 25 Mar 2020.
3. Avclass. 2016. https://github.com/malicialab/avclass. Accessed 05 Feb 2020.
4. Avira. Malware naming conventions. https://www.avira.com/en/support-malware-naming-conventions. Accessed 07 Feb 2020.
5. Azab, Ahmad, Mamoun Alazab, and Mahdi Aiash. 2016. Machine learning based botnet identification traffic. In *2016 IEEE trustcom/BigDataSE/ISPA*, 1788–1794, IEEE.
6. Azab, Ahmad, Robert Layton, Mamoun Alazab, and Jonathan Oliver. 2014. Mining malware to detect variants. In *2014 5th cybercrime and trustworthy computing conference*, 44–53, IEEE.
7. Bencsáth, Boldizsár. 2012. Duqu, flame, gauss: Followers of stuxnet. https://www.rsaconference.com/writable/presentations/file_upload/br-208_bencsath.pdf. Accessed 10 July 2016.
8. BitDefender. 2006. Virus naming. the "who's who?" dilemma. Technical report, BitDefender. http://download.bitdefender.com/resources/files/Main/file/Virus_Naming_Whitepaper.pdf. Accessed 10 Jan 2020.
9. Bontchev, Vesselin. 2015. Current status of the caro malware naming scheme. *Virus bulletin (VB2005)*, Dublin, Ireland. Accessed 07 Feb 2020.
10. Bragen, Simen Rune. 2015. Malware detection through opcode sequence analysis using machine learning. *Gjøvik University College*.
11. CARO. Naming scheme - caro - computer antivirus research organization. www.caro.org/naming/scheme.html. Accessed 07 Feb 2020.
12. Check Point. What is a cyber attack? https://www.checkpoint.com/definitions/what-is-cyber-attack/. Accessed 25 Mar 2020.
13. Chen, Chong-Kuan. 2015. Malware classification and detection. http://www.slideshare.net/Bletchley131/malware-classificationanddetection. Accessed 10 July 2016.
14. Chess, David M, and Steve R White. 2000. An undetectable computer virus. In *Proceedings of virus bulletin conference*, vol. 5.
15. Cisco. What are the most common cyber attacks? https://www.cisco.com/c/en/us/products/security/common-cyberattacks.html. Accessed 25 Mar 2020.
16. ClamAV. Potentially unwanted applications (pua). https://www.clamav.net/documents/potentially-unwanted-applications-pua. Accessed 06 Feb 2020.
17. Cohen, Fred. 1987. Computer viruses: Theory and experiments. *Computers & Security* 6 (1): 22–35.
18. Comodo. A short history of computer viruses. https://antivirus.comodo.com/blog/computer-safety/short-history-computer-viruses/. Accessed 11 Feb 2020.
19. Cvedetails.com - the ultimate security vulnerability datasource. 2020. https://www.cvedetails.com/. Accessed 17 Feb 2020.

20. Critical Infrastructure Cybersecurity. 2014. Framework for improving critical infrastructure cybersecurity. *Framework* 1: 11.
21. Damshenas, Mohsen, Ali Dehghantanha, and Ramlan Mahmoud. 2013. A survey on malware propagation, analysis, and detection. *International Journal of Cyber-Security and Digital Forensics (IJCSDF)* 2 (4): 10–29.
22. Distler, Dennis, and Charles Hornat. 2007. *Malware analysis: An introduction An introduction*. Sans Reading Room.
23. Egele, Manuel, Theodoor Scholte, Engin Kirda, and Christopher Kruegel. 2012. A survey on automated dynamic malware-analysis techniques and tools. *ACM Computing Surveys (CSUR)* 44 (2): 6.
24. Ericsson: Internet of things forecast. https://www.ericsson.com/en/mobility-report/internet-of-things-forecast. Accessed 11 Feb 2020.
25. F-secure. F-secure classifies threats. https://www.f-secure.com/v-descs/guides/classification_guide.shtml. Accessed 06 Feb 2020.
26. Grini, Lars Strande, Andrii Shalaginov, and Katrin Franke. 2016. Study of soft computing methods for large-scale multinomial malware types and families detection. In *Proceedings of the the 6th world conference on soft computing*.
27. Guarnieri, Claudio, Allessandro Tanasi, Jurriaan Bremer, and Mark Schloesser. 2012. The cuckoo sandbox.
28. Hardikar, A. 2008. Malware 101-viruses. *SANS Institute*.
29. Impe, Koen Van. 2018. How to choose the right malware classification scheme to improve incident response. https://securityintelligence.com/how-to-choose-the-right-malware-classification-scheme-to-improve-incident-response/. Accessed 07 Feb 2020.
30. Kaspersky. Types of malware. https://www.kaspersky.com/resource-center/threats/malware-classifications. Accessed 06 Feb 2020.
31. Kendall, Kris, and Chad McMillan. 2007. Practical malware analysis. In *Black hat conference*, USA.
32. Kirillov, Ivan, Desiree Beck, Penny Chase, and Robert Martin. 2011. Malware attribute enumeration and characterization. https://www.researchgate.net/profile/Robert_Martin10/publication/267691330_Malware_Attribute_Enumeration_and_Characterization/links/54bd188e0cf218d4a169ee0c/Malware-Attribute-Enumeration-and-Characterization.pdf. Accessed 07 Feb 2020.
33. Kolter, J.Zico, and A. Marcus, Maloof. 2006. Learning to detect and classify malicious executables in the wild. *Journal of Machine Learning Research* 7: 2721–2744.
34. Krebs, Brian. 2003. A short history of computer viruses and attacks. *Washingtonpost.com* 14.
35. Lee, Alan, Vijay Varadharajan, and Udaya Tupakula. 2013. On malware characterization and attack classification. *Proceedings of the 1st Australasian web conference* 144: 43–47.
36. Malware Bytes. What is malware? https://www.malwarebytes.com/malware/#what-is-the-history-of-malware. Accessed 11 Feb 2020.
37. Markel, Zane, and Michael Bilzor. 2014. Building a machine learning classifier for malware detection. In *2014 2nd workshop on anti-malware testing research (WATeR)*, 1–4, IEEE.
38. Martin, Lockheed. 2014. Cyber kill chain®. http://cyber.lockheedmartin.com/hubfs/GainingtheAdvantageCyberKillChain.pdf.
39. Mell, Peter, and Tim Grance. 2002. *Use of the common vulnerabilities and exposures (cve) vulnerability naming scheme*. National inst of standards and technology gaithersburg md computer security div: Technical report.
40. Micro, Trend. New threat detection naming scheme in trend micro. https://success.trendmicro.com/solution/1119738-new-threat-detection-naming-scheme-in-trend-micro. Accessed 10 Feb 2020.
41. Microsoft. Malware names. https://docs.microsoft.com/nb-no/windows/security/threat-protection/intelligence/malware-naming. Accessed 06 Feb 2020.
42. Microsoft. Understanding alert levels in microsoft security essentials. https://docs.microsoft.com/nb-no/archive/blogs/robmar/understanding-alert-levels-in-microsoft-security-essentials. Accessed 06 Feb 2020.

43. Microsoft. 2016. The microsoft windows malicious software removal tool helps remove specific, prevalent malicious software from computers that are running supported versions of windows. https://support.microsoft.com/en-us/kb/890830. Accessed 15 July 2016.
44. MITRE. Common vulnerabilities and exposures. https://cve.mitre.org/about/index.html. Accessed 05 Feb 2020.
45. MITRE. 2006. Common malware enumeration: reducing public confusion during malware outbreak. https://cme.mitre.org/about/index.html. Accessed 07 Feb 2020.
46. Mo, Jianpeng. 2015. What can we learn from anti-malware naming conventions? https://www.opswat.com/blog/what-can-we-learn-anti-malware-naming-conventions, 2015. Accessed 07 Feb 2020.
47. Mushtaq, Atif. 2010. World's top malware. https://www.fireeye.com/blog/threat-research/2010/07/worlds_top_modern_malware.html. Accessed 15 July 2016.
48. Netmarketshare - market share statistics for internet technologies. 2020. https://www.netmarketshare.com/. Accessed 06 Feb 2020.
49. NI Business Info. Cyber security for business: Reasons behind cyber attacks. https://www.nibusinessinfo.co.uk/content/reasons-behind-cyber-attacks. Accessed 05 Feb 2020.
50. MISP-Open Source Threat Intelligence Platform. Open standards for threat information sharing. http://www.misp-project.org/index.html. Accessed 07 Feb 2020.
51. Rankin, B. 2018. A brief history of malware—its evolution and impact. https://www.lastline.com/blog/history-of-malware-its-evolution-and-impact/.
52. Rieck, Konrad, Thorsten Holz, Carsten Willems, Patrick Düssel, and Pavel Laskov. 2008. Learning and classification of malware behavior. In *Proceedings of the 5th international conference on detection of intrusions and malware, and vulnerability assessment*, DIMVA'08, 108–125. Berlin: Springer.
53. S21Sec. 2013 Zeus timeline. https://www.s21sec.com/zeus-timeline-i/. Accessed 10 Jan 2020.
54. Saarinen, Juha. 2017. Malware authors camouflage code with russian terms. https://www.itnews.com.au/news/malware-authors-camouflage-code-with-russian-terms-452012.
55. Schiffman, Mike. 2010. A brief history of malware obfuscation. http://blogs.cisco.com/security/a_brief_history_of_malware_obfuscation_part_1_of_2. Accessed 13 July 2016.
56. Sebastián, Marcos, Richard Rivera, Platon Kotzias, and Juan Caballero. 2016. Avclass: A tool for massive malware labeling. In *International symposium on research in attacks, intrusions, and defenses*, 230–253, Springer.
57. Securosis. 2012. Measuring and optimizing malware analysis: An open model. Technical report, Securosis. https://cdn.securosis.com/assets/library/reports/Securosis-MAQuant-v1.4_FINAL.pdf. Accessed 10 Jan 2020.
58. Shalaginov, Andrii. 2017. Dynamic feature-based expansion of fuzzy sets in neuro-fuzzy for proactive malware detection. In *2017 20th international conference on information fusion (Fusion)*, 1–8, IEEE.
59. Shalaginov, Andrii. 2018. *Advancing neuro-fuzzy algorithm for automated classification in largescale forensic and cybercrime investigations: Adaptive machine learning for big data forensic*. PhD thesis, Norwegian University of Science and Technology.
60. Shalaginov, Andrii, Sergii Banin, Ali Dehghantanha, and Katrin Franke. 2017. Machine learning aided static malware analysis: A survey and tutorial. *Cyber Threat Intelligence 2017*.
61. Shalaginov, Andrii, and Katrin Franke. 2016. Automated intelligent multinomial classification of malware species using dynamic behavioural analysis. In *2016 14th annual conference on privacy, security and trust (PST)*, 70–77, IEEE.
62. Shalaginov, Andrii, Lars Strande Grini, and Katrin Franke. 2016. Understanding neuro-fuzzy on a class of multinomial malware detection problems. In *International joint conference on neural networks (IJCNN) 2016*, 684–691, Research Publishing Services.
63. Shankarapani, M, Kesav Kancherla, S Ramammoorthy, R Movva, and Srinivas Mukkamala. 2010. Kernel machines for malware classification and similarity analysis. In *The 2010 international joint conference on neural networks (IJCNN)*, 1–6, IEEE.
64. Sikorski, Michael, and Andrew Honig. 2012. *Practical malware analysis: The hands-on guide to dissecting malicious software*. San Francisco: No Starch Press.

65. Skulason, Fridrik, Alan Solomon, and Vesselin Bontchev. 1991. CARO naming scheme.
66. Sophos. 2020. Sophos 2020 threat report. we're covering your blind spots. Technical report, Sophos. Accessed 11 Feb 2020.
67. Stack Overflow. 2016. Stack overflow - developer survey results. http://stackoverflow.com/research/developer-survey-2016. Accessed 11 July 2016.
68. Symantec. Preparing for a cyber attack. https://www.symantec.com/content/en/us/enterprise/other_resources/b-preparing-for-a-cyber-attack-interactive-SYM285k_050913.pdf. Accessed 05 Feb 2020.
69. Symantec. 2011. Malware categories for policies. https://support.symantec.com/us/en/article.howto54185.html#v46370003. Accessed 06 Feb 2020.
70. Symantec. 2019. Malicious code classifications and threat types. https://support.symantec.com/us/en/article.tech226322.html. Accessed 06 Feb 2020.
71. Tehranipoor, Mohammad, and Farinaz Koushanfar. 2010. A survey of hardware trojan taxonomy and detection. *IEEE Design & Test of Computers* 27 (1): 10–25.
72. TrendMicro. Malware naming. https://docs.trendmicro.com/all/ent/tms/v2.5/en-us/tda_2.5_olh/malware_naming.htm. Accessed 07 Feb 2020.
73. UK National Cyber Security Center. 2016. How cyber attacks work. https://www.ncsc.gov.uk/information/how-cyber-attacks-work. Accessed 25 Mar 2020.
74. Uppal, Dolly, Roopak Sinha, Vishakha Mehra, and Vinesh Jain. 2014. Malware detection and classification based on extraction of api sequences. In *2014 international conference on advances in computing, communications and informatics (ICACCI)*, 2337–2342, IEEE.
75. Virusshare. https://www.VirusShare.com/. Accessed 17 Feb 2020.
76. VirusTotal. Report on trojandownloader zlob. https://www.virustotal.com/gui/file/e8331ed32e33ba0abb6a73c320552bd17d5fe7acd4189cbea5a72f933e2a09e9/detection. Accessed 10 Feb 2020.
77. VirusTotal. https://www.virustotal.com/. Accessed 17 Feb 2020.
78. Virustotal statistics. https://www.virustotal.com/en/statistics/. Accessed 04 Feb 2020.
79. VxHeaven.org website mirror. 2018. https://github.com/opsxcq/mirror-vxheaven.org.
80. Wichers, Dave. 2013. Owasp top-10 2013. *OWASP Foundation*.
81. Wikia. The malware database. http://malware.wikia.com/wiki/. Accessed 06 Feb 2020.
82. Wu, C.H., and J.D. Irwin. 2016. *Introduction to computer networks and cybersecurity*. Boca Raton: CRC Press.
83. Zabidi, M.N.A, M.A. Maarof, and A. Zainal. 2012. Malware analysis with multiple features. In *2012 UKSim 14th international conference on computer modelling and simulation (UKSim)*, 231–235.
84. Zelster, L. 2015. Mastering 4 stages of malware analysis. https://zeltser.com/mastering-4-stages-of-malware-analysis/.
85. Zelster, L. 2011. Assigning descriptive names to malware – why and how? https://zeltser.com/descriptive-names-for-malware/.
86. Zelster, L. 2011. How security companies assign names to malware specimens [web blog]. https://zeltser.com/malware-naming-approaches/.
87. Zhang, Boyun, Jianping Yin, Jingbo Hao, Dingxing Zhang, and Shulin Wang. 2007. Malicious codes detection based on ensemble learning. In *Proceedings of the 4th international conference on autonomic and trusted computing*, ATC'07, 468–477. Berlin: Springer.

Addressing Malware Attacks on Connected and Autonomous Vehicles: Recent Techniques and Challenges

Aiman Al-Sabaawi, Khamael Al-Dulaimi, Ernest Foo, and Mamoun Alazab

Abstract Part of the wider development and monitoring of smart environments for an intelligent cities approach is the building of an intelligent transportation system. Such a system involves the development of modern vehicles which significantly improve passenger safety and comfort, a trend that is expected to increase in the coming years. There are key factors relating to safety impacts and security vulnerabilities that may emerge during the increased deployment of automated vehicles and the security and privacy of connected and automated vehicle systems. They include ways of defining the security of malware-relevant system boundaries including electronic control units, silicon hardware, software, vehicle systems, infrastructure, network connectivity and more. In addition, vehicle industries are facing many problems with critical security and privacy issues, influenced by the smart environments for an intelligent cities approach. Such problems are related to hardware and software applications that allow the interfacing of Vehicle to Vehicle (V2V) and Vehicle to Infrastructure networks (V2I). In this chapter, we present connected car methods relating to the attack, defence and detection of malware in vehicles. Critical issues are introduced regarding the sharing of safety information and the verification of the integrity of this information from V2V and V2I networks. In particular, we discuss the challenges and review state-of-the-art intra–inter-vehicle communication. Hack-

A. Al-Sabaawi (✉) · K. Al-Dulaimi
Queensland University of Technology, Queensland, Australia
e-mail: a.alsabaawi@connect.qut.edu.au

K. Al-Dulaimi
e-mail: khamaelabbaskhudhair.aldulaimi@hdr.qut.edu.au

Department of Computer Science, Al-Nahrain University, Baghdad, Iraq

E. Foo
School of Information and Communication Technology, Griffith University, Queensland, Australia
e-mail: e.foo@griffith.edu.au

M. Alazab
College of Engineering, Information Technology and Environment, Charles Darwin University, Northern Australia, Australia
e-mail: alazab.m@ieee.org

M. Stamp et al. (eds.), *Malware Analysis Using Artificial Intelligence and Deep Learning*, https://doi.org/10.1007/978-3-030-62582-5_4

ers can access this information in V2V/V2I networks and broadcast fake messages and malware to break the security system by using weak points in vehicles and networks. We present important security approaches that are used in vehicles which can fully protect the vehicle security architecture by detecting the attempts made and the methods used by hackers to tackle malware and security problems in vehicles. We present a comprehensive overview of current research on advanced intra-inter-vehicle communication networks and identify outstanding research questions that may be used to achieve high levels of vehicle security and privacy in intelligent cities in the future.

1 Introduction

Smart city or smart building technology has been advancing in recent years due to the development of communication technologies and wiring which are associated with the fields of power, health, education, industries and transportation. Buildings are becoming more complex, with interconnected Internet of Things (IoT) systems offering technological equipment and cost-efficient buildings and energy. The infrastructure of the IoT is still developing and offers many benefits for humans including monitoring asset movements, turning lights on/off as needed, optimizing room occupancy, air conditioning systems, health systems, vehicular and road connected systems, security systems (monitoring camera systems) and location systems [38, 49].

The IoT enables cities to grow and expand. Officials in cities with the technology of the IoT can access valuable data to gain a better understanding of their city's operations. Those data enable the control of traffic, the empowerment of local law enforcement, allow improved security of connected vehicles, monitoring of the environment and enable city-wide connectivity and tracking of parking efficiency. These cities play a significant role in fostering creativity and innovation. The creation of customized IoT applications positions cities at the technological forefront which, in turn, attracts new residents and businesses. Intelligent transportation systems, monitoring the way people commute in metros and smart cities, are one benefit. An intelligent transportation system offers a novel approach to the provision of different transportation modes, advanced infrastructure and traffic and mobility management solutions. It uses a number of electronic, sensor, wireless and communication technologies to provide consumers with access to a smarter, safer and faster way of travelling [38, 49].

1.1 Important Technologies in Intelligent Transportation System in Smart Cities

1. Advanced Tracking System: modern vehicles are connected with in-vehicle GPS. The GPS system can offer two-way communication, helping traffic professionals to locate vehicles, check speeding vehicles and provide emergency services. Smartphones, mobile applications and Google Maps have become useful tools in tracking, understanding road quality, traffic density and locating different routes and places.
2. Advanced Sensing Technologies: These include intelligent sensors both in vehicles and road infrastructure. Radio Frequency Identification (RFI) and intelligent beacon sensing technologies are ensuring the safety of drivers in cities worldwide. Road reflectors and inductive loops are built into roads, assisting with traffic control and safe driving, especially at night. They can also provide information about vehicle density at particular times and can identify vehicles at both slow and high speeds.
3. Advanced Video Vehicle Detection: Video cameras or CCTV surveillance can solve many problems for traffic managers. Video footage of strategic places and prime junctions can help operators observe traffic flow and identify any emergency situation or road congestion. In-built vehicle sensors and automatic number plate detection help to check vehicles for security purposes.
4. Advanced Traffic Light Systems: Radio Frequency Identification (RFID) is used in traffic light systems. This technology can offer correct algorithms and databases even when applied to multiple lanes, road junctions and vehicles. These lights can adjust themselves during critical and peak hour traffic situations without any human presence.
5. Emergency E-Call Vehicle Service: During an emergency situation such as an accident or mishap, in-vehicle sensors can establish contact with a nearby emergency centre. An e-call will help a driver to connect to a trained operator and also transmit important information such as time, location, direction of vehicle and vehicle identification directly to the centre [38].

1.2 Benefits of Intelligent Transportation System

1. Minimizing Pollution: An intelligent transportation system aims to promote the use of public transport by the general public. If it provides single-point services and access to real-time information about the transport schedule, people will prefer to use an intelligent transportation system and reduce private vehicle usage, thereby lowering traffic congestion and lowering pollution levels.
2. Security and Safety: Advanced sensing technologies help to provide emergency and critical care services to drivers and people when required, such as real-time data analysis, including CCTV, GPS, internet connectivity and wireless and virus

and malware detection [12]. Surveillance of public transportation also helps to alert city managers to the risk of terror elements and to avoid mishaps or terror attacks.

3. Market for Mobile Applications: Recently, modern transportation has come to depend more on smartphones and mobile applications to identify parking spots, route guides, destination points, weather forecasts and arrival and departure details.
4. Smart Parking Solutions: Smart parking solutions, combined with appropriate infrastructure, internet connectivity and security cameras, can minimize parking problems. Many urban areas now have multi-layer parking systems. There are also many applications which provide users with information about free parking spaces available nearby [38].

1.3 Challenges of Intelligent Transportation System

To support these benefits and sophisticated features in an intelligent transportation system, modern vehicles are developed using software, an assortment of embedded computing devices, sensors, communication interfaces and actuators. However, these lead to many challenges that affect human life and safety [23].

Vehicle industries predict that as the cost of software and electronics fall, security-related incidents will become a serious threat. Exploiting vulnerabilities in the vehicle's electronics may allow the remote control of vehicle components. An attacker can turn off the vehicle's lights or even control the brakes while on the move [30]. More recently, attacks on production vehicles, for example, exploiting vulnerabilities in Fiat/Chrysler's Uconnect system, enabled hackers to control the vehicles, turning off the engine and controlling the steering over the Internet [24, 49], with the company urging owners to update their vehicles' software to patch the identified vulnerabilities [49].

Moreover, many of the enhanced services of these modern inter-connected vehicles rely on the location of the vehicles and their drivers, information that, by its nature, gives rise to significant privacy concerns. Securing the various heterogeneous hardware and software platforms and networks in the intelligent transformation system ecosystem is still a challenging task. While security is an important key in various aspects of smart vehicle-related Information and Communications Technology (ICT) deployments, many aspects of efficient intelligent transportation operations have safety issues and other Quality of Service (QoS) characteristics which may limit the applicability of complex security initiatives. Therefore, potential solutions should be considered for these limitations by identifying attack type, defence and detection [23].

Given the advantages of a connected vehicle, security in vehicle networks and their characteristics and issues, it is crucial to understand how current intelligent transportation systems can be adapted to work with smart environments for intelligent cities. In this chapter, we provide an overview of connected vehicle methods in intra-inter-vehicle communication. We then provide a survey of the recent history of three key features in-vehicle security and malware: attack, defence and detection. There has been extensive research in each area and many studies address intra–inter-vehicle communication, which is a critical problem in-vehicle technology. The following review critically describes the literature regarding the identification of attacks such as malware and their types and the defences and solutions in intra-inter-vehicle communication. The detection issues and challenges for each type of communication will be presented. The literature also includes the recent techniques and their challenges and issues regarding vehicle security from malware including attacks, defences and detection in intra-inter-communication networks. The objective of this chapter is to help researchers to address these challenges in future work and in the further investigation of attacks, defences and detection, as well as to make significant changes to the design of vehicle systems to improve automotive security and prevent any malware and cyber terrorists from attacking vehicles.

This chapter is organized as follows: the first section presents a comprehensive overview of the vehicle connected methods, including attacks, detection and defence. In the second section, recent techniques and their challenges are discussed. The third section comprises the conclusion of the chapter.

2 Literature Review

Because modern technology has introduced more intelligence and complexity into the car industry, researchers are required to take greater responsibility for both safety and security. Vehicle security is different from vehicle safety, which includes vehicle speed [10] and vehicle integrated design [21]. Vehicle security, however, is essential to delivering vehicle safety from malware [45]. With a connected environment, vehicles, infrastructure and pedestrians can exchange information, either through a peer-to-peer connectivity protocol or a centralized system via a 4G or more advanced telecommunication and security network. This technology has the potential to be one of the most disruptive technologies for urban and smart cities. The interaction and exchange of information regarding the use of malware may occur in V2V, V2I, pedestrian-to-infrastructure (P2I) or vehicle-to-pedestrian applications (V2P) [52].

Vehicle security covers many aspects including immobilizers, car-to-car communication, car-to-infrastructure communication, car-to-X communications, cloud and smartphone or smart device [13]. Recently, security analysis has been investigated in vehicle production and it was discovered that there are many reasons for security development. By accessing the in-vehicle 3G or Bluetooth, an attacker may tamper with the brakes while people are driving cars. In addition, car thieves have the ability to exploit security breaches in keyless-entry systems or to generate spare keys by using the on-board diagnostic system or using malware. Today, the weakness of security measures in vehicles can cause many financial problems, such as decreasing mileage to extend warranty claims by illegal chip adjusting [48]. Connected car methods are shown in Fig. 1.

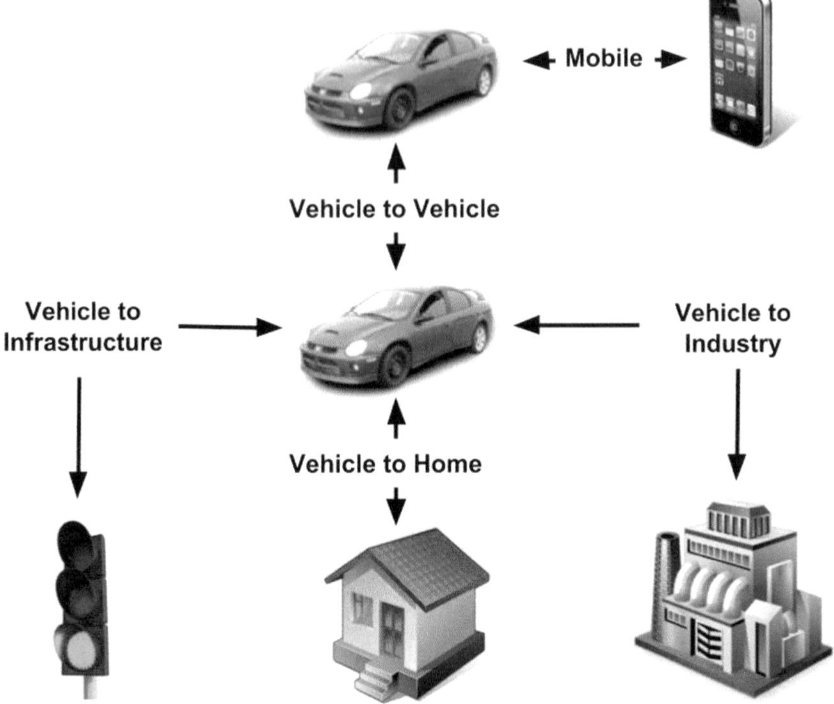

Fig. 1 The concept of the connected car

1. Car-to-Car communication: This term refers to inter-vehicle communication exchange between two cars, for example, to warn others of a change in the road surface, obstacles on the roadway or other dangers.
2. Car-to-infrastructure communication: This refers to communication between cars and components of the infrastructure using wireless communication. Components of the infrastructure include nodes in a cellular network or intelligent traffic signs that can be utilized to establish car-industry communication, infotainment platforms or the Internet.
3. Car-to-X communication: This term refers to the sending and receiving of data between cars, the infrastructure, other transport, traffic management systems and different Internet applications. While other communications receive and process information, cars can also exchange information.
4. Smartphone or smart device: Given the implementation of common modern technology, smartphones, tablets and smartwatch use is widespread and they become an obvious goal in the communication system, as shown in Fig. 2.
5. Cloud: Computing can exchange data from cars and data stored in the (Cloud) by using the Internet, as shown in Fig. 3.

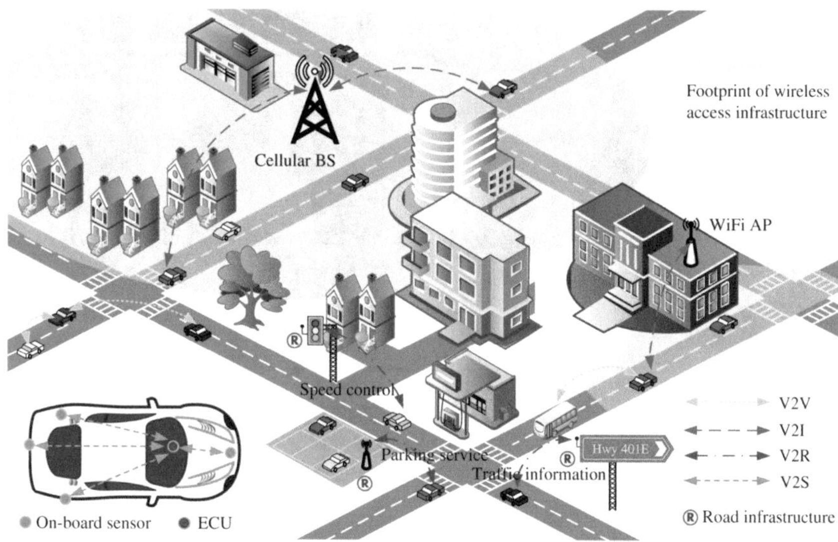

Fig. 2 Connection types [33]

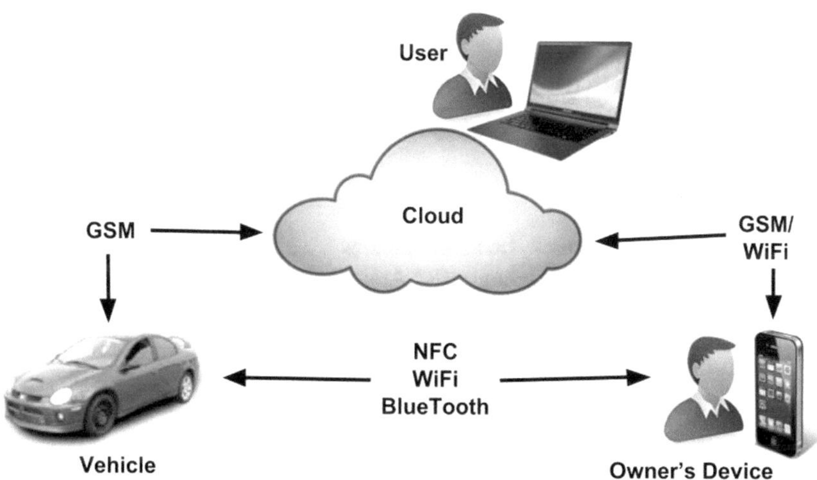

Fig. 3 Simple car connections

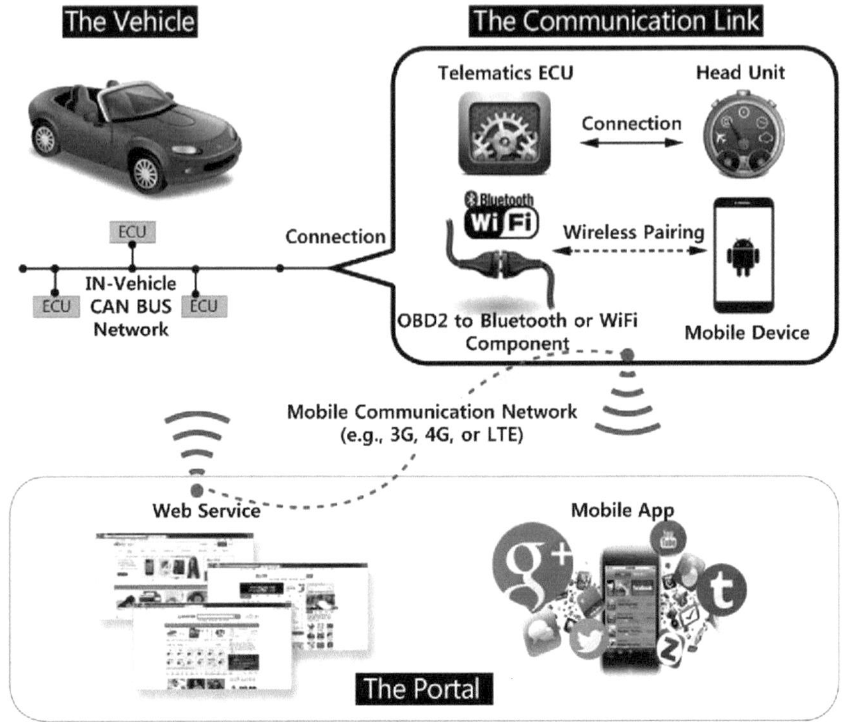

Fig. 4 Connected car environment [58]

In recent years, electronic systems in vehicles have been controlled by an Electronic Control Unit (ECU). The controller Area Network (CAN) uses an in-vehicle network to structure an effective network of ECUs [58].

The ECU is an important component in automotive application components that can control one or more of the electrical systems and subsystems in cars [57]. The on-board architecture of vehicles can contain more than 70 ECUs [17] that can interconnect via different networks such as Local Interconnect Network (LIN), CAN or FlexRay [1, 40]. In fact, CAN use has become widespread because it significantly reduces the number of communication lines and ensures the reliability of higher data transmission [28], as shown in Fig. 4.

Recently, due to increasing penetration of smartphones and advanced communication technologies, Global Positioning System (GPS) data [54, 56], media access control (MAC) addresses from Bluetooth and Wi-Fi components [14, 20] and mobile phone data [15, 16] are becoming available for the analysis of traffic conditions or even travel behaviour and security in vehicles. The data sources listed above are important in developing and monitoring smart environments for intelligent cities.

With such characteristics, more detailed analysis of attack, detect and defence of vehicle security could be conducted.

2.1 Attack

One of the effects of the extensive introduction of technology in vehicles is car hacking using malware [8]. Nowadays it could be conducted to exploit a new generation of vehicles that are even more connected to wireless networks, to the Internet, and with each other [41], as shown in Fig. 4. Vehicles in Vehicular Ad hoc Networks (VANETs) transmit self-information to fixed remote nodes such as their speed, direction, acceleration and traffic conditions. For example, Dedicated Short-Range Communications (DSRC) are emerging as a standard to support IEEE 802.11 in communications between vehicles. FCC has allocated a 75 MHz of DSRC spectrum at 5.9 GHz to be used in VANETs communications. There is also an IEEE P1609 working group which has proposed DSRC as the IEEE 802.11p standard which gives specifications for a wireless Medium Access Control (MAC) layer and a physical layer for Wireless Access in Vehicular Environments (WAVE) [39]. Attacks on VANETs create a large number of issues for all network users by using different types of attacks, such as malware [53]. In this chapter, two types of attacks using malware are addressed, namely attacks on Inter-vehicle communication (IVC) and attacks on Intra-vehicle communication. Car methods are shown in Fig. 1.

2.1.1 Attacks on Inter-Vehicle Communication (IVC)

Several years ago, work on Inter vehicle communication (IVC) started in industrial research labs and academic institutions. To date, some academic research teams have started addressing security issues in vehicles, however, some research projects are still highly theoretical and do not suggest realistic solutions [26]. In [45], various perspectives on IVC security were considered and the focus was on secure positioning and privacy problems. In this chapter, classifying and identifying the types of attacks in IVC aims to suggest practical solutions.

–Denial of Service Attack (DoS)
The purpose of (DoS) attacks is to prevent legal users from accessing services or data in computer networks. In vehicular networks, this attack jams and overflows the traffic with huge volumes of irrelevant messages that negatively affect communication among the nodes of the network, roadside units and on-board units. There is a huge number of high-powered computing facilities in close proximity to the target because the vehicle under attack is part of the vast infrastructure of the Social Internet of Vehicles (SIoV) embedded in a smart city environment. An attacker can use these for jamming attacks on the target's on-board sensory tool or use malware attacks, thereby countering the ability of the target vehicle to detect irregular messages dur-

Fig. 5 DOS attack in V2V communications

ing collaboration with its local information resources. A voting scheme can address the issues of DoS attacks [36]. However, if attackers can produce false identities to masquerade themselves, voting schemes may fail [32]. According to [34], DoS attacks have three levels:

First level (Basic Level): Overwhelm the Node Resources. The goal of the attacker is to overwhelm the node resources so that other important and necessary tasks cannot be performed by their nodes. These nodes become constantly busy and use all the resources to check the messages.

- Case 01: DOS Attack in V2V Communications. A warning message is sent by an attacker (Accident at location Y) and this message is received by a victim node behind the attacker node as shown in Fig. 5. The attacker continuously repeats the sending of the same message, so the victim node is kept busy and is completely denied access to the network [46].
- Case 02: Launch DOS Attack in V2I Communications as shown in Fig. 6, an attack is launched on a Road-Side Unit (RSU). Any other nodes that attempt to communicate with the RSU will be unable to get any response from the RSU, therefore, the service is unavailable when the RSU is continuously busy attempting to verify the messages. The key risk, in this example, is the inability to send critical life information [46].

Second Level (Extended Level): Jamming the Channel. The highest level of DOS attack involves Jamming the Channel, therefore, denying other users' access to the network. There are two possible cases:

- Case 01: A high-frequency channel, sent by an attacker, jams the communication among any nodes in a domain as shown in Fig. 7. Messages cannot be sent or received by these nodes in that domain (services are not available in that domain due to this attack). It can send and/or receive messages when a node leaves the domain of attack [46].
- Case 02: Jamming the communication channel between the nodes and the infrastructure. In Fig. 8, an attack is launched near the infrastructure to jam the channel.

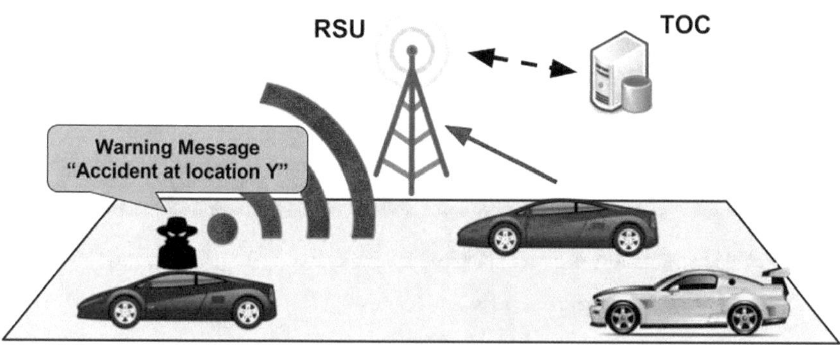

Fig. 6 Launch DOS attack in V2I communications

Fig. 7 A domain of jammed channel for vehicle-to-vehicle communications

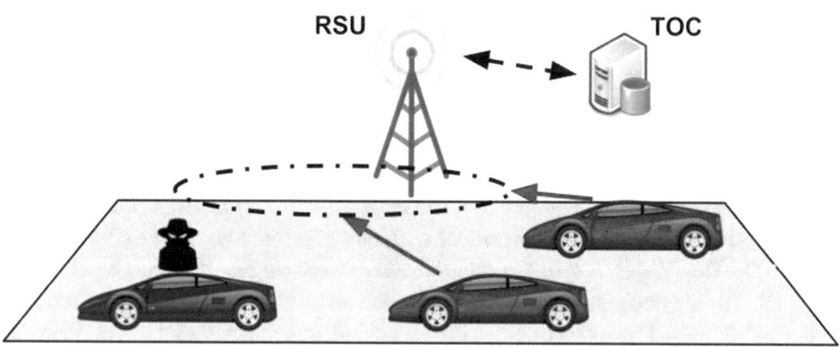

Fig. 8 *Source* denial of service (DOS) attack and its possible solutions in VANET

As a result, the network breaks down. In this way, because the network is unavailable, sending and/or receiving messages and/or malware to/from other nodes is not possible and would fail.

Third Level: Distributed Denial of Services (DDOS): DDOS attacks are more serious in the vehicular ad hoc network (VANET) because of the distribution of this

Fig. 9 DDOS in vehicle-to-vehicle communications

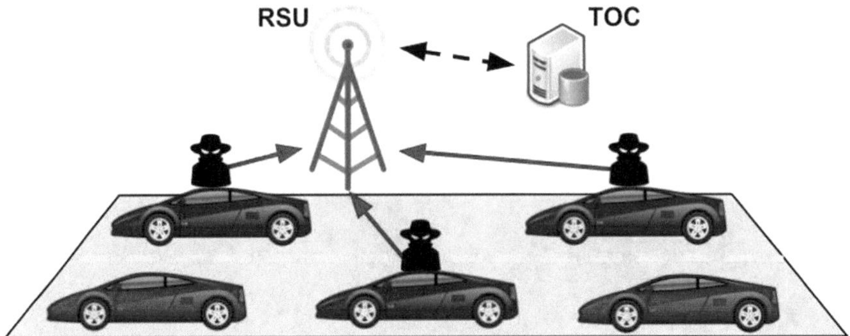

Fig. 10 DDOS in vehicle-to-infrastructure communications

attack which spreads over a wide area of the network. The attacker can launch attacks from various resources. The two possible cases [27] are

- Case 01: An attack is launched from various resources and different time slots may be used to send the messages. These messages and time slots may differ from node to node. The objective of this attack is to make a network unavailable by bringing the network down at a goal node. Figure 9 shows three attackers, black cars (nodes), sending messages to a red car (target node) in front. After a period of time, the goal node cannot connect with any other nodes in the network.
- Case 02: The VANET infrastructure (RSU) is the target of attack as shown in Fig. 10. Three attackers in the network launch an attack on the infrastructure from various sources. The infrastructure is overloaded, causing a denial of service when other nodes in the network want to access the network.

– GPS Spooling Attack

This attack tries to fake a GPS Receiver by broadcasting false GPS signals, structured to match a set of normal GPS signals, or by rebroadcasting genuine signals captured elsewhere or at a different time. The attacker may modify these spoofed signals in such a way by using malware to cause the receiver to estimate its location to be somewhere other than where it actually is, or to be located where it is but at a different time. GPS spooling detection requires swiftness and accuracy. In many GPS-based applications, it is critical to detect GPS spooling attacks as soon as possible, as shown in Fig. 11. GPS has been used in wide-area monitoring systems (WAMSs) in

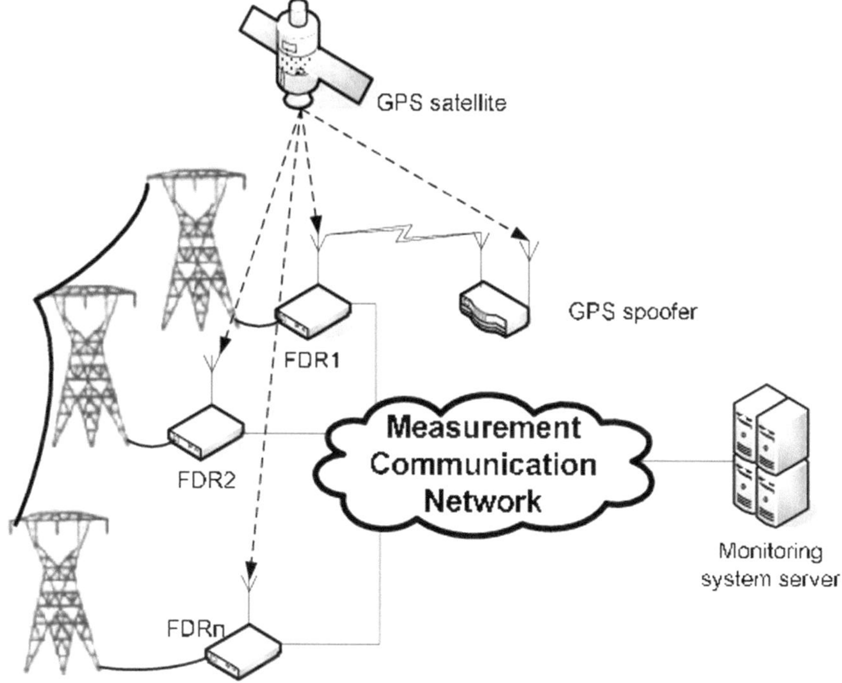

Fig. 11 Illustration of time-critical spooling detection in power grid system [33]

the power grid [63]. WAMS consists of frequency disturbance recorders (FDRs), a communication network and a monitoring system server. Each FDR is provided with a GPS receiver to obtain its position and accurate timing [50].

–Masquerading and Sybil

In a masquerading attack, a vehicle conceals its identity and appears to be legal in the vehicle network. Strangers can conduct attacks, such as injecting false messages or malware. In a Sybil attack, the attackers create several identities, appearing to be several legal vehicles at the same time. They can artificially damage a roadway and impact on the decision making of the other drivers during smart routing systems. In this attack mode, a vehicle can claim several locations concurrently, that can lead to traffic congestion [35].

– Impersonation Attack

The attackers steal the identity of a legal vehicle and can then broadcast security messages on the behaviour of that vehicle. These messages can affect the decision-making of other drivers and generate traffic issues. In [19], a method called Building Up secure Connection along with Key factors (BUCK) has been proposed to detect and separate the impersonation attack.

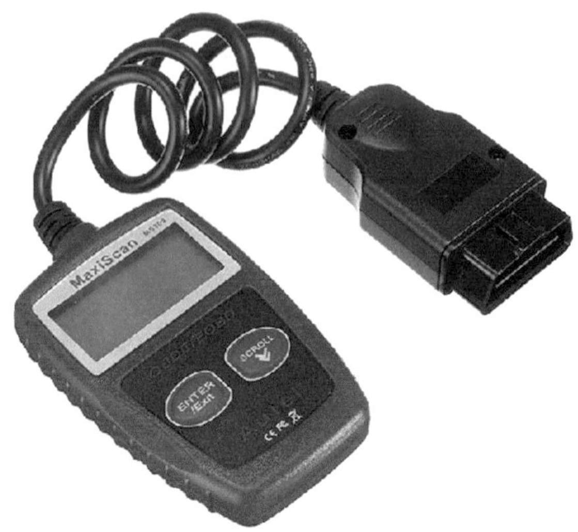

Fig. 12 OBD-II [25]

2.1.2 Attack on Intra Vehicle Communication

– Indirect physical access

In modern vehicles, the internal networks can be accessed either directly or indirectly by several physical interfaces:

- The OBD-II port, as shown Fig. 12, is the most significant automotive interface that can provide direct access to the vehicle's key CAN buses. It provides service personnel with sufficient access to the full range of automotive systems, allowing routine maintenance for both diagnostics and ECU programming [18]. Attackers can also access the in-car entertainment system, for example, introducing false code into MP3 files when playing the file and inserting malicious information and malware in the in-vehicle entertainment system without the owner's knowledge [61].
- Entertainment includes Disc, USB and iPod. A USB port or an iPod/iPhone docking port is external digital multimedia ports provided by vehicle manufacturers, allowing users to control their vehicle's media system by using their personal phone or audio player. Thus, an attacker can deliver malicious information and malware by using encoding algorithms as a song file on a CD and convincing the user to play it by using social engineering. Also, it may compromise an iPod or the mobile phone of the user and install software on them that can help to attack the media system in a vehicle when connected.

– Short-Range Wireless Access

There are many drawbacks in indirect physical access to the network, including challenges to precise targeting, the inability to control the time of compromise and its operational complexity. Therefore, the ability of an attacker to locate a vehicle's

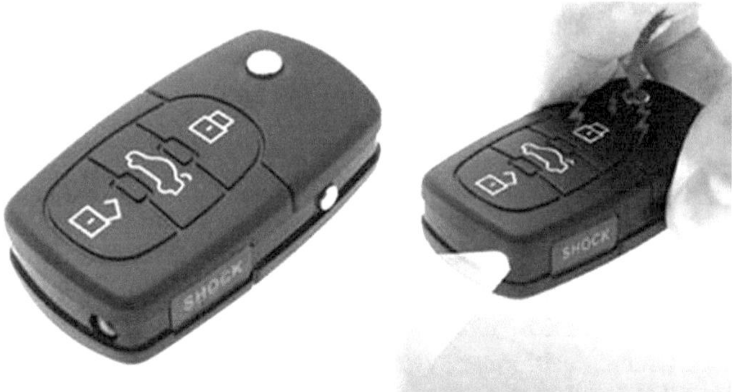

Fig. 13 Remote keyless entry [37]

Fig. 14 Tire pressure monitoring system [3]

wireless interface for devices is required to weaken that ability over a Short-Range Wireless Access [18]. Examples include the following:

- Bluetooth has been used to support hands-free calling in vehicles and it is sold by all vehicle manufacturers. Generally, Bluetooth devices used in vehicles have a range of 10 m. The management services component of the Bluetooth stack is often implemented in software, while the Bluetooth protocol is typically implemented in hardware [45].
  For example, the attacker can place a wireless transmitter close to the vehicle's receiver device. The hackers need to know the vehicle's Bluetooth MAC address to exploit the vehicle's vulnerability without physical contact [61].
- Remote Keyless Entry: automobiles have been equipped with RF-based remote keyless entry (RKE) systems to open doors from a distance, flashlights, switch on the engine of the vehicle and activate alarms, as shown in Fig. 13.
- Tire pressure: Modern vehicles have used a system to support a Tire Pressure Monitoring System (TPMS) to warn a driver about over- or under-inflated tires. It is called Direct TPMS and uses rotating sensors to transmit digital telemetry, as shown in Fig. 14.

– Long-Range Wireless

Modern vehicles include long-distance wireless digital access channels greater than 1 km. These comprise two categories (Adam, 2011):

- Broadcast channels are not specifically aimed at a given vehicle but can be (tuned into) by receivers on request to be a part of the external attack surface. Long-range broadcast media, such as control channels (to make attacks), can be attractive. Because they are difficult to detect, malware can control multiple receivers at once and does not need attackers to get an accurate address for their prey. There is a plethora of broadcast receivers for long-range signals in modern vehicles that include Global Positioning Systems (GPS) (Honda/Acura, GM, Toyota, Saab, Ford, Kia, BMW and Audi). Remote telematics systems are the most significant systems targeted in long-range wireless attacks, with companies, such as Ford (Sync), GM (OnStar) and Toyota (Safety Connect) supplying their vehicles with data networks and cellular voices that provide numerous features, such as (1) supporting safety (crash reporting); (2) convenience (hands-free data access such as driving directions or weather); (3) diagnostics (early alert of mechanical issues); and (4) anti-theft (remote track and disable).
- Cellular channels also have many features vulnerable to attack over considerable distances by using malware, in a mostly covert way, because of the wide coverage of the cellular data structure and its relatively high bandwidth. Moreover, they are two-way channels "supporting interactive control and data ex-filtration" (Adam, 2011) and are individually addressable.

2.2 *Defence*

In the last decade, vehicle industries have been faced with critical security and privacy issues when they developed telematics systems. These issues relate to everyday applications that allow interfacing between vehicles and humans and vehicles and infrastructure. The current risks that face vehicle architectures are wireless security break-ins and sensors, but future automotive architectures and systems will increase these risks and, therefore, they need to be mitigated. Vehicle systems can be protected from (hackers) and infectious viruses using malware that will, from the consumer's perspective, have a direct impact on trustworthiness, the vehicle's safety dynamics and quality. Hacking occurs by taking full advantage of the telematics and wireless features that have become an important part of the vehicle, performing the function of an electrical system brain in the vehicle. Therefore, this allows the module to become the open input to the world. There are two potential solutions to defend against an attack [47].

2.2.1 Inter-Vehicle Communication Solution

This solution combines cryptography and data security with the packet data session through (TCP/IP) and the voice service. A number of researchers have proposed trying to secure V2V networks, but these methods are not sufficient to efficiently provide safety and security [47]. Many attempts have addressed this technique. In [22], a new technique was proposed, namely Elliptic Curve cryptography and Digital Signature Algorithms (ECDSA), by using two parties (a remote agent and network embedded system) to create a 128-bit symmetric key and encrypting all transmitted data through the Advance Encryption Scheme (AES). An Identity-based Batch-Verification (IBV) technique that creates a private key for use in [62]. It does not require a certificate and it will verify each received signature within 300 ms. However, it relies on (the Dynamic Short-Range Communication (DSRC) protocols). The research of [44] investigated how much the Medium Access Control (MAC) protocol can acquire through both quality of Service and security necessity for vehicle network safety applications and how to design an efficient MAC protocol to acquire the safety-related vehicle networks.

2.2.2 Intra-Vehicle Communication Solution

As automotive industries began utilizing more and more electronics in vehicles, huge wire harnesses, that were expensive and heavy, were the result. Specific wiring was then replaced by in-vehicle networks, which reduced wiring weight, complexity and cost. CAN, a high integrity serial bus system for intelligent networking devices, emerged as the standard in-vehicle network (see Fig. 15). It requires data transmission security between the vehicle's ECU via a CAN Bus which is a protocol for an open and unsecured vehicle. Vehicle companies have no concerns about the security of this type of communication because of the low risk of remotely accessing the CAN Bus. The only way of accessing the CAN Bus is by using an On-Board Diagnostic (OBD) connector that can connect a diagnostic tool physically to the vehicle, so that problem analysis can be performed by authorized technicians [59]. However,

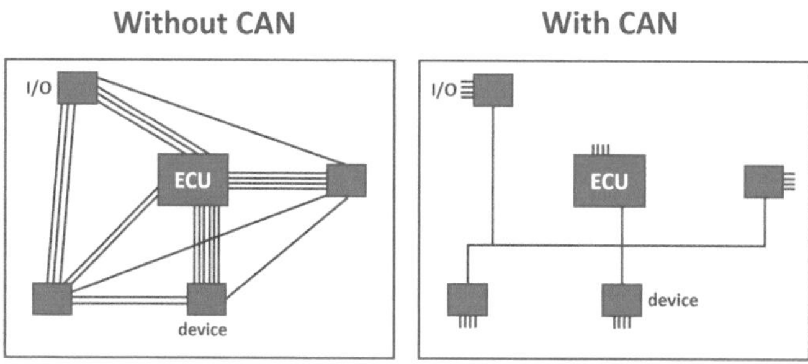

Fig. 15 A high-integrity serial bus system with or without CAN

automotive companies are able to easily develop hardware interfaces and software application layers that allow malware to access the CAN Bus directly through the telematics ECU by using Wi-Fi, BT and cellular networks. Today, technology has increased security risks to the point of allowing unauthorized systems and network access, audit ability and compliance, customer data breaches, internal and external sabotage and the theft of intellectual property and confidential business information [51].

2.3 Detect

2.3.1 Challenges of Inter-Vehicle Communication

There is a critical problem in securing vehicle to vehicle or vehicle to infrastructure communication. This is because all communication between vehicle and vehicle or vehicle and roadside units occurs using wireless technology, therefore, if security is not enforced, the probability of various attacks or viruses being injected into the unprotected system is high [47]. Inter-Vehicle Communication still faces challenges regarding the following issues: trust; real-time communication; quality of service; message dissemination; fault detection; efficient physical layer transmission schemes; wireless network access; secure protocols; information security mechanisms; network scalability; and robustness [42]. Therefore, automotive industries need to create a secure, reliable and effective system to avoid these problems [47].

2.3.2 Challenges of Intra-Vehicle Communication

Internal Vehicle Communication faces a range of issues [33]:

- The use of different generic wireless sensor networks possessing unique characteristics that provide the space for optimization.
- Sensors are stationary so that the network topology does not change over time.
- Sensors are typically connected to the ECU through one hop, which yields a simple star-topology.
- There is no energy constraint for sensors having a wired connection to the vehicle power system. The design and deployment of Internal Vehicle wireless sensor networks are still challenging.
- The Internal Vehicle Communication environment is difficult due to severe scattering in a very limited space and often with no line-of-sight. This is the major reason for the extensive effort to characterize the Internal Vehicle wireless channels.
- Data transmissions require low latency and high reliability to satisfy the stringent requirement of real-time Internal Vehicle control systems.

- Interference from neighbouring vehicles in a highly dense urban scenario may not be negligible.
- Security is critical to protect the in-vehicle network and control system from malicious attacks.

3 Recent Techniques and Challenges

Electrical wiring systems in vehicles have become increasingly sophisticated. They require more and more connectors, control units, relays and terminals to connect the ECU with other devices. Recently, due to developments in automotive technology, vehicles have become even more connected through wireless networks and have become more dependent on complex electronic systems. Therefore, vehicles can be attacked through wireless networks, smartphones, GPS and cameras [60]. Automotive industries, such as AVnu and OPENSIG, argue that Ethernet represents the standard of next-generation automotive networks because ethernet is wide-ranging and includes bandwidth improvements, improved implementation, flexibility and cost savings. Currently, it is not convenient to replace all in-vehicle devices with Ethernet-enabled replacements [43]. Thus, it is likely that Ethernet will function as a high-speed backbone network at first, coexisting with legacy technologies until such time it becomes cost-effective to migrate to a full end-to-end Ethernet solution. As automotive networks become more complicated, the standardization of approaches becomes more and more attractive to manufacturers. This is happening at all levels of the automotive communication stack and is gaining momentum, with organizations such as IEEE RTPGE, OPENSIG, the AVnu alliance and AUTOSAR coordinating an industry-led push towards extensible and cost-effective standards that will drive the development of in-vehicle networks, as shown in Fig. 16. Research in this field has been increased. For example, in [27], a method of V2I cybersecurity architecture, known CVGuard, can detect and prevent cyberattacks on V2I applications. A Stop Sign Gap Assist (SSGA) application has shown that CVGuard was effective in mitigating the adverse safety effects created by a DDoS attack.

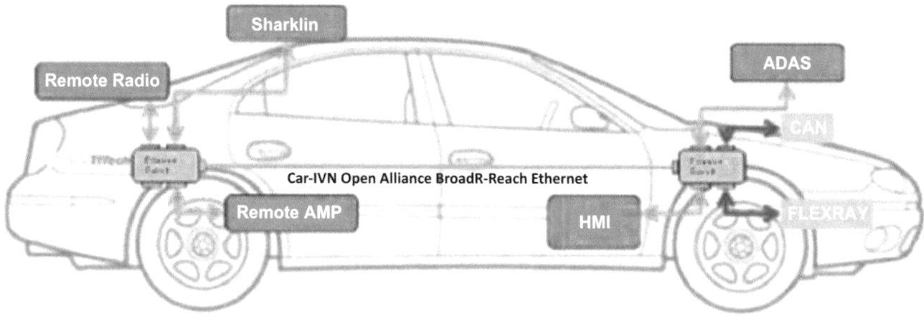

Fig. 16 Ethernet switch to connect vehicle's devices [29, 31]

The literature suggests that, as in-vehicle technology becomes more and more complex, there will be a drive to standardize approaches across the industry, allowing manufacturers to focus on improving the existing applications built on similar foundations. This provides an excellent structure for the future expansion and improvement of in-vehicle network systems and leads, ultimately, to greater driver comfort and, most importantly, safety [5, 55].

4 Conclusion

Developing security solutions compatible with the automotive ecosystem and smart cities is challenging and we believe it will require greater engagement between the computer security community and automotive manufacturers. This chapter provides an opportunity to reflect on the security and privacy risks and malware associated with modern automobiles. We synthesized concrete, pragmatic recommendations for future automotive security and identified fundamental challenges. Defending against known vulnerabilities does not mean the non-existence of other vulnerabilities, thus, many of the specific vulnerabilities identified will need to be addressed. In the future, it may be that the future of intelligent transportation systems and smart cities falls within the multiple layers of the connected environment including cybersecurity and forensics [6], artificial intelligence and machine learning in identification traffic [11], biometric recognition [2, 4, 9], traffic congestion control-based In-Memory Analytics [7] and connected networks of vehicles. These will lead to the development of future intelligent transportation systems and smart cities and vehicle industries that include the analysis of information regarding malware from cyber sources, CSP network modelling and flow models in a connected environment.

References

1. Saber A., F. Di Troia, and M. Stamp. 2020. Intrusion detection and can vehicle networks. *Digital Forensic Investigation of Internet of Things (IoT) Devices*.
2. Abbas, Khamael. 2011. Eye recognition technique based on eigeneyes method. *International Conference on Software and Computer Applications*, vol. 9, 212–219. Singapore: IACSIT Press.
3. Adam. Tire pressure, 2011. Retrieved from: http://www.bergenimports.com/tire-pressure.
4. Khamael Abbas Al-Dulaimi., and Aiman Abdul Razzak Al-Saba'awi. 2011. Handprint recognition technique based on image segmentation for recognize. *International Journal of Computer Information Systems* 2 (6): 7–12.
5. Al-Sabaawi, A., H.M. Ibrahim, M.A.B.M. Almalullah, J. Kaur, K. Al-Dulaimi, and A. Zwayen. 2019. Proposal specifications of building data centre for virtual globalnets. In *2019 IEEE Asia-Pacific Conference on Computer Science and Data Engineering (CSDE)*, 1–7.

6. Al-Sabaawi, Aiman, and Ernest Foo. 2019. A comparison study of android mobile forensics for retrieving files system. *International Journal of Computer Science and Security (IJCSS)* 13 (4): 148.
7. Al-Sabaawi, Aiman Abdul-Razzak Fatehi. 2017. Traffic congestion control based in-memory analytics: Challenges and advantages. *International Journal of Computer Applications* 975: 8887.
8. M. Alazab, S. Venkatraman, and P. Watters. 2013. *Information Security Governance: The Art of Detecting Hidden Malware*. IGI Global.
9. Ali, S.M., and Khamael A. AL-Phalahi. 2009. Face recognition technique based on eigenfaces method. In *3rd Scientific Conference of the College of Science-Baghdad University-Iraq*, 781–785.
10. Atombo, C., C. Wu, H. Zhang, and A.A. Agbo. 2017. Drivers speed selection behaviors, intention, and perception towards the use of advanced vehicle safety. *Advances in transportation studies* 42: 23–38.
11. Azab, A., M. Alazab, and M. Aiash. 2016. Machine learning based botnet identification traffic. In *2016 IEEE Trustcom/BigDataSE/ISPA*, 1788–1794.
12. Azab, A., R. Layton, M. Alazab, and J. Oliver. 2014. Mining malware to detect variants. In *2014 Fifth Cybercrime and Trustworthy Computing Conference*, 44–53.
13. Bécsi, Tamás, Szilárd Aradi, and Péter Gáspár. 2015. Security issues and vulnerabilities in connected car systems. In *2015 International Conference on Models and Technologies for Intelligent Transportation Systems (MT-ITS)*, 477–482. IEEE.
14. Bhaskar, Ashish, and Edward Chung. 2013. Fundamental understanding on the use of bluetooth scanner as a complementary transport data. *Transportation Research Part C: Emerging Technologies* 37: 42–72.
15. Caceres, N., J.P. Wideberg, and F.G. Benitez. 2007. Deriving origin-destination data from a mobile phone network. *IET Intelligent Transport Systems* 1 (1): 15–26.
16. Francesco Calabrese, Mi Diao, Giusy Di Lorenzo, Joseph Ferreira Jr, and Carlo Ratti. 2013. Understanding individual mobility patterns from urban sensing data: A mobile phone trace example. *Transportation Research Part C: Emerging Technologies*, 26:301–313.
17. Robert N Charette. 2009. This car runs on code. *IEEE spectrum* 46 (3):3.
18. Stephen Checkoway, Damon McCoy, Brian Kantor, Danny Anderson, Hovav Shacham, Stefan Savage, Karl Koscher, Alexei Czeskis, Franziska Roesner, Tadayoshi Kohno, et al. 2011. Comprehensive experimental analyses of automotive attack surfaces. In *USENIX Security Symposium*, vol. 4, 447–462. San Francisco.
19. Simranpreet Singh Chhatwal., and Manmohan Sharma. 2015. Detection of impersonation attack in vanets using buck filter and vanet content fragile watermarking (vcfw). In *2015 International Conference on Computer Communication and Informatics (ICCCI)*, 1–5. IEEE.
20. Danalet, Antonin, Bilal Farooq, and Michel Bierlaire. 2014. A bayesian approach to detect pedestrian destination-sequences from wifi signatures. *Transportation Research Part C: Emerging Technologies* 44: 146–170.
21. Dedes, G., S. Wolfe, D. Guenther, Byungkyu Brian Park, J.J. So, K. Mouskos, D. Grejner-Brzezinska, C. Toth, X. Wang, and G. Heydinger. 2011. A simulation design of an integrated gnss/inu, vehicle dynamics, and microscopic traffic flow simulator for automotive safety. *Advances in Transportation Studies*.
22. Duraisamy, Roshan, Zoran Salcic, Maurizio Adriano Strangio, and Miguel Morales-Sandoval. 2007. An elliptic curve key establishment protocol-on-chip. Supporting symmetric 128-bit aes in networked embedded systems. *EURASIP Journal on Embedded Systems* 2007: 1–9.
23. Fysarakis, Konstantinos, Ioannis Askoxylakis, Vasilios Katos, Sotiris Ioannidis, and Louis Marinos. 2017. Security concerns in co-operative intelligent transportation systems.
24. Greenberg, Andy. 2015. Hackers remotely kill a jeep on the highway–with me in it. *Wired* 7: 21.
25. Harborfreigh. OBD II and can code reader with multilingual menu, 2016. Retrieved from: http://www.harborfreight.com/can-obdii-code-reader-with-multilingual-menu98568.html.

26. Hubaux, Jean-Pierre, Srdjan Capkun, and Jun Luo. 2004. The security and privacy of smart vehicles. *IEEE Security and Privacy* 2 (3): 49–55.
27. Islam, Mhafuzul, Mashrur Chowdhury, Hongda Li, and Hu Hongxin. 2018. Cybersecurity attacks in vehicle-to-infrastructure applications and their prevention. *Transportation Research Record* 2672 (19): 66–78.
28. Henrik Johansson, Karl, Martin Törngren, and Lars Nielsen. 2005. Vehicle applications of controller area network. In *Handbook of Networked and Embedded Control Systems*, 741–765. Springer.
29. Koopman. S. 2015. Automotive advanced driver assistance systems ADAS market will reach \$ 18.2bn in 2014. According to a New Study on ASDReports, 2015. https://www.asdreports.com/news-5198/automotive-advanced-driver-assistance-systems-adas-market-will-reach-182bn-2014-according-new-study-asdreports.
30. Koscher, Karl, Alexei Czeskis, Franziska Roesner, Shwetak Patel, Tadayoshi Kohno, Stephen Checkoway, Damon McCoy, Brian Kantor, Danny Anderson, Hovav Shacham, et al. 2010. Experimental security analysis of a modern automobile. In *2010 IEEE Symposium on Security and Privacy*, 447–462. IEEE.
31. Reger, L. 2015. *Advances in automotive at ces* 2015. Retrieved fromhttps://blog.nxp.com/automotive/advances-in-automotive-at-ces-2015.
32. Lesser, Victor, Charles L Ortiz Jr, and Milind Tambe. 2012. *Distributed Sensor Networks: A Multiagent Perspective*, vol. 9. Springer Science & Business Media.
33. Lu, Ning, Nan Cheng, Ning Zhang, Xuemin Shen, and Jon W Mark. 2014. Connected vehicles: Solutions and challenges. *IEEE Internet of Things Journal* 1 (4):289–299.
34. Maglaras, Leandros A, Ali H Al-Bayatti, Ying He, Isabel Wagner, and Helge Janicke. Social internet of vehicles for smart cities. *Journal of Sensor and Actuator Networks* 5 (1):3.
35. Maglaras, Leandros A, Pavlos Basaras, and Dimitrios Katsaros. 2013. Exploiting vehicular communications for reducing co2 emissions in urban environments. In *2013 International Conference on Connected Vehicles and Expo (ICCVE)* 32–37. IEEE.
36. Malla, Adil Mudasir, and Ravi Kant Sahu. 2013. Security attacks with an effective solution for dos attacks in vanet. *International Journal of Computer Applications* 66 (22).
37. Maurizio. Hacking car security system and remote keyless entry, 2015. Retrieved from: http://dev.emcelettronica.com/hacking-car-security-system-and-remote-keyentry-rke.
38. Menouar, Hamid, Ismail Guvenc, Kemal Akkaya, A. Selcuk Uluagac, Abdullah Kadri, and Adem Tuncer. 2017. Uav-enabled intelligent transportation systems for the smart city: Applications and challenges. *IEEE Communications Magazine* 55 (3):22–28.
39. Mokhtar, Bassem, and Mohamed Azab. 2015. Survey on security issues in vehicular ad hoc networks. *Alexandria Engineering Journal* 54 (4): 1115–1126.
40. Nolte, Thomas, Hans Hansson, and Lucia Lo Bello. 2005. Automotive communications-past, current and future. In *2005 IEEE Conference on Emerging Technologies and Factory Automation*, vol. 1, 8. IEEE.
41. Paganini, P. 2013. Car hacking is today possible due the massive introduction of technology in our vehicles, 2013. Retrieved from: https://www.cyberdefensemagazine.com/car-hacking-is-today-possible-due-the-massive-introduction-of-technology-in-our-vehicles.
42. Paruchuri, Vineetha. 2011. Inter-vehicular communications: Security and reliability issues. In *ICTC 2011*, 737–741. IEEE.
43. Porter, Donovan. 2018. *100base-t1 ethernet: the evolution of automotive networking*. Texas Instruments: Techn. Ber.
44. Qian, Yi, Kejie Lu, and Nader Moayeri. 2008. Performance evaluation of a secure mac protocol for vehicular networks. In *MILCOM 2008-2008 IEEE Military Communications Conference*, 1–6. IEEE.
45. Raya, Maxim, and Jean-Pierre Hubaux. 2005. Security aspects of inter-vehicle communications. In *5th Swiss Transport Research Conference (STRC)*, number CONF.
46. Rizvi, Syed, Jonathan Willet, Donte Perino, Seth Marasco, and Chandler Condo. 2017. A threat to vehicular cyber security and the urgency for correction. *Procedia Computer Science* 114: 100–105.

47. Saed, Mustafa, Scott Bone, and John Robb. 2014. Security concepts and issues in intra-inter vehicle communication network. In *Proceedings of the International Conference on Security and Management (SAM)*, 1. The Steering Committee of The World Congress in Computer Science, Computer.
48. Sagstetter, Florian, Martin Lukasiewycz, Sebastian Steinhorst, Marko Wolf, Alexandre Bouard, William R Harris, Somesh Jha, Thomas Peyrin, Axel Poschmann, and Samarjit Chakraborty. 2013. Security challenges in automotive hardware/software architecture design. In *2013 Design, Automation and Test in Europe Conference and Exhibition (DATE)*, 458–463. IEEE.
49. Samuel. G. 2015. Jeep owners urged to update their cars after hackers take remote control. https://www.theguardian.com/technology/2015/jul/21/jeep-owners-urged-update-car-software-hackers-remote-control.
50. Sparwasser, Nils, Markus Stöbe, Hartmut Friedl, Thomas Krauß, and Robert Meisner. 2007. Simworld–automatic generation of realistic landscape models for real time simulation environments–a remote sensing and gis-data based processing chain. *Advances in Transportation Studies* 21.
51. Stallings, W. 2017. *Cryptography and network security: principles and practice*. Upper Saddle River: Pearson.
52. Sumalee, Agachai, and Hung Wai Ho. 2018. Smarter and more connected: Future intelligent transportation system. *IATSS Research* 42 (2): 67–71.
53. Ahmed Sumra, Irshad, Halabi Bin Hasbullah, Iftikhar Ahmad, Daniyal M Alghazzawi, et al. 2013. Classification of attacks in vehicular ad hoc network (vanet). *International Information Institute (Tokyo). Information* 16 (5):2995.
54. Sun, Dihua, Hong Luo, Fu Liping, Weining Liu, Xiaoyong Liao, and Min Zhao. 2007. Predicting bus arrival time on the basis of global positioning system data. *Transportation Research Record* 2034 (1): 62–72.
55. Tuohy, Shane, Martin Glavin, Ciarán Hughes, Edward Jones, Mohan Trivedi, and Liam Kilmartin. 2014. Intra-vehicle networks: A review. *IEEE Transactions on Intelligent Transportation Systems* 16 (2): 534–545.
56. Vanajakshi, Lelitha, Shankar C Subramanian, and R Sivanandan. 2009. Travel time prediction under heterogeneous traffic conditions using global positioning system data from buses. *IET Intelligent Transport Systems* 3 (1):1–9.
57. Wolf, Marko, André Weimerskirch, and Thomas Wollinger. 2007. State of the art: Embedding security in vehicles. *EURASIP Journal on Embedded Systems* 1–16: 2007.
58. Woo, Samuel, Hyo Jin Jo, and Dong Hoon Lee. 2014. A practical wireless attack on the connected car and security protocol for in-vehicle can. *IEEE Transactions on Intelligent Transportation Systems* 16 (2):993–1006.
59. Yadav, Aastha, Gaurav Bose, Radhika Bhange, Karan Kapoor, NCSN Iyengar, and Ronnie D Caytiles. 2016. Security, vulnerability and protection of vehicular on-board diagnostics. *International Journal of Security and Its Applications* 10 (4):405–422.
60. Yang, Teng, Frank Wolff, and Chris Papachristou. 2018. Connected car networking. In *NAECON 2018-IEEE National Aerospace and Electronics Conference*, 60–64. IEEE.
61. Yoshida, J. 2013. How hackers can take control over your car. EE Times. https://www.eetimes.com/document.asp?doc_id=1318838&page_number=2.
62. Zhang, Chenxi, Rongxing Lu, Xiaodong Lin, P-H Ho, and Xuemin Shen. 2008. An efficient identity-based batch verification scheme for vehicular sensor networks. In *IEEE INFOCOM 2008-The 27th Conference on Computer Communications*, 246–250. IEEE.
63. Zhang, Yingchen, Penn Markham, Tao Xia, Lang Chen, Yanzhu Ye, Wu Zhongyu, Zhiyong Yuan, Lei Wang, Jason Bank, Jon Burgett, et al. 2010. Wide-area frequency monitoring network (fnet) architecture and applications. *IEEE Transactions on Smart Grid* 1 (2): 159–167.

A Survey of Intelligent Techniques for Android Malware Detection

Rajesh Kumars, Mamoun Alazab, and WenYong Wang

Abstract The revolution of smart devices such as smartphones, smart washing machines, smart cars is increasing every year, as these devices are provided connected with the network and provide the online functionality and services available with the lowest cost. In this context, the Android operating system (OS) is very popular due to its openness. It has major stakeholder in the smart devices but has also become an attractive target for cyber-criminals. This chapter presents a systematic and detailed survey of the malware detection mechanisms using deep learning and machine learning techniques. Also, it classifies the Android malware detection techniques in three main categories including static, dynamic, and hybrid analysis. The main contribution of this chapter are (1) It briefly describing the background and feature extraction of the static, dynamic, and hybrid analysis. (2) This chapter discusses the basic methodology and frameworks which classify, cluster, or extract Android malware features. (3) Exploring the dataset, harmful features, and classification results. (4) Discussing the current challenges and issues. Moreover, it discusses the most important factors, data-mining algorithms, and processed frameworks.

1 Introduction

With the growth of smartphone and the services they provide such as online shopping, health monitoring system, money transaction, and many more. The android has largest global market in the world. The frequent use of mobile devices with those

R. Kumars (✉) · W. Wang
School of Computer Science and Engineering, University of Electronic Science and Technology of China, Chengdu, China
e-mail: rajakumarlohano@gmail.com

W. Wang
e-mail: wangwy@uestc.edu.cn

M. Alazab
College of Engineering, IT and Environment at Charles Darwin University, Darwin, Australia
e-mail: alazab.m@ieee.org

M. Stamp et al. (eds.), *Malware Analysis Using Artificial Intelligence and Deep Learning*, https://doi.org/10.1007/978-3-030-62582-5_5

facilities encourage people to store and share their personal and critical information through using mobile devices, and the wide use of devices with Android system makes Android-based mobile devices a target for malicious application developers [3, 4, 6, 7, 38, 41, 42, 70–72]. Therefore, the malicious activity can affect the working of many devices connected in a network. Malware is a program or a set of programs that can cause harm to financial forgery, identity, sensitive information or data, and resources. These malicious applications may leak the user's private information without their knowledge or consent.

Personal data leakage: People are not concerned with the security of data or personal information in mobile devices while they are normally very concerned for the same in PC environments [5, 6, 10, 11, 35, 40, 51]. Some apps steal personal information and at the same time demand payments. Such Trojan apps have been downloaded 9,252 times and 211 affected users paid a total of $250,000 to the malware developers [50]. Malware developers successfully stole personal data such as contacts, emails, SMS, and device information which can be used in identity theft and spamming [50].

Social: GPS location, call log, and contact lists can be captured by malware [50]. The contact list and location are user-sensitive information. This information can be captured by malware and can do harm by leaking social identity that can be used in various ways to threaten the security of a user's social image.

Business: Business organizations have their own apps to run their business. Malware can capture user information or business data which will put the business organization at a risk. The business owner will be at a risk of financial loss as well as reputation

Financial loss: The motive of malware development has changed and now focuses on financial gain [23]. Capital expenses related to malware average $ 6–7bn dollars in a fiscal year [23]. " Zeus in the Mobile" is a Trojan that captures the authentication code of the user in a banking application, which may cause financial losses to the user. It is also expensive to remove, where a security firm charged $21/s for the first detection in 2010 [51]. This type of malware can cause user financial losses as well as large financial losses to a business owner in detection fees. In some cases, a user may have to pay large phone bills for premium rate services because of the malicious activity of an app [50].

Every day has various new applications in the market. It is assessed that there will be roughly 6.1 billion smartphone clients by 2020 [55, 60]. Google, the manufacturers of the Free Phone Alliance, and the open-source community of Android developers have made great efforts to enhance security for Android. However, a major concern tends to be the proliferation and development of emerging security threats. Hence, in this context, we discuss the static, dynamic, and hybrid analysis detection Android malware features extraction techniques. After that, the most popular framework to detect malware is discussed. Then, the most popular and basic algorithm and techniques are discussed which is mostly an analysis of malware. Finally, some conclusions about Android malware detection techniques. Additionally, this chapter identifies many elements of security threats involved in using mobile phones and applications, and the user will feel confident in using these applications. The following are the main contributions of this survey:

1. Providing a summary of the current static, dynamic, and hybrid analysis related to Android malware detection using the machine and deep learning.
2. Presenting a current approach to detect Android malware.
3. Exploring the important features extraction methods and results of the machine learning and deep learning approach.
4. Discussing the challenges and open-source dataset of the Android malware detection.

The rest of the paper was structured as follows: Sect. 2 overviews the static, dynamic, and hybrid analysis approaches and discusses the features extraction methods. Section 3 discusses the current methodologies for the classification, clustering , and data mining for the feature extraction. Section 4 discuss the dataset and results of the current machine learning techniques. Section 6 discusses the challenges. Finally, Sect. 7 concludes the chapter.

2 Static, Dynamic, and Hybrid Analysis of Android Malware Background

In this chapter, we discuss the background of Android malware detection techniques. There are three basic techniques to detect Android malware. (i) Static analysis, (ii) Dynamic analysis, iii) Hybrid analysis. Firstly, we discuss the static analysis, which consists of two methods (i) Permission-based analysis (ii) API Call based analysis. Secondly, we elaborate on the dynamic analysis that is used to extract the training characteristics of the model. Also, we consider the hybrid analysis that combines static and dynamic analysis. Finally, we compare the static, dynamic, and hybrid analysis.

2.1 Static Analysis

The static analysis method refers to analyzing source code files or executable files without running applications. There are several features such as API call and permissions to analyze the static analysis. The feature extraction methods are shown in Fig. 1 (Table 1).

Furthermore, some static features detection methods are shown in Table 2. The k-nearest neighbors machine learning classifier achieve better performance and accuracy in the detection of the malware. However, it takes more processing time with a large amount of data. That's why most of the authors used Support Vector Machine and Random Forest classifiers. Therefore, we use and enhance the Random Forest algorithm for Android malware detection.

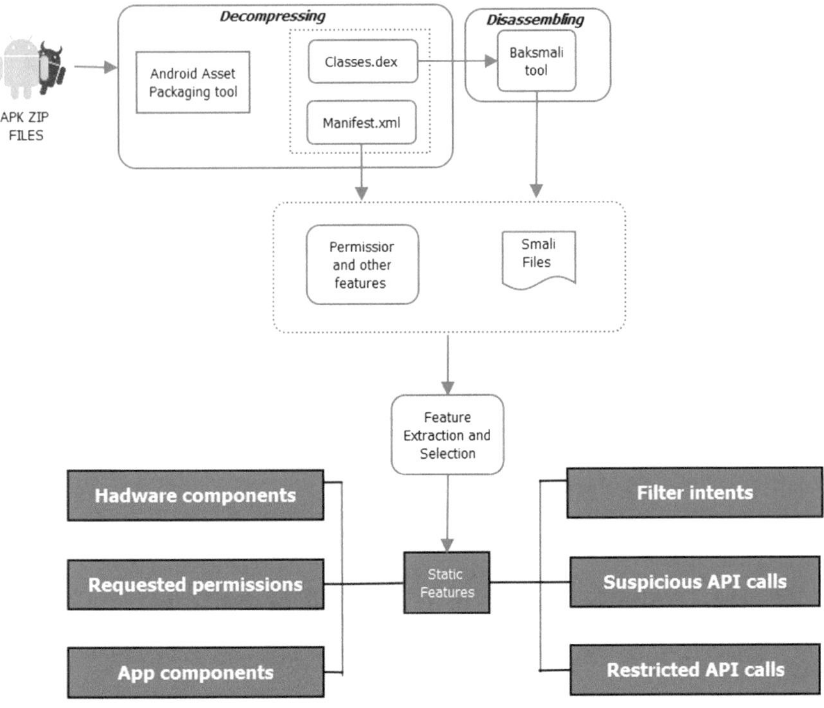

Fig. 1 Static feature extraction of method

Table 1 Overview of feature sets of Android APK decompiled files

Feature sets		
Manifest	S1	Hardware components
	S2	Requested permissions
	S3	Application components
	S4	Filtered intents
Dexcode	S5	Restricted API calls
	S6	Used permission
	S7	Suspicious API calls
	S8	Network addresses

2.1.1 Permission-Based Analysis

Permission-based access control mechanism is a major component of the Android platform security mechanism. On the Android platform, applications are separated from applications, and applications and systems are isolated. When applications perform certain operations or access certain data, they must apply for corresponding permissions. This means that permissions defined in the manifest file can indicate

Table 2 Static features detection methods

Ref	Features	Accuracy	Machine learning models	Contribution	Limitation
[9]	Permission	91.75%	Random Forest	Permission-based approach using KNN clustering	Risky permission not founded
[29]	Permission	81%	C4.5, SVM	The framework quick identify the malicious permission	It uses the limited number of malware. It requires the evidence
[58]	Permission	88.20%	HMNB	Probabilistic generative models for ranking the permission. It identifies ranging from the simple Naive Bayes, hierarchical mixture models	Susceptible to adversarial attack
[17]	Permission	–	AHP	a global threat score deriving set of permissions required by the app	Only depends on permissions with known limitations—susceptible to attack
[47]	Permission	98.6	J48	Build a framework for based on SIGPID. It extracts top 22 permissions.	Susceptible to impersonate attack
[39]	Permission	92.79%	Random Forest	Design a model which score the malicious permission	Susceptible to adversarial attack
[57]	Permission	94.90%	Random Forest	It uses the classification algorithm to detect the malware.	Susceptible to adversarial attack
[13]	Permission, API calls	92.36%	Random Forest		Susceptible to adversarial attack
[78]	Permission, API calls, intent	97.87%	k-nearest neighbors	Design a DroidMat Framework which is based on manifest and API call tracing	Susceptible to adversarial attack
[1]	API call	99%	k-nearest neighbors	It mitigates Android malware installation through providing lightweight classifiers	Susceptible to impersonate attack

(continued)

Table 2 (continued)

Ref	Features	Accuracy	Machine learning models	Contribution	Limitation
[16]	API call	93.04%	Signature matching	It measures the similarity of malware	Susceptible to impersonate attack
[15]	API call	96.69%	SVM	The paper uses malicious-preferred features and normal-preferred features for the detection of malware	Susceptible to impersonate attack
[79]	ICC related features	97.40%	SVM	Design a ICCDetector framework which classify the malware based on android intent filters	Susceptible to impersonate attack
[82]	Permission, command, API calls	98.60%	Parallel classifier	This paper combine the machine learning classifiers to classify the malware.	Susceptible to impersonate attack
[27]	Requested permissions-used permissions sensitive API calls-Actions-app components	F1 97.3 Prec. 98.2 Recall 98.4	DBN	DroidDeep for detection of malware using deep belief network	Susceptible to adversarial attack
[75]	Risky Permissions-dangerous API calls	F1-94.5 Recall-94.5 Prec-93.09	DBN	Proposed DroidDeepLearner combines risky permission and dangerous API calls to build a DBN classification model.	Susceptible to adversarial attack
[28]	API call blocks	ACC 96.66%	DBN	DroidDelver Detection system is used to identify malware using an API call block.	Susceptible to adversarial attack
[22]	Requested permission	Acc 93%	CNN-AlexNet	Proposed a detection system that converts the requested permissions into an image format and then uses CNN for classification	Only depends on permissions with known limitations—susceptible to attack

(continued)

Table 2 (continued)

Ref	Features	Accuracy	Machine learning models	Contribution	Limitation
[88]	323 features	F1 95.05	DBN	An identification system designed by FlowDroid uses data flow analysis to identify malware.	Susceptible to adversarial attack
[52]	Learn to detect sequences of opcode that indicate malware	ACC 98 Prec. 99 Recall 95 F1 97	CNN	Developed a detection system that uses automatic functions to learn from raw data and to treat the disassembled code as text	Although trained on a large dataset, performance dropped when tested on a new dataset—Susceptible
[54]	API call sequence	Acc 99.4 Prec. 100 Recall 98.3 Acc 97.7	CNN	The proposed method based on API call sequence that can use the multiple layers of CNN.	Susceptible to impersonate attack
[27]	Extract features from the transferred images		CNN	Proposed a RGB scheme based on color representation.	Results showed that human experts are still needed in the collection and updating of long-term samples. Susceptible to an attack
[46]	Dangerous API calls-risky permissions	Recall 94.28	DBN	DBN was used to create an automatic malware classifier	Susceptible to adversarial attack
[86]	API calls Permissions-Intent filters	Prec 96.6 Recall 98.3 ACC 97.4 F1 97.4	CNN	Presented system detection of malware DeepClassifyDroid Android based on CNN	Susceptible to impersonate attack
[65]	API calls	Acc 95.7	DBN	Suggested approach to image texture analysis for malware detection	Risky permission not founded

(continued)

Table 2 (continued)

Ref	Features	Accuracy	Machine learning models	Contribution	Limitation
[74]	Permissions requested permissions filtered intents restricted API calls-hardware features-code related features suspicious API calls	Acc 98.8 Recall 99.91 F1 99.82	CNN	A hybrid malware detection model has been developed using CNN and DAE	It uses the limited number of malware. It requires the evidence
[34]	API sequence calls	F1 96.29 Prec 96.29 Recall 96.29	CNN	MalDozer used natural language processing technique to detect Android malware that can identify the malware family attributes.	Susceptible to adversarial attack
[80]	The semantic structure of Android bytecode	Acc 97.74	CNN LSTM	DeepRfiner was proposed to identify the malware. The structure of method use the LSTM for semantic byte code	Only depends on permissions with known limitations—susceptible to attack
[44]	Permissions API Calls	Prec 97.15 Recall 94.18 F1 95.64	DNN	Implemented DNN—based malware detection engine	Susceptible to impersonate attack
[26]	Code analysis	Acc 95.4	CNN	The proposed method for analyzing a small portion of raw APK using 1-D CNN	Susceptible to adversarial attack

the behavior of the application. Developers can declare the permissions that need to be applied in the <uses-permission> tag or <permission> tag. The permissions in the <uses-permission> tag are predefined by android, and the permissions in the <permission> tag are customized by the developer and belong to third-party permissions. According to Android's official documentation, the level of protection of permissions implies the potential risks involved and points out the verification process that should be followed when the system decides whether to grant application permissions. The four protection levels are described as follows: Normal defines the low-risk permissions to access the system or other applications, which does not require user confirmation and is automatically authorized. Dangerous can access user data or control the device in some form, such as READ_SMS (allowing applications to read SMS). When granting such permissions, the system will pop up a confirmation dialog box and display the permission information requested by the application. The user can choose to agree or cancel the installation. Signature is the most severe permission level and requires an encryption key. It only grants applications that use the same certificate as the declared permissions. Therefore Signature usually only appears in applications that perform device management tasks, such as ACCESS_ALL_- EXTERNAL_STORAGE (access to external storage). System can be granted either partial applications of the system image or applications with the same signature key as the declaration permission.

2.1.2 Suspicious API Calls

The second solution is a static analysis of the source code of the app. Malicious codes usually use a combination of services, methods and API calls that is not common for non-malicious applications [12]. To differentiate malicious and non-malicious applications, Machine learning algorithms can learn common malware services such as combinations of APIs and system calls. Figure 2 shows the some of suspicious API calls, which are mostly used by malware applications. Figure 3 shows the extracted features from the APK file that contains the classes.dex file.

2.2 Dynamic Analysis

The dynamic analysis method is not affected by code transformation technologies, such as bytecode encryption, reflection, and native code execution, and can deeply analyze the malicious behaviors of the application. Therefore, it makes sense to collect dynamic features, which can effectively compensate for the limitations of static analysis. Figure 4 shows the feature extraction method and detection technique of the dynamic analysis. Many machine learning algorithm used for dynamic analysis, for instance, Logistic regression (LR), K-means Clustering, SVM, KNN_E,KNN, Bayesian network (BN), and NaÃ¯ve Bayes. Table 3 illustrates the accuracy level, dynamic features, and detection methods. For example, some malware may obtain malicious files through the network or other means during the running process, and then write them into the system files to perform malicious behaviors. These means can escape static detection and affect the accuracy of detection. DroidBox is an Android application sandbox that extends TaintDroid. It can perform dynamic strain analysis at the application framework level, and monitor various operations of the application, such as information leakage, network, file input / output, and encryption operations. DroidBox provides two scripts, startemu.sh and droidbox.sh. The former is used to start a simulator dedicated to the dynamic analysis of Android applications, and the

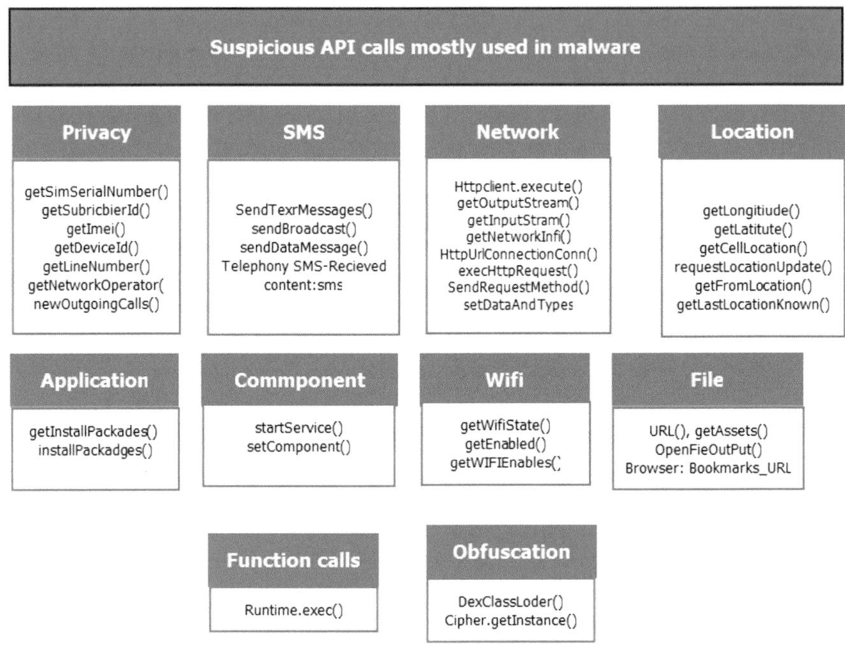

Fig. 2 Suspicious API calls

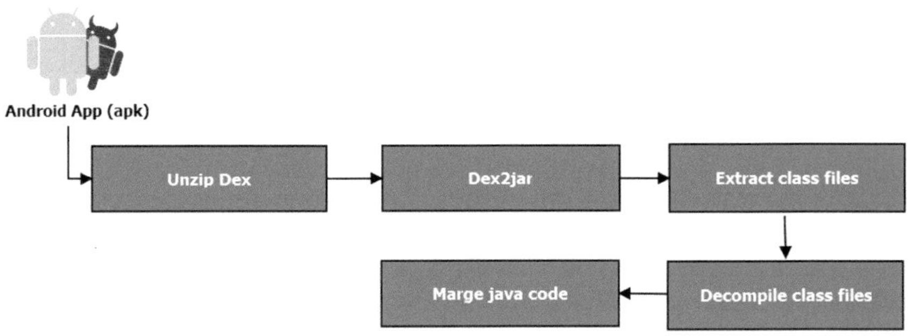

Fig. 3 Workflow of android file decompiling

latter is used to perform specific dynamic analysis. We obtain the dynamic operation log of each application by installing and running each application in DroidBox for 30s, and extract features from them (Table 4).

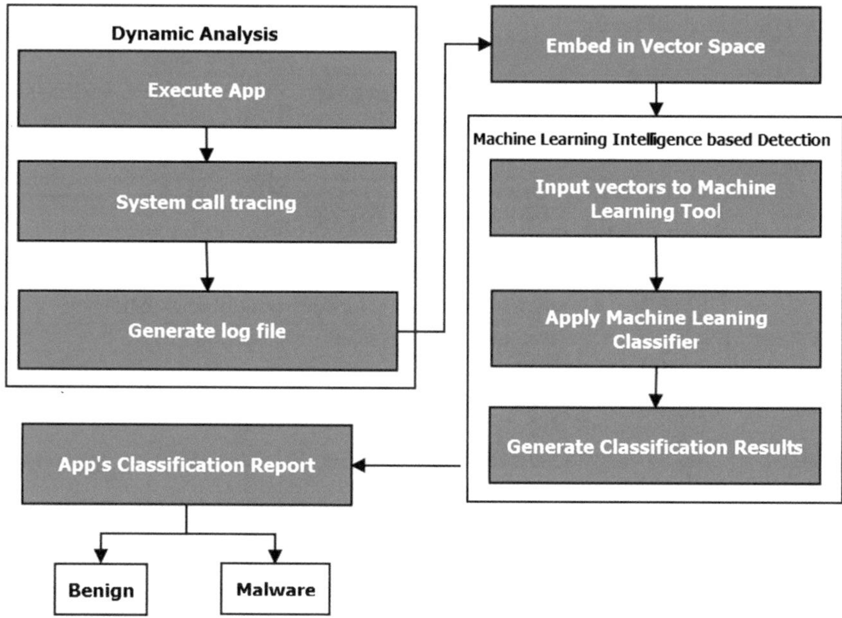

Fig. 4 Dynamic feature extraction and detection

Table 3 Dynamic features detection methods

Ref	Features	Accuracy	Machine learning models
[32]	System call	91.75%	Signature Matching
[12]	System call	81%	K-Means
[24]	System call	88.2%	Frequency
[25]	System call	–	Pattern matching
[77]	API call	97.6	KNN_M
[29]	Native size	99.9%	RF, SVM

Table 4 Suspicious API call

Name	Used in malicious	Used in benign
PTRACE	Most often utilized [24, 49]	Utilized in benign applications [24]
SIGPROCMASK	Most often utilized [24, 49]	Utilized in benign applications [24]
CLOCK	Most often utilized [49, 68]	–
CLOCK-GETTIME	Utilized in malicious applications [24]	Utilized in benign applications [24]
RECV	Most often utilized [24, 68]	Not Utilized [24]
RECVFROM	Most often utilized [25, 49, 68]	Not Utilized [24]
WRITE	Most often utilized [25, 49, 68]	Utilized in benign applications [24]
WRITEV	Most often utilized [24, 68]	Utilized in benign applications [24]
WAIT4	Most often utilized [49]	
SEND	Most often utilized [68]	
SENDTO	Most often utilized [49, 68]	
MPROJECT	Most often utilized [25, 49, 68]	Utilized in benign applications [24]
FUTEX	Most often utilized [24, 49]	Utilized in benign applications [24]
IOCTL	Most often utilized [24, 49]	Utilized in benign applications [24]
FCNTL64	Most often utilized [24]	Utilized in benign applications [24]
GETPID	Most often utilized [24, 49]	Utilized in benign applications [24]
GETUID32	Most often utilized [24, 49]	Utilized in benign applications [24]
EPOLL	Most often utilized [24]	Utilized in benign applications [24]
EPOLL-CTL	Most often utilized [24]	Utilized in benign applications [24]
EPOLL-WAIT	Most often utilized [25, 68]	Utilized in benign applications [24]
CACHEFLUS	–	–
READ	Most often utilized [49, 68]	Utilized in benign applications [24]
READV	Most often utilized [68]	–
STAT64	–	–

(continued)

Table 4 (continued)

Name	Used in malicious	Used in benign
GETTIMEEOFDAY	utilized in malicious applications [24]	Utilized in benign applications [24]
ACCESS	Most often utilized [25, 68]	Utilized in benign applications [24]
PREAD	–	–
UMASK	Most often utilized [24]	Not Utilized [24]
CLOSE	utilized in malicious applications [24]	Utilized in benign applications [24]
OPEN	Most often utilized [24, 68]	Utilized in benign applications [24]
MMAP2	utilized in malicious applications [24]	Utilized in benign applications [24]
MUNMAP	–	–
MADVISE	utilized in malicious applications [24]	Utilized in benign applications [24]
FCHOWN32	Most often utilized [24]	Not Utilized[24]
PRCTL	Not Utilized [24]	Utilized in benign applications [24]
BRK	Most often utilized [24]	Not Utilized[24]
LSEEK	Utilized in malicious applications [24]	Utilized in benign applications [24]
DUP	Utilized in malicious applications [24]	Utilized in benign applications [24]
GETPRIORTY	Utilized in malicious applications [24]	Utilized in benign applications [24]
PIPE		
CLONE	Utilized in malicious applications [24]	Utilized in benign applications [24]
FSYNC	Most often utilized in [24]	Not Utilized[24]
GETDENTS64	Utilized in malicious applications [24]	Utilized in benign applications [24]
GETTID	Utilized in malicious applications [24]	Utilized in benign applications [24]
LSTA64	Utilized in malicious applications [24]	Utilized in benign applications [24]
FORK	–	–
NANOSLEEP	Not Utilized [24]	Only Utilized in benign applications [24]
RECVMSG	–	–
CHMOD	Utilized in malicious applications [24]	Utilized in benign applications [24]
SENDMSG	Most widely Utilized[49]	–
FLOCK	Not Utilized [24]	Only Utilized in benign applications [24]

(continued)

Table 4 (continued)

Name	Used in malicious	Used in benign
MKDIR	Most often utilized [24]	Not Utilized [24]
CONNECT	Most often utilized [24]	Not Utilized [24]
POLL	Not Utilized [24]	Only Utilized in benign applications [24]
RENAME	Most widely Utilized [68]	Not Utilized [24]
SETPRIORITY	–	–
SETSOCKOPT	Most often utilized [24]	Not utilized [24]
SOCKET	Most often utilized [24]	Not utilized [24]
UNLINK	–	–

2.3 *Hybrid Analysis*

To improve the performance of learning algorithms, the hybrid analysis was developed, which utilizes the dynamic and static features as shown in Figure fig: Hybrid Analysis. Some researches proposed multi-classification techniques [20, 30] to obtain high accuracy in the hybrid analysis. Furthermore, The static features are Publisher ID, API call, Class structure, Java Package name, Crypto operations, Intent receivers Services, Receivers, and Permission, and dynamic are Crypto operations, File operations, Network activity. The APK file extracted static features from classes.dex files, and dynamic features from Androidmanifest.xml file. Hybrid Analysis combines static features and dynamic features. These features are used to detect malicious applications. In [48], the following features are selected form static (permission and APICall) and dynamic (SystemCall). Y. Liu, et al. [48] used the SVM and Naive Bayes machine learning classifier. The SVM classifier used for static analysis achieved 93.33 to 99.28 percent accuracy, while the Naive Bayes used for dynamic

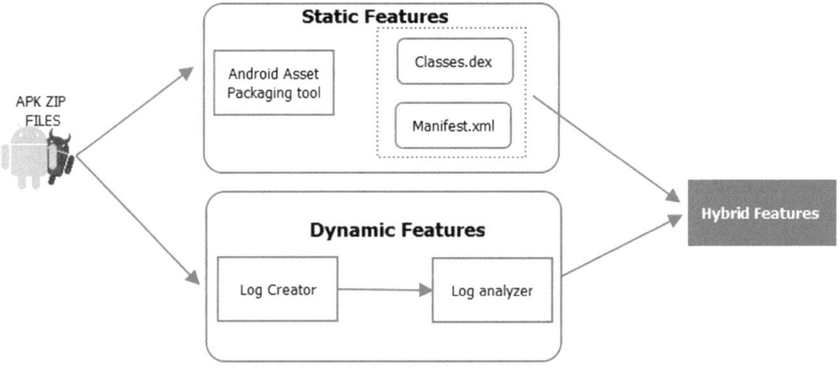

Fig. 5 Dynamic feature extraction and detection

Table 5 Hybrid analysis methods

Ref	Methodology	Tools	Achements	Limitations
[69]	Decompress and decompile the Android app using the tool Baksmali. Scans decompiled samli files to extract static patterns. Generate static behavior vector. Installs and executes the applications on emulator Runs monkey to give user inputs Hijacks system calls using LKM logs the system calls	Baksmali Monkey tool Emulator	can detect the malicious system calls at kernel space	Insufficient test results for malware detection No comparison of the system is provided against any other malware detection techniques. Not any classification results are available Increase in malware detection rate is not shown Incomplete evaluation system
[87]	Detects known malware samples by filtering and foot printing based on permission. Detects zero-day malware through heuristic filtering and dynamic monitoring of execution	–	Successfully detects 211 malicious apps among 204,040 apps. +Detect two zero-day malware Droid Dream light and Plankton Achieves 86.1 accuracy	This study is limited to two heuristics Permission-based filtering only considered the essential permission of 10 malware families
[2]	Pre-process the App through API Monitor to obtain static features such as API calls. Install the app on AVD. Uses APE_BOX, combination of DroidBox and APE, to collect the run-time activities and simulation of GUI-based event. Combines the static and dynamic features and applies SVM classification	API Monitor APE DroidBox LIBSVM	Achieves 86.1% accuracy	Time consuming due to use of emulators High resource consumption in log collection. Malware can easily evade anti-emulator techniques

(continued)

Table 5 (continued)

Ref	Methodology	Tools	Achements	Limitations
[37]	Extract the static features from manifest file and disassembled dex file using Aapt Extracts dynamic features using CuckooDroid Maps the features into vector space and performs vector selection. Uses LinearSVC classifier in Misuse detection to classify the application, if app is malware uses signature-based detection to identify the malware. Applies anomaly detection if App is not classified by misuse detection and uses signature-based detection to identify the family of malware	Android Asset Packaging Tool	Detects known malwares and their variants with 98.79% true positive rate. Detects the zero-day malwares real positive rate with 98.76 percent accuracy	Comparison of proposed scheme with other well-known malware detection schemes, e.g., RiskRanker, Drebin, Kirin, etc. is not provided
[56]	Parameters related to permissions, such as broadcast receivers, intents and services, are decompiled from the manifest file in the static analysis phase using Aapt. In the behavior analysis phase, the Android emulator app is executed and the functions related to user interactions, java based, and native function calls are extracted. Performs feature on the basis of information gain and records them in CSV file. Rule generation module uses CSV file to create rules and maps the permission against the function calls for classification	Android Asset Packaging Tool	Achieves 96.4% detection rate	High time for scanning. High electricity consumption. High consumption of resources/storage
[84]	Extracts sensitive API calls and permissions as static features. Logs dynamic action for dynamic analysis Applies deep learning model for classification	7ZIP, XML-printer2 Tinyxml, DropidBOX Baksmali	Detects 96.7 percent accurate malware	Unrealistic malware for dynamic analysis that does not display malicious behavior throughout the monitoring interval can evade the detection system

(continued)

Table 5 (continued)

Ref	Methodology	Tools	Achements	Limitations
[64]	Extracts PSI from binary code files as static features sort features according to the frequency of occurrence in each file. Selects feature with occurrence frequency above certain threshold value and creates static feature vector. For dynamic feature use cuckoo malware analyzer. For each file, create API call grams and analyze API call sequences based on the n-gram method. Selects grams of API call above a certain threshold value and creates a dynamic function vector. Concatenates both feature vector for each file and input them to Machine learning classifiers	WEKA	Classifies 98.7 percent accurate unknown applications	Comparison of proposed scheme with other well-known malware detection schemes, e.g., RiskRanker, Derbin, DroidRanger, etc. is not provided
[48]	Decompiles applications using Akptool and analyze the decompiled results. Automatically switches to static analysis if app is correctly decompiled. Performs extraction of static features, permission and API calls, from manifest and smali files. Inputs the feature vectors to machine learning classifiers, SVM, KNN, and Naive Bayes. If application does not correctly decompile then it performs dynamic analysis by operating the app with monkey tool and monitoring the app's actions using strace. Generates the feature vector of traced system call logs and applies the machine learning classifier on the feature vector for classification	APK tool Strace Monkey tool	Achieves 99% accuracy as a result of static analysis and 90% accuracy as a result of dynamic analysis	Only static or dynamic analysis can be performed on the application, so that the dynamically labeled data cannot be detected in an easy way for static analysis Only the executed code is analyzed when dynamic analysis is carried out. The non-executed code remains undetected

(continued)

Table 5 (continued)

Ref	Methodology	Tools	Achements	Limitations
[62]	Extracts features at four different levels: user level, application level, kernel level, and package level user activities at user level and market information and riskiness of application at package level Generates feature vectors consisting of 14 features and input the vector to KNN classifier. Notifies the user about malicious apps and helps the user to block and remove them through UI			Only runs on rooted devices with a carnal having module support due to which it has not been conceived for distribution in the mass market. Pre-installed apps are not analyzed by the app evaluator. Thus, will not be included in apps suspicious list and so will not be dejected against known malware behavior patterns. only the apps identified as risky or added to the apps suspicious list. 9.4% memory overhead because classifier requires the training data and memory
[61]	Feature collector collects static features of at the application at installation. GramDroid a web tool that extracts the features of applications and provides their visual representation in order to identify the threads posed by the application Local detector classifies the application as legitimate, malware, or risk using static features. Response manager gives control to use if app as detects as malware. Cloud detector performs detailed dynamic analysis at a remote server if app is detected is risk by local detector updates the database if app is detecting malware		From top 20 enlisted frequently requested permission	
[33]	The Android device's client application captures the application's specific information and sends it to the server. Detailed analysis and application execution based on emulation is carried out. Otherwise, the APK file will be sent from the client device to the server	Androgaurd	Detects 99% accurate malware applications	The malware can easily evade emulation-based detection

(continued)

Table 5 (continued)

Ref	Methodology	Tools	Achements	Limitations
[67]	User permission to detect malware behavior as static analysis. The signature data type contains all applications signature. Android user offers users a malware analysis service. The central server connects the Android client to the signature database		Archives 92.5% specificity	It lacks the advantages of dynamic analysis, as dynamic malicious payloads cannot be detected
[66]	Uses static functions, manifest file, and code files assembled. Uses system calls and binder transactions as dynamic behavior features. The user and the application monitor and signature are forwarded to the server which applies to generate the signature. The signature matching algorithm		Achieves 99% accuracy	Overall causes 7.4 percent overhead performance and 8.3 percent overhead memory

analysis achieved accuracy up to 90 percent. Furthermore, Kim et al. [36], used the J48 machine learning classier, the features are selected from static (permission) and dynamic (APICal l). A. Saracino el al. [62], achieved 96.9% accuracy based on KNN by selecting the static feature (permission) and dynamic (critical API, SMS, User activity System call) feature (Fig. 5 and Table 5).

2.4 A Comparison of Static, Dynamic, and Hybrid Analysis

Static Analysis:

1. Single Category features: The advantages of single category features are easy to extract, and low power computation. The limitations associated with this method are code obstruction, imitation attack, and low precision.
2. Multiple categories of Features: The advantages of multiple category features are easy to extract, and high accuracy. The limitations associated with this method are Mimicry attack, high computation, code obfuscation, and difficult to handle multiple features

Dynamic Analysis:

1. Single Category features: it poses a better accuracy and easy to recover code obfuscation as compared with static analysis. However, its feature extraction process is difficult, and it consumes high resources.
2. Multiple categories of Features: It gives better accuracy and easy to recover code obfuscation as compared with a static and dynamic single category. The limitations

of this approach are (1) difficult to handle multiple features, (2) high resources, and (3) more time computation.

Hybrid Analysis: The main benefits of hybrid analysis are to perform the highest accuracy as compared to static and dynamic analysis. The limitations are (1) highest complexity, (2) framework requirement to combine the static and dynamic features, (3) more resources utilization, and (4) time-consumption.

3 Android Malware Detection Approaches

3.1 Basic Proposed Framework to Detect Android Malware

In this section, we discuss the methodology to detect malicious codes detection techniques based on deep learning and machine learning. Kim et al. [38] proposed an multi-model malware detection-based malware analysis system to automatically analyze and classify malware behaviors. Figure 6 shows the overall architecture of the developed framework. The multimodal deep learning framework uses seven kinds of the feature; String feature, method opcode feature, method API feature, shared library function opcode feature, permission feature, component feature, and environmental feature. Using those features, the seven corresponding feature vectors are generated first, and then, among them, the permission/component/predefined setting feature vectors are merged into one feature vector. Finally, the five feature vectors are fed to the classification model for malware detection.

Moreover, Tao Lei et al. [43] proposed an Graph-based malware detection model based on three components: (1) call graph extraction; (2) event group building; and

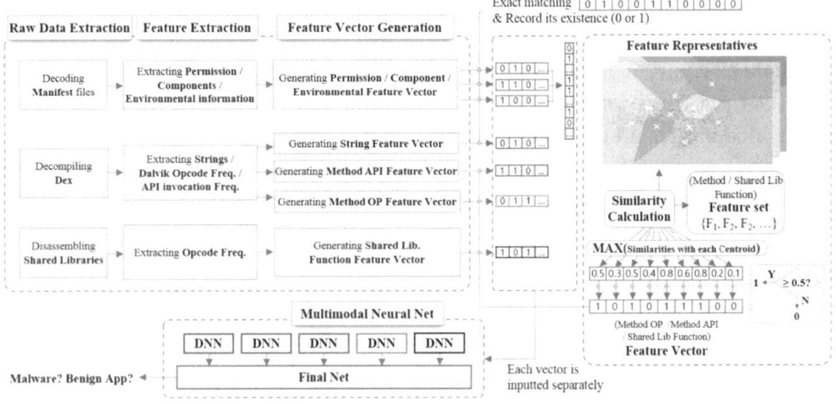

Fig. 6 A multimodal deep learning method for android malware detection using various features [38]

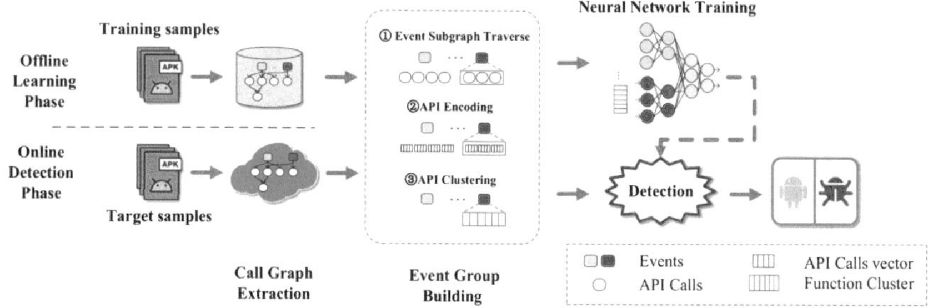

Fig. 7 EveDroid: event-aware android malware detection against model degrading for IoT devices [43]

(3) NN training. These three phases are shown in 7. In call graph phase it extracts the call graphs of every sample from the training samples by using the static analysis tools and then filters out repetitive API calls. The event group building component aims to build the event group for apps, which consists of event subgraph traverse, API calls encoding and clustering. Finally, the event group (clustering result) is fed into the NN to train the parameters.

Andrea Saracino et al. [62] detect malicious behavioral-patterns extracted from several categories of malware. The features at the four system levels, and to detect and prevent a misbehavior. It consists of four steps shown in Fig. 9. The first one is the App Risk Assessment, which includes the App Evaluator that implements an analysis of metadata of an app package (apk) (permission and market data), before the app is installed on the device. The second block is the Global Monitor, which monitors the device and OS features at three levels, i.e., kernel (SysCall Monitor), user (User Activity Monitor), and application (Message Monitor). The third block is the Per-App Monitor, which implements a set of known behavioral patterns to monitor the actions performed by the set of suspicious apps (App Suspicious List), generated by the App Risk Assessment, through the Signature-Based Detector (Fig. 8).

Huijuan Zhu et al. [89] raises a stacking ensemble framework SEDMDroid to identify Android malware. Specifically, to ensure individual's diversity, it adopts random feature subspaces and bootstrapping samples techniques to generate subset, and runs Principal Component Analysis (PCA) on each subset. The accuracy is probed by keeping all the principal components and using the whole dataset to train each base learner Multi-Layer Perception (MLP). Then, Support Vector Machine (SVM) is employed as the fusion classifier to learn the implicit supplementary information from the output of the ensemble members and yield the final prediction result. Figure 9 shows the overall proposed framework of the SEDMDroid (Fig. 10).

Jin Li, et al. [45] propose the malware detection framework based on static analysis for permission feature. The proposed framework consists of three-technique to collect risky permissions. (i) Permission Ranking With Negative Rate (ii) Support-Based Permission Ranking (iii) Permission Mining With Association Rules. It extracts

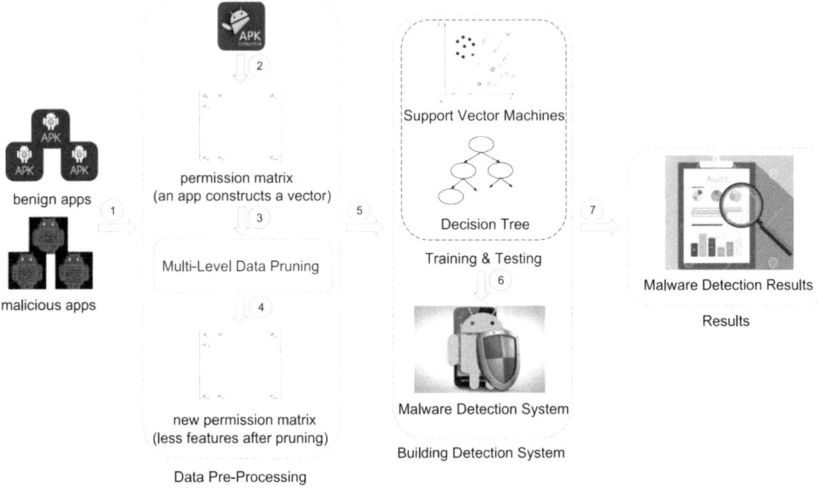

Fig. 8 Significant permission identification for machine-learning-based android malware detection [45]

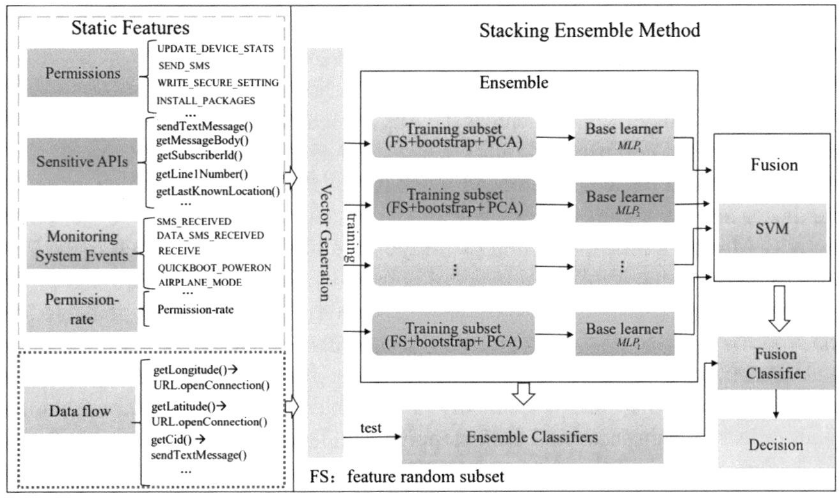

Fig. 9 SEDMDroid: an enhanced stacking ensemble framework for Android malware detection [89]

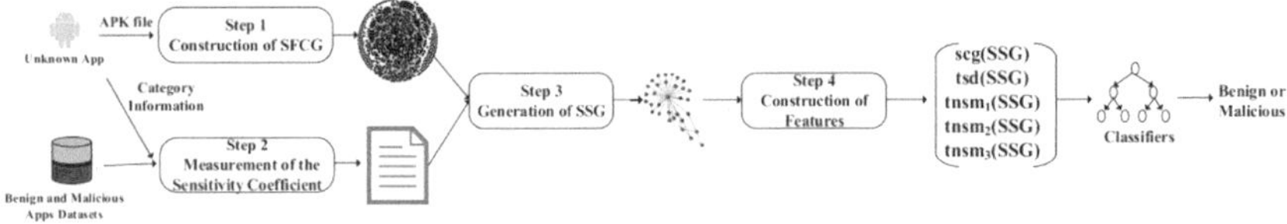

Fig. 10 DAPASA: detecting android piggybacked apps through sensitive subgraph analysis [18]

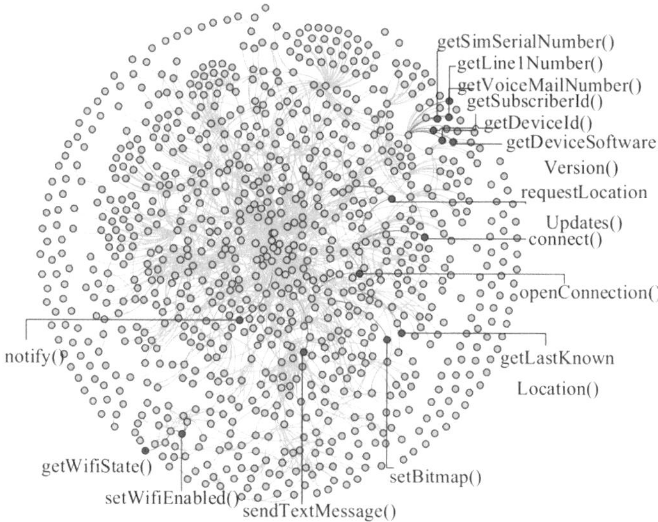

Fig. 11 DAPASA: SSG graph [18]

significant permissions from apps and uses the extracted information to effectively detect malware using a supervised learning algorithm (Fig. 11).

Kumar et al. [41] propose the malware detection framework which is based on three techniques, (i) Clustering Algorithm (ii) Naive Bayes Classifier for Multi-Feature (iii) Blockchain-based malware detection framework. Overall architecture of the proposed system shown in Fig. 12. A new blockchain-based framework was presented to evaluate the performance of malware detection. The newly proposed machine learning technique provides an efficient approach to train the model and then stores and exchanges the trained model results throughout the blockchain network for spreading the information of newly generated malware.

More precisely, the first method based on a clustering algorithm, which reduces the high dimensional data and removes unnecessary features. Secondly, we use a classification method based on naïve Bayes for multi-feature classification. Finally, a blockchain database store the malware information.

3.2 *Basic Proposed Algorithms for Android Malware Features*

This section discusses the basic algorithms and techniques which is used commonly.

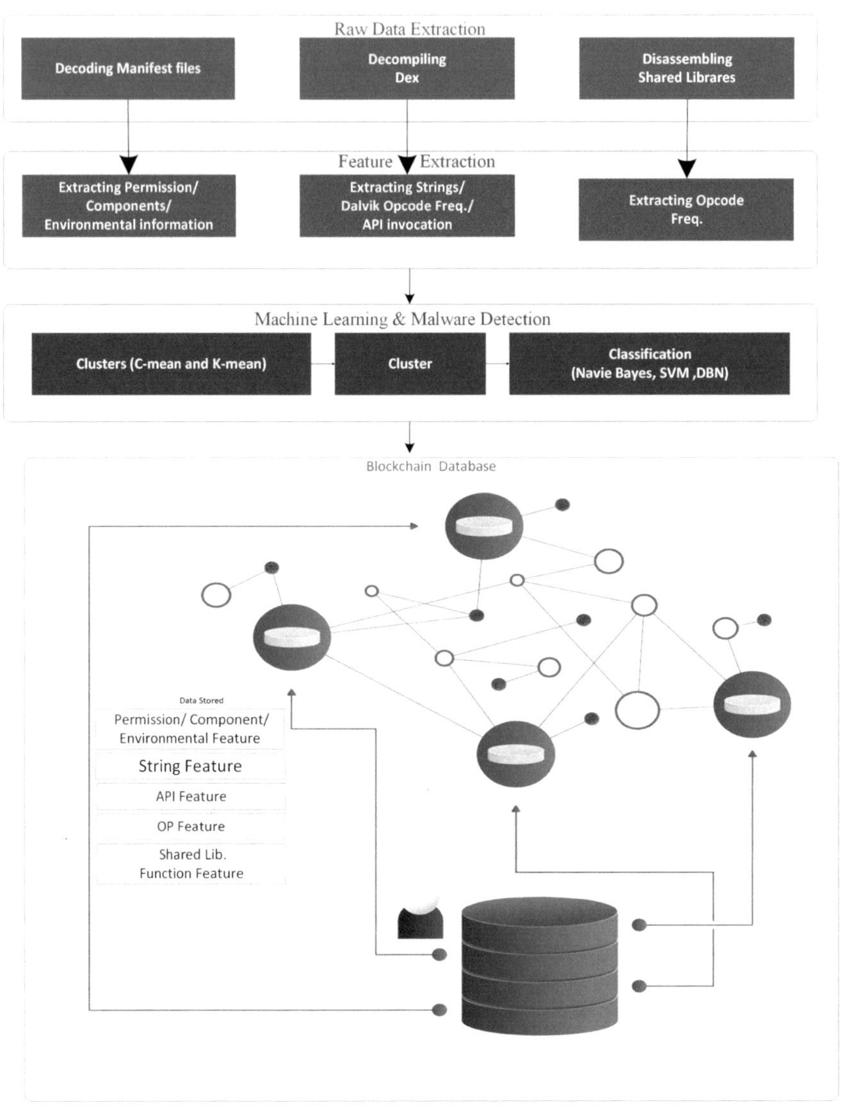

Fig. 12 A multimodal malware detection technique for Android IoT devices using various features [42]

3.2.1 Clustering Techniques to Classify the Malware

The centroids of the clusters which are calculated using the basic K-means [53] clustering algorithm shown in Algorithm 1. The process of future generation values in the malicious feature database corresponds to the elements of the feature vector, and every feature value is searched in the features extracted from input applications. If there is no certain feature value in the extracted features, its absence is represented as zero. Otherwise, the existence of the feature value is represented as one in the vector. The overall process of future generation is shown in Algorithm 2. Additionally, the similarity-based feature vectors are generated in Algorithm 3.

Algorithm 1 K-means algorithm [53]

1: Select K centroids arbitarity for each cluster Ci, i $\varepsilon[1, k]$
2: Assign each data point to the cluster whose centroid is closet to the data point
3: Calculate the centroid Ci of cluster Ci, i$\varepsilon[1, k]$
4: Repeat Steps 2 and 3 until no points change between clusters

Algorithm 2 Feature value clustering-based feature transformation [38]

$Centroids \leftarrow k\_$ means (k, F_db)
Feature vector $\leftarrow [\,00\ldots\ldots0\,]$
index $\leftarrow 0$
for $\forall c \in$ Centroids **do**
    min-sim $\leftarrow 0$
    for $\forall f \in F\_$ app **do**
        dist $\leftarrow$ get-euclidean-dist(c, f)
        min-sim $\leftarrow sim$
    end for
end for
$return \leftarrow$ feature vector

Algorithm 3 Feature value clustering-based feature transformation [38]

$input \leftarrow F = \{f_{ij}\}_{1\le i\le m, 1\le j\le n}$, G : number of clusters desired, Clu a clustering algorthim, $\oplus$ associative and communicative feature combination algorithm
Cluster the n basic features into G groups accordingly by considering each feature to be a column vector in F
$minConf \leftarrow$ a minimum threshold of confidence coefficient
for each sample APK i **do**
    for each feature group g **do**
        $f_{ig}^{FC} = \oplus\{i.f \mid f \in g\}$% combine values of APK i value of the feature f for each f in feature group g
    end for
    $\mathbf{f}_i^{FC} = \left(f_{i1}^{FC}, \cdots, f_{iG}^{FC}\right)$ %
    sample i
end for
$return \leftarrow F_{FC} = \left\{\mathbf{f}_i^{FC} \mid 1 \le i \le m\right\}$ (feature value clustering-based G-dimensional features vector for m sample APKs)

3.2.2 Feature Ranking-Based Algorithms

Average Accuracy-Based Ranking Scheme: The ranking is designed to be directly proportional to the average prediction accuracies across the classes.

Let P_{base} be the set of performance accuracies $P_{k,c} \in P_{base}$ of K base classifiers. If m denotes malware and b, benign then the average accuracy of the k-th base classifier is given by

$$a_k = 0.5 \times \sum_{c=m,b} P_{k,c} | k \in \{1, \ldots, K\}, 0 < P_{k,c} \leq 1. \quad (1)$$

Let $A \leftarrow a_k, \forall k \in \{1, \ldots, K\}$ be a set of the average predictive accuracies, to which a ranking function $Rank_{\text{desc}}$ (.) is applied

$$\bar{A} \leftarrow Rank_{\text{desc}}(A) \quad (2)$$

Thus, $\bar{A}$ contains an ordered ranking of the level-1 base classifiers average predictive accuracies in descending order. Next, the top Z rankings are utilized in weight assignments as follows:

$$\omega_1 = Z, \omega_2 = Z - 1, \ldots, \omega_Z = 1, Z \leq K \quad (3)$$

Class Differential-Based Ranking Scheme: let the average accuracy of each base classifier be given by a_k in (1) and define $\bar{D}$ with cardinality K as a set of ordered rankings in descending order of magnitude. Calculate d_k proportional to average accuracies and inversely proportional to absolute difference of interclass accuracies.

$$d_k = \frac{a_k}{\left| P_{k,m} - P_{k,b} \right|}, k \in \{1, \ldots, K\} \quad (4)$$

$$\bar{D} \leftarrow Rank_{\text{desc}}(D) \quad (5)$$

Ranked Aggregate of Per Class Accuracies-Based Scheme: With $\bar{F}$ defined as the set of ordered rankings with cardinality K, given the initial performance accuracies of $P_{p,c}$ of the K base classifiers.

$$\begin{cases} P_m \leftarrow P_{k,c} \text{ where } c \neq b \\ P_b \leftarrow P_{k,c} \text{ where } c \neq m \end{cases}, k \in \{1, \ldots, K\}, c \in \{m, b\} \quad (6)$$

$$\begin{cases} \bar{P}_m \leftarrow Rank_{\text{desc}}(P_m) \\ \bar{P}_b \leftarrow Rank(P_b) \end{cases} \quad (7)$$

$$\begin{cases} f_k \leftarrow \bar{P}_{k,m} + \bar{P}_{k,b} \\ F \leftarrow f_k \end{cases}, \forall k \in \{1, \ldots, K\}$$
$$\bar{F} \leftarrow Rank_{\text{desc}}(F) \tag{8}$$

3.3 Feature Selection-Based Algorithms

Feature selection is extremely important in static, dynamic, and hybrid analysis. The appropriate feature set is selected using different selection methods such as information gain, mutual information, fisher score, and similarity function.

Information gain (IG) feature ranking approach to rank the features and then selecting the top n features. IG evaluates the features by calculating the IG achieved by each feature. Specifically, given a feature X, IG is expressed as

$$\text{IG} = E(X) - E(X/Y) \tag{9}$$

where $E(X)$ and $E(X/Y)$ represent the entropy of the feature X before and after observing the feature Y, respectively. The entropy of feature X is given by

$$E(X) = -\sum_{x \in X} p(x) \log_2(p(x)) \tag{10}$$

where $p(x)$ is the marginal probability density function for the random variable X. Similarly, the entropy of X relative to Y

$$E(X/Y) = -\sum_{x \in X} p(x) \sum_{x \in X} p(x \mid y) \log_2(p(x \mid y)) \tag{11}$$

Similarity-based feature selection is shown in the below equation, B represents the benign and M represents the malware. X is the feature list and γ is the similarity between the features.

$$S_B(X_j) = e_p \sum_{i=1}^{n} \gamma^{S_B}(X_j^{sb}) \psi(X_j^{sb}), (X_j) \in X^{sb} \tag{12}$$

$$S_{score} = S_p + S_j \tag{13}$$

3.4 Association Rule-Based Algorithms

Association rule mining is used to discover meaningful relationships between variables in huge databases. For example, if events A and B always occur at the same time, then the two events are likely to be associated, for instance, we found that many permissions are always together, i.e., READ_CONTACTS and WRITE CONTACTS are always used together. These dangerous Android permissions belong to the permission Google's list. As we know that those permissions are always together. So we only need one of them to characterize certain behavior.

STEP1: Find out the frequent two permissions sets

STEP2: Diversity-based interestingness measures for association rule using frequent two itemsets that were developed by Piatetsky-Shapiro [21]

- When support($Y \bigcup Z$) $\approx support(Y)support(Z)$, the two-item sets($Y$, Z) are mutually independent. That is, the association rule $Y \Rightarrow Z$ is uninteresting.

$$interest(y, z) = \frac{support(Y \cup Z)}{support(Y \cup Z)} - 1$$
$$= \frac{P(Y|Z)}{P(Z)}$$

- if interest(Y, Z) > 0, Y, and$Z$ are correlated positively.
- if interest(Y, Z) ≈ 0, Y, and$Z$ are commonly independent, and the common two-item sets should be rejected.
- if interest(Y, Z) < 0, Y, and Z are negatively correlated.

STEP3: Create the association rule based on the permission (see Algorithm 4).

STEP4: Calculate the probability table of the association rules.

Algorithm 4 Association rule set R for permission based [41]

```
input ← 1 Associaion Rule Set R
minSub ← minimum thershold of support coefficient
minConf ← minimum thershold of confidence coefficient
for Z=D do
    r = null
    r.PushTail(Z)
    for Y in D do
        if Y ⇒ Z ∈ L2 and support(Y ⇒ Z) > minsup and confidence(Y ⇒ Z) > mincof then
            r.PushTail(Y)
        end if
        r.PushTail(r)
    end for
end for
output ← Association Rule R
```

3.5 *Model Evaluation Measures*

Python programming language contains tools for data pre-processing, classification, clustering, regression, association rules, and visualization, which make it the best tool for the data scientist to measure and test the performance of classifiers. There are various criteria for evaluating classifiers and criteria are set based on the selected goals. For the classification methods are evaluating such as True Positive Rate (TPR) and False Positive Rate (FPR) and classification accuracy. we used the following standard measurements: Given the number of true positives for malicious applications using the following formulas:

$$TPR = \frac{T_p}{T_p + F_n} \tag{14}$$

False Positive rate is the proportion of negative instance for the benign apps

$$FPR = \frac{F_p}{F_p + T_n} \tag{15}$$

The accuracy is defined as below equation

$$Accuracy = \frac{T_p + T_n}{T_p + T_n + F_p + F_n} \tag{16}$$

4 Experimental Analysis and Dataset Discussion

The proposed framework poses strong evidence over acquired experiments results. Here, we discuss major aspects for experimentation which include statistics and source of dataset, evaluation measures to understand the performance criteria for exploited machine learning algorithm, and result outcomes which give strong evidence towards the significance of our proposed model.

4.1 Publicly Available Most Popular Dataset

In order to excavate practical significance, we introduce 10 most popular dataset in Table 6. More description of the dataset are discussed in the provided links.

4.2 Dataset Other Research Work

The comparison of the number of benign and malware apps used in previous work is shown in Table 7.

Table 6 Publicly available most popular dataset

	Original label	Sources
1	Android Malware Genome Project	http://www.malgenomeproject.org
2	M0Droid Dataset	http://m0droid.netai.net/modroid/
3	The Drebin Dataset	http://user.informatik.uni-goettingen.de/~darp/drebin/
4	AndroMalShare	http://sanddroid.xjtu.edu.cn:8080/#home
5	Kharon Malware Dataset	http://kharon.gforge.inria.fr/dataset/
6	AMD Project	http://amd.arguslab.org
7	AAGM Dataset	http://www.unb.ca/cic/datasets/android-adware.html
8	Android PRAGuard Dataset	http://pralab.diee.unica.it/en/AndroidPRAGuardDataset
9	AndroZoo	https://androzoo.uni.lu/
10	A Dataset based on ContagioDump	http://cgi.cs.indiana.edu/~nhusted/dokuwiki/doku.php?id=datasets

Table 7 Compersion of dataset using benign and malware apps

Authors	Benign	Malware
[29]	480	124769
[31]	45	300
[79]	5264	12026
[58]	378	324658
[1]	3978	500
[13]	175	621
[57]	1446	2338
[8]	5560	123453
[82]	2925	3938
[16]	238	1500

5 Experimental Analysis

5.1 Permission-Based Experimental Analysis

Among the 145 permission set, 48 permission are risky permissions which are mentioned in previous literature [13, 63, 73] and Table 8. Moreover, Jin Li, et.al, [47], developed a SIGPID framework to detect the risky permission, the authors generate top 22 risky permission mentioned in Table 9. Furthermore Kumar et al. [41] used a data-mining technique to extract the risky permission, based on association rule set of risky permission shown in Table 10.

Table 8 Permission set mostly used in malware

Risky permissions	
ACCESS_WIFI_STATE	SEND_SMS
READ_LOGS	READ_CALL_LOG
CAMERA	DISABLE_KEYGUARD
CHANGE_NETWORK_STATE	RESTART_PACKAGES
WRITE_APN_SETTINGS	SET_WALLPAPER
CHANGE_WIFI_STATE	INSTALL_PACKAGES
READ_CONTACTS	WRITE_CONTACTS
WRITE_SETTINGS	GET_TASKS
RECEIVE_MMS	ACCESS_WIFI_STATE
WRITE_APN_SETTINGS	SYSTEM_ALERT_WINDOW
READ_HISTORY_BOOKMARKS	RECEIVE_BOOT_COMPLETED
ACCESS_NETWORK_STATE	CALL_PHONE
READ_EXTERNAL_STORAGE	ACCESS_FINE_LOCATION
EXPAND_STATUS_BAR	ADD_SYSTEM_SERVICE
PERSISTENT_ACTIVITY	INTERNET
GET_ACCOUNTS	WRITE_SMS
PROCESS_OUTGOING_CALLS	CHANGE_CONFIGURATION
READ_HISTORY_BOOKMARKS	GET_PACKAGE_SIZE
WAKE_LOG	ACCESS_MOCK_LOCATION
WRITE_CALL_LOG	WRITE_HISTORY_BOOKMARKS
READ_PHONE_STATE	RECEIVE_WAP_PUSH
SET_ALARAM	WRITE_SMS
RECEIVE_SMS	READ_SMS

Table 9 Top 22 permissions [45]

Top 22 Risky permission extract from SIGPID	
ACCESS_WIFI_STATE	SEND_SMS
READ_LOGS	READ_CALL_LOG
RESTART_PACKAGES	DISABLE_KEYGUARD
READ_EXTERNAL_STORAGE	CHANGE_NETWORK_STATE
WRITE_APN_SETTINGS	SET_WALLPAPER
CHANGE_WIFI_STATE	INSTALL_PACKAGES
READ_CONTACTS	WRITE_CONTACTS
CAMERA	GET_TASKS
READ_HISTORY_BOOKMARKS	ACCESS_WIFI_STATE
WRITE_APN_SETTINGS	SYSTEM_ALERT_WINDOW
WRITE_SETTINGS	RECEIVE_BOOT_COMPLETED

Table 10 Permission patterns Malware and Benign [41]

Permission patterns	Benign	Malware
Common android request permission		
READ_PHONE_STATE, ACCESS_WIFI_STATE	2.36	63.08
INTERNET, ACCESS_WIFI_STATE	5.05	63.49
READ_PHONE_STATE	31.87	93.4
ACCESS_WIFI_STATE	5.22	63.49
ACCESS_NETWORK_STATE, ACCESS_WIFI_STATE	3.99	60.31
INTERNET, WRITE_EXTERNAL_STORAGE, READ_PHONE_STATE	13.28	65.44
INTERNET, READ_PHONE_STATE, ACCESS_NETWORK_STATE	24.21	78.97
INTERNET, READ_PHONE_STATE	31.21	93.078
WRITE_EXTERNAL_STORAGE, READ_PHONE_STATE	13.37	65.53
READ_PHONE_STATE, ACCESS_NETWORK_STATE	24.21	79.05

(continued)

Table 10 (continued)

Permission patterns	Benign	Malware
Common android run-time permissions		
READ_PHONE_STATE, ACCESS_NETWORK_STATE	23.63	77.18
INTERNET, READ_LOGS	6.85	6.85
READ_PHONE_STATE	30.32	91.69
INTERNET, READ_PHONE_STATE, ACCESS _NETWORK_STATE	26.36	77.18
READ_PHONE_STATE,VIBRATE	21.92	65.28
INTERNET, READ_PHONE_STATE	29.9	91.52
READ_PHONE_STATE, READ_LOGS	5.38	46.86
READ_LOGS	6.93	47.6
INTERNET, READ_PHONE_STATE, VIBRATE	21.68	65.12
Unique android request permission		
READ_PHONE_STATE, WRITE_SMS	0	50.94
INTERNET, READ_PHONE_STATE, ACCESS_WIFI_STATE	0	63.09
ACCESS_NETWORK_STATE, RECEIVE_BOOT_COMPLETED	0	51.68
ACCESS_NETWORK_STATE, WRITE_SMS	0	49.64
RECEIVE_BOOT_COMPLETED, ACCESS_WIFI_STATE	0	42.63
INTERNET, RECEIVE_BOOT_COMPLETED	0	44.75
WRITE_EXTERNAL_STORAGE, ACCESS_NETWORK_STATE, ACCESS_WIFI_STATE	0	54.53
READ_PHONE_STATE, RECEIVE_BOOT_COMPLETED	0	43.12
INTERNET, SEND_SMS	0	43.12
INTERNET, ACCESS_NETWORK_STATE, ACCESS_WIFI_STATE	0	60.31
Unique android run-time permissions		
INTERNET, READ_PHONE_STATE, ACCESS_NETWORK_STATE, VIBRATE	0	55.42
ACCESS_NETWORK_STATE, VIBRATE, READ_LOGS	0	38.55
READ_PHONE_STATE, ACCESS_NETWORK_STATE, READ_LOGS	0	43.2
READ_LOGS, INTERNET, ACCESS_NETWORK_STATE	0	43.2
READ_PHONE_STATE, VIBRATE, READ_LOGS	0	41.33
INTERNET, VIBRATE, READ_LOGS	0	41.49
READ_LOGS, INTERNET, READ_PHONE_STATE,	0	46.87
ACCESS_FINE_LOCATION, READ_PHONE_STATE, VIBRATE,INTERNET	0	34.23
INTERNET, SEND_SMS	0	33.58
INTERNET, ACCESS_FINE_LOCATION, READ_LOGS	0	28.45

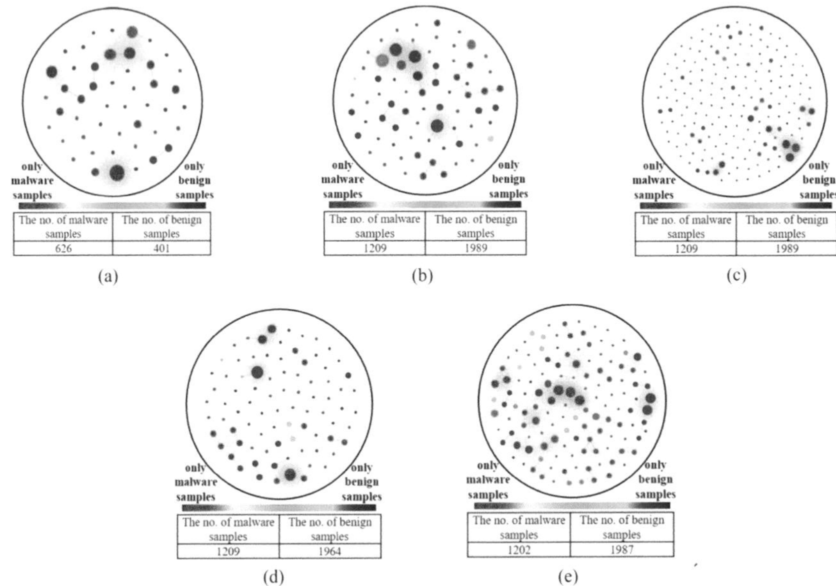

Fig. 13 Topological data analysis (TDA) result of each feature data. Density-based spatial clustering algorithm was utilized in the TDA. **a–e** the visualized result for each feature type. Malicious samples from Malgenome project were used [38]

5.2 Clustering-Based Experimental Analysis

Kim et al. [38], cluster the malware features based on frequency analysis. The red color shows the highest risk features. Figure 13 shows the clustering results obtained by [38].

5.3 Classification Experimental Analysis

From the machine learning-based methods to the general classification-based methods, various kinds of the Android malware detection methods were surveyed. As shown in Table 11, the detection accuracy or the F-measure values of our framework were higher than the other methods including the deep learning-based methods [30, 36, 47, 54].

Table 11 Classification results

Authors	Algorthim	Capicity for feature diversity	Accuracy	F-measure
[38]	Multimodal deep neural network	High	98%	0.99
[81]	Ranking based	High	98%	0.98
[42]	KNN & Navie Bayes	High	98%	0.98
[83]	DNN/RNN	Medium	90%	NA
[52]	CNN	Low	90%	NA
[19]	XGBoost	Low	97%	0.97
[29]	Adaboost/NB/DT	Low	NA	0.78
[85]	NB	Low	93%	NA
[8]	SVM	Low	93.9	NA
[76]	KNN+K-means	Low	NA	0.91
[14]	Bayesian	Low	92%	NA
[84]	SVM	Low	NA	0.98
[59]	RF	Low	97.5%	NA

6 Additional Challenges of Android Malware Detection

Mobile malware and account fraud have exploded around the world. Cybersecurity strategy that allows you to protect your digital assets from hackers. We observed that increasing cyber threats targeting Android mobile devices. Cyber Threat Actors and their use and monetization of stolen data. We discuss and analyze the current effort of monetizing mobile malware in detail below.

- **Premium Rate Number Billing:** In this case, the attacker sets and registers an additional rate number. Usually, these are "shortcodes" that are shorter than the usual phone numbers. The Android application can request permission to send SMS messages during installation. These SMS messages can be sent without user confirmation. Sending a text message to an advanced shortcode causes the phone owner to charge his phone bill and attacker to generate revenue.

- **Spyware:** Several Android apps allow someone to track and monitor a mobile phone user. These apps can record and export all SMS, emails, messages, call logs, microphone, and GPS locations. These applications typically require an attacker to buy a vendor application and then gain physical access to the phone. Although these apps may not generate an attacker's revenue, they generate revenue for the spyware application vendor. Table 8 shows the required permission and APIs used in an Android application to perform these tasks.
- **Search Engine Poisoning:** Some search engines recommend websites or change search engine rankings by monitoring user access rates. These recommendations can be further customized when using a mobile version of the search site and are explicitly monitored by mobile users. A malicious application can initiate multiple requests to these sites, thereby poisoning the hit rate monitored by the search engine. Artificially increasing their search rankings allows an attacker to increase the number of visits by potential customers, or generate revenue through pay-per-view or pay-per-click advertising displayed on the website.
- **Pay-Per-Click:** Each service (such as an ad network) pays for each time an affiliate member refers to a particular website (pay-per-click). Using malicious applications, an attacker can manually access these sites for a few cents per click. Mobile television in China is a wide range of value-added services, and content providers can participate in revenue-sharing programs with operators based on the payer's view. An attacker can create a video channel with the carrier and then register it, generating revenue each time a user views the video or channel. Malicious apps can generate revenue for downloading such video content
- **Pay-Per-Install:** In the mobile market, the pay-per-view scheme usually refers to a model that differs from the pay-per-install scheme in the PC malware space. The term usually refers to a legitimate distribution market in the mobile market, which hosts download applications and charges vendors based on the number of downloads and installations. The opposite is a pre-installation in the PC malware space; the reseller pays the affiliate each time they can install an app on a user's computer. Installing pay-per-install software on an infected computer allows an attacker to generate revenue. Although PC applications have many pay-as-you-go solutions, only a handful of mobile apps are available.
- **Adware:** Many ad networks pay for each view and click when the ad appears. Malicious apps can display ads by launching a browser. An attacker generates ad revenue each time the app is used and an ad is displayed.
- **mTAN Stealing:** Some banks must send additional credentials out of the band to prevent man-in-the-middle attacks when they make a transaction or log in to an online bank account. In particular, the bank will send a random number to the registered mobile phone number called a transaction authentication number (mTAN). They need malware on their phone to get this number for the attacker to succeed.

7 Conclusion

This chapter presented a systematic literature survey of the Android malware detection techniques using deep learning and machine learning. Te reviewed and papers were categorized as three categories of Android malware detection: (1) static analysis, (2) dynamic analysis, and (3) hybrid analysis approaches. The most popular and useful Android malware detection techniques were analyzed via classification approaches, clustering approaches, data-mining approaches, deep learning and, machine-based approaches. Moreover, this chapter discusses the all available dataset and experimental analysis of android malware detection. Furthermore, it assessed the effectiveness of current methods for analyzing malware and detection techniques. That's different from previous surveys that usually study mobile attacks only, this chapter introduces static, dynamic, and hybrid analysis techniques and proposed algorithms.

References

1. Aafer, Yousra, Wenliang Du, and Heng Yin. 2013. Droidapiminer: Mining api-level features for robust malware detection in android. In *International Conference on Security and Privacy in Communication Systems*, 86–103. Springer.
2. Afonso, Vitor Monte, Matheus Favero de Amorim, André Ricardo Abed Grégio, Glauco Barroso Junquera, and Paulo Lício de Geus. 2015. Identifying android malware using dynamically obtained features. *Journal of Computer Virology and Hacking Techniques* 11 (1):9–17.
3. Alazab, Mamoun, and Roderic Broadhurst. 2017. An analysis of the nature of spam as cybercrime. In *Cyber-Physical Security* 251–266. Springer.
4. Alazab, Mamoun, Shamsul Huda, Jemal Abawajy, Rafiqul Islam, John Yearwood, Sitalakshmi Venkatraman, and Roderic Broadhurst. 2014. A hybrid wrapper-filter approach for malware detection. *Journal of Networks* 9 (11): 2878–2891.
5. Alazab, Mamoun, Robert Layton, Roderic Broadhurst, and Brigitte Bouhours. 2013. Malicious spam emails developments and authorship attribution. In *2013 Fourth Cybercrime and Trustworthy Computing Workshop*, 58–68. IEEE.
6. Alazab, Mamoun, Sitalakshmi Venkatraman, Paul Watters, and Moutaz Alazab. Information security governance: the art of detecting hidden malware. In *IT Security Governance Innovations: Theory and Research*, 293–315. IGI Global.
7. Alazab, Moutaz, Mamoun Alazab, Andrii Shalaginov, Abdelwadood Mesleh, and Albara Awajan. 2020. Intelligent mobile malware detection using permission requests and api calls. *Future Generation Computer Systems* 107: 509–521.
8. Arp, Daniel, Michael Spreitzenbarth, Malte Hübner, Hugo Gascon, and Konrad Rieck. 2014. Drebin: Effective and explainable detection of android malware in your pocket. In *Proceedings 2014 Network and Distributed System Security Symposium*.
9. Aung, Zarni, and Win Zaw. 2013. Permission-based android malware detection. *International Journal of Scientific and Technology Research*.
10. Azab, Ahmad, Mamoun Alazab, and Mahdi Aiash. 2016. Machine learning based botnet identification traffic. In *2016 IEEE Trustcom/BigDataSE/ISPA*, 1788–1794. IEEE.
11. Azab, Ahmad, Robert Layton, Mamoun Alazab, and Jonathan Oliver. 2014. Mining malware to detect variants. In *2014 Fifth Cybercrime and Trustworthy Computing Conference*, 44–53. IEEE.

12. Burguera, Iker, Urko Zurutuza, and Simin Nadjm-Tehrani. 2011. Crowdroid: Behavior-based malware detection system for android. *Proceedings of the 1st ACM Workshop on Security and Privacy in Smartphones and Mobile Devices - SPSM '11*, 15.
13. Chan, Patrick P.K., and Wen Kai Song. 2014. Static detection of android malware by using permissions and API calls. In *Proceedings - International Conference on Machine Learning and Cybernetics*, vol. 1, 82–87.
14. Chen, Kai, Peng Wang, Yeonjoon Lee, XiaoFeng Wang, Nan Zhang, Heqing Huang, Wei Zou, and Peng Liu. 2015. Finding unknown malice in 10 seconds: Mass vetting for new threats at the google-play scale. In *24th {USENIX} Security Symposium ({USENIX} Security 15)*, 659–674.
15. Chuang, Hsin Yu, and Sheng De Wang. 2015. Machine learning based hybrid behavior models for android malware analysis. *Proceedings - 2015 IEEE International Conference on Software Quality*. Reliability and Security: QRS.
16. Desnos, Anthony, and Geoffroy Gueguen. 2011. Android: From reversing to decompilation. In *Proceeding of Black Hat*, Abu Dhabi.
17. Dini, Gianluca, Fabio Martinelli, Ilaria Matteucci, Marinella Petrocchi, and Andrea Saracino. 2018. *and Daniele Sgandurra. Risk analysis of Android applications: A user-centric solution. Future Generation Computer Systems.*
18. Fan, Ming, Jun Liu, Wei Wang, Haifei Li, Zhenzhou Tian, and Ting Liu. 2017. Dapasa: Detecting android piggybacked apps through sensitive subgraph analysis. *IEEE Transactions on Information Forensics and Security* 12 (8): 1772–1785.
19. Fereidooni, Hossein, Mauro Conti, Danfeng Yao, and Alessandro Sperduti. 2016. ANASTASIA: android malware detection using static analysis of applications. In *2016 8th IFIP International Conference on New Technologies, Mobility and Security (NTMS)*, 1–5. IEEE.
20. Ferrante, Alberto, Miroslaw Malek, Fabio Martinelli, Francesco Mercaldo, and Jelena Milosevic. 2018. Extinguishing ransomware - a hybrid approach to android ransomware detection. In *Lecture Notes in Computer Science (Including Subseries Lecture Notes in Artificial Intelligence and Lecture Notes in Bioinformatics).*
21. Piatetsky-Shapiro, G. 1991. Discovery, analysis and presentation of strong rules. *Knowledge Discovery in Databases.*
22. Ganesh, Meenu, Priyanka Pednekar, Pooja Prabhuswamy, Divyashri Sreedharan Nair, Younghee Park, and Hyeran Jeon. 2017. CNN-based android malware detection. In *2017 International Conference on Software Security and Assurance (ICSSA)*, 60–65. IEEE.
23. Gold, Steve. 2011. Android insecurity. *Network Security* 2011 (10): 5–7.
24. Ham, You Joung, and Hyung-Woo Lee. 2014. Detection of malicious android mobile applications based on aggregated system call events. *International Journal of Computer and Communication Engineering.*
25. Ham, You Joung, Daeyeol Moon, Hyung Woo Lee, Jae Deok Lim, and Jeong Nyeo Kim. 2014. Android mobile application system call event pattern analysis for determination of malicious attack. *International Journal of Security and its Applications.*
26. Hasegawa, Chihiro, and Hitoshi Iyatomi. 2018. One-dimensional convolutional neural networks for Android malware detection. In *Proceedings -, 2018. IEEE 14th International Colloquium on Signal Processing and its Application. CSPA 2018.*
27. Hou, Shifu, Aaron Saas, Lingwei Chen, Yanfang Ye, and Thirimachos Bourlai. 2017. Deep neural networks for automatic android malware detection. In *Proceedings of the 2017 IEEE/ACM International Conference on Advances in Social Networks Analysis and Mining 2017 - ASONAM '17.*
28. Hou, Shifu, Aaron Saas, Yanfang Ye, and Lifei Chen. 2016. Droiddelver: An android malware detection system using deep belief network based on api call blocks. In *International Conference on Web-Age Information Management*, 54–66. Springer.
29. Huang, Chun Ying, Yi Ting Tsai, and C.H. Hsu. 2013. Performance evaluation on permission-based detection for android malware. *Smart Innovation, Systems and Technologies.*
30. Huda, Shamsul, Rafiqul Islam, Jemal Abawajy, John Yearwood, Mohammad Mehedi Hassan, and Giancarlo Fortino. 2018. A hybrid-multi filter-wrapper framework to identify run-time behaviour for fast malware detection. *Future Generation Computer Systems* 83: 193–207.

31. Idrees, Fauzia, and Muttukrishnan Rajarajan. 2014. Investigating the android intents and permissions for malware detection. In *International Conference on Wireless and Mobile Computing, Networking and Communications*, 354–358.
32. Isohara, Takamasa, Keisuke Takemori, and Ayumu Kubota. 2011. Kernel-based behavior analysis for android malware detection. In *Proceedings - 2011 7th International Conference on Computational Intelligence and Security, CIS 2011*.
33. Jang, Jae Wook, Hyunjae Kang, Jiyoung Woo, Aziz Mohaisen, and Huy Kang Kim. 2016. Andro-Dumpsys: Anti-malware system based on the similarity of malware creator and malware centric information. *Computers and Security*.
34. Karbab, El Mouatez Billah, Mourad Debbabi, Abdelouahid Derhab, and Djedjiga Mouheb, 2018. MalDozer: Automatic framework for android malware detection using deep learning. *Digital Investigation*.
35. Khan, Riaz Ullah, Xiaosong Zhang, Rajesh Kumar, Abubakar Sharif, Noorbakhsh Amiri Golilarz, and Mamoun Alazab. 2019. An adaptive multi-layer botnet detection technique using machine learning classifiers. *Applied Sciences* 9 (11): 2375.
36. Kim, D., J. KIm, S. Kim. 2013. Proceedings: 3rd International, and undefined 2013. A malicious application detection framework using automatic feature extraction tool on android market. In *3rd International Conference on Computer Science and Information Technology (ICCSIT)*, 1–4.
37. Kim, Gisung, Seungmin Lee, and Sehun Kim. 2014. A novel hybrid intrusion detection method integrating anomaly detection with misuse detection. *Expert Systems with Applications*.
38. Kim, TaeGuen, BooJoong Kang, Mina Rho, Sakir Sezer, and Eul Gyu Im. A multimodal deep learning method for android malware detection using various features. *IEEE Transactions on Information Forensics and Security* 14 (3): 773–788.
39. Kumar, Ajit, K.S. Kuppusamy, and G. Aghila. 2018. FAMOUS: Forensic analysis of mobile devices using scoring of application permissions. *Future Generation Computer Systems*.
40. Kumar, Rajesh, Zhang Xiaosong, Riaz Ullah Khan, Ijaz Ahad, and Jay Kumar. 2018. Malicious code detection based on image processing using deep learning. In *Proceedings of the 2018 International Conference on Computing and Artificial Intelligence*, 81–85.
41. Kumar, Rajesh, Xiaosong Zhang, Riaz Khan, Abubakar Sharif, Rajesh Kumar, Xiaosong Zhang, Riaz Ullah Khan, and Abubakar Sharif. 2019. Research on data mining of permission-induced risk for android IoT devices. *Applied Sciences* 9 (2): 277.
42. Kumar, Rajesh, Xiaosong Zhang, Wenyong Wang, Riaz Ullah Khan, Jay Kumar, and Abubaker Sharif. 2019. A multimodal malware detection technique for android iot devices using various features. *IEEE Access* 7: 64411–64430.
43. Lei, Tao, Zhan Qin, Zhibo Wang, Qi Li, and Dengpan Ye. 2019. Evedroid: Event-aware android malware detection against model degrading for iot devices. *IEEE Internet of Things Journal* 6 (4): 6668–6680.
44. Li, Dongfang, Zhaoguo Wang, and Yibo Xue. 2018. Fine-grained android malware detection based on deep learning. In *2018 IEEE Conference on Communications and Network Security (CNS)*, 1–2. IEEE.
45. Li, Jin, Lichao Sun, Qiben Yan, Zhiqiang Li, Witawas Srisa-An, and Heng Ye. 2018. Significant permission identification for machine-learning-based android malware detection. *IEEE Transactions on Industrial Informatics* 14 (7): 3216–3225.
46. Li, Wenjia, Zi Wang, Juecong Cai, and Sihua Cheng. 2018. An android malware detection approach using weight-adjusted deep learning. In *2018 International Conference on Computing, 2018*. Networking and Communications: ICNC.
47. Li, Yuqi, Yanghao Li, Hongfei Yan, and Jiaying Liu. 2018. Deep joint discriminative learning for vehicle re-identification and retrieval. In: *Proceedings - International Conference on Image Processing, ICIP, 2017*, 395–399.
48. Liu, Yu, Yichi Zhang, Haibin Li, and Xu Chen. 2016. A hybrid malware detecting scheme for mobile android applications. In *2016 IEEE International Conference on Consumer Electronics, ICCE 2016*.

49. Malik, Sapna, and Kiran Khatter. 2016. System call analysis of android malware families. *Indian Journal of Science and Technology* 9 (21).
50. Mansfield-Devine, Steve. 2012. Android malware and mitigations. *Network Security* 2012 (11): 12–20.
51. Mansfield-Devine, Steve. 2012. Paranoid android: just how insecure is the most popular mobile platform? *Network Security* 2012 (9): 5–10.
52. McLaughlin, Niall, Adam Doupé, Gail Joon Ahn, Jesus Martinez del Rincon, BooJoong Kang, Suleiman Yerima, Paul Miller, Sakir Sezer, Yeganeh Safaei, Erik Trickel, and Ziming Zhao. 2017. Deep android malware detection. In *Proceedings of the Seventh ACM on Conference on Data and Application Security and Privacy - CODASPY '17*, 301–308.
53. Mohamad, Ismail Bin, and Dauda Usman. 2013. Standardization and its effects on k-means clustering algorithm. *Research Journal of Applied Sciences, Engineering and Technology* 6 (17): 3299–3303.
54. Nix, Robin, and Jian Zhang. 2017. Classification of android apps and malware using deep neural networks. In *Proceedings of the International Joint Conference on Neural Networks*.
55. Park, Ji Sun, Taek Young Youn, Hye Bin Kim, Kyung Hyune Rhee, and Sang Uk Shin. 2018. *Sensors (Switzerland): Smart Contract-Based Review System for an IoT Data Marketplace*.
56. Patel, Kanubhai, and Bharat Buddadev. 2015. Detection and mitigation of android malware through hybrid approach. In *International symposium on Security in Computing and Communication*.
57. Pehlivan, Ugur, Nuray Baltaci, Cengiz Acarturk, and Nazife Baykal. 2014. The analysis of feature selection methods and classification algorithms in permission based Android malware detection. In *IEEE SSCI 2014: 2014 IEEE Symposium Series on Computational Intelligence - CICS 2014: 2014 IEEE Symposium on Computational Intelligence in Cyber Security, Proceedings*.
58. Peng, Wuxu, Linda Huang, Julia Jia, and Emma Ingram. 2018. Enhancing the Naive Bayes spam filter through intelligent text modification detection. In *Proceedings - 17th IEEE International Conference on Trust, Security and Privacy in Computing and Communications and 12th IEEE International Conference on Big Data Science and Engineering, Trustcom/BigDataSE 2018*.
59. Rastogi, Vaibhav, Yan Chen, and Xuxian Jiang. 2013. DroidChameleon: Evaluating android anti-malware against transformation attacks. In *Proceedings of the 8th ACM SIGSAC symposium on Information, Computer and Communications Security - ASIA CCS '13*, 329. New York: ACM Press.
60. Biljana, L., Risteska Stojkoska and Kire V. Trivodaliev. 2017. *A Review of Internet of Things for Smart Home: Challenges and Solutions*.
61. Rodriguez-Mota, Abraham, Ponciano Jorge Escamilla-Ambrosio, Salvador Morales-Ortega, Moises Salinas-Rosales, and Eleazar Aguirre-Anaya. 2016. Towards a 2-hybrid android malware detection test framework. In *2016 International Conference on Electronics, Communications and Computers (CONIELECOMP)*, 54–61. IEEE.
62. Saracino, Andrea, Daniele Sgandurra, Gianluca Dini, and Fabio Martinelli. 2018. MADAM: Effective and efficient behavior-based android malware detection and prevention. *IEEE Transactions on Dependable and Secure Computing*.
63. Seo, Seung Hyun, Aditi Gupta, Asmaa Mohamed Sallam, Elisa Bertino, and Kangbin Yim. 2014. Detecting mobile malware threats to homeland security through static analysis. *Journal of Network and Computer Applications* 38 (1): 43–53.
64. Shijo, P. V., and A. Salim. 2015. Integrated static and dynamic analysis for malware detection. In *Procedia Computer Science*.
65. Shiqi, Luo, Tian Shengwei, Yu Long, Yu Jiong, and Sun Hua. 2018. Android malicious code classification using deep belief network. *KSII Transactions on Internet and Information Systems*.
66. Sun, Mingshen, Xiaolei Li, John C.S. Lui, Richard T.B. Ma, and Zhenkai Liang. 2017. Monet: A user-oriented behavior-based malware variants detection system for android. *IEEE Transactions on Information Forensics and Security* 12 (5).

67. Talha, Kabakus Abdullah, Dogru Ibrahim Alper, and Cetin Aydin. 2015. APK auditor: Permission-based Android malware detection system. *Digital Investigation*.
68. , Tchakounté, F., P. Dayang. 2013. International journal of science and undefined. System calls analysis of malwares on android. *International Journal of Science and Technology* 2 (9): 669–674.
69. Thomas, Blasing, Leonid Batyuk, Aubrey Derrick Schmidt, Seyit Ahmet Camtepe, and Sahin Albayrak. 2010. An android application sandbox system for suspicious software detection. In *Proceedings of the 5th IEEE International Conference on Malicious and Unwanted Software, Malware 2010*.
70. Tran, Khoi-Nguyen, Mamoun Alazab, Roderic Broadhurst, et al. 2014. Towards a feature rich model for predicting spam emails containing malicious attachments and urls. *Eleventh Australasian Data Mining Conference*.
71. Vasan, Danish, Mamoun Alazab, Sobia Wassan, Hamad Naeem, Babak Safaei, and Qin Zheng. 2020. Imcfn: Image-based malware classification using fine-tuned convolutional neural network architecture. *Computer Networks* 171: 107138.
72. Venkatraman, Sitalakshmi, and Mamoun Alazab. Use of data visualisation for zero-day malware detection. *Security and Communication Networks*.
73. Wang, Jingdong, Heng Tao Shen, Jingkuan Song, and Jianqiu Ji. 2014. Hashing for similarity search: A survey. *arXiv preprint* arXiv:1408.2927.
74. Wang, Panpan, and Bo Li. 2018. Vehicle re-identification based on coupled dictionary learning. In *2018 2nd International Conference on Robotics and Automation Sciences (ICRAS)*, 1–5. IEEE.
75. Wang, Zi, Juecong Cai, Sihua Cheng, and Wenjia Li. 2016. Droid deep learner: Identifying Android malware using deep learning. In *2016 IEEE 37th Sarnoff Symposium*, 160–165. IEEE.
76. Wu, Dong Jie, Ching Hao Mao, Te En Wei, Hahn Ming Lee, and Kuo Ping Wu. 2012. DroidMat: Android malware detection through manifest and API calls tracing. In *Proceedings of the 2012 7th Asia Joint Conference on Information Security, AsiaJCIS 2012*.
77. Songyang, Wu, Pan Wang, Xun Li, and Yong Zhang. 2016. Effective detection of android malware based on the usage of data flow APIs and machine learning. *Information and Software Technology* 75: 17–25.
78. Wu, W.-C., and S.-H. Hung. DroidDolphin: A dynamic android malware detection framework using big data and machine learning. In *2014 Conference on Research in Adaptive and Convergent Systems, RACS 2014*, 247–252.
79. Ke, Xu, Yingjiu Li, and Robert H. Deng. 2016. ICCDetector: ICC-based malware detection on android. *IEEE Transactions on Information Forensics and Security*.
80. Xu, Ke, Yingjiu Li, Robert H. Deng, and Kai Chen. 2018. DeepRefiner: Multi-layer android malware detection system applying deep neural networks. In *2018 IEEE European Symposium on Security and Privacy (EuroS&P)*, 473–487. IEEE.
81. Yerima, Suleiman Y., and Sakir Sezer. 2018. Droidfusion: A novel multilevel classifier fusion approach for android malware detection. *IEEE Transactions on Cybernetics* 49 (2): 453–466.
82. Yerima, Suleiman Y., Sakir Sezer, and Igor Muttik. 2014. Android malware detection using parallel machine learning classifiers. *Proceedings - 2014 8th International Conference on Next Generation Mobile Applications*, 2014. Services and Technologies: NGMAST.
83. Yu, Wei, Linqiang Ge, Guobin Xu, and Xinwen Fu. 2014. Towards neural network based malware detection on android mobile devices. In *Cybersecurity Systems for Human Cognition Augmentation*, 99–117. Springer.
84. Yuan, Zhenlong, Lu Yongqiang, and Yibo Xue. 2016. Droiddetector: Android malware characterization and detection using deep learning. *Tsinghua Science and Technology* 21 (1): 114–123.
85. Zhang, Mu, Yue Duan, Heng Yin, and Zhiruo Zhao. 2014. Semantics-aware android malware classification using weighted contextual api dependency graphs. In *Proceedings of the 2014 ACM SIGSAC Conference on Computer and Communications Security*, 1105–1116.
86. Zhang, Yi, Yuexiang Yang, and Xiaolei Wang. 2018. A novel android malware detection approach based on convolutional neural network. In *Proceedings of the 2nd International Conference on Cryptography, Security and Privacy - ICCSP 2018*, 144–149. New York: ACM Press.

87. Zhou, Yajin, Zhi Wang, Wu Zhou, and Xuxian Jiang. 2012. Hey, you, get off of my market: Detecting malicious apps in official and alternative android markets. In *NDSS (Network and Distributed System Security Symposium)*.
88. Zhu, Dali, Hao Jin, Ying Yang, Di Wu, and Weiyi Chen. 2017. DeepFlow: Deep learning-based malware detection by mining android application for abnormal usage of sensitive data. In *Proceedings - IEEE Symposium on Computers and Communications*.
89. Zhu, Huijuan, Yang Li, Ruidong Li, Jianqiang Li, Zhu-Hong You, and Houbing Song. 2020. Sedmdroid: An enhanced stacking ensemble of deep learning framework for android malware detection. *IEEE Transactions on Network Science and Engineering*.

Deep Learning in Malware Identification and Classification

Balram Yadav and Sanjiv Tokekar

Abstract Albeit the cyber world has become an essential part and the lifeline of the present day, there are threats associated with it. People access the cyber world for various services like networking, banking, communication, shopping, and for other uses. Malware is one of the primary and perilous threats among malevolent software for the decades in the cyber and the computing world. Due to its magnification in volume and in complexity, malware and its variant identification and classification are the most central and severe problems nowadays. Since malware inception, more and more malware is engendered and designed, as time passes; more intricate malware is designed enormously. Researchers and analysts are perpetually probing for a solution that is the most efficacious to fight back with malware. The most-famed methods utilized for malware analysis is signature-based detection, static, and dynamic analysis. In recent years, signature-based detection has been proven ineffective against the escalation of malware and its variants. Malware classification is attracting widespread interest due to its vast proliferation. In this chapter, we have chosen to discuss and explore another method of malware analysis that is image-based malware analysis utilizing deep learning. We are specifically discussing malware classification utilizing malware visualization and deep learning, one of the most widely implemented techniques in many real-world applications. To better understand the concept from a practical perspective, we additionally discussed and implemented a fundamental level malware classifier, for the reader's further research and study purpose. The main objective of this chapter is to avail readers a better and in-depth understanding of malware classification, visualization, deep learning algorithms and emerging challenges, open issues.

B. Yadav (✉)
I.E.T, D.A.V.V., Indore, India
e-mail: balram.dreamsworld@gmail.com

S. Tokekar
I.E.T., D.A.V.V., Indore, India
e-mail: stokekar@ietdavv.edu.in

M. Stamp et al. (eds.), *Malware Analysis Using Artificial Intelligence and Deep Learning*, https://doi.org/10.1007/978-3-030-62582-5_6

1 Malware and Malware Analysis

In this section, we are discussing what is malware, what is malware analysis, what is malware classification, how we visualize malware, etc. This section is prodigiously needed and avails readers to better understand the malware analysis and malware classification.

1.1 Malware

To efficaciously understand and analyze malware, you should be familiar with it. Malware is an abbreviation for malicious software (malicious software is an umbrella term used to refer a variety of forms of inimical software or programs) intending to access information, resources without the user's notification, and sanction. Malware is any code that performs inimical. Malware infections are among the most frequently encountered threats in the digital and computing world. Malware is additionally utilized for obtaining a password, obtaining confidential data; additionally, they are acclimated to trap the government. The malware is mall functioning software that is found on the computer systems. Malware and other threats are defined as specially indited programs to perform deleterious activities. An assailant designs malware to compromise computer services, access data, bypass access controls, and affects the functioning of a computer, its applications, or data.

The accelerated growth of devices in the cyber world has designated a massive obstruction in front of malware analysts, researchers, and additionally for antivirus companies. Assailants utilize the cyber world for illegitimate activities to commit financial frauds, to gain access to sensitive and personal information, to gain access for systems and networks. In recent years, there has been an expeditious increase in Internet attacks [7, 8]. The researchers and analysts customarily suggested security mechanisms and designed novel methods to fight malware and its variant attacks. There has been a great amendment in the design of malware. Afore the term malware was coined, all the malignant programs were considered under the term computer virus. Malware is an umbrella term for any program that contravenes the confidentiality, integrity, and availability of accommodations, contrivances, networks, or systems.

The list below provides an overview of variants of malware based on malware's behavior includes Trojans, viruses, worms, rootkits, botnets, phishing, spam, spyware, key loggers, logic bombs, etc.

Adware is kenned as advertisement software. Adware is the designation given to those programs which are designed to exhibit advertisements on your computer when you explore the cyber world, and then redirect your search requests to advertising websites and accumulate information about you and your interest. Adware is considered as malevolent because it amasses data without your consent or sanction. It is a type of malware that automatically distributes advertisements. Advertising-fortified software often comes bundled with software and applications and most of them serve as a revenue tool.

Virus A computer virus is a malevolent program that cyber attackers program to reproduce in massive amounts and affects the functioning of a computer and degrades its performance. It is also known as infectors. It conventionally does so by assailing and infecting subsisting files on the target system and from one host to another. Viruses must execute to do their deleterious task, so they target any type of file that the system can execute. A virus is a software program that modifies other programs and affixes itself to their code. A virus can run by itself; they perform intended malevolent activities when the infected program is executed.

Spyware Abbreviated for spy software (software that spies on a computer system). It is programmed to monitor and record browsing data as well as confidential information and other activities. It is a type of malware that spies and tracks utilizer activity without their erudition. The capabilities of spyware can include keystrokes accumulation, financial data harvesting, or activity monitoring.

Worm Functionally virus and worms are homogeneous. Worms are infectious and spreads. Assailers design worms to replicate themselves like a virus. However, a worm replicates without targeting and infecting specific files that are already present on a computer. They utilize a computer network to spread, relying on security failures on the target computer to access it, and steal or delete data. Worms are network viruses that can spread over the network by duplicating themselves. They do not transmute or ravage the user's files but they reside in main memory and duplicate themselves, and by this they make the system and network unresponsive.

Trojan A trojan or trojan horse is a maleficent program that represents its utilizer to be appearing utilizable and innocuous files or legitimate software. Attackers distribute trojans as routine software, game, or an implement that persuades a utilizer to install it on their computer. The denomination is derived from the antediluvian Greek story of the wooden horse that used to march into the city of Troy by stealth. Trojan horses are just as pernicious on computers and considered destructive. Cybersecurity experts consider trojans to be among the most hazardous types of malware, concretely trojans are designed to glom financial information from users.

Key logger A keystroke logger, or key logger, captures keystroke ingression made on a computer by the utilizer, often without the sanction or erudition of the utilizer. Key loggers have legitimate uses as a professional information technology monitoring tool . However, keystroke logging is commonly utilized for malefactor purposes, capturing sensitive information like usernames, passwords, answers to security questions, and financial information.

Rootkit A rootkit is a set of software tools, typically malevolent, which gives an unauthorized utilizer privileged access to a computer. Once a rootkit has been installed, the controller of the rootkit can remotely execute files and transmute system configurations on the host machine. Rootkits cannot self-propagate or replicate. They must be installed on a device.

Bots and Botnets Additionally kenned as robots. Bots are maleficent programs designed to infiltrate a computer and automatically respond to and carry out

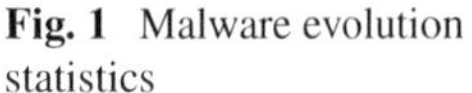
Fig. 1 Malware evolution statistics

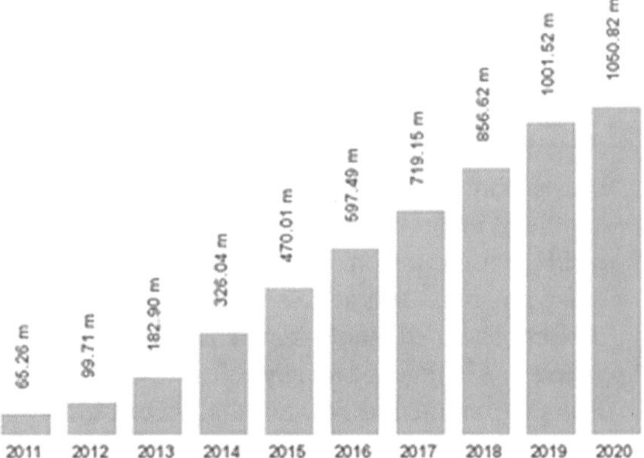

instructions received from a central command and control server. Bots can self-replicate (like worms) or replicate via user action (like viruses and trojans).

Ransomware Ransomware is a type of malware that locks the data on a victim's computer, typically by encryption. The cybercriminal behind the malware demands payment afore decrypting the ransomed data and returning access to the victim. The motive for ransomware attacks is proximately always monetary, and unlike other types of attacks, the victim is conventionally notified that an exploit has occurred and is given instructions for making payment to have the data renovated to normal. It is a type of malignant software that essentially restricts utilizer access to the computer by encrypting the files or locking down the system while injunctively authorizing a ransom. Users are forced to pay the malware author to remove the restrictions and gain access to their computers.

1.1.1 Current Scenario of Malware Magnification

This section deals with the current scenario of the magnification of malware and its variants. We can see from Fig. 1 that the number of attacks is growing every year. The number of malware found perpetual to increment because malware and its variants can be engendered utilizing automated tools and reusing code modules. Reports from different antivirus companies limpidly describe that number of malware, and its variants are incrementing expeditiously.

A report from the av-test institute verbalized that in the period 2011- august 2020, 1050.82 million malware were recorded [7] and 10.87 million new malware were reported in the month August 2020 Fig. 2.

One more report from McAffe antivirus company placidly describes the statistics of malware evolution, millions of malware and variants are discovered [8].

There are many more reports from different antivirus companies conspicuous the fact that malware and its variant assailments are incrementing every year and besides malware, reports additionally present the current scenario of attacks of Internet of

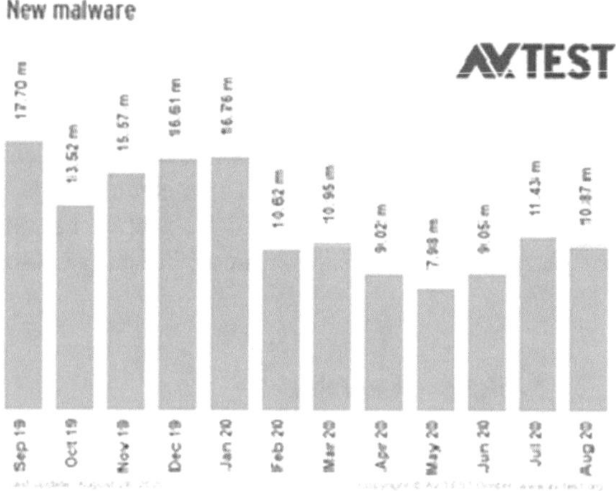

Fig. 2 New malware evolution statistics

Things (IoT) malware, mobile malware, and withal an expeditious increase is in ransomware recently. With these statistics, manual malware analysis is not feasible anymore; it does not scale to handle this enormous count of malware that's why the process of malware analysis needs to be automated. This is discovered or reported malware and its variants. It does not account for malware that has not been discovered or reported yet. There could be millions more out there that are still relishing the comfort of not being detected

1.1.2 Malware Family

A malware family is a group of malware that comports and functions in the same way. A family can be divided into different variants, especially if an incipient malware has different functionality and structure than the precedent ones. Malware family is the term utilized for the malware samples that belong to the same family designates they apportion their code or can have homogeneous code, capabilities, damage potential, inchoation, or behavior. Malware family betokens that incipient malware is designed by utilizing antecedent malware so we can group them in a single malware family.

For example, the Loylda family refer Table 1 of malware has four known variants: Loylda.AA1, Loylda.AA2, Loylda.AA3 and Loylda.AT, malware samples from malimg dataset [19].

1.1.3 Threats From Malware

The damage caused by malware depends upon it, whether it infected a computer, a business organization or whole network. The consequences of the damage caused by malware depend upon the type of malware. There are many threats associated with variants of malware, such as some malware interrupts the services of the system and operating system, some accesses file system without sanction, some access user's confidential data, some perform a denial of accommodation attacks, some minimize the space of a system, effacing, misplacement and corrupts files, some access systems resources, they additionally decelerate the process of the system, engender multiple shortcuts, automatically consumes an abundance of space in the system and truncating the recollection of the system. Malware greatly affects the functionality of computers and networks. Malware additionally causes hardware failure.

1.2 Malware Analysis

Malware analysis is the process of inspection and dissection of the functionality, purport, inception, and potential impact of malevolent code. In another way, it is the process of extracting cryptic information from malware code through static, dynamic, or hybrid inspection by utilizing tools, techniques, and methods. The data that is extracted from malware can be simple like its file type, strings to more perplexed information like malfeasance. Malware analysis denotes analyzing and inspecting binaries of malignant code to understand its working and finding methods for identification and classification of homogeneous files. Attributes or properties of data/samples, and these attributes are analyzed to engender paramount insights into the data under analysis. We accumulate features from malware binaries.

For example, in the facial detection system, the features would be shape, size, color, and structure of eye perceivers, nose perceiver and in malware analysis, features can be strings from the malware binaries, application programming interface (API) call sequences, n-grams, etc.

1.2.1 Traditional Approaches

The investigation of maleficent code is done traditionally mainly with static, dynamic, and hybrid analysis. Traditional approaches Fig. 3 such as static, dynamic, or hybrid analysis extract separate levels of features from malignant samples for identification and relegation, which cannot perform efficiently and accurately. The utilizations of deep learning for malware classification offers an expedient of building scalable machine learning models, which may handle any scale of data, without expending of resources such as memory. Deep learning marks malware depend on the general pattern, which directs the distinguishing of a variety of malware attacks and their variations. Furthermore, deep learning conducts a profound classification and improves its accuracy because deep learning identifies more features than conventional machine learning methods by passing through many calibers of feature extraction. This enables deep learning models to acquire an incipient pattern of malware after the fundamental training phase.

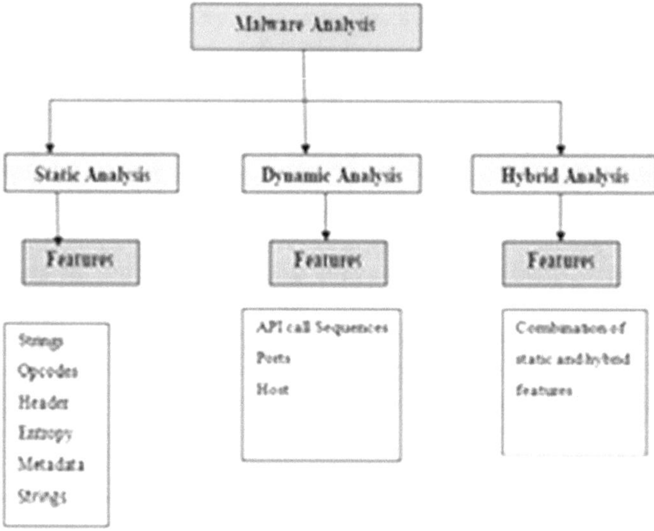

Fig. 3 Traditional malware analysis approaches

1.2.2 Features and Feature Engineering

The performance of any classification, prediction, and recognition system is closely dependent on feature. In machine learning, features are learned manually or we can say that hand-crafted features are used. They dominate the past on image- and video-based applications. There are many disadvantages associated with this feature learning like deep knowledge of data for feature extraction; feature extraction and classification were two different modules, where hundreds of features crafted for applications, feature dimension is high and to select optimized features from feature vector is a slow process.

Traditionally malware was identified and analyzed by utilizing the following approaches.

1.2.3 Static Analysis

It refers to the analysis or investigation of a malignant program without executing it. It is the process of extracting information from malware while it is not executing. Static analysis can be performed directly with the actual code (if present) and if not, can be applied to sundry representations of executables. Static analysis is considered the most facile, expeditious, and less precarious analysis process. It is the most facile and expeditious because there are no special conditions and requisites needed for the analysis process. The malware is simply subjected to analysis implements. It is less jeopardous because the malware is not executed during analysis; consequently, there is not at all any jeopardy of an infection yielding and spreading while analysis is going on, and we do not worry about engendering a safe environment for static analysis. The patterns detected in this kind of analysis include string signature, byte-sequence or operation codes (opcodes), frequency distribution, byte-sequence n-grams or opcodes n-grams, API calls, the structure of the disassembled program, etc. The terminus goal is to identify malware afore the program goes under assessment. Disassembly of malevolent programs is required to detect the patterns some prevalent disassembly implements are objdump, IDA Pro, etc. Static analysis is considered to be a less profit method of analysis as the data extracted from the static analysis is less promising because data is amassed when malware is in passive mode (not executing). Data extracted is constrained and not reveal much paramount information about malware. Prevalent techniques applied in the static analysis are flow analysis, string analysis and signature analysis.

1.2.4 Dynamic Analysis

Dynamic analysis is the process of extracting data from malware while it is executing. It refers to the analysis of the deportment of a malevolent program while it is being executed in a controlled environment (virtual machine, emulator, sandbox, etc) to identify inimical activities after the program executes. The demeanor is monitored by utilizing implements like process monitor, process explorer, wire shark, or capture bat. This kind of analysis endeavors to monitor system calls, injunctive authorization trace, function and API calls, the network, the flow of information, etc. Unlike the static analysis, which provides inhibited information from the malware being analyzed, the dynamic analysis offers an in-depth view into the malware's functions and comportment because it is accumulating information while the malware is executing. To conduct dynamic analysis we require two things, first is the environment where we can execute malware is in a controlled manner for the analysis purport and second is analysis implements that monitor and records the environment for any vicissitudes made by the malware to its target system. Unlike static analysis, dynamic analysis is considered to be highly jeopardous but paramount, or high-profit process. The peril of infection, spreading, or something inimical transpiring is high because the malware is executing; the profit is high because the data extracted from malware reveals more of itself during execution. In the dynamic analysis, we are probing for the following vicissitudes in registry activity, network traffic activity, process, and file activity. Some prevalent dynamic analysis implements are process monitor, wire shark, capture bat, anubis, etc.

1.2.5 Hybrid Analysis

The hybrid analysis technique includes consolidating static and dynamic features accumulated from examining the application and drawing data while the application is running, discretely. Nevertheless, it would boost the precision of the identification.

The principal drawback of hybrid analysis consumes the system resources and takes a long time to perform the analysis. The hybrid analysis amalgamates the traits of static and dynamic analysis for expeditious analysis and better results.

1.2.6 Comparison

Static analysis cannot detect unknown malware and its variants. Compared to static analysis, dynamic analysis is more efficacious and does not require the executable to be disassembled but on the other hand, it takes more time and consumes more resources than static analysis, being more arduous to scale. One more issue is as the controlled environment in which the malware is monitored is different from the genuine one, the program may comport differently because some deportment of malware might be triggered only under certain conditions such as via a concrete command or on the concrete system date and in consequence, cannot be detected in a virtual environment. In static analysis, data extraction is effective only if the malware is free from any type of encryption or obfuscation. Dynamic analysis is all about making the malware prosperously run in a controlled environment. Therefore, its circumscription is because of the different malware dependencies like time, event, program, etc. Static analysis can facilely be subjugated by a packed and encrypted file. This is why file unpacking and decryption are paramount in the fight against malware. Static analysis reveals some immediate information about malware but it is expeditious, exhaustive analysis more in-depth information but it is hard and time-consuming.

Malware analysis is a highly manual and laborious task, additionally requires analysts to have expertise in software internals and reverse engineering. Data mining and machine learning have shown promise in automating certain components of malware analysis, but these methods still rely heavily on extracting paramount features from the data, which is a nontrivial task that perpetuates to require practitioners with specialized skill sets. As the number of devices connected over the cyber world increases parallelly the attacks additionally increase exponentially. In reality, malware analysis does not reveal most of the information from the malware because of the known limitation of the malware analysis process.

1.3 Malware Classification

We now shift our discussion toward the main topic of this chapter that is malware classification and identification. In general, malware classification is defined as to group or classify malware together predicated on some mundane properties like they apportion homogeneous code, same potential damage, their inceptions, etc. In more simple words, classification is the process of assigning an object to a category or class. Classification refers to methods for presaging the likelihood that a given sample belongs to a predefined class or category, like whether a piece of email belongs to the class "spam" or a url is benign or malignant.

Malware can be classified in many ways such as depending on task, inception, authorship, damage potential, etc. In general, malware samples are grouped by family. Malware samples that show homogeneous functionality, structure with little differences are grouped under one roof and referred to as they belong to the same malware family. Classification is the prediction of incipient samples into its class whereas clustering is about discrimination of one group of samples from other groups. Classification is supervised whereas clustering is unsupervised. Classification examples are like to classify the taste of food as good or bad, to classify the thoughts or thinking as right and wrong, etc.

The prevalent term for non-maleficent files is a benign file. These are the examples of a binary classification problem one with only two output classes, "spam" and "not spam," "botnet" or "benign." By convention, samples that possess the attribute we are investigating (e.g., that an electronic mail is spam) are labeled as belonging to class "1" while samples that don't possess this attribute (e.g., mail that is not spam) are labeled as belonging to the class "0." These 1 and 0 class labels are often referred to as positive and negative cases, respectively.

Classification is a puissant and efficacious supervised learning model that can be applied productively to a broad range of security and other quandaries. The algorithms used to perform classification are referred to as "classifiers." There are numerous classifiers available to solve binary classification problems, each with its strengths and impotencies. By the definition of malware classification, one can be confused with the identification of any given file as malicious and non-malicious. One should be kept in mind that malware classification includes the identification and classification of malicious and non-malicious files. So we can conclude that a given arbitrary binary file identified or classified as benign or malware comes under malware classification. This classification is utilized to determine whether a binary is malicious or not.

1.3.1 Classification Steps

A classification typically proceeds through the following steps:

1. A training/learning phase: In this phase, an analyst builds a model and applies a classifier on the training inputs. Training data consists of two things, data or samples, and its associated labels/class.
2. A validation phase: This phase is applied to assess the training performance on validation data. The validation phase is optional but researchers and analysts vigorously suggest utilizing the validation phase. In this, training data is split into two sets, one is for training and the second is for validation. Training is done on training data and to assess the training performance (customarily accuracy) we apply validation data on training.
3. A testing phase: To assess the performance of the deep learning model, we apply testing data on the classifier and monitor the classifiers prognosticated labeled

with an authentic or ground-truth label of test samples. The test precision is the overall precision of the model. test data is not optically discerned afore data.

1.3.2 Why Malware Classification?

Malware and its variant detection and classification have become one of the most adverse quandaries in the field of cybersecurity and the digital world. The daily increase of malware and its variants is a rigorous quandary of malware analysis. The main quandary with malware analysis is that the number of attack files submitted to antivirus companies for the investigation purpose is enormous. It is virtually infeasible and arduous to analyze each file manually, so there is a desideratum for some automation system and implements to analyze these files efficiently with less human intervention and efforts.

In the cybersecurity domain, traffic classification as malicious and benign is considered the first step toward security. By classifying malware into their respective families is helpful to analyze samples of a given family by human experts and some defensive measures can be proposed to mitigate malware attacks. Features or characteristics are extracted from the malware binaries utilizing data extraction methods and implements. The attack of malware and its variants is not only inhibited to the cyber world, it is withal affecting the IoT networks, mobile networks, and contrivances. Researchers and analysts commenced to explore malware analysis utilizing deep learning and visualization techniques in IoT, mobile, and cloud infrastructure.

1.3.3 Why Malware Visualization?

Malware visualization is the process of visualizing malware binaries as images—examples are given in Fig. 4. Visualization avails to visualize kindred attributes and distinctions between two variants of the same family. Visualization is efficacious in the representation of internal structure kindred attribute of malware. Malware binaries are ready to run or executable programs referred to as binary files and has an extension of .bin or .exe .

As we can visually perceive from Fig. 4 that malware from the same malware family exhibits the same internal structure while malware from different malware families has a different internal structure. This is the prevalent advantage of visual malware as an image and it avails in classifying malware. The advantage of images utilized in visualization is that they can give more in-depth information about the internal structure of the malware binary code and could identify even small changes in code while retaining the whole structure of the code.

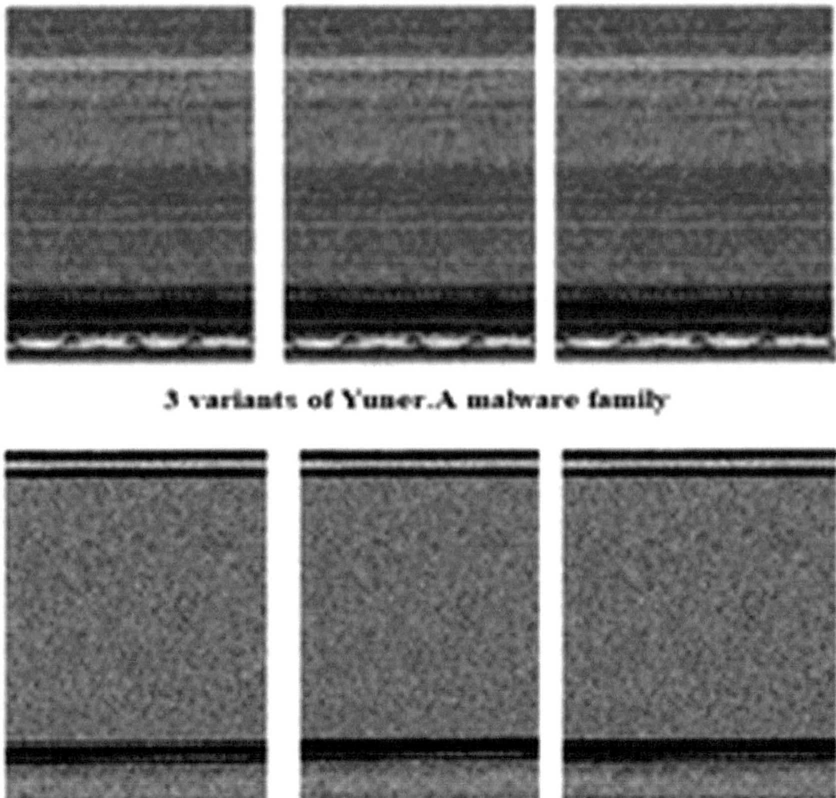

Fig. 4 Visual representation of malware

1.3.4 Challenges of Malware Classification

Here we are going to discuss the challenges that are reported during the study of malware analysis. One of the most sizably voluminous challenges is that everyday millions of malware are being designed and the complexity to detect this massive amplitude of sophisticated malware are very difficult to identify. Traditional approaches for malware analysis were very tedious and manual intervention was required for analysis. Obfuscation techniques present most immensely colossal hurdle and one of the major factors which affect the analysis of malware. Scalability is one of the major challenges in the malware defense system as the number and variety of malware are kept incrementing. Classification algorithms and models can engender precise results on propitious conditions but this case is not possible in the genuine world. To obtain a dataset for training and testing that is sizably voluminous and accurately labeled is arduous. The number of samples in each class additionally affects the relegation precision. The classifier's performance is highly dependent upon the ample quantity of labeled data. Overfitting and underfitting are two well-kenned quandaries associated with the classifier's performance. There is not a single performance measure that is used to assess the performance of the classifier; there are varieties of measures available like accuracy, precision, f1 score, roc curves, etc. Deep learning is about deep neural networks and neural networks have a variety of hyper parameters that affects the models or classifier's performance like the number of hidden layers, number of neurons per layer, learning rate, dropout, etc. Some features extracted from malware samples have high dimensionality, which denotes a more involutes system and incremented processing time. One of the latest emerging threat in malware analysis is the file less malware [10], it does not utilize the file system for its malevolent activities, thereby eschewing traditional approaches and became one of the hurdles in malware analysis.

2 Deep Learning

In this section, we are discussing what is deep learning, what are different deep learning algorithms, how is the deep learning model defined? The topic is briefly explicated to relate with the malware classification.

2.1 What is Deep Learning?

Deep learning is a sub-branch of machine learning and its functioning is inspired by the structure and function of the brain called neural networks. Deep learning refers to the set of techniques utilized for learning in neural networks. It refers to deep, or many-layered, neural networks withal kenned as deep neural network. Deep learning is about learning abstract representations of data or observations utilizing network layers that avail to make sense of some kind of hidden patterns, features of data like images, sound, and text. In pursuing malware analysis and lowering human intervention, deep learning has been introduced into malware analysis. Deep learning depends on studying various levels (from low level to higher level) of representations, where top-level features (for example, face) are tenacious from lower level ones (like edges, curve, etc.), and similarly lower- level features avail in determining numerous top-level features.

2.1.1 Machine Learning

Machine learning is defined as the subfield of artificial intelligence. The goal of machine learning is to understand the data and build a numerical model, fit that data into a model that can be understood and utilized by the user.

Antivirus companies commenced to utilize modern classification techniques dependent on data mining and machine learning methods. All the methods either data mining or machine learning approach dependent upon the extraction of features, applying more clever frameworks or classifiers for classification purposes. The disadvantage of machine learning is that it requires manual feature extraction. Many authors applied support vector machine (SVM) classifier, naÃ¯ve bayes classifier, or mixed classifiers to classify malware.

2.1.2 Shallow and Deep Learning

Deep learning is a subfield of machine learning, concerned with functionality and structure inspired by the human brain called artificial neural networks. The term "deep" in deep learning isn't a reference to any kind of in-depth understanding achieved by the approach; rather, it stands for conception and number of stacked layers of representations of the input. How many layers contribute to a deep learning model of the input data is called the depth or deepness of the model? The term shallow learning algorithms are normally referred to as traditional machine learning algorithms. It refers to algorithms that are not deep in architecture, e.g., decision trees, support vector machines, naive bayes classifier, etc.

Modern deep learning models often constitute tens or even hundreds of stacked layers of representations and they're all learned /extract features automatically from exposure to training data. Machine learning inclines to fixate on learning/extracting only one (mostly) or two layers of representations of the input data; hence, they're sometimes called shallow learning.

2.1.3 What Makes Deep Learning Different?

1. Deep learning algorithms offered better performance on many involutes real-world problems.
2. It makes problem-solving more facile.
3. It automates the most critical phase of machine learning that is optimized feature extraction.
4. With deep learning, we can acquire more refined transformations of complex problems.

2.1.4 Deep Learning Framework

Generally, a framework is a platform, interface, accumulation of libraries, and implements for developing applications. We have deep learning frameworks for building deep learning models facilely and without going into depth cognizance of algorithms. Some popular frameworks are tensor flow, keras, pytorch, caffe, deeplearning4j, etc.

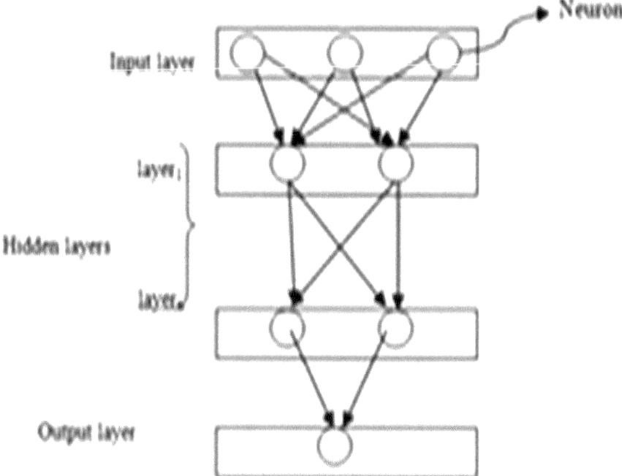

Fig. 5 Deep neural network

2.2 *Deep Learning Algorithms*

Deep learning can be considered as a subfield of machine learning. It is predicated on learning and improving on its own by examining algorithms. While machine learning uses simpler concepts, deep learning works with artificial neural networks, which are designed to be homogeneous to how humans think and learn. Artificial neural network (ANN) drive deep learning. Neural networks were restricted by computing power and thus were limited in complexity. However, deep learning (ANN with many layers) sanction computers to observe, learn, and react to intricate situations more expeditious than humans. Deep learning has availed image classification, language translation, and speech recognition. Deep learning can be acclimated to solve any pattern recognition problem, to classify images, for language translation, to recognize speech and without human intervention. Deep learning is to learn hierarchical representations of input data.

Commonly used deep learning algorithms are

2.2.1 Deep Neural Network (DNN)

Deep neural networks are the ANN with many layers Fig. 5. Typically deep neural networks are feed-forward networks in which input flows from the input layer to the output layer and hidden layers(two or more) and the sodalities between the layers are one way which is in the forward direction(input layer to output layer). The outputs are obtained by learning with datasets of labeled information predicated on backpropagation. The circumscription of deep neural networks is that they don’t have any memory unit.

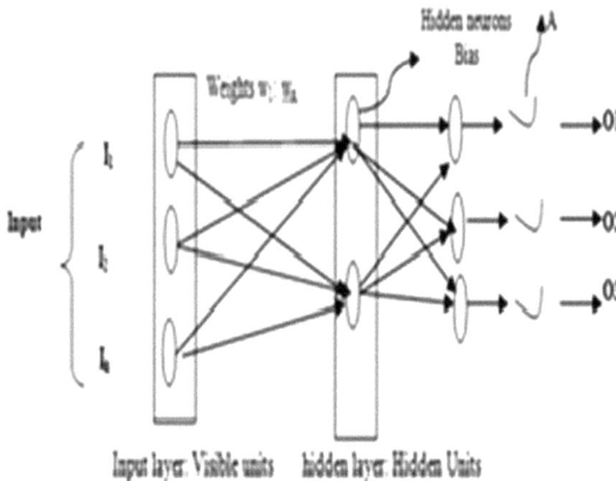

Fig. 6 Restricted boltzmann machine

2.2.2 Restricted Boltzmann Machine (RBM)

RBMs are a two-layered artificial neural network with generative capabilities Fig. 6. They can learn a probability distribution over its set of input. RBM can be utilized for dimensionality reduction, relegation, regression, collaborative filtering, feature learning, and topic modeling. RBMs are a special class of Boltzmann machines and they are restricted in terms of the connections between the visible and the hidden units. This makes it facile to implement them when compared to boltzmann machines. As stated earlier, they are a two-layered neural network (one being the visible layer and the other one being the hidden layer) and these two layers are connected by a fully bipartite graph. This denotes that every node in the visible layer is connected to every node in the hidden layer but no two nodes in the same group are connected. There are two other layers of bias units (hidden bias and visible bias) in a RBM. This is what makes RBMs different from auto encoders. The hidden bias RBM produces the activation on the forward pass and the visible bias avails RBM to reconstruct the input during a rearward pass. The reconstructed input is always different from the actual input as there are no connections among the visible units and therefore, there is no way of transferring information among them.

2.2.3 Convolutional Neural Network (CNN)

Convolutional neural networks are very subsidiary for images based processing, especially for image-based classification. A convolutional neural network Fig. 7 is a type of feed-forward neural network in which the connectivity pattern between its neurons is inspired by the organization of the animal visual cortex, whose individual neurons are arranged in such a way that they respond to overlapping regions tilling the visual field. Convolutional layers are the core of a convolutional neural network. Convolutional neural networks, like neural networks, are composed of neurons with

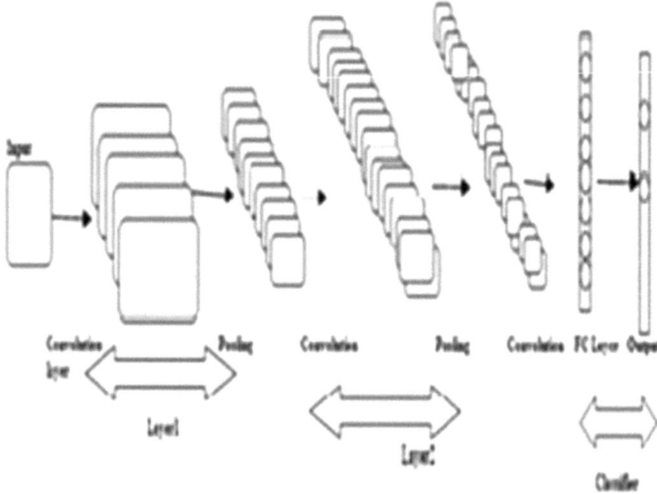

Fig. 7 Convolutional neural network

weights and biases (updates through learning). Each neuron receives inputs, applies a convolution operation (weighted sum of multiplication) over them, passes it through an activation function, and responds with an output. The network has a loss function and weights and biases are updated according to the loss function. CNN is composed of three types of layers: convolution layers, pooling/subsampling layers, fully-connected/dense layers.

2.2.4 Deep Belief Network (DBN)

A DBN is a class of deep neural network, a graphical model, composed of multiple layers of latent variables (hidden units utilized for detecting features), with connections between the layers but not between units within each layer and have direct and undirected connections Fig. 8. RBMs can be stacked and trained to compose so-called deep belief networks. Multiple RBMs can withal be stacked and learned through the process of gradient descent and backpropagation. Such a network is called a deep belief network. A deep belief network utilizes an unsupervised machine learning model to produce results. One of the mundane features of a deep belief network is that albeit layers have connections between them, the network does not include connections between units in a single layer. A DBN can work as a supervised learning algorithm (as a classifier) and additionally utilized as an unsupervised learning algorithm (to cluster data).

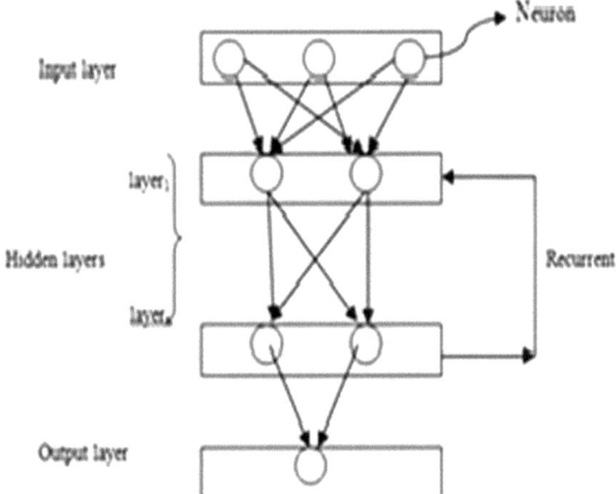

Fig. 8 Deep belief network

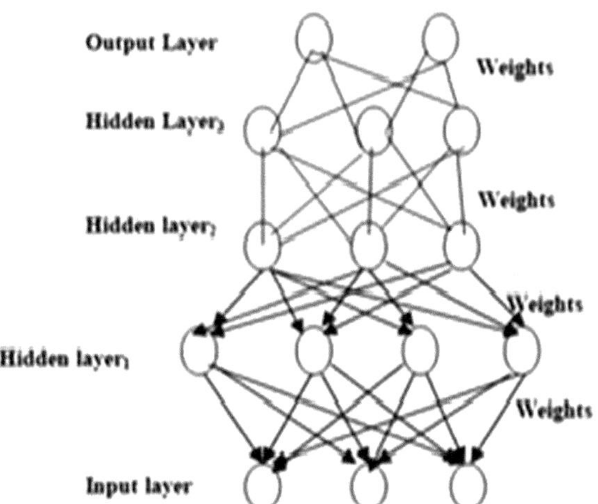

Fig. 9 Recurrent neural network

2.2.5 Recurrent Neural Network (RNN)

Recurrent neural networks are best to process sequences. A recurrent neural network Fig. 9 addresses the issue of the memory limitation of deep neural networks. Deep neural networks are stateless, but recurrent neural networks have connections between passes and connections through time. A recurrent neural network looks similar to a traditional artificial neural network except that it has a memory-state and is added to the neurons. With a recurrent neural network, this output is sent back to the previous layer number of times. RNNs can remember parts of the inputs and use them to make accurate predictions.

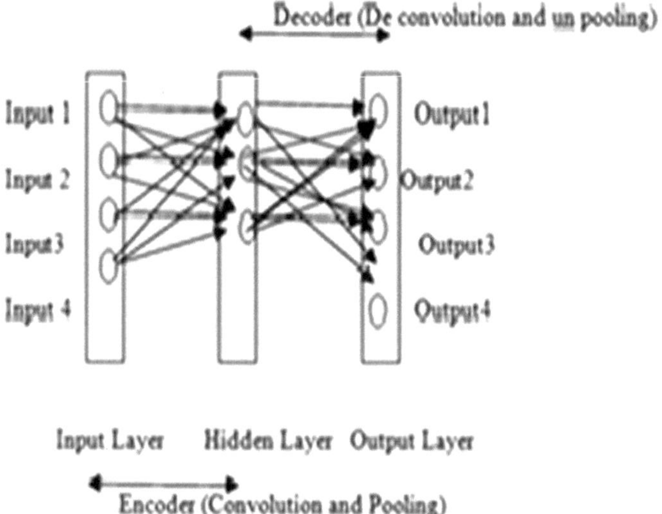

Fig. 10 Autoencoder

2.2.6 Deep Autoencoder (AE)

Autoencoder is a neural network Fig. 10 that utilizes unsupervised learning algorithms and backpropagation. It efficiently compresses and encodes input data then learns how to set output values identically to the input values. How to decode the data back from the minimized encoded representation to a representation that is as proximate to the original input as possible. Autoencoder, by design, transforms data into a hidden representation and then reconstructs data from that hidden representation inputs are high-dimensional data. It is compressed by the hidden layer and the output layer reconstructs the inputs. The main applications of the autoencoder are data denoising and dimensionality abbreviation.

2.3 Steps for Building a Deep Learning Model

The main advantage of deep learning systems for malware analysis is that they automate the work of feature extraction, and they have the potential to perform more accurately and efficiently than traditional approaches to malware analysis, especially we want to focus on malware classification especially on new, previously unseen malware. Essentially, the following steps Fig. 11 are used to build any deep learning model for malware classification.

1. Data/samples collection: To train the DL model, we require data (training data). For malware analysis, we require malware as well as benign (good wares) data.

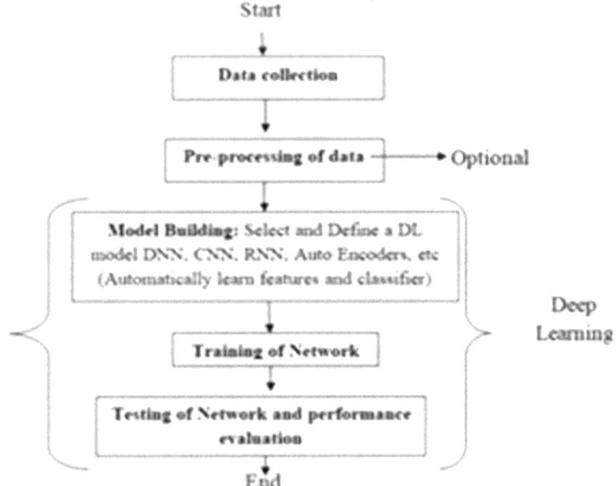

Fig. 11 Deep learning model building

The performance of the dl model depends profoundly on the quantity and quality of training examples, you provide for training. The quality of the training data is also important. If you want to apply the dl model for a multi-class classification problem you have to amass adequate data for each class. The general rule of thumb is that the more data (training data) you feed into your dl model for training, the more precise results you will get.

2. Model building: We have to define a deep learning model among various available deep learning models (DNN, CNN, RNN, Auto Encoders, etc.) as per the requirements. We first build the model then training and testing is applied on the defined model.
3. Training: Train the model for recognition of malware on the optimized features extracted automatically by the dl model. For training, we provide data /samples and associated labels of samples. Mundanely, training is considered to be an arduous task to perform because of the settings of hyper parameters. We feed the training images into different CNN model architectures (it varies with several layers, number of neurons in layers, learning rate, number of epochs, batch size, etc.) with different hyper parameters settings, several epochs, and batch size and probe for the model that fits our dataset.
4. Testing: Once you trained your model, we require to test the model on the data samples that were not included in the training to assess the model's performance or how precise the model is. Generally, testing is done by running the trained deep learning model on the data samples that were not included in the training denotes data that has been never seen by the model.

3 Malware Classification Based on Malware Visualization and Deep Learning

In the following section, we are going to present and discuss detailed procedures on the recent cognate work predicated on malware classification utilizing deep learning. The exhaustive study covers techniques of malware visualization predicated on different deep learning algorithms.

To visualize malware as an image [19] is the field of representing malware in the form of visual features. To analyze malware more deeply, malware has to be transformed into an image, refer Fig. 13. The main benefit of visualizing malware as an image is that different sections of a binary file can be facilely differentiated, Fig. 4. Many solutions have been proposed and implemented utilizing static and dynamic approaches but this work is predicated on the malevolent code and its variants detection and classification utilizing visualization techniques and deep learning. There has been extensive research and study done on analyzing malware, many papers are published which denotes static, dynamic, and signature-based malware analyzing techniques. A publication on image-based malware visualization is one of the preferred ways [19]. This section explicates how to compose an image out of binary malware files, how to visualize those images, and how these images are utilized for image-based classification.

Traditionally this task is done by signature matching. In signature matching, a database is prepared of properties, the behavior of previously seen malware; incipient binaries are compared by this dataset to compare previously stored data to determine that something visually perceived afore. Signature matching performs well as long as malware designers alter the behavior and properties of malware to evade detection. A malware designer continuously changes the properties and behavior of malware to avoid detection. By utilizing obfuscation techniques like metamorphic and polymorphic, authors of malware changes properties of code, behavior to avoid detection of malware by signature matching or malware identification implementations.

We have studied papers which utilize the same principles as [19] to classify the malware into their families. It has been observed that the deep learning model is efficient. We propose to utilize malware visualization technique, converts every malware bytes code to a grayscale image. In research and analysis, it was observed that malware from different families has kindred attributes in visual appearance presenting to us an opportunity to exploit this impotency where these images will be utilized for image-based classification. In image generation and classification technique, every byte of data is converted into a grayscale pixel; array of the byte stream was converted into an image. Image representation of the malware engenders very convincing images for analysis purposes.

3.1 Related Work: Recent Innovations in Malware Classification Using Deep Learning and Visualization

In this section, we are going to discuss and review the current state-of-the-art approaches that have been established to address the malware classification utilizing deep learning models and malware visualization.

The solution based on malware visualization by Nataraj et al. [19] in the year 2011 is considered as the first solution of this kind. Authors proposed a method to represent malware as a grayscale image and after that extracted gist (a texture-based feature), afterward, the grayscale malware images are classified utilizing a data mining algorithm knn (k-nearest neighbor). They experimented with the malimg dataset consisting of 9458 malware grayscale images belonging to 25 different families, amassed from the anubis system. For the experimental purpose, they converted malware into a grayscale image of dimension 64*64. They extracted 320-dimensional texture features from malware image predicated on gist. They divided the samples into a ratio of 90–10% for train and test ratio. They obtained the test accuracy of 97.18% which is very high as compared to traditional approaches. The results obtained evidence that visualization of malware is very efficacious and can relegate malware with more precision and expeditiously than subsisting static and dynamic approaches.

In the year 2013, Han k. et al. [5] proposed an incipient way of visualizing malware. They visualized malware as a color image utilizing binary values. The proposed method generates RGB colored pixels by utilizing binary information extracted through static analysis. First, the author disassembled malware binary files utilizing implements such as IDA pro or ollyDbg, after disassembling the extracted sequence of assembly codes are divided into blocks of opcodes (example of opcode sequence: pusmovaddsubmov), after block building, every block of opcode instruction sequence is processed by two hash functions to engender matrix of coordinate values and RGB color pixels information. To compute the homogeneous attribute between image matrices authors utilized a selective area matching algorithm. For experiment purposes, the authors utilized a color image of size 256 * 256, 2505 benign, and 8169 malware image matrices are engendered utilizing a visualization implement. 95% test accuracy is achieved by this method. Results deduced relegation efficaciously and the time spent to calculate homogeneous attribute was about 2.4ms.

In 2016, K. K. Pal and k. S. Sudeep [21] presented a data preprocessing technique for the malware relegation model utilizing a convolutional neural network and image representation. Authors proved that by applying preprocessing techniques on the data, classification accuracy can amend. Raw data applied to any deep neural network does not engender good results. The authors conducted three types of normalization on the dataset and showed how precision varies. They applied to mean normalization, standardization, zero component analysis on the dataset. For experimental purport authors used malware color images of size 32*32 and they utilized the cifar 10 dataset (dataset contains 60000 color images of size 32*32 belongs to 10 different classes). They obtained an accuracy of 64–68% when zero component analysis is applied, they got increased accuracy as compared to when no preprocessing applied.

In the year 2016 Ding y. et al. [2] have prosperously applied a deep belief network, one of the unsupervised learning algorithms for malware relegation. The authors represented malware as opcode sequences and then use deep belief network to detect malware. An opcode (operation code) additionally kenned as an instruction machine code that designates the operation to be performed. In a deep learning algorithm, the neural network is trained multiple times by the raw opcode sequences extracted from the decompiled file, so that the hidden feature information can be efficaciously learned and the malware can be detected efficiently and more accurately. Feature extractor measures different measures like information gain, document frequency to evaluate the relegation . Author used information gain to cull subsidiary n-gram. To accurately describe the opcode comportment, the author extracted opcode sequences from 3000 benign and 3000 malware samples. The extractor evaluates 10000 different n-grams with different information gain values. From these 10000 values author utilized the top 400 n-grams as the features of an executable. DBN architecture has 3 hidden units with 200,200 50 hidden neurons, respectively. Each layer is trained with 30 epochs. The authors obtained 96.1% accuracy.

In the year 2016 Hardy w. et al. [6] proposed an intelligent deep learning framework for malware detection. They applied auto encoder, it is one of the unsupervised deep learning algorithms used to detect generic features from malware to detect unknown malware. They utilized a greedy-based feature learning at each layer, followed by supervised tuning of weights and biases. The authors extracted windows-based API call sequences from the portable executable (PE) files. For experimental purpose, authors used the comodo cloud security center dataset (dataset contains 22500 malware samples and 22500 benign samples total of 50000) and the train and test ratio was 90–10%. The experiment is performed with a different number of neurons in the hidden layer but 100 neurons at each hidden layer and 3 hidden layers configuration yield the maximum accuracy that is 96.85% at training and 95.64% at testing.

Tobiyama s. Et al. [23], in the year 2016, proposed the fusion of deep learning models in malware analysis. The authors first applied a recurrent neural network to extract the features based on malware behavior and then applied CNN to classify malware feature images of size image 30*30. To capture the behavior of malicious application authors utilized API call sequences. The proposed malware detection framework is mainly using API call sequence extraction and deep learning technique for classification. A process behavior is defined as various activities and to perform each activity various operations are associated with activities. To record process behavior API call sequence is generated; the API call sequence represents activities and related operations. They extracted feature vector by training of recurrent neural network and then these extracted feature vectors are converted into an image and applied CNN for classification. For experimental purpose 81 malware process log files of 11 different malware families, 69 benign processes log files data collected by NTT secure platform laboratory. The architecture of recurrent neural network consists of an input layer, a hidden layer, 2 LSTM hidden layers, and an output layer. The architecture of CNN consists of 2 convolution and pooling layers with 10 and

20 filters, respectively. Used max pooling with stride 2, no of epochs: 5, batch size: 20. They obtained 96% accuracy.

Azab A. et al. [1] in the year 2016 proposed and addressed machine learning technique for identification of untrained botnets traffic. Authors applied the c4.5 learning algorithm (for building classifier with 10,20 and 30 FN costs and 1 FP cost) and correlation-predicated feature cull (cfs, applied to filter out duplicate, redundant and impertinent features form extracted features) algorithm on the communication traffic between compromised contrivances and botmaster, they extracted 511 different features coalescence from 9 different features categories from this communication that avails to relegate between botnet traffic and legitimate traffic. For botnet network traffic accumulation, Zeus (a botnet toolkit) was culled and it is considered one of the major threats, especially for attacks on online banking transmissions. Two separate datasets accumulated for experiment one are for training (the 432 botnet traffic engendered utilizing zeus builder version 1.x and 2774 HTTP traffic) and the second is for testing(the 144 botnet traffic engendered utilizing Zeus builder version 2.x and 2396 HTTP traffic). All the built classifiers were evaluated utilizing the K-10 cross-validation to optate the lenient classifiers. The built classifiers were evaluated utilizing the K-10 cross-validation to cull the rigorous classifier. The voting results from the three costs achieved 88 TP, 56 FN, and 1 FP results, providing 0.989 precision, 0.611 recall, and 0.755 F-Measure results. These results betoken that the utilization of the stringent classifier might affect the detection of the untrained version's flows that were included by the lenient classifier.

In the year 2018, Kalash M. et al. [11] proposed and implemented a deep CNN model for malware classification. They translate the malware classification problem into an image classification by following the approach used by Nataraj et al. [19], converting malware binaries to grayscale images of size 224*224 and then applied a convolutional neural network for classification. The proposed convolutional neural network model architecture is based on VGG-16. They applied the proposed method on two different datasets, namely, malimg(dataset consists of 9458 malware samples belonging to 25 different families) and Microsoft dataset (contains 21741 malware samples, each malware sample belongs to 9 different malware families). Train and test ratio used by the authors are 90–10% in the malimg dataset and 10868 samples for training and 10873 samples for testing on the Microsoft dataset. They utilized cross-entropy loss to train the network. The authors achieved 98.52% accuracy on the malimg dataset (with 25 epochs and a batch size of 6) and 98.99 and 99.97% on two different settings of Microsoft dataset (with 25 epochs and a batch size of 8).

In 2018, Ni S. et al. [20] proposed a malware classification algorithm that utilizes static features and convolutional neural network. They converted the disassembled malware codes into grayscale images based on simhash, and then classification is done by convolutional neural network. They extracted the opcode sequence from the code section as features then after extraction of the opcode sequence they calculated simhash for sequence similarity comparison. By using simhash and bipolar interpolation they converted the opcode sequence into a malware image then applied convolutional neural network for training and classification. Each input image needs to go through two convolutional layers, two subsampling layers, and three full connection layers. During the convolution process, they applied 32 filters of size 2*2 and during subsampling max pooling is used whose size is 2*2 to dimension reduction. The authors used the dataset for the experiment in Microsoft malware classification challenge on kaggle by Microsoft 2015. The dataset consists of 10868 labeled malware images from 9 families, from 10868, 80% of them used for training and the rest for testing. The classification accuracy they obtained was 99.260% with a 98.07% f1 Score and 2.34% false positive rate (FPR).

Kim C. H. et al. [13] in the year 2018 proposed a convolution gated neural network for the task of malware identification and classification. Proposed model comprised of convolutional neural network, gated recurrent unit (GRU), layer of deep neural network, and a sigmoid layer. Each convolutional neural network has a convolution layer, activation function, and pooling layer. All convolutional neural network produces a single output, and this output is applied to gated recurrent unit layers and treats this output of convolutional neural network as time-series data. Each gated recurrent unit produces a single output equal to the number of convolutional neural networks in the first layer. Output of GRU is input to deep neural network. Each deep neural network produces single output. The final layer of the network is the sigmoid layer and the result of this layer is the classification.

In the year 2019, Singh A. et al. [22] explored and implemented a new way to represent malware as color images as they used RGB representation of malware (RGB images of size 32*32) over grayscale images to classify malware. They experimented with 37374 binary samples belonging to 22 families collected from malshare, virusshare, and virustotal, and malimg dataset. They applied deep neural network architectures ResNet-50(residual network) architecture including a dense convolutional neural network for classifying images. With their implemented model they obtained 98.98% using convolutional neural network and 99.40% using ResNet-50 on the authors dataset and 96.08% using convolutional neural network and 98.10% using ResNet-50 on the malimg dataset. The authors introduced a novel approach to convert the binary file string of zeros and ones into rgb color images. They used a 15 layer convolutional neural network model (5 convolutional layers and 2 dense layers).

Yin Q. et al. [25] in the year 2019 presented a fused model of convolutional neural network and recurrent neural network for image classification. Authors extracted features using convolutional and recurrent neural network networks from the intermediate convolutional neural network network.

In the year 2019, Naeem H. [17] proposed a fast deep learning model to detect malware in the IoT network. IoT devices improved the user experience of the internet by smart devices to connect and information sharing. The author proposed the detection of malware by converting malware binaries into the color images of size 192*192 and then applied a deep convolutional model for efficient malware detection on the malimg dataset (dataset consists of 9458 malware samples belonging to 25 different families) and leopard mobile datasets(contains 14733 malware samples and 2486 benign samples of IoT applications.) The train and test ratio was utilized as 55–45% for the malimg dataset and 34–66% for the leopard dataset. The author obtained an accuracy of 98.18% on the malimg dataset and 97.34% on the leopard dataset. The author achieved better accuracy and response time.

In the year 2019, Khan U. R., et al. [12] defined an improved, more intelligent convolutional neural network model for intrusion detection. Authors mentioned that machine learning algorithms have a low detection rate, as well as manual extraction of features, which is a laborious and time-consuming task that's why they applied deep learning in intrusion detection. Deep convolutional neural network is used for training and classification and it automatically extracts optimized features from input samples. The dataset used for experiments is the KDD99 dataset; the dataset contains 494021 training samples and 311029 test samples, from 5 different categories (contains normal, DOS, R2L, U2R, probe). They obtained the accuracy on improved convolutional neural network that is 99.23 for 800 epochs, which is a promising result. CNN model architecture has two convolutional and two pooling layers.

Mourtaji Y. et al. [16] in the year 2019 proposed a deep learning framework for malware classification. Authors first converted malware binaries into grayscale images as used by Nataraj et al. [19] and then trained a convolutional neural network model for classification. Different parameters for experimental purpose used malimg and Microsoft datasets. Train-test ratio used 85–15% for the malimg dataset and 10868 samples for training and 10873 samples for testing, and they utilized the convolutional neural network architecture defined by K. Simonyan and used a cross-entropy to learn and train the model from the network after that utilize stochastic gradient descent (SGD) to optimize the learning parameters of the model, initialized the learning rate to be 0.001 and 25 epochs, batch size of 6 for malimg dataset and 25 epochs, batch size of 8 for Microsoft dataset. The authors obtained 97.02% on malimg and 98.72% and 99.881% on two different experiment settings on Microsoft dataset.

Jain M. et al. [9] in the year 2020 applied and compared CNN and extreme learning machines (ELM) for malware classification. Results are evident that ELMs required less time to train as compared to train a CNN and achieves higher accuracy on one-dimensional data processing. Authors also found that for two-dimensional data processing ELMs are faster than CNN. Authors experimented with different settings of the CNN model like they applied CNN with one hidden layer than with two hidden layers with different hyperparameters settings, and the best results they got with a two-layer configuration of CNN with input images of size 128 $\times$ 128 pixels with 32 and 64 filter maps. With ELMs, they have to perform very fewer experiment settings like only they tuned some neurons in the hidden layer, the chosen 50 neurons for the experiment. The authors utilized grayscale images of size 128*128 for CNN and grayscale images of size 64*64 for the ELM model. They used the malimg dataset for experiments and 80% for training, 10% for testing, and 10% for validation division is applied to the dataset. The configuration of CNN architecture: CNN with two convolutional layers, 128 $\times$ 128 images, and (32, 64) filters and ELM architecture has 50 neurons in the hidden layer. They obtained an accuracy of 96.3% on the CNN model and 97.7% on the ELM model.

In January 2020, Kumar G. S. and Bagane P. [3] presented a hybrid deep learning-based model for malware classification. They applied convolutional neural network with bi-directional long short-term memory(LSTM) to do the task. First, they applied convolutional neural network for feature extraction, and then in the last layer after flattening the output they applied the LSTM model for the classification.

In the year 2020, Vasan D. et al. [24] proposed a novel approach based on the ensemble CNN architecture model for effective detection and classification of malware images. Authors utilized the pre-trained models and combined different optimized features extracted to fine-tune the VGG 16 and ResNet50 and fused the extracted features from both models and classified the malware into their corresponding families. Results proved the effectiveness of the proposed method.

In the year 2020, Naeem et al. [18] proposed a deep learning model for malware detection in the android operating system. They transformed a raw android file into a color image (of dimension 224*224 and 229*229) and then applied a deep convolutional neural network model for android malware classification. The author designed a very deep convolutional neural network that has 4 convolutional layers each followed by an activation function and max pooling layer, followed by a dense layer and softmax layer. The author applied a deep convolutional neural network model on the leopard dataset (dataset contains 14,733 malware samples and 2486 benign samples of different IIoT applications) and the malimg dataset. They achieved 97.81% accuracy on a leopard mobile malware dataset (224*224 color image dimension), a well-known industrial Internet of Things (IIOT) dataset, and 98.79% on a malimg dataset (with 229*229 color image dimension).

3.1.1 Generative Adversarial Networks (GANs)

Generative adversarial networks provide a new way of addressing computer vision, detection and classification problems. One of the biggest problems with deep learning model is lacking of sufficient training samples as we know that good quality and sufficient data is the key of deep learning model. Any deep learning model heavily dependent on the number of samples providing for training. Many datasets available today face this problem. We can notice from Table 1 that malimg dataset is also a high imbalanced dataset Allaple. A malware family contains 2949 samples as compare to Wintrim.BX and Skintrim.N malware family contains 97 and 80 samples respectively. This imbalance affects the training process as well as the classification performance. To address this problem GANs we can utilize. GAN can be used to generate samples from the data.

GANs are types of deep learning technique for generative modeling and most recent development in machine learning. GANs are very incipient in the literature on deep learning, and they belong to unsupervised learning. The first paper published by Goodfellow et al. In 2014 [4] introduced the generative adversarial networks framework. A GAN is trained utilizing two neural network models. Generative modeling requires a model to engender incipient samples from a subsisting distribution of available samples, for example, engendering incipient images that are generally homogeneous but concretely different from available images in the dataset. GANs are mainly utilized with convolutional neural network which denotes GANs are specially utilized for image cognate applications. A GAN is trained utilizing two neural network models. One model is referred as the generator or generative network model, which learns to engender incipient likely samples. The other model is called the discriminator or discriminative network and learns to differentiate engendered samples from authentic samples, discriminator works like a classifier; it distinguishes authentic samples from the engendered samples.

We can apply GAN in cybersecurity field, and it is proving very promising. One of the quandaries with any deep learning model is data imbalance issue. As we all very well know that good quality and adequate magnitude of training data is the key for any deep learning models performance. So, we can surmount this issue of data imbalance utilizing GAN and can engender incipient samples from the genuine samples for training. In reality, all the real-world datasets are imbalance datasets and there is much variation in the number of samples in each family. So GANs can be proved very efficient to address this issue.

In the year 2019, Y. Lu and J. Li [15] applied GAN for malware classification/predication on deep learning model. Authors addressed the data imbalance issue and engender incipient samples for training. They applied GAN utilizing convolutional neural network and called this model as deep convolutional generative adversarial network (DCGAN) to engender malware samples from the available dataset. Experimental results are conspicuous that utilizing GAN accuracy of the proposed model is incremented by 6%. In their implementation, they utilized a 18-layers deep residual net as the malware classifier. Network learns from the trained data that engenders the potential distribution of the incipient genuine samples from the authentic samples, while the discriminator differentiates the incipient genuine samples with the genuine samples as accurately as possible. Multiple convolutional and convolutional-transpose layers are utilized in the discriminator and engenderer for training. They trained the GAN network for 10000 epochs to engender the authentic samples, starting from the 1000 training epochs, preserved 25 engendered samples for every 100 epochs for each class. So after the training is done. They have 2250 engendered synthetic samples for each class. They achieved the overall average testing accuracy of the deep residual network is 84% and the precision, recalls, and f1-scores of the classes with more samples size are supplementally incremented.

In the year 2017, Kim JY. et al. [14] proposed a transferred generative adversarial network (tGAN) for automatic malware relegation and detection of the zero-day attack. They surmount the constraint of GAN training to pre-train GAN with auto encoder structure. The proposed model gets the best performance compared to the conventional learning algorithms. To address the data imbalance issue and to engender incipient samples they proposed and applied tGAN model predicated on GAN. Their proposed architecture consists of three modules: pre-training module, engendering data module, and malware detecting module. First module pre-trains the second module which has an engenderer that engenders data kindred for training, and a discriminator that distinguishes genuine data from engendered data. The discriminator is trained to distinguish the authentic data from the engendered data, and the engenderer is trained to make the discriminator to classify the engendered data into the genuine data. They used malware data utilized in the kaggle Microsoft malware classification challenge. The accuracy of malware type detection is 96.39%. The entire data is divided into training and test data at a ratio of 90:10. It shows the best performance compared to other conventional models, and it enables to detect malware even with a minute of data.

A detailed review of generative adversarial networks and its application in cyber security is presented by Banjo Y. et al. [26]. They explained how GANs are very useful and applicable in cyber security field. They reviewed two very widely utilized GAN architectures the deep convolutional generative adversarial network (DCGAN), and wasserstein GAN. Their reviews are notable to study cyber security where the GAN plays a vital role in the design of a security system . This paper guide the scope of modern cyber security studies with generative adversarial networks.

Deep learning can help to solve problems caused by modern malware and the way they function. Using deep learning we can also automate the malware analysis process. The biggest advantage with deep learning is that the manual extraction of features or data is skipped, deep learning architectures automatically extracts features from samples, based on the extracted features from the training dataset, samples are distinguished by samples belonging to a particular class to other classes. Traditionally malware classification has been a manual process, involving experts having in-depth knowledge of malware, their working, properties in malware to design malware identification, or classification engines. Deep learning facilitates automatic extraction of optimized features from the training dataset, letting the automatic detection of features analysts can make effort for designing more efficient algorithms, and better results.

3.2 Performance Metrics: To Measure the Performance of the Deep Learning Model

To evaluate the performance of the developed system or solution following metrics are calculated. Using these metrics we can compare different techniques and can conclude which technique is better than others.

3.2.1 Confusion Matrix

It is utilized to visualize the performance of a technique. In general, a classifier is evaluated by a confusion matrix Fig. 12. Structure-wise confusion matrix is a table representation that is used to describe the performance of a classification model on the test datasets. All other performance metrics are calculated utilizing the confusion metric. In the confusion matrix, there are four possible states denominated true positive (TP), false positive (FP), true negative (TN), and false negative (FN) defined as follows

TP: when the sample is identified as an attack and the sample is an attack (Remark: identification of attack).

Fig. 12 Confusion matrix

Predicted Outcomes ⇨ / ⇩ Actual Outcomes	positive	Negative
Positive	TP	FP
Negative	FN	TN

FP: when the sample is identified as an attack and the sample is not an attack (false alarm).
TN: when the sample is not identified as an attack and the sample is not an attack.
FN: when the sample is not identified as an attack and the sample is an attack.

Accuracy indicates the proportion of all samples with correct predictions to the total sample size. The formula to calculate accuracy is

$$\text{Accuracy} = \frac{\text{TP} + \text{TN}}{\text{TP} + \text{TN} + \text{FP} + \text{FN}} \tag{1}$$

Precision describes the ratio of predictive positive samples positive. The formula to calculate precision is

$$\text{Precision} = \frac{\text{TP}}{\text{TP} + \text{FP}} \tag{2}$$

Recall is also known as True Positive Rate (TPR.)The formula to calculate recall is

$$\text{Recall} = \frac{\text{TP}}{\text{TP} + \text{FN}} \tag{3}$$

F1 is the harmonic mean of precision and recall. The formula to calculate the F1 score is

$$\text{F1 score} = 2 \times \frac{\text{Precision} \times \text{Recall}}{\text{Precision} + \text{Recall}} \tag{4}$$

Receiver operating characteristic curve (ROC) is a graph that is used to summarize the performance of a classifier over all possible thresholds. The graph is generated by plotting a graph between True Positive Rate (TPR) and False Positive Rate (FPR). the formulas for TPR and FPR are

$$\text{TPR} = \frac{\text{TP}}{\text{TP} + \text{FN}} \text{ and FPR} = \frac{\text{FP}}{\text{FP} + \text{TN}}.$$

We observed that accuracy (1) is a common measure used to judge the classifier's performance but it seems inadequate, and other measures like f1-score (4) and recall (3) are also important to evaluate the performance of the classification. High accuracy and recall with lower misclassification are required for an efficient model.

3.3 *A Practical Implementation of Malware Classification Using CNN and Malware Image Visualization*

In this section, we are going to discuss a practical example of malware classification utilizing a convolutional neural network, which is considered the most prosperous deep learning architecture in computer vision, pattern matching, and natural language processing. We also discuss an idea of how convolution works, convolutional neural network works, and the main operation types are used in building the convolutional neural network model. We can implement any other deep learning model withal but most advanced applications and models are currently being utilized by convolutional neural network so we decided to implement convolutional neural network for our practical implementation. After reading this section, you will have an early understanding of how deep neural networks work, and you will be able to move on to practical applications. Our goal is to give you an expeditious and facile tutorial on how to implement image relegation. Hopefully, you will be able to understand the main practical concepts and utilize this to build your applications and research.

A fundamental convolutional neural network model architecture Fig. 7 contains convolutional layer followed by an activation function (CONV), pooling layer (max or avg pooling based on the requisite) (POOL), dense/fully connected layer (FC).

3.3.1 Convolution Layer

To implement convolution operation kernels/filters are frequently utilized. The convolution operation (betokened by *) consists of multiplying the corresponding pixels with the kernel pixels, one pixel at a time, and summing up the values to assign that value to the central pixel. The same operation will then be applied, shifting the convolution matrix to the left until all possible pixels are visited. Kernels or filters are a matrix of values and the kernel slides over the input image and performs element-wise multiplication operation between the values in the filter with the pristine pixel values of the image. The multiplications are summed up engendering a single number for that particular receptive field. The input to the convolutional layer is an image that is resized to an optimal size (mundanely image size n*n) and fed as input to the convolutional layer. Let us consider image size is 32*32*1, where 32*32 is image dimension and 1 is the channel depth, it will be 1 for grayscale image and will take value 3 for color images.

3.3.2 Pooling Layer

Convolutional layers in a convolutional neural network methodically apply kernels/filters to images to extract optimized features and outputs feature maps. A quandary with feature maps is that they are sensitive to the location of the features in the input image which denotes a minuscule change (this can transpire due to shifting, cropping, rotation, or any other transformation) in the input image will yield different feature maps. A prevalent solution to this problem is downsampling. Mundane pooling methods are average pooling and max pooling. Max pooling is commonly utilized for the downsampling. Pooling layer operates on each feature map individually.

3.3.3 Dense/Fully Connected Layer

Fully connected layers or dense layers are a crucial layer of convolutional neural network, which are responsible for recognizing and classifying images or we can say that the final classification decision is taken by a fully connected layer. Fully connected layer takes the output from previous layers (convolutional and pooling layers of the defined convolutional neural network model) and predicts the class/label that best describes the input image.

3.3.4 Dataset

In this practical implementation, we will be working on one of the most extensively used datasets in malware classification that is the malimg dataset. The dataset details are given in Table 1. In this demonstration, we will build a simple convolutional neural network model to have an idea of the general structure of computations needed to tackle the multi-class classification problem.

First, let us understand the dataset. We are going to use malimg malware dataset [19] for practical purpose. Description of the dataset is as follows

1. The dataset contains 9339 malware images.
2. Malware images belong to 25 different malware families/classes.
3. Images are grayscale images.
4. All images of different sizes.
5. Dataset is highly imbalanced.

Table 1 Malimg dataset

Serial no.	Family/class	Family name	No. of variants
1	Worm	Allaple.A	2949
2	Worm	Allaple.L	1591
3	Worm	Yuner.A	800
4	Dialer	Instant access	431
5	Worm	VB.AT	408
6	Rogue	Fakerean	381
7	PWS	Lolyda.AA 1	213
8	Trojan	C2Lop.gen!G	200
9	Trojan	Alueron.gen!J	198
10	PWS	Lolyda.AA 2	184
11	Dialer	Dialplatform.B	177
12	Trojan-Downloader	Dontovo.A	162
13	PWS	Lolyda.AT	159
14	Backdoor	Rbot!gen	158
15	Trojan	C2Lop.P	146
16	Trojan-Downloader	Obfuscator.AD	142
17	Trojan	Malex.gen!J	136
18	Trojan-Downloader	Swizzor.gen!I	132
19	Trojan-Downloader	Swizzor.gen!E	128
20	PWS	Lolyda.AA 3	123
21	Dialer	Adialer.C	122
22	Backdoor	Agent.FYI	116
23	Worm:AutoIT	Autorun.K	106
24	Trojan-Downloader	Wintrim.BX	97
25	Trojan	Skintrim.N	80
—	—	Total	9339

3.3.5 Preprocessing

Our malimg dataset already contains malwarein the form of images (grayscale images), to demonstrate how a malware binary can be visualized as an image, we are going to use a random text file and we will show you how to convert the file into image Fig. 13. The ultimate goal of this step is to convert files into images and use them as the input of our convolutional neural network. We can convert any file using the following python code used by [19]. We have created a notepad file abc.txt with the contents of activeds.dll. Activeds.dll is the dynamic link library file of the windows operating system, which is stored in location c:/windows/system32/ activeds.dll.

The following python program is used to convert the abc.txt file into a grayscale image.

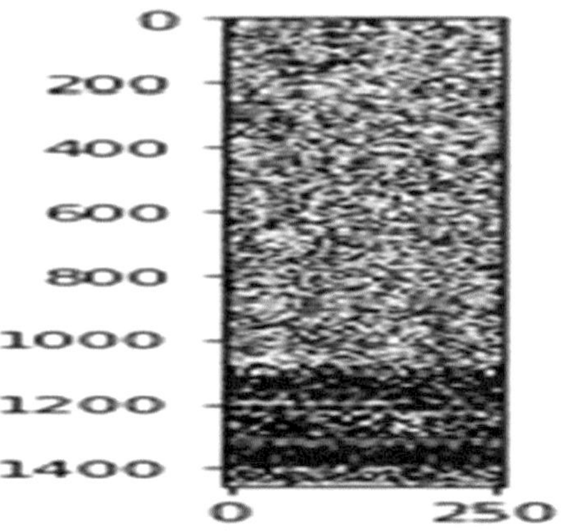

Fig. 13 Converted grayscale image

```
Python code to convert file into image
import numpy, os, array
import imageio
import matplotlib.pyplot as plt
from PIL import Image
filename = 'abc.txt'
f = open(filename,'rb')
print(f)
ln = os.path.getsize(filename)
width = 256
rem = ln%width a = array.array("B")
a.fromfile(f,ln-rem)
f.close()
b=(len(a)/width,width)
b=numpy.uint(b)
g = numpy.reshape(a,b)
imageio.imwrite('000f0e45f4f120838dd17500000de69a.png',g)
plt.imshow(g,cmap='gray')
```

...

After running the above program we got the following grayscale image Fig. 13 of the corresponding abc.txt.

3.3.6 Image Resizing

Let us proceed, so our malimg dataset already contains malware samples in grayscale image format. To input these images into convolutional neural network for training, we need all images of the same size, so there is need to resizes all images of the malimg dataset to the specified size you want. I chose the (48*48) image dimension.

As we can see from Fig. 4, some differences among malware images, However, it would be too complex to accurately classify malware into their corresponding families as we have 9339 total malware images.

3.3.7 Implementation Details

For a programming perspective, we performed the following experiment utilizing a personal laptop with i3, 2.40 GHz intel processor; a 64-bit system with 4GB of random access memory. We used python programming language, python packages, and libraries which are availed to experiment. Keras library is utilized to train and test the model which utilizes a convolutional neural network. We have utilized spyder 4.0.1 which is a scientific python development environment tool, python 3.7.5 on windows 10, 64-bit windows 10 operating system.

3.3.8 Architecture

Let us define the convolutional neural network model for training and classification. Our dataset is ready; we have built our model using keras. Here, we will define our model, which is a stacked layer of convolution and pooling operations, with a final flattened layer and a softmax activation function applied to determine the class probability of the malware samples. The following network architecture will be used for training and testing purpose, the chosen convolutional neural network architecture Table 2 will only be for study and understanding purpose, We have randomly chosen the number of filters, layers, filter size. Hyper parameters tuning is also a research topic. So basically, we don't know how it is going to perform, what will be the accuracy, and we do not need to worry about these things here.

1. Convolutional layer : 30 filters, (3 * 3) kernel size, activation=ReLU
2. Max pooling layer : (2 * 2) pool size
3. Convolutional layer : 48 filters, (3 * 3) kernel size, activation=ReLU
4. Max pooling layer : (2 * 2) pool size
5. Dropout layer: dropping 50 percent of neurons
6. Flatten layer
7. Dense/fully connected layer : 1024 neurons, ReLU activation function
8. Dropout layer: dropping 50 perecnt of neurons
9. Dense/fully connected layer : number of output class, softmax activation function

Table 2 summarizes our chosen convolutional neural network architecture. model.summary() is used to visualize defined model architecture.

The input for convolutional neural network training has a shape of [48 * 48 * 1]: [image width * image height * channel /depth]. In our case, each malware is a

Table 2 Summary of our chosen convolutional neural network architecture

S. no.	Layer (type)	Output shape	Parameters
1	conv2d_3(Conv2D)	(None, 46, 46, 30)	300
2	max_pooling2d_3(MaxPooling2)	(None, 23, 23, 30)	0
3	conv2d_4(Conv2D)	(None, 21, 21, 48)	13008
4	max_pooling2d_4(MaxPooling2)	(None, 10, 10, 48)	0
5	dropout_3(Dropout)	(None, 10, 10, 48)	0
6	flatten_2(Flatten)	(None, 4800)	0
7	dense_3(Dense)	(None, 1024)	4916224
8	dropout_4(Dropout)	(None, 1024)	0
9	dense_4(Dense)	(None, 25)	25625

grayscale image, so the image channel value will be 1, if we use color images we have to assign value 3 in the image channel.We used the train test split() function of scikit learn to split dataset images between train and test, following a (90-10) % ratio.

Here is the code used to define convolutional neural network architecture using keras.

```
#Python code to define convolutional neural network model architecture
model = Sequential ()
model.add (Conv2D (30, (3, 3), activation='relu', input_shape= (img_rows,
img_cols,img_channels)))
model.add (MaxPooling2D ((pool_size, pool_size)))
model.add (Conv2D (48, (3, 3), activation='relu'))
model.add (MaxPooling2D ((pool_size, pool_size)))
model.add (Dropout (0.5))
model.add (Flatten ())
model.add (Dense (1024, activation='relu'))
model.add (Dropout (0.5))
model.add (Dense (no_out_classes,activation='softmax'))
opti_mizer=Adam (lr=0.001)
model.compile (loss ='categorical_crossentropy', optimizer = opti_mizer,
metrics=['accuracy'])
```

We executed our program for 15 epochs. Epochs summary are as follows
Train on 7564 samples, validate on 841 samples
Epoch 1/15
7564/7564 [==============================] - 174s 23ms/step - loss:

```
0.9531 - acc: 0.7132 - val_loss: 0.2436 - val_acc: 0.9394
Epoch 2/15
7564/7564 [==============================] - 491s 65ms/step - loss:
0.2252 - acc: 0.9332 - val_loss: 0.1796 - val_acc: 0.9441
Epoch 3/15
7564/7564 [==============================] - 713s 94ms/step - loss:
0.1643 - acc: 0.9510 - val_loss: 0.1265 - val_acc: 0.9631
Epoch 4/15
7564/7564 [==============================] - 168s 22ms/step - loss:
0.1370 - acc: 0.9594 - val_loss: 0.1125 - val_acc: 0.9750
Epoch 5/15
7564/7564 [==============================] - 168s 22ms/step - loss:
0.1158 - acc: 0.9636 - val_loss: 0.1214 - val_acc: 0.9655
Epoch 6/15
7564/7564 [==============================] - 166s 22ms/step - loss:
0.1062 - acc: 0.9681 - val_loss: 0.0941 - val_acc: 0.9774
Epoch 7/15
7564/7564 [==============================] - 170s 22ms/step - loss:
0.0913 - acc: 0.9718 - val_loss: 0.1050 - val_acc: 0.9727
Epoch 8/15
7564/7564 [==============================] - 719s 95ms/step - loss:
0.0890 - acc: 0.9710 - val_loss: 0.1350 - val_acc: 0.9679
Epoch 9/15
7564/7564 [==============================] - 166s 22ms/step - loss:
0.0819 - acc: 0.9741 - val_loss: 0.0810 - val_acc: 0.9798
Epoch 10/15
7564/7564 [==============================] - 167s 22ms/step - loss:
0.0726 - acc: 0.9757 - val_loss: 0.1052 - val_acc: 0.9703
Epoch 11/15
7564/7564 [==============================] - 167s 22ms/step - loss:
0.0753 - acc: 0.9753 - val_loss: 0.0797 - val_acc: 0.9822
Epoch 12/15
7564/7564 [==============================] - 166s 22ms/step - loss:
0.0650 - acc: 0.9791 - val_loss: 0.1039 - val_acc: 0.9738
Epoch 13/15
7564/7564 [==============================] - 166s 22ms/step - loss:
0.0773 - acc: 0.9751 - val_loss: 0.0852 - val_acc: 0.9798
Epoch 14/15
7564/7564 [==============================] - 166s 22ms/step - loss:
0.0634 - acc: 0.9790 - val_loss: 0.0847 - val_acc: 0.9822
Epoch 15/15
7564/7564 [==============================] - 166s 22ms/step - loss:
0.0620 - acc: 0.9795 - val_loss: 0.0977 - val_acc: 0.9715
```

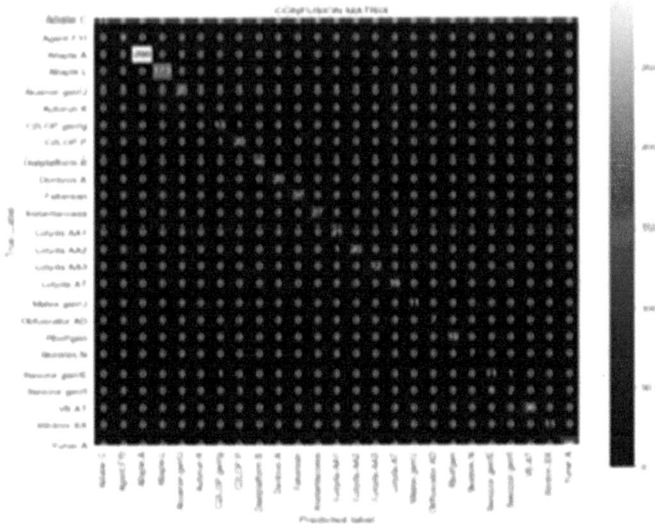

Fig. 14 Confusion matrix

3.3.9 Results

After training and testing our convolutional neural network model, we reached a final test accuracy of 97.537% which is very high! We got test time at 0.078. Here is the confusion matrix of our classification Fig. 14.

We can observe from the confusion matrix that most of the malware samples were well classified into its corresponding actual family, Autorun. K is always misclassified for Yuner. A, it is probably because we have only 80 samples of Autorun.K; this is very few in our dataset and that both are a component of a close worm type. Moreover, Swizzor.gen!E is often misclassified with Swizzor.gen!l, which can be explicated by the fact that they emanate from authentically close kind of families and types and thus could have homogeneous attributes in their code.We can also plot train and validation accuracy Fig. 15 and loss Fig. 16 during per epoch and analyze precision and losses, ups and downs during the whole journey. We can also calculate some more performance-based quantifications such as precision of the model which is 0.965, recall of the model is 0.975, and f1-score of the model is 0.968.

It is all about how to implement initial level malware image classification, and to further explore the results and analysis we can plot the confusion matrix, which give us some more statistics about the classification, some hints about what went erroneous during classification. It was the initial level understanding of how to implement

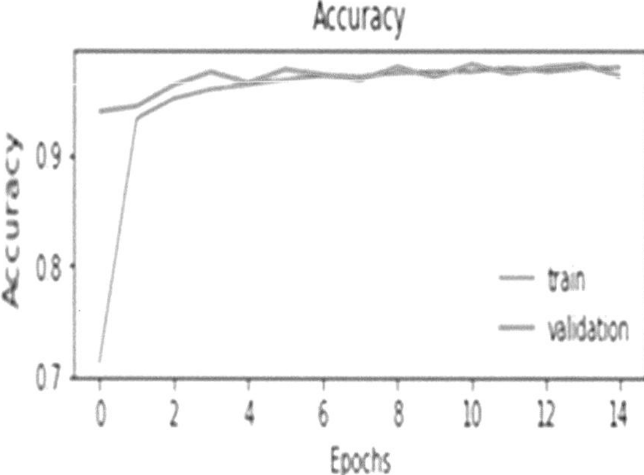

Fig. 15 Training and validation accuracy during 15 epochs

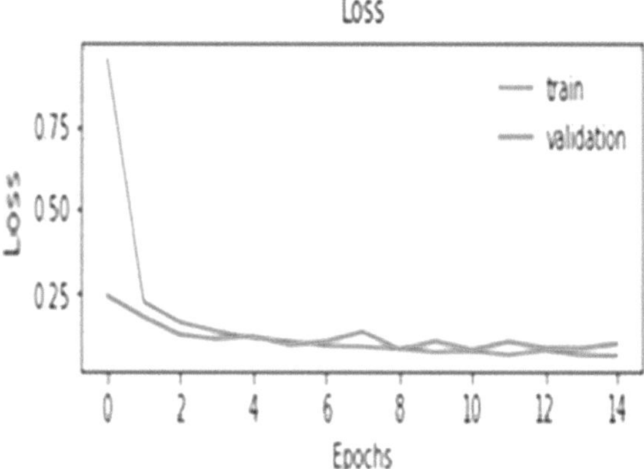

Fig. 16 Training and validation losses during 15 epochs

malware image classification utilizing convolutional neural network. Utilizing this basic understanding, you can further ameliorate classification results, perform more applications, and do further research.

3.3.10 Datasets for Malware Analysis

Here we are mentioning some popular datasets Table 3 available for practice and research. Datasets play a consequential role in training, testing, and validation of systems. Datasets of malware images consist of many images that belong to different families. Readers can make utilization of it for their research and projects. Some popular datasets are

Table 3 Summary of publicly available datasets for malware analysis

S. no.	Dataset	Dataset description
1	Malimg dataset	The malimg dataset consists of 9000 malware files belonging to 25 malware families and their variants
2	Malicia dataset	The dataset comprises 11,688 malware binaries collected from 500 drive-by download servers
3	Microsoft malware classification challenge dataset	This dataset contains 9 classes for training and testing purposes. It includes 21741 malware samples
4	Malshare	A free malware repository providing researchers access to samples, malicious feeds
5	IoT-23	A labeled dataset with malicious and benign IoT network traffic
6	AMD	Android malware dataset has 24,553 samples, it is integrated by 71 malware families ranging from 2010 to 2016
7	Android malware genome project	More than 1,200 malware samples that cover the majority of existing Android malware families, ranging from their debut in August 2010 to recent ones in October 2011
8	Drebin dataset	The dataset contains 5,560 applications from 179 different malware families. The samples have been collected from August 2010 to October 2012 and were made available to us by the mobile sandbox project

4 Challenges and Open Issues

This chapter and study is the first step toward enhancing our understanding of visualization and deep learning-based malware classification. During the study, many difficulties and challenges of malware classification were found; the present findings might have important implications for suggesting several courses of action to solve this problem. For a consistent and effective framework, it is important to address all the challenges and difficulties. Traditional malware classification approaches are very time-consuming and complex.

In our view, malware classification is very well handled by image visualization and deep learning approaches as compared to traditional approaches. Deep learning approaches efficiently perform learning but we found some limitations such as dl models requires all input images of the same size, which limits the training model. Work done by many researchers transform malicious binaries into grayscale images, so there is the scope of training and classification with color images. Different model architectures (differ in several layers, number of kernels/filters, size of strides, etc) show different results, so there is the scope of more intelligent architectures using deep learning to improve performance. Existing approaches which achieved high accuracy are often specific to a particular dataset, so still, there is a need for more generic deep learning architecture, which can be utilized for any type of dataset. The size of the dataset has significant importance on the accuracy and performance of the model. There is scope to address the data imbalance issue (as there is a difference in the number of samples in one family and rest). Available malware datasets constitute different formats and specifications containing both infected and non-infected files; inconsistencies may arise in the accuracy of the results, deep learning methods can be substantially influenced by adversarial attacks using the experience of the learning algorithm to avoid detection, or infuse harming instances into the training data. One of the latest emerging hurdles in malware analysis is the file less malware [10], which makes malware analysis more complicated. A combination of deep learning models for malware analysis can prove more intelligent and effective. To achieve good classification accuracy architecture alone is not only responsible it is also dependent on the dataset, so quality and enough data generate more accurate results, so preprocessing of the dataset is one of the important considerations for classification.

5 Conclusion

In this chapter, we provided a detailed study of the malware, malware analysis, deep learning, and its algorithms. The exponential development of the Internet, connected devices, services and applications, user's activity, and confidential information attracts cybercriminals. Although malware is not a new threat in the cyber world, but the device manufacturer, attacker's techniques to avoid detection and different service providers use different communication technologies creates a heterogeneous environment where malware analysis becomes a critical task. In this context, this chapter aimed to present an overview of the fundamental aspects of malware detection and classification using image visualization and deep learning techniques.

Within the next few years, malware classification and identification are likely to become important and inevitably be an issue that is going to be explored more. As can be concluded from the above-discussed information and study, the use of visualization techniques to represent malware and deep learning models in malware detection and classification proved to be efficient than traditional approaches. It is important to keep in mind that deep learning approaches prove to be a state-of-the-art approaches for malware detection and classification in some cases, but they are always the possibility of better to do. Deep learning methods also have some limitations such as a limited number of samples for analysis, to increase hidden layers which also increases complexity in the model and increases training time.

One biggest advantage of deep learning is automation. There is a need for automating the detection and classification process is still an important issue as most of the traditional malware analysis in the identification and classification of new threats continues to be a human task. Deep learning also has human interaction but limited sense. Until now, this methodology has only been applied to very less in literature and also in practice, so this chapter will encourage readers to do further research in this vast and important topic, and malware analysis is also connected to our lives directly. Malware classification is a fundamental and vital issue for future research and we have mentioned some state-of-the-art researcher's approaches, scope and emerging challenges for malware classification using deep learning for the reader's further studies. Finally, it is expected that the information presented and discussed in this chapter would help readers, analysts, and researchers to obtain a general and practical view of the malware analysis especially malware identification and classification from where they can visualize and explore new avenues of research.

References

1. Azab, A., M. Alazab, and M. Aiash. 2016. Machine learning based botnet identification traffic. In *2016 IEEE Trustcom/BigDataSE/ISPA*, 1788–1794.
2. Ding, Y., S. Chen, and J. Xu. 2016. Application of deep belief networks for opcode based malware detection. In *2016 International Joint Conference on Neural Networks (IJCNN)*, 3901–3908.
3. Bagane Pooja, Garminla Sampath Kumar. 2020. Detection of malware using deep learning techniques. *International Journal of Scientific and Technology Research* 9: 1688–1691.
4. Goodfellow, Ian, Jean Pouget-Abadie, Mehdi Mirza, Bing Xu, David Warde-Farley, Sherjil Ozair, Aaron Courville, and Yoshua Bengio. 2014. Generative adversarial nets. In *Advances in Neural Information Processing Systems*, ed. Z. Ghahramani, M. Welling, C. Cortes, N.D. Lawrence, and K.Q. Weinberger, vol. 27, 2672–2680. Curran Associates, Inc.
5. KyoungSoo Han, Jae Hyun Lim, and Eul Gyu Im. 2013. Malware analysis method using visualization of binary files. In *Proceedings of the 2013 Research in Adaptive and Convergent Systems, RACS '13*, 317–321. New York: Association for Computing Machinery.
6. Hardy, W., Lingwei Chen, Shifu Hou, Yanfang Ye, and X. Li. 2016. Dl 4 md : A deep learning framework for intelligent malware detection.
7. AV-TEST The Independent IT-Security Institute. Malware statistics and trends report [online] by av-test institute, 2020.
8. McAfee LLC is an American global computer security software company. Mcafee labs threats reports [online] by mcafee, 2019.
9. Jain, Mugdha, William Andreopoulos, and Mark Stamp. 2020. Convolutional neural networks and extreme learning machines for malware classification. *Journal of Computer Virology and Hacking Techniques*, vol. 04.
10. Sudhakar, K., and K. Sushil. 2019. An emerging threat fileless malware: a survey and research challenges 3: 1, 12.
11. Kalash, M., M. Rochan, N. Mohammed, N. D. B. Bruce, Y. Wang, and F. Iqbal. 2018. Malware classification with deep convolutional neural networks. In *2018 9th IFIP International Conference on New Technologies, Mobility and Security (NTMS)*, 1–5.
12. Khan, R.U., X. Zhang, M. Alazab, and R. Kumar. 2019. An improved convolutional neural network model for intrusion detection in networks. In *2019 Cybersecurity and Cyberforensics Conference (CCC)*, 74–77.

13. Kim, C.H., E.K. Kabanga, and S. Kang. 2018. Classifying malware using convolutional gated neural network. In *2018 20th International Conference on Advanced Communication Technology (ICACT)*, 40–44.
14. Kim, Jin-Young, Seok-Jun Bu, and Sung-Bae Cho. 2017. Malware detection using deep transferred generative adversarial networks. In *Neural Information Processing*, ed. Derong Liu, Shengli Xie, Yuanqing Li, Dongbin Zhao, and El-Sayed M. El-Alfy, 556–564. Cham: Springer International Publishing.
15. Lu, Y., and J. Li. 2019. Generative adversarial network for improving deep learning based malware classification. In *2019 Winter Simulation Conference (WSC)*, 584–593.
16. Mourtaji, Youness, Mohammed Bouhorma, and Daniyal Alghazzawi. 2019. *Intelligent Framework for Malware Detection with Convolutional Neural Network. NISS19*. New York: Association for Computing Machinery.
17. Naeem, Hamad. 2019. Detection of malicious activities in internet of things environment based on binary visualization and machine intelligence. *Wireless Personal Communications*, 1–21.
18. Naeem, Hamad, Farhan Ullah, Muhammad Rashid Naeem, Shehzad Khalid, Danish Vasan, Sohail Jabbar, and Saqib Saeed. 2020. Malware detection in industrial internet of things based on hybrid image visualization and deep learning model. *Ad Hoc Networks* 105: 102154.
19. Nataraj, L., S. Karthikeyan, G. Jacob, and B.S. Manjunath. 2011. Malware images: Visualization and automatic classification. In *Proceedings of the 8th International Symposium on Visualization for Cyber Security, VizSec '11*. New York: Association for Computing Machinery.
20. Ni, Sang, Quan Qian, and Rui Zhang. 2018. Malware identification using visualization images and deep learning. *Computers and Security* 77: 04.
21. Pal, K.K., and Sudeep, K.S. (2016). Preprocessing for image classification by convolutional neural networks. In *2016 IEEE International Conference on Recent Trends in Electronics, Information Communication Technology (RTEICT)*, 1778–1781.
22. Singh, Ajay, Anand Handa, Nitesh Kumar, and Sandeep Kumar Shukla. 2019. Malware classification using image representation. In *Cyber Security Cryptography and Machine Learning*, ed. Shlomi Dolev, Danny Hendler, Sachin Lodha, and Moti Yung, 75–92, Cham: Springer International Publishing.
23. Tobiyama, S., Y. Yamaguchi, H. Shimada, T. Ikuse, and T. Yagi. 2016. Malware detection with deep neural network using process behavior. In *2016 IEEE 40th Annual Computer Software and Applications Conference (COMPSAC)*, vol. 2, 577–582.
24. Vasan, Danish, Mamoun Alazab, Sobia Wassan, Babak Safaei, and Qin Zheng. 2020. Image-based malware classification using ensemble of cnn architectures (imcec). *Computers and Security* 92: 101748, 05.
25. Yin, Qiwei, Ruixun Zhang, and XiuLi Shao. 2019. Cnn and rnn mixed model for image classification. *MATEC Web of Conferences*, 277: 02001, 01.
26. Yinka-Banjo, Chika, and Ogban-Asuquo Ugot. 2019. A review of generative adversarial networks and its application in cybersecurity. *Artificial Intelligence Review* 53: 06.

Review of Artificial Intelligence Cyber Threat Assessment Techniques for Increased System Survivability

Nikolaos Doukas, Peter Stavroulakis, and Nikolaos Bardis

Abstract This chapter presents an overview of the problem of survivability of information systems, along with solutions that are currently available to designers of such systems. The notion of survivability in the context of cybersecurity over multi-user distributed information systems is defined, which is set as the target of cyber defense to prevent the adversary from successfully completing their mission. The cyber-attackers' kill chain is explained. Artificial Intelligence (AI) techniques that may be employed in order to promote information system survivability are outlined and the technical issues toward which each technique can contribute are listed. Following that, schemes for increased cyber survivability are presented, which focus on solving particular problems that commonly appear by employing artificial intelligence techniques. First, the problem of email message filtering in the context of breaking the cyber kill chain is analyzed and a typical AI-assisted technical solution is given. Following that, the effect of malware in survivability is presented and an approach to its solution based on the static analysis and detection of patterns is presented. Subsequently, the collusion attack, an attack where multiple malware programs collaborate in order to achieve malicious goals, is presented and an AI-powered solution is outlined based on currently available technology. A three-level anomaly detection system is presented that employs AI primitives and detects problematic behavior in network traffic, packed files, and SQL statements in order to produce cybersecurity defense actions and warnings. Dynamic analysis of potentially harmful programs is analyzed and a technique that performs such analysis is presented that examines the executed machine-level instruction opcodes and utilizes AI in order to circumvent efforts of malware creators to obfuscate the actions and intents of their code. A recently proposed comprehensive cooperative infrastructure defense system is briefly

N. Doukas · N. Bardis
Hellenic Army Academy, Varis - Koropiou Avenue, P.O. 16673, Vari, Athens, Greece
e-mail: nd@ieee.org

N. Bardis
e-mail: bardis@ieee.org

P. Stavroulakis (✉)
Telecommunications Research Institute of Crete, Technical University of Crete, Chania, Greece
e-mail: pete_tsi@yahoo.gr

M. Stamp et al. (eds.), *Malware Analysis Using Artificial Intelligence and Deep Learning*, https://doi.org/10.1007/978-3-030-62582-5_7

presented that is based on the artificial intelligence ant colony paradigm. The system aims to coordinate human and automated efforts to protect the integrity of large-scale information systems. It uses multiple AI principles in order to utilize existing information and obtain novel knowledge, adapting to new threats and user expectations. Finally, survivability promoting countermeasures are presented that act as additional fail-safe mechanisms to impair the cyber-attackers mission.

1 Introduction

Survivability is a notion applicable to both military and civilian or commercial systems and describes the ability of such systems to remain operational or gracefully degrade after an attack against them [17]. The range of possible attacks is easy to determine in the case of physical systems, but presents a challenge of itself in the case of cyber systems. In the context of the cyberspace, maintaining an information system operational principally involves protecting the CIA triad of information security, namely confidentiality, integrity, and availability [4]. Some researchers further expand this description with three further specialized cybersecurity targets, authorization, authentication, and non-repudiation [12]. Defense efforts against cyberattacks are facilitated by understanding the cyber kill chain, a widely accepted model describing the steps taken by attackers in order to achieve their goals. These steps start involve (i) reconnaissance, whereby a would-be attacker collects information about specific targets and tactics of the target information system, (ii) weaponization, which describes the development by the attacker of target-specific weapons or exploits based on the identified weaknesses, (iii) delivery, when the actual delivery of the malicious software to the target takes place, (iv) exploitation, where the weaknesses identified are used in order to execute the malicious application, (v) installation of the malware, (vi) command and control, when the attacker acquires communications and management channels with the target and finally, and (vii) actions, when the attackers utilize the target to achieve their objectives of illegitimate data access, data integrity violation, or denial of service [15, 17]. Engineers aiming to protect information systems against cyberattacks distribute their efforts in four fronts, namely the ability of a system to (i) remain invisible from the attacker, (ii) remain inaccessible to the attacker's weapons, (iii) withstand the hit, and (iv) maintain partial or full functionality despite possible damages, while offering capabilities for containing the problems arising and following recovery procedures [17]. The survivability approach to cyber defense involves making assumptions that conventional defenses in each phase have failed and designing techniques to minimize the effect of possible attacker actions. Such techniques include cyber deception to confuse the adversary, analysis of outbound traffic to detect data theft, white lists for software, limitations on use of non-essential applications, use of redundant computational units, etc. Designing and implementing cyber defense for survivability is a process relying on intuition, requires analysis of system data and activities and prioritization of threats, all for the purpose of deriving the context and allowing human operators to make real-time informed decisions. Defending cyber systems against attacks to this effect, artificial intelligence has been successfully used for the purpose of supporting the decision-making process. This chapter reviews emergent research results concerning AI tools that contribute to effective and timely threat prioritization and mitigation, for the purpose of developing components of AI cyber threat assessment decision support systems. AI primitives that contribute to these aims include pattern recognition, anomaly detection, predictive analytics, and natural language processing. Particular topics to be reviewed are anomaly-based network intrusion detection, detection of insider threats, spam and phishing detection, behavior-based malicious software identification, malicious activity detection based on low-level instruction analysis, suspicious network activity identification, data provenance tracking, etc. The actual AI techniques involved include classification, clustering, statistical decision models, fuzzy logic, stream data analytics, and data visualization. A review of these techniques will be covered in this chapter regarding the use of AI for cyber survivability. Special emphasis will be given to the use of AI for cyber survivability decision support including machine learning for software analysis, cyber-physical system integration, blind testing of programs without specifications, machine learning and data mining techniques for enhanced database security, malicious executable detection via ML, etc. A particular challenge in the problem of classification in the context of cyber security is class imbalance or the “one class classification” issue. This arises from the necessity to train the detector using a limited training set of positive examples, since the behavior of malware is unknown and rapidly evolving.

2 AI Support to Survivability

AI techniques are being applied to cybersecurity problems, despite the fact that the foundation technology may still be considered as under development or experimental [13], since it is producing promising results in a significant range of possible cyber threats belonging to the cyber kill chain explained earlier. The application of AI to the problem of identifying and alleviating cyber threats provides a defense tool capable of adapting to the continuously changing tactics of the attackers [13, 18]. AI includes a variety of techniques and algorithms, several of which become useful in the fight against cyber threats [17, 18]. Pattern recognition algorithms are commonly used for identifying phishing or spam e-mail messages, malware programs, untrusted sites, and other threats. Anomaly detection techniques are employed for spotting unusual computer process activity or questionable data. Natural language processing schemes may be used to process texts that have been created for the purpose of obfuscating the attackers aims, into structured intelligence information useable in the context of defense. Predictive analytics process large quantities of data in search of patterns and outliers that may reveal upcoming threats. The use of AI is, despite fears to the contrary, not aiming to replace humans, but to provide decision support tools in order to assist human efforts to make optimal and rapid decisions [13, 17, 18]. It is not expected to solve all possible cybersecurity problems. The aim is to enable the timely and adequate response to emerging threats, either by quickly filtering simplistic attacks or by identifying the suspicious cases in order to bring them to the attention of the expert security analysts. AI techniques have been successfully applied in the context of some more abstract challenges concerning the ability of information systems to withstand and survive attacks, hence supporting their survivability, namely [17]

- Compliance. The pre-deployment verification of the consistency of an information system's implementation with its specifications and the relevant security standards of its conformance with the rules of the network to which it is going to connect, and of its compatibility to the architecture of the other systems in conjunction with which it is going to operate.
- Intrusion Detection. The run-time operation of identifying incoming threats. It may be based on signatures, in the case of attacks that have already been observed at least once, or with anomaly detection techniques after long-term observations that have permitted the establishment of the baseline normality of network traffic. Determining what is normal is a particularly challenging task in environments presenting rich cyber activity.
- Patching. The act of identifying vulnerabilities discovered by the community that affect the system of interest and for which remedies, or patches, are already available and can be readily applied.
- Digital Forensics. The post-attack process of discovering what went wrong, how, when, where, and why, in order to avoid the occurrence of similar events in the future.
- Defense in Depth. The run-time process of identifying ongoing attacks and impeding their progress within the system until suitable countermeasures and defensive actions may be employed.

2.1 Security Threat Detection for Preventive Survivability

Survivability as an engineering design target does not imply the prevention of every possible attack and the elimination of the probability of an intrusion [14]. On the contrary, survivability analyses need to consider multiple "what if" scenaria determining the behavior and availability of the information system after different types of security breach events [14, 17]. The ultimate aim is not to prevent the attack, but prevent the attackers to complete their mission [17]. The mission is the final link of the kill chain described and consists of one or more of the three subtasks of data theft, system integrity compromise, and denial of service. Therefore, in designing for survivability, the AI algorithms to be used may be selected according to the expected level of threat, the anticipated types of attacks and the amount of processing capabilities available to be assigned for this purpose. The survivability subsystem of the design hence becomes configurable and scalable according to the necessities of the application. The success of the defense can be quantified as the extent to which the mission of the attacker failed. According to this observation, AI techniques suitable for the survivability in the event of different types of attacks will be outlined in the order of the progression of the attack within the cyber kill chain described earlier and in increasing level of sophistication.

2.2 *Email Message Filtering by Linear Classifiers*

Email messages are commonly used as an attack vector and depending on their type, e.g., spam, phishing, malware, etc., may present a significant cyber threat. Email messages can be the vehicle used by the attackers in multiple links of the kill chain, as for example a phishing message may be part of a reconnaissance effort or a spam message may be used for the delivery of the malicious payload [13, 17]. Consider a simplistic approach, whereby undesired messages are recognized by the presence and frequency of appearance of words or patterns from a pre-determined set of suspicious ones within their body [13]. A generic classifier may then be designed, which based the decision on a score value for the suspicious words calculated for each message. The score could be a simple weighted sum of the frequencies of appearance of each of these words and the weight for each word is a design parameter [13]. The final decision is based on a threshold; if the score for a message exceeds a certain threshold, which is another design parameter, the message is classified as spam, otherwise it is considered normal. In realistic applications, however, given the variety of spam sources and the evolution of the practices of the spammers, who will adapt in order to circumvent the filter, the design parameters for this classifier would have to be regularly or continuously adjusted. If this is organized in an iterative manner, in interaction with the users, the use of AI-supervised learning techniques becomes a naturally occurring choice [13]. The weighted sum approach represents a linear classifier in a space whose dimensions are determined by the size of the word-set. The AI instrument suitable for solving this problem is the perceptron. An initial set of weights is determined via an arbitrary principle and a desired result is defined for the weighted sum function described above that requires it to give e.g., an answer of +1 when a message is spam and -1 when it is not. The perceptron, which is realized as an implementation of this function, can then be trained, using a training set of emails. Each training message is presented to the perceptron, the result produced is observed, compared to the known result and the weights of the function are adjusted depending on the difference from the desired value. The correction functions for the weights are the object of many independent research efforts, but a simple and robust approach may be found in [13], along with further details about the implementation. The corrections become smaller after a number of iterations and the training converges. The principal and significant limitation of this class of algorithms is that they can only be effective if the actual data is separable by a multidimensional line in the space defined by the data. In fact, if the data is not linearly separable, then the training algorithm will not converge, i.e., the corrections will cause the values of the weights to oscillate indefinitely, irrespective of the number of iterations [13]. Better classification for spam messages can be achieved if the separation is attempted by a hyperplane function not limited to a linear form. Support Vector Machines (SVMs) are also supervised learning algorithms that attempt to determine the hyperplane that separates the data and are not limited to linear models [13]. Another major difference from perceptrons is that they attempt to maximize the distance of the training data from the separation hyperplane, rather than minimizing the number of misclassifications. SVMs are a powerful tool and can also detect spam that is encoded in the form of images [13]. A completely different approach for spam e-mail classification was proposed in [1]. Instead of trying to detect spam messages based on the content, messages are clustered according to the author to whom they are attributed. Incoming messages undergo a two-phase process. In the first phase, messages are grouped together using an authorship-based clustering algorithm. Finer categorization is achieved in the second step, whereby linguistic, syntactical, and other structural analyses are performed [1]. The results are clusters of e-mails with high probability of being attributed to the same author. This approach, as reported by its authors in [1], does not produce an actual solution to the problem of detecting spam, as it is easy for spammers to occasionally modify their style in order to evade detection. It should be seen as a means of reducing the volume of processing required in order to process all messages, and therefore cancelling the advantage spammers acquire by launching massive numbers of spam messages. Additionally, the method has been successful [1] in detecting links between seemingly different spam campaigns.

2.3 Malware Detection

Malware, or MALicious softWARE, is software that is employed by cyber-attackers for the purposes of completing their mission with the kill chain. Depending on the method of delivery, malware could be very easy or extremely difficult to recognize. An executable file arriving as an attachment to an e-mail message can be readily detected by the spam detectors described earlier on. However, malware hidden in a downloaded file, a non-authorized installation medium or a shared storage unit may present a greater challenge. Additionally, there exist collusion attacks, when multiple, seemingly innocent applications combine AI algorithms can again be employed to facilitate the detection and promote survivability. Detection can be broadly divided into two categories in this case, namely static and dynamic analysis. Static analysis involves examining the file, without executing it. Recognition may be achieved based on a variety of factors, ranging from the name of the file to specialized signatures of particular malware that have been identified by the community. This process is essentially a feature extraction process: large volumes of data (e.g., the executable files) are processed in order to extract features (the signatures, the filename, other distinguishing strings) which are then used for recognition either by humans or in this case by means of AI algorithms. Hash functions are mathematical functions with several properties that render them suitable for fast processing of large quantities of data [9, 10], among which is the ability to summarize. Suitable many-to-one hash functions are applied on a large number of bytes to produce a relatively small-sized result used as a signature, i.e., a quick way to identify identical data items. For well-designed hash functions, the signature is unique, in that slightly dissimilar items produce completely different results [9], while collisions are possible, since the function is many-to-one. Widely used signature functions are SHA256 and MD5, with MD5 being susceptible to collisions [16]. Community-identified hashes of the executables of known malware can be used as features to identify a particular file as an instance of that malware. Other elements of the file, such as the magic numbers encountered in some file types (pdf, image files) or the portable executable header in Windows, are also used as recognition features. A toolbox and database for such file feature extraction is Yara (https://yara.readthedocs.io/en/stable/). Commonly encountered malware in the form of viruses, trojans, etc., is successfully detected by commercially available antivirus and other similar programs. There exist, however, cases where this process is not as straightforward as it seems, and the assistance of AI is indispensable.

2.4 Collusion Attacks

In recent years, an everexpanding use of smart phones has been observed. Even though such devices use operating systems, and therefore software isolation and other information security principles can be applied, the need for openness has created several ways that applications running on such devices can communicate or share data between them [3]. This means that multiple, seemingly unrelated, apps collaborating between them have the possibility to achieve malicious intents, such as data theft. This type of attack is called a collusion attack. Examining applications individually, e.g., in a sandbox environment, would not reveal such behavior. Machine learning and other AI techniques may be used as detection tools for colluding behavior. Even though the Android operating system incorporates many of the strict Linux security primitives isolating running applications, both open and secret means of communication between apps have been identified [3]. An open, or overt means could be file sharing, while a hidden or covert one might be a special setting of a shared resource, such as the system volume, which passes a special message to the collaborating application [3]. Since there exist legitimate applications that need to perform similar actions as those observed in malware, the principal difficulty of the detection is that the intent of the application, a purely subjective and qualitative notion, needs to be detected. Suitable features for detecting colluding behavior need to be identified before any AI technique can be employed for detecting colluding behavior. Such features can include actions that applications take [3], e.g.,

- accessing sensitive information (contacts, e-mail messages),
- accessing cost-bearing services (calling, SMS sending),
- accessing sensitive hardware (camera, microphone),

- sending data over the Internet, and
- permissions it requires.

The above actions are actions that can be legitimately used by authorized application in order to create services desired by the users. The fact that an application takes one or more of these actions is by no means an indication of malicious intent. The Naive Bayes classifier has been successfully employed as an AI tool in order to estimate the likelihood of sets of two or more applications are serving malicious purposes, given the set of actions that each one of them uses [3]. The classifier was trained over a set of known for their malicious intent applications. A probabilistic filter was hence created that operates in two stages:

- An initial, fast filter examines applications and application sets based on the easily detectable features outlined above
- The second, more thorough filter performs the static code analysis described in the previous section, along with intent determination rules, in order to discover candidates with increased likelihood for colluding behavior.

On the identified candidates a software model checking procedure is used in order to detect sequences of actions that are likely to be part of collusion of malicious intent [3]. This approach presents both false positives and false negatives, especially since given the fact that the analysis is of probabilistic character. This same fact, however, also implies that the approach promotes survivability. Calibrating the likelihood of false positives, for example, can be viewed as a means of calibrating the survivability of the system in relation to its capability of running applications that may perform risky but desirable operations. Additionally, scores of the intermediate detection phases may be used in order to create breaking points at which to block the attacker's mission from being completed. For example, even though data transfer operations by applications could be allowed, data transfers after a sequence of actions are classified as likely to be malicious.

2.5 Anomaly Detection

An artificial intelligence technique for survivability has been presented that is based on the method of operation of biological beings [8]. Three different machine learning systems are used in order to construct a hybrid framework for detecting anomalies and can be viewed as a multi-agent system. Each agent operates in a different sector of the information system and synchronization is achieved by cooperation or negotiation, since no single agent has a complete perspective of all the available information and there exists no central coordinator [8]. Temporal programming is used in order to phase contradiction of intentions and contradiction of resource management concluding about the extent of the threat and risk. Automatic actions may be taken, such as termination of network connections and reports are sent to human administrators for taking more complicated risk mitigation actions. The first agent uses neural network classifiers trained on a database of information system intrusion information with recordings of the total flow of a network simulating the US air-force military network [8]. Events are analyzed including the connection between the IP addresses, the TCP packages exchanged, the protocols used, and the operation time. The features extracted include content features, traffic features, time-based traffic features, and host-based traffic features, totaling 41 features. The attacks noted include Denial of Service, Remote to User, Probing, and User to Root. A three-step procedure is developed, with each step using an increasing number of features, in order to filter out irrelevant events and raise the alarm with high confidence for more dangerous occurrences. The second agent also employs a neural network classifier in order to detect packed or unpacked executables of possible malware [8]. A genetic algorithm is hence used in order to reduce the probability of false positives and false negatives. The agent is trained based on an available dataset of patterns used as features that are extracted from packed malware, and benign executables [8]. A three-step procedure is employed:

- The neural network classifies binaries as packed files or benign executables.
- The executables are sent to the regular antivirus software for further assessment. The Evolving Classification Function technique is then used upon the packed files in order to again classify those files as benign or malignant. A genetic algorithm is used in order to increase the process of the integrity.
- Benign files are sent to the antivirus software, while suspect ones are unpacked in sandbox, the diagnosis is verified by antivirus software and are finally quarantined for treatment by human administrators.

The third agent uses a neural network and a genetic algorithm in order to train the neural network's weights for minimum error in classifying SQL injection attacks [8]. A dataset of 13384 SQL statements is used, including both legitimate and malicious ones. Various features are extracted from the available SQL statements including length, symbols present, linguistic constructs in the SQL syntax, and the correlation with known malicious SQL commands. Particularly interesting is the entropy feature measuring the amount of information the SQL command is expected to extract [8]. The classification is two-phase, firstly eliminating outlier data and then employing the classifier in order to examine the instances that are more likely to be malicious SQL injections. The three-agent system's functionality is consistent with the aims of survivability. The first agent uses AI to detect network anomalies that evade the firewall and intrusion detection systems. The second agent enhances survivability in virus attacks by employing AI techniques in order to filter out regular threats that can be successfully handled by the antivirus software and employing advanced and computationally intensive AI detection procedures for more involved cases.

The use of network traffic analysis and irregularity detection has been shown in [5] to be capable of detecting new botnets before they launch their attack, i.e., being able to predict the attack. The detection is based on extracted features from network traffic in the phase where the botnet is establising or reconfiguring their command and control (CC) structure. Machine learning is employed that is trained on observations before and after the deployment of the botnet CC traffic. There exist elements of botnet CC traffic that are similar between different botnets and deviate from the corresponding observations of legitimate traffic, which make the training of machine learning algorithms effective [5]. This approach is, therefore, extremely helpful in detecting zero-day versions of botnets.

2.6 Dynamic Analysis of Malware

Static analysis of malware, as explained in a previous section, aims to identify malware threats based on features such as the structure of the code, data structures encountered, patterns that can be observed within, and signatures. The structure of the code may be parametrized as call graphs. Patterns within files may reveal information such as IP addresses or URLs, command line options, passwords and windows portable execution files utilized by the malware for attaining its aims. The process is considered safe, since it is not necessary to execute the file. However, it is not always feasible to examine alternative call paths, since attackers can easily insert redundant code in order to obfuscate the path for the true malicious actions [7]. Dynamic analysis of executables involves observing the file under investigation while it is being executed. Information about memory accesses during the execution of the program is recorded, thereby eliminating many of the possibilities of attackers to hide their intentions. The execution may take place in the actual operating system, in an emulated environment that is monitored by the controllers' host, or a purely virtual environment that completely isolates the target system from possible effects of the malware. Countermeasures taken by malware creators to evade dynamic analysis include

- the detection of emulated or virtual environments,
- insertion of complicated mathematical or other complex code that serves no purpose,
- re-ordering of instructions within their code so that it does not resemble the known patterns,
- variable use of registers,
- altering code appearance by using alternative instruction constructs to achieve the same results, and
- packing software to obfuscate the appearance of the Windows portable executable files.

Additional techniques employed by attackers are polymorphic approaches that encrypt malicious code until it is time to attempt its malicious assignment and metamorphic malware where the malicious code is rewritten at every iteration [7]. Opcode analysis has been proposed as a counter-measure against these practices of the attackers to mask their actions. Opcoderefers to the machine-level instructions executed by the malware in the target system. It is hence possible to observe the actual actions of the malware and circumvent obfuscation efforts. A scheme for dynamic malware analysis was proposed in [7]. The training and testing are based on the VirusTotal database of malware. The opcodes corresponding to the stored executables were extracted via an automated procedure. The Intel architecture commonly used in most current personal computer incorporates 610 different opcodes. From these opcode sequences, run-lengths of various lengths were extracted. N-grams are a concept used in solving natural language processing problems. A similar approach was used for analyzing the malware opcode language. Hence, the extracted opcode data was processed for bigrams and trigrams, and run-lengths were determined [7]. The proposed method uses the Random Forest (RF) classification algorithm. The RF is an ensemble learner, combining decisions from multiple small decision trees, each one with a random number of features at each node. This technique presents increased immunity to noise and reduces overfitting, while it is possible to be parallelized. For better training accuracy, the features were reduced by selecting features with the largest information gain [7]. Accuracy in detection for this dynamic analysis technique is reported at more than 99%. The technique promotes the survivability target of detecting unknown attacks based on the knowledge of known attacks, via the transformation of malware commands into the opcode space.

A scheme for calculating the similarity between obfuscated versions of malware binaries is proposed in [6]. This is achieved using the Trend Locality Sensitive Hashing (TLSH) and k Nearest Neighbors in order to detect similarities between obfuscated versions of the same binaries. Hashing techniques that are common in document similarity detection [9, 10] are used. A sliding window is used to populate an array of buckets. The quartile points of each bucket is then estimated. A digest is hence created that also includes a checksum. The digest body is further quantized and each bucket count is converted to a two-bit value. The output digest created from the last two steps is used for the decision-making. The method is shown [6] to be able to efficiently detect and being resilient to obfuscations commonly used by cybercriminals.

3 Cooperative Infrastructure Defense

The cyber survivability systems reviewed so far involve an automatic process, with an optional asynchronous intervention by human operators. A cyber defense scheme has been proposed that organizes humans and digital agents into a cooperative scheme, where the initiative is mixed [11]. The Cooperative Infrastructure Defense (CID) implements a hierarchical framework of humans and digital agents called ants. The notion of digital agents that are parallelized to ants is well established in AI. The CID rapidly adapts to respond to unknown attacks, with the high-level supervision and guidance by human operators—administrators [11]. The software agents (ants) are also hierarchically organized, in order to better organize the flow of guidance instructions from the human to the ants and also provide a concise and accurate representation of the current situation as feedback to the administrator. Each of the agents possesses a level of rationality which is a combination of human instructions, logic programming rules, and machine learning results. The system is capable of collecting information and using it to update its knowledge and recognizing unknown states as safe or potentially problematic. It is, therefore, inherently continuously seeking survivable states, thus enhancing its survivability. The software agents take real-time decisions at low level, while permitting human intervention at all levels. Agents have the ability to spawn more agents or terminate their operation. In normal operation, agents stop operating only when they encounter a difficult problem that requires intervention by humans. Human-initiated intervention is permitted at any time, but is discouraged as it disrupts the real-time response and adaptation to the events happening within the system. Agents are organized in three levels of hierarchy, enclave-level, host-level and swarm-level [11].

- At enclave level, the agents interface with humans and other agents at the same level in order to coordinate the establishment and application of business drivers and human-defined policies. Their role involves creating and enforcing executable policies for the entire information system (enclave). A supervisor may be responsible for one or more information systems and could be members of regulatory bodies. They interact with the top-level software agents via natural language or graphical commands. They oversee the overall system survivability. They translate human instructions to actionable policies to be applied over the entire system. Using supervised learning, they adapt their behavior in order to better interact with the human operators and create responses that are closer to what the human will eventually choose. They can broker agreements with agents at the same level of other information systems. Their actions are uniquely identifiable within the system for attribution and problem detection. Since such actions may incur physical costs, especially when involving external systems, special rules impose limits upon the actions they can take.
- At host level, the agents protect and configure a single host or a set of similar hosts and interact with the administrators to obtain clarifications about ambiguous evidence from the swarm. The subsystem under the control of such agents could be a single server or router, a storage area network or a set of user workstations. They are responsible for implementing the policies they receive from higher level agents. They are also responsible for collecting the information gathered by the lower level agents (the sensors). They give sensors local connectivity information and control their spawning. They combine the information from the sensors, their own previously obtained information and previous human input and policies from higher level agents in order to detect problems and derive solutions. They are essentially the principal survivability seeking architecture of the system. Host-level agents give positive feedback to sensors that provided information that turned out to be useful, in order to encourage them to collect more similar information.
- At swarm level, the agents that can also be seen as sensors, continuously scan machines for problems and reporting them to the sentinel. Sensors are specialized and possess classifiers capable of detecting a particular set of problem indicators. They are independent processes and communicate via messaging primitives. Sensors may move from machine to machine, by suitable remote procedure calls

and message passing. They employ ant colony algorithms and swarm intelligence as documented in literature [11]. They must contain simple logic so that it can be possible to spawn large numbers of such agents and give them high mobility. They move according to a predefined geography within their information system, but their motion is random similar to that of ants when moving on the ground. They employ a classifier and search the hosts they visit for patterns of known problems or anomalous conditions. There also exist differential sensors that detect differences between similar hosts that they visit, such as network traffic volume, number of open files, etc. They communicate with other sensors via messages. Sensors that have contributed a significant amount of information that turns out useful are spawned by higher level agents so as to employ more identical instances.

Trust is maintained at the host-level agents, as they have been shown to be capable of hence protecting and promoting trust at lower levels. Results show that this system for safeguarding system survivability can match the performance of an equivalent intrusion detection system with the number of operational sensors of each type being about 50% of the number of the existing devices. In rough terms, this means that the information system level survivability is achieved with approximately half the computational load overhead required for the equivalent intrusion detection system.

A fully automated system that incorporates techniques to detect unknown and hidden malware, thereby providing a solution for the zer-day malware problem is proposed in [2]. Both the systems described in this section demonstrate how AI-based tools are capable of easing the burden of forensic analysis on information systems for the purpose of detecting hidden malware threats and providing a real-time command and control center for defending cyber infrastructure.

4 Post Attack Survivability

The survivability analysis is based on the assumption that the attack has already occurred [17], and therefore, when trying to mitigate the effects of the attack, one or more links of the cyber kill chain have already been materialized. The ultimate goal of the attacker is to achieve their mission, which is normally qualified into one of three primary attack objectives which are denial of service, information theft, and deception attacks [17]. The ultimate target of survivability design involves preventing the attacker from completing their mission, i.e., reaching the final link of the kill chain. In the previous sections, a series of algorithms that detect or give warning about the possibility that a security breach has occurred. In this section, it will be demonstrated with two scenaria how AI and other techniques can be used to implement countermeasures that are capable of interrupting the progress of an attack that has not been detected. The first scenario involves the establishment of a secure communications link to disseminate critical mission messages to remotely operating military personnel. Assuming that the communication breach will inevitably occur, the following preventive countermeasures are capable of limiting or completely eliminating the effects of this breach [17].

- Use of deceptive messages which are randomly mixed with the proper ones. The attacker will then have to devise a means of distinguishing between the two.
- Re-establishment of the session after each mission has been accomplished, or at random instances. The attacker will need to detect the change and repeat the breach successfully.
- Multiple channel creation, hence dividing the maximum possible gain of any single breach.
- Agreement of session keys between the legitimate participants before they are separated in order to arrive at their remote positions. The attacker will hence be forced to attempt to deduce the session keys by data observation, a process that requires more time and processing capabilities.

The second scenario involves protecting the position information for a fleet of UAVs when the base control station has been compromised [17]. In this case, the following preventive actions could be taken:

- Indications of compromise coming from the attack detection AI algorithms of possible security breaches cause a secure reboot of the system to occur. The attacker will hence have to overcome the secure operator authentication process. This process could be triggered, even if the quantitative indices show a low probability of breach.
- Disk encryption at both operating system level and at data level. The adversary will hence have to cryptanalyze both encryption schemes.
- Indications of compromise coming from the attack detection AI algorithms of possible security breaches trigger the destruction of all data stored in the base station. The attacker will have to try to recover the data without triggering the destruction mechanism. Provided that the data required by the legitimate users is regularly transferred to secure storage, this process could also be triggered, even if the quantitative indices show a low probability of breach.
- Regular verification of the integrity of the base station is required, e.g., by having a legitimate user or the administrator authenticate locally. If the deadline passes without this event occurring, data destruction is initiated.

The above scenaria demonstrate the cyber survivability principle of designing information systems that have as principle aim to prevent the adversary from completing their mission after they have completed parts of their attack (the cyber kill chain). The countermeasures described are complementary to current cybersecurity schemes, but have a significant contribution toward this very important aim.

5 Conclusions

Survivability is a concept that is well known in the context of military operations and is also understood in other contexts, such as business, and medicine. The notion of survivability in the context of cybersecurity over multi-user distributed information systems was defined. Rather than setting as the target of cyber defense the prevention of the adversary from breaching the system, the aim becomes to stop them from bringing their mission to an end, which can involve data theft, denial of service, or integrity violation. An overview of the problem of survivability in information systems and solutions that are currently available to system designers was presented. The cyber attacker's kill chain was explained. Artificial Intelligence (AI) techniques that may be employed in order to promote system survivability were outlined and the technical issues toward which they can contribute were listed. Following that, schemes for increased cyber survivability were presented, which focused on solving particular problems that commonly appear, by employing artificial intelligence techniques. First, the problem of email message filtering in the context of breaking the cyber kill chain was analyzed and a typical AI-assisted technical solution was given. Following that, the effect of malware in survivability was presented and an approach to its solution based on static analysis and detection of patterns was presented. Subsequently, the collusion attack, an attack where multiple malware programs collaborate in order to achieve malicious goals, was presented and an AI-powered solution was outlined based on current technology. A three-level anomaly detection system was presented that employs AI primitives and detects problematic behavior in network traffic, packed files, and SQL statements in order to produce cybersecurity warnings. Dynamic analysis of potentially harmful programs was analyzed and a technique that performs such analysis was presented that examined the executed machine-level instruction opcodes and utilized AI primitives in order to circumvent efforts of malware creators to obfuscate the actions and intents of their code. The final section presented a comprehensive cooperative infrastructure defense system that was based on the ant colony paradigm of artificial intelligence theory. The system aimed to coordinate human and automated efforts to protect the integrity of large-scale information systems. It used multiple AI principles in order to utilize existing information and obtain novel knowledge, adapting to new threats and user expectations. A set of countermeasures that inhibit the cyber adversary's mission and promote survivability against cyber attacks were finally outlined and evaluated in the context of cyberattack scenaria. The countermeasures acted as fail-safe mechanisms to break the cyber kill chain when the attack has not been detected.

References

1. Alazab, M., R. Layton, R. Broadhurst, and B. Bouhours. 2013. Malicious spam emails developments and authorship attribution. In *Fourth cybercrime and trustworthy computing workshop*, 58–68.

2. Alazab, M., S. Venkatraman, P. Watters, and M. Alazab. 2013. Information security governance: The art of detecting hidden malware. In *IT security governance innovations: theory and research*, 293–315, IGI Global, Harrisburg.
3. Asăvoae, Irina Măriuca, Jorge Blasco, Thomas M. Chen, Harsha Kumara Kalutarage, Igor Muttik, Hoang Nga Nguyen, Markus Roggenbach, and Siraj Ahmed Shaikh. 2017. Detecting malicious collusion between mobile software applications: The Android™ case. https://www.jorgeblascoalis.com/assets/papers/Asavoae2017.pdf.
4. Atighetchi, Michael, and Joseph Loyall. 2010. Meaningful and flexible survivability assessments: Approach and practice. In *CrossTalk: The journal of defense software engineering*, 12–18.
5. Azab, A., M. Alazab, and M. Aiash. 2016. Machine learning based botnet identification traffic. In *IEEE Trustcom/BigDataSE/ISPA*, 1788–1794.
6. Azab, A., R. Layton, M. Alazab, and J. Oliver. 2014. Mining malware to detect variants. In *Fifth cybercrime and trustworthy computing conference*, 44–53.
7. Carlin, Domhnall, Philip O'Kane, and Sakir Sezer. 2017. Dynamic analysis of malware using run-time opcodes. In *Data analytics and decision support for cybersecurity*, 99–125, Springer, Berlin.
8. Demertzis, Konstantinos, and Lazaros Iliadis. 2015. A bio-inspired hybrid artificial intelligence framework for cyber security. In *Computation, cryptography, and network security*, 161–193, Springer, Berlin.
9. Doukas, Nikolaos. 2017. Technologies for greener internet of things systems. In *Green IT engineering: components, networks and systems implementation*, 23–42. Springer, Berlin.
10. Doukas, Nikolaos, Oleksandr P. Markovskyi, and Nikolaos G. Bardis. 2019. Hash function design for cloud storage data auditing. *Theoretical Computer Science* 800: 42–51.
11. Haack, J.N., G.A. Fink, W.M. Maiden, A.D. McKinnon, S.J. Templeton, and E.W. Fulp. 2011. Ant-based cyber security. In *Eighth international conference on information technology: new generations*, 918–926.
12. Asif Khan, M., and Mureed Hussain. 2010. Cyber security quantification model. In *Proceedings of the 3rd international conference on security of information and networks*.
13. Parisi, Alessandro. 2019. *Hands-On Artificial Intelligence for Cybersecurity: Implement Smart AI Systems for Preventing Cyber-Attacks and Detecting Threats and Network Anomalies*. Birmingham: Packt Publishing Ltd.
14. Stavroulakis, P., M. Kolisnyk, V. Kharchenko, N. Doukas, O. P. Markovskyi, and N. G. Bardis. 2017. Reliability, fault tolerance and other critical components for survivability in information warfare. In *International Conference on E-Business and Telecommunications*, 346–370, Springer, Berlin.
15. Tarnowski, Ireneusz. 2017. How to use cyber kill chain model to build cybersecurity. *European Journal of Higher Education IT*.
16. Tsukerman, Emmanuel. 2019. *Machine Learning for Cybersecurity Cookbook*. Birmingham: Packt Publishing Ltd.
17. Wilson, Duane. Cyber survivability—Keeping mission systems survivable in the event of a mission-based cyberattack. https://www.survice.com/media/technology-spotlight/cyber-survivability-keeping-mission-systems-survivable-event-mission.
18. Wolfgang, Ertel. 2011. *Introduction to Artificial Intelligence*. London: Springer. Translated by Nathanael Black with illustrations by Florian Mast.

On Ensemble Learning

Mark Stamp, Aniket Chandak, Gavin Wong, and Allen Ye

Abstract In this chapter, we consider ensemble classifiers, that is, machine learning based classifiers that utilize a combination of scoring functions. We provide a framework for categorizing such classifiers, and we outline several ensemble techniques, discussing how each fits into our framework. From this general introduction, we then pivot to the topic of ensemble learning within the context of malware analysis. We present a brief survey of some of the ensemble techniques that have been used in malware (and related) research. We conclude with an extensive set of experiments, where we apply ensemble techniques to a large and challenging malware dataset. While many of these ensemble techniques have appeared in the malware literature, previously there has been no way to directly compare results such as these, as different datasets and different measures of success are typically used. Our common framework and empirical results are an effort to bring some sense of order to the chaos that is evident in the evolving field of ensemble learning—both within the narrow confines of the malware analysis problem, and in the larger realm of machine learning in general.

1 Introduction

In ensemble learning, multiple learning algorithms are combined, with the goal of improved accuracy as compared to the individual algorithms. Ensemble techniques are widely used, and as a testament to their strength, ensembles have won numerous

M. Stamp (✉) · A. Chandak
San Jose State University, San Jose, CA, USA
e-mail: mark.stamp@sjsu.edu

A. Chandak
e-mail: aniket.chandak@sjsu.edu

G. Wong · A. Ye
Lynbrook High School, San Jose, CA, USA
e-mail: gavinmwong@gmail.com

A. Ye
e-mail: allenye66@gmail.com

M. Stamp et al. (eds.), *Malware Analysis Using Artificial Intelligence and Deep Learning*, https://doi.org/10.1007/978-3-030-62582-5_8

machine learning contests in recent years, including the KDD Cup [15], the Kaggle competition [14], and the Netflix prize [26].

Many such ensembles resemble Frankenstein's monster [33], in the sense that they are an agglomeration of disparate components, with some of the components being of questionable value—an "everything and the kitchen sink" approach clearly prevails. This effect can be clearly observed in the aforementioned machine learning contests, where there is little (if any) incentive to make systems that are efficient or practical, as accuracy is typically the only criteria for success. In the case of the Netflix prize, the winning team was awarded $1,000,000, yet Netflix never implement the winning scheme, since the improvements in accuracy "did not seem to justify the engineering effort needed to bring them into a production environment" [3]. In real-world systems, practicality and efficiency are necessarily crucial factors.

In this chapter, we provide a straightforward framework for categorizing ensemble techniques. We then consider specific (and relatively simple) examples of various categories of such ensembles, and we show how these fit into our framework. For various examples of ensembles, we also provide experimental results, based on a large and diverse malware dataset.

While many of the techniques that we consider have previously appeared in the malware literature, we are not aware of any comparable study focused on the effectiveness of various ensembles using a common dataset and common measures of success. While we believe that these examples are interesting in their own right, they also provide a basis for discussing various tradeoffs between measures of accuracy and practical considerations.

The remainder of this chapter is organized as follows. In Sect. 2, we discuss ensemble classifiers, including our framework for categorizing such classifiers. Section 3 contains our experimental results and some discussion of these results. This section also includes a discussion of our dataset, scoring metrics, software used, and so on. Finally, Sect. 4 concludes the paper and includes suggestions for future work.

2 Ensemble Classifiers

In this section, we first give a selective survey of some examples of malware (and closely related) research involving ensemble learning. Then we provide a framework for discussing ensemble classifiers in general.

2.1 Examples of Related Work

The paper [18] discusses various ways to combine classifiers and provides a theoretical framework for such combinations. The focus is on straightforward combinations, such as a maximum, sum, product, majority vote, and so on. The work in [18] has

Table 1 Security research papers using ensemble classifiers

Authors	Application	Features	Ensemble
Alazab et al. [2]	Detection	API calls	Neural networks
Comar et al. [8]	Detection	Network traffic	Random forest
Dimjašević et al. [9]	Android	System calls	RF and SVM
Guo et al. [10]	Detection	API calls	BKS
Idrees et al. [12]	Android	Permissions, intents	RF and others
Jain & Meena [13]	Detection	Byte n-grams	AdaBoost
Khan et al. [17]	Detection	Network based	Boosting
Kong & Yan [19]	Classification	Function call graph	Boosting
Morales et al. [24]	Android	Permissions	Several
Narouei et al. [25]	Detection	DLL dependency	Random forest
Shahzad et al. [31]	Detection	Opcodes	Voting
Sheen et al. [32]	Various	Detection efficiency	Pruning
Singh et al. [34]	Detection	Opcodes	SVM
Smutz & Stavrou [36]	Malicious PDF	Metadata	Random forest
Toolan & Carthy [40]	Phishing	Various	C5.0, boosting
Ye et al. [58]	Detection	API calls, strings	SVM, bagging
Ye et al. [59]	Categorization	Opcodes	Clustering
Yerima et al. [60]	Zero day	179 features	RF, regression
Zhang et al. [61]	Detection	n-grams	Dempster-Shafer

clearly been influential, but it seems somewhat dated, given the wide variety of ensemble methods that are used today.

The book [20] presents the topic of ensemble learning from a similar perspective as [18] but in much more detail. Perhaps not surprisingly, the more recent book [62] seems to have a somewhat more modern perspective with respect to ensemble methods, but retains the theoretical flavor of [18, 20]. The brief blog at [35] provides a highly readable (if highly selective) summary of some of the topics covered in the books [20, 62].

Here, we take an approach that is, in some sense, more concrete than that in [18, 20, 62]. Our objective is to provide a relatively straightforward framework for categorizing and discussing ensemble techniques. We then use this framework as a frame of reference for experimental results based on a variety of ensemble methods.

Table 1 provides a summary of several research papers where ensemble techniques have been applied to security-related problems. The emphasis here is on malware, but we have also included a few closely related topics. In any case, this represents a small sample of the many papers that have been published, and is only intended to provide an indication as to the types and variety of ensemble strategies that have been considered to date. On this list, we see examples of ensemble methods based on bagging, boosting, and stacking, as discussed below in Sect. 2.3.

2.2 A Framework for Ensemble Classifiers

In this section, we consider various means of constructing ensemble classifiers, as viewed from a high-level perspective. We then provide an equally high-level framework that we find useful in our subsequent discussion of ensemble classifiers in Sect. 2.3 and, especially, in Sect. 2.4.

We consider ensemble learners that are based on combinations of scoring functions. In the general case, we assume the scoring functions are real valued, but the more restricted case of zero-one valued "scoring" functions (i.e., classifiers) easily fits into our framework. We place no additional restrictions on the scoring functions and, in particular, they do not necessarily represent "learning" algorithms, per se. Hence, we are dealing with ensemble methods broadly speaking, rather than ensemble learners in a strict sense. We assume that the ensemble method itself—as opposed to the scoring functions that comprise the ensemble—is for classification, and hence ensemble functions are zero-one valued.

Let $\omega_1, \omega_2, \ldots, \omega_n$ be training samples, and let v_i be a feature vector of length m, where the features that comprise v_i are extracted from sample ω_i. We collect the feature vectors for all n training samples into an $m \times n$ matrix that we denote as

$$V = \left(\, v_1 v_2 \cdots v_n \, \right), \tag{1}$$

where each v_i is a column of the matrix V. Note that each row of V corresponds to a specific feature type, while column i of V corresponds to the features extracted from the training sample ω_i.

Let $S : \mathbb{R}^m \to \mathbb{R}$ be a scoring function. Such a scoring function will be determined based on training data, where this training data is given by a feature matrix V, as in Eq. (1). A scoring function S will generally also depend on a set of k parameters that we denote as

$$\Lambda = \left(\, \lambda_1 \lambda_2 \ldots \lambda_k \, \right). \tag{2}$$

The score generated by the scoring function S when applied to sample x is given by

$$S(x; V, \Lambda),$$

where we have explicitly included the dependence on the training data V and the function parameters Λ.

For any scoring function S, there is a corresponding classification function that we denote as $\widehat{S} : \mathbb{R}^m \to \{0, 1\}$. That is, once we determine a threshold to apply to the scoring function S, it provides a binary classification function that we denote as $\widehat{S}$. As with S, we explicitly indicate the dependence on training data V and the function parameters Λ by writing

$$\widehat{S}(x; V, \Lambda).$$

For example, each training sample ω_i could be a malware executable file, where all of the ω_i belong to the same malware family. Then an example of an extracted feature v_i would be the opcode histogram, that is, the relative frequencies of the mnemonic opcodes that are obtained when ω_i is disassembled. The scoring function S could, for example, be based on a hidden Markov model that is trained on the feature matrix V as given in Eq. (1), with the parameters Λ in Eq. (2) being the initial values that are selected when training the HMM.

In its most general form, an ensemble method for a binary classification problem can be viewed as a function $F : \mathbb{R}^\ell \rightarrow \{0, 1\}$ of the form

$$F\big(S_1(x; V_1, \Lambda_1), S_2(x; V_2, \Lambda_2), \dots, S_\ell(x; V_\ell, \Lambda_\ell)\big). \tag{3}$$

That is, the ensemble method defined by the function F produces a classification based on the scores $S_1, S_2, \dots, S_\ell$, where scoring function S_i is trained using the data V_i and parameters Λ_i.

2.3 *Classifying Ensemble Classifiers*

From a high-level perspective, ensemble classifiers can be categorized as bagging, boosting, stacking, or some combination thereof [20, 35, 62]. In this section, we briefly introduce each of these general classes of ensemble methods and give their generic formulation in terms of Eq. (3).

2.3.1 Bagging

In bootstrap aggregation (i.e., bagging), different subsets of the data or features (or both) are used to generate different scores. The results are then combined in some way, such as a sum of the scores, or a majority vote of the corresponding classifications. For bagging, we assume that the same scoring method is used for all scores in the ensemble. For example, bagging is used when generating a random forest, where each individual scoring function is based on a decision tree structure. One benefit of bagging is that it reduces overfitting, which is a particular problem for decision trees.

For bagging, the general Eq. (3) is restricted to

$$F\big(S(x; V_1, \Lambda), S(x; V_2, \Lambda), \dots, S(x; V_\ell, \Lambda)\big) \tag{4}$$

That is, in bagging, each scoring function is essentially the same, but each is trained on a different feature set. For example, suppose that we collect all available feature vectors into a matrix V as in Eq. (1). Then bagging based on subsets of samples would correspond to generating V_i by deleting a subset of the columns of V. On the other hand, bagging based on features would correspond to generating V_i by deleting a subset of the rows of V. Of course, we can easily extend this to bagging based on both the data and features simultaneously, as in a random forest. In Sect. 2.4, we discuss specific examples of bagging.

2.3.2 Boosting

Boosting is a process whereby distinct classifiers are combined to produce a stronger classifier. Generally, boosting deals with weak classifiers that are combined in an adaptive or iterative manner so as to improve the overall classifier. We restrict our definition of boosting to cases where the classifiers are closely related, in the sense that they differ only in terms of parameters. From this perspective, boosting can be viewed as "bagging" based on classifiers, rather than data or features. That is, all of the scoring functions are re-parameterized versions of the same scoring technique. Under this definition of boosting, the general Eq. (3) becomes

$$F\big(S(x; V, \Lambda_1), S(x; V, \Lambda_2), \ldots, S(x; V, \Lambda_\ell)\big). \tag{5}$$

That is, the scoring functions differ only by re-parameterization, while the scoring data and features do not change.

Below, in Sect. 2.4, we discuss specific examples of boosting; in particular, we discuss the most popular method of boosting, AdaBoost. In addition, we show that some other popular techniques fit our definition of boosting.

2.3.3 Stacking

Stacking is an ensemble method that combines disparate scores using a meta-classifier [35]. In this generic form, stacking is defined by the general case in Eq. (3), where the scoring functions can be (and typically are) significantly different. Note that from this perspective, stacking is easily seen to be a generalization of both bagging and boosting.

Because stacking generalizes both bagging and boosting, it is not surprising that stacking-based ensemble methods can outperform bagging and boosting methods, as evidenced by recent machine learning competitions, including the KDD Cup [15], the Kaggle competition [14], as well as the infamous Netflix prize [26]. However, this is not the end of the story, as efficiency and practicality are often ignored in such competitions, whereas in practice, it is virtually always necessary to consider such issues. Of course, the appropriate tradeoffs will depend on the specifics of the problem at hand. Our empirical results in Sect. 3 provide some insights into these tradeoff issues within the malware analysis domain.

In the next section, we discuss concrete examples of bagging, boosting, and stacking techniques. Then in Sect. 3, we present our experimental results, which include selected bagging, boosting, and stacking architectures.

2.4 Ensemble Classifier Examples

Here, we consider a variety of ensemble methods and discuss how each fits into the general framework presented above. We begin with a few fairly generic examples, and then discuss several more specific examples.

2.4.1 Maximum

In this case, we have

$$F\big(S_1(x; V_1, \Lambda_1), S_2(x; V_2, \Lambda_2), \ldots, S_\ell(x; V_\ell, \Lambda_\ell)\big) = \max\{S_i(x; V_i, \Lambda_i)\}. \quad (6)$$

2.4.2 Averaging

Averaging is defined by

$$F\big(S_1(x; V_1, \Lambda_1), S_2(x; V_2, \Lambda_2), \ldots, S_\ell(x; V_\ell, \Lambda_\ell)\big) = \frac{1}{\ell}\sum_{i=1}^{\ell} S_i(x; V_i, \Lambda_i). \quad (7)$$

2.4.3 Voting

Voting could be used as a form of boosting, provided that no bagging is involved (i.e., the same data and features are used in each case). Voting is also applicable to stacking, and is generally applied in such a mode, or at least with significant diversity in the scoring functions, since we want limited correlation when voting.

In the case of stacking, a simple majority vote is of the form

$$\begin{aligned} &F\big(\widehat{S}_1(x; V_1, \Lambda_1), \widehat{S}_2(x; V_2, \Lambda_2), \ldots, \widehat{S}_\ell(x; V_\ell, \Lambda_\ell)\big) \\ &\qquad = \mathrm{maj}\big(\widehat{S}_1(x; V_1, \Lambda_1), \widehat{S}_2(x; V_2, \Lambda_2), \ldots, \widehat{S}_\ell(x; V_\ell, \Lambda_\ell)\big), \end{aligned}$$

where "maj" is the majority vote function. Note that the majority vote is well defined in this case, provided that ℓ is odd—if ℓ is even, we can simply flip a coin in case of a tie.

As an aside, we note that it is easy to see why we want to avoid correlation when voting is used as a combining function. Consider the following example from [47]. Suppose that we have the three highly correlated scores

$$\begin{pmatrix} \widehat{S}_1 \\ \widehat{S}_2 \\ \widehat{S}_3 \end{pmatrix} = \begin{pmatrix} 1\,1\,1\,1\,1\,1\,1\,1\,0\,0 \\ 1\,1\,1\,1\,1\,1\,1\,1\,0\,0 \\ 1\,0\,1\,1\,1\,1\,1\,1\,0\,0 \end{pmatrix},$$

where each 1 indicates correct classification, and each 0 is an incorrect classification. Then, both $\widehat{S}_1$ and $\widehat{S}_2$ are 80% accurate, and $\widehat{S}_3$ is 70% accurate. If we use a simple majority vote, then we obtain the classifier

$$C = (1\ 1\ 1\ 1\ 1\ 1\ 1\ 1\ 0\ 0)$$

which is 80% accurate. On the other hand, the less correlated classifiers

$$\begin{pmatrix} \widehat{S}_1{}' \\ \widehat{S}_2{}' \\ \widehat{S}_3{}' \end{pmatrix} = \begin{pmatrix} 1\ 1\ 1\ 1\ 1\ 1\ 1\ 1\ 0\ 0 \\ 0\ 1\ 1\ 1\ 0\ 1\ 1\ 1\ 0\ 1 \\ 1\ 0\ 0\ 0\ 1\ 0\ 1\ 1\ 1\ 1 \end{pmatrix}$$

are only 80%, 70% and 60% accurate, respectively, but the majority vote in this case gives us

$$C' = (1\ 1\ 1\ 1\ 1\ 1\ 1\ 1\ 0\ 1)$$

which is 90% accurate.

2.4.4 ML-Based Combination

Recall that the most general formulation of an ensemble classifier is given in Eq. (3). In this formulation, we can select the function F based on a machine learning technique, which is applied to the individual scores $S(x; V_i, \Lambda_i)$. In the remainder of this section, we consider specific ensemble examples involving machine learning techniques.

2.4.5 AdaBoost

Given a collection of (weak) classifiers $c_1, c_2, \ldots, c_\ell$, AdaBoost is an iterative algorithm that generates a series of (generally, stronger) classifiers, $C_1, C_2, \ldots, C_M$ based on the classifiers c_i. Each classifier is determined from the previous classifier by the simple linear extension

$$C_m(x) = C_{m-1}(x) + \alpha_m c_i(x)$$

and the final classifier is given by $C = C_M$. Note that at each iteration, we include a previously unused c_i from the set of (weak) classifiers and determine a new weight α_i. A greedy approach is used when selecting c_i, but it is not a hill climb, so that results might get worse at any step in the AdaBoost process.

From this description, we see that the AdaBoost algorithm fits the form in Eq. (5), with $\widehat{S}(x; V, \Lambda_i) = C_i(x)$, and

$$F\big(\widehat{S}(x; V, \Lambda_1), \widehat{S}(x; V, \Lambda_2), \ldots, \widehat{S}(x; V, \Lambda_M)\big) = \widehat{S}(x; V, \Lambda_M) = C_M(x)$$

2.4.6 SVM as Meta-Classifier

It is natural to use an SVM as a meta-classifier to combine scores [38]. For example, in [34], an SVM is used to generate a malware classifier based on several machine learning and statistical-based malware scores. In [34], it is shown that the resulting SVM classifier consistently outperforms any of the component scores, and the differences are most pronounced in the most challenging cases.

The use of SVM in this meta-classifier mode can be viewed as a general stacking method. Thus, this SVM technique is equivalent to Eq. (3), where the function F is simply an SVM classifier based on the component scores $S_i(x; V_i, \Lambda_i)$, for $i = 1, 2, \ldots, \ell$.

2.4.7 HMM with Random Restarts

A hidden Markov model can be viewed as a discrete hill climb technique [37, 38]. As with any hill climb, when training an HMM we are only assured of a local maximum, and we can often significantly improve our results by executing the hill climb multiple times with different initial values, selecting the best of the resulting models. For example, in [51] it is shown that an HMM can be highly effective for breaking classic substitution ciphers and, furthermore, by using a large number of random restarts, we can significantly increase the success rate in the most difficult cases. The work in [51] is closely related to that in [7], where such an approach is used to analyze the unsolved Zodiac 340 cipher.

From the perspective considered in this paper, an HMM with random restarts can be seen as a special case of boosting. If we simply select the best model, then the "combining" function is particularly simple, and is given by

$$F\big(S(x; V, \Lambda_1), S(x; V, \Lambda_2), \ldots, S(x; V, \Lambda_\ell)\big) = \max\{S(x; V, \Lambda_i)\}. \quad (8)$$

Here, each scoring function is an HMM, where the trained models differ based only on different initial values. We see that Eq. (8) is a special case of Eq. (6). However, the "max" in Eq. (8) is the maximum over the HMM model scores, not the maximum over any particular set of input values. That is, we select the highest scoring model and use it for scoring. Of course, we could use other combining functions, such as an average or majority vote of the corresponding classifiers. In any case, since there is a score associated with each model generated by an HMM, any such combining function is well-defined.

2.4.8 Bagged Perceptron

Like a linear SVM, a perceptron will separate linearly separable data. However, unlike an SVM, a perceptron will not necessarily produce the optimal separation, in the sense of maximizing the margin. If we generate multiple perceptrons, each with different random initial weights, and then average these models, the resulting classifier will tend to be nearer to optimal, in the sense of maximizing the margin [21, 47]. That is, we construct a classifier

$$F\big(S(x;V,\Lambda_1),S(x;V,\Lambda_2),\ldots,S(x;V,\Lambda_\ell)\big)=\frac{1}{\ell}\sum_{i=1}^{\ell}S(x;V,\Lambda_i), \quad (9)$$

where S is a perceptron and each P_i represents a set of initial values. We see that Eq. (9) is a special case of the averaging example given in Eq. (7). Also, we note that in this sum, we are averaging the perceptron models, not the classifications generated by the models.

Although this technique is sometimes referred to as "bagged" perceptrons [47], by our criteria, it is a boosting scheme. That is, the "bagging" here is done with respect to parameters of the scoring functions, which is our working definition of boosting.

2.4.9 Bagged Hidden Markov Model

Like the HMM with random restarts example given above, in this case, we generate multiple HMMs. However, here we leave the model parameters unchanged, and simply train each on a subset of the data. We could then average the model scores (for example) as a way of combining the HMMs into a single score, from which we can easily construct a classifier.

2.4.10 Bagged and Boosted Hidden Markov Model

Of course, we could combine both the HMM with random restarts discussed in Sect. 2.4.7 with the bagging approach discussed in the previous section. This process would yield an HMM-based ensemble technique that combines both bagging and boosting.

3 Experiments and Results

In this section, we consider a variety of experiments that illustrate various ensemble techniques. These experiments involve malware classification based on a challenging dataset that includes a large number of samples from a significant number of malware families.

3.1 Dataset and Features

Our dataset consists of samples from the 21 malware families listed in Table 2. These families are from various different types of malware, including Trojans, worms, backdoors, password stealers, so-called VirTools, and so on.

Each of the malware families in Table 2 is summarized below.

Adload downloads an executable file, stores it remotely, executes the file, and disables proxy settings [41].
Agent downloads Trojans or other software from a remote server [42].
Allaple is a worm that can be used as part of a denial of service (DoS) attack [52].
BHO can perform a variety of actions, guided by an attacker [45].
Bifrose is a backdoor Trojan that enables a variety of attacks [4].
CeeInject uses advanced obfuscation to avoid being detected by antivirus software [48].
Cycbot connects to a remote server, exploits vulnerabilities, and spreads through backdoor ports [5].
FakeRean pretends to scan the system, notifies the user of supposed issues, and asks the user to pay to clean the system [53].
Hotbar is adware that shows ads on webpages and installs additional adware [1].
Injector loads other processes to perform attacks on its behalf [49].
OnLineGames steals login information of online games and tracks user keystroke activity [28].

Table 2 Type of each malware family

Index	Family	Type	Index	Family	Type
1	Adload [41]	Trojan downloader	12	Renos [43]	Trojan downloader
2	Agent [42]	Trojan	13	Rimecud [54]	Worm
3	Allaple [52]	Worm	14	Small [44]	Trojan downloader
4	BHO [45]	Trojan	15	Toga [46]	Trojan
5	Bifrose [4]	Backdoor	16	VB [6]	Backdoor
6	CeeInject [48]	VirTool	17	VBinject [50]	VirTool
7	Cycbot [5]	Backdoor	18	Vobfus [55]	Worm
8	FakeRean [53]	Rogue	19	Vundo [56]	Trojan downloader
9	Hotbar [1]	Adware	20	Winwebsec [22]	Rogue
10	Injector [49]	VirTool	21	Zbot [23]	Password stealer
11	OnLineGames [28]	Password stealer	–	–	–

Renos downloads software that claims the system has spyware and asks for a payment to remove the nonexistent spyware [43].

Rimecud is a sophisticated family of worms that perform a variety of activities and can spread through instant messaging [54].

Small is a family of Trojans that downloads unwanted software. This downloaded software can perform a variety of actions, such as a fake security application [44].

Toga is a Trojan that can perform a variety of actions of the attacker's choice [46].

VB is a backdoor that enables an attacker to gain access to a computer [6].

VBinject is a generic description of malicious files that are obfuscated in a specific manner [50].

Vobfus is a worm that downloads malware and spreads through USB drives or other removable devices [55].

Vundo displays pop-up ads and may download files. It uses advanced techniques to defeat detection [56].

Winwebsec displays alerts that ask the user for money to fix supposed issues [22].

Zbot is installed through email and shares a user's personal information with attackers. In addition, Zbot can disable a firewall [23].

From each available malware sample, we extract the first 1000 mnemonic opcodes using the reversing tool Radare2 (also know as R2) [29]. We discard any malware executable that yields less than 1000 opcodes, as well as a number of executables that were found to be corrupted. The resulting opcode sequences, each of length 1000, serve as the feature vectors for our machine learning experiments.

Table 3 gives the number of samples (per family) from which we successfully obtained opcode feature vectors. Note that our dataset contains a total of 9725 samples from the 21 malware families and that the dataset is highly imbalanced—the number of samples per family varies from a low of 129 to a high of nearly 1000.

Table 3 Type of each malware family

Index	Family	Samples	Index	Family	Samples
1	Adload	162	12	Renos	532
2	Agent	184	13	Rimecud	153
3	Allaple	986	14	Small	180
4	BHO	332	15	Toga	406
5	Bifrose	156	16	VB	346
6	CeeInject	873	17	VBinject	937
7	Cycbot	597	18	Vobfus	929
8	FakeRean	553	19	Vundo	762
9	Hotbar	129	20	Winwebsec	837
10	Injector	158	21	Zbot	303
11	OnLineGames	210		Total	9725

3.2 Metrics

The metrics used to quantify the success of our experiments are accuracy, balanced accuracy, precision, recall, and the F1 score. Accuracy is simply the ratio of correct classifications to the total number of classifications. In contrast, the balanced accuracy is the average accuracy per family.

Precision, which is also known as the positive predictive value, is the number of true positives divided by the sum of the true positives and false positives. That is, the precision is the ratio of samples classified as positives that are actually positive to all samples that are classified as positive. Recall, which is also known as the true positive rate or sensitivity, is computed by dividing the number of true positives by the number true positives plus the number of false negatives. That is, the recall is the fraction of positive samples that are classified as such. The F1 score is computed as

$$\mathrm{F1} = 2 \cdot \frac{\text{precision} \cdot \text{recall}}{\text{precision} + \text{recall}},$$

which is the harmonic mean of the precision and recall.

3.3 Software

The software packages used in our experiments include `hmmlearn` [11], `XGBoost` [57], `Keras` [16], `TensorFlow` [39], and `scikit-learn` [30], as indicated in Table 4. In addition, we use `Numpy` [27] for linear algebra and various tools available in the package `scikit-learn` (also known as `sklearn`) for general data processing. These packages are all widely used in machine learning.

Table 4 Software used in experiments

Technique	Software
HMM	`hmmlearn`
XGBoost	`XGBoost`
AdaBoost	`sklearn`
CNN	`Keras`, `TensorFlow`
LSTM	`Keras`, `TensorFlow`
Random Forest	`sklearn`

3.4 Overview of Experiments

For all of our experiments, we use opcode sequences of length 1000 as features. For CNNs, the sequences are interpreted as images.

We consider three broad categories of experiments. First, we apply "standard" machine learning techniques. These experiments, serve as a baseline for comparison for our subsequent experiments. Among other things, these standard experiments show that the malware classification problem that we are dealing with is challenging.

We also conduct bagging and boosting experiments based on a subset of the techniques considered in our baseline standard experiments. These results demonstrate that both bagging and boosting can provide some improvement over our baseline techniques.

Finally, we consider a set of stacking experiments, where we restrict our attention to simple voting schemes, all of which are based on architectures previously considered in this paper. Although these are very basic stacking architectures, they clearly show the potential benefit of stacking multiple techniques.

3.5 Standard Techniques

For our "standard" techniques, we test several machine learning methods that are typically used individually. Specifically, we consider hidden Markov models (HMM), convolutional neural networks (CNN), random forest, and long short-term memory (LSTM). The parameters that we have tested in each of these cases are listed in Table 5, with those that gave the best results in boldface.

From Table 5, we note that a significant number of parameter combinations were tested in each case. For example, in the case of our random forest model, we tested

$$5^3 \cdot 3 \cdot 6 = 2250$$

different combinations of parameters.

The confusion matrices for all of the experiments in this section can be found in the Appendix in Fig. 2a through Fig. 2d. We present the results of all of these experiments—in terms of the metrics discussed previously (i.e., accuracy, balanced accuracy, precision, recall, and F1 score)—in Sect. 3.9.

3.6 Bagging Experiments

Recall from our discussion above that we use the term bagging to mean a multi-model approach where the individual models are trained with the same technique and essentially the same parameters, but different subsets of the data or features. In

Table 5 Parameters for standard techniques

Technique	Parameters	Values tested
HMM	n_components	[1,2,5,**10**]
	n_iter	[50,100,**200**,300,500]
	tol	[0.01,0.5]
CNN	learning_rate	[**0.001**,0.0001]
	batch_size	[**32**,64,128]
	epochs	[50,75,**100**
Random Forest	n_estimators	[100,200,300,500,**800**]
	min_samples_split	[**2**,5,10,15,20]
	min_samples_leaf	[**1**,2,5,10,15]
	max_features	[**auto**,sqrt,$\log_2$]
	max_depth	[30,**40**,50,60,70,80]
LSTM	layers	[**1**,3]
	directional	[uni-dir,**bi-dir**]
	learning_rate	[**0.01**]
	batch_size	[**1**,16,32]
	epochs	[**20**]

contrast, we use boosting to refer to multi-model cases where the data and features are essentially the same and the models are of the same type, with the model parameters varied.

We will use AdaBoost and XGBoost results to serve as representative examples of boosting. We also consider bagging experiments (in the sense described in the previous paragraph) involving each of the HMM, CNN, and LSTM architectures. The results of these three distinct bagging experiments—in the form of confusion matrices—are given in Fig. 3 in the Appendix. In terms of the metrics discussed above, the results of these experiments are summarized in Sect. 3.9.

3.7 *Boosting Experiments*

As representative examples of boosting techniques, we consider AdaBoost and XGBoost. In each case, we experiment with a variety of parameters as listed in Table 6. The parameter selection that yielded the best results are highlighted in boldface.

Confusion matrices for these two boosting experiments are given in Fig. 4 in the Appendix. The results of these experiments are summarized in Sect. 3.9, in terms of accuracy, balanced accuracy, and so on.

Table 6 Parameters for boosting techniques

Technique	Parameters	Values tested
AdaBoost	n_estimators	[100,200,300,500,800,**1000**]
	learning_rate	[0.5,1.0,1.5,2.0]
	algorithm	[**SAMME**,SAMME.R]
XGBoost	eta	[0.05,0.1,0.2,**0.3**,0.5]
	max_depth	[1,2,**3**,4]
	objective	[**multi:softprob**,binary:logistic]
	steps	[1,5,10,**20**,50]

3.8 Voting Experiments

Since there exists an essentially unlimited number of possible stacking architectures, we have limited our attention to one of the simplest, namely, voting. These results serve as a lower bound on the results that can be obtained with stacking architectures.

We consider six different stacking architectures. These stacking experiments can be summarized as follows.

CNN consists of the plain and bagged CNN models discussed above. The confusion matrix for this experiment is given in Fig. 5a.

LSTM consists of the plain and bagged LSTM models discussed above. The confusion matrix for this experiment is given in Fig. 5b.

Bagged neural networks combines our bagged CNN and bagged LSTM models. The confusion matrix for this experiment is given in Fig. 5c.

Classic techniques combines (via voting) all of the classic models considered above, namely, HMM, bagged HMM, random forest, AdaBoost, and XGBoost. The confusion matrix for this experiment is given in Fig. 5d.

All neural networks consists of all of the CNN and LSTM models, bagged and plain. The confusion matrix for this experiment is given in Fig. 5e.

All models combines all of the classic and neural network models into one voting scheme. The confusion matrix for this experiment is given in Fig. 5f.

In the next section, we present the results for each of the voting experiments discussed in this section in terms of our various metrics. These metrics enable us to directly compare all of our experimental results.

3.9 Discussion

Table 7 summarizes the results of all of the experiments discussed above, in term of the following metrics: accuracy, balanced accuracy, precision, recall, and F1 score. These metrics have been introduced in Sect. 3.1.

Table 7 Comparison of experimental results

Experiments	Case	Accuracy	Balanced accuracy	Precision	Recall	F1 score
Standard	HMM	0.6717	0.6336	0.7325	0.6717	0.6848
	CNN	0.8211	**0.7245**	**0.8364**	**0.8211**	0.8104
	Random Forest	0.7549	0.6610	0.7545	0.7523	0.7448
	LSTM	**0.8410**	0.7185	0.7543	0.7185	**0.8145**
Bagging	Bagged HMM	0.7168	0.6462	0.7484	0.7168	0.7165
	Bagged CNN	**0.8910**	**0.8105**	**0.9032**	**0.8910**	**0.8838**
	Bagged LSTM	0.8602	0.7754	0.8571	0.8602	0.8549
Boosting	AdaBoost	0.5378	0.4060	0.5231	0.5378	0.5113
	XGBoost	**0.7472**	**0.6636**	**0.7371**	**0.7472**	**0.7285**
Voting	Classic	0.8766	0.8079	0.8747	0.8766	0.8719
	CNN	0.9260	0.8705	0.9321	0.9260	0.9231
	LSTM	0.8560	0.7470	0.8511	0.8560	0.8408
	Bagged neural networks	**0.9337**	**0.8816**	**0.9384**	**0.9337**	**0.9313**
	All neural networks	0.9208	0.8613	0.9284	0.9208	0.9171
	All models	0.9188	0.8573	0.9249	0.9188	0.9154

In Table 7, the best result for each type of experiment is in boldface, with the best results overall also being boxed. We see that a voting strategy based on all of the bagged neural network techniques gives us the best result for each of the five statistics that we have computed.

Since our dataset is highly imbalanced, we consider the balanced accuracy as the best measure of success. The balanced accuracy results in Table 7 are given in the form of a bar graph in Fig. 1.

Note that the results in Fig. 1 clearly show that stacking techniques are beneficial, as compared to the corresponding "standard" techniques. Stacking not only yields the best results, but it dominates in all categories. We note that five of the six stacking experiments perform better than any of the standard, bagging, or boosting experiments. This is particularly noteworthy since we only considered a simple stacking approach. As a result, our stacking experiments likely provide a poor lower bound on stacking in general, and more advanced stacking techniques may improve significantly over the results that we have obtained.

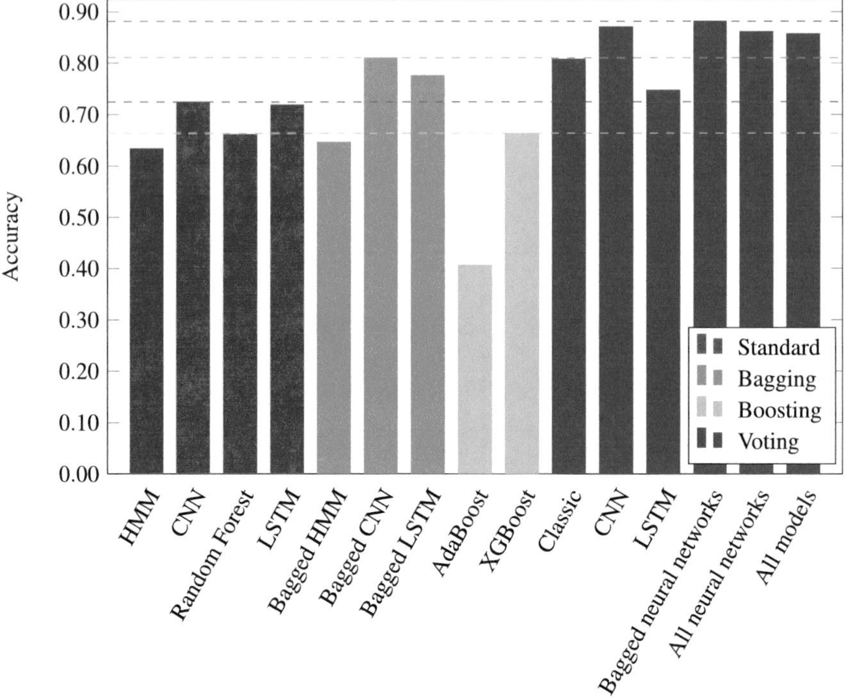

Fig. 1 Balanced accuracy results

4 Conclusion and Future Work

In this chapter, we have attempted to impose some structure on the field of ensemble learning. We showed that combination architectures can be classified as either bagging, boosting, or in the more general case, stacking. We then provided experimental results involving a challenging malware dataset to illustrate the potential benefits of ensemble architectures. Our results clearly show that ensembles improve on standard techniques, with respect to our specific dataset. Of course, in principle, we expect such combination architectures to outperform standard techniques, but it is instructive to confirm this empirically, and to show that the improvement can be substantial. These results make it clear that there is a reason why complex stacking architectures win machine learning competitions.

However, stacking models are not without potential pitfalls. As the architectures become more involved, training can become impractical. Furthermore, scoring can also become prohibitively costly, especially if large numbers of features are used in complex schemes involving extensive use of bagging or boosting.

For future work, it would be useful to quantify the tradeoff between accuracy and model complexity. While stacking will generally improve results, marginal improvements in accuracy that come at great additional cost in training and scoring are

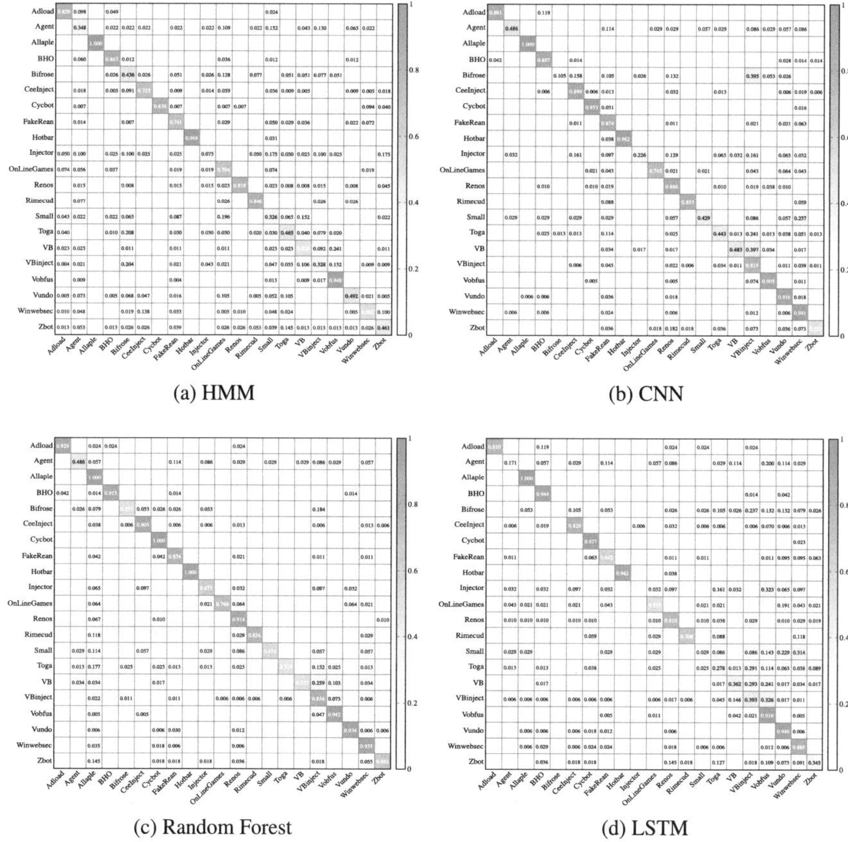

Fig. 2 Confusion matrices for standard techniques

unlikely to be of any value in real-world applications. More concretely, future work involving additional features would be very interesting, as it would allow for a more thorough analysis of bagging, and it would enable us to draw firmer conclusions regarding the relative merits of bagging and boosting. Of course, more and more complex classes of stacking techniques could be considered.

Appendix: Confusion Matrices

See Figs. 2, 3, 4, 5

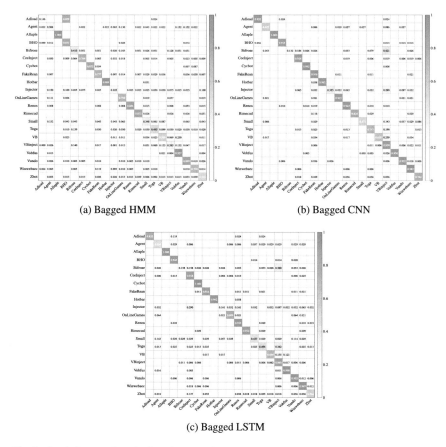

(a) Bagged HMM

(b) Bagged CNN

(c) Bagged LSTM

Fig. 3 Confusion matrices for bagging experiments

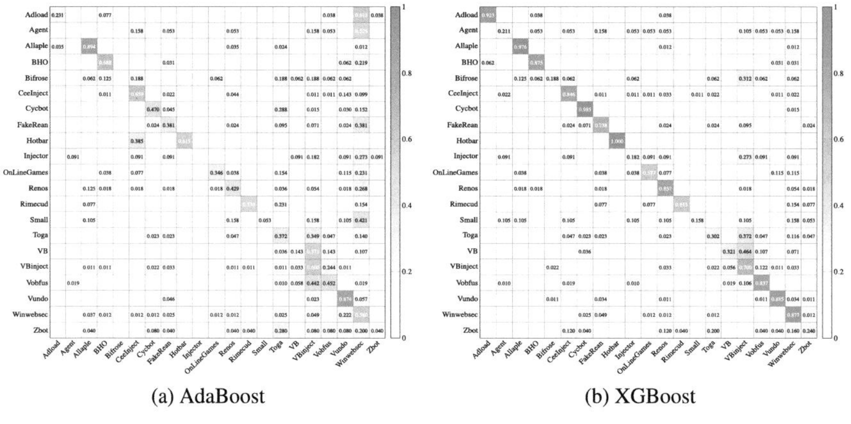

(a) AdaBoost

(b) XGBoost

Fig. 4 Confusion matrices for boosting techniques

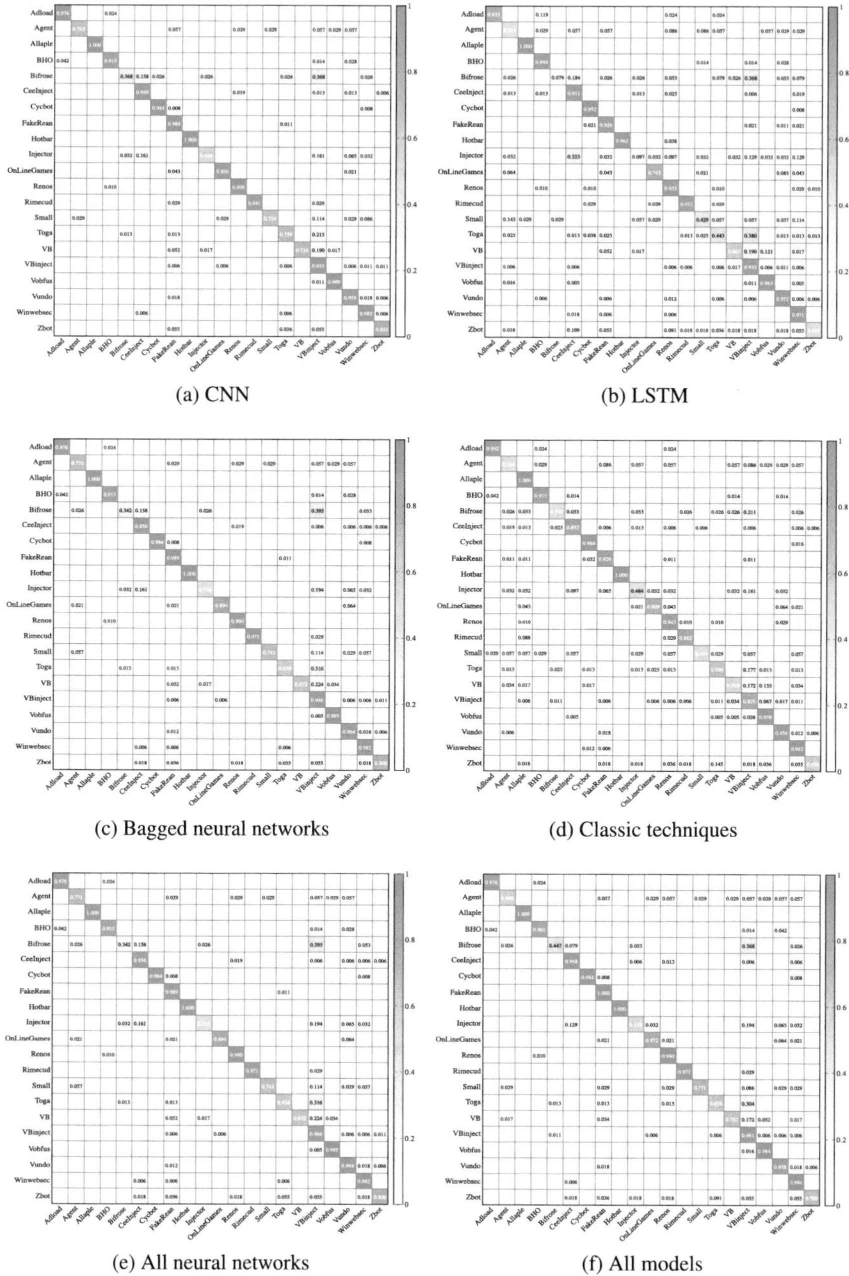

(a) CNN
(b) LSTM
(c) Bagged neural networks
(d) Classic techniques
(e) All neural networks
(f) All models

Fig. 5 Confusion matrices for voting ensembles

References

1. Adware:win32/hotbar. https://www.microsoft.com/en-us/wdsi/threats/malware-encyclopedia-description?Name=Adware:Win32/Hotbar&threatId=6204.
2. Alazab, Mamoun, Sitalakshmi Venkatraman, Paul Watters, and Moutaz Alazab. 2011. Zero-day malware detection based on supervised learning algorithms of API call signatures. In *Proceedings of the Ninth Australasian Data Mining Conference, volume 121 of AusDM '11*, 171–182, Australian Computer Society, Darlinghurst.
3. Amatriain, Xavier, and Justin Basilico. 2012. Netflix recommendations: Beyond the 5 stars (part 1). https://medium.com/netflix-techblog/netflix-recommendations-beyond-the-5-stars-part-1-55838468f429.
4. Backdoor:win32/bifrose. https://www.microsoft.com/en-us/wdsi/threats/malware-encyclopedia-description?Name=Backdoor:Win32/Bifrose&threatId=-2147479537.
5. Backdoor:win32/cycbot.g. https://www.microsoft.com/en-us/wdsi/threats/malware-encyclopedia-description?Name=Backdoor:Win32/Cycbot.G.
6. Backdoor:win32/vb. https://www.microsoft.com/en-us/wdsi/threats/malware-encyclopedia-description?Name=Backdoor:Win32/VB&threatId=7275.
7. Berg-Kirkpatrick, Taylor, and Dan Klein. 2013. Decipherment with a million random restarts. *Proceedings of the Conference on Empirical Methods in Natural Language Processing, EMNLP*, 874–878.
8. Comar, Prakash Mandayam, Lei Liu, Sabyasachi Saha, Pang-Ning Tan, and Antonio Nucci. 2013. Combining supervised and unsupervised learning for zero-day malware detection. *2013 Proceedings IEEE INFOCOM*, 2022–2030. IEEE: Piscataway.
9. Dimjaševic, Marko, Simone Atzeni, Ivo Ugrina, and Zvonimir Rakamaric. 2015. Android malware detection based on system calls. Technical Report UUCS-15-003, School of Computing, University of Utah, Salt Lake City, Utah.
10. Guo, Shanqing, Qixia Yuan, Fengbo Lin, Fengyu Wang, and Tao Ban. 2010. A malware detection algorithm based on multi-view fusion. *International Conference on Neural Information Processing, ICONIP 2010*, 259–266. Berlin: Springer.
11. hmmlearn. https://hmmlearn.readthedocs.io/en/latest/.
12. Idrees, Fauzia, Muttukrishnan Rajarajan, Mauro Conti, Thomas M. Chen, and Yogachandran Rahulamathavan. 2017. Pindroid: A novel android malware detection system using ensemble learning methods. *Computers & Security* 68: 36–46.
13. Jain, Sachin, and Yogesh Kumar Meena. 2011. Byte level n-gram analysis for malware detection. *Computer Networks and Intelligent Computing*, 51–59. Berlin: Springer.
14. Kaggle. 2018. Welcome to Kaggle competitions. https://www.kaggle.com/competitions.
15. KDD Cup of fresh air. 2018. https://biendata.com/competition/kdd_2018/.
16. Keras: The Python deep learning API. https://keras.io/.
17. Khan, Muhammad Salman, Sana Siddiqui, Robert D. McLeod, Ken Ferens, and Witold Kinsner. 2016. Fractal based adaptive boosting algorithm for cognitive detection of computer malware. *15th International Conference on Cognitive Informatics & Cognitive Computing, ICCI*CC*, 50–59. IEEE: Piscataway.
18. Kittler, Josef, Mohamad Hatef, Robert P.W. Duin, and Jiri Matas. 1998. On combining classifiers. *IEEE Transactions on Pattern Analysis and Machine Intelligence* 20 (3): 226–239.
19. Deguang, Kong, and Guanhua Yan. 2013. Discriminant malware distance learning on structural information for automated malware classification. In *Proceedings of the 19th ACM SIGKDD International Conference on Knowledge Discovery and Data Mining*, KDD '13, 1357–1365, ACM, New York.
20. Kuncheva, Ludmila I. 2004. *Combining Pattern Classifiers: Methods and Algorithms*. Wiley, Hoboken, NJ. 2004. https://pdfs.semanticscholar.org/453c/2b407c57d7512fdbe19fa1cefa08dd22614a.pdf.
21. Michailidis, Marios. 2017. *Investigating machine learning methods in recommender systems*. Thesis: University College London.

22. Microsoft malware protection center, winwebsec. https://www.microsoft.com/security/portal/threat/encyclopedia/entry.aspx?Name=Win32
23. Symantec security response, zbot. http://www.symantec.com/security_response/writeup.jsp?docid=2010-011016-3514-99.
24. Morales-Ortega, Salvador, Ponciano Jorge, Abraham Escamilla-Ambrosio, and Rodriguez-Mota, and Lilian D, Coronado-De-Alba. 2016. Native malware detection in smartphones with Android OS using static analysis, feature selection and ensemble classifiers. *11th International Conference on Malicious and Unwanted Software, MALWARE 2016*, 1–8. IEEE: Piscataway.
25. Narouei, Masoud, Mansour Ahmadi, Giorgio Giacinto, Hassan Takabi, and Ashkan Sami. 2015. Dllminer: Structural mining for malware detection. *Security and Communication Networks* 8 (18): 3311–3322.
26. Netflix Prize. 2009. https://www.netflixprize.com.
27. Numpy. https://numpy.org/.
28. Pws:win32/onlinegames. https://www.microsoft.com/en-us/wdsi/threats/malware-encyclopedia-description?Name=PWS%3AWin32%2FOnLineGames.
29. Radare2: Libre and portable reverse engineering framework. https://rada.re/n/.
30. scikit-learn: Machine learning in Python. https://scikit-learn.org/stable/.
31. Raja Khurram Shahzad and Niklas Lavesson. 2013. Comparative analysis of voting schemes for ensemble-based malware detection. *Journal of Wireless Mobile Networks, Ubiquitous Computing, and Dependable Applications* 4 (1): 98–117.
32. Sheen, Shina, R. Anitha, and P. Sirisha. 2013. Malware detection by pruning of parallel ensembles using harmony search. *Pattern Recognition Letters* 34 (14): 1679–1686.
33. Mary Wollstonecraft Shelley. 1869. Frankenstein or The Modern Prometheus. *Dent*.
34. Singh, Tanuvir, Fabio Di Troia, Visaggio Aaron Corrado, Thomas H. Austin, and Mark Stamp. 2016. Support vector machines and malware detection. *Journal of Computer Virology and Hacking Techniques* 12 (4): 203–212.
35. Smolyakov, Vadim. 2017. Ensemble learning to improve machine learning results. https://blog.statsbot.co/ensemble-learning-d1dcd548e936.
36. Smutz, Charles, and Angelos Stavrou. 2012. Malicious pdf detection using metadata and structural features. In *Proceedings of the 28th Annual Computer Security Applications Conference*, ACSAC 2012, 239–248, ACM, New York.
37. Stamp, Mark. 2004. A revealing introduction to hidden Markov models. https://www.cs.sjsu.edu/~stamp/RUA/HMM.pdf.
38. Stamp, Mark. 2017. *Introduction to Machine Learning with Applications in Information Security*. Boca Raton: Chapman and Hall/CRC.
39. TensorFlow: An end-to-end open source machine learning platform. https://www.tensorflow.org/.
40. Toolan, Fergus, and Carthy Joe. 2009. Phishing detection using classifier ensembles. *eCrime Researchers Summit, 2009, eCRIME '09*, 1–9. IEEE: Piscataway.
41. Trojandownloader:win32/adload. https://www.microsoft.com/en-us/wdsi/threats/malware-encyclopedia-description?Name=TrojanDownloader%3AWin32%2FAdload.
42. Trojandownloader:win32/agent. https://www.microsoft.com/en-us/wdsi/threats/malware-encyclopedia-description?Name=TrojanDownloader:Win32/Agent&ThreatID=14992.
43. Trojandownloader:win32/renos. https://www.microsoft.com/en-us/wdsi/threats/malware-encyclopedia-description?Name=TrojanDownloader:Win32/Renos&threatId=16054.
44. Trojandownloader:win32/small. https://www.microsoft.com/en-us/wdsi/threats/malware-encyclopedia-description?Name=TrojanDownloader:Win32/Small&threatId=15508.
45. Trojan:win32/bho. https://www.microsoft.com/en-us/wdsi/threats/malware-encyclopedia-description?Name=Trojan:Win32/BHO&threatId=-2147364778.
46. Trojan:win32/toga. https://www.microsoft.com/en-us/wdsi/threats/malware-encyclopedia-description?Name=Trojan:Win32/Toga&threatId=-2147259798.
47. van Veen, Hendrik Jacob, Le Nguyen The Dat, and Armando Segnini. 2015. Kaggle ensembling guide. https://mlwave.com/kaggle-ensembling-guide/.

48. Virtool:win32/ceeinject. https://www.microsoft.com/en-us/wdsi/threats/malware-encyclopedia-description?Name=VirTool%3AWin32%2FCeeInject.
49. Virtool:win32/injector. https://www.microsoft.com/en-us/wdsi/threats/malware-encyclopedia-description?Name=VirTool:Win32/Injector&threatId=-2147401697.
50. Virtool:win32/vbinject. https://www.microsoft.com/en-us/wdsi/threats/malware-encyclopedia-description?Name=VirTool:Win32/VBInject&threatId=-2147367171.
51. Vobbilisetty, Rohit, Fabio Di Troia, Richard M. Low, Corrado Aaron Visaggio, and Mark Stamp. 2017. Classic cryptanalysis using hidden Markov models. *Cryptologia* 41 (1): 1–28.
52. Win32/allaple. https://www.microsoft.com/en-us/wdsi/threats/malware-encyclopedia-description?Name=Win32/Allaple&threatId=.
53. Win32/fakerean. https://www.microsoft.com/en-us/wdsi/threats/malware-encyclopedia-description?Name=Win32/FakeRean.
54. Win32/rimecud. https://www.microsoft.com/en-us/wdsi/threats/malware-encyclopedia-description?Name=Win32/Rimecud&threatId=.
55. Win32/vobfus. https://www.microsoft.com/en-us/wdsi/threats/malware-encyclopedia-description?Name=Win32/Vobfus&threatId=.
56. Win32/vundo. https://www.microsoft.com/en-us/wdsi/threats/malware-encyclopedia-description?Name=Win32/Vundo&threatId=.
57. XGBoost documentation. https://xgboost.readthedocs.io/en/latest/.
58. Ye, Yanfang, Lifei Chen, Dingding Wang, Tao Li, Qingshan Jiang, and Min Zhao. 2009. Sbmds: An interpretable string based malware detection system using svm ensemble with bagging. *Journal in Computer Virology* 5 (4): 283.
59. Ye, Yanfang, Tao Li, Yong Chen, and Qingshan Jiang. 2010. Automatic malware categorization using cluster ensemble. In *Proceedings of the 16th ACM SIGKDD International Conference on Knowledge Discovery and Data Mining*, KDD '10, 95–104, ACM, New York.
60. Suleiman, Y., Sakir Sezer Yerima, and Igor Muttik. 2015. High accuracy android malware detection using ensemble learning. *IET Information Security* 9 (6): 313–320.
61. Zhang, Boyun, Jianping Yin, Jingbo Hao, Dingxing Zhang, and Wang Shulin. 2007. Malicious codes detection based on ensemble learning. *International Conference on Autonomic and Trusted Computing, ATC 2007*, 468–477. Berlin: Springer.
62. Zhou, Zhi-Hua. 2012. *Ensemble Methods: Foundations and Algorithms*. Boca Raton, FL: CRC Press. http://www2.islab.ntua.gr/attachments/article/86/Ensemble

Malware Analysis

Optimizing Multi-class Classification of Binaries Based on Static Features

Lasse Øverlier

Abstract Classification of binaries is often done with limited resources spent on pre-processing the input, assuming that the resource-intensive machine learning techniques will find the optimal results. In this paper, we identify pre-processing methods to perform faster malware multi-class classification of high accuracy, and we also use the same techniques to classify author (programmer) identification from executables. One method is via eight different types of code simplifications of the disassembled code to reduce storage and calculation time. Another is through visual analysis from running TFIDF N-gram analysis using both Random Forest and SVM, for a large range of different N-grams. The results show interesting features from our classification of executables which we base solely on the analysis of the disassembled code. We have in addition looked at using different training data sizes, compiler optimized code, and both ELF and PE-files and demonstrate methods for optimizing storage and computational complexity when classifying executable files. Our findings show that a higher size N-gram is only preferable for some code simplifications, and that some code simplifications can give a very high accuracy (99.2%) based on only a fraction of the code. In addition, the amount of training data can be quite low and still yield an accuracy of over 95%.

1 Introduction

When we move away from malware detection [10, 23] and binary classification to malware multi-class classification, there are many ways to group the malware binaries.

It is useful to classify the malware with other malware with similar functionality. It is also useful to classify the programmers who wrote the malware assuming that they leave some sort of identifiable signatures through their programming practices. The malware programmer may be anonymous, but it is useful to know which malware

L. Øverlier (✉)
Norwegian Defence Research Establishment, Kjeller, Norway
e-mail: lasse.overlier@ffi.no

M. Stamp et al. (eds.), *Malware Analysis Using Artificial Intelligence and Deep Learning*, https://doi.org/10.1007/978-3-030-62582-5_9

most likely originates from the same programmer or programmer group. These two methods of classification do seem like two completely different challenges since we might assume that any author identifiable texts are removed and code converted when the source codes are compiled into binaries.

So when we have access to the originating source code of the malware, this latter research problem resembles the author identification problem for identifying the origin of, or classification of, anonymous texts [1]. But for processed text, in our case binary software files without access to the source code, this is a different kind of problem. In our case, we have little or no control over how the software's source code has been parsed, analyzed, and optimized by the compiler to produce the binary software file. We may also assume that different types of compilers might leave identifiable markers. And in addition most malware programmers make an effort into being anonymous for obvious reasons, and this may also affect the resulting binary. After compilation, we may also assume that the binary also can be edited to hide potential identifying markers before being used as a malware.

Our original challenge was to start with the binaries and see if and how a classification of the source code may be optimized by using and extracting as little as possible from the binary files. This created two dimensions of experiments. First, we look at how accuracy of correct classification varies with different methods of machine learning techniques and variables. And second, we attempt to minimize the data that must be extracted from the binaries in order to create and maximize classification results and still maintain acceptable accuracy.

In one of our experiments, we examine whether malware classifications into family classes of similar functionality can be done and how the results vary with our optimizations. In the other experiment, we attempt to see whether an author can be identified to belong to a classification of binaries based on the authors (programmers).

This paper is structured as follows. We will in section two look at related work, and in section three look at the theory behind our classification experiment. In section four, we explain the experiment setup, and we continue in section five with the results. The discussion of the results is in chapter six, and in chapter seven we conclude.

2 Related Work and Background

Classifying programmer (group) when we have access to the source code, this research problem resembles the author identification problem with many earlier publications. In [5] deVel et al. used a support vector machines (SVM) classifier to identify authorship of multi-topic e-mails. Others have looked at different uses of N-gram features for author identification using histograms of character-based N-grams [6], and also trying to avoid language dependency and smaller profiles [13]. N-grams can be used on different types of entities, characters, bytes, opcodes, words, etc., depending on what kind of patterns we are looking for.

Using character-based techniques may not be best in all types of analyses and classifications. Houvardas et al. [9] use variable length word-based N-gram features to make author identification with original text as input. They experimented with various length (character) n-grams, *3-gram* up to *5-gram*, and showed that there is a variation in accuracy when comparing different lengths. Language independence was also shown in programmer identification where Peng et al. [18] present graphs for accuracy of the various lengths of N-grams from *1-gram* to *7-gram* for English, Chinese, and Greek.

Classifying file types is another similar problem where earlier N-gram analysis [22] have shown to be promising and produced "fingerprints" for the file types using unigrams ("*1-gram*"). Later research have improved this by using support vector machines and multiple various length N-grams for the same file type classification problem [2]. Most earlier N-gram analysis on binary files are made character-by-character.

Analyzing programmer authorship may involve other things more specific for programmers and code structure than authors' language and structures. Krsul and Spafford [15] looked at the programming style through measurable predetermined parameters in the code, like program line length, variable name length, function names, percentage of global variables, etc. Others have researched similar techniques later [14]. Caliskan-Islam et al. [4] use syntactic structures in the source code to extract selected features and run these features through a random forest classification. Burrows and Tahaghoghi [3] have made similar research using N-gram source code analysis.

This raises the question of which of the different lengths of *n-gram* that will contribute the most in our case where we only base the analysis on the binary files. Even for binary files, there are multiple methods of how the binary code is produced. Some binaries may include debug information and thus may include names and identifiers used in the source code. Most binaries remove this debug information, and some compile with a high degree of optimization which tries to make the most effective binary code from the source code files.

Other types of binary data analysis have been performed on network traffic. Analysis of different N-gram techniques for classifying and detecting binary protocols ("not text-based protocols") like RPC, Samba, and RDP can be found in [7].

Different types of compilers might leave identifiable markers in how they compile or maybe how they optimize the binary executable. In [12], Kalbhor et al. demonstrates successful results by using hidden Markov models for identifying the compiler used.

Spafford and Weeber [21] addressed the challenge of tracking both source code and binary code back to its authors already in 1992. Among other things, they discussed the use of compiler optimization, different source code languages, data structures and system calls, all which are relevant features to use in today's newer analysis methods. Research into identifying programmer(s) from binaries have received an extra boost with the current machine learning tools to simplify large-scale analysis. In Meng [16] and Meng et al. [17], extract features from basic blocks in the binaries and use this to attribute the different basic blocks to authors using support vector machines and random forest classification models. Rosemblum et al. [20] use features from the control flow graph, and Caliskan et al. [4] extracts both instruction features, lexical features, syntactic features, and flow features in their de-anonymization of programmers.

3 Methodology

In author identity analysis there are multiple identification indicators that will vary and may help identify an author, like language, vocabulary, word frequency, term frequency, sentence buildup, order of sentences, style of writing, etc. There are many more indicators which may be specific for any individual author. We will in this paper only use a few of these identifiers, but let us first review some of the additional challenges we have to address when we start with binary code to attempt to classify malware. We use a disassembler to give us a textual representation of the malware, but this contains information like memory addresses which will vary with compiler and operating system. We try to structure this disassembled information as explained in Chap. 4 and we will only look at some of the indicators: words, terms, term lengths, and term frequency. Then we will visualize how the prediction accuracy varies with the length of the terms in different setups.

3.1 *Selecting the Dictionary*

One term in disassembled code can be interpreted to be one of many variations. Figure 1 shows a tiny fraction of one such file from the experiment. The code can be found in the lower right half of the figure.

We may look at each word (or set of characters separated by one or more whitespace characters) as they are found in the code, like "mov," "edi," "[ecx-0DCh]," etc.,

```
.text:00401000 00 00 00 00        dword_401000    dd 0                ; DATA XREF: HEADER:0040006C^Xo
.text:00401000                                                        ; HEADER:00400144^Xo
.text:00401004
.text:00401004                    ; =============== S U B R O U T I N E =======================================
.text:00401004
.text:00401004                    ; Attributes: bp-based frame
.text:00401004
.text:00401004                    sub_401004      proc near           ; CODE XREF: sub_4010D9+E^Yp
.text:00401004 8B FF                              mov     edi, edi
.text:00401006 55                                 push    ebp
.text:00401007 8B EC                              mov     ebp, esp
.text:00401009 83 EC 1C                           sub     esp, 1Ch
.text:0040100C 8D 8A 40 01 00 00                  lea     ecx, [edx+140h]
.text:00401012 03 B9 24 FF FF FF                  add     edi, [ecx-0DCh]
.text:00401018 51                                 push    ecx
.text:00401019 50                                 push    eax
.text:0040101A 56                                 push    esi
.text:0040101B 52                                 push    edx
t(DOS)---  A75mw4S9QdqEua2sDL1n.asm    1% L44    (Assembler)
```

Fig. 1 Example of code from one of the data sets [19]

or we may look at one full line of code at a time, like "push ebp," "mov ebp, esp," etc., to be one single term in the dictionary.

Many earlier machine learning experiments have fed the binary files with limited pre-processing into the classifiers, often using character-based analysis, and including all the disassembled output, like addresses, comments, opcodes (machine code values) and have had very successful results [8, 11].

Since we wanted to build our analysis solely on the disassembled code and different variations of this code as found from the binary malware files, we have extracted the code without all the extra information shown in Fig. 1. The code lines with the disassembly code shown in the lower right part of the figure are everything we base our analysis on. This way we ignored the addresses, comments, and opcodes/machine code, and only used our parsing, pre-processing, and analysis on the textual representation of the code part.

The difference between using word frequency and term frequency can easily be observed in the figure as the word count for "push" is five, but if we use the term "push <REGISTER>" we see that there is only one occurrence of each term. We can also assume that a longer more inclusive term will give a lower occurrence rate/count of each term and a larger dictionary, compared to single word "terms." More complex terms will usually give a lower term count and a larger dictionary. To make the text easier to parse we combined words into selected "terms" (chosen by us) through removing spaces to build new words in the text to be analyzed.

We wanted to analyze the difference between the terms included in the dictionary to see whether we could make good predictions in classifying malware by using only a fraction of the information given from the disassembler and binary analysis tools. We have run our analysis on the following different simplifications of disassembled code:

1. Single letter—only the first letter from the first word (first letter of mnemonic).
2. First word—only the first word of each code line (the mnemonic)
3. First two words connected—as (4), but as one word in the dictionary ("2 words connected")
4. First two words—often mnemonic and first parameter, as separate words in the dictionary ("2 words")
5. All words connected—as (6), but as one word in the dictionary for each code line
6. All words—removed number values, type information, and special characters, except colon(':'), and kept all words—often mnemonic with all registers
7. All code connected—as (8), but as one word in the dictionary for each code line
8. All code—the complete code line with the whitespace output from the disassembler as the word separator. Meaning that commas, brackets, etc., will be present in the words in the dictionary.

An example of this code simplification is shown in Table 1. Here we have used the disassembled code from Fig. 1 to exemplify how the different dictionaries will become. The input disassembled code is shown in the "All code" column. Note that when using large *N-gram* we will include words from multiple lines, as a line separator will function as a normal whitespace.

Table 1 Code simplification in the different analysis formats

Single letter	First word	2 words connected	2 words	All words connected	All words	All code connected	All code
m	mov	movedi	mov edi	movediedi	mov edi edi	movedi,edi	mov edi, edi
p	push	pushebp	push ebp	pushebp	push ebp	pushebp	push ebp
m	mov	movebp	mov ebp	movebpesp	mov ebp esp	movebp,esp	mov ebp, esp
s	sub	subesp	sub esp	subesp	sub esp	subesp,1Ch	sub esp, 1Ch
l	lea	leaecx	lea ecx	leaecxedx	lea ecx edx	leaecx,[edx+140h]	lea ecx, [edx+140h]
a	add	addedi	add edi	addediecx	add edi ecx	addedi,[ecx-0DCh]	add edi, [ecx-0DCh]
p	push	pushecx	push ecx	pushecx	push ecx	pushecx	push ecx
p	push	pusheax	push eax	pusheax	push eax	pusheax	push eax

Table 2 shows the first 80 characters of the resulting documents representing the simplification from one file in the experiment. The full code can be seen/extracted when combining "All code" and "All code connected." All these different types of combining the disassembled code gives completely different dictionaries when run through a word-based term frequency-inverse document frequency (TFIDF) counter.[1]

So the list of words in the dictionaries based on the code from Table 1 will look something like this

1-gram, Single letter : ("m", "p", "s", "l", "a")
2-gram, Single letter : ("m p", "p m", "m s", "s l", "l a", "a p", "p p")
1-gram, 2 words : ("mov", "edi", "push", "ebp", "sub", "esp", "lea", "ecx", "add", "eax")
2-gram, 2 words : ("mov edi", "edi push", "push ebp", "ebp mov", "mov ebp", "ebp sub", "sub esp", "esp lea", ...)
3-gram, 2 words : ("mov edi push", "edi push ebp", "push ebp mov", "ebp mov ebp", "mov ebp sub", "ebp sub esp", ...)
1-gram, All code connected : ("movedi,edi", "pushebp", "movebp,esp", "subesp, 1Ch", "leaecx,[edx+140h]", ...)
2-gram, All code connected : ("movedi,edi pushebp", "pushebp movebp,esp", "movebp,esp subesp,1Ch", "subesp,1Ch leaecx,[edx+140h]", "leaecx,[edx+ 140h]", ...)
1-gram, All code : ("mov", "edi,", "edi", "push", "ebp", "ebp,", "esp", "sub", "esp,", "1Ch", "lea", "ecx,", "[edx+140h]", ...)

These are just a few examples to understand how our simplification of code is being built into "words" that will appear in the dictionaries and be used in the analyses.

[1]We used the SKLearn (https://scikit-learn.org/) TFIDF vectorizer.

Table 2 First 80 characters in one of the disassembled files in all simplification variations

Simplification	First 80 characters of "Document"
Single letter	j d i a a a a a a a a a a x a x a j a a a x a o s p a p a a a x a x a a a a a a
First word	jg dec inc add add add add add add add add add add xor add xor add je add add ad
First two words connected	jg decesp incesi addecx addeax addeax addeax addeax addeax addeax addeax addeax
First two words	jg dec esp inc esi add ecx add eax add eax add eax add eax add eax add eax add e
All words connected	jg decesp incesi addecxeax addeaxeax addeaxal addeaxal addeaxal addeaxal addeaxe
All words	jg dec esp inc esi add ecx eax add eax eax add eax al add eax al add eax al add
All code connected	jg0x47 decesp incesi adddword[ecx],eax adddword[eax],eax addbyte[eax],al addbyte
All code	jg 0x47 dec esp inc esi add dword [ecx], eax add dword [eax], eax add byte [eax]

Some special situations to be aware of

1. In "2 words," "All words," and "All code" each word in every code line separated by a whitespace (or new line) will be a word for the dictionary. This especially affects the large *N-gram* experiments with regards to dictionary size and memory usage.
2. In "All code," there will be special characters in the different words, like "ebp," (with a comma included) which makes the dictionary add this entry in addition to "ebp."
3. When using multiple words, and the *-connected words the resulting dictionary will grow extremely large and this have caused problems with the largest data set as it required up toward 1TB of memory when using standard ML-libraries for higher *N-gram* values. We have, therefore, reduced the maximum N-grams to 8-gram for the experiment using the biggest data set.

4 Experiments

For learning how to perform multi-class classification, there exists a few data sets available online, but not many (public) large ones. We have chosen two quite different data sets. One set from the Microsoft Malware Classification Challenge (MMCC) and one set extracted from the Google Code Jam (GCJ) competition. These two data sets are different in many ways. In MMCC, we have a classification of binaries which is made into malware families which consists of many variants of same functionality.

In GCJ, we have classified, by the programmer, and have different programs written by the same author with highly different functionality and purpose.

The similarities of the experiments is that we try to make the same type of dictionary variants as explained in Sect. 3, and plot the accuracy of support vector machines and random forest classification for all N-gram values from "*1-gram*" up to either "*8-gram*" or "*18-gram*" based on the data size.

But first we will explain the special circumstances for running our experiment on these two quite different type of data sets.

4.1 Microsoft Malware Classification Challenge

One of the largest public available data sets with malware can be found in the Microsoft Malware Classification Challenge [19][2] (MMCC). It consists of over 400 GB of data, with both binary and disassembled code from the use of the *IDA disassembler and debugger*.[3] The binary malware has been stripped of the PE-header to be made non-executable for security reasons. This does limit the value of the data set, but they have prioritized the potential security implications with hundreds of gigabytes of executable malware available to anyone. The MMCC data set consists of around 200 GB of training data, where the classification is known, and another 200 GB of test data to be classified as a part of the competition evaluation. The real classification of this test data set is still unknown at the time of our experiment—four years after the competition was completed. We, therefore, used only the training data for both training and testing our results.

The data classification is given on all the training data—10868 of them in nine different classes: Ramnit, Lollipop, Kelihos_ver3, Vundo, Simda, Tracur, Kelihos_ver1, Obfuscator.ACY, and Gatak. These classes were predefined and found in a separate file classifying each malware file. We wanted to see if the results varied with the amount of data used from each code line. Therefore, we made eight experiments on this data set where each line of disassembled code was converted into documents of the same simplification classes shown in Table 1, from single letter, first word, etc. to "all code."

There were 10859 files useful for classification based on disassembled code. Nine files were ignored simply because they only consisted of data without any code, and with the PE-header removed, these were so different from the other files, and therefore not used in the analysis.

We performed some simplifications before applying this conversion. We found that some disassembly instructions were over-represented in many files and these lines have been removed. "align" and especially the data indicator lines "dd" which is just listing unstructured (unrecognized) data in the binary. These data blocks are

[2]https://www.kaggle.com/c/malware-classification.

[3]https://www.hex-rays.com/products/ida/.

Table 3 MMCC code size in gigabytes (GB) after simplification in the different analysis formats, uncompressed (top) and Zlib-compressed (bottom)

Single letter	First word	2 words	2 words connected	All words	All words connected	All code	All code connected	Raw input
0.52	0.97	4.3	4.5	4.9	5.6	6.4	7.6	140
0.039	0.061	1.3	1.3	1.4	1.4	1.9	1.9	17

often encrypted, not very informative until the malware is executed, and therefore not used in our static analysis.

With our simplifications, the code base was reduced from the around 200 GB (binary and compressed disassembled code) down to 500 MB for "single letter" data and even down to 39MB for the compressed version. Sizes of the different code versions are shown in Table 3 with all sizes in gigabytes. So if we can get acceptable results with "single letter" or "first word," we can reduce storage needed for the classification from 140 GB to 39 MB/61 MB.

4.2 Google Code Jam (GCJ) Data

Another set of software was taken from the Google Code jam library which has a publicly available source code repository at https://www.go-hero.net/jam/ where the source code files from every competition can be downloaded (if scripting). The programs entered here are answers to programming challenges and authors submit their solutions into this open competition. All solutions are sorted by challenge, year, programming language, and author. The source files submitted are made in multiple programming languages, but we have only extracted those that used C/C++ for our experiment. These source code files had to be compiled to Linux ELF binaries for our experiment setup.

As with all machine learning algorithms, you need a minimum amount of training and testing data to have significance and see if identifying patterns can be discovered. There were many contributors with just a few entries of programs, so we start by using the top ten code writers (based on a number of contributions only). This gave us approximately the same number of classes as in the MMCC data set, but the number of files in total went down from 10800 in MMCC to 413 in GCJ. This makes GCJ a quite different classification challenge as we now only have 37 samples from the "least productive" programmer in our set up to 49 files from the "most productive" programmer as the content in this multi-class classification.

Table 4 GCJ code size in megabytes (MB) after simplification in the different analysis formats, uncompressed (top) and Zlib-compressed (bottom)

Single letter	First word	2 words	2 words connected	All words	All words connected	All code	All code connected	Raw input
2.1	4.6	7.1	8.0	8.3	9.8	14	16	18
0.33	0.48	0.72	0.74	0.86	0.92	1.8	1.9	2.7

We used a script made by F. Seehusen at FFI[4] to download the GCJ source code from the 2010–2017 competitions. The code was downloaded and sorted by username and this script also included how to produce both a 64-bit version with the optimization flag set ("g++ -w -m64 -O"), and a 32-bit version without any optimization ("g++ -w -m32 -O0") from each of the source codes. Compile time of these small programs were in the order of seconds.

So why did we not just use the source code? Because that would be a similar problem as described by others under Chap. 3. There have been many classification experiments on original source code(s). So even if we had the source code, our experiment was to see if we could classify disassembled code correctly. First the disassembled code was extracted using Radare2[5] and all the functions in the executable (as defined by Radare) were extracted together with their corresponding disassembled code. As with MMCC, we only used the code parts in our analysis. We do assume that data areas may contain useful information that will enhance the analysis and classification, but this experiment was narrowed down to only look at the code sections.

We converted the GCJ-code into the eight different code simplifications and the sizes of the analysis material and the resulting eight versions are shown in Table 4.

5 Results

Here, we will first describe the results from the MMCC malware experiments and then from the Google Code Jam data set. All accuracy values are made from running the classification with 80% of the files in training and 20% in the prediction set. Every test was run 10 times and the average of these runs is shown in the graphs. The average accuracy for correct classifications are shown in Fig. 2 as given from the *scikit-learn→metrics→ accuracy_score* library.

---

[4]Norwegian Defence Research Establishment.

[5]Radare2 can be found at https://rada.re/ and is one of the most used open-source disassembly tools today.

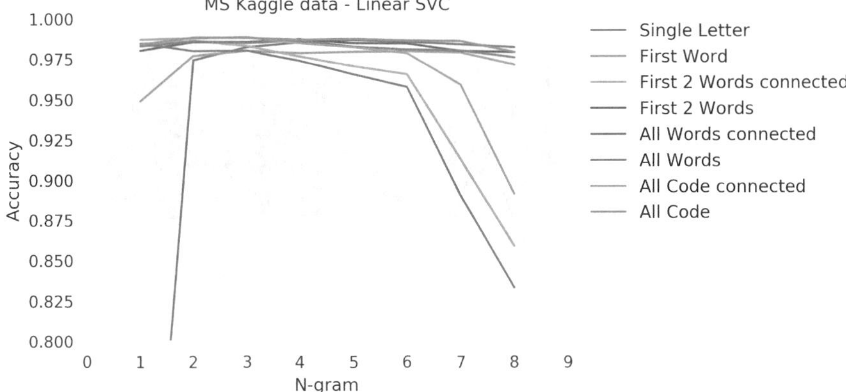

Fig. 2 Classification using SVM on MMCC malware

5.1 MMCC Malware

5.1.1 Using SVM

Using Support Vector Machines (SVM), we run the experiment 10 times for each n-gram from 1 to 8, using 80% random selected files for training, and the remaining 20% for testing.

To get some more details on the variation of the data we can shorten the scale on the y-axis as shown in Fig. 3.

We can see from the results that single letter have the least accurate result starting at 1-gram, but stay within top-5 and better than 98% accuracy from 3-gram and

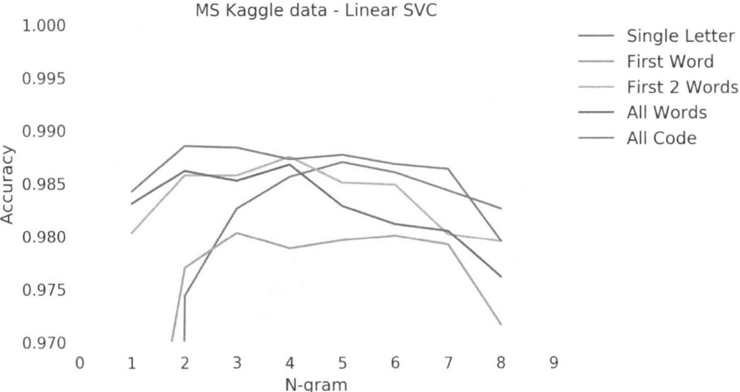

Fig. 3 Classification using SVM on MMCC malware—shortened y-axis

Table 5 Heatmap for best and worst accuracy using SVM

Confusion Matrix for All Code, 2-gram, LinearSVC

Actual \ Predicted	Ramnit	Lollipop	Kelihos_ver3	Vundo	Simda	Tracur	Kelihos_ver1	Obfuscator.ACY	Gatak
Ramnit	2037	0	0	4	10	0	0	1	0
Lollipop	0	758	0	3	0	5	0	0	0
Kelihos_ver3	0	10	5818	0	0	0	0	0	0
Vundo	0	0	0	5014	3	9	0	5	0
Simda	5	16	2	6	2306	75	1	7	5
Tracur	0	0	2	2	14	3083	0	0	2
Kelihos_ver1	1	5	0	1	4	5	64	0	2
Obfuscator.ACY	3	0	0	7	7	6	0	1448	0
Gatak	2	2	6	0	5	4	0	2	943

Confusion Matrix for Single Letter, 1-gram, LinearSVC

Actual \ Predicted	Ramnit	Lollipop	Kelihos_ver3	Vundo	Simda	Tracur	Kelihos_ver1	Obfuscator.ACY	Gatak
Ramnit	0	0	1672	97	170	38	0	0	2
Lollipop	0	80	629	22	66	3	0	0	6
Kelihos_ver3	0	0	5993	23	0	12	0	0	4
Vundo	0	0	1402	3474	3	49	0	0	9
Simda	0	5	273	449	1336	206	0	0	139
Tracur	0	1	766	959	56	1223	0	0	35
Kelihos_ver1	0	6	43	4	33	2	0	0	0
Obfuscator.ACY	0	0	607	299	476	42	0	0	54
Gatak	0	0	173	468	57	15	0	0	239

larger. Another interesting result is that all the manually connected dictionaries have their accuracy start high, but drops off significantly after 6-gram.

If we look at the accuracy more in detail in Fig. 3, we can observe that the "All code" keeps a slight advantage almost for all N-grams. But we also can see that the accuracy "Single letter" is actually (nearly) as good as the maximum accuracy value at 5-gram.

To visualize better and see which binary classes are easier to predict than others we can print a heatmap for the scenario of the best and another lower accuracy when using Linear SVC. We can see from Fig. 3 that the assumed worst accuracy is the "Single letter," *1-gram* result, and that the best accuracy (by a small margin) is the "All code," *2-gram* result. These heatmaps are shown in Table 5 where we can see that under the *1-gram/single letter*, some features seem to be over-represented to be predicted, like "Kelihos_ver3," but others are not occuring at all, like "Kelihos_ver1," "Ramnit," and "Obfuscator_ACY." But this improves a lot when using all code were close to everything that is predicted to be "Kelihos_ver1" actually is correct, and all other categories have a very high accuracy as well.

Looking at the results it seems like *2 words*, *all words* and *all code* are consistently better than 98% accuracy. And we can also see that *first word* and *single letter* reaches the 98% accuracy level already around *3-gram*.

5.1.2 Using Random Forest

If we use the same data in a random forest classification we get the results shown in Fig. 4. In order to see any differences between the top five code types we reduce the scale on the accuracy axis to observe any patterns. This is shown in Fig. 5.

As we can see every run for the top five code types gives an accuracy better than 98.5% from 2-gram and higher, and that a 99% accuracy is achieved already at 1-gram of "2 words," "all words," and "all code" and the need for building much higher n-grams seem unnecessary.

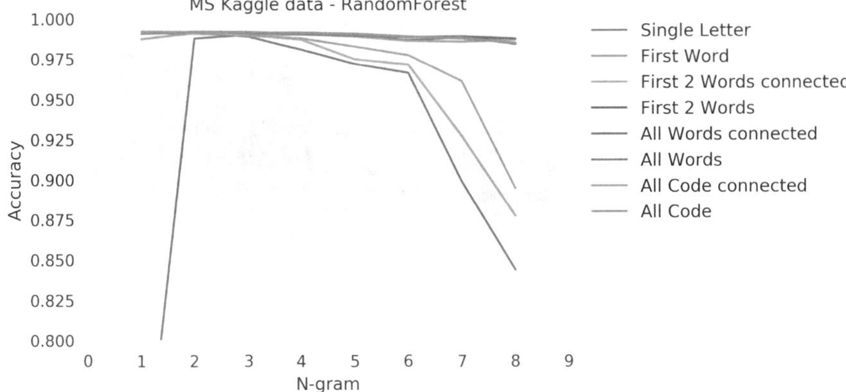

Fig. 4 Classification using Random Forest on MMCC malware

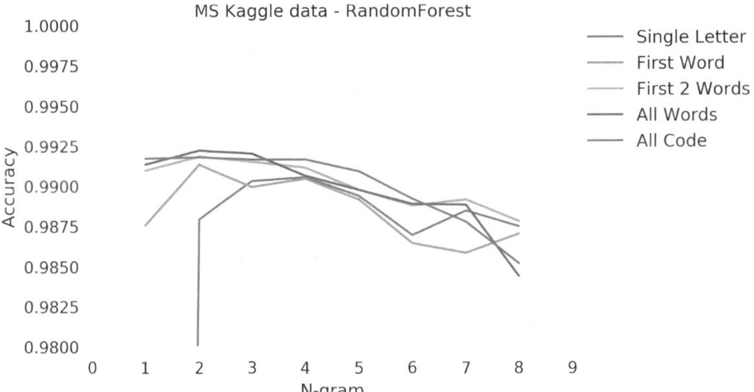

Fig. 5 Classification using Random Forest on MMCC malware—re-scaled y-axis

Best accuracy was found with *All words, 2-gram* at 99.2% accuracy with only one parameter taken into account. As expected the worst accuracy was found in *single letter, 1-gram*.

Table 6 shows the heatmap for best and worst accuracy for random forest, even if the differences and/or challenges are not easily identifiable from the heatmap. The hardest category to detect seems here to be "Simda" for our best result as these are more often than other classes wrongfully predicted to be either "Tracur" or "Obfuscator.ACY."

Table 6 Heatmap for best and worst accuracy using Random Forest

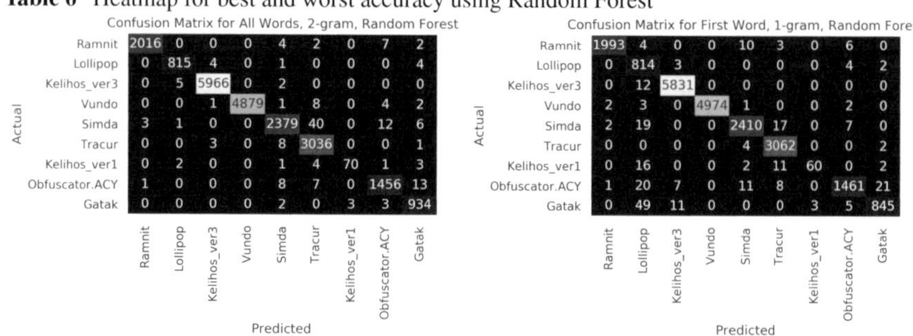

Confusion Matrix for All Words, 2-gram, Random Forest
Actual
Ramnit 2016 0 0 0 4 2 0 7 2
Lollipop 0 815 4 0 1 0 0 0 4
Kelihos_ver3 0 5 5966 0 2 0 0 0 0
Vundo 0 0 1 4879 1 8 0 4 2
Simda 3 1 0 0 2379 40 0 12 6
Tracur 0 0 3 0 8 3036 0 0 1
Kelihos_ver1 0 2 0 0 1 4 70 1 3
Obfuscator.ACY 1 0 0 0 8 7 0 1456 13
Gatak 0 0 0 0 2 0 3 3 934
Ramnit Lollipop Kelihos_ver3 Vundo Simda Tracur Kelihos_ver1 Obfuscator.ACY Gatak
Predicted
Confusion Matrix for First Word, 1-gram, Random Forest
Actual
Ramnit 1993 4 0 0 10 3 0 6 0
Lollipop 0 814 3 0 0 0 0 4 2
Kelihos_ver3 0 12 5831 0 0 0 0 0 0
Vundo 2 3 0 4974 1 0 0 2 0
Simda 2 19 0 0 2410 17 0 7 0
Tracur 0 0 0 0 4 3062 0 0 2
Kelihos_ver1 0 16 0 0 2 11 60 0 2
Obfuscator.ACY 1 20 7 0 11 8 0 1461 21
Gatak 0 49 11 0 0 0 3 5 845
Ramnit Lollipop Kelihos_ver3 Vundo Simda Tracur Kelihos_ver1 Obfuscator.ACY Gatak
Predicted

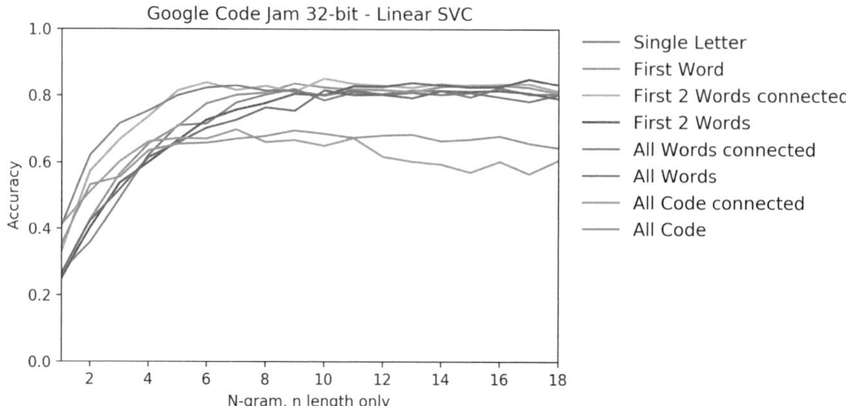

Fig. 6 Classification using Linear SVC on GCJ 32-bit binaries

5.2 *Google Code Jam (GCJ) Results*

In the Google Code Jam (GCJ) data set, we look at two different binary versions: Linux 32-bit and Linux 64-bit. In addition, we look at the same eight different code parsing methods.

5.2.1 Google Code Jam—32-Bit

For the GCJ 32-bit using Linear SVC, shown in Fig. 6, we see that the results are lower than for the MMCC experiments. The maximum accuracy can be found at 5-gram and higher and in the first six code categories. We see here that both "All code" and "All code connected" never reaches more than around 60% accuracy.

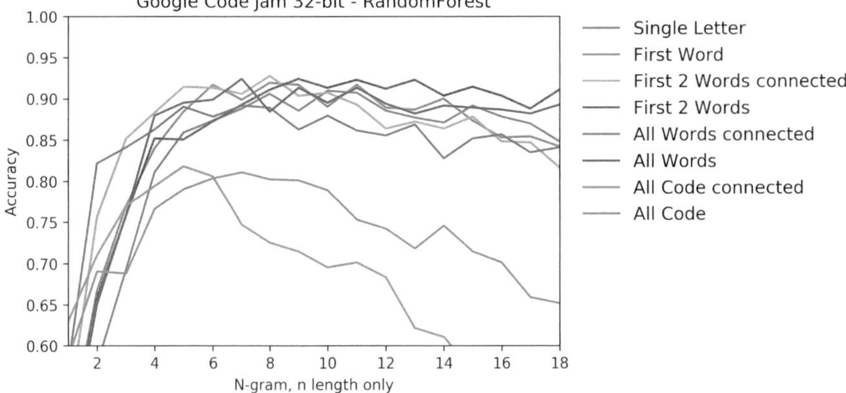

Fig. 7 Classification using Random Forest on GCJ 32-bit binaries, 60–100% accuracy

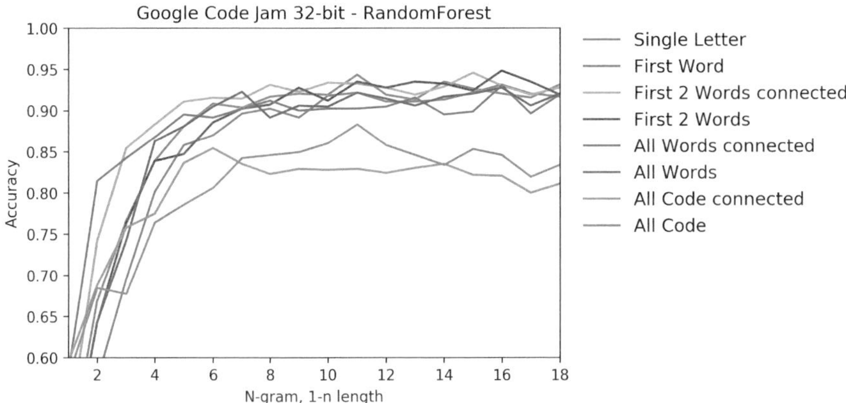

Fig. 8 Classification using Random Forest on GCJ 32-bit binaries, 1-gram through n-gram included, 60–100% accuracy

Using random forest classification, we get the results of Fig. 7. The scale of the figure is reduced and starts at 60% accuracy. We see that here the accuracy of the top six code types are around 90% from 5-gram and higher.

But Figs. 6 and 7 are only showing the results when the specific N-gram is included in the classification. Since we have a significantly smaller amount of code in this experiment compared to in the MMCC-experiment, we can try to build the dictionary from all combinations from 1-gram to 18-gram where all lower n-grams are included. Meaning that the 3-gram result will include 1-grams, 2-grams, and 3-grams. This result is shown in Fig. 8.

We can see that the only major difference is that the drop off disappears and every code class seems to more or less stabilize at around a maximum accuracy of 90% for the top six, and around 80% at the two "all code" types.

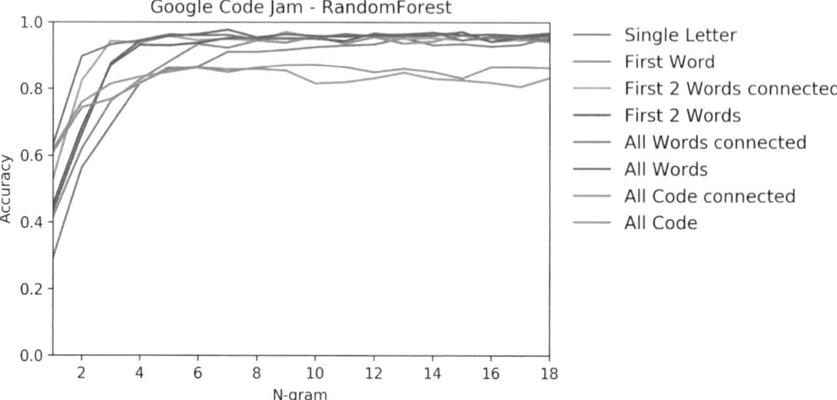

Fig. 9 Classification using Random Forest on GCJ 64-bit binaries

5.2.2 Google Code Jam—64-Bit

The 64-bit versions of the GCJ codes were compiled with the optimization flag set and including all lower n-grams when building classification vectors as explained above. Figure 9 shows the Random Forest classification of the 64-bit codes.

The top six code types for accuracy is enhanced in Fig. 10 with the scale showing 80–100% accuracy. We see that most of these six code types reach 95% accuracy for higher N-grams.

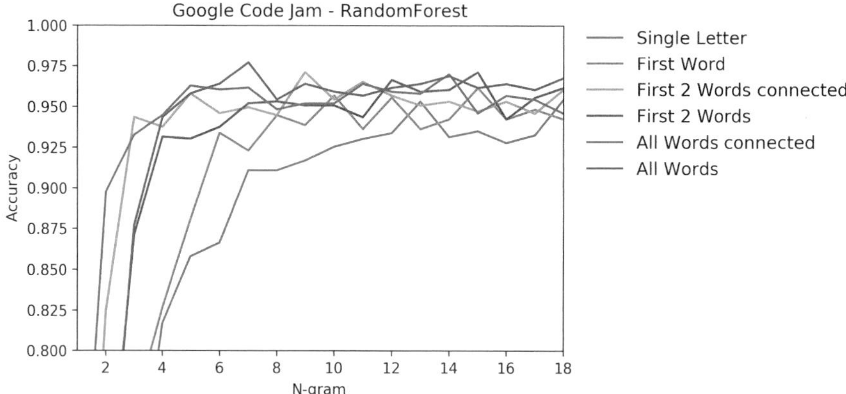

Fig. 10 Classification using Random Forest on GCJ 64-bit binaries, 80–100% accuracy, top six

6 Discussion

We have analyzed the classification of binaries through disassembly in several different dimensions. The paper set out to use only two variants: length of N-grams used in classification and different types of simplification of disassembled code for storage reduction. Through the experiment phase other dimensions were added as they were found to be interesting for optimizing the classification algorithms: the size of training material, 64-bit optimized compilation vs. 32-bit no optimization, classification algorithm, and accumulation of n-gram lengths. We will discuss each of these points here.

6.1 Length of N-Gram

The length of the N-grams is essential with regards to accuracy. Increase in length also means a significant increase in dictionary size, unless we make a strict cutoff in the TFIDF construction. We had problems with the implementation in our machine learning toolkit, sklearn, when both the training size and the dictionary size were large. With large N-grams the dictionary could be up millions of entries if we did not enforce cutoff. We have not had time to experiment with best cutoff values, but after some initial testing we chose 5% for the low boundary, and 75% for the high boundary. After this the dictionary was in most cases below 10000, and we were able to run the experiment on 1-gram–8-grams for the MMCC data set within reasonable time (< 4h for each setting) and keep sklearn below our 192 GB of RAM.

For the GCJ data set, we only had around 440 files, and therefore we could include all lengths up to 18-gram, including the accumulation of lower N-grams. Without accumulation we had some of the code simplifications demonstrating a "drop off." These were the "*-connected" simplifications where the dictionary usually become very large in higher N-grams. For accumulated results this "drop off" is likely the reason for the lower horizontal accuracy graph. We could also observe that the dictionary was about twice as big for the accumulated n-grams as for the single n-grams.

6.2 Simplification of Code

Looking at the accuracy graphs for each of the different code simplifications we observe that almost all variants occur in the highest levels. If this is to be used as an indication only with >98% accuracy, we would suggest just using "single letter/3-gram," "first word/2-gram" or "2words/1-gram." All these results will yield a significant indication upon the classification, and be both fast and demand less resources.

We also observe that in five of the code simplifications, the accuracy in the MMCC data analysis is kept above 98% in all N-gram lengths (ref Fig. 5).

6.3 Size of Training Data

As expected we have a higher accuracy when we have a high amount of training material (MMCC data, 99.2% accuracy) compared to a small training data size (GCJ data, 95% accuracy). Classifying a programmer based on only 440 files of training data with an accuracy of 95% might be a good indication and good enough for many analysis scenarios.

As we can see from Tables 3 and 4 we may save significant amounts of storage if we are comfortable with the "single letter" accuracy. The reduction is around 10:1 and almost as good with.

6.4 64-Bit Optimized Binaries Versus 32-Bit Non-optimized Binaries

One surprising result came with the analysis of optimized vs. non-optimized binaries. From the graphs in Figs. 10 and 8 we observe that the accuracy of the GCJ classification was better than 95% from the optimized binaries and around 92% from the 32-bin plain compilation. We also note that the difference between 64-bit and 32-bit code may play a part here, but we did not have time to include analysis of this variable as well.

6.5 Classification Algorithm

We see that the LinearSVC classification algorithm is almost identical in both the "single n-gram analysis" compared to the "accumulated n-gram analysis." But comparing the graphs for Random Forest classification, they show a clear "drop off" in increasing N-gram values for the largest dictionaries when using "single n-gram". This is compensated by including the use of lower length n-grams in "accumulated n-gram."

LinearSVC also demonstrates a lower accuracy in all classification results, but it was significantly faster (5-10x) in large dictionaries and slower (1-2x) in small dictionaries.

7 Conclusion

We have observed interesting features in multi-class classification of binaries based on analyzing only the disassembled code. We have looked at various methods of constructing simplification of the code, using different length N-grams for analysis, different classification algorithms, and different training data sizes.

Our findings show that a higher size N-gram is only preferable for some code simplifications, and that some code simplifications can give a very high accuracy (99.2%) based on only a fraction of the code. In addition, we found that the amount of training data storage can be reduced by over 97% from input to compressed first word, with corresponding reduced computing resources for *1-grams* in dictionary size, and still yield an accuracy of over 95%.

In future work, we hope to extend to other platforms and other types of malware and maybe be able to include more programmer specific code. In addition ROC-curves will be analyzed to optimize how to best visualize elements from a multi-class classification.

References

1. Argamon, Shlomo, Moshe Koppel, James W Pennebaker, and Jonathan Schler. 2009. Automatically profiling the author of an anonymous text. *Communications of the ACM*, 52(2): 119–123.
2. Beebe, N.L., L.A. Maddox, L. Liu, and M. Sun. 2013. Sceadan: Using concatenated n-gram vectors for improved file and data type classification. *IEEE Transactions on Information Forensics and Security* 8 (9): 1519–1530.
3. Burrows, Steven, and Seyed MM Tahaghoghi. 2007. Source code authorship attribution using n-grams. In *Proceedings of the twelth Australasian document computing symposium, Melbourne, Australia, RMIT University*, 32–39, Citeseer.
4. Caliskan-Islam, Aylin, Harang Richard, Liu Andrew, Narayanan Arvind, Voss Clare, Yamaguchi Fabian, and Greenstadt Rachel. 2015. De-anonymizing programmers via code stylometry. In *24th USENIX security symposium (USENIX Security 15)*, 255–270, USENIX Association, Washington, D.C.
5. de Vel, O., A. Anderson, M. Corney, and G. Mohay. 2001. Mining e-mail content for author identification forensics. *SIGMOD Record* 30 (4): 55–64.
6. Escalante, Hugo Jair, Thamar Solorio, and Manuel Montes-y Gómez. 2011. Local histograms of character n-grams for authorship attribution. In *Proceedings of the 49th annual meeting of the association for computational linguistics: human language technologies - volume 1*, HLT '11, USA, 288–298, Association for Computational Linguistics, Stroudsburg.
7. Hadžiosmanović, Dina, Lorenzo Simionato, Damiano Bolzoni, Emmanuele Zambon, and Sandro Etalle. 2012. N-gram against the machine: On the feasibility of the n-gram network analysis for binary protocols. In *Research in Attacks, Intrusions, and Defenses*, ed. Davide Balzarotti, Salvatore J. Stolfo, and Marco Cova, 354–373. Berlin, Heidelberg: Springer.
8. Hardy, William, Lingwei Chen, Shifu Hou, Yanfang Ye, and Xin Li. 2016. Dl4md: A deep learning framework for intelligent malware detection. In *Proceedings of the international conference on data mining (DMIN)*, 61. The Steering Committee of The World Congress in Computer Science, Computer.
9. Houvardas, John, and Efstathios Stamatatos. 2006. N-gram feature selection for authorship identification. In *Proceedings of the 12th international conference on artificial intelligence: methodology, systems, and applications*, AIMSA'06, 77–86. Berlin, Heidelberg: Springer.

10. Idika, Nwokedi, and Aditya P Mathur. A survey of malware detection techniques. 2007. *Purdue University*, 48: 2007–2.
11. Kalash, M., M. Rochan, N. Mohammed, N.D.B. Bruce, Y. Wang, and F. Iqbal. 2018. Malware classification with deep convolutional neural networks. In *2018 9th IFIP international conference on new technologies, mobility and security (NTMS)*, 1–5.
12. Kalbhor, Ashwin, Thomas H. Austin, Eric Filiol, Sébastien Josse, and Mark Stamp. 2015. Dueling hidden markov models for virus analysis. *Journal of Computer Virology and Hacking Techniques* 11 (2): 103–118.
13. Keselj, Vlado, Peng Fuchun, Cercone Nick, and Thomas Calvin. 2003. N-gram-based author profiles for authorship attribution.
14. Kothari, Jay, Shevertalov Maxim, Stehle Edward, and Mancoridis Spiros. 2007. A probabilistic approach to source code authorship identification. In *Fourth international conference on information technology (ITNG'07)*, 243–248. Piscataway: IEEE.
15. Krsul, Ivan, and Eugene H. Spafford. 1996. Authorship analysis: Identifying the author of a program. Technical report, Computers and Security.
16. Meng, Xiaozhu. 2016. Fine-grained binary code authorship identification. In *Proceedings of the 2016 24th ACM SIGSOFT international symposium on foundations of software engineering*, FSE 2016, 1097–1099, Association for Computing Machinery, New York, NY, USA.
17. Meng, Xiaozhu, Barton P. Miller, and Kwang-Sung Jun. 2017. Identifying multiple authors in a binary program. In *Computer Security – ESORICS 2017*, 286–304, ed. Simon N. Foley, Dieter Gollmann, and Einar Snekkenes. Cham: Springer International Publishing.
18. Peng, Fuchun, Dale Schuurmans, Shaojun Wang, and Vlado Keselj. 2003. Language independent authorship attribution using character level language models. In *Proceedings of the Tenth Conference on European Chapter of the Association for Computational Linguistics - Volume 1*, EACL '03, 267–274, USA, Association for Computational Linguistics, New York, NY.
19. Ronen, Royi, Marian Radu, Corina Feuerstein, Elad Yom-Tov, and Mansour Ahmadi. 2018. Microsoft malware classification challenge. *CoRR*, arXiv:abs/1802.10135.
20. Rosenblum, Nathan, Xiaojin Zhu, and Barton P. Miller. 2011. Who wrote this code? Identifying the authors of program binaries. In *Computer Security – ESORICS 2011*, ed. Vijay Atluri, and Claudia Diaz, 172–189. Berlin, Heidelberg: Springer.
21. Spafford, Eugene H., and Stephen A. Weeber. 1993. Software forensics: Can we track code to its authors? *Computers and Security* 12 (6): 585–595.
22. Li, Wei-Jen, Ke Wang, S.J. Stolfo, and B. Herzog. Fileprints: Identifying file types by n-gram analysis. 2005. In *Proceedings from the sixth annual IEEE SMC information assurance workshop*, 64–71.
23. Ye, Yanfang, Tao Li, Donald Adjeroh, and S. Sitharama Iyengar. 2017. A survey on malware detection using data mining techniques. *ACM Computing Surveys*, 50(3): 3073559.

Deep Learning Techniques for Behavioral Malware Analysis in Cloud IaaS

Andrew McDole, Maanak Gupta, Mahmoud Abdelsalam, Sudip Mittal, and Mamoun Alazab

Abstract This chapter focuses on online malware detection techniques in cloud IaaS using machine learning and discusses comparative analysis on the performance metrics of various deep learning models.

1 Introduction and Motivation

Cloud has become a popular platform due to its characteristics of on-demand services, infinite resources, ubiquitous availability and pay-as-you-go business model [22]. Infrastructure as a Service (IaaS) is a popular service model for many data centers. In IaaS, resources of a data center may be purchased by clients to perform their own personal tasks but require either the strength or availability of the data centers resources. This model allows customers to save money by removing the need for every customer to set up their own computers or computing cluster and it allows data centers to efficiently utilize their computing resources. Clients can purchase access to any number of virtual machines, which can include a few machines or thousands of virtual machines. With the scale of resource usage, there must be

A. McDole · M. Gupta (✉)
Tennessee Technological University, Cookeville, TN, USA
e-mail: mgupta@tntech.edu

A. McDole
e-mail: amcdole42@students.tntech.edu

M. Abdelsalam
Manhattan College, Riverdale, NY, USA
e-mail: mabdelsalam01@manhattan.edu

S. Mittal
University of North Carolina at Wilmington, Wilmington, NC, USA
e-mail: mittals@uncw.edu

M. Alazab
Charles Darwin University, Casuarina, NT, Australia
e-mail: alazab.m@ieee.org

M. Stamp et al. (eds.), *Malware Analysis Using Artificial Intelligence and Deep Learning*, https://doi.org/10.1007/978-3-030-62582-5_10

automatic monitoring of these virtual machines to provide security for the cloud provider and its clients. With many clients having their resources in a shared virtual space, there is a risk that one client's virtual machine becoming infected could mean more virtual machines within the data center become infected. Not only would the cloud providers hardware be at risk, any users who have virtual resources in this data center could also be at risk. As cloud providers grow and increase the size of their consumer base, the responsibility of cloud providers to ensure the protection and security of their customers also increases. Cloud providers must seek the best possible security mechanisms to employ in the defense of their clients. In IaaS, the risk of attack and infection is increased due to common configurations and automatic provisioning of virtual machines in data centers. When configurations are similar among virtual machines, malware attacks are able to be repeated across those similar virtual machines.

As cloud infrastructure grows and develops, it presents a larger attack vector for malicious actors to launch attacks and inject malware. Customers who utilize cloud resources in data centers are called cloud tenants. Those tenant's virtual machines (VMs) need to be protected and secured against any variety of attacks that may take place against their resources. Preventing these attacks is a critical task, but equally important is detecting when a novel exploit succeeds and a portion of a data center's resources become infected. These exploits can take the form of a system vulnerability, a configuration vulnerability, an insider threat, or credentials stolen from an external source. With all of the possible methods of malware entry, cloud presents the opportunity for large amounts of malware to infect data centers globally.

1.1 Relevance in Cloud IaaS

Cloud infrastructure is unique because of the flexibility of services that can be offered in a cloud format. The flexibility of cloud attracts many customers who can utilize IaaS for their own benefit. With this flexibility also brings more risk with the various types of malware that can affect the data center. With varying services hosted within a data center, there are more variations of malware and attacks that can be performed as opposed to a system where a single type of service is offered. With more viable attacks, cloud infrastructure becomes more likely to experience malware infections. Cloud IaaS would benefit from highly accurate malware analysis and detection. Deep learning techniques have proven to be highly effective in malware detection and can be used to improve the security of cloud IaaS as well.

Cloud has some essential characteristics such as on-demand self-service, rapid elasticity, migration, resource pooling, and controlled measured service. These characteristics are necessary to support the on-demand delivery of computing power and the pay-as-you-go service model of cloud computing. These characteristics also incentivize attackers to target cloud infrastructure. For example, attackers could make use of the rapid-elasticity characteristic to create bot-nets quickly. Reacting to emerging threats brought on by the adoption of cloud is necessary to increasing cloud security.

Research in [1–3, 9, 15–18, 23, 31] discuss vulnerabilities that involve the essential cloud properties. The largest threat to cloud infrastructure is malware infection. Cloud malware injection is an attack where malware is injected into pre-existing process. While this attack is not unique to cloud environments, it can affect cloud more drastically. This is due to automatic provisioning and ability for cloud to spin up more VMs on demand and those VMs are configured in a similar manner. Such VMs are vulnerable to alike attacks if they are from the same template, increasing the chances that malware will spread from one VM to the next. In such a case, an attacker is able to quickly gain control of large number of VMs to execute large-scale attacks.

2 Machine Learning-Based Malware Detection

In many cases, malware detection methods for a single machine will work for cloud systems as well. Figure 1 shows an overview of machine learning-based malware detection techniques and their commonly associated features.

2.1 *File Classification*

In file classification, the goal is to examine a binary file and classify it as either malicious or not. This is usually accomplished by executing the malware in an observable isolated environment. The environment is designed to prevent malware from spreading outside of the intended scope. If an executable is identified as benign, then it is

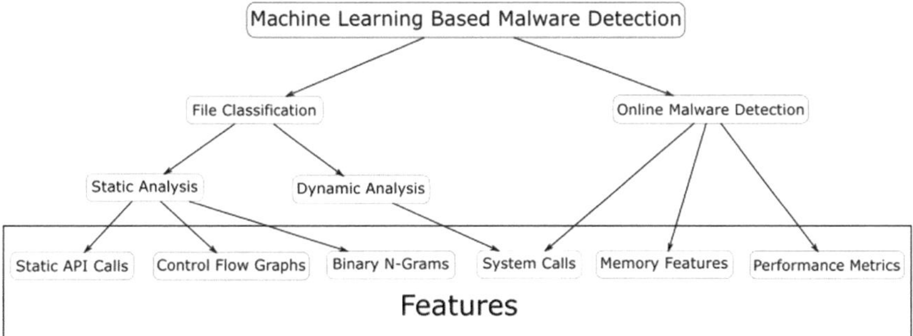

Fig. 1 Classification diagram of machine learning-based malware detection methods and associated features

allowed to be executed without further interference. File classification methods are split into static and dynamic analysis.

Static malware is the process of scanning files before they are executed to determine if they are malicious or not. If a file is being statically analyzed, then the malware executable is disassembled by dissemblers. These dissemblers produce an approximate source code of the malware which can be examined with various tools. More complex dissemblers can produce more accurate source codes than simpler ones. Certain dissemblers can also analyze control flow of a program to produce even more accurate source code. This technique is commonly used because it is simple, quick, and the tools needed are readily available. This technique can be defeated by malware writers who can craft binaries which cause the dissemblers to generate incorrect code. This is usually accomplished by the malware writers inserting in flawed code, which when executed, follow an obfuscated execution path that the dissemblers could not generate properly on their own. One example of this type of static malware analysis is extracting n-grams of a binary file. These n-grams can then be used as features to be used in a machine learning technique to uncover malicious patterns.

Even if dissemblers could generate the proper source code for a particular malicious binary, then the malware could be injected into an already running application. Since many files in static analysis are checked only once, usually before executing the file, the benign application would be scanned and deemed as such. If malware is injected into this already running process, then the malware's source code would not be scanned and go undetected. In cloud IaaS, this attack is referred to as a cloud malware injection [17]. If a cloud malware injection attacked is performed successfully, then the malware will go undetected and allowed to act without interference. Therefore, there is an essential need to constantly monitor applications running in cloud environments to maintain the services that have not been infected.

Where static analysis will fail to capture such malware attacks taking place, dynamic analysis can detect such an attack. Dynamic analysis works by recording the behavior of an executable and analyzing it to determine if malicious behavior is taking place. In dynamic analysis, the executable is executed and external software is recording its behavior. Typically, this is conducted within a sandbox where the executable would not be able to affect anything important. Information gathered during execution may include system calls, memory access, or network communications.

2.2 Online Malware Detection

In contrast, online detection methods involved a system that is being continuously monitored for malware. The features associated with online detection methods such as system calls, memory features, and performance metrics are costly to collect. These methods make up for this cost with the ability to detect malware that has infected an already running process that was initially determined to be benign.

Figure 1 shows a classification diagram of various machine learning-based malware detection techniques. Several works have been done in malware detection which focuses on different feature aspects. Most algorithms for detecting malware focus on a single machine. Some examples of this include support vector machines (SVMs) [30], all-nearest-neighbor classifier [13], and naïve bayes [4, 14]. While cloud environments generally involve more than a single machine acting on its own, there is not a large difference between a single virtual machine and a standalone host except for the hypervisor in a cloud environment. The hypervisor is critical in collecting information about the virtual machines running in a cloud environment. Due to this restriction, most works [4, 10–12, 19, 20, 24, 26, 32] are limited to using features which can be obtained through the hypervisor. Dawson et al. [10] utilize system calls for features and are concerned with rootkits. The work uses a non-linear phase-space algorithm to detect anomalies found in the system calls. The phase-space graph dissimilarities are used to evaluate the results.

Entropy-based Anomaly Testing (EbAT) was introduced in [29]. EbAT used multiple metrics such as memory utilization as well as CPU utilization. The work analyzed these metrics for anomaly detection based upon distribution instead of a flat threshold. Accurate results were generated from this approach for detection and the ability to scale up to meet metric processing demand. This work was limited in usefulness for practical and realistic cloud environment scenarios. Azmandian et al. [8] utilize performance metrics gathered from the hypervisor to form a new anomaly detection approach. These metrics included disk and network input–output. This work also uses KNN and Local Outlier Factor.

Abdelsalam et al. [3] show that malware detection can be conducted using a black-box approach. The metrics used in this work included VM-level performance and resource utilization. Highly active malware that made a large footprint in resource utilization records were detected by this approach. This approach was not as effective in detecting malware with low amounts of activity. These malwares likely attempted to hide themselves and reduced their activity to avoid leaving evidence in the resource utilization records.

3 Literature Review

Tables 1 and 2 summarize state-of-the-art research, challenges, and contributions with respect to online malware detection literature.

4 Cloud Security Monitoring Overview

Cloud security monitoring takes place at various levels. The levels include the Physical layer which contains computer hardware. The next level is the Infrastructure as a Service level which contains the cloud infrastructure made of virtual machines

Table 1 Online malware detection literature

Paper title	Focus/objective	Contribution	Limitation
Malware detection in cloud computing infrastructures [30]	Introduces an online cloud anomaly detection approach	• Effective anomaly detection using one-class Support Vector Machines • Assesses VM-Based feaures on detection performance	• Gathering features per process is computationally expensive • Uses highly active malware that is easier to detect
Malicious sequential pattern mining for automatic malware detection [13]	Proposes an effective sequence mining operation to discover malicious patters	• Proposes effective framework using sequence mining techniques • New nearest neighbor classifier to identify unknown malware	• Unable to perform malware classification, only detection
Analysis of machine learning techniques used in behavior-based malware detection [14]	Provide proof-of-concept on automatic behavior-based malware analysis	• Utilizes 220 unique Indonesian malware for Windows • Proof of concept for using ML in behavior-based malware detection	• Limited malware dataset • Limited feature selection
Zero-Day Malware Detection Based on Supervised Learning Algorithms of API Call Signatures [4]	Propose and evaluate a novel method of detecting and classifying zero-day malware	• Proposes machine learning framework to detect unknown malware with high accuracy, high efficiency, and signature free	• Only uses Windows API Calls as features
Towards understanding malware behavior by the extraction of API calls [6]	Provides automated method of extracting API call features	• Provides automated approach for API call feature extraction • Combines API call features with anomaly detection to analyze overall behavior of binaries	• Only applies to malware which uses Windows API calls
Malware detection based on structural and behavioral features of API calls [5]	Detecting obfuscated malware involving structural and behavioral features of API Calls	• Provides automated system to reverse-engineer program codes and apply feature extraction • Behavior features of API calls • Applied n-gram statistical model on executables for n-values	• If there are a lack of tools that can unpack certain malware, then it poses a challenge to the automated system

Table 2 Online malware detection literature (Continued)

Paper title	Focus/objective	Contribution	Limitation
Machine learning-based botnet identification traffic [7]	Botnet Identification using machine learning.	• Introduces CONIFA, a ML framework for botnet detection	• CONFIA relies on ML which can be affected by new data deviating from the data used during training.
Phase-space detection of virtual machine cyber events through hypervisor-level system call analysis [10]	Validating Oak Ridge National Lab's (ORNL) Beholder project is applicable to rootkit detection in virtual machines	• Implements system for detecting malware in a running VM. • Validates accuracy of ORNL's Beholder project algorithm • Builds on ORNL's Beholder project by applying new data	• Only tested with Beholder's parameter set, could have different results with other parameters • Future work left to experiment with other execution environments
On the feasibility of online malware detection with performance counters [11]	Building a malware detector in hardware using performance counters	• Tests the efficacy of using dynamic performance data to characterize and detect malware • Applies standard machine learning techniques such as KNN and Decision Trees to detect malware	• Detector accuracy can be improved in futher work • Unable to determine if new approach provides a significant advantage in malware detection
Deep learning approach for intelligent intrusion detection system [27]	Incorporating deep learning to introduce a new scalable intrusion detection system	• Proposes deep learning approach to detect cyberattacks proactively • Explores host-level events using natural language processing • Use multiple datasets in comparative analysis due to underlying flaws • Proposes a scalable hybrid intrusion detection system SHIA	• Further work includes enhancing accuracy of the proposed framework by adding the ability to monitor DNS and BGP events in the network. • Complex deep neural networks were not trained due to the computational cost
Robust intelligent malware detection using deep learning [28]	Evaluating classical machine learning architectures and deep learning architectures for malware detection	• Proposes scalable framework named ScaleMalNet to collect malware samples from distributed sources • Novel image processing technique • Independent performance evaluation of classical machine learning and deep learning architectures	• The deep learning architectures are vulnerable to adversarial environments and the robustness of the deep learning architectures against this vulnerability is not discussed.

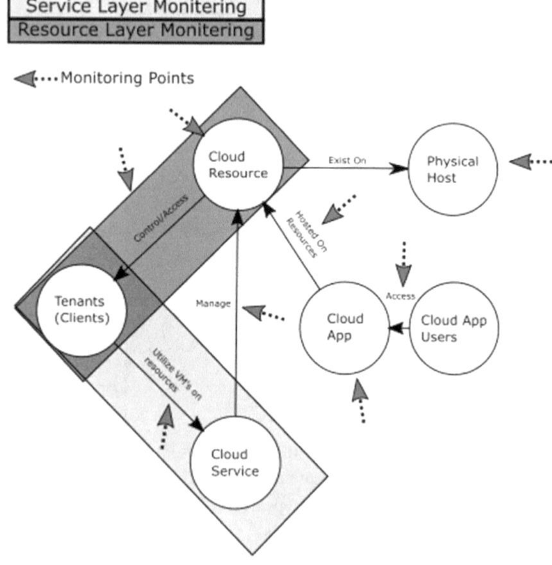

Fig. 2 Cloud monitoring points

and storage. The next level is the Platform as a Service level which is made of run-time environments. The last level is the Software as a Service level where cloud applications are used over the Internet.

With cloud systems becoming more complex and servicing more customers, the need volume of virtual machines that need to be monitored is increasing. Monitoring all of these virtual machines is resource intensive and failure to protect the tenants can result in service downtime. Tenants can also infect other tenants via co-resident attacks. All of the scenarios in the cloud systems where attacks could be performed need to be monitored. Figure 2 shows where cloud monitoring endpoints could be employed in a cloud IaaS scenario.

There are two categories of cloud security described by Fig. 2: Resource Layer Monitoring and Service Layer Monitoring. The figure illustrates interactions between components in a cloud environment and which components should be monitored. There is a security risk associated with every interaction and malware can spread through these interactions to infect unrelated parts of the ecosystem. Customers interact with various cloud services hosted on cloud storage. Users typically can update and manage this information. Clients can also host services such as web sites which have their own set of end users. If an end user of a hosted website is able to infect the Cloud App component, then the Cloud Storage component is at risk which endangers the rest of the environment and its entities.

Figures 3 and 4 show close up views of the Resource Layer Monitoring and Service Layer Monitoring, respectively. Figure 3 shows the tenants interacting with various cloud resource components. Figure 4 shows the tenants interacting in various ways with cloud services. These cloud services are also communicating between one

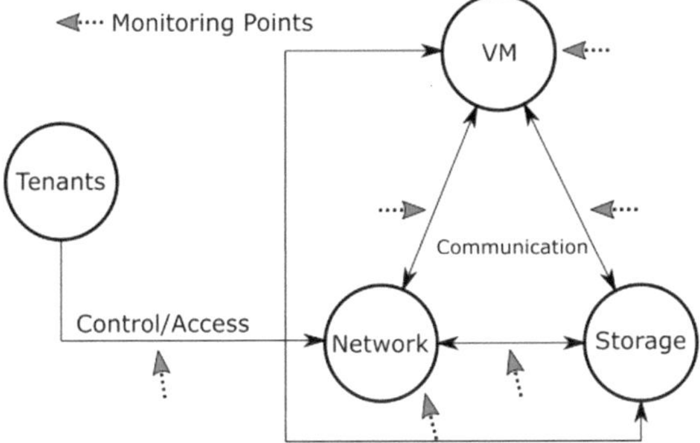

Fig. 3 Resource layer monitoring

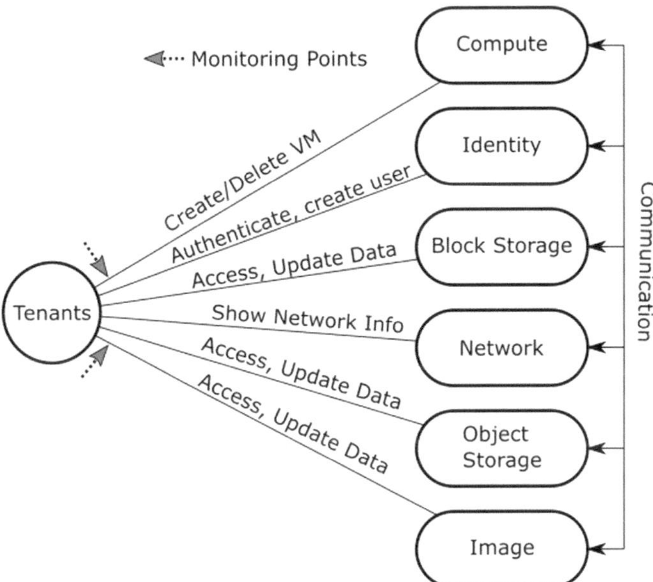

Fig. 4 Service layer monitoring

another in the background. While monitoring the behavior of the tenants' interactions may provide evidence for malicious activity, it may also be worth monitoring the background communications between services.

5 Behavioral Features and Characteristics

When building a deep learning model for malware analysis, features must be determined and collected from experiments or existing data. They should represent information that reflects the behavior of active malware in a particular system. Existing work such as [4, 24, 26] utilize API calls whereas [10, 12, 20] focus on system calls. Other features can include performance counters [11] and memory features [19, 32]. Table 3 shows an example set of features which could be used for behavioral malware detection.

For the collection of data, developing an ecosystem which allows malware to execute without issue is essential. Modern malware commonly has mechanisms to detect when it should or should not be active. This can be done to avoid detection and reduce the chance that a system's malware detection and prevention methods will catch the malware. If a malware detects that it is running inside of a "sandbox" or being monitored by an anti-virus application, the malware might decide to remain idle. If a malware is not active and features of the system are recorded, then those features may reflect normal activity of the system and should not be labeled as malicious.

An example experimental setup may look like a set of machines set up with Internet access and limited anti-virus measures, if any. Such machines need Internet access to allow most malware to conduct their malicious activities without interference. These may be part of a larger network that simulates some service and traffic to that service. An example service relevant to cloud IaaS is a web server that responds to various HTTP requests. Simulated traffic represents normal users interacting with this web service. Allow the simulated traffic to run for a period of time and record the selected features for the machine that is intended to be infected with malware.

Table 3 Sample virtual machine features

Metric	Description
CPU utilization	Average CPU utilization
Memory usage	Amount of memory in use by the VM that is allocated to it
Memory allocation	Amount of memory allocated to the VM by the hypervisor
Disk read requests	Amount of times the VM requested read access to the disk
Disk read amount	Number of bytes read from the disk by the VM
Disk write requests	Amount of times the VM requested write access to the disk
Disk write amount	Number of bytes write from the disk by the VM
Network incoming bytes	Number of bytes received by the VM
Network outgoing bytes	Number of bytes sent by the VM

This machine will generate benign samples that will be used later to establish what is normal behavior. After the period of time is over, malware should be injected into the machine and the simulated traffic should be allowed to continue. This represents a successful cloud malware injection attack in the real world, where users are unlikely to know that the service they are using has been compromised. If users notice no change in their service, their behavior should remain the same and it should be the same traffic behavior as before. Once the malware has been injected into the target machine, further recording should take place of the same selected features. This phase of recording should last as long as the benign phase to maintain a balanced amount of data. Once all of the data has been collected across the time samples, the network infrastructure should be wiped clean so that any changes made by the malware are reverted. This experiment should be run multiple times with different malware or network setups to generate a large volume of data which can be used in deep learning techniques.

Once the data has been collected, the data must be pre-processed and transformed into a format that is able to be input into deep learning algorithms. This usually means transferring all relevant information in a vector with floating point values and encoding all strings using one-hot encoding. Once the data is in the proper format, a set of data must be set aside for training and validation. A deep learning model requires data to train on to learn behaviors but there needs to be some way of comparing performances across models. The validation data is used for this comparison. Dataset aside as validation data should not be used during the training of the model, and instead be used as a measurement of a model's accuracy. An example of using validation data is to split the full dataset into 80% training and 20% validation. The model trains on the training dataset and every so often, the model is tasked with predicting the validation data samples. The accuracy of the model on the validation dataset should be the standard by which the model is judged. Various techniques exist when applying this dataset split including using a third test dataset when splitting the full dataset as well as cross-fold validation. It is necessary to generate enough data so that even after the split, there is enough data to train the deep learning model and enough data to test a varied set of examples.

6 Experimental Setup and Methodology

The dataset used for experimental analysis of deep learning techniques [21] was collected from an OpenStack testbed that simulated a 3-tier web architecture. The testbed utilized a database (MySQL), application server (Wordpress), and a web server (Apache) to create the architecture. In a typical 3-tier web architecture, a client makes a request to a web server which can then either return a static page to the client or access the application server. If the application server requires data access, then it can access the database. In any case, the response is returned back to the client. In this experimental setup, the web server and application server were allowed to scale up or down based on demand while the database was not. Two separate load balancers were used, one to distribute requests among web-servers from clients, and the other to distribute web server requests to application servers. Each computes node, acting as a cloud service, had network monitoring agents as well as agents collecting samples of the virtual machine itself.

Traffic was needed to accurately represent the intended use case: 3-tier web service. The traffic was simulated using two traffic different generation models: *Poisson Process* and *ON/OFF Pareto*. A program was built to act as multiple clients. This program sent requests to the web-servers which would go through the load balancer first. The parameters used for the simulation are

- Generation algorithm: Poisson or On/Off Pareto
- Concurrent Clients: 50
- Requests sent per hour: 3600
- Request Types: GET and POST (Randomly Generated)

The parameters for the On/Off Pareto algorithm used the NS2[1] tool defaults. The traffic volume was chosen to test the scalability policy by stressing the load balancers. The policy for scaling up was set to scale up when CPU utilization of either the app-servers or web-servers increased above %. Each experiment lasted for one hour.

The following four metrics were used to evaluate the effectiveness of the models:

$$\text{Accuracy} = \frac{\text{TP} + \text{TN}}{\text{TP} + \text{TN} + \text{FP} + \text{FN}}$$

$$\text{Precision} = \frac{\text{TP}}{\text{TP} + \text{FP}}$$

$$\text{Recall} = \frac{\text{TP}}{\text{TP} + \text{FN}}$$

$$\text{F1 Score} = 2 \times \frac{\text{Precision} \times \text{Recall}}{\text{Precision} + \text{Recall}}$$

A positive is when the system correctly detects a present malware and a negative is when the system fails to detect an existing malware. Therefore:

- TP: Malware successfully detected
- FP: System detected malware but no malware was present
- TN: System correctly identified no malware was present
- FN: System failed to detect present malware

In the experiments, a single virtual machine was randomly chosen to be injected with malware halfway through the experiment. The chosen features of this machine were recorded every 10 s. All machines were erased and rebuilt after every experiment due to the infection of live malware.

[1]NS2 tool manual. http://www.isi.edu/nsnam/ns/doc/node509.html.

Table 4 Sample values for metrics collected in [21]

Metric	Value	Metric	Value
sample_no	5672254	mem_swap	0
exp_no	23	mem_lib	0
vm_id	178	mem_text	217088
pid	1036	mem_uss	1105920
ppid	1	mem_dirty	0
sample_time	6/6/2018 19:32	mem_shared	3334144
process_creation_time	6/6/2018 19:32	mem_data	585728
status	sleeping	mem_vms	43921408
num_threads	1	mem_rss	3751936
kb_received	0	io_write_bytes	0
kb_sent	0	io_write_chars	76
num_fds	14	io_write_count	9
cpu_children_sys	0	io_read_bytes	958464
cpu_children_user	0	io_read_chars	61088
cpu_user	0.01	io_read_count	77
cpu_sys	0	ctx_switches_involuntary	43
cpu_percent	0	ctx_switches_voluntary	182
cpu_num	0	nice	0
name	dbus-daemon	ionice_ioclass	0
gid_real	111	ionice_value	0
gid_saved	111	label	0
gid_effective	111	–	–

Table 4 shows the metrics collected for the experiments in [21]. The example column represents a single process with the features collected about that process at a given time slice. This is the raw data that was collected by the polling agents. Further preparation of the data must be done before it is ready to be used in a neural network such as using one-hot encoding to encode the string values. All of the preparation will turn this raw data into a feature vector.

7 Deep Learning Techniques

Deep learning is a subset of machine learning, which revolves around using artificial neural networks (ANNs). ANNs are made of layers of neurons that activate in response to its input from the previous layers. Input in the form of tensors is passed through the layers and eventually makes a prediction. The prediction is then used to calculate the loss of the model and the weights of the model is then updated

Table 5 Comparison for evaluation metrics

Model	Accuracy	Precision	Recall	F1
LeNet-5	89.2	94.7	80.9	87.2
ResNet-50	88.4	86.0	88.9	87.4
ResNet-101	86.6	82.3	89.7	85.9
ResNet-152	89.5	89.0	87.8	88.4
DenseNet-121	92.9	100	84.6	91.5
DenseNet-169	92.8	99.7	84.4	91.4
DenseNet-201	92.8	99.5	84.6	91.5

to increase accuracy. Two of the most used types of artificial neural networks are convolutional neural networks (CNNs) and recurrent neural networks (RNNs).

CNNs are used in many applications where data can be visualized in the form of images. CNNs utilize a series of convolutional layers and pooling layers to break down an image into smaller mappings, and then begin to recognize smaller patterns until it builds up to predicting to full image. In the case of malware detection, features can be gathered from a virtual machine and used to create two-dimensional arrays of processes and features which can then be interpreted as an image. A CNN can work on this image to "recognize" and infected VM.

RNNs are used for applications where there is a strong time relationship between data samples that need to be captured. RNNs utilize either Long Short-Term Memory (LSTM) or Gate Recurrent Units (GRUs) cells to simulate memory. These cells are designed to solve the vanishing gradient where backpropagation fails to update the early layers within a deep network. Such LSTM and GRU cells have internal gates which regulate the flow of information. For example, the authors in [25] utilize an RNN to perform early detection of malware. They were able to detect malware within the first 5 s of execution with 94% accuracy.

7.1 Comparative Analysis

In this subsection, a detailed [21] comparison and contrasts of different state-of-the-art CNN models have been discussed used for detecing malwares in cloud IaaS. The models used were LeNet-5, ResNet-50, ResNet-101, ResNet-152, DenseNet-121, DenseNet-169, and DenseNet-201.

The results for the seven models tested are shown in Table 5. The baseline model LeNet-5 represents a shallow, simple CNN. The model lacks the depth necessary to capture complex and minute features. LeNet-5 reached an accuracy of $\simeq$87% while the best performing model, DenseNet-121 reached the highest accuracy of $\simeq$93%. The difference between DenseNet-121, DenseNet-169, and DenseNet-201 in accuracy was negligible. These results suggest that adding more layers to create deeper networks did not improve accuracy past a certain point within dense networks.

ResNet-152 performed slightly better than LeNet-5 in accuracy but fell behind in precision. One impact of this low precision is the ResNet models that were classifying benign samples as malicious. This could prove to be detrimental if regular use was being misidentified as malicious and a cloud provider was receiving false alarms. This may cause loss of availability to clients if the cloud provider were taking steps against these false alarms such as isolating virtual resources. Once the considerably longer training time of ResNet-152 is taken into account, the minor accuracy increase over LeNet-5 is likely not a worthwhile tradeoff. The higher recall of the residual networks may lend themselves to other scenarios where catching all of the malware incidents is more important than some false alarms. These may be cases where the resources being protected are extremely sensitive in nature and false positives are preferable to false negatives. ResNet-50 and ResNet-101 achieved the lowest accuracy overall however so they are not recommended over the other models tested.

All of the models underwent training and validation to find the point at which their validation accuracy was greatest. For all seven models, their highest performing version was used in a testing phase to generate the results in Table 5. The DenseNet models took much longer to train and therefore longer to reach their highest performing states than LeNet-5. DenseNet-121 took almost 10x longer to train than LeNet-5 and the other DenseNet models took longer than DenseNet-121 while providing marginally better results. It suffices to say that DenseNet-121 provided the best performance improvement in terms of accuracy with respect to training time from the baseline LeNet-5.

8 Conclusion

In this chapter, we discuss the importance of analyzing malware detection methods in cloud IaaS and the ability to utilize deep learning methods in those detection efforts. Cloud providers have an increasing responsibility to provide security mechanisms which will protect their data centers and the customers they serve. One such security mechanism is deep learning. Deep learning techniques provide highly accurate models which can detect malware. These models are useful for cloud environments especially when many virtual machines may be configured similarly and therefore could be susceptible to repeated attacks using the same malware. To utilize deep learning in cloud malware detection, data must first be found or generated. The data should be gathered from a simulated environment that closely represents real-world services and network infrastructure. If a new dataset is being generated, then the features being collected must be selected first. These features can include CPU usage, memory usage, system calls, or even network operations. Once data is generated and collected, it is ready to be fed into a deep learning model for predictions. A deep learning model should be designed to fit the dataset and the hyperparameters of the model will need to be tuned to increase accuracy. Once a model has been trained, it can then be used to perform predictions on new malware.

Acknowledgements This work is partially supported by National Science Foundation awards 1565562, 2025682, 2025685, and 2025686.

References

1. Abdelsalam, Mahmoud, et al. 2019. Online malware detection in cloud auto-scaling systems using shallow convolutional neural networks. In *Proceedings of IFIP annual conference on data and applications security and privacy*. Berlin: Springer.
2. Abdelsalam, Mahmoud, Ram Krishnan, Yufei Huang, and Ravi Sandhu. 2018. Malware detection in cloud infrastructures using convolutional neural networks. In *Proceedings of IEEE international conference on cloud computing*, 162–169.
3. Abdelsalam, Mahmoud, Ram Krishnan, and Ravi Sandhu. 2017. Clustering-based IaaS cloud monitoring. In *Proceedings of IEEE international conference on cloud computing (CLOUD)*, 672–679.
4. Alazab, Mamoun, et al. 2011. Zero-day malware detection based on supervised learning algorithms of API call signatures. In *Proceedings of the Australasian data mining conference*, 171–182, AUS. Australian Computer Society, Inc.
5. Alazab, Mamoun, Robert Layton, Sitalakshmi Venkatraman, and Paul Watters. 2010. Malware detection based on structural and behavioural features of API calls. In *Proceedings of the 1st International cyber resilience conference* ed. Craig Valli, 1–10. Edith Cowan University.
6. Alazab, Mamoun, Sitalakshmi Venkataraman, and Paul Watters. 2010. Towards understanding malware behaviour by the extraction of API calls. In *2010 second cybercrime and trustworthy computing workshop*, 52–59. IEEE.
7. Azab, Ahmad, Mamoun Alazab, and Mahdi Aiash. 2016. Machine learning based botnet identification traffic. In *2016 IEEE Trustcom/BigDataSE/ISPA*, 1788–1794. IEEE.
8. Azmandian, Fatemeh, et al. 2011. Virtual machine monitor-based lightweight intrusion detection. *ACM SIGOPS Operating Systems Review* 45 (2): 38–53.
9. Dahbur, Kamal, Bassil Mohammad, and Ahmad Bisher Tarakji. 2011. A survey of risks, threats and vulnerabilities in cloud computing. In *Proceedings of the 2011 international conference on intelligent semantic web-services and applications*, 1–6.
10. Dawson, Joel A., et al. 2018. Phase space detection of virtual machine cyber events through hypervisor-level system call analysis. In *Proceedings of IEEE international conference on data intelligence and security*, ICDIS, 159–167.
11. Demme, John, et al. 2013. On the feasibility of online malware detection with performance counters. *ACM SIGARCH Computer Architecture News* 41 (3): 559–570.
12. Dini, Gianluca, et al. 2012. Madam: A multi-level anomaly detector for android malware. In *Computer Network Security*, ed. Igor Kotenko, and Victor Skormin, 240–253. Berlin: Springer.
13. Fan, Yujie, Yanfang Ye, and Lifei Chen. 2016. Malicious sequential pattern mining for automatic malware detection. *Expert Systems with Applications* 52: 16–25.
14. Firdausi, Ivan, et al. 2010. Analysis of machine learning techniques used in behavior-based malware detection. In *Proceedings of IEEE international conference on advances in computing, control, and telecommunication technologies*, 201–203.
15. Gholami, Ali, and Erwin Laure. 2016. Security and privacy of sensitive data in cloud computing: A survey of recent developments. arXiv:1601.01498.
16. Grobauer, Bernd, Tobias Walloschek, and Elmar Stocker. 2010. Understanding cloud computing vulnerabilities. *IEEE Security & Privacy* 9 (2): 50–57.
17. Gruschka, Nils, et al. 2010. Attack surfaces: A taxonomy for attacks on cloud services. In *Proceedings of IEEE international conference on cloud computing*, 276–279.

18. Jensen, Meiko, Jörg Schwenk, Nils Gruschka, and Luigi Lo Iacono. 2009. On technical security issues in cloud computing. In *2009 IEEE international conference on cloud computing*, 109–116. IEEE.
19. Khasawneh, Khaled N., et al. 2015. Ensemble learning for low-level hardware-supported malware detection. In *Proceedings of international symposium on recent advances in intrusion detection*, 3–25. Berlin: Springer.
20. Luckett, P., et al. 2016. Neural network analysis of system call timing for rootkit detection, 1–6. In *Proceedings of Cybersecurity symposium, CYBERSEC*, April.
21. McDole, Andrew, Mahmoud Abdelsalam, Maanak Gupta, and Sudip Mittal. 2020. Analyzing CNN Based Behavioural Malware Detection Techniques on Cloud IaaS. arXiv:2002.06383.
22. Mell, Peter, and Tim Grance. 2011. The NIST definition of cloud computing. https://csrc.nist.gov/publications/detail/sp/800-145/final.
23. Piplai, Aritran, Sudip Mittal, Mahmoud Abdelsalam, Maanak Gupta, Anupam Joshi, and Tim Finin. 2020. Knowledge enrichment by fusing representations for malware threat intelligence and behavior. Technical report, UMBC, October
24. Pirscoveanu, Radu S., et al. 2015. Analysis of malware behavior: Type classification using machine learning. In *Proceedings of IEEE international conference on cyber situational awareness, data analytics and assessment*, 1–7.
25. Rhode, Matilda, Pete Burnap, and Kevin Jones. 2018. Early-stage malware prediction using recurrent neural networks. *Computers & Security* 77: 578–594.
26. Tobiyama, Shun, et al. 2016. Malware detection with deep neural network using process behavior. In *Proceedings of IEEE annual computer software and applications conference* vol. 2, 577–582.
27. Vinayakumar, R., K.P. Mamoun Alazab, Prabaharan Poornachandran Soman, Ameer Al-Nemrat, and Sitalakshmi Venkatraman. 2019. Deep learning approach for intelligent intrusion detection system. *IEEE Access* 7: 41525–41550.
28. Vinayakumar, R., K.P. Mamoun Alazab, Prabaharan Poornachandran Soman, and Sitalakshmi Venkatraman. 2019. Robust intelligent malware detection using deep learning. *IEEE Access* 7: 46717–46738.
29. Wang, Chengwei. 2009. Ebat: Online methods for detecting utility cloud anomalies. In *Proceedings of the middleware doctoral symposium*, 1–6.
30. Watson, Michael R., et al. 2015. Malware detection in cloud computing infrastructures. *IEEE Transactions on Dependable and Secure Computing* 13 (2): 192–205.
31. Xiao, Zhifeng, and Yang Xiao. 2012. Security and privacy in cloud computing. *IEEE Communications Surveys & Tutorials* 15 (2): 843–859.
32. Xu, Zhixing, et al. 2017. Malware detection using machine learning based analysis of virtual memory access patterns. In *Proceedings of IEEE design, automation & test in europe conference & exhibition*, 169–174.

A Comparison of Word2Vec, HMM2Vec, and PCA2Vec for Malware Classification

Aniket Chandak, Wendy Lee, and Mark Stamp

Abstract Word embeddings are often used in natural language processing as a means to quantify relationships between words. More generally, these same word embedding techniques can be used to quantify relationships between features. In this paper, we first consider multiple different word embedding techniques within the context of malware classification. We use hidden Markov models to obtain embedding vectors in an approach that we refer to as HMM2Vec, and we generate vector embeddings based on principal component analysis. We also consider the popular neural network-based word embedding technique known as Word2Vec. In each case, we derive feature embeddings based on opcode sequences for malware samples from a variety of different families. We show that we can obtain better classification accuracy based on these feature embeddings, as compared to HMM experiments that directly use the opcode sequences, and serve to establish a baseline. These results show that word embeddings can be a useful feature engineering step in the field of malware analysis.

1 Introduction

Malware detection and analysis are critical aspects of information security. The 2019 Internet Threat Security Report [46] claims an increase of 25% in 1 year in the number of attack groups using malware to disrupt businesses and organizations. According to the 2016 California Data Breach Report [13], malware contributed to 54% of all breaches and 90% of total records breached, with a staggering 44 million records

A. Chandak · W. Lee · M. Stamp (✉)
San Jose State University, San Jose, CA, USA
e-mail: mark.stamp@sjsu.edu

A. Chandak
e-mail: aniket.chandak@sjsu.edu

W. Lee
e-mail: wendy.lee@sjsu.edu

M. Stamp et al. (eds.), *Malware Analysis Using Artificial Intelligence and Deep Learning*, https://doi.org/10.1007/978-3-030-62582-5_11

breached due to malware in the years 2012–2016. Statistics such as these imply that malware is an increasing threat.

In this paper, we apply machine learning classification techniques to engineered features that are derived from malware samples. This feature engineering involves machine learning techniques. In effect, we apply machine learning to higher level features, where these features are themselves obtained using machine learning models. The motivation is that machine learning can serve to distill useful information from training samples, and hence the classification techniques may perform better on such data. In this research, we consider the effectiveness of using these derived features in the context of malware classification.

Specifically, we use word embeddings based on opcodes to derive features for subsequent classification. We consider three distinct word embedding techniques. First, we derive word embeddings from trained hidden Markov models (HMM). We refer to this technique as HMM2Vec. We then consider an analogous technique based on principal component analysis (PCA), which we refer to as PCA2Vec. And, as a third approach, we experiment with the popular neural network-based word embedding technique known as Word2Vec. In each case, we generate word embeddings for a significant number of samples from a variety of malware families. We then use several classification techniques to determine how well we can classify these samples using word embeddings as features.

The remainder of this paper is organized as follows. We provide a selective survey of relevant related work in Sect. 2. Section 3 contains an extensive and wide-ranging discussion of machine learning topics that play a role in this research. In Sect. 4, we provide details on the word embedding techniques that form the basis of our experiments. Section 5 gives our experiments and results, while Sect. 6 provides our conclusion and some paths for future work.

2 Related Work

Malware analysis and detection are challenging problems due to a variety of factors, including the large volume of malware and obfuscation techniques [10]. Every day, thousands of new malware are generated—manual analysis techniques cannot keep pace. Obfuscation is widely used by malware developers to make it difficult to analyze their malicious code.

Signature-based malware detection methods rely on pattern matching with known signatures [47]. Signature detection is relatively fast, and it is effective against "traditional" malware. However, extracting signatures is a labor-intensive process, and obfuscation techniques can defeat signature scanning.

Anomaly-based techniques are based on "unusual" or "virus-like" behavior or characteristics. An example of anomaly detection is behavior-based analysis, which can be used to analyze a sample when executed or under emulation [47]. When an executable file performs any action that does not fit its expected behavior, an alarm can be triggered. Such a method can detect obfuscated and zero-day malware, but it is slow, and generally subject to excessive false positives.

Recently, machine learning techniques have proven extremely useful for malware detection. The effectiveness of machine learning algorithms depends on the characteristics of the features used by such models. In malware detection and classification, a sample can be represented by a wide variety of features, including mnemonic opcodes, raw bytes, API calls, permissions, header information, etc. Opcodes are a popular feature that form the basis of the analysis considered in this paper.

In [7], the author experiments with opcodes and determines that such features can be successfully used to detect malware. The paper [11] achieves good results using API calls as a feature. Such features can be somewhat more difficult for malware writers to obfuscate, since API calls relate to the essential activity of software. However, extracting API calls from an executable is more costly than extracting opcodes.

Another example of malware research involving opcodes can be found in [33]. This paper features opcode n-grams, with a Markov blanket used to select from the large set of available n-gram. Classification is based on hidden Markov models, and experiments are based on five malware families.

In [3], malware opcodes are treated as a language, with Word2Vec used to quantify contextual information. Classification relies on k-nearest neighbors (k-NN). The research in [34] also uses Word2Vec to generate feature vectors based on opcode sequences, with a deep neural network employed for malware classification. In this latter research, the number of opcodes is in the range of 50–200, and the length of the Word2Vec embeddings range from 250 to 750.

Word2Vec embeddings are used as features to train bi-directional LSTMs in [20]. The experiments achieve good accuracy for malware detection, but training is costly. In [14], the author proposed a word embedding method based on opcode graphs—the graph is projected into vector space, which yields word embeddings. This technique is also computationally expensive.

In comparison to previous research, we consider additional vector embedding techniques, we experiment with a variety of classification algorithms, we use a smaller number of opcodes, and we generate short embedding vectors. Since we use a relatively small number of opcodes and short embedding vectors, our techniques are all highly efficient and practical. In addition, our experiments are based on a recently collected and challenging malware dataset.

3 Background

In this section, we present background information on the various learning techniques that are used in the experiments discussed in Sect. 5. Specifically, we introduce neural networks, beginning with some historical background and moving on to a modern context. We also introduce HMMs and PCA, which form the basis for the word embedding techniques that we refer to as HMM2Vec and PCS2Vec, respectively.

Finally, we introduce four classification techniques, which are used in our experiments.

In Sect. 4, we discuss HMM2Vec, PCA2Vec, and the neural network-based word embedding technique, Word2Vec, in detail. For our experiments in Sect. 5, we use these three word embedding techniques to generate features to classify malware samples.

3.1 Neural Networks

The concept of an artificial neuron [12, 49] is not new, as the idea was first proposed by McCulloch and Pitts in the 1940s [22]. However, modern computational neural networks begins with the perceptron, as introduced by Rosenblatt in the late 1950s [37].

3.1.1 McCulloch–Pitts Artificial Neuron

An artificial neuron with three inputs is illustrated in Fig. 1. In the original McCulloch–Pitts formulation, the inputs $X_i \in \{0, 1\}$, the weights $w_i \in \{+1, -1\}$, and the output $Y \in \{0, 1\}$. The output Y is 0 (inactive) or 1 (active), based on whether or not the linear function $\sum w_i X_i$ exceeds the specified threshold T. This form of an artificial neuron was modeled on neurons in the brain, which either fire or it do not (thus $Y \in \{0, 1\}$), and have input that comes from other neurons (thus each $X_i \in \{0, 1\}$). The weights w_i specify whether an input is excitatory (increasing the chance of the neuron firing) or inhibitory (decreasing the chance of the neuron firing). Whenever $\sum w_i X_i > T$, the excitatory response wins, and the neuron fires—otherwise the inhibitory response wins and the neuron does not fire.

Fig. 1 Artificial neuron

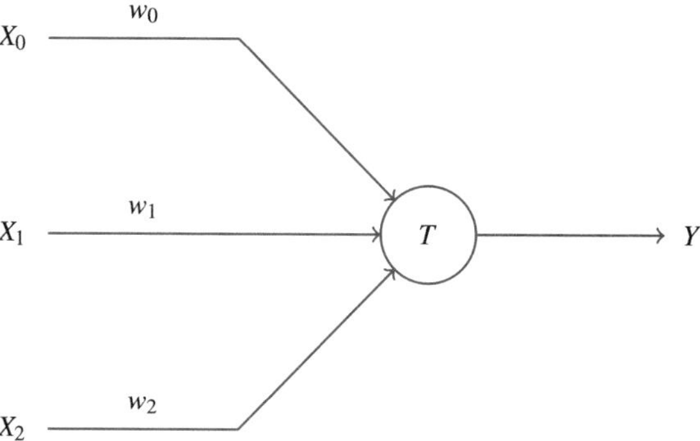

3.1.2 Perceptron

A *perceptron* is less restrictive than a McCulloch–Pitts artificial neuron. With a perceptron, both the inputs X_i and the weights w_i can be real valued, as opposed to the binary restrictions of McCulloch–Pitts. As with the McCulloch–Pitts formulation, the output Y of a perceptron is generally taken to be binary.

Given a real-valued input vector $X = (X_0, X_1, \ldots, X_{n-1})$, a perceptron can be viewed as an instantiation of a function of the form

$$f(X) = \sum_{i=0}^{n-1} w_i X_i + b,$$

that is, a perceptron computes a weighted sum of the input components. Based on a threshold, a single perceptron can define a binary classifier. That is, we can classify a sample X as "type 1" provided that $f(X) > T$, for some specified threshold T, and otherwise we classify X as "type 0."

In the case of two-dimensional input, the decision boundary of a is of the form

$$f(x, y) = w_0 x + w_1 y + b \tag{1}$$

which is the equation of a line. In general, the decision boundary of a perceptron is a hyperplane. Hence, a perceptron can only provide ideal separation in cases where the data itself is linearly separable.

As the name suggests, a multilayer perceptron (MLP) is an ANN that includes multiple (hidden) layers in the form of perceptrons. An example of an MLP with two hidden layers is given in Fig. 2, where each edge represent a weight that is to be determined via training. Unlike a single layer perceptron, MLPs are not restricted to linear decision boundaries, and hence an MLP can accurately model more complex functions. For example, the XOR function—which cannot be modeled by a single layer perceptron—can be modeled by an MLP.

To train a single layer perceptron, simple heuristics will suffice, assuming that the data is actually linearly separable. From a high-level perspective, training a single layer perceptron is somewhat analogous to training a linear support vector machine (SVM), except that for a perceptron, we do not require that the margin (i.e., minimum separation between the classes) be maximized. But training an MLP is clearly far more challenging, since we have hidden layers between the input and output, and it is not obvious how changes to the weights in these hidden layers will affect each other or the output.

As an aside, it is interesting to note that for SVMs, we deal with data that is not linearly separable by use of the "kernel trick," where the input data is mapped to a higher dimensional "feature space" via a (nonlinear) kernel function. In contrast, perceptrons (in the form of MLPs) overcome the limitation of linear separability by the use of multiple layers. With an MLP, it is as if a nonlinear kernel function has

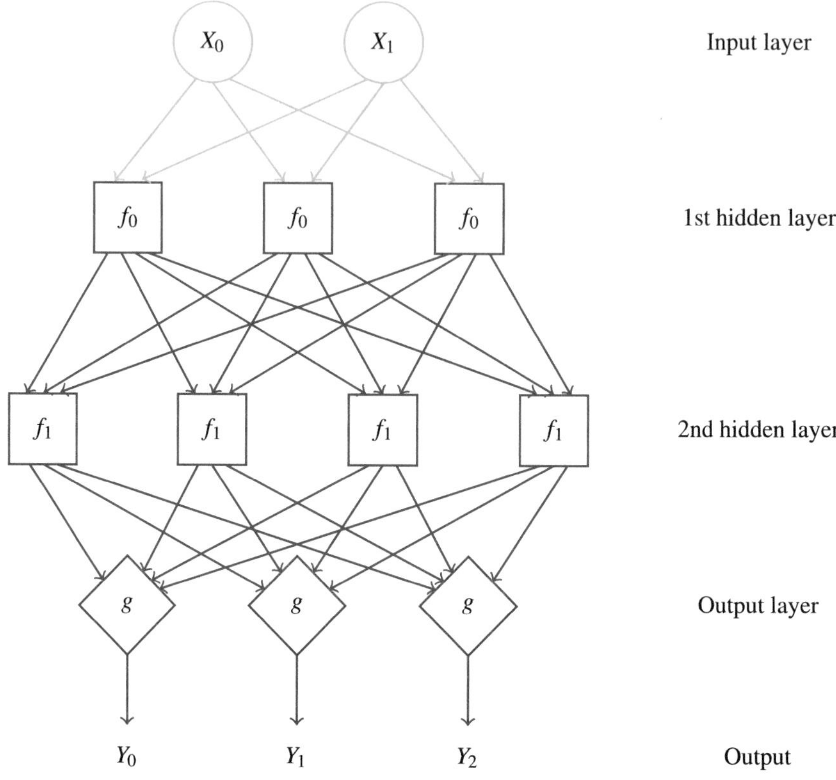

Fig. 2 MLP with two hidden layers

been embedded directly into the model itself through the use of hidden layers, as opposed to a user-specified explicit kernel function, which is the case for an SVM.

We can view the relationship between ANNs and deep learning as being somewhat akin to that of Markov chains and hidden Markov models (HMM). That is, ANNs serve as a basic technology that can be used to build powerful machine learning techniques, analogous to the way that an HMM is built on the foundation of an elementary Markov chain.

3.2 Hidden Markov Models

A generic hidden Markov model is illustrated in Fig. 3, where the X_i represent the hidden states and all other notations are shown in Table 1. The state of the Markov process, which we can be viewed as being hidden behind a "curtain" (the dashed line in Fig. 3), is determined by the current state and the A matrix. We are only able to

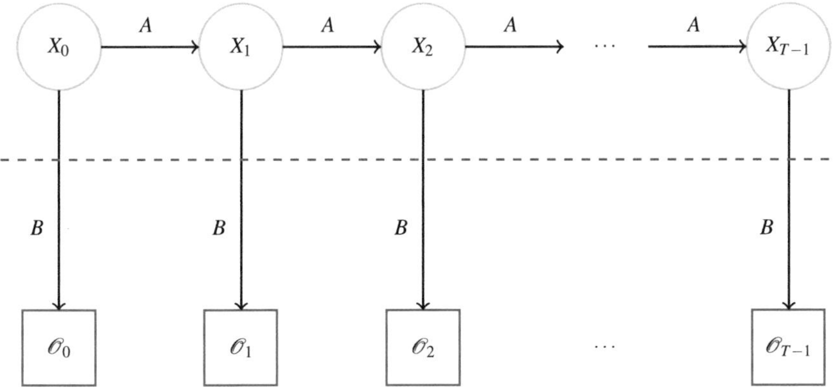

Fig. 3 Hidden Markov model

Table 1 HMM notation

Notation	Explanation
T	Length of the observation sequence
N	Number of states in the model
M	Number of observation symbols
Q	Distinct states of the Markov process, $q_0, q_1, \ldots, q_{N-1}$
V	Possible observations, assumed to be $0, 1, \ldots, M-1$
A	State transition probabilities
B	Observation probability matrix
π	Initial state distribution
$\mathcal{O}$	Observation sequence, $\mathcal{O}_0, \mathcal{O}_1, \ldots, \mathcal{O}_{T-1}$

observe the observations $\mathcal{O}_i$, which are related to the (hidden) states of the Markov process by the matrix B.

3.2.1 Notation and Basics

The notation used in an HMM is summarized in Table 1. Note that the observations are assumed to come from the set $\{0, 1, \ldots, M-1\}$, which simplifies the notation with no loss of generality. That is, we simply associate each of the M distinct observations with one of the elements $0, 1, \ldots, M-1$, so that we have $\mathcal{O}_i \in V = \{0, 1, \ldots, M-1\}$ for $i = 0, 1, \ldots, T-1$.

The matrix $A = \{a_{ij}\}$ is $N \times N$ with

$$a_{ij} = P(\text{state } q_j \text{ at } t+1 \mid \text{state } q_i \text{ at } t).$$

The matrix A is row stochastic, that is, each row satisfies the properties of a discrete probability distribution. Also, the probabilities a_{ij} are independent of t, and hence the A matrix does not vary with t. The matrix $B = \{b_j(k)\}$ is of size $N \times M$, with

$$b_j(k) = P(\text{observation } k \text{ at } t \mid \text{state } q_j \text{ at } t).$$

As with the A matrix, B is row stochastic, and the probabilities $b_j(k)$ are independent of t. The somewhat unusual notation $b_j(k)$ is convenient when specifying the HMM algorithms.

An HMM is defined by A, B, and π (and, implicitly, by the dimensions N and M). Thus, we denote an HMM as $\lambda = (A, B, \pi)$.

Suppose that we are given an observation sequence of length four, that is,

$$\mathcal{O} = (\mathcal{O}_0, \mathcal{O}_1, \mathcal{O}_2, \mathcal{O}_3).$$

Then the corresponding (hidden) state sequence is denoted as

$$X = (X_0, X_1, X_2, X_3).$$

We let π_{X_0} denote the probability of starting in state X_0, and $b_{X_0}(\mathcal{O}_0)$ denotes the probability of initially observing $\mathcal{O}_0$, while a_{X_0,X_1} is the probability of transiting from state X_0 to state X_1. Continuing, we see that the probability of a given state sequence X of length four is

$$P(X, \mathcal{O}) = \pi_{X_0} b_{X_0}(\mathcal{O}_0) a_{X_0,X_1} b_{X_1}(\mathcal{O}_1) a_{X_1,X_2} b_{X_2}(\mathcal{O}_2) a_{X_2,X_3} b_{X_3}(\mathcal{O}_3). \quad (2)$$

Note that in this expression, the X_i represent indices in the A and B matrices, not the names of the corresponding states.

To find the optimal state sequence in the dynamic programming (DP) sense, we simply choose the sequence (of length four, in this case) with the highest probability. In contrast, to find the optimal state sequence in the HMM sense, we choose the most probable symbol at each position. The optimal DP sequence and the optimal HMM sequence can differ.

3.2.2 The Three Problems

There are three fundamental problems that we can solve using HMMs. Here, we briefly describe each of these problems.

Problem 1 Given the model $\lambda = (A, B, \pi)$ and a sequence of observations $\mathcal{O}$, determine $P(\mathcal{O} \mid \lambda)$. That is, we want to compute a score for the observed sequence $\mathcal{O}$ with respect to the given model λ.

Problem 2 Given $\lambda = (A, B, \pi)$ and an observation sequence $\mathcal{O}$, find an optimal state sequence for the underlying Markov process. In other words, we want to uncover the hidden part of the hidden Markov model.

Problem 3 Given an observation sequence $\mathcal{O}$ and the parameter N, determine a model $\lambda = (A, B, \pi)$ that maximizes the probability of $\mathcal{O}$. This can be viewed as training a model to best fit the observed data. This problem is generally solved using Baum–Welch re-estimation [35, 43], which is a discrete hill climb on the parameter space represented by A, B, and π. There is also an alternative gradient ascent technique for HMM training [4, 45].

Since the technique we use to train an HMM (Problem 3) is a hill climb, in general, we obtain a local maximum. Training with different initial conditions can result in different local maxima, and hence it is often beneficial to train multiple HMMs with different initial conditions, and select the highest scoring model.

3.2.3 Example

Consider, for example, the problem of speech recognition which, not coincidentally, is one of the earliest and best-known successes of HMMs. In speech problems, the hidden states can be viewed as corresponding to movements of the vocal cords, which are not directly observed. Instead, we observe the sounds that are produced, and extract training features from these sounds. In this scenario, we can use the solution to HMM Problem 3 to train an HMM λ to, for example, recognize the spoken word "yes." Then, given an unknown spoken word, we can use the solution to Problem 1 to score the word against the trained model λ and determine the likelihood that the word is "yes." In this case, we do not need to solve Problem 2, but it is possible that such a solution (i.e., uncovering the hidden states) might provide additional insight into the underlying speech model.

English text analysis is another classic application of HMMs, which appears to have been first considered by Cave and Neuwirth [9]. This application nicely illustrates the strength of HMMs and it requires no background in any specialized field, such as speech processing or information security.

Given a length of English text, we remove all punctuation, numbers, etc., and converts all letters to lower case. This leaves 26 distinct letters and word-space, for a total of 27 symbols. We assume that there is an underlying Markov process (of order one) with two hidden states. For each of these two hidden states, we assume that the 27 symbols are observed according to fixed probability distributions.

This defines an HMM with $N = 2$ and $M = 27$, where the state transition probabilities of the A matrix and the observation probabilities of the B matrix are unknown, while the observations $\mathcal{O}_t$ consist of the series of characters we have extracted from the given text. To determine the A and B matrices, we must solve HMM Problem 3, as discussed above.

We have trained such an HMM, using the first $T = 50{,}000$ observations from the Brown Corpus,[1] which is available at [8]. We initialized each element of π and A randomly to approximately $1/2$, taking care to sure that the matrices are row stochastic. For one specific iteration of this experiment, the precise values used were

$$\pi = \begin{pmatrix} 0.51316 & 0.48684 \end{pmatrix}$$

and

$$A = \begin{pmatrix} 0.47468 & 0.52532 \\ 0.51656 & 0.48344 \end{pmatrix}.$$

Each element of B was initialized to approximately $1/27$, again, under the constraint that B must be row stochastic. The values in the initial B matrix (more precisely, the transpose of B) appear in the second and third columns of Table 2.

After the initial iteration, we find $\log\big(P(\mathcal{O}\,|\,\lambda)\big) = -165097.29$ and after 100 iterations, we have $\log\big(P(\mathcal{O}\mid\lambda)\big) = -137305.28$. These model scores indicate that training has improved the model significantly over the 100 iterations.

In this particular experiment, after 100 iterations, the model $\lambda = (A, B, \pi)$ has converged to

$$\pi = \begin{pmatrix} 0.00000 & 1.00000 \end{pmatrix} \text{ and } A = \begin{pmatrix} 0.25596 & 0.74404 \\ 0.71571 & 0.28429 \end{pmatrix}$$

with the converged B^T appearing in the last two columns of Table 2.

The most interesting part of an HMM is generally the B matrix. Without having made any assumption about the two hidden states, the B matrix in Table 2 shows us that one hidden state consists of vowels while the other hidden state consists of consonants. Curiously, from this perspective, word-space acts more like a vowel, while `y` is not even sometimes a vowel.

Of course, anyone familiar with English would not be surprised that there is a significant distinction between vowels and consonants. But, the crucial point here is that the HMM has automatically extracted this statistically important distinction for us—it has "learned" to distinguish between consonants and vowels. And, thanks to HMMs, this feature of English text could be easily discovered by someone who previously had no knowledge whatsoever of the language.

Cave and Neuwirth [9] obtain additional results when considering HMMs with more than two hidden states. In fact, they are able to sensibly interpret the results for models with up to $N = 12$ hidden states.

For more information on HMMs, see [43], which includes detailed algorithms including scaling or Rabiner's classic paper [35].

[1] Officially, it is the Brown University Standard Corpus of Present-Day American English, which includes various texts totaling about 1,000,000 words. Here, "Present-Day" means 1961.

Table 2 Initial and final B^T

Observation	Initial		Final	
a	0.03735	0.03909	0.13845	0.00075
b	0.03408	0.03537	0.00000	0.02311
c	0.03455	0.03537	0.00062	0.05614
d	0.03828	0.03909	0.00000	0.06937
e	0.03782	0.03583	0.21404	0.00000
f	0.03922	0.03630	0.00000	0.03559
g	0.03688	0.04048	0.00081	0.02724
h	0.03408	0.03537	0.00066	0.07278
i	0.03875	0.03816	0.12275	0.00000
j	0.04062	0.03909	0.00000	0.00365
k	0.03735	0.03490	0.00182	0.00703
l	0.03968	0.03723	0.00049	0.07231
m	0.03548	0.03537	0.00000	0.03889
n	0.03735	0.03909	0.00000	0.11461
o	0.04062	0.03397	0.13156	0.00000
p	0.03595	0.03397	0.00040	0.03674
q	0.03641	0.03816	0.00000	0.00153
r	0.03408	0.03676	0.00000	0.10225
s	0.04062	0.04048	0.00000	0.11042
t	0.03548	0.03443	0.01102	0.14392
u	0.03922	0.03537	0.04508	0.00000
v	0.04062	0.03955	0.00000	0.01621
w	0.03455	0.03816	0.00000	0.02303
x	0.03595	0.03723	0.00000	0.00447
y	0.03408	0.03769	0.00019	0.02587
z	0.03408	0.03955	0.00000	0.00110
Space	0.03688	0.03397	0.33211	0.01298

3.3 Principal Component Analysis

Principal component analysis (PCA) is a linear algebraic technique that provides a powerful tool for dimensionality reduction. Here, we provide a very brief introduction to the topic; for more details, Shlens' tutorial is highly recommended [40], while a good source for the math behind PCA is [39]. The discussion at [42] provides a brief, intuitive, and fun introduction to the subject.

Geometrically, PCA aligns a basis with the (orthogonal) directions having the largest variances. These directions are defined to be the principal components. A simple illustration of such a change of basis appears in Fig. 4.

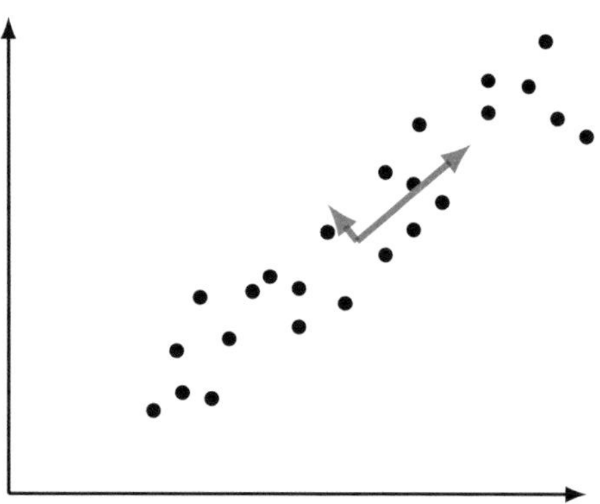

Fig. 4 A better basis

Intuitively, larger variances correspond to more informative data—if the variance is small, the training data is clumped tightly around the mean and we have limited ability to distinguish between samples. In contrast, if the variance is large, there is a much better chance of separating the samples based on the characteristic (or characteristics) under consideration. Consequently, once we have aligned the basis with the variances, we can ignore those directions that correspond to small variances without losing significant information. In fact, small variances often contribute only noise, in which cases we can actually improve our results by neglecting those directions that correspond to small variances.

The linear algebra behind PCA training (i.e., deriving a new-and-improved basis) is fairly deep, involving eigenvalue analysis. Yet, the scoring phase is simplicity itself, requiring little more than the computation of a few dot products, which makes scoring extremely efficient and practical.

Note that we treat singular value decomposition (SVD) as a special case of PCA, in the sense that SVD provides a method for determining the principal components. It is possible to take the opposite perspective, where PCA is viewed as a special case of the general change of basis technique provided by SVD. In any case, for our purposes, PCA and SVD can be considered as essentially synonymous.

3.4 Classifiers

In the research presented in this paper, we consider four different classifiers, namely, k-nearest neighbors (k-NN), multilayer perceptron (MLP), random forest (RF), and support vector machine (SVM). We have already discussed MLPs above, so in this section, we give a brief overview of k-NN, RF, and SVM.

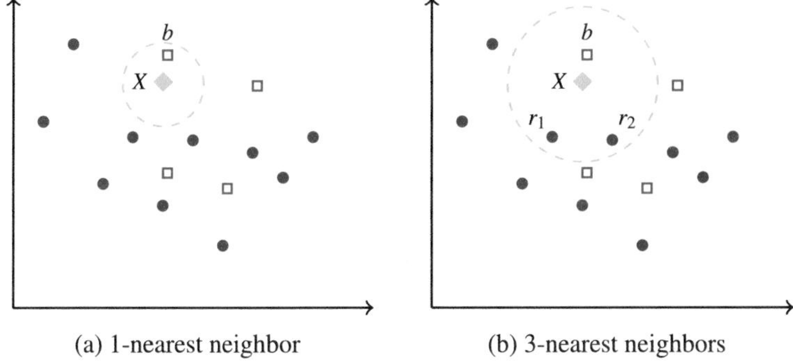

Fig. 5 Examples of k-NN classification [44]

3.4.1 k-Nearest Neighbors

Perhaps the simplest possible machine learning algorithm is k-nearest neighbors (k-NN). In the scoring phase, k-NN consists of classifying based on the k-nearest samples in the training set, typically using a simple majority vote. Since all computation is deferred to the scoring phase, k-NN is considered to be a "lazy learner."

Figure 5 shows examples of k-NN, where the training data consists of two classes, represented by the open blue squares and the solid red circles, with the green diamond (the point labeled X) being a point that we want to classify. Figure 5a shows that if we use the 1-nearest neighbor, we would classify the green diamond as being of same type as the open blue squares, whereas Fig. 5b shows that X would be classified as the solid red circle type if using the 3-nearest neighbors.

3.4.2 Random Forest

A random forest (RF) generalizes a simple decision tree algorithm. A decision tree is constructed by building a tree, based on features from the training data. It is easy to construct such trees, and trivial to classify samples once a tree has been constructed. However, decision trees tend to overfit the input data.

An RF combines multiple decision trees to generalize the training data. To do so, RFs use different subsets of the training data as well as different subsets of features, a process known as bagging [44]. A simple majority vote of the decision trees comprising the RF is typically used for classification [18].

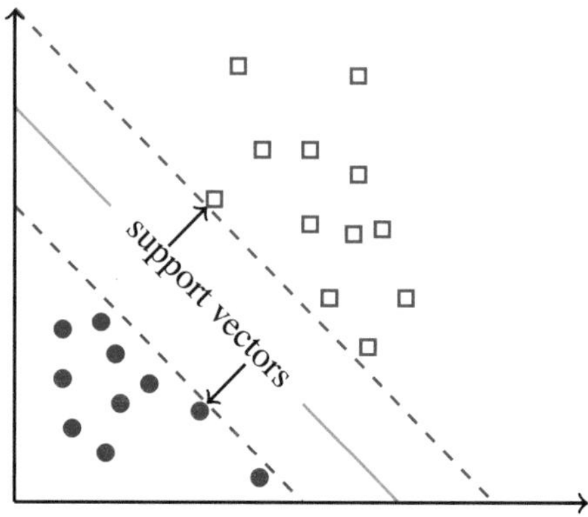

Fig. 6 Support vectors in SVM [44]

3.4.3 Support Vector Machine

Support vector machines (SVM) are a class of supervised learning methods that are based on four major ideas, namely, a separating hyperplane, maximizing the "margin" (i.e., separation between classes), working in a higher dimensional space, and the so-called kernel trick. The goal in SVM is to use a hyperplane to separate labeled data into two classes. If it exists, such a hyperplane is chosen to maximize the margin [44].

An example of a trained SVM is illustrated in Fig. 6. Note that the points that actually minimize the distance to the separating hyperplane correspond to support vectors. In general, the number of support vectors will be small relative to the number of training data points, and this is the key to the efficiency of SVM in the classification phase.

Of course, there is no assurance that the training data will be linearly separable. In such cases, a nonlinear kernel function can be embedded into the SVM process in such a way that the input data is, in effect, transformed to a higher dimensional "feature space." In this higher dimensional space, it is far more likely that the transformed data will be linearly separable. This is the essence of the kernel trick—an example of which is illustrated in Fig. 7. That we can transform our training data in such a manner is not surprising, but the fact that we can do so without paying any significant penalty in terms of computational efficiency makes the kernel trick a very powerful "trick" indeed. However, the kernel function must be specified by the user, and selecting an (near) optimal kernel can be challenging.

3.4.4 Last Word on Classification Techniques

We note in passing that MLP and SVM are related techniques, as both of these approaches generate nonlinear decision boundaries (assuming a nonlinear kernel). For SVM, the nonlinear boundary is based on a user-specified kernel function,

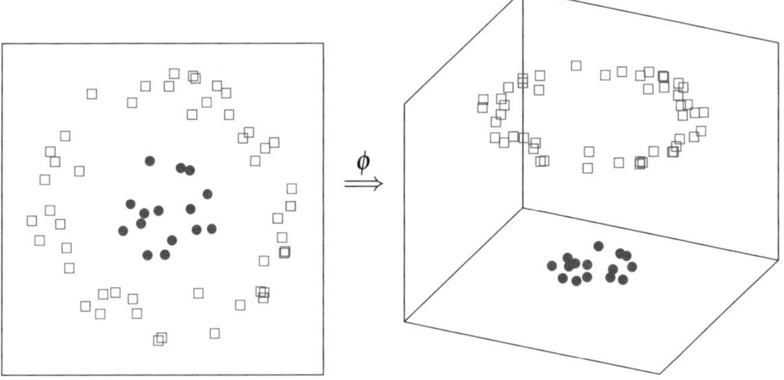

Fig. 7 A function ϕ illustrating the kernel trick [44]

whereas the equivalent aspect of an MLP is learned as part of the training process—in effect, the "kernel" is learned when training an MLP. This suggests that MLPs have an advantage, since there are limitations on SVM kernels, and selecting an optimal kernel is more art than science. However, the trade-off is that more data and more computation will generally be required to train a comparable MLP, since the MLP has more to learn, in comparison to an SVM.

It is also the case that k-NN and RF are closely related. In fact, both are neighborhood-based algorithms, but with neighborhood structures that are somewhat different [19].

Thus, we generally expect that the results obtained using SVM and MLP will be qualitatively similar, and the same is true when comparing results obtained using k-NN and RF. By using these four classifiers, we obtain a "sanity check" on the results. If, for example, our SVM and MLP results differ dramatically, this would indicate that we should investigate further. On the other hand, if, say, our MLP and RF results differ significantly, this would not raise the same level of concern.

4 Word Embedding Techniques

Word embeddings are often used in natural language processing as they provide a way to quantify relationships between words. Here, we use word embeddings to generate higher level features for malware classification.

In this section, we discuss three distinct word embedding techniques. First, we consider word embeddings derived from trained HMMs, which we refer to as HMM2Vec. Then we consider a word embedding technique based on PCA, which we refer to as PCA2Vec. Finally, we discuss the popular neural network-based technique known as Word2Vec.

4.1 HMM2Vec

Before discussing the basic ideas behind Word2Vec, we consider a somewhat analogous approach to generating vector representations based on hidden Markov models. To begin with we consider individual letters, as opposed to words—we call this simpler version Letter2Vec.

Recall that an HMM is defined by the three matrices A, B, and π, and is denoted as $\lambda = (A, B, \pi)$. The π matrix contains the initial state probabilities, A contains the hidden state transition probabilities, and B consists of the observation probability distributions corresponding to the hidden states. Each of these matrices is row stochastic, that is, each row satisfies the requirements of a discrete probability distribution. Notation-wise, N is the number of hidden states, M is the number of distinct observation symbols, and T is the length of the observation (i.e., training) sequence. Note that M and T are determined by the training data, while N is a user-defined parameter.

Suppose that we train an HMM on a sequence of letters extracted from English text, where we convert all uppercase letters to lowercase and discard any character that is not an alphabetic letter or word-space. Then $M = 27$, and we select $N = 2$ hidden states, and we use $T = 50{,}000$ observations for training. Note that each observation is one of the $M = 27$ symbols (letters plus word-space). For the example discussed below, the sequence of $T = 50{,}000$ observations was obtained from the Brown corpus of English [8]. Of course, any source of English text could be used.

In one specific case, an HMM trained with the parameters listed in the previous paragraph yields the B matrix in Table 2. Observe that this B matrix gives us two probability distributions over the observation symbols—one for each of the hidden states. We observe that one hidden state essentially corresponds to vowels, while the other corresponds to consonants. This simple example nicely illustrates the concept of machine learning, as no assumption was made a priori concerning consonants and vowels, and the only parameter we selected was the number of hidden states N. Thanks to this training process, the model has learned a crucial aspect of English directly from the data.

Suppose that for a given letter ℓ, we define its Letter2Vec representation $V(\ell)$ to be the corresponding row of the converged matrix B^T in the last two columns of Table 2. Then, for example,

$$\begin{aligned} V(\text{a}) &= \big(0.13845\ 0.00075\big) \quad & V(\text{e}) &= \big(0.21404\ 0.00000\big) \\ V(\text{s}) &= \big(0.00000\ 0.11042\big) \quad & V(\text{t}) &= \big(0.01102\ 0.14392\big). \end{aligned} \tag{3}$$

Next, we consider the distance between these Letter2Vec embeddings. However, instead of using Euclidean distance, we measure distance based on cosine similarity.

The cosine similarity of vectors X and Y is the cosine of the angle between the two vectors. Let $X = (X_0, X_1, \ldots, X_{n-1})$ and $Y = (Y_0, Y_1, \ldots, Y_{n-1})$. Then the cosine similarity is given by

$$\cos_\theta(X, Y) = \frac{\displaystyle\sum_{i=0}^{n-1} X_i Y_i}{\sqrt{\displaystyle\sum_{i=0}^{n-1} X_i^2}\sqrt{\displaystyle\sum_{i=0}^{n-1} Y_i^2}}.$$

In general, $-1 \le \cos_\theta(X, Y) \le 1$, but since our Letter2Vec encoding vectors consist of probabilities—and hence are non-negative—we have $0 \le \cos_\theta(X, Y) \le 1$ for the X and Y under consideration.

For the vector encodings in (3), we find that for the vowels "a" and "e," the cosine similarity is $\cos_\theta(V(\text{a}), V(\text{e})) = 0.9999$. In contrast, the cosine similarity between the vowel "a" and the consonant "t" is $\cos_\theta(V(\text{a}), V(\text{t})) = 0.0817$. These results indicate that these Letter2Vec embeddings—which are derived from a trained HMM—provide useful information on the similarity (or not) of pairs of letters.

Analogous to our Letter2Vec embeddings, we could train an HMM on words (or other features) and then use the columns of the resulting B matrix (equivalently, the rows of B^T) to define word (feature) embeddings.

The state of the art for Word2Vec based on words from English text is trained on a dataset corresponding to $M = 10{,}000$, $N = 300$ and $T = 10^9$. Training an HMM with such parameters would be decidedly non-trivial, as the work factor for Baum–Welch re-estimation is on the order of N^2T.

While the word embedding technique discussed in the previous paragraph—we call it HMM2Vec—is plausible, it has some potential limitations. Perhaps the biggest issue with HMM2Vec is that we typically train an HMM based on a Markov model of order one. That is, the current state only depends on the immediately preceding state. By basing our word embeddings on such a model, the resulting vectors would likely provide only a very limited sense of context. While we can train HMMs using models of higher order, the work factor would be prohibitive.

4.2 PCA2Vec

Another option for generating embedding vectors is to apply PCA to a matrix of pointwise mutual information (PMI). To construct a PMI matrix, based on a specified window size W, we compute $P(w_i, w_j)$ for all pairs of words (w_i, w_j) that occur within a window W of each other within our dataset, and we also compute $P(w_i)$ for each individual word w_i. Then we define the PMI matrix as

$$X = \{x_{ij}\} = \log\frac{P(w_j, w_i)}{P(w_i)P(w_j)}.$$

We treat column i of X, denoted X_i, as the feature vector for word w_i. Next, we perform PCA (using a singular value decomposition) based on these X_i feature vectors, and we project the feature vectors X_i onto the resulting eigenspace. Finally, by choosing the N dominant eigenvalues for this projection, we obtain embedding vectors of length N.

It is shown in [32] that these embedding vectors have many similar properties as Word2Vec embeddings, with the author providing examples analogous to those we give in the next section. Interestingly, it may be beneficial in certain applications to omit some of the dominant eigenvectors when determining the PCA2Vec embedding vectors [17].

For more details on using PCA to generate word embeddings, see [17]. The aforecited blog [32] gives an intuitive introduction to the topic.

4.3 Word2Vec

Word2Vec is a technique for embedding "words"—or more generally, any features—into a high-dimensional space. In Word2Vec, the embeddings are obtained by training a shallow neural network. After the training process, words that are more similar in context will tend to be closer together in the Word2Vec space.

Perhaps surprisingly, certain algebraic properties also hold for Word2Vec embeddings. For example, according to [30], if we let

$$w_0 = \text{“king”},\ w_1 = \text{“man”},\ w_2 = \text{“woman”},\ w_3 = \text{“queen”}$$

and we define $V(w_i)$ to be the Word2Vec embedding of w_i, then $V(w_3)$ is the vector that is closest to

$$V(w_0) - V(w_1) + V(w_2),$$

where "closest" is in terms of cosine similarity. Results such as this indicate that Word2Vec embeddings capture meaningful aspects of the semantics of the language.

Word2Vec uses a similar approach as the HMM2Vec concept outlined above. But, instead of using an HMM, Word2Vec embeddings are obtained from shallow (one hidden layer) neural network. Analogous to HMM2Vec, in Word2Vec, we are not interested in the resulting model itself, but instead we make use the learning that is represented by the trained model to define word embeddings. Next, we consider the basic ideas behind Word2Vec. Our approach is similar to that found in the excellent tutorial [21]. Here, we describe the process in terms of words, but these "words" can be general features.

Suppose that we have a vocabulary of size M. We encode each word as a "one-hot" vector of length M. For example, suppose that our vocabulary consists of the set of $M = 8$ words

$$\begin{aligned} W &= (w_0, w_1, w_2, w_3, w_4, w_5, w_6, w_7) \\ &= (\text{“for”}, \text{“giant”}, \text{“leap”}, \text{“man”}, \text{“mankind”}, \text{“one”}, \text{“small”}, \text{“step”}). \end{aligned}$$

Table 3 Training data

Offset	Training pairs
" one small step ..."	(one, small), (one, step)
"one small step for ..."	(small, one), (small, step), (small, for)
"one small step for man ..."	(step, one), (step, small), (step, for), (step, man)
"... small step for man one ..."	(for, small), (for, step), (for, man), (for, one)
"... step for man one giant ..."	(man, step), (man, for), (man, one), (man, giant)
"... for man one giant leap ..."	(one, for), (one, man), (one, giant), (one, leap)
"... man one giant leap for ..."	(giant, man), (giant, one), (giant, leap), (giant, for)
"... one giant leap for mankind"	(leap, one), (leap, giant), (leap, for), (leap, mankind)
"... giant leap for mankind"	(for, giant), (for, leap), (for, mankind)
"... leap for mankind "	(mankind, leap), (mankind, for)

Then we encode "for" and "man" as

$$E(w_0) = E(\text{“for”}) = 10000000 \text{ and } E(w_6) = E(\text{“man”}) = 00010000,$$

respectively.

Now, suppose that our training data consists of the phrase

$$\text{“one small step for man one giant leap for mankind.”} \tag{4}$$

To obtain our training samples, we specify a window size W, and for each offset we consider pairs of words within the specified window. For this example, we select $W = 2$, so that we consider words at a distance of one or two, in either direction. For the sentence in (4), a window size of two gives us the training pairs in Table 3.

Consider the pair "(for,man)" from the fourth row in Table 3. As one-hot vectors, this training pair corresponds to the input vector 10000000 and output vector 00010000.

A neural network similar to that illustrated in Fig. 8 is used to generate Word2Vec embeddings. The input is a one-hot vector of length M representing the first element of a training pair, such as those in Table 3. The network is trained to output the second element of each ordered pair which, again, is represented as a one-hot vector. The hidden layer consists of N linear neurons and the output layer uses a softmax function to generate M probabilities, where p_i is the probability of the output vector corresponding to w_i for the given input.

Observe that the Word2Vec network in Fig. 8 has NM weights that are to be determined via training, and these weights are represented by the blue lines from the

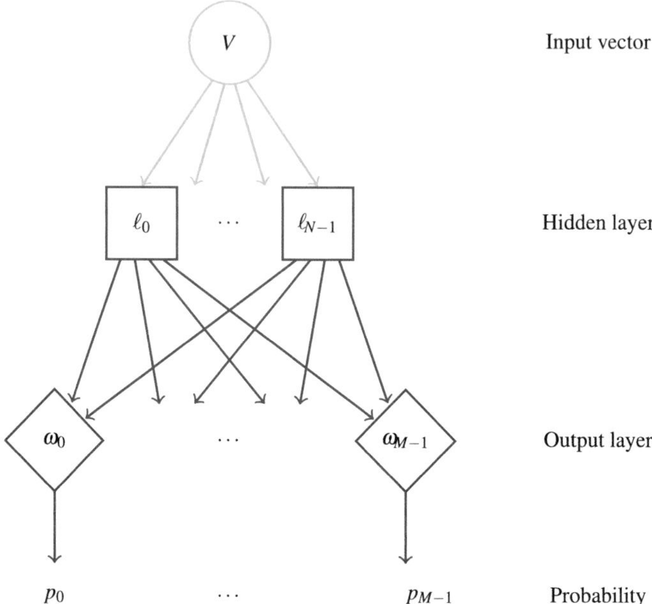

Fig. 8 Neural network for Word2Vec embeddings

hidden layer to the output layer. For each output node ω_i, there are N edges (i.e., weights) from the hidden layer. The N weights that connect to output node ω_i form the Word2Vec embedding $V(w_i)$ of the word w_i.

The state of the art in Word2Vec for English text is trained on a vocabulary of some $M = 10{,}000$ words, and embedding vectors of length $N = 300$, training on about 10^9 samples. Clearly, training a model of this magnitude is an extremely challenging computational task, as there are 3×10^6 weights to be determined, not to mention a huge number of training samples to deal with. Most of the complexity of Word2Vec comes from tricks that are used to make it feasible to train such a large network with such a massive amount of data.

One trick that is used to speed training in Word2Vec is "subsampling" of frequent words. Common words such as "a" and "the" contribute little to the model, so these words can appear in training pairs at a much lower rate than they are present in the training text.

Another key trick that is used in Word2Vec is "negative sampling." When training a neural network, each training sample potentially affects all of the weights of the model. Instead of adjusting all of the weights, in Word2Vec, only a small number of "negative" samples have their weights modified per training sample. For example, suppose that the output vector of a training pair corresponds to word w_0. Then the "positive" weights are those connected to the output node ω_0, and these weights are modified. In addition, a small subset of the $M - 1$ "negative" words (i.e., every word in the dataset except w_0) are selected and their corresponding weights are adjusted. The distribution used to select negative cases is biased toward more frequent words.

A general discussion of Word2Vec can be found in [5], while an intuitive—yet reasonably detailed—introduction is given in [21]. The original paper describing Word2Vec is [30] and an immediate follow-up paper discusses a variety of improvements that mostly serve to make training practical for large datasets [31].

5 Experiments and Results

In this section, we summarize our experimental results. These results are based on HMM2Vec, PCA2Vec, and Word2Vec experiments. But, first we discuss the dataset that we have used for all of the experiments reported in this section.

5.1 Dataset

The experimental results discussed in this section are based on the families in Table 4, with the number of available samples listed. In order to keep the test set balanced, from each of these families, we randomly selected 1000 samples, for a total of 7000 samples in our classification experiments. These families have been used in many recent studies, including [6, 48], for example.

The malware families in Table 4 are of a wide variety of different types. Next, we briefly discuss each of these families.

BHO can perform a wide variety of malicious actions, as specified by an attacker [25].

CeeInject is designed to conceal itself from detection, and hence various families use it as a shield to prevent detection. For example, CeeInject can obfuscate a

Table 4 Malware families and the number of samples

Family	Type	Samples
BHO	Trojan	1396
CeeInject	VirTool	1077
FakeRean	Rogue	1017
OnLineGames	Password stealer	1508
Renos	Trojan downloader	1567
Vobfus	Worm	1107
Winwebsec	Rogue	2302
Total	–	9974

Table 5 Classifier hyperparameters tested

Classifier	Hyperparameter	Tested values
MLP	learning_rate	constant, invscaling, adaptive
	hidden_layer_size	[(30, 30, 30), (10, 10, 10)]
	solver	sgd, adam
	activation	relu, logistic, tanh
	max_iter	[10000]
SVM	kernel	rbf, linear
	C	[1, 10, 100, 1000]
	gamma (rbf only)	[0.001, 0.0001]
k-NN	n_neighbors	[3, 5, 11, 19]
	weights	uniform, distance
	p	[1, 2, 3]
RF	n_estimators	[30, 100, 500, 1000]
	max_depth	[5, 8, 15, 25, 30]
	min_samples_split	[2, 5, 10, 15, 100]
	min_samples_leaf	[1, 2, 5, 10]

bitcoin mining client, which might have been installed on a system without the user's knowledge or consent [24].

FakeRean pretends to scan the system, notifies the user of nonexistent issues, and asks the user to pay to clean the system [29].

OnLineGames steals login information of online games and tracks user keystroke activity [26].

Renos will claim that the system has spyware and ask for a payment to remove the supposed spyware [23].

Vobfus is a family that downloads other malware onto a user's computer and makes changes to the device configuration that cannot be restored by simply removing the downloaded malware [27].

Winwebsec is a trojan that presents itself as antivirus software—it displays misleading messages stating that the device has been infected and attempts to persuade the user to pay a fee to free the system of malware [28].

In the remainder of this section, we present our experimental results. First, we discuss the selection of parameters for the various classifiers. Then we give results from a series of experiments for malware classification, based on each of the three word embedding techniques discussed in Sect. 4, namely, HMM2Vec, PCA2Vec, and Word2Vec. Note that all of our experiments were performed using scikit-learn [38].

5.2 Classifier Parameters

For each of our word embedding classification experiments, we test the three classifiers discussed in Sect. 3.4, namely, k-nearest neighbors (k-NN), random forest (RF), and support vector machine (SVM), along with the multilayer perceptron (MLP), which is discussed in Sect. 3.1.2. The features considered are the word embeddings from HMM2Vec, PCA2Vec, and Word2Vec. Note that this gives us a total of 12 distinct experiments.

For each case, we performed a grid search over a set of hyperparameters using `GridSearchCV` [41] in `scikit-learn`. `GridSearchCV` performs fivefold cross validation to determine the best parameters for each embedding technique. The parameters tested are listed in Table 5. Observe that for each of the three different word embedding techniques, we tested 36 combinations of parameters for MLP, we tested 12 combinations for SVM, we tested 16 combinations for k-NN, and we tested 400 RF combinations. Overall, we conducted

$$3 \cdot (36 + 12 + 16 + 400) = 1392$$

experiments to determine the parameters for the remaining experiments.

The optimal parameters selected for each classifier and for each embedding technique are listed in Table 6. We note that overall there is considerable agreement between the parameters for the different word embedding techniques, but in two cases (`learning_rate` and `n_estimators`), a different parameter is selected for each of the three embedding techniques.

Table 6 Classifier hyperparameters selected

Classifier	Hyperparameter	HMM2Vec	Word2Vec	PCA2Vec	Baseline HMM
MLP	`learning_rate`	`invscaling`	`constant`	`adaptive`	`constant`
	`hidden_layer_size`	(30, 30, 30)	(30, 30, 30)	(30, 30, 30)	(30, 30, 30)
	`solver`	`adam`	`adam`	`sgd`	`adam`
	`activation`	`relu`	`relu`	`relu`	`relu`
	`max_iter`	10000	10000	10000	10000
SVM	`kernel`	`linear`	`rbf`	`rbf`	`rbf`
	`C`	1000	1000	1000	10
	`gamma`	NA	0.001	0.001	0.0001
k-NN	`n_neighbors`	3	3	3	3
	`weights`	`distance`	`distance`	`distance`	`distance`
	`p`	1	2	1	3
RF	`n_estimators`	100	500	1000	1000
	`max_depth`	25	30	30	30
	`min_samples_split`	2	2	2	2
	`min_samples_leaf`	1	1	1	1

5.3 Baseline Results

First, we consider experiments based on opcode sequences and HMMs. These results serve as a baseline for comparison with the vector embedding techniques that are the primary focus of this research. We choose these HMM-based experiments for the baseline, as HMM trained on opcode features have proven popular and highly successful in the field of malware analysis [1, 2, 16, 36, 50].

Specifically, we train an HMM for each of the seven families in our dataset, using $N = 2$ hidden states in each case. For classification, we score a sample against all seven of these HMMs, and the resulting score vector (of length seven) serves as our feature vector. We use the same classification algorithms as in our word embedding experiments, namely, k-NN, MLP, RF, and SVM.

Note that we use the same opcode sequences here as in our vector embedding experiments. Specifically, the top 20 most frequent opcodes are used, with all remaining opcodes deleted.

The confusion matrices for these baseline HMM experiments are given in Fig. 9. The accuracies obtained for k-NN, MLP, RF, and SVM are 0.92, 0.44, 0.91, and 0.78, respectively. We see that MLP and SVM both perform poorly, whereas the neighborhood-based techniques, namely, k-NN and RF, are both strong, considering that we have seven classes. In addition, k-NN and RF give very similar results.

5.4 HMM2Vec Results

For these experiments, we train an HMM on each sample in our dataset. Recall that our dataset consists of 1000 samples from each of the seven families listed in Table 4. We train each of these 7000 models with $N = 2$ hidden states, using the $M = 20$ most frequent opcodes over all malware samples. Opcodes outside the top 20 are ignored.

As mentioned in Sect. 3.2.2, we often train multiple HMMs with different initial conditions, and select the best scoring model. This becomes more important as the length of the observation sequence decreases. Hence, when training our HMMs, we perform multiple random restarts—the number of restarts is determined by the length of the training sequence, as indicated in Table 7.

Each B matrix is 2×20, where each row corresponds to one of the hidden states of the model. From each of these matrices, we construct a vector of length 40 by appending the two rows. Since the order of the hidden states can vary between models, we select the order of the rows so as to obtain a consistency with respect to the most common opcode. That is, the row corresponding to the state that accumulates the highest probability for `MOV` is the first half of the feature vector, with the other row of the B matrix becoming the last 20 elements of the feature vector. This accounts for any cases where the hidden states differ.

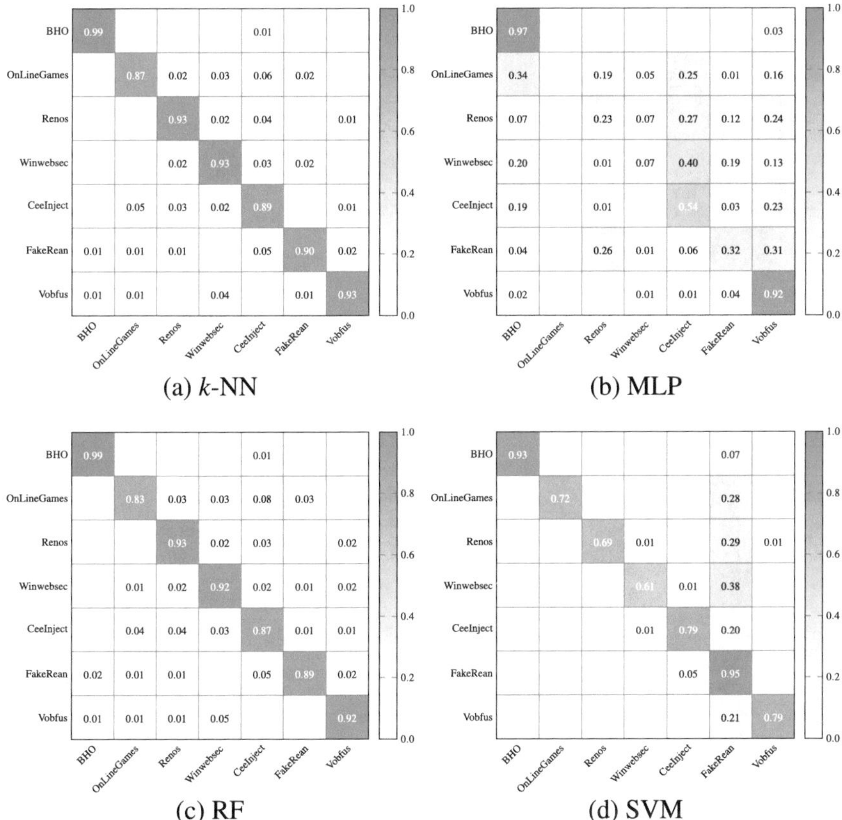

Fig. 9 Confusion matrices for baseline HMM experiments

Table 7 Number of random restarts

Observations	Restarts
Greater than 30,000	10
10,000–30,000	30
5000–10,000	100
Less than 500	500

Based on the resulting feature vectors, we use the parameters in the HMM2Vec column of Table 6 to classify the samples using k-NN, MLP, RF, and SVM. The confusion matrices for each of these cases is give in Fig. 10.

The accuracies obtained for k-NN, MLP, RF, and SVM based on HMM2Vec features are 0.93, 0.91, 0.93, and 0.89, respectively. From the confusion matrices in Fig. 10, we see that the greatest source of misclassifications is between FakeRean

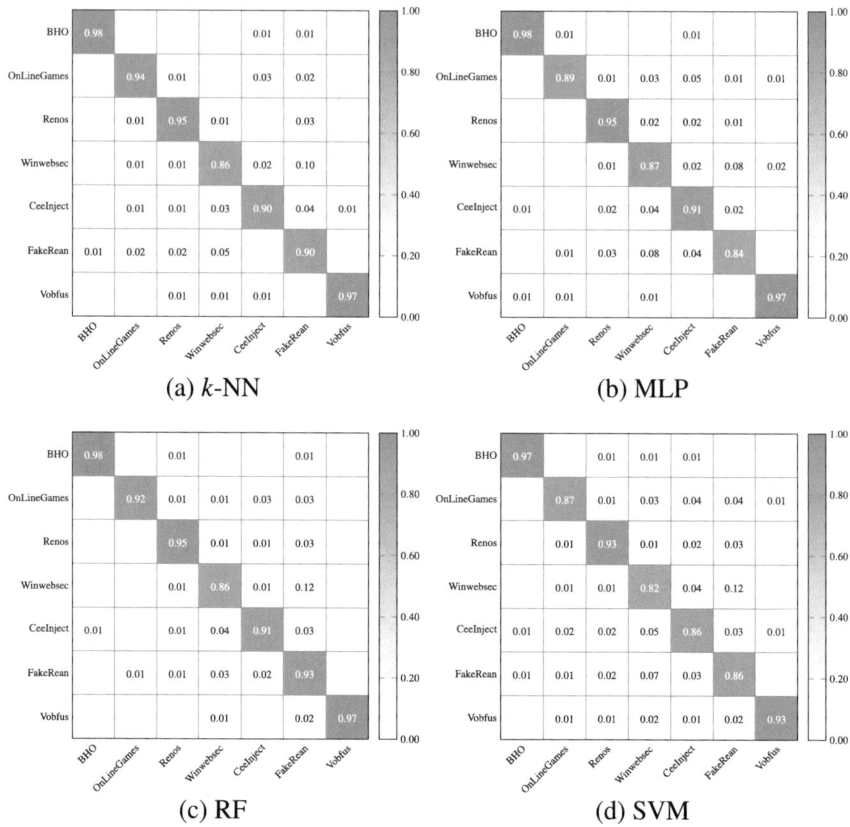

Fig. 10 Confusion matrices for HMM2Vec experiments

and Winwebsec families. In most—but not all—of our subsequent experiments, these two families will prove to be the most challenging to distinguish.

5.5 PCA2Vec Results

For our PCA2Vec experiments, we generate embedding vectors for each of the 7000 samples in our training set, as discussed in Sect. 4.2. We then train and classify the 7000 malware samples using these PCA2Vec feature vectors. The confusion matrices for these experiments are summarized in Fig. 11.

As above, each model is based on the 20 most frequent opcodes, which gives us a 20×20 PMI matrix. For consistency with the HMM2Vec experiments discussed above, we consider the two most dominant eigenvectors, and for consistency with the Word2Vec models discussed below, we use a window size of $W = 10$ when

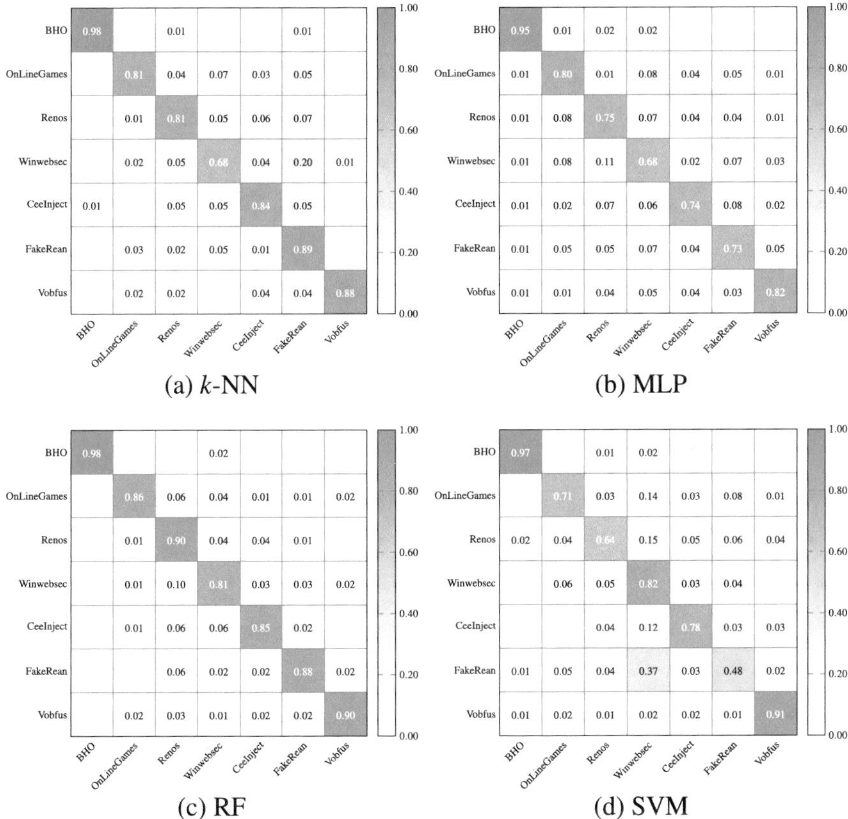

Fig. 11 Confusion matrices for PCA2Vec experiments

constructing the PMI matrix. The resulting projection into the eigenspace is 2×20, which we vectorize to obtain a feature vector of length 40.

The accuracies obtained for k-NN, MLP, RF, and SVM based on PCA2Vec features are 0.84, 0.78, 0.88, and 0.76, respectively. From these numbers, we see that PCA2Vec performed poorly for each of the classifiers considered, as compared to HMM2Vec.

5.6 Word2Vec Results

Analogous to the HMM2Vec and PCA2Vec experiments above, we classify the samples using the same four classifiers but with Word2Vec embeddings as features. The confusion matrices for these experiments are given in Fig. 12.

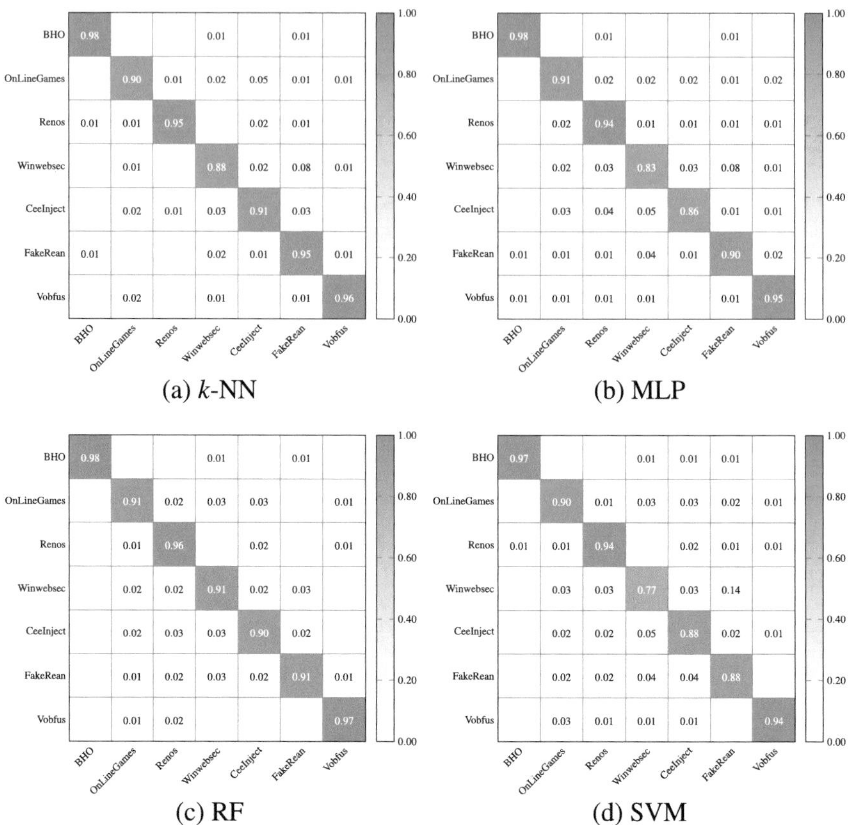

Fig. 12 Confusion matrices for Word2Vec experiments

As with the PCA2Vec experiments above, to generate our Word2Vec models, we use a window size of $W = 10$. And, to be consistent with both the HMM2Vec and PCA2Vec models discussed above, we use a vector length of two, giving us feature vectors of length 40. We use the so-called continuous-bag-of-words (CBOW) model, which is the model that we described in Sect. 4.3.

The accuracies obtained for k-NN, MLP, RF, and SVM based on Word2Vec features are 0.93, 0.91, 0.93, and 0.89, respectively. These results match those obtained using HMM2Vec.

In Sect. 5.8, we compare the accuracies obtained in our baseline HMM, HMM2Vec, PCA2Vec, and Word2Vec experiments. But first we discuss possible overfitting issues with respect to the k-NN and RF classifiers discussed above.

5.7 Overfitting

As discussed above in Sect. 3.4.4, both k-NN and random forest are neighborhood-based classification algorithms, but with different neighborhood structure. Thus, we expect that these two classification algorithms will generally perform in a somewhat similar manner, at least in a qualitative sense.

For k-NN, small values of k tend to result in overfitting. To avoid overfitting, the rule of thumb is that we should choose $k \approx \sqrt{N}$, where N is the number of samples in the training set [15]. Since we use an 80-20 split for training-testing and we have 7000 samples, for our k-NN experiments, this rule of thumb gives us $k = \sqrt{5600} \approx 75$. However, for each feature set considered, our grid search yielded an optimal value of $k \leq 3$.

In Fig. 13, we graph the accuracy of k-NN as a function of k for the baseline HMM, HMM2Vec, and Word2Vec feature sets. We see that all of these techniques perform more poorly as k increases. In particular, for $k \approx 75$, the performance of each is poor in comparison to $k \leq 3$, and this effect is particularly pronounced in the case of the baseline HMM. This provides strong evidence that small values of k in k-NN result in overfitting for each feature set, and the overfitting is especially pronounced for the baseline HMM.

For a random forest, the overfitting that is inherent in decision trees is mitigated by using more trees. In contrast, if the depth of the trees in the random forest is too large, the effect is analogous to choosing k too small in k-NN, and overfitting is likely to occur.

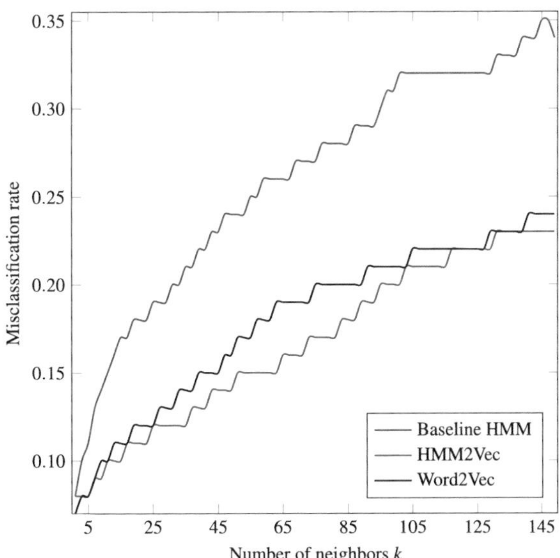

Fig. 13 k-NN results as a function of k

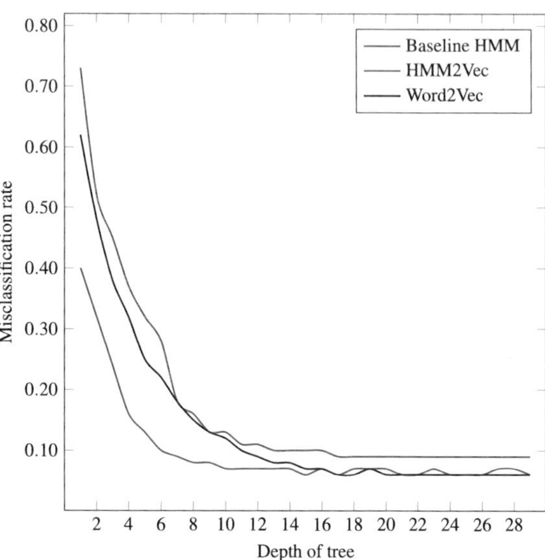

Fig. 14 Random forest results as a function of tree depth

To explore overfitting in our RF experiments, in Fig. 14, we give the misclassification results for the baseline HMM, HMM2Vec, and Word2Vec features, as a function of the maximum depth of the trees. In this case, Word2Vec performs best for smaller (maximum) depths, which indicates that the baseline HMM and HMM2Vec features are more prone to overfitting.

In Fig. 15a and b, we give misclassification results as a function of both the maximum depth and the number of trees for the baseline HMM and for HMM2Vec features, respectively. From these results, we see that the baseline HMM performs similarly as a function of the maximum depth, regardless of the number of trees. In contrast, the HMM2Vec features yield consistently better results than the baseline

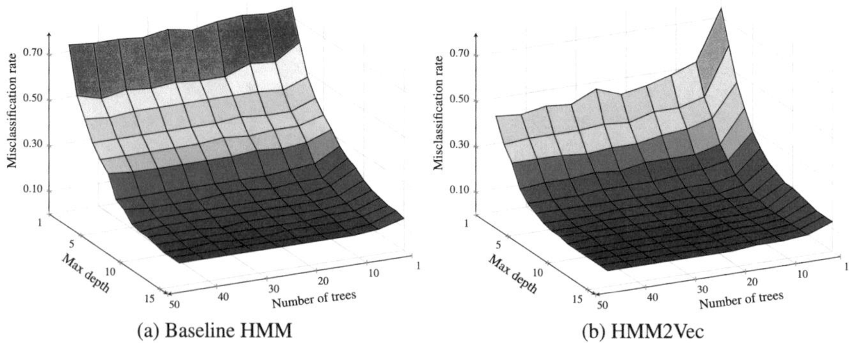

Fig. 15 Random forest maximum depth vs number of trees

HMM (as a function of the maximum depth), except when the number of trees is very small. This indicates that, with respect to the maximum depth, overfitting is significantly worse for the baseline HMM, since the overfitting cannot be overcome by increasing the number of trees.

From the discussion in this section, we see that all of our k-NN experimental results suffer from some degree of overfitting, with this effect being most significant in the case of the baseline HMM. For our RF results, overfitting is a relatively minor issue for the HMM2Vec- and Word2Vec-engineered features but, as with k-NN, it is a significant problem for the baseline HMM. Consequently, both the k-NN and RF results we have reported for the baseline HMM are overly optimistic, as these represent cases where significant overfitting has occurred.

5.8 Discussion

Figure 16 gives the overall accuracy for each of our multiclass experiments using k-NN, MLP, RF, and SVM classifiers, for our baseline HMM opcode experiments, and for each of the HMM2Vec-, PCA2Vec-, and Word2Vec-engineered feature experiments. In general, we expect RF and k-NN to perform somewhat similarly, since both are neighborhood-based algorithms. We also expect that in most cases, SVM and MLP will perform in a qualitatively similar manner to each other, since these techniques are closely related. We find that these expectations are generally met in our experiments, which can be viewed as a confirmation of the validity of the results.

From our 16 distinct experiments, we see that HMM2Vec and Word2Vec perform best, with PCA2Vec lagging far behind. The baseline HMM does well with respect to the neighborhood-based classifiers, namely, RF and k-NN. However, as discussed in

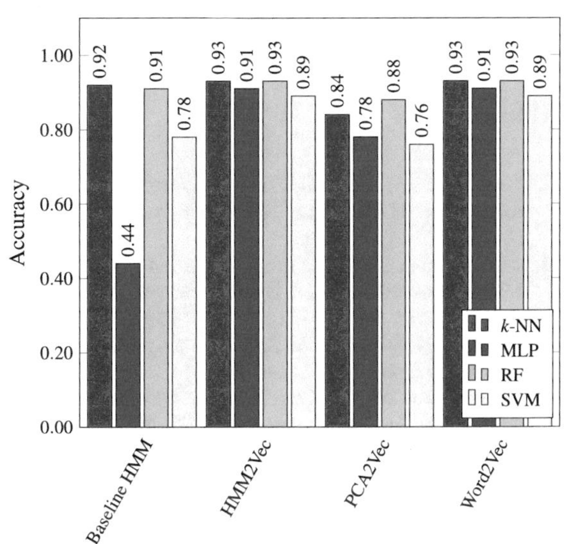

Fig. 16 Accuracies for combinations of features and classifiers

the previous section, these neighborhood-based techniques overfit the training data in the baseline HMM experiments. Neglecting these overfit results, we see that using the HMM2Vec- and Word2Vec-engineered features with SVM and MLP classifiers, give us the best results. Furthermore, these HMM2Vec and Word2Vec results are substantially better than either of the reliable results obtained for the baseline HMM, that is, the baseline HMM results using SVM and MLP classifiers.

6 Conclusion and Future Work

In this paper, we have presented results for a number of experiments involving word embedding techniques in malware classification. We have applied machine learning techniques to raw features to generate engineered features that are used for classification. Such a concept is not entirely unprecedented as, for example, PCA is often used to reduce the dimensionality of data before applying other machine learning techniques. However, the authors are not aware of previous work involving the use word embedding techniques in the same manner considered in this paper.

Our results show that word embedding techniques can be used to generate features that are more informative than the original data. This process of distilling useful information from the data before classifying samples is potentially useful, not only in the field of malware analysis, but also in other fields where learning plays a prominent role.

For future work, it would be interesting to consider other families and other types of malware. It would also be worthwhile to consider more complex and higher dimensional data—as with dimensionality-reduction techniques, such data would tend to offer more scope for improvement using the word embedding strategies considered in this paper.

References

1. Annachhatre, Chinmayee, Thomas H. Austin, and Mark Stamp. 2015. Hidden Markov models for malware classification. *Journal of Computer Virology and Hacking Techniques* 11 (2): 59–73.
2. Austin, Thomas H., Eric Filiol, Sébastien Josse, and Mark Stamp. 2013. Exploring hidden Markov models for virus analysis: A semantic approach. In *46th Hawaii international conference on system sciences HICSS 2013*, 5039–5048.
3. Awad, Y., M. Nassar, and H. Safa. Modeling malware as a language. In *2018 IEEE international conference on communications*, ICC, 1–6.
4. Baldi, Pierre, and Yves Chavin. 1994. Smooth on-line learning algorithms for hidden Markov models. *Neural Computation* 6: 307–318. https://core.ac.uk/download/pdf/4881023.pdf.
5. Banerjee, Suvro. 2018. Word2Vec — A baby step in deep learning but a giant leap towards natural language processing. https://medium.com/explore-artificial-intelligence/word2vec-a-baby-step-in-deep-learning-but-a-giant-leap-towards-natural-language-processing-40fe4e8602ba.

6. Basole, Samanvitha, Fabio Di Troia, and Mark Stamp. 2020. Multifamily malware models. *Journal of Computer Virology and Hacking Techniques*.
7. Bilar, Daniel. 2007. Opcodes as predictor for malware. *International Journal of Electronic Security and Digital Forensics* 1 (2): 156–168.
8. The Brown corpus of standard American English. http://www.cs.toronto.edu/~gpenn/csc401/a1res.html.
9. Cave, Robert L., and Lee P. Neuwirth. 1980. Hidden Markov models for English. In *Hidden Markov models for speech*, 16–56, IDA-CRD. New Jersey: Princeton. https://www.cs.sjsu.edu/~stamp/RUA/CaveNeuwirth/index.html.
10. Dhammi, Arshi, and Maninder Singh. 2015. Behavior analysis of malware using machine learning. In *Eighth international conference on contemporary computing*, IC3 2015, 481–486.
11. Hachinyan, Olga. 2017. Detection of malicious software on based on multiple equations of API-calls sequences. In *2017 IEEE conference of Russian roung researchers in electrical and electronic engineering*, EIConRus, 415–418.
12. Hardesty, Larry. 2017. Explained: Neural networks. http://news.mit.edu/2017/explained-neural-networks-deep-learning-0414.
13. Harris, Kamala. 2016. California data breach report. https://oag.ca.gov/sites/all/files/agweb/pdfs/dbr/2016-data-breach-report.pdf.
14. Hashemi, Hashem, Amin Azmoodeh, Ali Hamzeh, and Sattar Hashemi. 2016. Graph embedding as a new approach for unknown malware detection. *Journal of Computer Virology and Hacking Techniques* 13: 153–166.
15. Jirina, Marcel, and Marcel Jirina Jr. Using singularity exponent in distance based classifier. In *10th International Conference on Intelligent Systems Design and Applications*, ISDA 2010, 220–224.
16. Kalbhor, Ashwin, Thomas H. Austin, Eric Filiol, Sébastien Josse, and Mark Stamp. 2015. Dueling hidden Markov models for virus analysis. *Journal of Computer Virology and Hacking Techniques* 11 (2): 103–118.
17. Levy, Omer, Yoav Goldberg, and Ido Dagan. 2015. Improving distributional similarity with lessons learned from word embeddings. *Transactions of the Association for Computational Linguistics* 3: 211–225. https://levyomer.files.wordpress.com/2015/03/improving-distributional-similarity-tacl-2015.pdf.
18. Liaw, Andy, and Matthew Wiener. 2002. Classification and regression by randomForest. *R news* 2 (3): 18–22.
19. Lin, Yi, and Yongho Jeon. 2006. Random forests and adaptive nearest neighbors. *Journal of the American Statistical Association* 101 (474): 578–590.
20. Liu, Yingying, and Yiwei Wang. 2019. A robust malware detection system using deep learning on API calls. In *2019 IEEE 3rd information technology, networking, electronic and automation control conference*, ITNEC, 1456–1460.
21. McCormick, Chris. 2016. Word2vec tutorial — The skip-gram model. http://mccormickml.com/2016/04/19/word2vec-tutorial-the-skip-gram-model/.
22. McCulloch, Warren S., and Walter Pitts. 1943. A logical calculus of the ideas immanent in nervous activity. *Bulletin of Mathematical Biophysics* 5. https://pdfs.semanticscholar.org/5272/8a99829792c3272043842455f3a110e841b1.pdf.
23. Microsoft Security Intelligence. Renos. 2006. https://www.microsoft.com/en-us/wdsi/threats/malware-encyclopedia-description?Name=TrojanDownloader:Win32/Renos&threatId=16054.
24. Microsoft Security Intelligence. CeeInject. 2007. https://www.microsoft.com/en-us/wdsi/threats/malware-encyclopedia-description?Name=VirTool%3AWin32%2FCeeInject.
25. Microsoft Security Intelligence. BHO. 2008. https://www.microsoft.com/en-us/wdsi/threats/malware-encyclopedia-description?Name=Trojan:Win32/BHO&threatId=-2147364778.
26. Microsoft Security Intelligence. OnLineGames. 2008. https://www.microsoft.com/en-us/wdsi/threats/malware-encyclopedia-description?Name=PWS%3AWin32%2FOnLineGames.
27. Microsoft Security Intelligence. Vobfus. 2010. https://www.microsoft.com/en-us/wdsi/threats/malware-encyclopedia-description?name=win32%2Fvobfus.

28. Microsoft Security Intelligence. Winwebsec. 2010. https://www.microsoft.com/security/portal/threat/encyclopedia/entry.aspx?Name=Win32%2fWinwebsec.
29. Microsoft Security Intelligence. FakeRean. 2011. https://www.microsoft.com/en-us/wdsi/threats/malware-encyclopedia-description?Name=Win32/FakeRean.
30. Mikolov, Tomas, Kai Chen, Greg Corrado, and Jeffrey Dean. 2013. Efficient estimation of word representations in vector space. https://arxiv.org/abs/1301.3781.
31. Mikolov, Tomas, Ilya Sutskever, Kai Chen, Greg Corrado, and Jeffrey Dean. 2013. Distributed representations of words and phrases and their compositionality. https://papers.nips.cc/paper/5021-distributed-representations-of-words-and-phrases-and-their-compositionality.pdf.
32. Moody, Chris. Stop using word2vec. https://multithreaded.stitchfix.com/blog/2017/10/18/stop-using-word2vec/.
33. Pechaz, B., M.V. Jahan, and M. Jalali. 2015. Malware detection using hidden Markov model based on Markov blanket feature selection method. In *2015 International congress on technology, communication and knowledge*, ICTCK, 558–563.
34. Popov, Igor. 2017. Malware detection using machine learning based on Word2Vec embeddings of machine code instructions. In *2017 siberian symposium on data science and engineering*, SSDSE, 1–4.
35. Rabiner, Lawrence R. 1989. A tutorial on hidden Markov models and selected applications in speech recognition. *Proceedings of the IEEE* 77 (2): 257–286. https://www.cs.sjsu.edu/~stamp/RUA/Rabiner.pdf.
36. Raghavan, Aditya, Fabio Di Troia, and Mark Stamp. 2019. Hidden Markov models with random restarts versus boosting for malware detection. *Journal of Computer Virology and Hacking Techniques* 15 (2): 97–107.
37. Rosenblatt, Frank. 1961. Principles of neurodynamics: Perceptrons and the theory of brain mechanisms. http://www.dtic.mil/dtic/tr/fulltext/u2/256582.pdf.
38. `scikit-learn`: Machine learning in Python. https://scikit-learn.org/stable/.
39. Shalizi, Cosma. Principal component analysis. https://www.stat.cmu.edu/~cshalizi/uADA/12/lectures/ch18.pdf.
40. Shlens, Jonathon. 2005. A tutorial on principal component analysis. http://www.cs.cmu.edu/~elaw/papers/pca.pdf.
41. `sklearn.model_selection.GridSearchCV`. https://scikit-learn.org/stable/modules/generated/sklearn.model_selection.GridSearchCV.html.
42. Stack Exchange. 2015. Making sense of principal component analysis. https://stats.stackexchange.com/questions/2691/making-sense-of-principal-component-analysis-eigenvectors-eigenvalues.
43. Stamp, Mark. 2004. A revealing introduction to hidden Markov models. https://www.cs.sjsu.edu/~stamp/RUA/HMM.pdf.
44. Stamp, Mark. 2017. *Introduction to machine learning with applications in information security*. Boca Raton: Chapman and Hall/CRC.
45. Stamp, Mark. 2019. Deep thoughts on deep learning. https://www.cs.sjsu.edu/~stamp/RUA/ann.pdf.
46. Symantec. 2019. Internet security threat report: Malware. https://interactive.symantec.com/istr24-web.
47. Vinod, P., R. Jaipur, V. Laxmi, and M. Gaur. 2009. Survey on malware detection methods. In *Proceedings of the 3rd Hackers' workshop on computer and internet security*, IITKHACK'09, 74–79.
48. Wadkar, Mayuri, Fabio Di Troia, and Mark Stamp. 2020. Detecting malware evolution using support vector machines. *Expert Systems with Applications* 143.
49. Wallis, Charles. 2017. History of the perceptron. https://web.csulb.edu/~cwallis/artificialn/History.htm.
50. Wong, Wing, and Mark Stamp. 2006. Hunting for metamorphic engines. *Journal in Computer Virology* 2 (3): 211–229.

Word Embedding Techniques for Malware Evolution Detection

Sunhera Paul and Mark Stamp

Abstract Malware detection is a critical aspect of information security. One difficulty that arises is that malware often evolves over time. To maintain effective malware detection, it is necessary to determine when malware evolution has occurred so that appropriate countermeasures can be taken. We perform a variety of experiments aimed at detecting points in time where a malware family has likely evolved, and we consider secondary tests designed to confirm that evolution has actually occurred. Several malware families are analyzed, each of which includes a number of samples collected over an extended period of time. Our experiments indicate that improved results are obtained using feature engineering based on word embedding techniques. All of our experiments are based on machine learning models, and hence our evolution detection strategies require minimal human intervention and can easily be automated.

1 Introduction

Malware is a malicious software that causes disruption in normal activity, allows access to unapproved resources, gathers private data of users, or performs other improper activity [1]. Developing measures to detect malware is a critical aspect of information security.

Malware often evolves due to changing goals of malware developers, advances in detection, and so on [3]. This evolution can occur through a wide variety of modifications to the code. It is essential to detect and analyze malware evolution so that appropriate measures can be taken to maintain and improve the effectiveness of detection techniques [2].

An obvious technique for analyzing malware evolution consists of reverse engineering a large number of samples over an extended period of time, which is a highly labor-intensive process. Other approaches to malware evolution include graph prun-

S. Paul · M. Stamp (✉)
San Jose State University, San Jose, CA, USA
e-mail: mark.stamp@sjsu.edu

M. Stamp et al. (eds.), *Malware Analysis Using Artificial Intelligence and Deep Learning*, https://doi.org/10.1007/978-3-030-62582-5_12

ing techniques [7] and analysis of PE file features using support vector machines (SVM) [29]. This latter research shows considerable promise, and has the advantage of being fully automated, with no reverse engineering or other time-consuming analysis required. Our proposed research can be viewed as an extension of—and improvement on—the groundbreaking work in [29].

We consider several experiments that are designed to detect points in time where a malware family has likely evolved significantly. We then perform further experiments to confirm that such evolution has actually occurred. All of our experiments have been conducted using a significant number of malware families, most of which include a large number of samples collected over an extended period of time. Furthermore, all of our experiments are based on machine learning, and hence fully automatable.

For a given malware family, we first separate the available samples based on windows of time. We have extracted opcode sequences from every sample and we use these opcodes as features for detecting malware evolution. We experiment with a variety of feature engineering techniques, and in each case we train linear SVMs over sliding windows of time. The SVM weights of these models are compared based on a χ^2 distance measure, which enables us to detect changes in the SVM models over time. A point in time where a spike is observed in the χ^2 graph shows a substantial change in SVM models—which indicates a possible evolutionary branch in the malware family under consideration. To confirm that such evolution has actually occurred, we train hidden Markov models (HMM) on either side of a significant spike in the χ^2 graph. If a clear distinction between these HMMs is observed, it serves as a confirmation that significant evolution has been detected. The primary objective of this research is to implement and analyze different variants of this proposed malware evolution detection technique.

The remainder of this paper is organized as follows. In Sect. 2, we discuss relevant related work in the area of malware evolution. Section 3 provides an overview of the dataset that we use, as well as brief introductions to the various machine learning models and techniques that we use in this research. We present our experimental results in Sect. 4. The paper concludes with Sect. 5, where we also outline possible avenues for future work.

2 Related Work

Relative to the vast malware research literature, comparatively little has been done in the area of malware evolution. In this section, we provide a selective review of research related to malware evolution.

The malware evolution research in [7] is based on large and diverse malware dataset that spans for nearly two decades. This work focuses on the inheritance properties of malware, and the technique is based on graph pruning. The authors claim that many specific traits of various families in their dataset have been "inherited" from other families. However, it is not entirely clear that these "inherited" traits are actually inherited, as opposed to having been developed independently. In addition, the graph-based analysis in [7] requires "extensive manual investigation," which is in stark contrast to the automated techniques that are considered in this paper.

The authors of [9] extract a variety of features from Android malware samples and determine trends based on standard software quality metrics. These results are compared to a similar analysis of trends in Android non-malware, or goodware. This work shows that the trends in Android malware and goodware are fairly similar, indicating that the "improvement" in this type of malware has followed a similar path as that of goodware.

The paper [24] is focused on detecting new malware variants, which is closely related to an evolution problem. The authors considered malware variants that would typically defeat machine learning based detectors. Their approach relies as an extensive feature set and employs semi-supervised learning. In comparison, the approach in this paper relies entirely on unsupervised techniques, and we are able to detect less drastic code modifications.

The work in [4] is nominally focused on malware taxonomy. However, this work also provides insight into malware evolution, in the form of "genealogical trajectories." The work relies on a variety of features and uses support vector machines (SVM) for classification.

We note in passing that machine learning models are trained on features. Thus, extracting appropriate features from a dataset is a crucial step in any malware analysis technique that is based on machine learning. We can broadly classify features as static and dynamic—features that can be obtained without executing the code are said to be static, while those that require code execution or emulation are known as dynamic. In general, static features are more efficient to collect, whereas dynamic features can be more informative and are typically more robust [5].

The author in [29] use static PE file features of malware samples as the basis for their malware evolution research. Based on these features, linear SVMs are trained over various time windows and the resulting model weights are compared using a χ^2 distance. A spike in the χ^2 distance graph is shown to be indicative of an evolutionary change in a malware family. Note that this approach is easily automated, with no reverse engineering required.

The research presented in this paper extends and expands on the work in [29]. As in [29], we use SVMs together with χ^2 distance as a means of detecting evolutionary change. We make several important contributions that greatly increase the utility of this basic approach. The novelty of our work includes the use of more sensitive static features—we use opcodes as compared to derived PE file features—and we employ various feature engineering techniques. In addition, we develop an HMM-based secondary test to verify the putative evolutionary changes obtained from the SVM together with a χ^2 distance.

3 Implementation

In this section, we give a broad summary of the malware families that comprise the dataset used in the research. We also discuss the features and machine learning techniques used in our experiments. These features and techniques form the basis of our evolutionary experiments in Sect. 4.

3.1 Dataset

A malware family represents a collection of samples that have major traits in common. Over time, successful malware families will tend to evolve, as malware writers develop new features and find different applications for the code base.

The research in this paper is based on a malware dataset consisting of Windows portable executable (PE) files. From a large dataset, we have extracted 11,037 samples belonging to 15 distinct malware families. Table 1 lists these malware families and the number of samples per family that we use in our experiments.

Our Winwebsec and Zbot malware samples were acquired from the Malicia dataset [23], while the remaining 13 families were extracted from a vast malware dataset that was collected as part of the work reported in [8]. This latter dataset is

Table 1 Number of samples used in experiments

Family	Samples	Years
Adload	791	2009–2011
BHO	1,116	2007–2011
Bifrose	577	2009–2011
CeeInject	742	2009–2012
DelfInject	401	2009–2012
Dorkbot	222	2005–2012
Hupigon	449	2009–2011
IRCBot	59	2009–2012
Obfuscator	670	2004–2017
Rbot	127	2001–2012
VBInject	2,331	2009–2018
Vobfus	700	2009–2011
Winwebsec	1,511	2008–2012
Zbot	835	2009–2012
Zegost	506	2008–2011
Total	11,037	–

greater than half a terabyte in size and contains on the order of 500,000 malware executables. Our datasets are available from the authors, upon request.

Most of the malware families that were chosen for this research have a substantial number of samples available over an extended time period. The smaller families (e.g., IRCBot and Rbot) were chosen to test our analysis techniques in cases where the training data is severely limited.

As a pre-processing step, we have organized all the malware samples in each family according to their creation date. During this initial data-wrangling phase, any sample having an altered compilation or creation date was discarded.

Next, we briefly discuss each of the malware families in our dataset. Note that these families represent a wide variety of types of malware, including Trojan, worm, adware, backdoor, and so on.

Bifrose is a backdoor Trojan that allows an attacker to connect to a remote IP using a random port number. Some variants of Bifrose have the capability to hide files and processes from the user. Bifrose enables an attacker to view system information, retrieve passwords, or execute files by gaining remote control of an infected system [22].

CeeInject serves to shielding nefarious activity from detection. For example, CeeInject can obfuscate a bitcoin mining client, which might be installed on a system to mine bitcoins without the user's knowledge [13].

DelfInject is a worm that enters a system from a file passed by other malware, or as a file downloaded accidentally by a client when visiting malignant sites. DelfInject drops itself onto the system using an arbitrary document name (e.g., `xpdsae.exe`) and alters the relevant registry entry so that it runs at each system start. The malware then injects code into `svchost.exe` so that it can create a connection with specific servers and download files [14].

Dorkbot is a worm that steals user names and passwords by tracking online activities. It blocks security update websites and can launch denial of service (DoS) attacks. Dorkbot is spread via instant messaging applications, social networks, and flash drives [21].

Hotbar is an adware program that may be unintentionally downloaded by a user from a malicious website. Being adware, Hotbar displays advertisements as the user browses the web [11].

Hupigon is a family of backdoor Trojans. This malware opens a backdoor server enabling other remote computers to control a compromised system [12].

Obfuscator hides its purpose through obfuscation. The underlying malware can have virtually any payload [18].

Rbot is a backdoor Trojan that enables an attacker to control an infected computer using an IRC channel. It then spreads to other computers by scanning for network shares and exploiting vulnerabilities in the system. Rbot includes many advanced features and it has been used to launch DoS attacks [10].

VBInject primarily serves to disguise other malware. VBInject is a packaged malware, i.e., malware that utilizes techniques of encryption and compression to obscure its contents. This makes it difficult to recognize malware that it is concealing. VBInject was first seen in 2009 and appeared again in 2010 [15].

Vobfus is a malware family that downloads other malware onto a user's computer. It uses the Windows `autorun` feature to spread to other devices such as flash drives. Vobfus makes long-lasting changes to the device configuration that cannot be restored simply by removing the malware from the system [16].

Winwebsec is a Trojan that presents itself as an antivirus software. It shows misleading messages to the users stating that the device has been infected and attempts to persuade the user to pay to remove these non-existent threats [17].

Zbot is a Trojan that attempts to steal confidential information from a compromised computer. It explicitly targets system data, online sensitive data, and banking information, and it can also be easily modified to accumulate other kinds of data. The Trojan itself is generally disseminated through drive-by downloads and spam campaigns. Zbot was originally discovered in January 2010 [19].

Zegost is a backdoor Trojan that injects itself into `svchost.exe`, thus allowing an attacker to execute files on the compromised system [20].

3.2 Feature Extraction

Mnemonic opcodes are machine-level language instructions that specify a particular operation that is to be performed [26]. For this research, our dataset consists of malware samples in the form of portable executable (PE) files. The primary feature that we consider are opcode sequences extracted from these executable files. We have also segregated the malware samples in each family according to their creation date.

3.3 Classification Techniques

In this section, we will provide an overview of each machine learning technique that we employ in this research. Additional pointers to the literature are provided.

3.3.1 Support Vector Machines

Support vector machines (SVM) are one of the most popular classes of machine learning techniques. An SVM attempts to find a separating hyperplane between two labeled classes of data [27]. By utilizing the so-called "kernel trick," an SVM can map the input data to a high-dimensional space where the additional space can afford a greater opportunity to find a separating hyperplane. The "trick" of the kernel trick is that this mapping does not result in any significant increase in the work factor.

A linear SVM assigns a well-defined weight to each feature in the training vector. These weights specify the relative importance that the SVM places on each feature

Table 2 Sliding time window example

Time window	Class +1	Class −1
Jan. 2011–Jan. 2012	Jan. 2011–Dec. 2011	Jan. 2012
Feb. 2011–Feb. 2012	Feb. 2011–Jan. 2012	Feb. 2012
Mar. 2011–Mar. 2012	Mar. 2011–Feb. 2012	Mar. 2012

which can serve as useful when ranking the importance of features. In our experiments, we rely heavily on this aspect of linear SVMs.

Analogos to [29], in our experiments, we define the two classes of an SVM as follows. All the samples within the most recent one-year time window are class "+1," while all samples from the current month are defined as class "−1." For example, in Table 2, we give three consecutive time windows, along with the time frames corresponding to the two classes in each case.

3.3.2 χ^2 Statistic

The χ^2 statistic is a normalized sum of square deviation between the observed and expected frequency distributions. This statistic is calculated as

$$\chi^2 = \sum_{i=1}^{n} \frac{(o_i - e_i)^2}{e_i}$$

where n denotes the number of features or observations, o_i is the observed value of the ith instance, and e_i is expected value of the ith instance.

For our experiments, this statistic is used to quantify the differences between SVM feature weights of different models, where these models were trained over different time windows. As mentioned above, we use a time period of one year for one class, and a time window of the following month as the other class. We compute this χ^2 "distance" between pairs of models that are trained on overlapping time windows. Any points in the resulting χ^2 graph where a substantial change (i.e., a "spike") occurs indicates a point where adjacent SVM models differ significantly. These are points of interest, since they indicate the times at which the code has likely been substantially modified.

3.3.3 Word2Vec

Word2Vec is a "word" embedding technique that can be applied more generally to features. Word2Vec is extremely popular in language modeling. The embedding vectors produced by state-of-the-art implementations of Word2Vec capture a surprising level of the semantics of a language. That is, words that are similar in meaning are

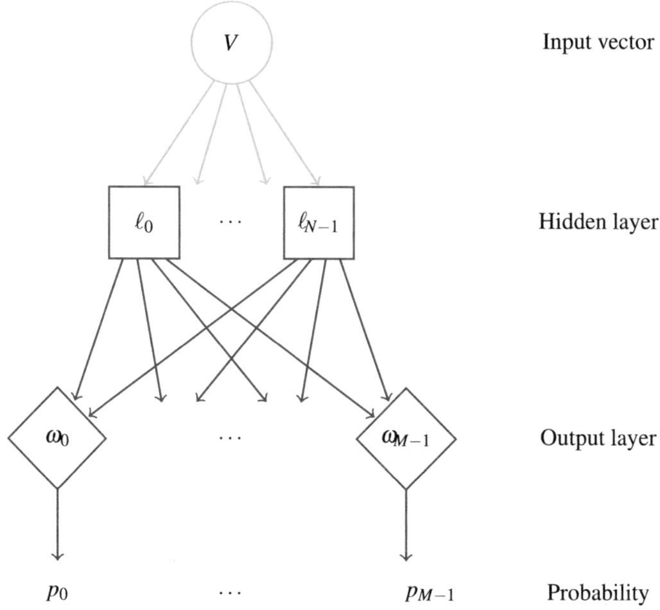

Fig. 1 Neural network to obtain Word2Vec embeddings

"close" in the Word2Vec embedding space [25]. An oft-cited example of the strength of Word2Vec is the following. If we let

$$w_0 = \text{"king"},\ w_1 = \text{"man"},\ w_2 = \text{"woman"},\ w_3 = \text{"queen"}$$

and $V(w_i)$ is the Word2Vec embedding of the word w_i, then $V(w_3)$ is the vector that is closest—in terms of cosine similarity—to

$$V(w_0) - V(w_1) + V(w_2)$$

Word2Vec is based on a shallow, two-layer neural networks, as illustrated in Fig. 1. Training such a model consists of determining the weights w_i based on a large training corpus [6]. These weights yield the Word2Vec embedding vectors.

In this paper, we compute Word2Vec embeddings based on extracted opcode sequence from malware samples. These word embeddings are then used as features in SVM classifiers. In this context, we can view the use of Word2Vec as a form of feature engineering.

One of the great strengths of Word2Vec is that training is extremely efficient. The key tricks that enable efficient training of such models are subsampling of frequent words, and so-called negative sampling, whereby only a subset of the weights that are affected by a training pair are adjusted at each iteration. For additional details on Word2Vec, see, for example [28].

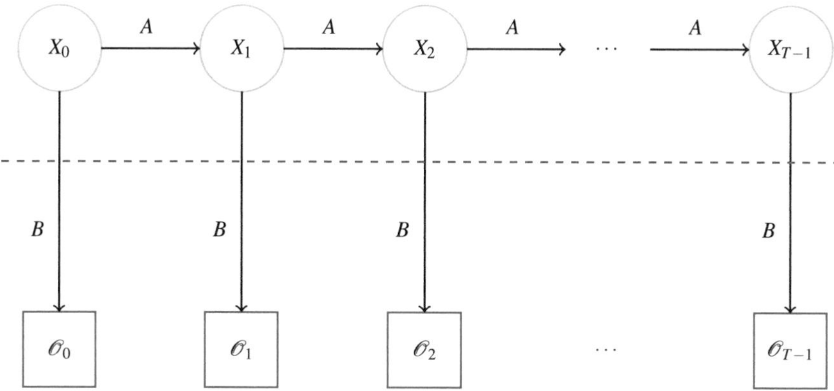

Fig. 2 Hidden Markov model

3.3.4 Hidden Markov Models

A Markov process is a statistical model that has states with known and fixed probabilities of state transitions. A hidden Markov model (HMM) extends this concept to the case where the states are "hidden," in the sense that they are not directly observable.

Figure 2 provides a generic view of an HMM. Here, the states X_i are determined by the row stochastic $N \times N$ matrix A. The states X_i are not directly observable, but as the name implies, the observations $\mathcal{O}_i$ can be observed. The hidden states are probabilistically related to the observations via the $N \times M$ row stochastic matrix B. Here, N is the number of hidden states of the model, and M is the number of distinct observation symbols, and T in Fig. 2 is the length of the observation sequence. There is also a row stochastic initial state distribution matrix, which is denoted as π. The three matrices, A, B, and π define an HMM, and we adopt the notation $\lambda = (A, B, \pi)$.

The following three HMM problems can be solved efficiently:

Problem 1 Given a model $\lambda = (A, B, \pi)$ and an observation sequence $\mathcal{O}$, we need to find $P(\mathcal{O} \mid \lambda)$. That is, an observation sequence can be scored to see how well it fits a given model.

Problem 2 Given a model $\lambda = (A, B, \pi)$ and an observation sequence $\mathcal{O}$, we can determine an "optimal" hidden state sequence. In the HMM sense, optimal means that we maximize the expected number of correct states. This is an expectation maximization (EM) algorithm.

Problem 3 Given $\mathcal{O}$ and a specified N, we can determine a model $\lambda = (A, B, \pi)$ that maximizes $P(\mathcal{O} \mid \lambda)$. This is, we can train a model to fit a given observation sequence.

In this research, we employ the algorithms for Problems 1 and 3 above. That is, we train HMMs, and we use trained HMMs to score samples.

3.3.5 Experimental Approach

To automatically determine points in time where significant evolutionary changes occur in malware families, we tag each sample in the family according to the date on which it was compiled. We also extract the opcode sequences from each of the malware samples.

As a first set of experiments, we train a series of linear SVMs directly on the extracted opcodes, as discussed above. We then attempt to improve on these results by considering several feature engineering techniques, in all cases using linear SVM weights and χ^2 graphs.

For our first attempt at feature engineering, we consider opcode n-grams. We then experiment using Word2Vec embeddings of the opocdes. Finally, we repeat the word embedding experiments, but based on the B matrices obtained from trained HMMs, instead of Word2Vec embeddings. We refer to this HMM-based word embedding technique as HMM2Vec.

As discussed above, when training, one class consists of all samples belonging to a specific family within a one-year time window, while the other class consists of the samples from the subsequent one-month time window. Such a model contrasts the family characteristics over a one month period to the characteristics of the previous one-year time interval. From each such model, we obtain a vector of linear SVM weights. Then we shift our time window one month ahead, and again train an SVM and obtain another vector of SVM weights.

For each set of experiments, we obtain a series of snapshots of the samples—in the form of linear SVM weights—based on overlapping sliding windows, where each SVM is trained over a one-year time-frame. Adjacent SVM weight vectors are based on one-month offsets. We use these SVM weight vectors as a basis for tracking changes in the underlying models, and we quantify those changes using the χ^2 statistics, as discussed above. Potential evolutionary points appear as spikes in the resulting χ^2 graph.

As a secondary test, for each significant spike in the χ^2 graph, we train two HMMs, one on either side of the spike. We then score samples on both sides of the spike using both HMMs. If the sample scores are observably different for each of these HMMs on each side of the spike, this serves to confirm that significant evolutionary change in the malware family has occurred.

For this secondary test, we can quantify the evolutionary effect by computing the χ^2-like evolution score

$$E = \frac{1}{n}\sum_{i=1}^{n}\frac{\left(\widehat{S}(x_i) - S(x_i)\right)^2}{S(x_i)}$$

where $S(x_i)$ is the HMM score of the sample x_i using the "correct" model and $\widehat{S}(x_i)$ is the score of x_i using the "incorrect" model. For example, if sample x_i occurs before the spike, then $S(x_i)$ is the score obtained using the model that was trained on data before the spike, and $\widehat{S}(x_i)$ is the score of x_i using the model trained after the spike. The larger the evolution score E, the stronger the evidence of evolution. Note that the $1/n$ factor is needed since the number of samples available differs for different families, and the number of samples might also differ for different spike computations within the same family.

4 Experiments and Results

In this section, we present and discuss the results of the experiments outlined in the previous section. First, we provide a graphical illustration of our HMM-based secondary test. Then we consider opcode-SVM experiments, followed by analogous SVM experiments based on opcode n-gram features. Neither of these techniques produce particularly strong results, and hence we then turn our attention to additional opcode-based feature engineering. Specifically, we apply word embedding techniques to opcode sequence and train SVM classifiers based on these engineered features. These experiments prove to be highly successful.

4.1 HMM-Based Secondary Test

The previous work in [29] is based on PE file features and uses linear SVM analysis to detect evolutionary changes in a malware family. We perform similar analysis in this paper, but based on opcode features and using word embedding techniques for feature engineering. In this paper, we also employ hidden Markov models as a secondary test to confirm suspected evolutionary changes.

As discussed above, once distinct spikes have been obtained from the χ^2 similarity graph, we train an HMM model on both sides such spikes using extracted opcode sequences. Both models are then used for scoring malware samples on either side of the spike. Here, we illustrate this secondary test for a specific case.

Figure 3a depicts the scores obtained when scoring samples before a χ^2 graph spike using both HMMs, while Fig. 3b gives the corresponding result for samples after the spike. In all cases, the scores are log likelihood per opcode (LLPO), that is, the scores have been normalized so as to be independent of the length of the opcode sequence.

In both Fig. 3a and b, we see that the scores are distinct for the two models on each sample tested. These results demonstrate that the model trained before and the model trained after the spike are significantly different which, in turn, indicates that the samples used to train the models differ significantly. This is a clear sign of an evolutionary branch point in the malware family.

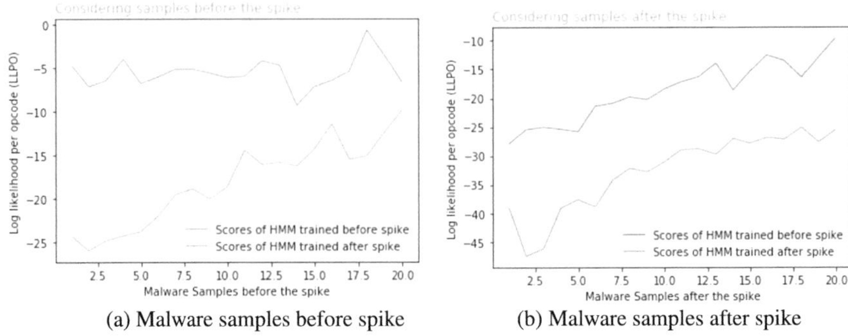

(a) Malware samples before spike (b) Malware samples after spike

Fig. 3 Hidden Markov model trained before and after the spike

4.2 *Opcode-SVM Results*

In this section, we discuss experiments on 15 malware families, based on opcode sequences and SVMs. That is, opcodes sequences are used directly as features in linear SVM models, with the resulting model weights used to compute χ^2 graphs.

Figure 4 gives such a χ^2 graph for the Zegost family, which has 506 malware samples from the years 2008 through 2011. We observe multiple spikes in the graph but the secondary HMM test does not yield impressive results for any of these spikes. Hence, we conclude that this particular test does not reveal any strong evolution result for the Zegost family.

We also computed opcode-SVM χ^2 graphs for the remaining 14 malware families, nine of which appear in Fig. 5. For Adload and BHO, we observe that the graphs

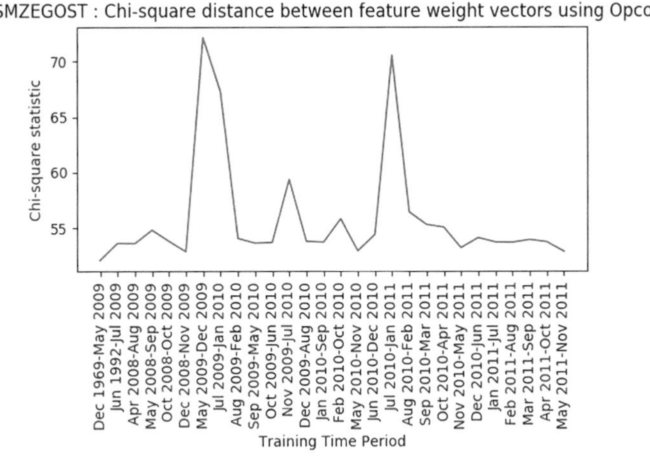

Fig. 4 Opcode-SVM χ^2 similarity graph for Zegost

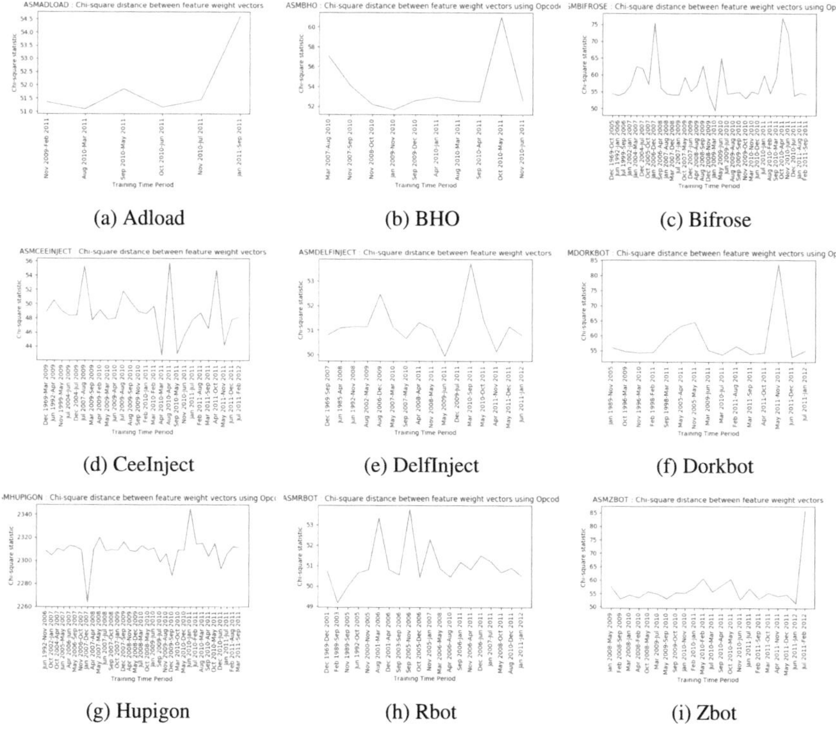

(a) Adload (b) BHO (c) Bifrose

(d) CeeInject (e) DelfInject (f) Dorkbot

(g) Hupigon (h) Rbot (i) Zbot

Fig. 5 Opcode-SVM χ^2 graphs for selected families

do not show any significant spikes except at the last time period. We are not able to perform the secondary HMM test at the extreme endpoint, so we are not able to confirm or deny these as evolution points.

From Fig. 5, we observe considerable fluctuation in the graphs for Bifrose, CeeInject, Hupigon, and Rbot, but none of these fluctuations stand out as clear points of possible evolutionary change in the families. That is, for these families, it appears that we only observe background noise.

4.3 Opcode n-Gram-SVM Results

Next, we consider analogous experiments to those of the previous section, but based on opcode n-grams. This can be viewed as a first attempt at feature engineering. As with the previous experiments, we train linear SVMs on these features and construct χ^2 graphs. We consider overlapping n-grams, and we experimented with $n = 2, 3, 5, 10$ on each of the 15 families.

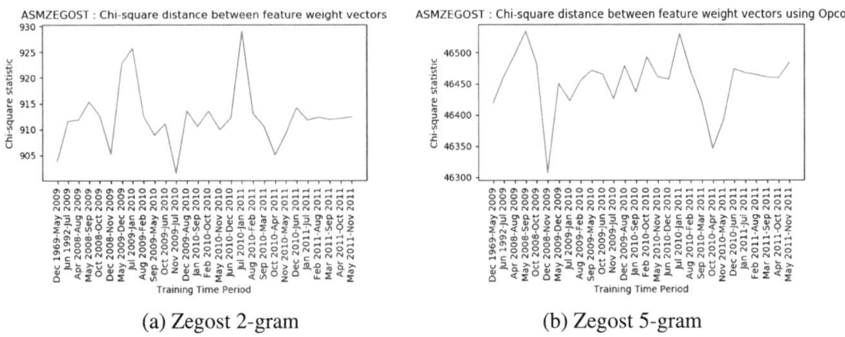

(a) Zegost 2-gram (b) Zegost 5-gram

Fig. 6 Opcode-n-gram-SVM χ^2 graphs for Zegost

Examples of the results of these n-gram experiments are given in Fig. 6, which shows 2-gram and 5-gram χ^2 graphs for the Zegost family. For both of these cases, we see only noisy results, with no clear evolutionary points. These results are typical of our n-gram experiments and we conclude that opcode n-grams are not useful for our purposes.

Next, we consider Word2Vec and HMM2Vec word embeddings These feature engineering techniques prove to be more effective for detecting evolutionary changes, with HMM2Vec giving us our strongest results.

4.4 Opcode-Word2Vec-SVM Results

Here, we generate Word2Vec models based on opcode sequences. We then train linear SVMs over each time window, based on these Word2Vec embeddings, and we compute χ^2 graphs of the SVM weights. As above, spikes in this graph indicate points in time where evolution might have occurred.

Figure 7a and b gives the χ^2 graphs for the Zegost family with Word2Vec embeddings of length 2 and 3, respectively. From Fig. 7, we observe that feature weights in certain time windows diverge significantly from their average values. Specifically, these time periods are November 2010 and May 2011, and these are the points in time where significant evolution in the family may have taken place. We also note the similarity between the results for embedding vector lengths 2 and 3. This can be viewed as a sign of the stability of the underlying approach, and serves to provide additional confidence in the putative evolution points.

Applying our secondary HMM verification technique to the spike in Fig. 7, we obtain the results in Fig. 8, which confirm that the malware family has evolved at this point. Since vector lengths of 2 and 3 give us consistent results for Zegost, for our remaining Word2Vec experiments, we use embedding vectors of length 2 in all cases.

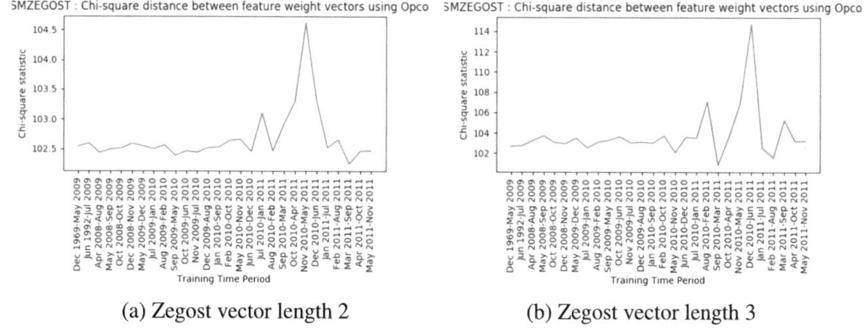

(a) Zegost vector length 2

(b) Zegost vector length 3

Fig. 7 Opcode-Word2Vec-SVM χ^2 graphs for Zegost

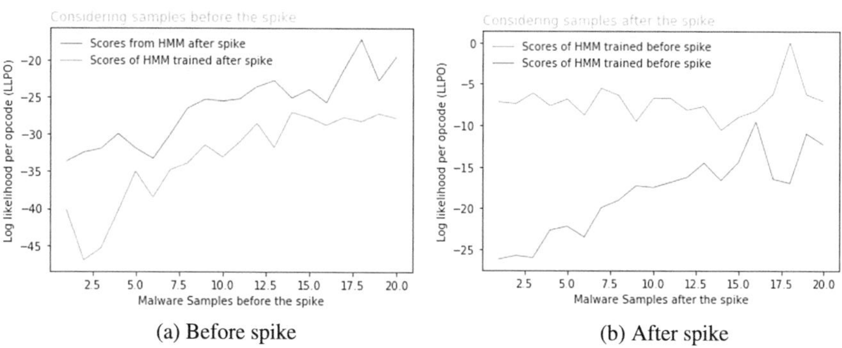

(a) Before spike

(b) After spike

Fig. 8 Zegost HMM secondary test for opcode-Word2Vec-SVM

Figure 9 shows our Word2Vec based χ^2 graphs for eight additional malware families. Four of these families—BHO, Bifrose, Adload, and Vobfus—perform well with this approach, in the sense that we detect clear spikes in their graphs.

For the remaining four families in Fig. 9, namely, CeeInject, Dorkbot, Hupigon, and Rbot, we do not detect any significant spikes in their χ^2 graphs. Additional secondary HMM tests showing evolution appear in Fig. 10, while Fig. 11 gives an example of a secondary test that shows no evolution. It is evident that the results of these opcode-Word2Vec-SVM experiments are a major improvement over the experiments considered above.

Of course, it is possible that there is no evolution to be detected, in some of these families. But, over the extended time periods under consideration, we believe it likely that evolution has occurred, which suggests that the Word2Vec features are simply not sufficiently sensitive to detect changes in all cases. In the next section, we consider another word embedding technique.

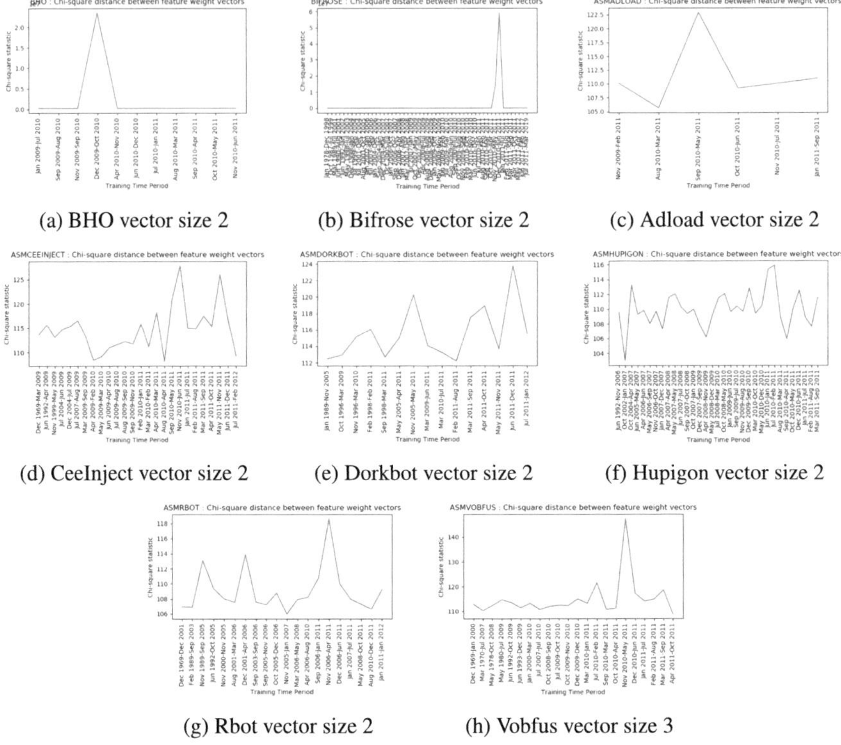

(a) BHO vector size 2 (b) Bifrose vector size 2 (c) Adload vector size 2

(d) CeeInject vector size 2 (e) Dorkbot vector size 2 (f) Hupigon vector size 2

(g) Rbot vector size 2 (h) Vobfus vector size 3

Fig. 9 Opcode-Word2Vec-SVM χ^2 graphs for selected families

4.5 *Opcode-HMM2Vec-SVM Results*

The experiments in this section are essentially the same as those in the previous section, except that we use HMM2Vec embeddings in place of Word2Vec embeddings. As discussed above, HMM2Vec embeddings use obtained from the B matrix of a trained HMM.

Figure 12a and b give HMM2Vec based χ^2 graphs for Zegost, using one random start and 10 random restarts, respectively. Note that these results are based on HMMs with $N = 2$ hidden states, which gives us embedding vectors of length 2. Since our HMM training algorithm is a hill climb technique, multiple random restarts often enable us to find a stronger model. For the Zegost results in Fig. 12, random restarts appear to offer little, if any, advantage. Consequently, for the remaining experiments in this section, we train a single HMM model, and we do not perform any random restarts.

We experimented with $N = 2$ and $N = 3$ hidden states (giving us embedding vectors of length 2 and 3, respectively), but we did not find any improvement

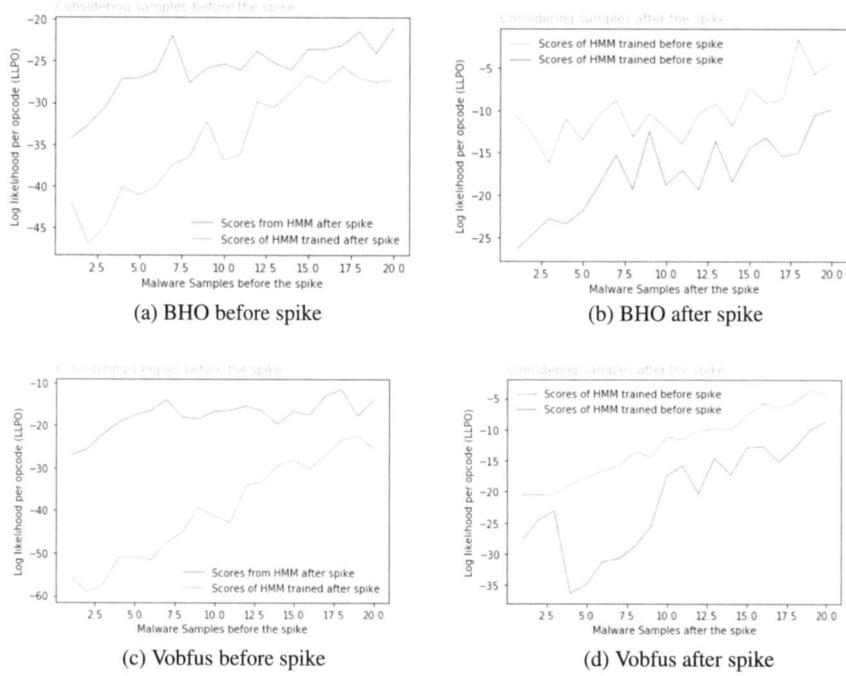

(a) BHO before spike

(b) BHO after spike

(c) Vobfus before spike

(d) Vobfus after spike

Fig. 10 HMM secondary tests for opcode-Word2Vec-SVM showing evolution

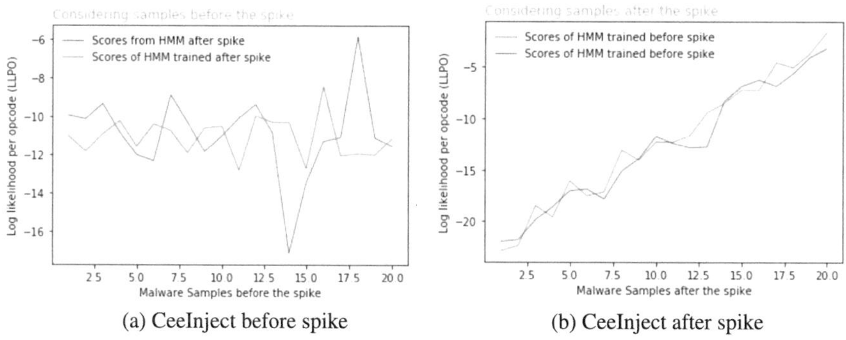

(a) CeeInject before spike

(b) CeeInject after spike

Fig. 11 HMM secondary tests for opcode-Word2Vec-SVM showing no evolution

using $N = 3$. Hence, we use HMM2Vec embedding vectors of length 2 in all experiments below.

Figure 13 shows the χ^2 graphs for 8 additional families based on HMM2Vec embedding vectors. Overall, this HMM2Vec-SVM approach seems to provide better results than the Word2Vec-SVM technique in the previous section, as we can detect more malware evolution using the HMM2Vec feature engineering.

Based on the graphs in Fig. 13, we make the following observations.

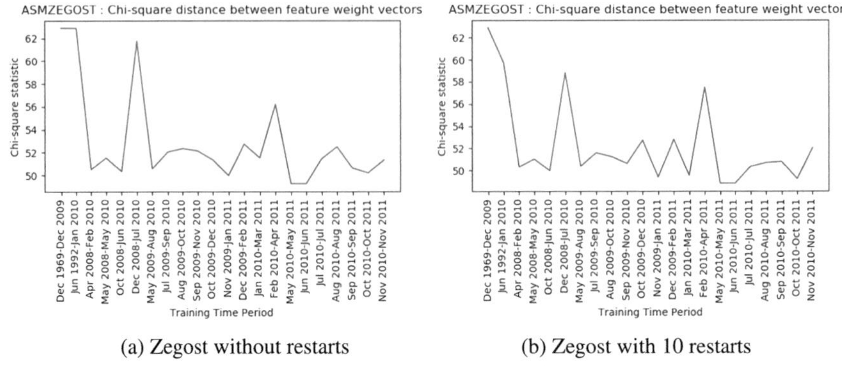

(a) Zegost without restarts

(b) Zegost with 10 restarts

Fig. 12 Opcode-HMM2Vec-SVM χ^2 graphs for Zegost

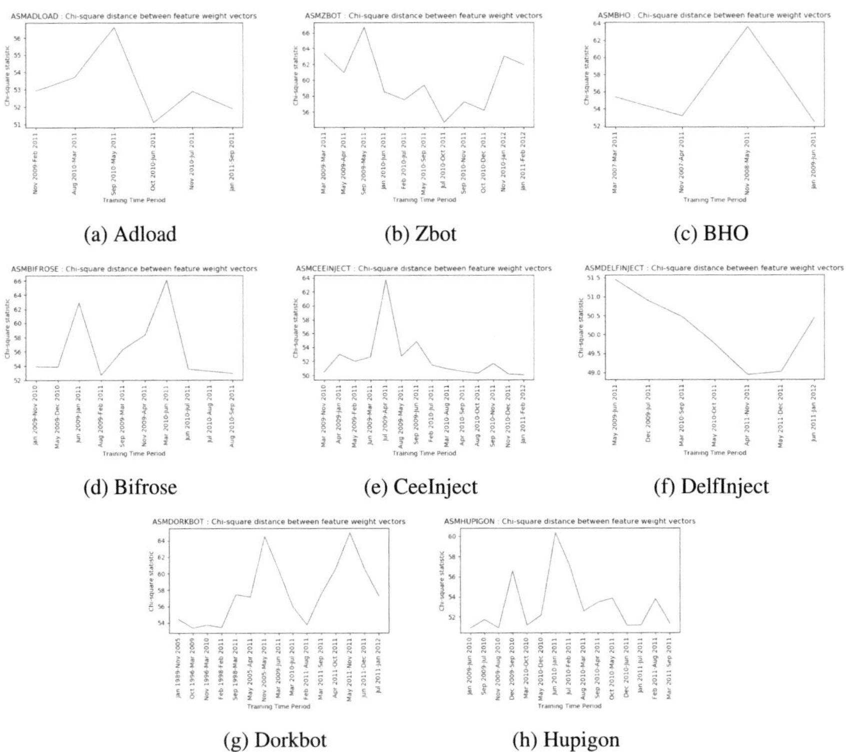

(a) Adload

(b) Zbot

(c) BHO

(d) Bifrose

(e) CeeInject

(f) DelfInject

(g) Dorkbot

(h) Hupigon

Fig. 13 Opcode-HMM2Vec-SVM χ^2 for selected families

Adload — An evolutionary event takes place in the time window Sep 2010–May 2011.

Zbot — No significant spike is observed in the χ^2 graph.

BHO — Significant evolution takes place in the time window Nov 2008–May 2011

Bifrose — Malware evolution takes place in the time window March 2010–June 2011. The other spike in the period June 2009–Jan 2011 is a part of noise in the data. This was confirmed by training HMMs on both sides.

CeeInject — Evolution occurs in the time window June 2009–April 2011.

DelfInject — No significant spike is observed in the χ^2 graph.

Dorkbot — No significant spike is observed in the χ^2 graph.

Hupigon — A significant spike in the time period June 2010–Jan2011 is observed in the graph.

From Fig. 13, we conclude that we can observe significant spikes in almost all families using HMM2Vec-SVM analysis. For the families Adload, BHO, Bifrose, CeeInject, and Hupigon we observe significant spikes in the χ^2 distribution graph. HMM secondary test confirming evolution for some of these cases appear in Fig. 14. In Fig. 15, we give results of HMM secondary tests that do not reveal evolution.

Comparing the families in which we could detect evolutionary changes using Word2Vec with those detected using HMM2Vec, we observe that the evolutionary points obtained using Word2Vec are also found using HMM2Vec. Yet, the HMM2Vec technique provides additional evolutionary points, indicating that it is more sensitive to change than Word2Vec embeddings. Overall, HMM2Vec performed better than any of the other approaches that we considered in this paper.

5 Conclusion and Future Work

Previous research has shown that analysis based on PE file features and linear SVM models can be useful in detecting malware evolution [29]. In this paper, we expanded on—and improved upon—this previous work in several ways. First, we considered opcode features, rather than PE file features. Our intuition was that opcode-based features would be more sensitive to the types of changes that we would like to detect, and our results support this intuition. Second, we experimented with various feature engineering techniques, and we found that vector embeddings increase the sensitivity of the SVM analysis. Thirdly, we showed that a secondary test using HMM techniques can be used to verify that suspected evolutionary points in the timeline.

We experimented with a variety of techniques, and our best results were obtained using an approach that we refer to as opcode-HMM2Vec-SVM. In this technique, we use mnemonic opcodes as raw features, then generate HMM2Vec encodings of the opcodes, which serve as features for linear SVMs, with the SVMs trained over sliding windows of time. The resulting SVM weights are compared using a χ^2 statistic, and we graph this statistic over the available timeline. Spikes in the χ^2

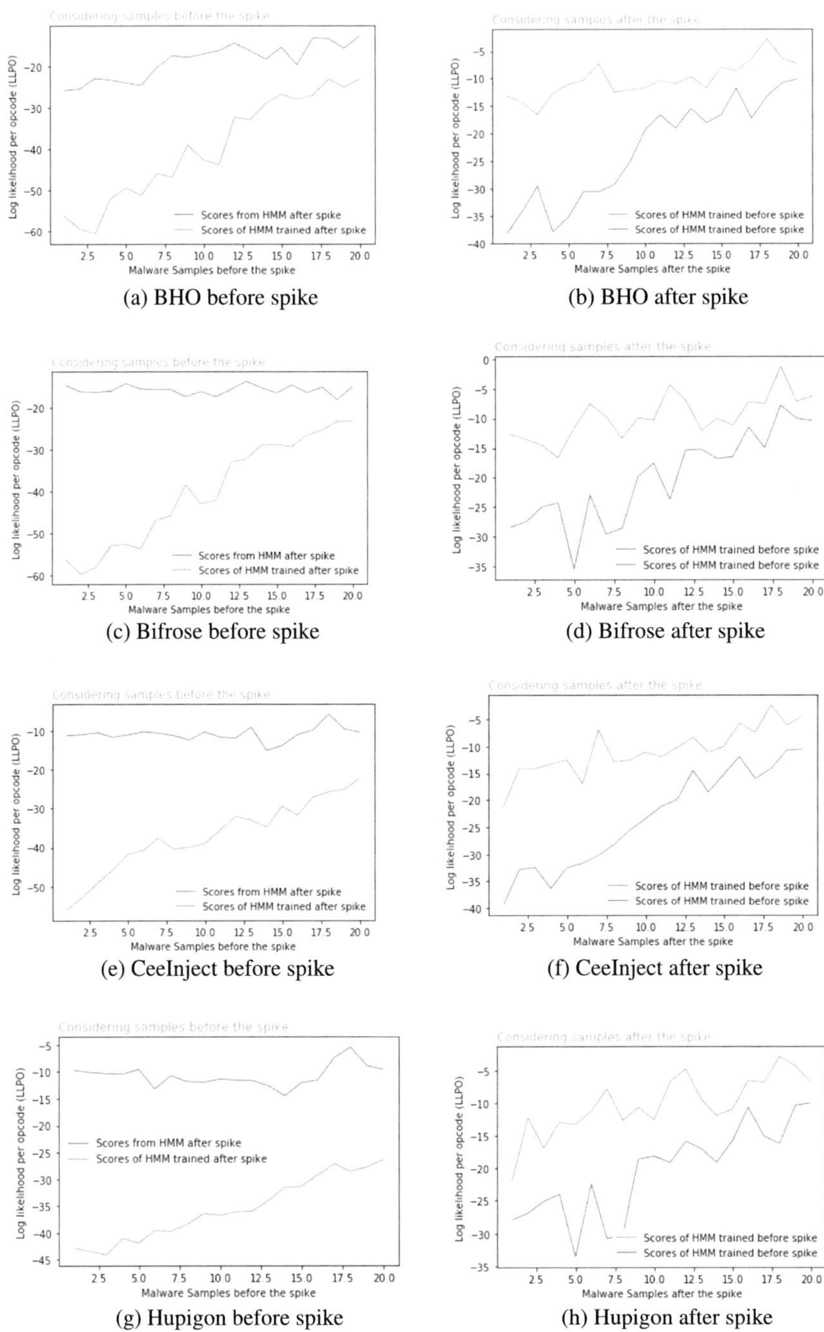

(a) BHO before spike

(b) BHO after spike

(c) Bifrose before spike

(d) Bifrose after spike

(e) CeeInject before spike

(f) CeeInject after spike

(g) Hupigon before spike

(h) Hupigon after spike

Fig. 14 HMM secondary tests for opcode-HMM2Vec-SVM showing evolution

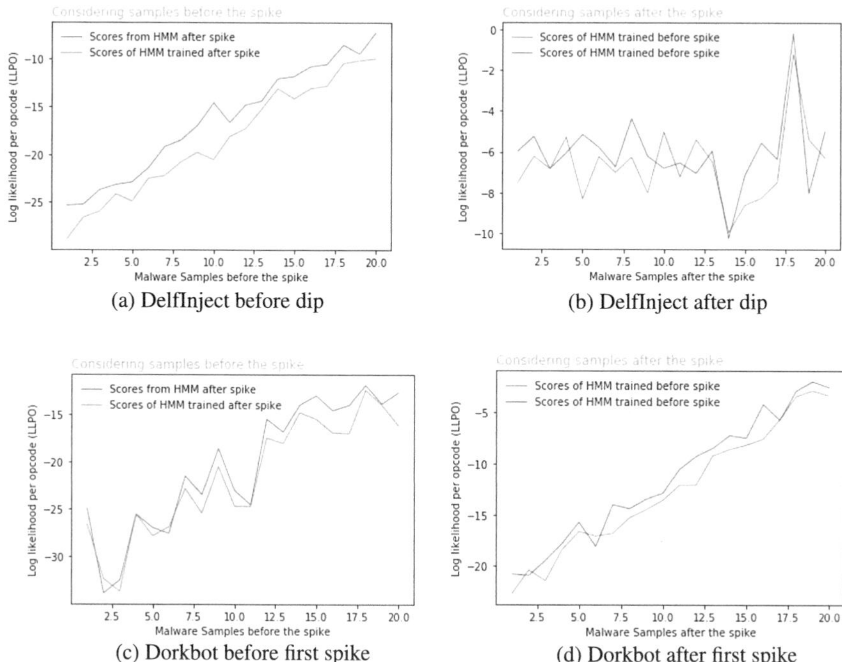

(a) DelfInject before dip

(b) DelfInject after dip

(c) Dorkbot before first spike

(d) Dorkbot after first spike

Fig. 15 HMM secondary tests for opcode-HMM2Vec-SVM showing no evolution

graph serve as indicators of likely evolutionary change. We were then able to further confirm evolutionary changes using a secondary test based on training HMMs on either side of a spike. This overall approach was more sensitive than previous work, in the sense that we were able to detect additional changes in the codebase of various families, and it was more precise, since we have a secondary test available to confirm (or deny) putative evolutionary changes.

In the realm of future work, additional machine learning techniques and additional features and engineering strategies could be considered. For example, neural network techniques could be used in place of SVMs, with multiclass output probabilities playing the role of the linear SVM weights. Another option would be to elevate our HMM-based secondary test to the role of the primary test. This might enable a more fine-grained analysis of the timeline, as relatively little data is needed for HMM training. With respect to feature engineering, dimensionality reduction techniques would be a natural topic to consider. The use of dynamic features might add value as well, although the additional complexity involved with collecting such features might be a concern.

References

1. Aycock, John. 2010. *Computer viruses and malware*. Berlin: Springer.
2. Bai, Jin, Shi Mu, and Guo Zou. 2014. The application of machine learning to study malware evolution. *Applied Mechanics and Materials* 530–531(Advances in Measurements and Information Technologies): 875–878.
3. Barat, Marius, Dumitru-Bogdan Prelipcean, and Drago Gavrilu. 2013. A study on common malware families evolution in 2012. *Journal of Computer Virology and Hacking Techniques* 9 (4): 171–178.
4. Chen, Zhongqiang, Mema Roussopoulos, Zhanyan Liang, Yuan Zhang, Zhongrong Chen, and Alex Delis. 2012. Malware characteristics and threats on the internet ecosystem. *The Journal of Systems & Software* 85 (7): 1650–1672.
5. Damodaran, Anusha, Fabio Di Troia, Corrado Visaggio, Thomas Austin, and Mark Stamp. 2017. A comparison of static, dynamic, and hybrid analysis for malware detection. *Journal of Computer Virology and Hacking Techniques* 13 (1): 1–12.
6. Gilyadev, Julian. 2017. Word2vec explained. https://israelg99.github.io/2017-03-23-Word2Vec-Explained/.
7. Gupta, A., P. Kuppili, A. Akella, and P. Barford. 2009. An empirical study of malware evolution. In *2009 first international communication systems and networks and workshops*, 1–10. IEEE.
8. Kim, Samuel. 2018. PE header analysis for malware detection. Master's thesis, San Jose State University, Department of Computer Science.
9. Mercaldo, Francesco, Andrea Di Sorbo, Corrado Aaron Visaggio, Aniello Cimitile, and Fabio Martinelli. 2018. An exploratory study on the evolution of android malware quality. *Journal of Software: Evolution and Process* 30 (11): n/a–n/a.
10. Microsoft. Win32 rbot detected with windows defender antivirus. 2005. https://www.microsoft.com/en-us/wdsi/threats/malware-encyclopedia-description?Name=Win32%2FRbot.
11. Microsoft. Adware: Win32 hotbar detected with windows defender antivirus. 2006. https://www.microsoft.com/en-us/wdsi/threats/malware-encyclopedia-description?Name=Adware%3AWin32%2FHotbar.
12. Microsoft. Backdoor: Win32 hupigon detected with windows defender antivirus. 2006. https://www.microsoft.com/en-us/wdsi/threats/malware-encyclopedia-description?Name=Backdoor%3AWin32%2FHupigon.
13. Microsoft. Virtool: Win32 ceeinject detected with windows defender antivirus. 2007. https://www.microsoft.com/en-us/wdsi/threats/malware-encyclopedia-description?Name=VirTool%3AWin32%2FCeeInject.
14. Microsoft. Virtool: Win32 delfinject detected with windows defender antivirus. 2007. https://www.microsoft.com/en-us/wdsi/threats/malware-encyclopedia-description?Name=VirTool:Win32/DelfInject&ThreatID=-2147369465.
15. Microsoft. Virtool: Win32 vbinject detected with windows defender antivirus. 2010. https://www.microsoft.com/en-us/wdsi/threats/malware-encyclopedia-description?Name=VirTool:Win32/VBInject&ThreatID=-2147367171.
16. Microsoft. Win32 vobfus detected with windows defender antivirus. 2010. https://www.microsoft.com/en-us/wdsi/threats/malware-encyclopediadescription?name=win32%2Fvobfus.
17. Microsoft. Win32 winwebsec detected with windows defender antivirus. 2010. https://www.microsoft.com/security/portal/threat/encyclopedia/entry.aspx?Name=Win32%2fWinwebsec.
18. Microsoft. Win32 obfuscator detected with windows defender antivirus. 2011. https://www.microsoft.com/en-us/wdsi/threats/malware-encyclopedia-description?Name=Win32%2FObfuscator.
19. Microsoft. Win32 zbot detected with windows defender antivirus. 2011. http://www.symantec.com/securityresponse/writeup.jsp?docid=2010-011016-3514-99.
20. Microsoft. Win32 zegost detected with windows defender antivirus. 2011. https://www.symantec.com/security-center/writeup/2011-060215-2826-99.

21. Microsoft. Worm: Win32 dorkbot detected with windows defender antivirus. 2011. https://www.microsoft.com/en-us/wdsi/threats/malware-encyclopedia-description?Name=Worm%3AWin32/Dorkbot.
22. Microsoft. Win32 bifrose detected with windows defender antivirus. 2012. https://www.trendmicro.com/vinfo/us/threat-encyclopedia/malware/bifrose.
23. Nappa, Antonio, M. Rafique, and Juan Caballero. 2015. The malicia dataset: Identification and analysis of drive-by download operations. *International Journal of Information Security* 14 (1): 15–33.
24. Ouellette, J., A. Pfeffer, and A. Lakhotia. 2013. Countering malware evolution using cloud-based learning. In *Proceedings of the 8th international conference on malicious and unwanted software*, 85–94.
25. Popov, Igor. 2017. Malware detection using machine learning based on word2vec embeddings of machine code instructions. In *2017 Siberian symposium on data science and engineering (SSDSE)*, 1–4. IEEE.
26. Rezaei, Saeid, Fereidoon Rezaei, Ali Afraz, and Mohammad Reza Shamani. 2016. Malware detection using opcodes statistical features. In *2016 8th international symposium on telecommunications (IST)*, 151–155. IEEE.
27. Stamp, Mark. 2018. *Introduction to machine learning with applications in information security*. Boca Raton: CRC Press, Taylor & Francis Group.
28. Stamp, Mark. 2019. Alphabet soup of deep learning topics. https://www.cs.sjsu.edu/~stamp/RUA/alpha.pdf.
29. Wadkar, Mayuri, Fabio Di Troia, and Mark Stamp. 2020. Detecting malware evolution using support vector machines. *Expert Systems with Applications* 143.

Reanimating Historic Malware Samples

Paul Black, Iqbal Gondal, Peter Vamplew, and Arun Lakhotia

Abstract Many types of malicious software are controlled from an attacker's command and control (C2) servers. Anti-virus organizations seek to defeat malware attacks by requesting removal of C2 server Domain Name Server (DNS) records. As a result, the life span of most malware samples is relatively short. Large datasets of historical malware samples are available for countermeasures research. However, due to the age of these malware samples, their C2 servers are no longer available. To cope with high volumes of malware production, malware analysis is increasingly performed using machine learning techniques. Dynamic analysis is commonly used for feature extraction. However, due to the absence of their C2 servers, after initialization, malware samples may exit or loop attempting to establish C2 server connections and, as a result, no longer exhibit their original capabilities. Therefore, partial execution of historical malware samples in a sandbox results in features that differ from those that would be extracted in-the-wild, thus invalidating the results of any machine learning research based on these features. One approach to extracting accurate features is to build an emulated C2 server to provide an environment that allows control of the full capabilities of the malware in an isolated environment. To illustrate the benefits of building C2 server emulators, this chapter provides examples of techniques for the creation of C2 server emulators for three malware families (Zeus, CryptoWall, and CryptoLocker) using manual reverse engineering techniques and a review of semi-automated techniques for the construction of C2 server emulators.

P. Black (✉) · I. Gondal
ICSL, Federation University, Ballarat, Australia
e-mail: p.black@federation.edu.au

I. Gondal
e-mail: iqbal.gondal@federation.edu.au

P. Vamplew
SEIT, Federation University, Ballarat, Australia
e-mail: p.vamplew@federation.edu.au

A. Lakhotia
CMIX, University of Louisiana at Lafayette, Lafayette, USA
e-mail: arun@louisiana.edu

M. Stamp et al. (eds.), *Malware Analysis Using Artificial Intelligence and Deep Learning*, https://doi.org/10.1007/978-3-030-62582-5_13

1 Introduction

Many types of malware, including information-stealing malware, ransomware, and Remote Access Trojans (RATs) are controlled from an attacker's Command and Control (C2) servers [3]. Anti-virus organizations seek to defeat malware attacks by requesting removal of C2 server Domain Name Server (DNS) records. For discussion, a historical malware sample is defined as a malware sample that has had its C2 servers removed. Large datasets of historical malware samples are available for academic experiments. However, due to the age of these malware samples, their C2 servers are no longer available. To cope with high volumes of malware production, malware analysis is increasingly performed using machine learning techniques [4]. Dynamic analysis is commonly used for feature extraction. However, due to the absence of their C2 servers, after initialization, malware execution may exit, or loop attempting to establish C2 server connections. As a result, the command interface of historic malware samples is no longer controlled. This results in the extraction of features that differ from those that would be extracted in-the-wild, thus invalidating the results of machine learning research based on these features.

It is noted that research techniques exist for automatic protocol analysis of malware [13]. However, these techniques depend on malware communications with live C2 servers. The usage of the malware capabilities is determined by the malware operators, and live testing may not reveal the full extent of the malware's capability. Other issues related to performing research with live malware include difficulties in obtaining a consistent supply of live malware, unknown configuration, unknown triggering conditions, detection of the analysis IP address (mitigated by the use of an anonymizing proxy), or the malware operators gaining access to the analysis VM via malware provided interfaces.

Researchers have recognized the need to prevent malware experiments from causing harm on the Internet. Research systems have been built to provide containment of malware research [23]. However, these systems do not address the C2 server problems faced when performing experiments with historical datasets. Internet simulator programs [20] may be used as part of a malware analysis environment and can provide generic responses to requests for common Internet services. A malware process may request a connection to a common website to perform a connectivity check, and an Internet simulator may be able to satisfy this request. However, if a connection to a C2 server or other attacker-controlled infrastructure is requested, an Internet simulator will not be able to respond with the protocol required by the malware.

The Botnet Evaluation Environment (BEE) provides an isolated environment for botnet research with emulated C2 servers for execution of the Agobot, SDBot, GTBot, Phatbot, and Spybot malware [6]. An isolated Waledac botnet was created by reverse engineering the Waledac malware and identifying the Waledac botnet protocol. An emulated C2 server was built to support this protocol, and a 3000 node Waledac botnet was built. This isolated botnet was used to research security vulnerabilities that could be used to take down the Waledac botnet [12].

To illustrate the benefits of building C2 server emulators for machine learning purposes, this chapter provides examples of techniques for the creation of C2 server emulators for three malware families (Zeus, CryptoWall, and CryptoLocker) using manual reverse engineering techniques.[1] This chapter also provides a review of semi-automated techniques for the construction of C2 server emulators.

1.1 Motivation

At a high level, the need to build a C2 server emulator will be the result of the following requirements:

- The need to perform research using historical malware samples,
- The need to control the full capabilities of a malware sample,
- The need to perform the research in an isolated environment.

The construction of C2 server emulators has the following benefits: the ability to control the malware through its network interface allows the execution of the full capabilities of the malware and the extraction of features that would otherwise not be possible using historical malware samples. Using an emulated C2 server allows the testing of malware samples in isolation from the Internet, which prevents criminal groups from becoming aware of the research.

1.2 Emulator Architecture

The architecture of a C2 server emulator will be similar irrespective of whether a manual or semi-automated process is used to construct the emulator. A representative C2 server emulator consists of an isolated network using two or more virtual machines (VMs). One VM (VM1) is configured to run the C2 server emulator script and a DNS simulator, while another VM (VM2) is configured to run the selected malware samples and any related programs that will be attacked by this malware. The DNS simulator resolves requests from the malware VM, and malware protocol requests are read by the C2 emulator. An illustration of this architecture is shown in Fig. 1. C2 server emulators may be created using either manual or semi-automated construction techniques. Sections 2–4 discuss manual construction, and Sect. 5 provides a review of semi-automated construction techniques.

---

[1]The datasets and code related to this research are available on request from the corresponding author.

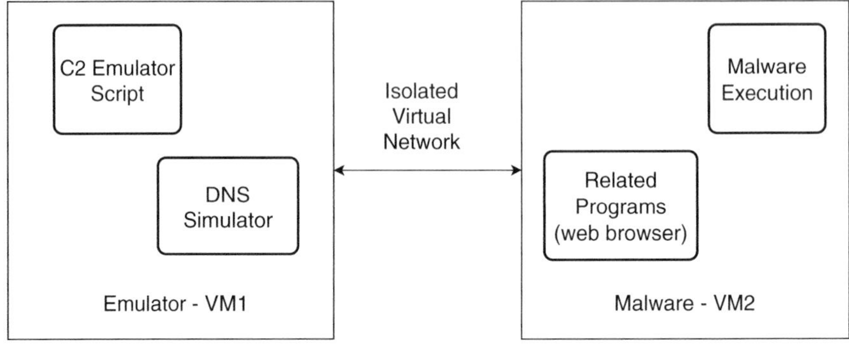

Fig. 1 C2 emulator architecture

2 Manual Construction

The creation of C2 emulators provides a test harness that allows the full capability of historic malware samples to be controlled in an isolated network. The process for the manual construction of C2 emulators can be described in an abstract manner as a process of the guided discovery of the communication and command processing paths of a malware sample using a debugger and the corresponding iterative development of a script to generate network traffic to control this execution path. A difficulty with this high-level description of the C2 server emulator construction process is that there may be difficulties in fully understanding how to implement this process. A malware analysis environment using a manually constructed C2 emulator is described in [18]. To illustrate the manual construction process, Sects. 3 and 4 provide examples of the manual construction of C2 server emulators for a common information-stealing malware and two ransomware families.

The requirement for the emulation of a malware C2 server arose from a research project using machine learning for the detection of webinjects. Webinjects are malicious HTML that are injected into web browser sessions and are used to steal banking credentials and to illegally transfer funds [16]. Information stealing malware targeting banking infrastructure (banking malware, banking trojans) contain facilities for intercepting credentials prior to encryption and injecting content into Internet banking sessions. This is performed by injecting malware into the browser process and gaining control of networking Application Programming Interface (API) functions with the use of user-mode hooking techniques [7]. Three options were considered for webinject generation:

- The use of live malware to perform webinjects, Zarathustra, and Prometheus performed webinject detection using live malware samples [9, 14],
- The use of Java-Script methods to inject code into the browser session,
- The creation an emulated C2 server that can be used in conjunction with a historical malware sample to perform user specified webinjects.

The problems associated with using live malware for research purposes have been discussed previously. While Java-Script methods can be used to inject HTML code into a browser session, this injected HTML may differ from webinjects created by malware. The time required to perform the reverse engineering work is a significant consideration when deciding whether to build an emulated C2 server. However, this may be offset by the significant benefits of being able to control the full capabilities of the malware on an isolated network and the collection of more representative features.

The following sections provide the details of the construction of the C2 server emulators for the Zeus V2, CryptoLocker, and CryptoWall malware.

3 Zeus C2 Server Emulator

The Zeus v2 malware was selected for this research due to familiarity with this malware from previous research. The Zeus C2 server emulator provides the capability to create custom webinjects on an isolated network and to capture the modified webpages for use in a webinject detection machine learning system [24]. In the following description, class and function names (e.g., Core::GetBaseConfig) are taken from the leaked Zeus source code [32]. A Zeus v2.1.0.1 malware sample with an MD5 hash of `a2a21d66f72ee53cfbc2dcfe929ffaba` was used in this research. This malware was unpacked using a custom static unpacker. The unpacked Zeus sample was loaded into the Interactive Disassembler (IDA). This IDA database was used to record the malware execution and to determine suitable API calls for breakpoints.

The Zeus v2 malware has an anti-analysis mechanism known as hardware locking. When Zeus malware infects a computer, a copy of the malware is installed in the filesystem, and a block of encrypted binary data is embedded into this installed malware. This encrypted data includes the malware's installation directory and a Globally Unique Identifier (GUID) that was generated from the computer's hard disk [31]. When a previously installed Zeus sample is executed in an analysis environment, execution on a new computer will be detected, and the malware will exit.

The hardware locking test was disabled by editing the machine code in the Core::EntryPoint function, and the jump instruction controlling the call to the Exit-Process API was overwritten with no operation (NOP) instructions. This was the only change that was made to the Zeus malware.

The guiding principle in building the Zeus C2 server emulator was to perform the minimum amount of reverse engineering needed to produce a C2 server emulator. Two VMs were used where VM1 was running the python C2 server emulator and the Internet simulator, and VM2 was running the Zeus malware sample and Internet Explorer version 8 for the webinjects testing. When the Zeus sample is executed, a copy of this malware is injected into the Explorer process, and the injected malware attempts to connect to the C2 server.

```
data_before
<title>Savings</title>
end\_before

data_inject
<p><font color="red"><b>ICSL Web Inject test framework</b></font></p>
end_inject

url_target
http://redacted.com.au/check.aspx?p=52
end_target
```

Fig. 2 Simplified Zeus configuration

The Zeus configuration is an encrypted binary data structure containing text specifying the target URL, injection start pattern, and the corresponding webinjects. The Zeus configuration is created by the Zeus configuration builder, subject to the malware author's security controls. To simplify the researcher workflow, the emulated C2 server uses a simplified text based configuration containing the targeted URLs, injection start patterns, and the corresponding webinjects. An example of the simplified configuration used by the emulated Zeus C2 server is given in Fig. 2. The C2 server emulator injects additional JavaScript to dump the DOM of the injected webpage into a shared host directory for analysis.

An initial C2 server emulator returning a response of 256 null bytes was created using the python BaseHttpServer class. On VM2, a debugger was attached to the explorer.exe process, a breakpoint was set at the InternetReadFile API, and the Zeus malware sample was executed. The emulated C2 server returned a response, and the breakpoint on the InternetReadFile API was hit in the Wininet::DownloadData function. Single stepping in the debugger was continued until the BinStorage::Unpack function was called where the following operations were observed:

- RC4 decryption of the received configuration,
- Recursive XOR decoding of the decrypted configuration Crypt::_visualEncrypt,
- Checking of the MD5 signature stored in the header of the decoded configuration,
- The writing of the encrypted Zeus configuration to the Windows registry.

Based on these observations, the C2 server emulator was updated to use RC4 encryption, recursive XOR encoding, and MD5 signing of the configuration data. A flowchart showing the steps involved in the creation of the encrypted Zeus configuration is given in Fig. 3.

A full explanation of the Zeus configuration and webinjects processing would require excessive detail. The following provides a high-level view of the structure of the Zeus configuration and the operation of the malware in the browser. The Zeus configuration file consists of the following sections: header, filters, a number of webinject sections, and an injects list containing the targeted URLs.

Using a debugger to follow the execution of the injected Zeus code in the web browser was necessary in order to debug the processing of the encrypted Zeus configuration created by the C2 server emulator and to determine the minimum configuration sections required to allow successful webinjects. To gain control of the Zeus

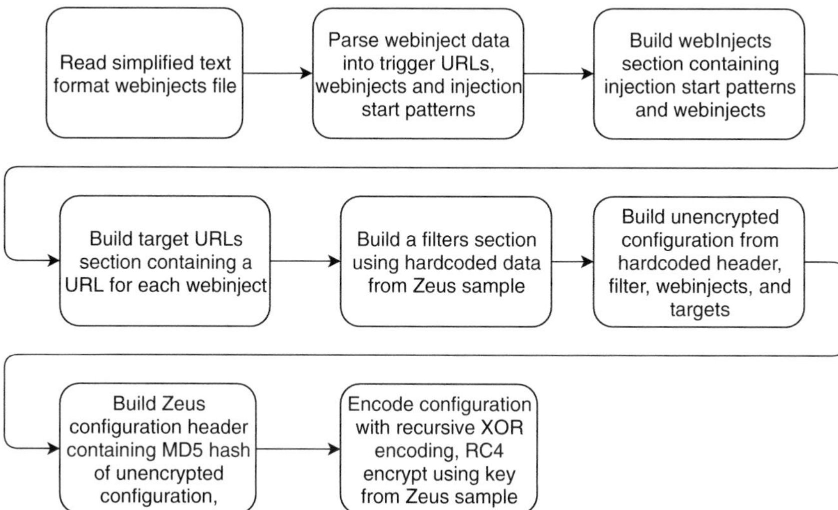

Fig. 3 Creation of encrypted Zeus configuration

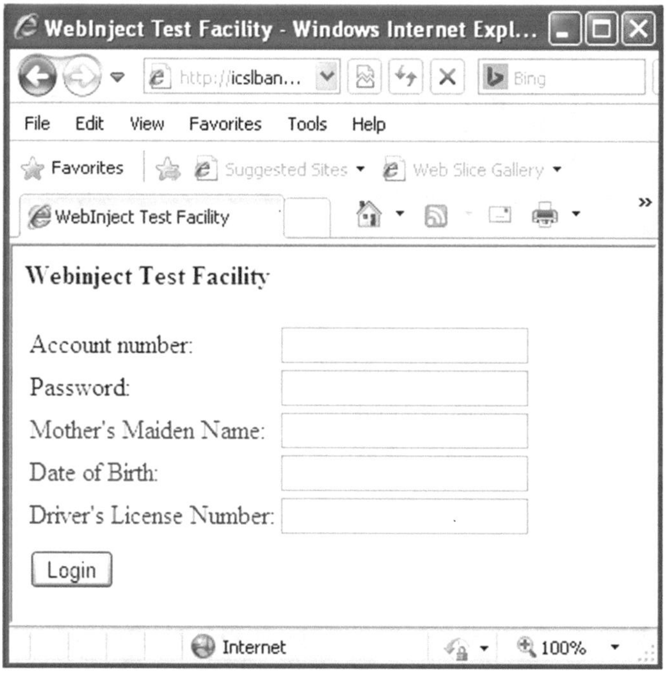

WebInject Test Facility - Windows Internet Expl...
http://icslban...
Bing
File Edit View Favorites Tools Help
Favorites
Suggested Sites
Web Slice Gallery
WebInject Test Facility
Webinject Test Facility
Account number:
Password:
Mother's Maiden Name:
Date of Birth:
Driver's License Number:
Login
Internet
100%

Fig. 4 Example Zeus webinject

code in Internet Explorer, a debugger was attached to the Internet Explorer 8 parent process, and a breakpoint was set on the GetModuleHandleW API. Following the validation of the Zeus configuration in the Explorer process, Zeus malware is injected into the browser process to monitor the current webpages, to detect triggering URLs, and to perform injection of the webinjects.

When executing in the context of a web browser, Zeus hooks the browser's HttpSendRequest and InternetReadFile APIs. When the HttpSendRequest API is called, this results in a call to WininetHook::OnHttpSendRequest to check the Zeus configuration filter actions. If the filter action is not "ignore," the HTTP request is added to an HTTP connections tracking table. When the InternetReadFile API is called, this results in a call to WininetHook::OnInternetReadFile to check the HTTP response. If this connection is in the tracked HTTP connections table, then the HttpGrabber::ExecuteInjects function is called to determine whether the URL is in the targeted URLs section, and if required injects the webinject into the HTTP response data. An example of a Zeus webinject, injected text in red, is shown in Fig. 4.

4 Ransomware C2 Server Emulators

The following section provides details of the construction of a C2 server emulator for the CryptoLocker and CryptoWall ransomware. The reverse engineering of the CryptoLocker malware was straightforward, and the construction of the C2 server emulator was simple. However, the reverse engineering of the CryptoWall ransomware was complicated by injection into multiple processes and API obfuscation.

4.1 CryptoLocker C2 Server Emulator

CryptoLocker ransomware was identified in 2013, and the number of infected computers is not known. The MD5 hash of the CryptoLocker sample used in this research is `fec5a0d4dea87955c124f2eaa1f759f5` [15]. This sample was obtained from Malpedia [26] and includes an unpacked version of the malware. CryptoLocker uses the Microsoft CryptoAPI, which simplifies the identification of cryptographic operations. CryptoLocker encryption of communications and files uses a randomly generated AES key. This AES key is then RSA encrypted and is embedded into each encrypted object. CryptoLocker communications encryption makes use of a public key embedded in the malware and a private key stored in the C2 server. CryptoLocker file encryption uses a public key provided by the C2 server. The private key needed to decrypt the files is only provided after the ransom is paid [21, 25].

Running the unpacked malware in a debugger showed that a second malware process was started, and the first process terminated. Examination of the malware in IDA showed that the function controlling C2 server communications and user file encryption was located at address `0x40B2A1`. A shortcut to gaining control of this malware was performed by editing the first two bytes of this function in the unpacked malware to `0xEBFE`. This is a two byte loop that will cause any process executing this function to loop and will stop the malware from progressing. The looping process was identified by its high CPU usage using the task manager. The debugger was then attached to gain control of the malware. Stepping through the malware with the debugger showed that the following (before encryption) data was sent to the C2 server `"version=1&id=1&name=USERNAME-06752E85&group=sel103-10&lid=en-US"`.

The response from the C2 server is intended to be encrypted with a private key contained in the C2 server. However, the C2 servers are no longer active, and the private key is not available. Two approaches to address the missing private key are to edit the CryptoLocker malware and replace the hard-coded RSA public key with a generated public key, and use the corresponding private key in the C2 emulator, or create unencrypted responses in the C2 emulator and modify the CryptoLocker malware to no longer check the decryption status by replacing a conditional jump with NOP instructions. The latter option was selected as it was easier to implement.

The unencrypted C2 server response was read using the InternetReadFile API and was decrypted using the CryptDecrypt API. The conditional jump instruction testing the return code from the CryptDecrypt API was overwritten with a NOP instruction, allowing unencrypted C2 server responses to be processed by the CryptoLocker malware. The malware was observed to test that the last byte of the decrypted response is zero, and the C2 server emulator was updated to send a null terminated unencrypted response.

Further use of the debugger showed that a value of 1 in the first byte of the response results in a call to a function that calls the CryptDecodeObjectEx API to decode a Privacy Enhanced Mail (PEM) format public key. This public key is located at byte 3 of the response. The completed C2 server emulator reads the initial message from the malware and returns a response of 0x01, 0x00, 0x00, followed by a null terminated PEM format public key. A screenshot of the CryptoLocker ransom demand screen displayed after the user files were encrypted is shown in Fig. 5.

4.2 *CryptoWall C2 Server Emulator*

CryptoWall ransomware was identified in 2014. The MD5 hash of the CryptoWall version 4 sample used in this research is `d9993ab7397f5d2a34f786b54fc55b2c`. This sample was obtained from Malpedia [26] and included an unpacked version of the malware. Descriptions of the CryptoWall protocol were provided by industry blogs [1, 15, 29], and this information significantly reduced the amount of reverse engineering required to build the CryptoWall C2 emulator.

Early versions of CryptoWall copied CryptoLocker's appearance, and the malware authors adopted the name CryptoWall in May 2014. CryptoWall was primarily distributed through malicious spam attachments. CryptoWall deletes volume shadow copies, and the Windows System Restore feature is disabled. CryptoWall version 4

CryptoLocker
Your personal files are encrypted!
Your important files encryption produced on this computer: photos, videos, documents, etc. Here is a complete list of encrypted files, and you can personally verify this.
Encryption was produced using a unique public key RSA-2048 generated for this computer. To decrypt files you need to obtain the private key.
The single copy of the private key, which will allow you to decrypt the files, located on a secret server on the Internet; the server will destroy the key after a time specified in this window. After that, nobody and never will be able to restore files...
To obtain the private key for this computer, which will automatically decrypt files, you need to pay 300 USD / 300 EUR / similar amount in another currency.
Click «Next» to select the method of payment.
Any attempt to remove or damage this software will lead to the immediate destruction of the private key by server.
Private key will be destroyed on
2/11/2020
12:36 PM
Time left
71 : 59 : 04
Next >>

Fig. 5 CryptoLocker ransom demand

uses a locally generated AES key to encrypt user files and filenames, RSA encryption is used to protect the AES key. CryptoWall communications are RC4 encrypted, and the RC4 cipher is passed to the C2 server in the URL of the infection announcement message [1, 29].

CryptoWall version 4 malware injects itself into a newly started `explorer` process [15]. The injected malware creates a new `svchost` process, which is injected with a copy of the malware [29]. The new `svchost` process performs ransomware operations. To gain control of the ransomware, run the analysis VM before the C2 server emulator is started. This will prevent the C2 server connection from being established and will keep the ransomware in its initialization state. Before running the CryptoWall ransomware, record a list of the process identifiers of the svchost processes. Start the CryptoWall malware and identify the new svchost process, connect to this process and set a breakpoint at the InternetConnectA API, next start the CryptoWall C2 emulator and allow the debugger to run the CryptoWall malware.

When the ransomware is executed, an HTTP POST is sent to the C2 server. The C2 server uses the sorted URL parameter as ciphertext to create an RC4 key [11]. The data passed by the HTTP POST is ASCII encoded binary data, which is decoded using the Python binascii.unhexlify function. The decoded data is decrypted using the RC4 key. The decrypted request is `"{1|crypt13001|32DC0066DCE410C9285635F121811FB99|1|2|1}"`.

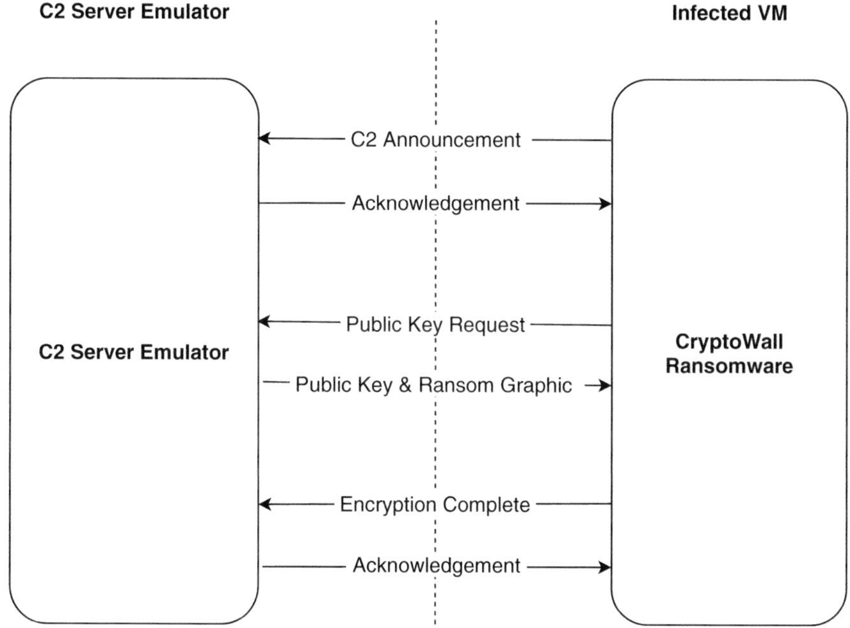

Fig. 6 CryptoWall messages

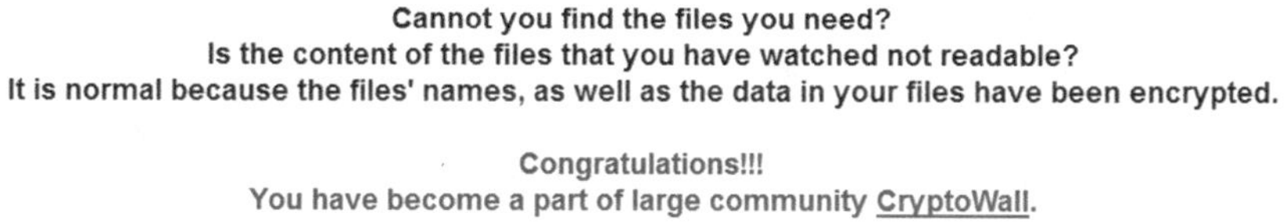

Cannot you find the files you need?
Is the content of the files that you have watched not readable?
It is normal because the files' names, as well as the data in your files have been encrypted.

Congratulations!!!
You have become a part of large community CryptoWall.

Fig. 7 CryptoWall ransom demand

The C2 server responds by sending an RC4 encrypted response e.g., `"{204|1}"` to the infected computer. The CryptoWall ransomware responds by sending a public key request to the C2 server. The C2 server responds with a message containing a public key and a base64 encoded ransom demand graphic. When this C2 response is received, the ransomware process scans the infected computer's storage and encrypts user files. When the user files have been encrypted, an infection notification message e.g., `"{260|1}"` is sent to the C2 server. Finally, a window is displayed on the infected computer to demand payment [1, 15]. The messages exchanged between the ransomware and the C2 server emulation are shown in Fig. 6.

The CryptoWall C2 emulator implements the CryptoWall protocol that allows the ransomware to exercise its full capabilities. A screenshot of a section of the CryptoWall ransom demand screen displayed after the user files were encrypted is shown in Fig. 7.

5 Semi-automated Generation of C2 Server Emulators

While the ability to automatically generate C2 server emulators for arbitrary malware families would be useful, this is not currently feasible, and the recent work in literature is a semi-manual construction process.

The Imaginary C2 program [30] converts captured network traffic into request definitions that allow C2 HTTP response to be replayed. However, this traffic replay approach is not suitable for situations where initial network traffic samples are not available.

One automation approach for the creation of C2 server emulators is provided in [2]. This research refers to C2 server emulators as Custom Impersonators. Malware samples are executed on a QEMU VM, and instruction traces are collected using DECAF [19], and the instruction traces are translated into VINE intermediate language [28]. Symbolic execution [22] is performed on the instruction traces, and symbolic variables are assigned to network input. A Simple Theorem Prover (STP) constraint solver [17] is used to determine the values that determine the outcome of the control flow tests. These values can be used to identify malware control dependencies controlled by values in the network input [2]. The malware control flow graph and control dependencies are provided to assist analysts with the manual construction of C2 server emulators.

Research using ANGR [27], an open-source symbolic execution framework, creates a technique that employs static analysis to determine the C2 command protocol and associated commands implemented in a common RAT. The top-level command processing function of the RAT is analyzed, and for each explored path, a list of the malware API calls and their arguments, function call relationships, and the network data required to trigger the path's execution are provided [5]. Windows API models and support for the `stdcall` calling convention were added to ANGR in order to support the analysis of Windows malware. Heuristics were created to limit the number of paths explored by the symbolic execution in order to prevent potential path explosion problems. Symbolic execution commences at the manually selected Symbolic Execution Point (SEP), and an execution context is needed to provide precondition values that are generated in malware initialization. In this research, the execution context was generated using two different techniques: by performing concrete execution, setting a breakpoint at the SEP, taking a memory dump, and extracting the necessary parts of the execution context, or by moving the SEP backward, allowing initialization of execution context values. Symbolic execution was used to explore the command processing loop. The report produced by this technique showed the API calls, and the functions called in processing each command, as well as the network data required to trigger the processing of each command. This research targets analysis for a single RAT, requires manual SEP identification, does not support analysis of encrypted protocols and does not support mining of the analysis report from the tool output [5].

The S2E symbolic execution engine is used as the basis of research that constructs C2 servers for RATs [8]. The S2E engine performs symbolic execution of instructions and forks execution when branches are taken. An SMT solver is used to evaluate expressions and obtain concrete values. To prevent performance problems and scalability issues due to path explosion, an analyst must provide the location of the command processing loop and details of how to reach this address. The process used in this research can be summarized as Trace Generation, Trace Analysis, Speculation, Validation, and C2 Server Generation. Trace generation uses symbolic execution to explore execution paths and to maximize code coverage. A number of the recorded traces will cover the RAT command processing code. The branches taken and API execution details are recorded. Trace analysis builds Augmented Prefix Tree Acceptors (APTA) that captures API execution and branches taken along the explored paths. APTA's are Deterministic Finite Automata (DFA) that have been used in the protocol reverse engineering [10]. The goal of speculation is to generate a small number of paths that cover all of the commands. Speculative edges are added to the APTAs in an attempt to combine symbolically executed command fragments into paths containing multiple commands. The symbolic execution engine is then used to validate speculatively generated paths, and when speculative edges are validated, the branches and API calls are recorded. C2 server generation is performed for each validated path that contains multiple commands. This research generates a C2 server from the code of a small RAT created for research purposes [8]. Due to the requirement for manual analysis to provide the location of the command processing loop as a starting point, this research is classified as semi-automated.

6 Limitations

Irrespective of the C2 emulator construction technique, some malware samples require minor modification before they can be executed with a C2 server emulator. Examples of the modifications required to allow the Zeus V2 and CryptoLocker malware to run with C2 server emulators are given below.

Zeus v2 is a self-modifying malware with a hardware locking feature that only allows the installed Zeus malware to execute on the computer that it was installed on. In the Zeus C2 server emulator, the Zeus malware was unpacked using a static unpacker, and the jump instruction that controls the hardware locking test was overwritten with NOP instructions to prevent the malware from terminating.

The CryptoLocker malware contains a hard-coded RSA public key, and the C2 server emulator is expected to respond with communications encrypted with the corresponding private key. Due to the removal of the original C2 servers, this private key is no longer available. The C2 server emulator was developed to return unencrypted responses, and the unpacked CryptoLocker malware sample was modified to skip the successful decryption check. The modified CryptoLocker sample operates in the same manner as the original ransomware, it connected to the emulated C2 server, scanned the hard disk for user files, performed file encryption, and displayed the ransom demand window.

In both of these cases, a modification of the malware's machine code allowed a historical malware sample to operate at a high level of fidelity with an emulated C2 server, allowing the collection of feature sets that are comparable with malware execution in-the-wild. It is acknowledged that the technique of manually building an emulated C2 server cannot currently be performed at scale. However, cases exist where manually building a C2 server emulator allows academic research projects to be performed that would not otherwise be possible.

A limitation in the manual construction of C2 emulators is the need for a skilled analyst to perform manual reverse engineering. Techniques for the semi-automated generation of C2 server emulators do exist. However, the fully automatic generation of C2 emulators is not currently feasible due to current limitations in symbolic execution techniques.

7 Conclusion

Academic malware datasets consist of collections of historic malware samples. The C2 servers of these malware samples no longer exist, and when executed on a VM, these malware samples perform their initialization functions and then wait for C2 server connections that no longer exist. This initialization-only behavior of historic malware samples provides more limited features than would be collected when the malware was running in-the-wild. Historic malware samples running without an emulated environment cannot perform many of the malware's original capabilities.

Live malware samples with active C2 servers have been used for research [14]. This approach is feasible but uncertain due to problems associated with the short life span of malware C2 servers, unknown malware configuration, the malware being controlled by the malware operator, and the possibility of the malware operator becoming aware of the research.

The creation of C2 server emulators allows the full capabilities of malware samples to be fully controlled by researchers in an isolated network. In the case of historic malware samples, the use of C2 server emulators allows control of malware capabilities that would no longer be available without emulation. This chapter discussed methods for the manual reverse engineering of the malware sample's command protocol and created an emulated C2 server that can control the full command interface of the malware. Three examples of methods used for the construction of emulated command and control servers were provided. Apart from the generation of C2 server emulators, some malware samples require minor modifications to bypass anti-analysis systems or to compensate for lost encryption keys. Examples of these modifications were provided for two malware families.

A review of the literature related to the creation of emulated C2 servers was undertaken. This review showed that the use of C2 server emulators and the automated generation of C2 server emulators is a new research topic with research in the early stages. Existing research provides the semi-automated generation of C2 server emulators based on individual samples.

References

1. Allievi, Andrea, Holger Unterbrink, and Warren Mercer. 2015. Cryptowall 4 - the evolution continues.
2. Alwabel, Abdulla, Hao Shi, Genevieve Bartlett, and Jelena Mirkovic. 2014. Safe and automated live malware experimentation on public testbeds. In *7th Workshop on cyber security experimentation and test (CSET 14)*.
3. Azab, Ahmad, Mamoun Alazab, and Mahdi Aiash. 2016. Machine learning based botnet identification traffic. In *2016 IEEE Trustcom/BigDataSE/ISPA*, 1788–1794. IEEE.
4. Azab, Ahmad, Robert Layton, Mamoun Alazab, and Jonathan Oliver. 2014. Mining malware to detect variants. In *2014 Fifth Cybercrime and Trustworthy Computing Conference*, 44–53. IEEE.
5. Baldoni, Roberto, Emilio Coppa, Daniele Cono D'Elia, and Camil Demetrescu. 2017. Assisting malware analysis with symbolic execution: A case study. In *International conference on cyber security cryptography and machine learning*, 171–188. Springer.
6. Barford, Paul, and Mike Blodgett. 2007. Toward botnet mesocosms. *HotBots* 7: 6–6.
7. Black, Paul, Iqbal Gondal, and Robert Layton. 2018. A survey of similarities in banking malware behaviours. *Computers & Security* 77: 756–772.
8. Borzacchiello, Luca, Emilio Coppa, Daniele Cono D'Elia, and Camil Demetrescu. 2019. Reconstructing c2 servers for remote access trojans with symbolic execution. In *International symposium on cyber security cryptography and machine learning*, 121–140. Springer.
9. Bosatelli, Fabio. 2013. Zarathustra: detecting banking trojans via automatic, platform independent webinjects extraction.
10. Bugalho, Miguel, and Arlindo L. Oliveira. 2005. Inference of regular languages using state merging algorithms with search. *Pattern Recognition* 38 (9): 1457–1467.
11. Cabaj, Krzysztof, and Wojciech Mazurczyk. 2016. Using software-defined networking for ransomware mitigation: The case of cryptowall. *IEEE Network* 30 (6): 14–20.
12. Calvet, Joan, Carlton R. Davis, José M. Fernandez, Jean-Yves Marion, Pier-Luc St-Onge, Wadie Guizani, Pierre-Marc Bureau, and Anil Somayaji. 2010. The case for in-the-lab botnet experimentation: creating and taking down a 3000-node botnet. In *Proceedings of the 26th annual computer security applications conference*, 141–150.
13. Cho, Chia Yuan, Domagoj Babi ć, Eui Chul Richard Shin, and Dawn Song. 2010. Inference and analysis of formal models of botnet command and control protocols. In *Proceedings of the 17th ACM conference on Computer and communications security*, 426–439.
14. Continella, Andrea, Michele Carminati, Mario Polino, Andrea Lanzi, Stefano Zanero, and Federico Maggi. 2017. Prometheus: Analyzing webinject-based information stealers. *Journal of Computer Security* 25 (2): 117–137.
15. Dell Secureworks Counter Threat Unit. 2014. Cryptowall ransomware threat analysis.
16. Forrester, Jock Ingram. 2014. An exploration into the use of webinjects by financial malware. PhD thesis, Rhodes University.
17. Ganesh, Vijay, and David L. Dill. 2007. A decision procedure for bit-vectors and arrays. In *International conference on computer aided verification*, 519–531. Springer.
18. Hakkarainen, Jani. 2015. Malware analysis environment for windows targeted malware.
19. Henderson, Andrew, Lok Kwong Yan, Xunchao Hu, Aravind Prakash, Heng Yin, and Stephen McCamant. 2016. Decaf: A platform-neutral whole-system dynamic binary analysis platform. *IEEE Transactions on Software Engineering* 43 (2): 164–184.
20. www.Inetsim.org. 2020. Inetsim: Internet services simulation suite.
21. Jarvis, Keith. 2013. Cryptolocker ransomware.
22. King, James C. 1976. Symbolic execution and program testing. *Communications of the ACM* 19 (7): 385–394.
23. Kreibich, Christian, Nicholas Weaver, Chris Kanich, Weidong Cui, and Vern Paxson. 2011. Gq: Practical containment for measuring modern malware systems. In *Proceedings of the 2011 ACM SIGCOMM conference on Internet measurement conference*, 397–412.

24. Moniruzzaman, Md, Adil Bagirov, Iqbal Gondal, and Simon Brown. 2018. A server side solution for detecting webinject: A machine learning approach. In *Pacific-Asia conference on knowledge discovery and data mining*, 162–167. Springer.
25. Panda Security. 2015. Cryptolocker: What is and how to avoid it.
26. Plohmann, Daniel, Martin Clauss, Steffen Enders, and Elmar Padilla. 2018. Malpedia: A collaborative effort to inventorize the malware landscape. *The Journal on Cybercrime & Digital Investigations* 3 (1).
27. Shoshitaishvili, Yan, Ruoyu Wang, Christopher Salls, Nick Stephens, Mario Polino, Andrew Dutcher, John Grosen, Siji Feng, Christophe Hauser, Christopher Kruegel, et al. 2016. Sok:(state of) the art of war: Offensive techniques in binary analysis. In *2016 IEEE symposium on security and privacy (SP)*, 138–157. IEEE.
28. Song, Dawn, David Brumley, Heng Yin, Juan Caballero, Ivan Jager, Min Gyung Kang, Zhenkai Liang, James Newsome, Pongsin Poosankam, and Prateek Saxena. 2008. Bitblaze: A new approach to computer security via binary analysis. In *International conference on information systems security*, 1–25. Springer.
29. Sophos. 2015. The current state of ransomware: Cryptowall.
30. Weyne, Felix. 2020. Imaginary c2.
31. Wyke, James. 2011. What is zeus? Technical report, Sophos Labs.
32. Zeus Author. 2011. Zeus source code.

Cluster Analysis of Malware Family Relationships

Samanvitha Basole and Mark Stamp

Abstract In this chapter, we use K-means clustering to analyze various relationships between malware samples. We consider a dataset comprising 20 malware families with 1000 samples per family. These families can be categorized into seven different types of malware. We perform clustering based on pairs of families and use the results to determine relationships between families. We perform a similar cluster analysis based on malware type. Our results indicate that K-means clustering can be a powerful tool for data exploration of malware family relationships.

1 Introduction

Previous research has demonstrated that it is possible in some cases to train a machine learning model to detect multiple malware families [3]. Specifically, neighborhood-based techniques are relatively effective in such a situation. Although support vector machines (SVM) did not perform well in this previous research, both k-nearest neighbors (k-NN) and random forest (RF) were able to distinguish malware from benign with good accuracy, even when several malware families were combined to form the malware class.

In this research, we consider the same dataset used in [3], which includes 20 malware families. Here, we apply cluster analysis to these families. Our goal is to determine whether we can discover interesting connections, relationships, and differences between these various families, based on elementary clustering techniques. While [3] considers binary classification experiments to distinguish malware from benign, the research in this chapter is focused on a data exploration problem. The features we use are byte n-gram frequencies, while the clustering method we consider is the well-known K-means algorithm.

S. Basole (✉) · M. Stamp
San Jose State University, San Jose, CA, USA
e-mail: s97basole@gmail.com

M. Stamp
e-mail: mark.stamp@sjsu.edu

M. Stamp et al. (eds.), *Malware Analysis Using Artificial Intelligence and Deep Learning*, https://doi.org/10.1007/978-3-030-62582-5_14

The remainder of this paper is organized as follows. In Sect. 2, we provide relevant background information, including a brief discussion of related work. Then in Sect. 3, we give our experimental results and analysis. Finally, Sect. 4 concludes this chapter, where we have included suggestions for future work.

2 Background

In this section, we first consider relevant related work. Then we discuss our malware dataset, and we provide information on each family in the dataset. We also present the various metrics that we use in our clustering experiments. Finally, we provide an introduction to the K-means clustering algorithm.

2.1 Related Work

In [9], the authors use API calls to classify malware based on their types. They consider a random forest (RF) classifier and achieve an average area under the ROC curve (AUC) of 0.98. In contrast, we use n-grams and a clustering approach, and we are considering a data exploration problem, rather than a straightforward classification problem.

The authors in [4] propose a clustering approach for malware, wherein the goal is to cluster samples based on similar behavior. These authors use features such as system calls and network activity, to cluster malicious code based on its behavior. In contrast to this previous work, our cluster analysis is based on simpler and easier to collect features and, again, we are in a data exploration mode of operation.

The research in [10] uses BIRCH clustering based on static and dynamic features. These authors consider 18 families, but 12 of those families contain less than 100 samples each. Our approach uses a larger and balanced dataset for clustering.

2.2 Dataset

The same dataset as used in [3] is considered in this research. This dataset includes 20 families, which we categorize into malware types, as listed in Table 1.

Each of the malware families in Table 1 is summarized below.

Adload downloads an executable file, stores it remotely, executes the file, and disables proxy settings [18].

Agent downloads Trojans or other software from a remote server [19].

Alureon exfiltrates usernames, passwords, credit card data, and other confidential data from an infected system [25].

Table 1 Type of each malware family

Index	Family	Type	Index	Family	Type
0	Adload [18]	Trojan Downloader	10	Obfuscator [27]	VirTool
1	Agent [19]	Trojan	11	OnLineGames [13]	Password Stealer
2	Alureon [25]	Trojan	12	Rbot [28]	Backdoor
3	BHO [21]	Trojan	13	Renos [20]	Trojan Downloader
4	CeeInject [24]	VirTool	14	Startpage [22]	Trojan
5	Cycbot.G [2]	Backdoor	15	Vobfus [29]	Worm
6	DelfInject [11]	VirTool	16	Vundo [30]	Trojan Downloader
7	FakeRean [26]	Rogue	17	Winwebsec [31]	Rogue
8	Hotbar [1]	Adware	18	Zbot [14]	Password Stealer
9	Lolyda.BF [12]	Password Stealer	19	Zeroaccess [23]	Trojan

BHO can perform a variety of actions, guided by an attacker [21].

CeeInject uses advanced obfuscation to avoid being detected by antivirus software [24].

Cycbot.G connects to a remote server, exploits vulnerabilities, and spreads through backdoor ports [2].

DelfInject sends usernames, passwords, and other personal and private information to an attacker [11].

FakeRean pretends to scan the system, notifies the user of supposed issues, and asks the user to pay to clean the system [26].

Hotbar is adware that shows ads on webpages and installs additional adware [1].

Lolyda.BF sends information from an infected computer and monitors the system. It can share user credentials and network activity with an attacker [12].

Obfuscator tries to obfuscate or hide itself to defeat malware detectors [27].

OnLineGames steals login information of online games and tracks user keystroke activity [13].

Rbot gives control to attackers via a backdoor that can be used to access information or launch attacks, and serves as a gateway to infect additional sites [28].

Renos downloads software that claims the system has spyware and asks for a payment to remove the nonexistent spyware [20].

Startpage changes the default browser homepage and may perform other malicious activities [22].

Vobfus is a worm that downloads malware and spreads through USB drives or other removable devices [29].

Table 2 Number of samples of each type

Malware type	Samples
VirTool	3000
Password Stealer	3000
Backdoor	2000
Trojan	8000
Worm	1000
Rogue	2000
Adware	1000

Vundo displays pop-up ads and may download files. It uses advanced techniques to defeat detection [30].

Winwebsec displays alerts that ask the user for money to fix supposed issues [31].

Zbot is installed through email and shares a user's personal information with attackers. In addition, Zbot can disable a firewall [14].

Zeroaccess is a Trojan horse that downloads applications that click on ads, thereby making money for the attacker [23].

The features we use for clustering are based on byte n-gram frequencies. Specifically, we choose the top 20 byte n-grams, with $n = 2$, from our malware class. These frequency vectors are then normalized, so that each vector can be viewed as a discrete probability distribution. The resulting normalized bigram frequency vectors (of length 20) form our feature set.

Our experiments include clustering based on pairs of families, clustering selected families belonging to different malware types, and clustering families belonging to the same malware type. The number of samples of each of the seven malware types found in our dataset is given in Table 2. Note that we categorize "Trojan Downloader" as a type of Trojan, giving us the seven distinct types listed in Table 2.

2.3 Metrics

In this section, we discuss the metrics used to numerically evaluate our clustering results. Note that we do not use accuracy, due to the label-switching problem that occurs when we attempt to apply this metric to clustering results [17].

One popular choice for clustering is the so-called v-measure, which is a robust metric for cluster evaluation—robust in the sense that a permutation of the cluster labels does not affect the score. The v-measure is defined as the harmonic mean between homogeneity (i.e., the case where each cluster contains all points from a single class) and completeness (i.e., the case where all points from the same class are in one cluster) [8]. Another nice feature of this metric is that it is symmetric, in that it yields the same score if the predicted classes and the true classes are switched. The v-measure ranges from 0 to 1.

Although v-measure is a robust evaluation metric with many advantages, it is not the best choice for the research considered in this chapter. The v-measure is not normalized for random cluster results, hence it would tend to produce a higher score for random cluster assignments when a "large" number of clusters K is chosen, say, $K > 10$. In contrast, adjusted mutual information (AMI) results in random label assignments having a score close to 0, regardless of the size of the dataset or the number of clusters.

Another useful metric for clustering is the adjusted Rand index (ARI). The ARI is similar to AMI, in the sense that it is adjusted to account for chance. The Rand index is a similarity measure that considers all pairs of samples and uses the number of pairwise agreements in the true and predicted clusters. Specifically, the Rand index is calculated as [16]

$$\text{RI} = \frac{a+d}{a+b+c+d} = \frac{a+d}{\binom{n}{2}}$$

where a, b, c, and d are defined as follows: If U and V are two different partitions or clusterings of the same data, then let a, b, c, and d be the number of objects determined by

$$\begin{aligned}
a &= \text{in the same cluster in } U \text{ and in the same cluster in } V \\
b &= \text{in the same cluster in } U \text{ but in different clusters in } V \\
c &= \text{in the same cluster in } V \text{ but in different clusters in } U \\
d &= \text{in different clusters in } U \text{ and in different clusters in } V
\end{aligned}$$

The formula for ARI is calculated using the raw Rand index RI as [8]

$$\text{ARI} = \frac{\text{RI} - \text{E(RI)}}{\max(\text{RI}) - \text{E(RI)}}$$

where E is the expected value operator.

The authors in [15] state that AMI should be used when the true clusters are unbalanced in size, while ARI should be used when the true clusters are large and roughly equal-sized. In our research, the size of the ground truth for family labels is precisely balanced with 1000 samples in each family. Thus, we use ARI to evaluate our clustering predictions.

2.4 K-Means

In this section, we first discuss a generic approach to clustering. We then consider how to implement such an approach, which leads directly to the K-means algorithm.

Suppose that we are given the n data points $X_1, X_2, \ldots, X_n$, where each of the X_i is a vector of m real numbers. For example, we could analyze a set of malware samples based on, say, five distinct scores, denoted $s_1, s_2, \ldots, s_5$. Then each data point would be of the form

$$X_i = (s_1, s_2, s_3, \ldots, s_5).$$

We assume that the desired number of clusters K is specified in advance and that we want to partition our n data points $X_1, X_2, \ldots, X_n$ into K clusters. We also assume that we have a distance function $d(X_i, X_j)$ that is defined for all pairs of data points.

We associate a centroid with each cluster, where the centroid can be viewed as representative of its cluster. Intuitively, a centroid is the center of mass of a cluster. We denote cluster j as C_j with the corresponding centroid denoted as c_j. Note that in K-means, centroids need not be actual data points.

Now, suppose that we have clustered our n data points. Then we have a set of K centroids,

$$c_1, c_2, c_3, \ldots, c_K$$

and each data point is associated with exactly one centroid. Let $\text{centroid}(X_i)$ denote the (unique) centroid of the cluster to which X_i belongs. The centroids determine the clusters, in the sense that whenever we have

$$c_j = \text{centroid}(X_i),$$

then X_i belongs to cluster C_j.

Before we can cluster data based on the process outlined above, we need to address the following two questions:

1. How do we determine the centroids c_j?
2. How do we determine the clusters? That is, we need to specify the function $\text{centroid}(X_i)$, which assigns data points to centroids. This has the effect of determining the clusters.

There are many reasonable ways to answer these questions. Next, we consider an intuitively appealing approach that leads directly to the K-means algorithm.

Intuitively, it seems clear that the more "compact" a cluster is, the better. Of course, this will depend on the data points X_i and the number of clusters K. Since the data is given, and we are assuming that K has been specified, we have no control over the X_i or K. But, we do have control over the selection of the centroids c_j and the assignment of points to centroids via the function $\text{centroid}(X_i)$. The choice

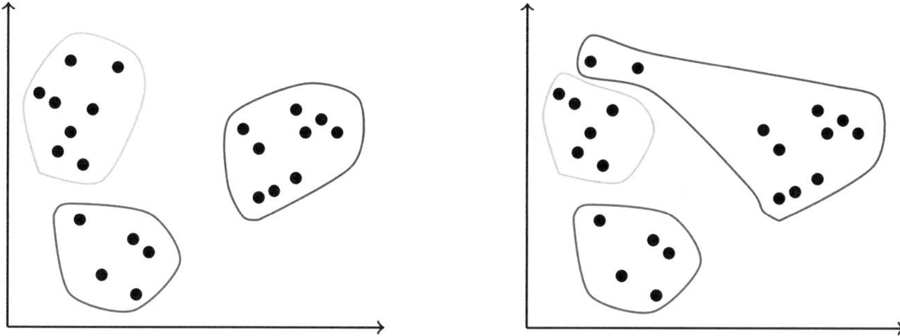

Fig. 1 Smaller and larger distortion for the same dataset

of centroids and the assignment of points to centroids will clearly influence the compactness (i.e., "shape") of the resulting clusters.

Let

$$\text{distortion} = \sum_{i=1}^{n} d\big(X_i, \text{centroid}(X_i)\big). \tag{1}$$

Intuitively, the smaller the distortion, the better, since a smaller distortion implies that individual clusters are more compact.[1]

For example, consider the data in Fig. 1, where the same data points are clustered in two different ways. It is clear that the clustering on the left-hand side in Fig. 1 has a smaller distortion than that on the right-hand side. Therefore, we would say that the left-hand clustering is superior, at least with respect to the measure of distortion.

Suppose that we try to minimize the distortion. First, we observe that distortion depends on K, since more clusters imply more centroids—in the limit, we could let $K = n$, and make each data point a centroid, in which case the distortion is 0. To emphasize this dependence on K, we write distortion_K. As mentioned above, we assume that K is specified in advance.

The problem we want to solve can be stated precisely as

$$\begin{aligned} &\text{Given: } K \text{ and data points } X_1, X_2, \ldots, X_n \\ &\text{Minimize: } \text{distortion}_K = \sum_{i=1}^{n} d\big(X_i, \text{centroid}(X_i)\big). \end{aligned} \tag{2}$$

Finding an exact solution to this problem is computationally infeasible. But, there is a simple iterative approximation that works well in practice.

We claim that a solution to (2) must satisfy the following two conditions.

[1] In addition to having compact clusters, we might also want a large separation between clusters. However, such separation is not (directly) accounted for in K-means.

Condition 1 Each X_i is clustered according to its nearest centroid. That is, if the data point X_i belongs to cluster C_j, then $d(X_i, c_j) \leq d(X_i, c_\ell)$ for all $\ell \in \{1, 2, \ldots, K\}$, where the c_ℓ are the centroids.
Condition 2 Each centroid is located at the center of its cluster.

To verify the necessity of Condition 1, suppose that X_i is in cluster C_j and that we have $d(X_i, c_\ell) < d(X_i, c_j)$ for some ℓ. Then by simply reassigning X_i to cluster ℓ, we will reduce distortion$_K$. Condition 2 also seems intuitively clear, and it is a straightforward calculus exercise to prove the necessity of this condition as well.

Condition 1 tells us that given any clustering for which there are points that are not assigned to their nearest centroid, we can improve the clustering by simply reassigning such data points to their nearest centroid. Condition 2 implies that we always want a centroid to be at the center of its cluster. Therefore, given any clustering, we may improve it—and we cannot make it worse—with respect to distortion$_K$ by performing either of the following two steps:

Step 1 Assign each data point to its nearest centroid.
Step 2 Recompute the centroids so that each lies at the center of its respective cluster.

It is clear that nothing can be gained by applying Step 1 more than once in succession, and the same holds true for Step 2. However, by alternating between these two steps, we obtain an iterative process that yields a series of solutions that will generally tend to improve and can never get worse with respect to distortion$_K$. This is precisely the K-means algorithm [7], which we state somewhat more precisely as Algorithm 1.

Algorithm 1 K-means clustering

```
Given:
    Data points X_1, X_2, ..., X_n to cluster
    Number of clusters K
Initialize:
    Partition X_1, X_2, ..., X_n into clusters C_1, C_2, ..., C_K
while stopping criteria is not met do
    for j = 1 to K do
        Let centroid c_j be the center of cluster C_j
    end for
    for i = 1 to n do
        Assign X_i to cluster C_j so that d(X_i, c_j) ≤ d(X_i, c_ℓ)
                for all ℓ ∈ {1, 2, ..., K}
    end for
end while
```

The stopping criteria in Algorithm 1 could be that distortion$_K$ improves (i.e., decreases) by less than a set threshold, or that the centroids do not change by much, or we could simply run the algorithm for a fixed number of iterations.

Note that Algorithm 1 is a hill climb, and hence K-means is only assured of finding a local maximum. And, as with any hill climb, the maximum we obtain will depend on our choice for the initial conditions. For K-means, the initial conditions correspond to the initial selection of centroids. Therefore, it can be beneficial to repeat the algorithm multiple times with different initializations of the centroids.

In the experiments below, we use the K-means clustering algorithm to explore the natural formation of malware clusters. We employ elbow plots as a tool to discern structure from the 20 malware families based on pairwise clusters. Next, we discuss elbow plots in this context.

2.5 Elbow Plots

Suppose that we graph the clustering error as a function of the number of clusters, K. Then an "elbow" in this graph indicates the point where adding another cluster does not significantly improve the clustering results [5]. Such an elbow can be used to determine the (near) optimal number of clusters.

We choose distortion and inertia for our elbow plots. Distortion is calculated as the average of the squared Euclidean distances from each point to the nearest centroid, whereas inertia is calculated as the sum of these same distances. For our experiments, elbow plots using distortion and inertia indicate that the clusters are not well formed, and thus, the number of clusters is somewhat subjective. From Fig. 2, it appears that $K \in \{4, 5, 6\}$ should be good values for the number of clusters, as the inertia and distortion only slightly decrease from that point onward. In any case, these elbow plots clearly indicate that the optimal number of clusters is less than 10, which is somewhat surprising, given that we are dealing with 20 families. This is a strong indication that there is significant similarity between some of the families in our dataset.

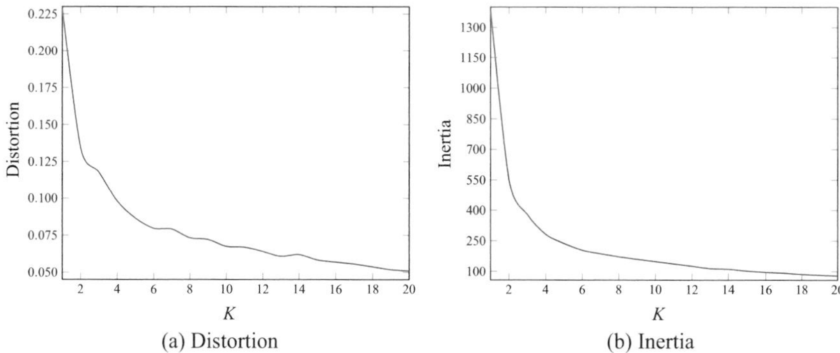

(a) Distortion (b) Inertia

Fig. 2 Elbow plots

3 Experiments and Results

In this section, we present results of three different sets of clustering experiments. First, we cluster each pair of malware families and show that we can draw meaningful conclusions based on these clustering results. Then we consider clustering experiments where we restrict our attention to one family of each malware type under consideration. Finally, we consider clustering multiple families from the same malware type.

3.1 Clustering by Family

In this set of experiments, we perform clustering for each pair of families. Since there are 20 families, we have $\binom{20}{2} = 190$ such clustering experiments. In each case, the top 20 n-grams is extracted to form the features under consideration. Every sample in the two families under consideration is then converted to a normalized vector of n-gram frequencies. The resulting data is clustered using K-means, with $K = 2$.

The results of these experiments consist of 190 ARI scores and 190 confusion matrices. Representative examples of the resulting confusion matrices are given in Fig. 3.

Each of these 190 clustering experiments provides information on how closely one family is related to another. From such results, we can deduce weak and strong links between malware family pairs. The 190 ARI similarity scores are given in the form of a heatmap in Table 3. Note that the diagonal elements are 1 in every case, since the similarity between a family and itself is always 1. Also, the heatmap is symmetric, since the ARI similarity score is itself symmetric.

Figure 4 gives the total pairwise ARI for each family in the form of a bar graph. That is, each bar represents the sum of the 19 ARI scores of a given family with all other families in our dataset. We refer to this sum as the total ARI.

In Fig. 5 we give the average ARI for all pairwise clusters formed with a given family. Based on the horizontal line at $y = 0.5$, we see that there are four families with a high average ARI, that is, an ARI that exceeds the $y = 0.5$ line. This implies that when each of these four families is clustered against the other families, the ARI is, on average, particularly high. The four high-ARI families are BHO, Adload, Hotbar, and Vobfus. Note that these strong ARI results are also apparent from the total ARI scores in Fig. 4 and from the heatmap in Table 3.

From Table 1, we see that Hotbar is the only adware in the dataset and Vobfus is the only worm. It is intuitive that these malware families would tend to stand out more from the other families, due to their being of unique types, and would thus be easier to cluster. This is clearly indicated by the high-ARI results for Hotbar and Vobfus. On the other hand, BHO and Adload are both Trojans, which is the most common type in our dataset. This result indicates that in spite of Adload and BHO being Trojans, they contain byte bigram features that are significantly different from

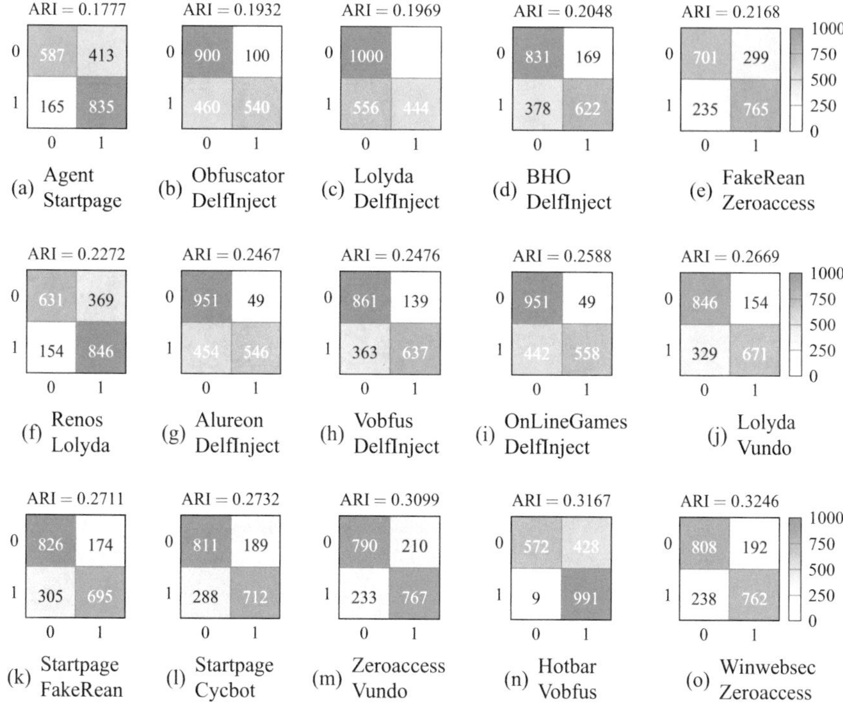

Fig. 3 Selected examples from the 190 pairwise confusion matrices

the other Trojans in the dataset, namely, Agent, Alureon, Renos, Startpage, Vundo, and Zeroaccess. It is also interesting that Adload and BHO are similar to each other, in the sense that their pairwise clustering result is poor, as can be observed from the heatmap in Table 3.

To further explore these high-ARI families, we provide graphs showing the relationship strength of each with respect to all other families. To generate these graphs, we use the NetworkX library in Python. The resulting graphs are given in Figs. 6, 7, 8 and 9, where each node represents a family, with the node numbers corresponding to the "index" column in Table 1. In each of these figures, the darkened node corresponds to the family mentioned in the caption. Also, a dotted edge between two nodes indicates an ARI score of 0.5 or less, while a solid line represents an ARI score greater than 0.5. The nodes are positioned by simulating a force-directed representation, based on the Fruchterman–Reingold force-directed algorithm [6].

These graphs help visualize how other families are related to the four most-distinct families in our dataset. For example, Hotbar is almost equally distinguishable from all other families. On the other hand, Adload is distinguishable from all families except Vobfus and BHO. This means that Adload, BHO, and Vobfus are mostly similar to each other, but highly distinguishable when clustered with other families in the dataset.

Table 3 Heatmap of pairwise clustering ARI scores

Family	Adload	Agent	Alureon	BHO	CeeInject	Cycbot.G	DelfInject	FakeRean	Hotbar	Lolyda	Obfuscator	OnLineGames	Rbot	Renos	Startpage	Vobfus	Vundo	Winwebsec	Zbot	Zeroaccess
Adload	1.00	0.67	0.88	0.01	0.68	0.72	0.27	0.76	0.75	0.88	0.83	0.88	0.74	0.85	0.77	0.01	0.81	0.95	0.64	0.74
Agent	0.67	1.00	0.11	0.51	0.04	0.01	0.04	0.01	0.66	0.15	0.04	0.13	0.06	0.05	0.18	0.55	0.02	0.03	0.01	0.17
Alureon	0.88	0.11	1.00	0.64	0.02	0.06	0.25	0.08	0.71	0.59	0.02	0.00	0.01	0.07	0.59	0.71	0.05	0.03	0.01	0.58
BHO	0.01	0.51	0.64	1.00	0.63	0.55	0.20	0.55	0.34	0.69	0.63	0.66	0.56	0.64	0.60	0.00	0.63	0.70	0.48	0.68
CeeInject	0.68	0.04	0.02	0.63	1.00	0.02	0.12	0.01	0.72	0.36	0.00	0.03	0.00	0.01	0.39	0.56	0.00	0.02	0.01	0.34
Cycbot	0.72	0.01	0.06	0.55	0.02	1.00	0.10	0.00	0.59	0.02	0.02	0.02	0.01	0.01	0.27	0.62	0.02	0.00	0.00	0.17
DelfInject	0.27	0.04	0.25	0.20	0.12	0.10	1.00	0.07	0.48	0.20	0.19	0.26	0.17	0.18	0.06	0.25	0.17	0.13	0.08	0.04
FakeRean	0.76	0.01	0.08	0.55	0.01	0.00	0.07	1.00	0.68	0.14	0.02	0.08	0.02	0.00	0.27	0.61	0.01	0.01	0.01	0.22
Hotbar	0.75	0.66	0.71	0.34	0.72	0.59	0.48	0.68	1.00	0.76	0.73	0.71	0.66	0.68	0.72	0.32	0.72	0.76	0.77	0.76
Lolyda	0.88	0.15	0.59	0.69	0.36	0.02	0.20	0.14	0.76	1.00	0.36	0.58	0.40	0.23	0.56	0.72	0.27	0.04	0.01	0.48
Obfuscator	0.83	0.04	0.02	0.63	0.00	0.02	0.19	0.02	0.73	0.36	1.00	0.04	0.00	0.00	0.41	0.70	0.00	0.02	0.01	0.44
OnLineGames	0.88	0.13	0.00	0.66	0.03	0.02	0.26	0.08	0.71	0.58	0.04	1.00	0.01	0.01	0.60	0.71	0.05	0.02	0.01	0.59
Rbot	0.74	0.06	0.01	0.56	0.00	0.01	0.17	0.02	0.66	0.40	0.00	0.01	1.00	0.00	0.42	0.58	0.00	0.03	0.01	0.41
Renos	0.85	0.05	0.07	0.64	0.01	0.01	0.18	0.00	0.68	0.23	0.00	0.01	0.00	1.00	0.42	0.67	0.01	0.01	0.01	0.40
Startpage	0.77	0.18	0.59	0.60	0.39	0.27	0.06	0.27	0.72	0.56	0.41	0.60	0.42	0.42	1.00	0.66	0.34	0.39	0.01	0.00
Vobfus	0.01	0.55	0.71	0.00	0.56	0.62	0.25	0.61	0.32	0.72	0.70	0.71	0.58	0.67	0.66	1.00	0.66	0.77	0.53	0.56
Vundo	0.81	0.02	0.05	0.63	0.00	0.02	0.17	0.01	0.72	0.27	0.00	0.05	0.00	0.01	0.34	0.66	1.00	0.01	0.01	0.31
Winwebsec	0.95	0.03	0.03	0.70	0.02	0.00	0.13	0.01	0.76	0.04	0.02	0.02	0.03	0.01	0.39	0.77	0.01	1.00	0.00	0.32
Zbot	0.64	0.01	0.01	0.48	0.01	0.00	0.08	0.01	0.77	0.01	0.01	0.01	0.01	0.01	0.01	0.53	0.01	0.00	1.00	0.01
Zeroaccess	0.74	0.17	0.58	0.68	0.34	0.17	0.04	0.22	0.76	0.48	0.44	0.59	0.41	0.40	0.00	0.56	0.31	0.32	0.01	1.00

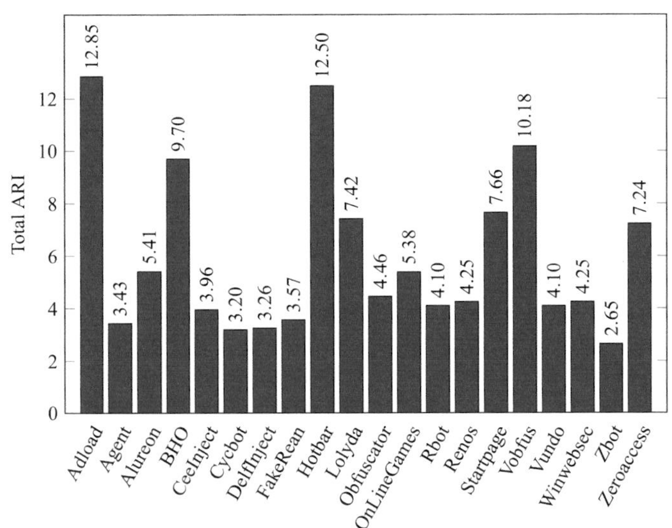

Fig. 4 Total ARI score (sum of 19 ARI scores) for each family

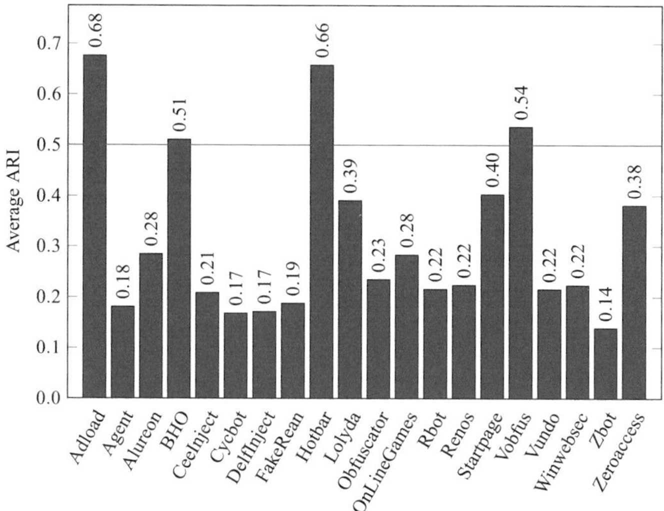

Fig. 5 Average ARI score for each family

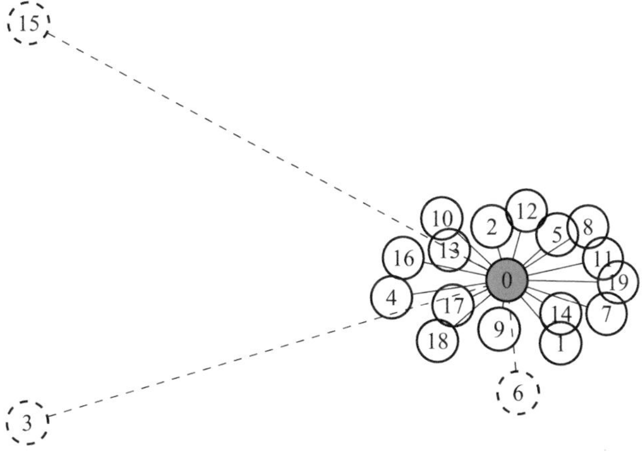

Fig. 6 Adload relationship with its paired families

It is also interesting to note that there are many families our dataset with extremely poor pairwise clustering results. We see that Agent, CeeInject, Cycbot, DelfInject, FakeRean, Obfuscator, Rbot, Renos, Vundo, Winwebsec, and Zbot all have average ARI scores below 0.23. This indicates that there is a large subset of the families that are virtually indistinguishable from each other.

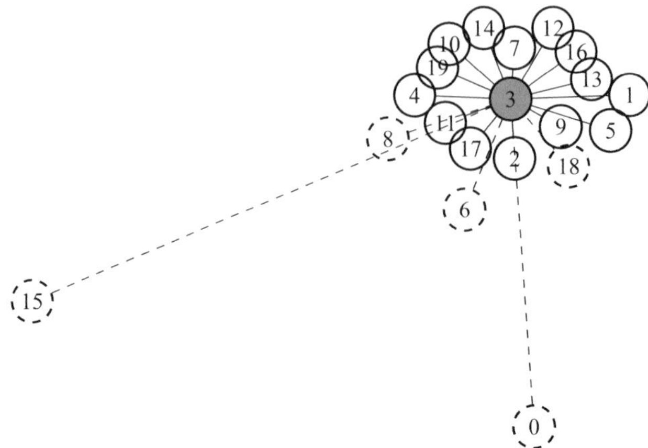

Fig. 7 BHO relationship with its paired families

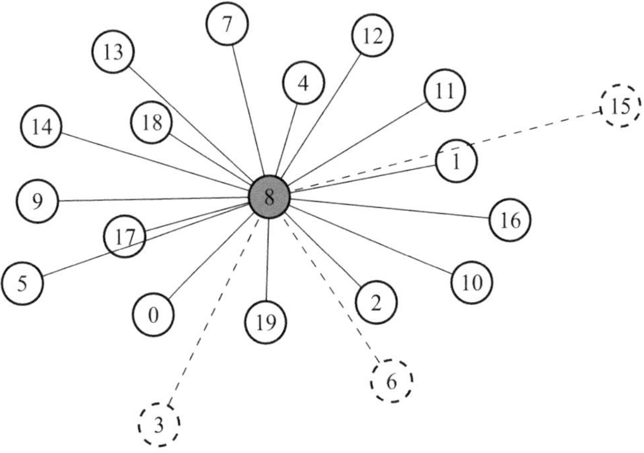

Fig. 8 Hotbar relationship with its paired families

3.2 Clustering Families of Different Type

In this set of clustering experiments, we consider seven families, each of which is of a different malware type. Specifically, the seven families considered, and their type, are the following:

Agent—Trojan
Ceeinject—VirTool
Cycbot—Backdoor
FakeRean—Rogue

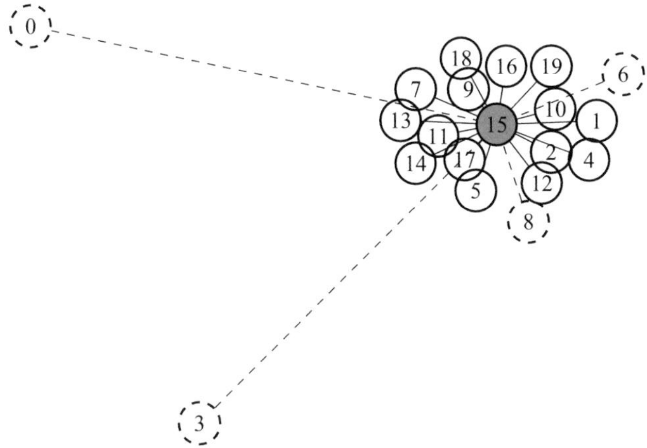

Fig. 9 Vobfus relationship with its paired families

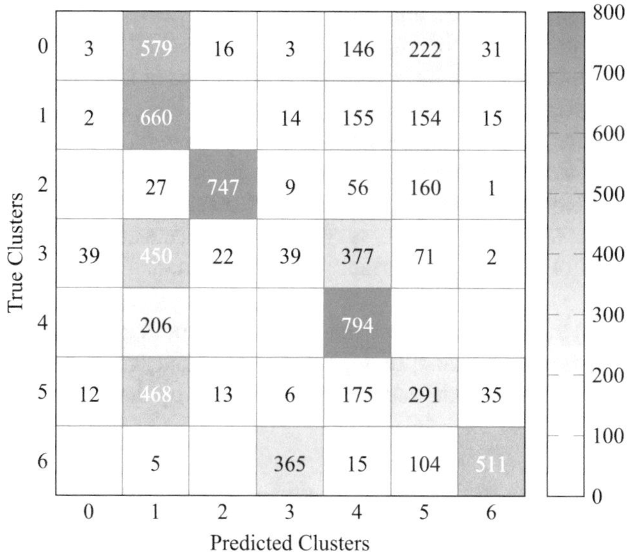

Fig. 10 Clustering using seven families from different malware types

Hotbar—Adware
Lolyda—Password Stealer
Vobfus—Worm

Figure 10 shows the results of clustering these seven families, each of which belongs to a different malware type. We might expect well-defined clusters in this case, but the ARI score is only 0.23, suggesting that a few families are still very similar, in spite of belonging to different malware types.

The results in Fig. 10 indicate that "type" is not a strong feature of malware. More specifically, we can say that the characteristics of bigrams that distinguish one malware family from another are not strongly type-dependent. This somewhat surprising result is useful, since it shows that attempts to identify malware by generic type are, in general, unlikely to be successful, at least when the analysis is based on byte bigram features. However, we note in passing that the authors in [9] use API calls as features and appear to have successfully classified selected malware by its type. Hence, it may be possible to obtain better results for malware type by using other features.

3.3 Clustering Families of the Same Type

In this section, we conducted two experiments to examine how well K-means clustering can distinguish between families belonging to the same malware type. Figure 11 illustrates the results of clustering four families, all of which are Trojans—the specific families considered in this case are Agent, Alureon, BHO, and Startpage. This result suggests that there are three well-defined clusters among these four families. We obtain an ARI of 0.35 in this case, which, interestingly, is much higher than the result obtained for malware samples of different types in the previous section.

Next, we cluster the three VirTool families in our dataset, namely, CeeInject, DelfInject, and Obfuscator. In this, we obtain the results in Fig. 12, which give us an ARI score of just 0.07. This number suggests a random clustering result and implies that these VirTool families are virtually indistinguishable, based on byte bigram features.

The results in this section indicate that the Trojan type is generic, in the sense that Trojan families can (and generally do) differ significantly from each other. This is not surprising, as Trojan code tends to be dominated by the non-malicious part of

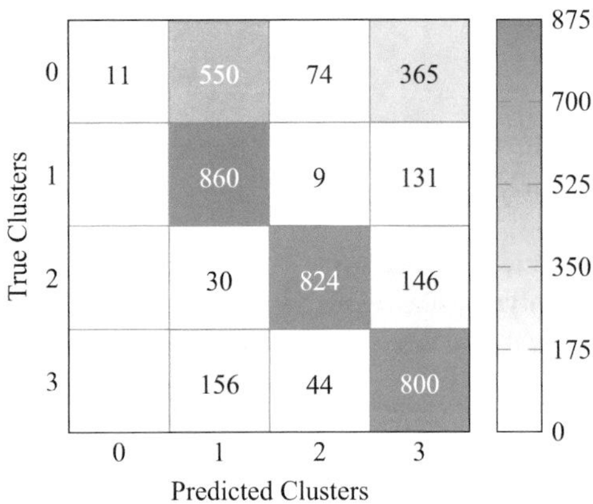

Fig. 11 Clustering four Trojan families (Agent, Alureon, BHO, and Startpage)

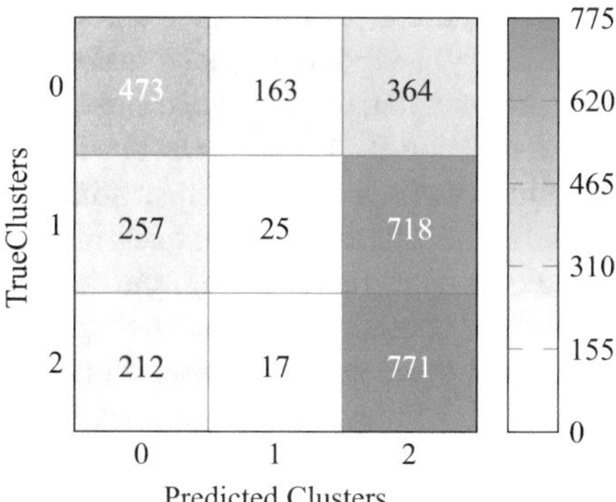

Fig. 12 Clustering VirTool families (DelfInject, CeeInject, and Obfuscator)

the application, which would be expected to vary widely between different Trojan families.

On the other hand, the VirTool type is highly specific, which results in an inability to distinguish between these families. That is, the VirTool type is relatively homogeneous, making such samples difficult to distinguish from each other, even when they are from different families.

4 Conclusion and Future Work

The goal of this research was to analyze malware clustering, with respect to families and types, based on elementary features and clustering techniques. We considered three different sets of experiments. In our first set of experiments, we clustered all families in pairs. In our second set of experiments, we clustered seven families, with one family from each of the distinct types in our dataset. Finally, we conducted experiments where the clustered families all belong to the same malware type. All of our experiments were based on K-means clustering using byte bigram features.

Our findings indicate that the relationship between malware families and malware type is somewhat complex. This is not entirely unexpected, since some malware types, such as Trojans, are only very loosely related, while other types, such as VirTool, are much more specific. Indeed, we did find that families of the Trojan type were far easier to distinguish from each other based on clustering, as compared to VirTool families.

More generally, our pairwise clustering results—and, in particular, the heatmap of ARI scores generated from these pairwise clusters—enabled us to draw many conclusions concerning similarities and differences between families. We were able to clearly see which families were most distinct from all other families, and which subsets of families were the most similar to each other. These results show that elementary cluster analysis is extremely useful for exploring relationships between malware families and that such analysis could serve as a guide for additional (and more costly) analysis of a given malware dataset.

In this chapter, we performed cluster analysis to examine the relationship between malware families. Our focus was clustering using K-means and byte bigram features. For future work, it would be interesting to consider larger numbers of clusters and to explore other clustering techniques, including Gaussian mixture models, hierarchical techniques, spectral clustering, and density-based clustering. While K-means can be viewed as generating "circular" or "spherical" clusters, other techniques can produce clusters of more general shapes. In addition, it would be interesting to experiment with other features, such as opcodes and API call sequences.

References

1. Adware:win32/hotbar. https://www.microsoft.com/en-us/wdsi/threats/malware-encyclopedia-description?Name=Adware:Win32/Hotbar&threatId=6204.
2. Backdoor:win32/cycbot.g. https://www.microsoft.com/en-us/wdsi/threats/malware-encyclopedia-description?Name=Backdoor:Win32/Cycbot.G.
3. Basole, Samanvitha, Fabio Di Troia, and Mark Stamp. 2020. Multifamily malware models. *Journal of Computer Virology and Hacking Techniques* 16 (1): 79–92.
4. Bayer, Ulrich, Paolo Milani Comparetti, Clemens Hlauschek, Christopher Kruegel, and Engin Kirda. 2009. Scalable, behavior-based malware clustering. In *NDSS*, vol. 9, 8–11.
5. Bholowalia, Purnima, and Arvind Kumar. 2014. EBK-means: A clustering technique based on elbow method and k-means in WSN. *International Journal of Computer Applications* 105 (9).
6. Hagberg, Aric A., Daniel A. Schult, and Pieter J. Swart. 2008. Exploring network structure, dynamics, and function using networkx. In *Proceedings of the 7th python in science conference, SciPy 2008*, 11–15. http://citeseer.ist.psu.edu/viewdoc/download;jsessionid=045872D50E1F472150E79500F79F4B93?doi=10.1.1.522.2540&rep=rep1&type=pdf.
7. Moore, Andrew W. k-means and hierarchical clustering. https://www.cs.cmu.edu/~cga/ai-course/kmeans.pdf.
8. Pedregosa, F., G. Varoquaux, A. Gramfort, V. Michel, B. Thirion, O. Grisel, M. Blondel, P. Prettenhofer, R. Weiss, V. Dubourg, J. Vanderplas, A. Passos, D. Cournapeau, M. Brucher, M. Perrot, and E. Duchesnay. 2011. Scikit-learn: Machine learning in python. *Journal of Machine Learning Research* 12: 2825–2830.
9. Pirscoveanu, Radu S., Steven S. Hansen, Thor M.T. Larsen, Matija Stevanovic, Jens Myrup Pedersen, and Alexandre Czech. 2015. Analysis of malware behavior: Type classification using machine learning. In *2015 International conference on cyber situational awareness, data analytics and assessment (CyberSA)*, 1–7.
10. Pitolli, Gregorio, Leonardo Aniello, Giuseppe Laurenza, Leonardo Querzoni, and Roberto Baldoni. 2017. Malware family identification with birch clustering. In *2017 International Carnahan Conference on Security Technology, ICCST*, 1–6.
11. Pws:win32/delfinject. https://www.microsoft.com/en-us/wdsi/threats/malware-encyclopedia-description?Name=PWS:Win32/DelfInject&threatId=-2147241365.
12. Pws:win32/lolyda.bf. https://www.microsoft.com/en-us/wdsi/threats/malware-encyclopedia-description?Name=PWS%3AWin32%2FLolyda.BF.
13. Pws:win32/onlinegames. https://www.microsoft.com/en-us/wdsi/threats/malware-encyclopedia-description?Name=PWS%3AWin32%2FOnLineGames.
14. Pws:win32/zbot. https://www.microsoft.com/en-us/wdsi/threats/malware-encyclopedia-description?Name=PWS:Win32/Zbot&threatId=-2147368817.

15. Romano, Simone, Nguyen Xuan Vinh, James Bailey, and Karin Verspoor. 2016. Adjusting for chance clustering comparison measures. *The Journal of Machine Learning Research* 17 (1): 4635–4666.
16. Santos, Jorge M., and Mark Embrechts. 2009. On the use of the adjusted rand index as a metric for evaluating supervised classification. In *International conference on artificial neural networks*, 175–184.
17. Stephens, Matthew. 2000. Dealing with label switching in mixture models. *Journal of the Royal Statistical Society: Series B (Statistical Methodology)* 62 (4): 795–809.
18. Trojandownloader:win32/adload. https://www.microsoft.com/en-us/wdsi/threats/malware-encyclopedia-description?Name=TrojanDownloader%3AWin32%2FAdload.
19. Trojandownloader:win32/agent. https://www.microsoft.com/en-us/wdsi/threats/malware-encyclopedia-description?Name=TrojanDownloader:Win32/Agent&ThreatID=14992.
20. Trojandownloader:win32/renos. https://www.microsoft.com/en-us/wdsi/threats/malware-encyclopedia-description?Name=TrojanDownloader:Win32/Renos&threatId=16054.
21. Trojan:win32/bho. https://www.microsoft.com/en-us/wdsi/threats/malware-encyclopedia-description?Name=Trojan:Win32/BHO&threatId=-2147364778.
22. Trojan:win32/startpage. https://www.microsoft.com/en-us/wdsi/threats/malware-encyclopedia-description?Name=Trojan:Win32/Startpage&threatId=15435.
23. Trojan.zeroaccess. https://www.symantec.com/security-center/writeup/2011-071314-0410-99.
24. Virtool:win32/ceeinject. https://www.microsoft.com/en-us/wdsi/threats/malware-encyclopedia-description?Name=VirTool%3AWin32%2FCeeInject.
25. Win32/alureon. https://www.microsoft.com/en-us/wdsi/threats/malware-encyclopedia-description?Name=Win32/Alureon.
26. Win32/fakerean. https://www.microsoft.com/en-us/wdsi/threats/malware-encyclopedia-description?Name=Win32/FakeRean.
27. Win32/obfuscator. https://www.microsoft.com/en-us/wdsi/threats/malware-encyclopedia-description?Name=Win32/Obfuscator&threatId=.
28. Win32/rbot. https://www.microsoft.com/en-us/wdsi/threats/malware-encyclopedia-description?Name=Win32/Rbot&threatId=.
29. Win32/vobfus. https://www.microsoft.com/en-us/wdsi/threats/malware-encyclopedia-description?Name=Win32/Vobfus&threatId=.
30. Win32/vundo. https://www.microsoft.com/en-us/wdsi/threats/malware-encyclopedia-description?Name=Win32/Vundo&threatId=.
31. Win32/winwebsec. https://www.microsoft.com/en-us/wdsi/threats/malware-encyclopedia-description?Name=Win32/Winwebsec.

Beyond Labeling: Using Clustering to Build Network Behavioral Profiles of Malware Families

Azqa Nadeem, Christian Hammerschmidt, Carlos H. Gañán, and Sicco Verwer

Abstract Malware family labels are known to be inconsistent. They are also black-box since they do not represent the capabilities of malware. The current state of the art in malware capability assessment includes mostly manual approaches, which are infeasible due to the ever-increasing volume of discovered malware samples. We propose a novel unsupervised machine learning-based method called MalPaCA, which automates capability assessment by clustering the temporal behavior in malware's network traces. MalPaCA provides meaningful behavioral clusters using only 20 packet headers. Behavioral profiles are generated based on the cluster membership of malware's network traces. A Directed Acyclic Graph shows the relationship between malwares according to their overlapping behaviors. The behavioral profiles together with the DAG provide more insightful characterization of malware than current family designations. We also propose a visualization-based evaluation method for the obtained clusters to assist practitioners in understanding the clustering results. We apply MalPaCA on a financial malware dataset collected in the wild that comprises 1.1 k malware samples resulting in 3.6 M packets. Our experiments show that (i) MalPaCA successfully identifies capabilities, such as port scans and reuse of Command and Control servers; (ii) It uncovers multiple discrepancies between behavioral clusters and malware family labels; and (iii) It demonstrates the effectiveness of clustering traces using temporal features by producing an error rate of 8.3%, compared to 57.5% obtained from statistical features.

A. Nadeem (✉) · C. Hammerschmidt · C. H. Gañán · S. Verwer
Delft University of Technology, Delft, The Netherlands
e-mail: azqa.nadeem@tudelft.nl

C. Hammerschmidt
e-mail: c.a.hammerschmidt@tudelft.nl

C. H. Gañán
e-mail: c.hernandezganan@tudelft.nl

S. Verwer
e-mail: s.e.verwer@tudelft.nl

M. Stamp et al. (eds.), *Malware Analysis Using Artificial Intelligence and Deep Learning*, https://doi.org/10.1007/978-3-030-62582-5_15

1 Introduction

The first malware was discovered over thirty years ago. Yet, it is still one of the leading threats in cybersecurity.[1] AV-test, a security research institute, reported detecting over 1000 Million malware samples in 2019.[2] Anti-Virus (AV) companies play a pivotal role in analyzing malware by assigning labels to newly discovered samples. However, there are several shortcomings of malware family labels: (i) Each vendor has its own way of determining a malware family. Labels obtained from different vendors are often inconsistent [29]. (ii) The precise methods used by each vendor are proprietary and unstandardized [49]. (iii) The current labels are heavily based on static and system-level activity analysis. The problem is that malware family labels do not represent the capabilities of malware samples. The black-box (unexplainable) nature of the labeling methods also makes it impossible to verify assigned family labels, causing the evaluation of newer detection methods to depend on unreliable ground truth [33]. Moreover, network traffic is rarely used to determine family labels because of noisy ground truth and non-stationary data distribution [3]. As a result, malware samples that exhibit identical network behavior but have different code attributes end up in different families, see, e.g., Perdisci et al. [44].

In this chapter, we address the limited interpretability of malware family labels by proposing white-box[3] behavioral profiles for malware samples. Existing research suggests that network traffic shows malware's core behavior by capturing direct interactions with the attacker or C&C server [14]. Network traffic analysis can also be performed remotely, which presents a lower overhead than many popular system-activity solutions. Therefore, we place emphasis in building network behavioral profiles. To this end, we propose MalPaCA (Malware Packet Sequence Clustering and Analysis) for automated capability assessment of malware samples. The goal of *Capability Assessment* is to discover the behaviors a malware sample can exhibit. We investigate the usage of unsupervised machine learning for intelligent capability assessment to tackle the ever-increasing volume of newly discovered malware.

Until now, malware capability assessment has primarily been a manual effort [11, 40, 50], resulting in behavioral profiles that are quickly outdated. Although machine learning-based behavioral analysis approaches exist, they construct a single model that describes either the whole network or each protocol usage individually [47]. However, the network traffic originating from even a single host can be so complex that these models fail to correctly represent malicious behaviors [23]. This is why MalPaCA splits the network traffic between hosts into *uni-directional connections* and considers them as discrete behaviors (or *capabilities*).

MalPaCA clusters similar connections based on their temporal similarity, where each cluster represents a unique capability. A malware sample is then represented by its *Behavioral Profile*—a list of cluster membership of its connections. We represent

[1]https://www.cybersecurity-insiders.com/top-15-cyber-threats-for-2019/.

[2]https://www.av-test.org/en/statistics/malware/.

[3]In white-box ML, all steps are explainable—the input, output and how the output was generated. In contrast, only the input and output are known in black-box ML, e.g., Neural Networks.

malware's behavioral profiles in a Directed Acyclic Graph that shows different samples' overlapping behaviors. The graph also shows malware samples from different families behaving identically, showing potentially incorrect family labels. MalPaCA is novel as it adopts sequential features that keep the temporal nature of the traffic intact. It uses a combination of Dynamic Time Warping and Ngrams to measure the distance between network connections. MalPaCA utilizes only 20 packets to identify the network behavior shown by any given connection. It also utilizes only the packet header features that are available even when traffic is encrypted.

The last step of MalPaCA's pipeline is assigning capability labels to clusters. Each discovered cluster is visualized using *temporal heatmaps* to determine which capability it captures. The temporal heatmaps provide a goal- and data-driven approach to investigate the performance of MalPaCA's clustering, by clearly showing the network connections that are grouped together. This eliminates the need to manually investigate thousands of network traces. Security analysts can also fine-tune MalPaCA's parameters by visualizing the temporal heatmaps. The key advantage of this methodology is its white-box and explainable nature: it provides a visual representation to investigate MalPaCA's rationale for finding behavioral similarity. In doing so, we address the interpretability problem of typical black-box analysis methods, which is an important stepping stone towards better detection methods.

We evaluate MalPaCA's performance on 1.1 k malware samples (resulting in 3.6 M packets) coming from 15 families collected in the wild. We also compare the effectiveness of sequence clustering by comparing with an existing method based on frequently-used statistical (aggregate) features [54].

Results. The results are very promising: (i) MalPaCA's capability assessment works on low quality datasets with as low as 20 packets in each trace, though additional traces result in more thorough profiles; (ii) It successfully discovers several attacking capabilities, such as port scans and reuse of C&C servers; (iii) MalPaCA demonstrates the effectiveness of sequence clustering by producing an error rate of 8.3% compared to 57.5% obtained from statistical features; and (iv) MalPaCA uncovers multiple discrepancies between behavioral clusters and family labels. We believe this happens either because the labels are incorrect or because the overlapping families share significant behavior.

Contributions. We summarize our contributions as follows:

1. We show that short sequences of packet header features are capable of characterizing network behavior;
2. We build *MalPaCA*[4]—a tool to automatically build network behavioral profiles of malware samples collected in the wild;
3. We introduce *temporal heatmaps*—a data-driven and visualization-based cluster evaluation method that requires no ground truth;
4. We show the behavioral relationships between malwares using a Directed Acyclic Graph, which also uncovers discrepancies between behavioral clusters and traditional family labels;

---

[4]https://github.com/azqa/malpaca-pub.

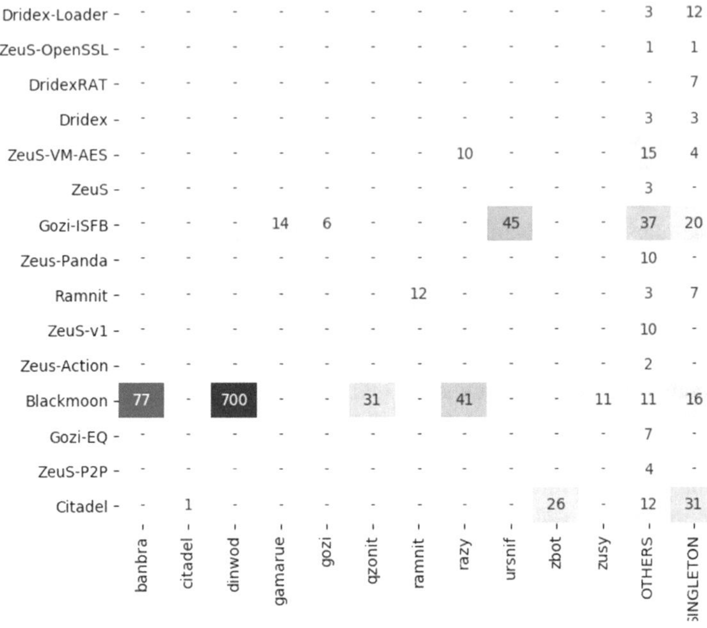

Fig. 1 Disagreements between AV vendors. Rows: YARA labels, Columns: AVClass labels, Counts: # malware binaries

5. We demonstrate the effectiveness of sequence clustering, which shows less errors than an existing solution based on statistical features.

2 The Problem with AV Labels

This section presents an analysis of our experimental dataset to emphasize the problem of inconsistent AV labels and motivates the need for explainable behavioral profiles. We compare the *agreement rate* of two popular malware labeling practices, i.e., YARA rules[5] and VirusTotal[6] labels. The malware collection process is given in Sect. 5.1. Table 2 shows the number of binaries in each malware family.

The malware binaries in the dataset are labeled using YARA rules. Each malware binary also has a Virus Total (VT) scan report. On average, there are 61 AV vendors for each malware sample, out of which 25.8% vendors per malware sample return a `null` detection, i.e., unable to detect it as malicious. The rest assign various labels to each malware binary.

[5]https://virustotal.github.io/yara/.

[6]https://www.virustotal.com/.

Since each AV vendor has its own vocabulary, a trivial filtering attempt on a VT report cannot identify the true underlying family label. Sebastian et al. [49] have developed an open source tool, called AVClass, that takes VT reports as input and returns the most likely family label. If, after all the filtering steps, AVClass is unable to identify the family name, it declares the malware as a "SINGLETON". We use AVClass to reduce a VT report into its representative VT family label. In the experimental dataset, AVClass returns "SINGLETON" for 101/1196 (8.4%) VT reports, while assigning 42 unique family labels to the rest 1095 malware binaries.

Figure 1 shows the label agreement rate between the YARA and VT labels. The y-axis shows the YARA labels. The x-axis shows the VT labels as aggregated by AVClass. For brevity, "OTHERS" category contains all samples for which $counts < 10$. Only 3 family names co-exist in both YARA and VT labels, i.e., `Citadel`, `Gozi`, and `Ramnit`. Also, although `Ramnit` is detected under the same name by both YARA and VT, 10 malware samples are still labeled differently. In fact, YARA family labels are assigned 4.2 distinct VT labels on average, while VT labels are assigned 1.5 distinct YARA labels on average. One example demonstrating this is: `YARA`: `Zeus-VM-AES` (29 samples) are predicted as `VT`: `razy` (10 samples), `gamarue` (6 samples), `cerber` (3 samples), `upatre` (3 samples), `farfli` (1 samples), `locky` (1 samples), `hpcerber` (1 samples), and `SINGLETON` (4 samples). This makes it very hard to understand the collected malware. One fair conclusion is that some VT labels can be considered as sub-families of the popular YARA malware family. For example, `Dinwod` and `Banbra` seem to be sub-families of `Blackmoon`, but the names alone do not explain which attributes set them apart from each other.

3 Related Work

The field of malware analysis has existed since the first malware was discovered over 30 years ago. Since then, multiple machine learning-based approaches have been proposed to automate malware detection and analysis. In this section, we present a brief survey of the major research challenges targeted by prior work. In doing so, we highlight how our work fills the gaps across various research themes.

3.1 Challenges in Malware Labeling

Existing research has repeatedly shown that malware family labels are noisy and inconsistent. Popular tools, such as VirusTotal, run multiple AV scanners and return an array of labels predicted by each scanner, without any indication as to which is correct. There is also an absence of a common vocabulary that all security companies can follow to label malware samples. Maggi et al. [37] propose a method to find inconsistencies in malware family labels generated by Anti-Virus (AV) scanners. Mohaisen et al. [38] are the first to measure the accuracy, consistency, and completeness of AV scanners. Their results show that AV vendors produce inconsistent labels 50% of the time, on average. These findings resulted in research that found ways to deal with the inconsistencies in the family labels. Kantchelian et al. [29] proposed an algorithm based on Expectation Maximization and Bayesian models that assign weights to each vendor's trustworthiness. Sebastián et al. [49] developed a useful open source tool, called AVClass that determines the likely family name after performing heavy filtering on all the predicted labels. However, these methods do not address the key underlying issue—malware family labels are black-box with limited interpretability.

Behavioral profiles complement family names in that they also describe the behavior of a sample. *Capability assessment* is done to characterize a malware family, which has primarily been a manual effort resulting in behavioral profiles that are quickly outdated. Also, most of the prior works in capability assessment utilize information extracted from the static analysis of malware executables: Black et al. [11] bridge the semantic gap between low-level API calls and high-level behaviors in order to build a taxonomy of banking malware. They extract API calls by statically analyzing a banking malware dataset, and map them to high-level behaviors manually with the help of domain experts. Sharma et al. [50] recently proposed a method to automatically build behavioral profiles. They select a few high-level capabilities possessed by malware by investigating the literature, and map them to low-level behaviors extracted from the static analysis of 56 malware samples. *In contrast, we propose MalPaCA that automatically builds dynamic (network) behavioral profiles.*

3.2 Research Objectives: Detection Versus Analysis

Existing research on malware comes in two strains: detection-based and analysis-based. Malware detection and signature generation dominates existing literature, with the end-goal of optimizing metrics [1, 2, 7, 10, 17, 24, 35, 36, 39, 44, 46, 54, 60], while only a few of these works also help the readers understand and analyze the obtained results [23, 43]. Recently, however, several malware analysis approaches have been proposed that aim to improve malware understandability rather than optimizing detection rates. These methods provide essential insights that can improve malware detection methods. Black et al. [11] perform an in-depth analysis of the key behaviors of banking malware families and how they have evolved over time. Moubarak et al. [40] discuss malware evolution and the structural relationship between several potentially state-sponsored malware. In [51], the authors cluster Android malware samples and build a dendrogram of the malware families showing overlapping code snippets. Sharma et al. [50] build behavioral profiles of malware samples using static analysis. *In this chapter, we follow a similar approach and build an analysis tool, MalPaCA. MalPaCA uses unsupervised clustering to group network connections that behave similarly and uses them to construct malware's behavioral profiles.*

Although clustering is an unsupervised technique, existing literature has often used some notion of ground truth (family labels) to evaluate the cluster quality. Bayer et al. [7] evaluate their malware clustering approach using labels obtained by the majority voting of 6 AV vendors. Perdisci et al. [44] evaluate their malware clustering approach by introducing a notion of AV graphs that depict the agreement between AV vendors as a measure of cluster cohesion and separation. In [35], the authors report the precision and recall of higher than 0.95 of their malware clustering approach. They use the majority voted family labels from 25 AV vendors as their ground truth. Li et al. [33] have examined the challenges of evaluating malware clustering and have advised caution when deciphering highly accurate clustering results as they can be impacted by spatial bias: performing majority voting on AV-provided labels is hazardous, since if most of the AV vendors are in agreement, it typically indicates that the families are already easy to detect. *In this chapter, we propose a data-driven and visualization-based method to evaluate clusters, without using family labels. Instead of optimizing clustering accuracy, our emphasis is on explainability of the results.*

3.3 Challenges in Malware Behavior Modeling

Modeling software behavior is a challenging task, but modeling malware's behavior is even more challenging since malware authors specifically try to evade detection [15]. Static analysis of malware binaries and disassembled code has been a popular malware analysis approach in the literature [6, 11, 21, 35, 39]. Increasingly more malware uses obfuscation techniques to evade analysis, causing difficulties for statically analyzing malware. The obfuscation attempts gave rise to dynamic analysis of malware that executes a malware sample in a sandbox and collects execution traces from it. Dynamic analysis is generally divided into two strains: System activity and Network traffic analysis. Network traffic analysis collects traces of malware samples remotely using existing network monitoring infrastructures [44], making it much easier to apply. However, the behavioral analysis and signature generation literature is heavily focused on system activity analysis, e.g., see [7, 16, 50, 52]. Research suggests that network traffic shows the core behavior of malware [14]. Although sometimes encrypted, network traffic contains the direct interaction with the attacker. In this section, we discuss three major challenges of modeling malware behavior via traffic analyses.

Feature selection. Network traffic analysis is generally applied when designing Network Intrusion Detection Systems (NIDS), which either detect anomalous traffic [24] or generate signatures for malware families [22, 26, 55]. Deep Packet Inspection (DPI) is one commonly used approach in NIDS to extract information from packet payloads. For example, Rafique et al. [46] use DPI for automatic signature generation of malware families. Although effective, downsides to DPI-based approaches are that they are privacy-intrusive, operationally expensive, and do not work out-of-the-box for encrypted traffic. There are also approaches that detect specific attacks. For example, HTTP-based malware can be detected using specific features from the Application header [44]. Similar approaches exist for DNS-based malware [32, 45], and HTTPs-based malware [4]. In the absence of the HTTP, DNS, and TLS headers, these approaches seize to work.

Several works use *coarse* or high-level features that are protocol-agnostic and work out-of-the-box even with encrypted traffic. For example, Conti et al. [17] use sequences of packet sizes to characterize the network behaviors generated by Android applications. Aiolli et al. [2] use various statistical features computed over packet sizes to detect bitcoin wallet application functionality. Acar et al. [1] use network traffic direction and packet lengths to identify commands issued to smart home IoT devices. These works aim to characterize benign network behaviors. In the malware domain, Tegeler et al. [54] use average packet size, average packet inter-arrival time, average connection duration, and the FFT of C&C communication to detect bot-infected hosts. Garcia [23] builds a behavioral Intrusion Detection System by using the size, duration, and periodicity of Netflows. *In this chapter, we also use high-level features from packet headers to characterize malware's network behavior. To the best of our knowledge, network traffic analysis has not been used in capability assessment or for generating behavioral profiles of malware samples.*

Feature representation. Machine learning methods take a feature vector as input, which can represent anything ranging from a single behavior to a complete malware sample. Multiple observations for a single feature are aggregated into *statistical features*, e.g., mean packet size of a netflow. Existing literature is filled with approaches that use such statistical features, e.g., see [5, 10, 23, 54]. Although they are computationally efficient, they lose local behavioral details, which can be a problem when the goal is to characterize that behavior.

Another approach that is gaining momentum is the use of *sequential features*. Numeric sequential features are typically used in two ways: *Discretized* and *Raw sequences*. A raw sequence (or a continuous sequence) is composed of the original observations, while a discretized sequence encodes the observations into a finite set of bins. Discretizing sequences is typically faster and makes measuring distances easier. Pellegrino et al. [43] learn state machines from discretized netflow data in order to detect bot-infected traffic, while Hammerschmidt et al. [27] use it to cluster host behavior over time. Lin et al. [36] detect anomalies in industrial water treatment plant by using discretized sequences from sensor readings. In practice, malware-related data is often scarce and noisy. In this case, discretization can lose important information.

Raw sequences are rarely used for modeling network traffic because it is non-stationary and contains noise (e.g., empty acknowledgment packets or retransmissions), and delays (due to varying network latency) [3]. Ntlangu et al. [41] provide a brief overview of time-series approaches to model network traffic. As noted in [41], due to the nature of network traffic and their distributions, (auto-)regressive models struggle to accurately capture them. Kim et al. [30] use a multi-variate time-series regression model on host-based resource consumption, such as CPU and memory usage (not network traffic) to identify Android malware. Conti et al. [17] propose a method to detect the action performed by Android applications using raw sequential features. *To the best our knowledge, MalPaCA is the first method that successfully uses short raw sequential features to characterize malware network behavior.*

Distance measure. The notion of behavioral similarity requires the means to be able to measure distance between two objects. The choice of the distance measure is directly dependent on the data type of the feature set (e.g., numeric or categorical) and the way the features are represented (e.g., statistical or sequential). For statistical features, Euclidean distance is most commonly used. For instance, Chan et al. [16] use Euclidean distance to determine similar Android processes.

Calculating the distance between sequential features is more challenging because they may not always be properly aligned. For categorical (or discretized) sequences, there exist Bioinformatics inspired solutions using sequence alignment [57]. They require pre-computed substitution matrices, which currently do not exist for malware. There also exist String matching solutions frequently used in the Natural Language Processing domain. Baysa et al. [8] use Levenshtein, or edit distance, to measure the similarity between two malware binary files. A sequence can also be broken down into sub-sequences, represented as Ngrams, which have been used to model genomic sequences [58] and to match files [34]. They have also been used to classify malware families in [13]. Longest Common Subsequence (LCS) with `k`-gaps can also be used to measure distances between sequences. The gaps account for the occasional noise. Chan et al. [16] use LCS to group similar resource-access-patterns (not network traffic) in Android applications.

A few distance measures exist for raw or continuous sequences. Verwer et al. [56] have used Kullback–Leibler divergence to measure the distance between two sequences while learning probabilistic automata. However, it requires substantial amount of data to measure the similarity with a high confidence, which is not always available for malware. Another promising distance measure is Dynamic Time Warping (DTW). DTW has been used in fingerprint verification [31], characterizing DDoS attack dynamics [59], and measuring similarity in android application behavior [17]. *MalPaCA uses a combination of DTW and Ngrams to measure the distance between network connections.*

4 MalPaCA: Malware Packet Sequence Clustering and Analysis

The ultimate goal of MalPaCA is to construct a behavioral profile for each malware sample that is more descriptive than its family label. Research shows that malware belonging to the same family exhibits similar behaviors since malware authors often share code and resources [53]. To this end, MalPaCA automatically identifies the various network behaviors exhibited by malware samples, and groups samples that share common behavior. MalPaCA does not assume any a priori knowledge about

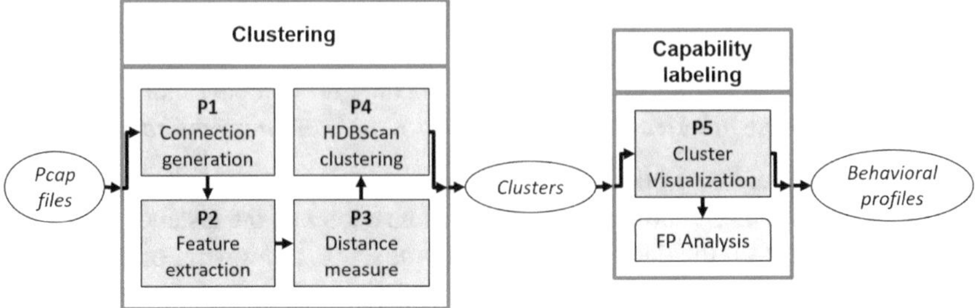

Fig. 2 MalPaCA: Connections clustered on behavioral similarity; malware described using connections' cluster membership

the malware's family name or its capabilities, and hence can be used out-of-the-box for other malware datasets. The profiles are built using observed behavior since only the executed functionality is relevant for behavioral profiling. Profiles for individual families can be enriched further by observing additional traffic. We release MalPaCA to the public.[7]

Figure 2 illustrates the architecture of MalPaCA with its five phases (P1 to P5). Network traces (Pcap files) are given as input to the system, which are split into uni-directional packet streams (or *connections*) that are clustered based on temporal similarities. Each cluster is assigned a capability label by visualizing temporal heatmaps showing connections' feature values. Each malware sample (and its associated Pcap file) is then described by a *Cluster Membership String*, forming a descriptive behavioral profile.

4.1 Connection Generation (P1)

A *connection* is defined as an uninterrupted uni-directional list of all packets sent from source IP to destination IP address. This means 8.8.8.8 → 123.123.123.123 is a different connection than 123.123.123.123 → 8.8.8.8. We refer to these as *Outgoing* and *Incoming* connections based on their direction with respect to the `localhost`. Note that we do not use IP address as a feature, except to create connections.

Ideally, a connection captures one complete capability. The connection length can vary significantly depending upon the behavior and network delays. Since the network delay is an artifact of the network, not of the malware, it is important to reduce its impact when measuring behavioral similarity. MalPaCA does so by capping the sequence length to a fixed threshold, avoiding artifacts that are due to connection length.

[7]https://github.com/azqa/malpaca-pub.

Existing research suggests that it is possible to identify behavioral differences from a *handshake*.[8] Wang et al. [61] use the first 3 to 12 bytes of packet headers in order to identify the different so-called Protocol Format Messages. MalPaCA builds upon this idea and utilizes the first few packets of a connection to identify the capability. This is a fixed threshold denoted by the tunable parameter len. It should be large enough to allow the handshake to be modeled, the length of which is often unknown in network traffic analysis. Larger values of len not only include noise artifacts but also increase the computational resources required to process longer connections.

4.2 Feature-Set Extraction (P2)

The choice of feature-set is crucial for determining the kind of behaviors that are identified by MalPaCA. Two considerations motivate our choice: (1) MalPaCA should be generalizable to more than one type of malware; (2) The feature set is small and easy to extract. Hence, we do not use features extracted from the packet payload itself as they limit the applicability of the method. We also do not use IP addresses as they are easy to spoof and are considered Personally Identifiable Information[9] in countries like the Netherlands. We use four sequential features: (i) packet size, (ii) time interval, (iii) source port, (iv) destination port. All four features are independent of the protocol type, making them available for every connection. Each feature is represented as a sequence of raw observations for subsequent packets. Although these features are simplistic, we demonstrate that their sequential nature captures malware behavior effectively.

Packet size (f_{ps}) measures the size of the *IP datagram* of each packet in bytes. *Time interval* (f_{in}) captures the inter packet arrival time in milliseconds. We use time interval because malware tends to show a periodic behavior, e.g., bots send periodic heartbeat packets[10] to inform the C&C server about the infected host. MalPaCA is meant to be used on a single network at a time since using inter-arrival time makes connections collected on different latency networks incomparable.

We use both *source* (f_{sp}) and *destination* (f_{dp}) *port numbers* because the connections are uni-directional. We particularly use source port so the analysts can limit the use of problematic ports in case of outgoing connections. The usage of certain vulnerable ports can also indicate suspicious activity. Each connection is represented by four sequences, one per feature, $C = (f_{ps}, f_{in}, f_{sp}, f_{dp})$.

---

[8] *Handshake traffic refers to the introductory few packets of a connection.*

[9] https://www.enterprisetimes.co.uk/2016/10/20/ecj-rules-ip-address-is-pii/.

[10] https://www.ixiacom.com/company/blog/mirai-botnet-things.

4.3 Distance Measure (P3)

Three considerations motivate our choice of distance measure: (1) Different distance measures are applicable on numeric and categorical data types; (2) The distance measure should be intuitive to help understand the results; (3) It must produce results that are resilient to delays and noise, which are common characteristics of network traces. The last consideration was added after observing distance measures producing results that were artifacts of network delays. MalPaCA uses a combination of Dynamic Time Warping (DTW) and Ngram analysis to measure distance between two connections.

Dynamic Time Warping. DTW [9] is used to measure distances between numeric sequences (packet size and time interval) due to its robustness to delays and noise. It aligns two time-series that may contain distortions (or warps) in the time-axis. It maps local substructures in one sequence to those of the other sequence. For two sequences $a = [a_0, a_1, \ldots, a_n]$ and $b = [b_0, b_1, \ldots, b_m]$ the DTW distance $d_{dtw}(a, b)$ is

$$d_{dtw}(a,b) = \sum_{i=1}^{n+1}\sum_{j=1}^{m+1} ||a_i - b_j|| + \min \begin{cases} d(a_{i-1}, b_j), \\ d(a_i, b_{j-1}), \\ d(a_{i-1}, b_{j-1})) \end{cases} \quad (1)$$

The output is a *similarity* score, which we normalize using:

$$d_{ndtw}(a,b) = \frac{d_{dtw}(a,b) - \min_{x,y}\{d_{dtw}(x,y)\}}{\max_{x,y}\{d_{dtw}(x,y)\} - \min_{x,y}\{d_{dtw}(x,y)\}} \quad (2)$$

Ngram analysis. An *Ngram* is defined as the set of n (called $order$) consecutive items in a given sequence. The larger the value of $order$, the more sequence structure is captured. A sequence of port numbers is converted into a set of Ngrams, called its *Ngram profile* using a sliding window of length $order$. An example for $order = 2$ is shown in Table 1, where A, B, C, D are hypothetical port numbers. Let G be the set of all unique Ngrams occurring in the dataset. For each packet sequence a, a vector $a_g = [f(g_1, a), f(g_2, a), \ldots, f(g_{|G|}, a)]$ is generated, containing the occurrence frequencies $f(g_i, a)$ in a of each Ngram $g_i \in G$.

We measure the distance between two Ngram profiles using Cosine distance. Other distance measures exist for Ngrams, but Cosine has shown promise in measuring

Table 1 Example—Distance measurement using Ngram analysis

Input	Ngram profiles	G = [AB, BC, CB, DA, CA]	Cosine distance
$ABCBC$	AB, BC, CB, BC	[1, 2, 1, 0, 0]	0.3876
$DABCA$	DA, AB, BC, CA	[1, 1, 0, 1, 1]	

similarity between categorical sequences [63]. It is determined by the angle between two non-zero vectors. The similarity value lies between 0 and 1, where 1 means that the two vectors are the same (parallel to each other) and 0 means they are completely different (orthogonal to each other). For two sequences in their vector representations $a = [v_1, \ldots, v_{|G|}]$ and $b = [v'_1, \ldots, v'_{|G|}]$, the Cosine distance $d_{cos}(a, b)$ is

$$d_{cos}(a, b) = 1 - \frac{\sum_{i=1}^{|G|} a_i \times b_i}{\sqrt{\sum_{i=1}^{|G|} a_i^2} \times \sqrt{\sum_{i=1}^{|G|} b_i^2}} \tag{3}$$

Finally, the DTW and cosine distances are combined to calculate the final distance between two connections:

$$d_{conn}(A, B) = \frac{d_{ndtw}(a_{ps}, b_{ps}) + d_{ndtw}(a_{in}, b_{in}) + d_{cos}(a_{sp}, b_{sp}) + d_{cos}(a_{dp}, b_{dp})}{4} \tag{4}$$

where $A = (a_{ps}, a_{in}, a_{sp}, a_{dp})$ and $B = (b_{ps}, b_{in}, b_{sp}, b_{dp})$ are connections and their features: packet sizes $\{a|b\}_{ps}$, intervals $\{a|b\}_{in}$, source port Ngram profiles $\{a|b\}_{ps}$, and destination port Ngram profiles $\{a|b\}_{dp}$.

4.4 HDBScan Clustering (P4)

A key strength of MalPaCA is the clustering algorithm it uses. There exists a familial structure among malware behaviors [51, 55]. Therefore, it makes sense to use hierarchical clustering to model the relationships between them. We have used Hierarchical Density-Based Spatial Clustering of Applications with Noise (HDBScan) [12] for this purpose. The key strengths of HDBScan are twofold: it automatically determines the optimal number of clusters, and it generates high-quality clusters that remain stable over time. It also has minimal tunable parameters, which allow configurations to be generalizable.

HDBScan requires a pairwise distance matrix as input. It does not force data points to become part of clusters—all data points whose membership to a cluster cannot be determined are considered to be *noise*. In our context, *noise* refers to behaviors that are either too different from all the others or cannot be clearly assigned to one cluster. An ideal dataset with clear cluster boundaries will have no noise. Hence, in the presence of a less ideal dataset, noise is discarded to extract high-quality clusters. Keep in mind that discarding excessive connections as noise can also be counterproductive. We discuss this limitation in Sect. 8.

4.5 Cluster Visualization (P5)

Formalizing cluster quality without ground truth is a fundamental challenge in clustering. Although some metrics exist that capture cluster quality (i.e., Silhouette index [48] and DB Index [18]), they require a notion of distance from a cluster centroid, which is difficult to obtain for sequences. In MalPaCA, each connection is represented by four sequences and collapsing these into a single cluster quality measure loses important local behavior. Instead, we define the following properties to be indicative of good clustering: (1) Cluster homogeneity is high—a cluster contains only similar connections. (2) Cluster separation is high—each cluster captures a unique capability. (3) Clusters are small and specific so they only capture the core capability. The first two properties ensure that we obtain meaningful capability-based clusters, the third ensures that only the core capabilities are captured.

We use *temporal heatmaps* for a white-box cluster analysis. We graphically show the connection features and rely on human visualization skills to determine cluster quality. Analysts can inspect heatmaps to determine which behavior is captured in a cluster. This gives them control over the clustering results. We leave the automation of this process as future work.

Four temporal heatmaps are associated with each cluster, one corresponding to each feature. Each row in a heatmap shows the corresponding feature sequence of the first len packets in a connection. Figure 3 shows example temporal heatmaps. The figure highlights one dissimilar connection among the eight in the cluster, clearly highlighted in red.

Clustering Error Analysis. Visualizing the cluster content helps to identify which connections belong in a cluster. A *Clustering Error* (CE) is defined as a connection that is placed in cluster `X` despite half of its features being different from the remaining connections in the cluster. Since each feature holds equal weight, we only consider a connection as CE if more than two features *differ*. We consider two features *different* if more than 50% of their sequences differ so significantly that a different color appears on the temporal heatmap. This is where human visualization skills play a key role in determining feature similarity. Figure 3 shows a cluster containing one CE, highlighted in red. It shows that three out of four feature values of this connection are different from other connections in the same cluster. The clustering error rate is

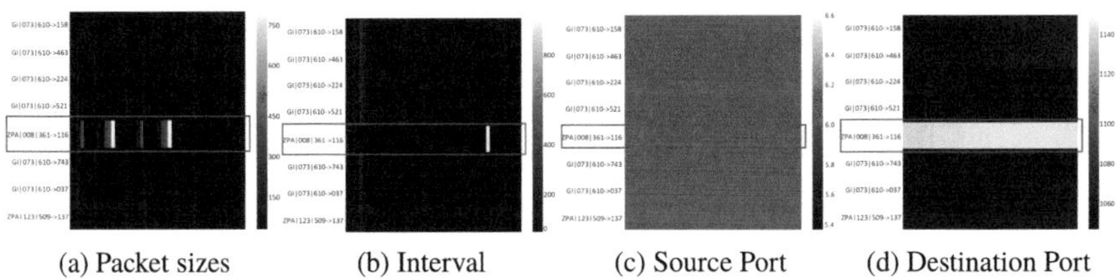

(a) Packet sizes (b) Interval (c) Source Port (d) Destination Port

Fig. 3 A clustering error: one connection does not belong in the cluster it is assigned

calculated as $\frac{CEs}{Clustersize}$, i.e., $\frac{1}{8}$. We measure the error rate of each cluster similarly and calculate the *average percentage of errors per cluster* as a notion of clustering quality.

In practice, we first establish the common majority by finding two or more connections that are most similar to one another, i.e., the ones that have the least mutual distance. The pairwise distance matrix computed during clustering is used as a lookup table for finding such connections. Figure 3 shows a simple case where the *rightful owners of a cluster* are easily visible since 7 out of 8 connections are very similar. The rest of the connections are compared with the rightful owners and are either considered as true positives or clustering errors, depending on how many feature sequences differ.

5 Experimental Setup

In this section, we describe the dataset used for the experiments and the configuration details of MalPaCA's parameters.

5.1 Experimental Dataset

MalPaCA was evaluated on financial malware samples collected in the wild. We worked in collaboration with a security company that specializes in malware analysis and threat intelligence. They collected the dataset independently. The dataset contained 1196 malware samples that were collected over one year. Each malware sample was executed in a sandboxed environment containing several virtual machines. The resulting network traffic was stored in a Pcap file. Some samples showed sandbox evasion. They were re-executed in a VM with different settings. This resulted in a total of 1196 Pcap files. Uni-directional connections were extracted, resulting a total of 8997 connections containing 3.6 M packets.

The dataset contains 15 famous financial malware families. They were labeled by the security company using their proprietary YARA rules. Additionally, each sample was submitted to VirusTotal (VT), which hosts 68 AV vendors. For each sample, VT returns a report containing detection results from each vendor. Table 2 summarizes the dataset.

5.2 MalPaCA Parameters

MalPaCA has four parameters, i.e., $order$ of the Ngrams used for port numbers, len of packet sequences for features, and the two parameters of HDBScan clustering algorithm: $Minimum\_Cluster\_Size$ and $K\_nearest\_neighbors$. In our experi-

Table 2 Experimental dataset: malware binaries and their associated YARA family labels

Family name (YARA)	# Malware binaries
Blackmoon (B)	887 (74.10%)
Gozi-ISFB (GI)	122 (10.19%)
Citadel (C)	70 (5.85%)
Zeus-VM-AES (ZVA)	29 (2.42%)
Ramnit (R)	22 (1.83%)
Dridex-Loader (DL)	15 (1.25%)
Zeus-v1 (Zv1)	10 (0.83%)
Zeus-Panda (ZPa)	10 (0.83%)
Gozi-EQ (GE)	7 (0.58%)
Dridex RAT Fake Pin	7 (0.58%)
Dridex (D)	6 (0.50%)
Zeus-P2P (ZP)	4 (0.33%)
Zeus (Z)	3 (0.25%)
Zeus OpenSSL	2 (0.17%)
Zeus Action	2 (0.16%)
Total	**1,196** (100%)

ments, we have used trigrams ($order = 3$) for port numbers, because they form a good trade-off between performance and data sparsity [28]. In the experimental dataset, the length of connections is highly skewed towards shorter sequences, with a mean of 20 packets. We use this mean as len.[11] Out of 8997 connections in the dataset, 733 connections are longer than len. The HDBScan algorithm uses $Minimum\_Cluster\_Size = 7$ and $K\_nearest\_neighbors = 7$. These parameters were selected by tuning MalPaCA on a configuration dataset (5% of the usable data). The experiments were run on a machine with Intel Xeon E3-12xx v2 processor, 8 cores and 64 GB RAM.

The specificity of the identified behaviors is highly dependent on the length of sequences, i.e., len. Based on preliminary experiments with $len = \{5, 10, 20, 50\}$, we found that $len = 20$ provided the optimal trade-off between behavior characterization and the amount of connections that were discarded. For smaller values, the connections were too generic. For larger values, connections with slight behavioral differences were considered very different. For example, at $len = 50$ several clusters capture slightly different variations of port scans, while at $len = 20$ those variations merge to form a few strong clusters.

[11] *len can be adjusted based on the required behavioral specificity.*

6 Malware Capability Assessment

MalPaCA produces 18 clusters from the dataset. There are, on average, 25 connections in each cluster. The algorithm discards 284 connections as noise. The remaining 449 connections originate from 216 Pcap files. Each cluster captures a unique behavior, listed in Table 3 along with the malware families that show that behavior. We describe a few of the interesting behaviors obtained by MalPaCA. We also discuss how host-based blacklisting [25, 54], which is a very common practice in security companies, will fail to detect these behaviors.

1. **Connection Direction Identification.** MalPaCA successfully identifies the direction of traffic flow even though no such feature is used. The clusters and their traffic direction are listed in Table 3. Interestingly, we continue to see this pattern even when port-related features are removed from the clustering. Hence, the sequence of packet sizes and their inter-arrival time are collectively indicative of the flow direction. This important trait identifies whether the suspicious behavior is originating from inside the network or from outside it.

Table 3 For each cluster, (i) # connections, (ii) # malware families, (iii) Capability label, and (iv) Traffic direction

Cluster	# Conns	# Families	Behavior	Direction
c1	39	***9 (Common)***	SSDP traffic	Out
c2	90	***9 (Common)***	Broadcast traffic	Out
c3	9	4	LLMNR traffic	Out
c4	49	5	Systematic port scan	In
c5	56	5	Randomized port scan	Out
c6	25	***1 (Rare)***	Connection spam	In
c7	23	***1 (Rare)***	Connection spam	Out
c8	16	***1 (Rare)***	Malicious subnet	Out
c9	11	***1 (Rare)***	Connection spam	Out
c10	9	2	HTTPs traffic	Out
c11	8	2	C&C Reuse	In
c12	18	4	HTTPs traffic	In
c13	25	5	Misc.	In
c14	10	3	Misc.	In
c15	20	3	Misc.	In
c16	12	3	Misc.	Out
c17	19	3	Misc.	Out
c18	10	4	Misc.	Out

2. **Device Probing.** Some clusters capture connections that connect to the same host. For example, one cluster contains all connections broadcasting to 239.255.255.250, which is used by the SSDP protocol to find Plug and Play devices. Another cluster captures all connections broadcasting to 224.0.0.252, which is used by the Link-Local Multicast Name Resolution (LLMNR) protocol to find local network computers. These clusters could easily have been obtained by using IP-based blacklist, but they would not have clustered behaviorally similar hosts with different IP addresses.
3. **Split-personality C&C Servers.** In several instances, an infected host was observed responding differently to the same request, so much so that the resulting connections ended up in different clusters. For example, two connections of `Gozi-ISFB` contact 46.38.238.XX, which has been reported as a malicious server located in Germany. The outgoing connections are identical as they both request for the same resource. However, the responses received are very different—the first response contains a small packet followed by a series of 1200-byte packets, while the second one contains a periodic list of small and large packets in the range of 600–1800 bytes. This insight portrays a better picture of the behavior of said C&C server. In contrast, a blacklist would have grouped these connections since they belong to the same host.
4. **Port Scan Detection.** Some clusters capture a *Port Scan*,[12] which is a method for determining open ports on a device in a network. Port scans are usually a part of the reconnaissance phase in the attack kill chain [62]. Utilizing sequences of port numbers enables us to detect any suspicious temporal behavior before an attack happens. The clusters identify two types of port scans: (i) *Systematic port scan* where ports are swept incrementally, which is seen as a gradient in the corresponding temporal heatmap; and (ii) *Randomized port scan* where ports are contacted randomly, which shows up in the heatmap as a checkered pattern. See Fig. 4. Port scans carried out by different connections are clustered together if they contact the same range of port numbers, which increases their mutual similarity. This result is in direct contrast with Mohaisen et al. [39] who conclude that port numbers are the least useful features in distinguishing malware families.
5. **C&C Reuse by Multiple Families.** One cluster contains connections from different families that contact the same C&C server, and their temporal heatmaps look behaviorally identical. The cluster includes three `Zeus-Panda` (ZPA) connections and one `Blackmoon` (B) connection who contact a single IP address (encoded as 009), which has been reported as malicious. Figure 5 shows the temporal heatmaps of this cluster. The said connections are highlighted in green. This result suggests that either the YARA rules mislabeled one of the samples or that authors share C&C servers.
6. **Malicious Subnet Identification.** In some instances, several connections contact IP addresses that fall in the same subnet. For example, two `Zeus-VM-AES` connections contact one host from 62.113.203.XX subnet, while another connection detected 15 days later contacts another host in the said subnet. Similarly, two

[12] https://whatismyipaddress.com/port-scan.

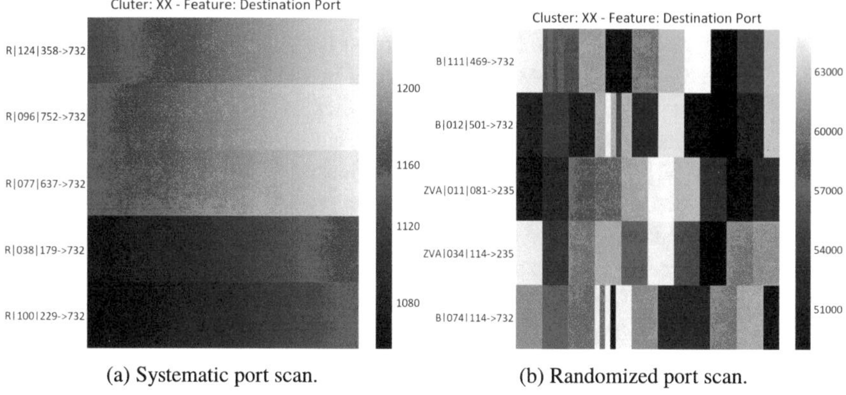

(a) Systematic port scan. (b) Randomized port scan.

Fig. 4 Clusters showing systematic and randomized port scans

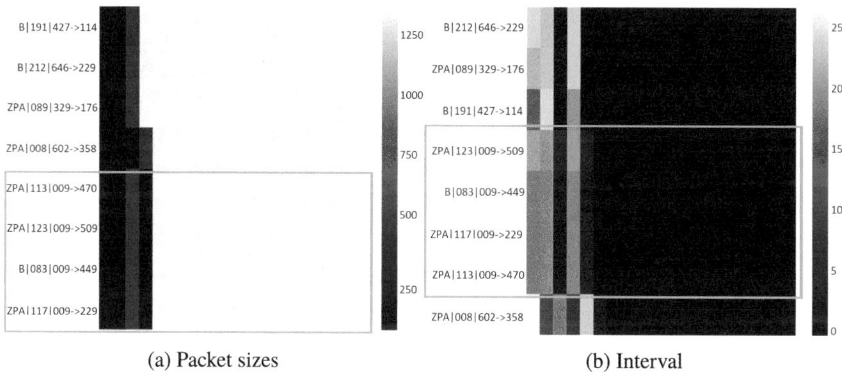

(a) Packet sizes (b) Interval

Fig. 5 Similar Zeus-Panda and Blackmoon connections

`Zeus-Panda` connections and one Blackmoon connection contact two hosts in 88.221.14.XX subnet. This gives actionable intelligence to ISPs to investigate if other IPs in these subnets are also hosting C&C servers.

6.1 Cluster Characterization

We analyze the temporal heatmaps for the behavioral trend of each cluster in order to label it. MalPaCA's goal is to identify different behaviors in the network traffic and it does so regardless of their maliciousness and origin. Hence, the resulting clusters contain both, benign and malicious behaviors. The common clusters can be discarded if they contain known-benign behaviors, drastically reducing the number of connections to analyze.

We successfully assigned labels to 12 clusters. For example, in the case of connection spam, the whole cluster is filled with almost identical connections originating from the same host. We validate this observation by specifically looking at the network traffic of these connections to see exactly what behavior is shown. Six clusters were left unlabeled since we could not identify the captured capability simply by exploring temporal heatmaps. These particular clusters were also the source of clustering errors. Table 3 shows that *SSDP* and *Broadcast traffic* are the most common behaviors and are both specific to Windows OS. Since the dataset is composed of Windows-based malware, it explains why 9 out of 12 families have connections in these two clusters. On the contrary, *Connection Spam* and *Malicious Subnet* are the rarest behaviors. *Malicious Subnet* only captures `Zeus-VM-AES`. `Gozi-ISFB` opens numerous connections, creating a *Connection Spam*. The incoming connections are stored in one cluster, while the outgoing traffic is split into two clusters due to the difference in the type of requests. This detailed behavioral analysis enables the identification of interesting clusters to analyze further.

Performance Analysis. The temporal heatmaps show that on average, 8.3% connections per cluster are CEs—their feature sequences are different from their fellow connections in a cluster. The majority of the errors originate from the last six clusters. Note that this error rate is low for an unsupervised setting since not all connections require manual revision.

6.2 Constructing Behavioral Profiles

MalPaCA identifies 18 distinct behaviors in the dataset. Hence, each malware sample (and its associated Pcap file) can be described as a binary string of 18 characters, known as *Cluster Membership String (CMS)*, where each character signifies whether the Pcap's connections were found in that cluster. Precisely, for a malware sample x, $CMS_x = b^n$, where $b \in \{0, 1\}$, n is the number of behavioral clusters, and b^i indicates whether x's connections are present in the ith cluster. The Cluster Membership String can be regarded as the behavioral profile of a given malware sample. In this work, we consider binary CMSs because we are only interested in the behavior overlap of different malware samples. Non-binary $CMS_x = z^n$, for connection counts $z \in \mathbb{Z}$, is an interesting avenue to investigate.

Table 4 lists the composite behavioral profiles for each YARA malware family in the dataset—each YARA family is represented as the union of all its samples' CMSs. `Dridex`, `Gozi-EQ`, `Zeus-P2P` and `Zeus-v1` only generate either *SSDP* or *Broadcast traffic*. Since this traffic is obtained from standard Windows services, it is likely that the malware was not activated when the associated Pcap files were recorded. Hence, the only connections observed from these families seem benign. `Gozi-ISFB` has the most diverse profile, with connection in 16 out of 18 clusters, which exhibit attacking capabilities such as *Port Scans* and *Connection Spamming*. Specifically, the *Connection Spamming* behavior is never exhibited by any other

Table 4 Composite behavioral profiles of malware families. Columns: YARA labels, Rows: Cluster labels by MalPaCA

	B	C	D	DL	GE	GI	R	Z	ZP	ZPa	Zv1	ZVA
SSDP traffic	✓	✓	✓	✓	✓	✓	✓	✓	–	✓	–	✓
Broadcast traffic	✓	✓	–	✓	–	✓	✓	–	✓	–	✓	✓
LLMNR traffic	✓	✓	–	✓	–	✓	–	–	–	–	–	–
System. port scan	✓	✓	–	–	–	✓	✓	–	–	–	–	✓
Random. port scan	✓	✓	–	–	–	✓	✓	–	–	–	–	✓
In conn spam	–	–	–	–	–	✓	–	–	–	–	–	–
Out conn spam	–	–	–	–	–	✓	–	–	–	–	–	–
Malicious Subnet	–	–	–	–	–	–	–	–	–	–	–	✓
In HTTPs	–	✓	–	✓	–	✓	–	–	–	✓	–	–
Out HTTPs	–	–	–	–	–	✓	–	–	–	✓	–	–
C&C reuse	✓	–	–	–	–	–	–	–	–	✓	–	–
Misc.	✓	✓	–	✓	–	✓	–	✓	–	✓	–	✓
# Clusters	7	11	*1*	8	*1*	*16*	4	2	*1*	7	*1*	7

malware family in the dataset. There are two reasons for `Gozi-ISFB`'s diversity: (i) `Gozi-ISFB` is the largest family under consideration, so many of its behavioral aspects are captured; and (ii) `Gozi-ISFB` opens more connections per sample compared to other families. For example, one sample of `Gozi-ISFB` opens 111 connections, while the average number of connections for other malware samples is 3.

6.3 Showing Relationships Using DAG

We extract the behavioral relationships between the 216 Cluster Membership Strings by considering it a *Set Membership* problem. It dictates that, e.g., `Set A= {0,1,1}` is a *subset* of `Set B={1,1,1}` because `Set B` encapsulates all of `Set A`'s behaviors and more. Similarly, `Set C= {0,0,0}` is a subset of every other set in this domain. `Set C` represents Pcaps where all connections were discarded as Noise due to significant differences in behavior.

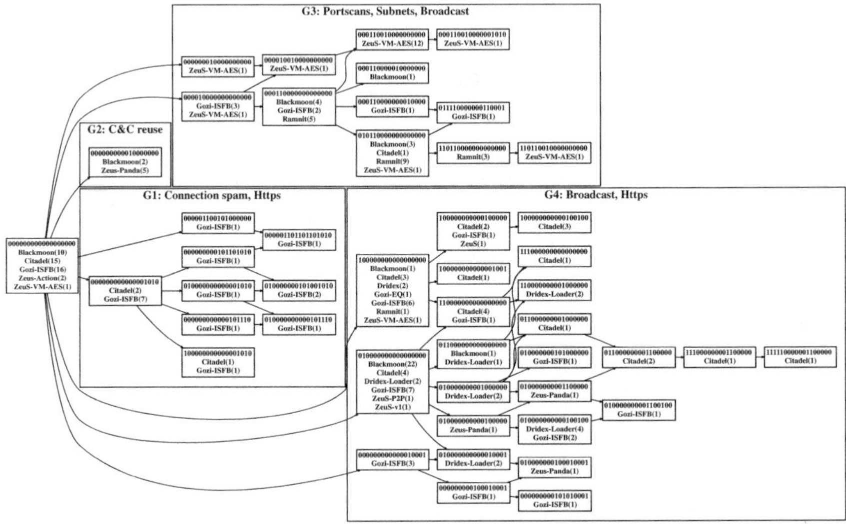

Fig. 6 MalPaCA's behavioral profiles: The DAG shows the behavioral relationships between malware samples. Each node shows a CMS and compares with YARA family labels (+ # Pcaps)

We represent the relationships between Pcap files using a Directed Acyclic Graph (DAG), shown in Fig. 6. Each node represents a unique Cluster Membership String. Multiple Pcaps can share a single CMS IFF their behaviors overlap. The nodes with minimum Hamming distance are connected using edges. This method allows multiple parents, i.e., a CMS of `"111"` may be reached by both `"110"` and `"101"`. Note that this graph is constructed purely from a data-driven approach without using any knowledge of family labels. In combination with human intelligence, we believe that it can serve as a powerful tool in understanding malware's network behavior.

7 Comparative Analysis

We show MalPaCA's results in relation to existing work by conducting two comparative analyses: (i) Comparing MalPaCA's behavioral profiles with YARA family labels, and (ii) Comparing MalPaCA's cluster quality with an existing approach that uses statistical features.

7.1 Comparison with Traditional Family Labels

We use the DAG from Fig. 6 to contrast between YARA labels and MalPaCA's behavioral profiles. Each node shows a unique CMS, and the number of malware families that share it. For example, the node with the CMS of "00000000000001010" is labeled as "Citadel(2), Gozi-ISFB(7)" because 2 Citadel Pcaps and 7 Gozi-ISFB Pcaps show the same behavior—their connections are co-located in the clusters 15 and 17. The root (on the left most side) contains the Pcaps for which all connections were discarded as Noise. Pcaps showing subsequently more behaviors are placed towards the right of the graph, with the right most node "11111000001100000 Citadel(1)" containing one Citadel Pcap that shows the most diverse number of behaviors. Note that observing additional network traffic will enrich this graph even further.

The graph shows four major partitions (denoted by G1-G4), indicating that there are four high-level behavioral sub-groups present in the dataset. The G2 group containing only one node stands out. It contains Pcaps from Zeus-Panda and Blackmoon, and are the only malware samples that share a C&C server. This observation makes a strong case that these particular Pcap files, albeit originating from two families, are behaviorally alike. The G3 group contains Pcaps from various families that are observed doing port scans and broadcasting behaviors. Some servers from this group also form malicious subnets. The G4 group, on the other hand, is the largest group that uses HTTPs traffic along with broadcasting behaviors. The G1 group is highly dominated by Gozi-ISFB and is observed doing Connection spamming, along with using HTTPs traffic. Some connections from these Gozi-ISFB Pcaps were placed in the behavioral clusters that we failed to identify (c13-c18).

The node location for some malware families is intriguing. For example, most of the Zeus-VM-AES Pcaps that are associated with malicious subnets are located in the G3 group, together with Ramnit files that are associated with port scans. Dridex-Loader is only observed in group G4, while most of the Citadel Pcaps are also seen in the same. Blackmoon and Gozi-ISFB have Pcaps that are distributed over all of the behavioral sub-groups. However, Gozi-ISFB is seen dominating the G1 group, while Backmoon dominates the G4 group. Furthermore, as observed from Table 4, Gozi-ISFB's Pcaps collectively show 18 discrete behaviors and Citadel's Pcaps show 11 behaviors. However, Citadel shows more discrete behaviors in a single Pcap compared to Gozi-ISFB, as Gozi-ISFB's Pcaps contain more (behaviorally similar) connections on average. Also, each of Gozi-ISFB's Pcaps is more behaviorally dissimilar than Citadel's Pcaps.

Zeus-Panda's Pcaps are clearly divided into two behavioral sub-groups—one in G2 group with Blackmoon samples and the other in the G4 group. Zeus-v1, Zeus-P2P, ZeuS, Gozi-EQ, and Dridex are only seen at the left side of the graph, indicating that none of their distinguishing behaviors were present in the dataset.

To conclude, the DAG clearly identifies the discrepancies in the malware's behavioral profiles and their traditional family names. A significant portion of the analysis pipeline is automated and unsupervised. The temporal heatmaps together with the DAG are intended for human-in-the-loop exploration—they actively support malware behavior analysis and provide more insightful characterization of malware than current family labels.

7.2 Comparison with Statistical Features

Baseline Setup. We compare the cluster quality of using sequential versus statistical features. We use the existing method by Tegeler et al. [54] (called baseline, henceforth) to compare our results since they not only use statistical features, but also incorporate periodic behavior using Fourier transform to detect bot-infected network traffic. Although the goal of their study diverges from ours, their feature selection approach is aligned with ours. For objectivity, we keep the rest of the pipeline as explained in Sect. 4. Taking guidelines from Tegeler et al. [54] and adapting them to our problem statement, each connection in the baseline is characterized by (1) average packet size, (2) average interval between packets, (3) average duration of a connection, and (4) the maximum Power Spectral Density (PSD) of the FFT obtained by the binary sampling approach by Tegeler et al. [54]—the signal is 1 when a packet is present in the connection and is 0 in between.

Cluster quality comparison. The baseline method results in 22 clusters, with an average of 21.2 connections per cluster. 265 connections are discarded as noise. These results are in comparison with sequence clustering—18 clusters; on average 25 connections per cluster; 284 connections discarded as noise.

Baseline seems to perform better with smaller cluster size on average and discarding fewer connections as noise. However, a deeper analysis shows the obtained clusters lack quality.

1. With statistical features, connections present in most clusters appear very different from their fellow connections. On average, 57.5% connections per cluster have visually different temporal heatmaps, compared to 8.3% for sequential features. Figure 7 shows a cluster from the baseline. It has nine connections, out of which six are errors based on their behavior. The *rightful owners* of the cluster are the connections that have the least mutual distance, i.e., `GI|090|178` →`021`, `GI|073|610`→`131`, `GI|073|610`→`346`. The other six connections have minor differences in all features, except the source port which is 6 for all. They were clustered together because their statistical features had the least mutual distance, i.e., $average\_time\_interval = 19.77 \pm 3.11$; $fft = 0.07 \pm 0.05$; $average\_duration = 397.7 \pm 61.7$; $average\_bytes = 573.3 \pm 113.8$. The temporal heatmaps clearly show behavioral differences in nearly all clusters.
2. Statistical features are also unable to identify the direction of network traffic. In the cluster shown in Fig. 7, there is one incoming connection in the cluster along with eight outgoing ones. A similar trend is observed for 19 out of 22 clusters. In contrast, sequences of packet size and inter-arrival time are enough to identify traffic direction in sequence clustering.

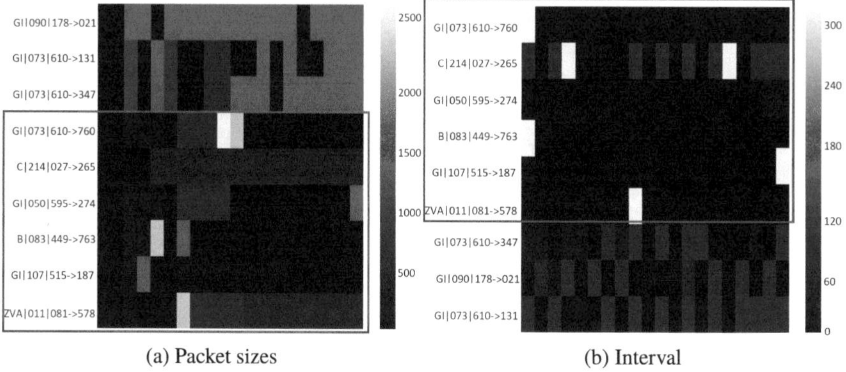

(a) Packet sizes (b) Interval

Fig. 7 Baseline clusters: Six out of nine behaviorally different connections clustered together in baseline version

In summary, while statistical features may be simple to use, they lose behavioral information that plays a crucial role in accurately determining similarities in network behavior. Sequence clustering obtains significantly better clusters. Given that modeling behavioral profiles is already challenging for short sequences, it is remarkable that MalPaCA can identify network behaviors using only 20 packets and 4 coarse features.

8 Limitations and Future Work

Limitations. Performance optimizations are needed to make sequence clustering more efficient and scalable. In MalPaCA, DTW forms the main bottleneck as the length of sequences grows longer. There exist streaming versions of DTW that compute results in real time. One such technique is presented by Oregi et al. [42]. Moreover, using Locality Sensitive Hashing [6, 7] can make MalPaCA more scalable.

Density-based clustering discards rare events as noise. This makes sense if the dataset is noisy. However, in the presence of a purely malicious dataset, the connections that lie in lower density regions may represent rare attacking capabilities, which may be discarded in the current implementation.

Malware authors can try to evade detection by modifying malware's code. A common assumption is that malware can easily evade detection by adding random delays and padding to packets. However, there is a limit to what an attacker can change. For example, a TCP handshake needs to happen in a certain way because this is how the protocol dictates it. Also, padding-related provisions are already standardized by some commonly used protocols, such as TLS making it difficult to hide "coarse" features like packet sizes and inter-arrival times [19]. We expect that MalPaCA is evasion resilient, e.g., since MalPaCA only uses coarse features, evading it is not a trivial task. Moreover, the usage of Dynamic Time Warping distance makes it resilient to random delays [20] and due to the relative distance measures used in HDBScan, randomized port numbers are already clustered together, as shown in Sect. 6. If, after all this, attackers still manage to evade MalPaCA, the malware sample will end up with a new behavioral profile, making analysts more prone to analyze it. More study is needed to strengthen these claims.

Future work. There are several research directions this work can take: (i) We will work on fully automating the capability assessment of malware by building a directory of observed behaviors, which will be used for cluster labeling. (ii) We will test and improve MalPaCA's adversarial evasion resilience. (iii) We will integrate additional behavioral data sources in MalPaCA so the profiles are based on all static, system-level, and network behavior. (iv) Since MalPaCA is a generic technique, we will test its applicability in building behavioral profiles for everyday-use software.

9 Conclusions

In this chapter, we propose MalPaCA, an intuitive network traffic-based tool to perform malware capability assessment: It groups capabilities using sequence clustering and uses the cluster membership to build network behavioral profiles. We also propose a visualization-based cluster evaluation method whose key advantage is its white-box nature, allowing malware analysts to investigate, understand, and even correct labels, if necessary. We implement MalPaCA and evaluate it on real-world financial malware samples collected in the wild. MalPaCA independently identifies attacking capabilities. We build a DAG to show overlapping malware behaviors and discover a number of samples that do not adhere to their family names, either because of incorrect labeling by black-box solutions or extensive overlap in the families' behavior. We also show that sequence clustering outperforms existing statistical features-based methods by making only 8.3% errors, as opposed to 57.5%. MalPaCA, with its visualizations and capability assessment, can actively support the understanding of malware samples. The resulting behavioral profiles give malware researchers a more informative and actionable characterization of malware than current family designations.

References

1. Acar, Abbas, Hossein Fereidooni, Tigist Abera, Amit Kumar Sikder, Markus Miettinen, Hidayet Aksu, Mauro Conti, Ahmad-Reza Sadeghi, and A. Selcuk Uluagac. 2018. Peek-a-boo: I see your smart home activities, even encrypted! *arXiv*.
2. Aiolli, Fabio, Mauro Conti, Ankit Gangwal, and Mirko Polato. 2019. Mind your wallet's privacy: Identifying bitcoin wallet apps and user's actions through network traffic analysis. In *SIGAPP*, 1484–1491. ACM.

3. Anderson, Blake, and David McGrew. 2017. Machine learning for encrypted malware traffic classification: Accounting for noisy labels and non-stationarity. In *Proceedings of the 23rd ACM SIGKDD*, 1723–1732.
4. Anderson, Blake, Subharthi Paul, and David McGrew. 2017. Deciphering malware's use of TLS (without decryption). *CVHT Journal* 14 (3).
5. Azab, Ahmad, Mamoun Alazab, and Mahdi Aiash. 2016. Machine learning based botnet identification traffic. In *IEEE Trustcom/BigDataSE/ISPA*, 1788–1794. IEEE.
6. Azab, Ahmad Robert Layton, Mamoun Alazab, and Jonathan Oliver. 2014. Mining malware to detect variants. In *Cybercrime and trustworthy computing conference*, 44–53. IEEE.
7. Bayer, Ulrich, Paolo Milani Comparetti, Clemens Hlauschek, Christopher Kruegel, and Engin Kirda. 2009. Scalable, behavior-based malware clustering. In *NDSS*, vol. 9, 8–11. Citeseer.
8. Baysa, Donabelle, Richard M. Low, and Mark Stamp. 2013. Structural entropy and metamorphic malware. *CVHT Journal* 9 (4): 179–192.
9. Berndt, Donald J., and James Clifford. 1994. Using dynamic time warping to find patterns in time series. *KDD* 10: 359–370
10. Bilge, Leyla, Davide Balzarotti, William Robertson, Engin Kirda, and Christopher Kruegel. 2012. Disclosure: Detecting botnet command and control servers through large-scale netflow analysis. In *ACSAC*, 129–138. ACM.
11. Black, Paul, Iqbal Gondal, and Robert Layton. 2017. A survey of similarities in banking malware behaviours. *Computers and Security*.
12. Campello, Ricardo J.G.B., Davoud Moulavi, and Jörg Sander. 2013. Density-based clustering based on hierarchical density estimates. In *PAKDD*, 160–172. Springer
13. Canfora, Gerardo, Andrea De Lorenzo, Eric Medvet, Francesco Mercaldo, and Corrado Aaron Visaggio. 2015. Effectiveness of opcode ngrams for detection of multi family android malware. In *ARES*, 333–340. IEEE.
14. Cavallaro, Lorenzo, Christopher Kruegel, Giovanni Vigna, Fang Yu, Muath Alkhalaf, Tevfik Bultan, Lili Cao, Lei Yang, Heather Zheng, Christopher C. Cipriano, et al. 2009. Mining the network behavior of bots. Technical report 2009-12.
15. Chakkaravarthy, S. Sibi, D. Sangeetha, and V. Vaidehi. 2019. A survey on malware analysis and mitigation techniques. *Computer Science Review* 32: 1–23.
16. Chan, Neil Wong Hon, and Shanchieh Jay Yang. 2017. Scanner: Sequence clustering of android resource accesses. In *IEEE DSC 2017*.
17. Conti, Mauro, Luigi V. Mancini, Riccardo Spolaor, and Nino Vincenzo Verde. 2015. Can't you hear me knocking: Identification of user actions on android apps via traffic analysis. In *CODASPY*, 297–304. ACM.
18. Davies, David L. and Donald W. Bouldin. 1979. A cluster separation measure. In *TPAMI 1979*.
19. Dyer, Kevin P., Scott E. Coull, Thomas Ristenpart, and Thomas Shrimpton. 2012. Peek-a-boo, i still see you: Why efficient traffic analysis countermeasures fail. In *S&P*, 332–346. IEEE.
20. Elfeky, Mohamed G., Walid G. Aref, and Ahmed K. Elmagarmid. 2005. Warp: Time warping for periodicity detection. In *Data Mining*, 8–pp. IEEE.
21. Feng, Yu, Saswat Anand, Isil Dillig, and Alex Aiken. 2014. Apposcopy: Semantics-based detection of android malware through static analysis. In *SIGSOFT*, 576–587. ACM.
22. Gandotra, Ekta, Divya Bansal, and Sanjeev Sofat. 2014. Malware analysis and classification: A survey. *Information Security Journal* 5 (02): 56.
23. Garcia, Sebastian. 2015. Modelling the network behaviour of malware to block malicious patterns. the stratosphere project: A behavioural IPS. *VB*.
24. Garcia-Teodoro, Pedro, Jesus Diaz-Verdejo, Gabriel Maciá-Fernández, and Enrique Vázquez. 2009. Anomaly-based network intrusion detection: Techniques, systems and challenges. *Computers and Security* 28 (1–2): 18–28.
25. Ghafir, Ibrahim and Vaclav Prenosil. 2015. Blacklist-based malicious IP traffic detection. In *GCCT*, 229–233. IEEE.
26. Ghorbani, Ali A., and Saeed Nari. 2013. Automated malware classification based on network behavior. In *ICNC*, 642–647. IEEE.

27. Hammerschmidt, Christian, Samuel Marchal, Radu State, and Sicco Verwer. 2016. Behavioral clustering of non-stationary IP flow record data. In *CNSM*, 297–301. IEEE.
28. Kalgutkar, Vaibhavi, Natalia Stakhanova, Paul Cook, and Alina Matyukhina. 2018. Android authorship attribution through string analysis. In *ARES*, 4. ACM.
29. Kantchelian, Alex, Michael Carl Tschantz, Sadia Afroz, Brad Miller, Vaishaal Shankar, Rekha Bachwani, Anthony D. Joseph, and J Doug Tygar. 2015. Better malware ground truth: Techniques for weighting anti-virus vendor labels. In *AISec*.
30. Kim, Ki-Hyeon and Mi-Jung Choi. 2015. Android malware detection using multivariate time-series technique. In *APNOMS*, 198–202.
31. Kovacs-Vajna, Zsolt Miklos. 2000. A fingerprint verification system based on triangular matching and dynamic time warping. *TPAMI* 22 (11): 1266–1276.
32. Lee, Jehyun, and Heejo Lee. 2014. Gmad: Graph-based malware activity detection by DNS traffic analysis. *Computer Communications* 49.
33. Li, Peng, Limin Liu, Debin Gao, and Michael K. Reiter. 2010. On challenges in evaluating malware clustering. In *RAID*, 238–255. Springer.
34. Li, Wei-Jen, Ke Wang, Salvatore J. Stolfo, and Benjamin Herzog. 2005. Fileprints: Identifying file types by n-gram analysis. In *IEEE SMC information assurance workshop*, 64–71. IEEE.
35. Li, Yuping, Jiyong Jang, Xin Hu, and Xinming Ou. 2017. Android malware clustering through malicious payload mining. In *RAID*, 192–214. Springer.
36. Lin, Qin, Sridha Adepu, Sicco Verwer, and Aditya Mathur. 2018. Tabor: a graphical model-based approach for anomaly detection in industrial control systems. In *Asia CCS*, 525–536. ACM.
37. Maggi, Federico, Andrea Bellini, Guido Salvaneschi, and Stefano Zanero. 2011. Finding non-trivial malware naming inconsistencies. In *ICISS*, 144–159
38. Mohaisen, Aziz, Omar Alrawi, Matt Larson, and Danny McPherson. 2013. Towards a methodical evaluation of antivirus scans and labels. In *ISA workshop*, 231–241. Springer.
39. Mohaisen, Aziz, Omar Alrawi, and Manar Mohaisen. 2015. Amal: High-fidelity, behavior-based automated malware analysis and classification. *Computers and Security* 52.
40. Moubarak, Joanna, Maroun Chamoun, and Eric Filiol. 2017. Comparative study of recent mea malware phylogeny. In *ICCCS*, 16–20. IEEE.
41. Ntlangu, Mbulelo Brenwen, and Alireza Baghai-Wadji. 2017. Modelling network traffic using time series analysis: A review. In *IoTBDS*, 209–215.
42. Oregi, Izaskun, Aritz Pérez, Javier Del Ser, and José A Lozano. 2017. On-line dynamic time warping for streaming time series. In *ECML-PKDD*, 591–605. Springer.
43. Pellegrino, Gaetano, Qin Lin, Christian Hammerschmidt, and Sicco Verwer. 2017. Learning behavioral fingerprints from netflows using timed automata. In *IFIP*, 308–316. IEEE.
44. Perdisci, Roberto, Wenke Lee, and Nick Feamster. 2010. Behavioral clustering of http-based malware and signature generation using malicious network traces. In *NSDI*, vol. 10.
45. Pomorova, Oksana, Oleg Savenko, Sergii Lysenko, Andrii Kryshchuk, and Kira Bobrovnikova. 2015. A technique for the botnet detection based on DNS-traffic analysis. In *CN*, 127–138. Springer.
46. Rafique, M. Zubair, and Juan Caballero. 2013. Firma: Malware clustering and network signature generation with mixed network behaviors. In *RAID*, 144–163. Springer.
47. Rieck, Konrad, Philipp Trinius, Carsten Willems, and Thorsten Holz. 2011. Automatic analysis of malware behavior using machine learning. *Journal of Computer Security* 19 (4): 639–668.
48. Rousseeuw, Peter J. 1987. Silhouettes: a graphical aid to the interpretation and validation of cluster analysis. *CAM Journal* 20.
49. Sebastián, Marcos, Richard Rivera, Platon Kotzias, and Juan Caballero. 2016. Avclass: A tool for massive malware labeling. In *RAID*, 230–253. Springer.
50. Sharma, Arushi, Ekta Gandotra, Divya Bansal, and Deepak Gupta. 2019. Malware capability assessment using fuzzy logic. *Cybernetics and Systems* 1–16.
51. Suarez-Tangil, Guillermo, Juan E. Tapiador, Pedro Peris-Lopez, and Jorge Blasco. 2014. Dendroid: A text mining approach to analyzing and classifying code structures in android malware families. *Expert Systems with Applications* 41 (4).

52. Sun, Mingshen, Xiaolei Li, John C.S. Lui, Richard T.B. Ma, and Zhenkai Liang. 2017. Monet: a user-oriented behavior-based malware variants detection system for android. *TIFS* 12 (5).
53. Tajalizadehkhoob, S.T., Hadi Asghari, Carlos Gañán, and M.J.G. Van Eeten. 2014. Why them? extracting intelligence about target selection from zeus financial malware. In *WEIS*.
54. Tegeler, Florian, Xiaoming Fu, Giovanni Vigna, and Christopher Kruegel. 2012. Botfinder: Finding bots in network traffic without deep packet inspection. In *CoNEXT*, 349–360. ACM.
55. Tian, Ronghua, Lynn Batten, Rafiqul Islam, and Steve Versteeg. 2009. An automated classification system based on the strings of trojan and virus families. In *MALWARE*. IEEE.
56. Verwer, Sicco, Rémi Eyraud, and Colin De La Higuera. 2014. Pautomac: A probabilistic automata and hidden Markov models learning competition. *Machine Learning* 96 (1–2): 129–154.
57. Vinod, P., V. Laxmi, M.S. Gaur, and Grijesh Chauhan. 2012. Momentum: Metamorphic malware exploration techniques using MSA signatures. In *IIT*, 232–237. IEEE.
58. Volis, George, Christos Makris, and Andreas Kanavos. 2016. Two novel techniques for space compaction on biological sequences. *WEBIST*.
59. Wang, An, Aziz Mohaisen, Wentao Chang, and Songqing Chen. 2015. Capturing DDoS attack dynamics behind the scenes. In *DIMVA*, 205–215. Springer.
60. Wang, Wei, Ming Zhu, Xuewen Zeng, Xiaozhou Ye, and Yiqiang Sheng. 2017. Malware traffic classification using convolutional neural network for representation learning. In *ICOIN*, 712–717.
61. Wang, Yipeng, Zhibin Zhang, Danfeng Daphne Yao, Buyun Qu, and Li Guo. 2011. Inferring protocol state machine from network traces: a probabilistic approach. In *ACNS*, 1–18. Springer.
62. Yadav, Tarun and Arvind Mallari Rao. 2015. Technical aspects of cyber kill chain. In *SSCC*.
63. Zahrotun, Lisna. 2016. Comparison jaccard similarity, cosine similarity and combined both of the data clustering with shared nearest neighbor method. *CE&AJ* 5 (1): 11–18.

An Empirical Analysis of Image-Based Learning Techniques for Malware Classification

Pratikkumar Prajapati and Mark Stamp

Abstract In this chapter, we consider malware classification using deep learning techniques and image-based features. We employ a wide variety of deep learning techniques, including multilayer perceptrons (MLP), convolutional neural networks (CNN), long short-term memory (LSTM), and gated recurrent units (GRU). Among our CNN experiments, transfer learning plays a prominent role—specifically, we test the VGG-19 and ResNet152 models. As compared to previous work, the results presented in this chapter are based on a larger and more diverse malware dataset, we consider a wider array of features, and we experiment with a much greater variety of learning techniques. Consequently, our results are the most comprehensive and complete that have yet been published.

1 Introduction

Traditionally, malware detection and classification has relied on pattern matching against signatures extracted from specific malware samples. While simple and efficient, signature scanning is easily defeated by a number of well-known evasive strategies. This fact has given rise to statistical and machine learning-based techniques, which are more robust to code modification. In response, malware writers have developed advanced forms of malware that alter statistical and structural properties of their code, which can cause statistical models to fail.

In this chapter, we compare deep learning (DL) models for malware classification. For most of our deep learning models, we use image-based features, but we also experiment with opcode features. The DL models consider include a wide variety of neural networking techniques, including multilayer perceptrons (MLP), several variants of convolutional neural networks (CNN), and vanilla recurrent neural net-

P. Prajapati · M. Stamp (✉)
San Jose State University, San Jose, CA, USA
e-mail: mark.stamp@sjsu.edu

P. Prajapati
e-mail: pratikkumar.prajapati@sjsu.edu

M. Stamp et al. (eds.), *Malware Analysis Using Artificial Intelligence and Deep Learning*, https://doi.org/10.1007/978-3-030-62582-5_16

works (RNN), as well as the advanced RNN architectures known as long short-term memory (LSTM) and gated recurrent units (GRU). We also experiment with a complex stacked model that combines both LSTM and GRU. In addition, we consider transfer learning, in the form of the ResNet152 and VGG-19 architectures.

The remainder of this chapter is organized as follows. In Sect. 2 we provide relevant background information, including a discussion of related work, an overview of the various learning techniques considered, and we introduce the dataset used in this research. Section 3 is the heart of the chapter, with detailed results from a wide variety of malware classification experiments. Section 4 concludes the chapter and provides possible directions for future work.

2 Background

In this section, we discuss related work and we introduce the various learning techniques that are considered in this research. We also discuss the dataset that we use in our malware classification experiments. In addition, we provide the specifications of the hardware and software that we use to conduct the extensive set of experiments that are summarized in Sect. 3.

2.1 Related Work

To the best of our knowledge, image-based analysis was first applied to the malware problem in [16], where high-level "gist" descriptors are used as features. More recently, [44] confirmed the results in [16] and presented an alternative deep learning approach that produces equally good—if not slightly better—results, without the extra work required to extract gist descriptors.

Transfer learning, where the output layer of an existing pre-trained DL model is retrained for a specific task, is often used in image analysis. Such an approach allows for efficient training, as a new model can take advantage of a vast amount of learning that is embedded in the pre-trained model. Leveraging the power of transfer learning has been shown to yield strong image-based malware detection and classification results [44].

There is a vast malware analysis literature involving classic machine learning techniques. Representative examples include [2, 5, 8, 25, 28, 42]. Intuitively, we might expect models based on image analysis to be somewhat stronger and more robust, as compared to models that rely on opcodes, byte n-grams, or similar statistical features that are commonly used in malware research.

The work presented in this chapter can be considered an extension of the work in [6], where image-based transfer learning is applied to the malware classification problem. We have extended this previous work in multiple dimensions, including a larger, more challenging, and more realistic dataset. In addition, we perform much more experimentation with a much wider variety of techniques, and we consider a large range of hyperparameters in each case.

2.2 *Learning Techniques*

In this section, we provide a brief introduction to each of the learning techniques considered in this paper. Additional details on most of the learning techniques discussed here can be found in [27], which includes examples of relevant applications of the techniques. We provide additional references for the techniques discussed below that are not considered in [27].

2.2.1 Multilayer Perceptron

A perceptron computes a weighted sum of its components in the form of a hyperplane, and based on a threshold, a perceptron can be used to define a classifier. It follows that a perceptron cannot provide ideal separation in cases where the data itself is not linearly separable. This is a severe limitation, as something as elementary as the XOR function is not linearly separable.

A multilayer perceptron (MLP) is an artificial neural network that includes multiple (hidden) layers in the form of perceptrons. Unlike a single layer perceptron, MLPs are not restricted to linear decision boundaries, and hence an MLP can accurately model more complex functions. The relationship between perceptrons and MLPs is very much analogous to the relationship between linear support vector machines (SVM) and SVMs based on nonlinear kernel functions.

Training an MLP would appear to be challenging since we have hidden layers between the input and output, and it is not clear how changes to the weights in these hidden layers will affect each other or the output. Today, MLPs are generally trained using backpropagation. The discovery that backpropagation can be used for training neural networks was a major breakthrough that made deep learning practical.

2.2.2 Convolutional Neural Network

Generically, artificial neural networks use fully connected layers. The advantage of a fully connected layer is that it can deal effectively with correlations between any points within training vectors. However, for large training vectors, fully connected layers are infeasible, due to the vast number of weights that must be learned.

In contrast, a convolutional neural network (CNN) is designed to deal with local structure. A convolutional layer cannot be expected to perform well when significant information is not local. The benefit of CNNs is that convolutional layers can be trained much more efficiently than fully connected layers, due to the reduced number of weights.

For images, most of the important structure (edges and gradients, for example) is local. Hence, CNNs are an ideal tool for image analysis and, in fact, CNNs were developed precisely for image classification. However, CNNs have performed well in a variety of other problem domains. In general, any problem for which local structure predominates is a candidate for CNNs.

2.2.3 Recurrent Neural Network

MLPs and CNNs are feedforward neural networks, that is, the data feeds directly through the network, with no "memory" of previous feature vectors. In a feedforward network, each input vector is treated independently of other input vectors. While feedforward networks are appropriate for many problems, they are not well suited for dealing with sequential data.

In some cases, it is necessary for a classifier to have memory. Suppose that we want to tag parts of speech in English text (i.e., noun, verb, etc.), this is not feasible if we only look at words in isolation. For example, the word "all" can be an adjective, adverb, noun, or pronoun, and this can only be determined by considering its context. A recurrent neural network (RNN) provides a way to add memory (or context) to a feedforward neural network.

RNNs are trained using a variant of backpropagation known as backpropagation through time (BPTT). A problem that is particularly acute in BPTT is that the gradient calculation tends to be become unstable, resulting in "vanishing" or "exploding" gradients. To overcome these problems, we can limit the number of time steps, but this also serves to limit the utility of RNNs. Alternatively, we can use specialized RNN architectures that enable the gradient to flow over long time periods. Both long short-term memory and gated recurrent units are examples of such specialized RNN architectures. We discuss these two RNN architectures next.

2.2.4 Long Short-Term Memory

Long short-term memory (LSTM) networks are a class of RNN architectures that are designed to deal with long-range dependencies. That is, LSTM can deal with extended "gaps" between the appearance of a feature and the point at which it is needed by the model. In plain vanilla RNNs this is generally not possible, due to vanishing gradients.

The key difference between an LSTM and a generic vanilla RNN is that an LSTM includes an additional path for information flow. That is, in addition to the hidden state, there is a so-called cell state that can be used to, in effect, store information from previous steps. The cell state is designed to serve as a gradient "highway" during backpropagation. In this way, the gradient can "flow" much further back with less chance that it will vanish (or explode) along the way.

As an aside, we note that the LSTM architecture has been one of the most commercially successful learning techniques ever developed. Among many other applications, LSTMs play a critical role in Google Allo [11], Google Translate [43], Apple's Siri [13], and Amazon Alexa [9].

2.2.5 Gated Recurrent Unit

Due to its wide success, many variants on the LSTM architecture have been considered. Most such variants are slight, with only minor changes from a standard LSTM. However, a gated recurrent unit (GRU) is a fairly radical departure from an LSTM. Although the internal state of a GRU is somewhat complex and less intuitive than that of an LSTM, there are fewer parameters in a GRU. As a result, it is easier to train a GRU than an LSTM, and consequently less training data is required.

2.2.6 ResNet152

Whereas LSTM uses a complex gating structure to ease gradient flow, a residual network (ResNet) defines additional connections that correspond to identity layers. These identity layers allow a ResNet model to, in effect, skip over layers during training, which serves to effectively reduce the depth when training and thereby mitigate gradient pathologies. Intuitively, ResNet is able to train deeper networks by training over a considerably shallower network in the initial stages, with later stages of training serving to flesh out the intermediate connections. This approach was inspired by pyramidal cells in the brain, which have a similar characteristic, in the sense that they bridge "layers" of neurons [26].

ResNet152 is a specific deep ResNet architecture that has been pre-trained on a vast image dataset. As one of our two examples of transfer learning, we use this architecture, which includes an astounding 152 layers. That is, we use the ResNet152 model, where we only retrain the output layer specifically for our malware classification problem.

2.2.7 VGG-19

VGG-19 is a 19-layer convolutional neural network that has been pre-trained on a dataset containing more than 10^6 images [24]. This architecture has performed well in many contests, and it has been generalized to a variety of image-based problems. Here, we use the VGG-19 architecture and pre-trained model as one of our two examples of transfer learning for image-based malware classification.

Table 1 Type of each malware family

Family	Type	Family	Type
Adload [29]	Trojan downloader	Obfuscator [37]	VirTool
Agent [30]	Trojan	Onlinegames [22]	Password stealer
Alureon [35]	Trojan	Rbot [38]	Backdoor
BHO [32]	Trojan	Renos [31]	Trojan downloader
CeeInject [34]	VirTool	Startpage [33]	Trojan
Cycbot [3]	Backdoor	Vobfus [39]	Worm
DelfInject [20]	VirTool	Vundo [40]	Trojan downloader
FakeRean [36]	Rogue	Winwebsec [41]	Rogue
Hotbar [1]	Adware	Zbot [23]	Password stealer
Lolyda [21]	Password stealer	Zegost [4]	Backdoor

2.3 Dataset

Our dataset consists of 20 malware families. Three of these malware families, namely, Winwebsec, Zeroaccess, and Zbot, are from the Malicia dataset [15], while the remaining 17 families are taken from the massive malware dataset discussed in [12]. This latter dataset is almost half a terabyte and contains more than 500,000 malware samples in the form of labeled executable files.

Table 1 lists the 20 families used in this research, along with the type of malware present in each family. Next, we briefly discuss each of these 20 malware families.

Adload downloads an executable file, stores it remotely, executes the file, and disables proxy settings [29].

Agent downloads trojans or other software from a remote server [30].

Alureon exfiltrates usernames, passwords, credit card information, and other confidential data from an infected system [35].

BHO can perform a variety of actions, guided by an attacker [32].

CeeInject uses advanced obfuscation to avoid being detected by antivirus software [34].

Cycbot.G connects to a remote server, exploits vulnerabilities, and spreads through a backdoor [3].

DelfInject sends usernames, passwords, and other personal and private information to an attacker [20].

FakeRean pretends to scan the system, notifies the user of supposed issues, and asks the user to pay to clean the system [36].

Hotbar is adware that shows ads on webpages and installs additional adware [1].

Lolyda sends information from an infected system and monitors the system. It can share user credentials and network activity with an attacker [21].

Obfuscator tries to obfuscate or hide itself to defeat malware detectors [37].

Onlinegames steals login information and tracks user keystroke activity [22].

Rbot gives control to attackers via a backdoor that can be used to access information or launch attacks, and it serves as a gateway to infect additional sites [38].

Renos downloads software that claims the system has spyware and asks for a payment to remove the nonexistent spyware [31].

Startpage changes the default browser homepage and can perform other malicious activities [33].

Vobfus is a worm that downloads malware and spreads through USB drives or other removable drives [39].

Vundo displays pop-up ads and it can download files. It uses advanced techniques to defeat detection [40].

Winwebsec displays alerts that ask the user for money to fix nonexistent security issues [41].

Zbot is installed through email and shares a user's personal information with attackers. In addition, Zbot can disable a firewall [23].

Zegost creates a backdoor on an infected machine [4].

The number of samples per malware family for the various features is given in Table 2. The "Binaries" lists the number of binary executable files available, the "Images" column lists the number of binaries that were successfully converted to images, and the "Opcodes" column lists the number of samples from which a sufficient number of opcodes were extracted. From the table we see that 26,413 samples are used in our image-based experiments, and 25,901 samples are used in our opcode-based experiments.

2.4 Hardware

Table 3 lists the hardware configuration of the machine used for the experiments reported in this chapter. This machine was assembled for the purpose of training deep learning models and it is highly optimized for this task.

2.5 Software

For our deep learning neural network experiments, we have used PyTorch [18]. In addition, for general data processing and related operations, we employ both Numpy [17] and Pandas [14]. In addition, all code that was developed as part of this project is available at [19].

Table 2 Samples per malware family

Family	Samples		
	Binaries	Images	Opcodes
Adload	1050	1050	1044
Agent	842	842	817
Alureon	1328	1328	1327
BHO	1176	1176	1159
CeeInject	894	894	886
Cycbot	1029	1029	1029
DelfInject	1146	1146	1097
Fakerean	1063	1063	1063
Hotbar	1491	1491	1476
Lolyda	915	915	915
Obfuscator	1445	1445	1331
Onlinegames	1293	1293	1284
Rbot	1017	1017	817
Renos	1312	1312	1309
Startpage	1136	1136	1084
Vobfus	926	926	924
Vundo	1793	1793	1784
Winwebsec	3651	3651	3651
Zbot	1786	1786	1785
Zeroaccess	1120	1120	1119
Total	26,413	26,413	25,901

3 Deep Learning Experiments and Results

In this section, we present results of a wide variety of neural network-based experiments. First, we consider MLP experiments, followed by CNN experiments, and then RNN experiments. We consider a large number of CNN and RNN cases. We conclude this section with a pair of models based on transfer learning. The MLP, CNN, and transfer learning models are based on image features, while the RNN experiments use opcode sequences.

We consider various different sizes for images, in each case using square images. To generate a square image from an executable, we first specify a width N, with the height determined by the size of the sample. We then resize the image so that it is $N \times N$, which has the effect of stretching or shrinking the height, as required.

Table 3 Hardware characteristics

Feature	Description	Details
CPU	Brand and model	Intel i9-9940X
	Clock frequency	3.30 GHz
	Number of threads	28
	Cache	19.25 MB Intel Smart Cache
CPU liquid cooling	Brand and model	Corsair Hydro Series H115i PRO RGB
	Fan speed	1200 RPM
	Fan size	140 mm
	Radiator size	280 mm
DRAM	Brand and model	Corsair CMK32GX4M2A2666C16
	Speed	2666 MHz
	Capacity	$16 \times 8 = 128$ GB
Motherboard	Brand and model	ASUS WS x299 Sage
GPU	Brand and model	Nvidia Titan RTX
	Total video memory	24 GB GDDR6
	Tensor cores	576
	CUDA cores	4608
	Base clock (MHz)	1350 MHz
	Single-precision performance	16.3 TFLOPS
	Tensor performance	130 TFLOPS
Storage	Brand and model	Sabrent 2TB Rocket NVMe
	Read speed	3400 MB/s
	Write speed	2750 MB/s

3.1 Multilayer Perceptron Experiments

We experimented with various perceptron-based neural networks. The model we present here uses square input image and has four hidden layers, each using the popular rectified linear unit (relu) activation function. The output from the final hidden layer is passed to a fully connected output layer. The output layer is used to classify the sample—since we have 20 classes of malware in our dataset, the output vector is 20-dimensional. The hyperparameters used for these MLP experiments are given in Table 4.

Figure 2 gives the confusion matrix for the best results obtained in our MLP experiments. The hyperparameters used for this best case are those shown in boldface in Table 4. In this case, the DelfInject and Obfuscator families have the lowest detection rates, with both only slightly above 50% accuracy. The overall accuracy is 0.8644.

Table 4 MLP model parameters

Classifier	Hyperparameter	Tested values	Accuracy	
			Train	Test
MLP	`image_dim`	[64, **128**]	0.9529	0.8644
	`learning_rate`	[0.001, **0.0001**]		
	`batch_size`	256		
	`epochs`	50		

3.2 *Convolutional Neural Network Experiments*

We have conducted a large number of convolutional neural network (CNN) experiments. In this section we first discuss CNN experiments based on two-dimensional images. Then we consider one-dimensional CNN experiments, where the malware images are vectorized. We also present results for CNN experiments using opcodes extracted from PE files, as opposed to forming images based on the raw byte values in the executable files. The opcodes were extracted using objdump, and we use the resulting mnemonic opcode sequence (eliminating operands, labels, etc.) as features. The hyperparameters tested for all of these CNN experiments are given in Table 5.

3.2.1 Two-Dimensional Image CNNs

Based on two-dimensional image features, we test the CNN model hyperparameters listed under "CNN 2-d" in Table 5. All of these 2-d CNN experiments use two convolutional layers and three fully connected layers. The first convolutional layer takes as input a square gray-scale image with one channel and outputs data with 12 channels using a kernel size of three, padding of two, and a stride of one. A relu activation and max pooling is applied to the result before passing it to the second convolutional layer. This second layer outputs data with 16 channels, with the other parameters being the same as the first convolutional layer. Again, relu activation and max pooling is applied before passing data to the first fully connected layer. This first fully connected layer outputs a vector of dimension 120. After applying relu activation, the data is passed to the second fully connected layer, which reduces the output to a 90-dimensional vector. Finally, relu activation is again applied and the data passes to the last fully connected layer, which is used to classify the sample, and hence is 20-dimensional. For all image sizes less than 1024, we execute our CNN 2-d models for 50 epochs; for the case of 1024×1024 images, we use 8 epochs due to the costliness of training on these large images.

The best overall accuracy obtained for our CNN 2-d experiments is 0.8955. Figure 3 gives the confusion matrix for the best case. We note that the Obfuscator family is again the most difficult to distinguish.

Table 5 CNN model parameters

Classifier	Hyperparameter	Tested values	Accuracy	
			Train	Test
CNN 2-d	image_dim	[64, 128, 256, **1024**]	0.9294	0.8955
	learning_rate	[**0.001**, 0.0001]		
	batch_size	256		
	epochs	50		
CNN 1-d	image_dim	[**1024**, 2048, 4096, 8192]	0.8445	0.8664
	learning_rate	[**0.001**, 0.0001]		
	batch_size	256		
	epochs	20		
CNN 1-d refined	conv1d_1_out_channel	[64, **128**]	0.8538	0.8932
	conv1d_1_kernel_size	[**16**, 32]		
	conv1d_1_stride	[2, **8**]		
	conv1d_2_out_channel	[**32**, 64, 128]		
	conv1d_2_kernel_size	[**8**, 16]		
	conv1d_2_stride	[**2**, 4]		
	image_dim	4096		
	learning_rate	0.001		
	batch_size	512		
	epochs	15		
CNN opcode	opcode_length	[500, **5000**]	0.8418	0.8282
	num_filters	[3, 6, **9**]		
	filter_size	[[12, 6], [6, 12], **[12, 24]**]		
	embedding_dim	[128, **512**]		
	learning_rate	0.001		
	batch_size	256		
	epochs	50		

3.2.2 Vectorized Image CNNs

Recent work has shown promising results for malware classification using one-dimensional CNNs on "image" data [10]. Consequently, we experiment with flattened images, that is, we use images that are one pixel in height. A possible advantage of this approach is that two-dimensional results can depend on the width chosen for the images. We perform two sets of such experiments, which we denote as CNN 1-d and CNN 1-d refined, the latter of which considers additional fine-tuning parameters. The hyperparameters tested for these two cases are given in Table 5.

Our CNN 1-d model uses two one-dimensional convolutional layers, followed by three fully connected layers. The first convolution layer takes in an image with one channel and outputs data with 28 channels based on a kernel size of three. The second convolutional layer outputs data with 16 channels and again uses a kernel of size three. The first fully connected layer outputs a vector of 120 dimensions, which is reduced to 90 dimensions by the second fully connected which, in turn, is reduced to 20 dimensions by the third (and last) fully connected layer. We have applied relu activations in all layers.

The confusion matrix for our best CNN 1-d case is given in Fig. 4. The overall accuracy in this case is 0.8664. A handful of families (Agnet, Alureon, DelfInject, Obfuscator, and Rbot) have accuracies below 80%, which represents the majority of the loss of accuracy.

The CNN 1-d refined tests use the same basic setup as our CNN 1-d experiments, but includes different selections of hyperparameters. As expected, these additional parameters improved on the CNN 1-d case, as the best overall accuracy attained for our CNN 1-d refined experiments is 0.8932. Qualitatively, the CNN 1-d refined results are similar (per family) to the CNN 1-d experiments, so we have omitted the confusion matrix for this case.

3.2.3 Opcode-Based CNNs

We also apply 2-d CNNs to opcode features. For each malware sample, we use the first N opcodes from each binary file, where $N \in \{500, 5000\}$. We also experiment with various other parameters, as indicated in Table 5.

The results for the best choice of parameters for our opcode-based CNN experiments are summarized in the confusion matrix in Fig. 5. Perhaps not surprisingly, the results in this case are relatively weak, with an overall accuracy of 0.8282. However, it is interesting to note from the confusion matrix that some of the families that are consistently misclassified at high rates by image-based CNN models are classified with high accuracy by this opcode-based approach. For example, DelfInject is classified at no better than about 71% in our previous CNN experimetns, but it is classified with greater than 90% accuracy using the opcode-based features.

3.3 Recurrent Neural Networks

Next, we consider a variety of experiments based on various recurrent neural network (RNN) architectures. Specifically, we employ plain vanilla RNN, LSTM, and GRU models. We also consider a complex LSTM-GRU stacked model. The hyperparameters tested in these experiments are summarized in Table 6.

Table 6 RNN model parameters

Classifier	Hyperparameter	Tested values	Accuracy	
			Train	Test
RNN	embedding_dim	[**256**, 1024]	0.7710	0.7294
	hidden_dim	[**256**, 1024]		
	num_layers	[**1**, 3]		
	directional	[**uni-dir**, bi-dir]		
	learning_rate	0.001		
	batch_size	128		
	epochs	20		
LSTM	embedding_dim	[**256**, 1024]	0.9362	0.8916
	hidden_dim	[**256**, 1024]		
	num_layers	[**1**, 3]		
	directional	[**uni-dir**, bi-dir]		
	learning_rate	0.001		
	batch_size	128		
	epochs	20		
GRU	embedding_dim	[**256**, 1024]	0.9411	0.9003
	hidden_dim	[**256**, 1024]		
	num_layers	[**1**, 3]		
	directional	[**uni-dir**, bi-dir]		
	learning_rate	0.001		
	batch_size	128		
	epochs	20		
Stacked	embedding_dim	[256, **1024**]	0.9525	0.8990
	hidden_dim	[256, **1024**]		
	num_layers	[**1**, 3]		
	directional	[**uni-dir**, bi-dir]		
	LG	[**True**, False]		
	learning_rate	0.001		
	batch_size	128		
	epochs	20		

3.3.1 Vanilla RNN, LSTM, and GRU

We have trained our plain vanilla RNNs, LSTMs, and GRU-based models using 20 epochs in each case, with a learning rate of 0.001, a batch size of 128, and based on the first 500 opcodes from each malware sample. We performed multiple experiments with various other parameters, as given in Table 6. In addition, we have applied a dropout layer with 0.3 probability for all models with more than one layer.

The vanilla RNN experiments performed poorly, with an overall accuracy of just 0.7294, and hence we omit the confusion matrix for this case. On the other hand, both the LSTM and GRU models perform well, with accuracies of 0.8916 and 0.9003, respectively. The confusion matrix for the GRU case is given in Fig. 6. Since the LSTM results are so similar, we omit the LSTM confusion matrix. From Fig. 6, we see that, qualitatively, the results of our GRU experiments more closely match those of the CNN opcode-based experiments than the CNN image-based experiments. However, quantitatively, our GRU opcode-based experiments yield significantly better results than our CNN opcode-based experiments.

3.3.2 Stacked LSTM-GRU Model

As in [7], we have also experimented with stacked LSTM and GRU layers. The experiments in this chapter test more parameters and we use a larger dataset, as compared to [7]. A configuration option, which we refer to as LG, is used to decide whether the LSTM is stacked on top of the GRU (LG = false in this case) or GRU is stacked on top of the LSTM (LG = true). For example, when LG is "true," opcode inputs are first passed to LSTM layers, with the output of the LSTM (i.e., the hidden cells) becoming input to the GRU layers. The output of the GRU is then passed to fully connected layers that are used to classify the input data. We have applied a dropout layer with 0.3 probability for models with more than one layer.

The best overall accuracy we obtain for our stacked LSTM-GRU experiments is 0.8990; the confusion matrix for this case is given in Fig. 7. This is somewhat disappointing, as it is in between the results obtained for our LSTM and GRU models.

3.4 Transfer Learning

Finally, we have considered two popular image-based transfer learning models, namely RestNet152 and VGG-19. These are models that have been pre-trained on large image datasets, and we simply retrain the last few layers for the malware dataset under consideration, while the earlier layers are frozen during training. The parameters used in these experiments are summarized in Table 7.

For ResNet152, the model parameters for layer four were unfrozen for training. We also added two more layers of fully connected neurons for training. Resnet152 is pre-trained based on 1000 classes and hence its last fully connected layer has output dimensions of 1000. We reduce this output dimension to 500 via another fully connected layer, and an additional fully connected layer further reduces the output dimension to 20, which is the number of classes in our dataset.

For VGG-19, we froze all layers except 34, 35, and 36. As with ResNet152, we added two more layers of fully connected neurons to reduce the output dimension from 1000 to 20.

For all of our transfer learning experiments, we use a batch size of 256 and trained each model for 20 epochs with learning rates of 0.001 and 0.0001. Both

Table 7 Transfer learning model parameters

Classifier	Hyperparameter	Tested values	Accuracy	
			Train	Test
ResNet152	image_dim	256	0.9811	0.9150
	learning_rate	[0.001, **0.0001**]		
	batch_size	256		
	epochs	20		
VGG-19	image_dim	256	0.9690	0.9216
	learning_rate	[0.001, **0.0001**]		
	batch_size	256		
	epochs	20		

ResNet152 and VGG-19 expect image dimensions of 224×224 and hence we resize our 256×256 images to 224×224.

The performance of these transfer learning models was the best of our deep learning experiments, with ResNet152 achieving an overall accuracy of 0.9150 and VGG-19 doing slightly better at 0.9216. The confusion matrix for VGG-19 is given in Fig. 8; we omit the confusion matrix for ResNet152 since it is similar, but marginally worse. As compared to the other image-based deep learning models we have considered, we see marked improvement in the classification accuracy of the most challenging families, such as Obfuscator.

3.5 Discussion

The results of the malware classification experiments discussed in this section are summarized in Fig. 1. We see that among the deep learning techniques, the image-based pre-trained models, namely, ResNet152 and VGG-19, perform best, with VGG-19 classifying more than 92% of the samples correctly. The best of our other (i.e., not pre-trained) image-based models achieved slightly less than 90% accuracy.

Although the opcode-based results performed relatively poorly overall, it is interesting to note that they were able to classify some families with higher accuracy than any of the image-based models. This suggests that a model that combines both image features and opcode features might be more effective than either approach individually.

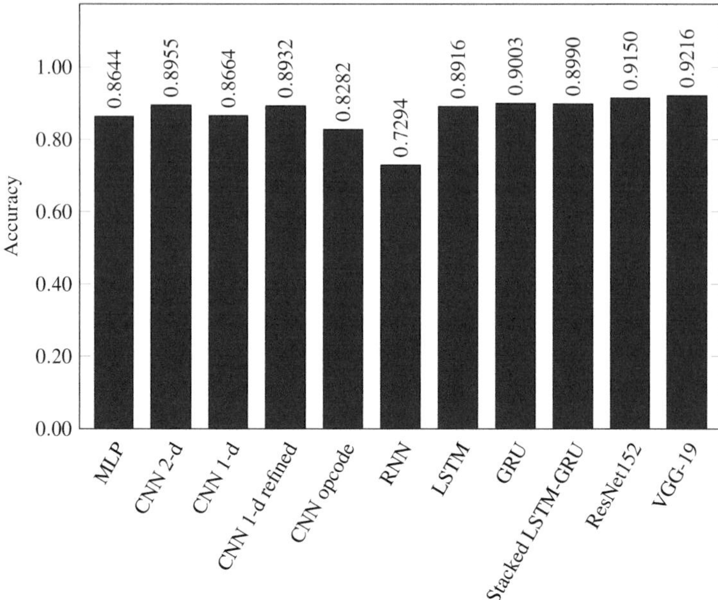

Fig. 1 Comparison of results

4 Conclusions and Future Work

Malware classification is a fundamental and challenging problem in information security. Previous work has indicated that treating malware executables as images and applying image-based techniques can yield strong classification results.

In this chapter, we provided results from a vast number of learning experiments, comparing deep learning techniques using image-based features to some cases involving opcode features. For our deep learning techniques, we focused on multilayer perceptrons (MLP), convolutional neural networks (CNN), and recurrent neural networks (RNN), including long short-term memory (LSTM) and gated recurrent units (GRU). We also experimented with the image-based transfer learning techniques ResNet152 and VGG-19. Among these techniques, the image-based transfer learning models performed the best, with the best classification accuracy exceeding 92%.

For future work, additional transfer learning experiments would be worthwhile, as there are many more parameters that could be tested. Larger and more diverse datasets could be considered. In addition, it would be interesting to consider both image-based and opcode features as part of a combined classification technique. As noted above, the opcode-based techniques perform worse overall, but they do provide better results for some families that are particularly challenging to distinguish based only on image features.

Appendix: Confusion Matrices

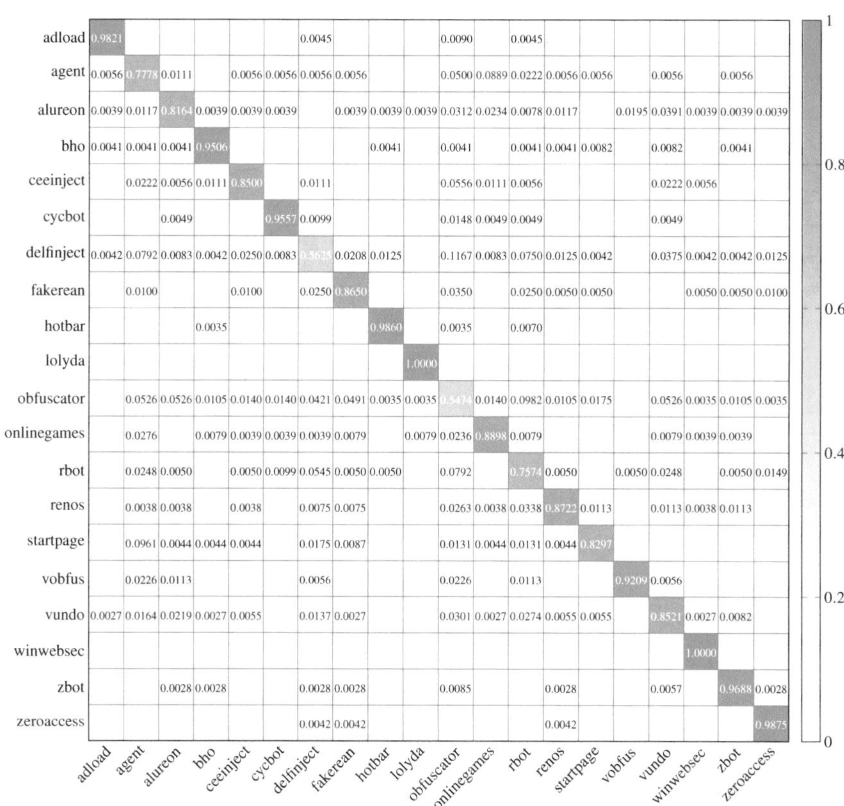

Fig. 2 Confusion matrix for MLP experiment

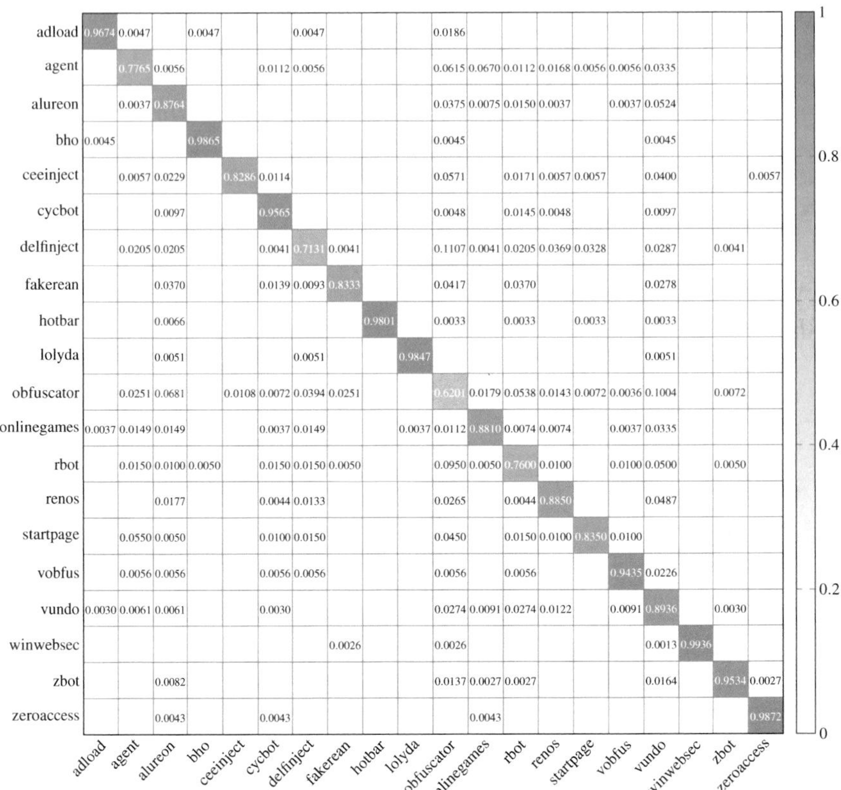

Fig. 3 Confusion matrix for CNN 2-d experiment

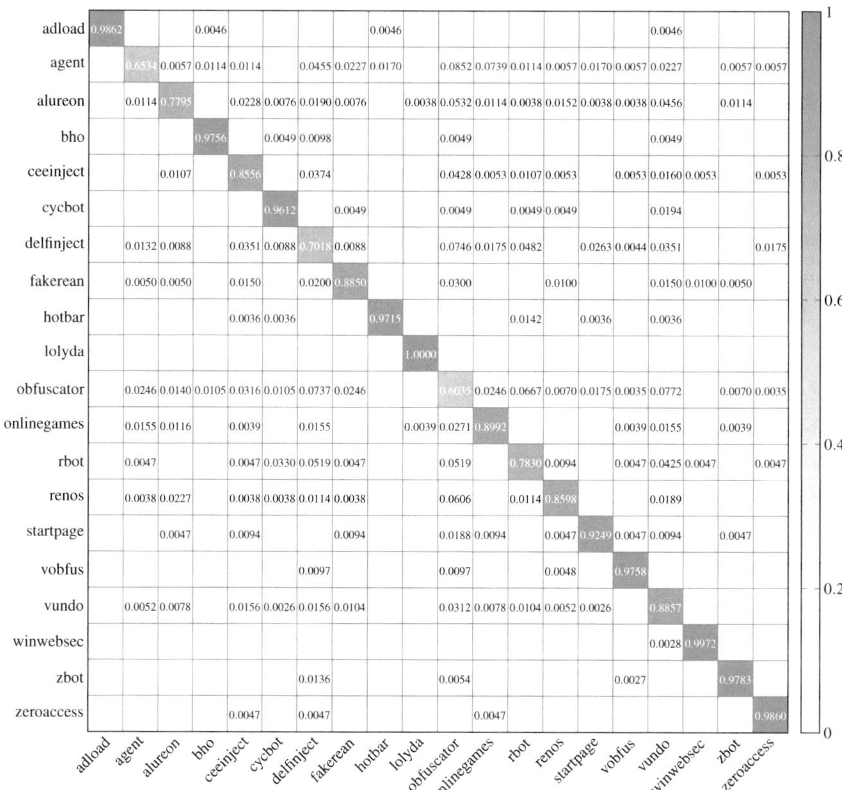

Fig. 4 Confusion matrix for CNN 1-d experiment

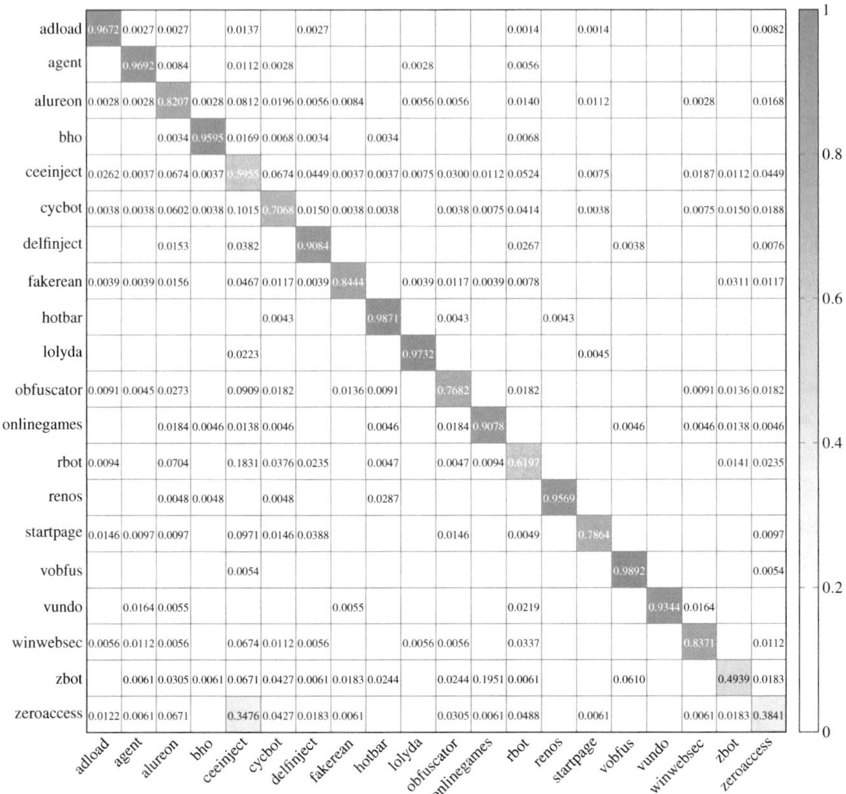

Fig. 5 Confusion matrix for opcode-based CNN experiment

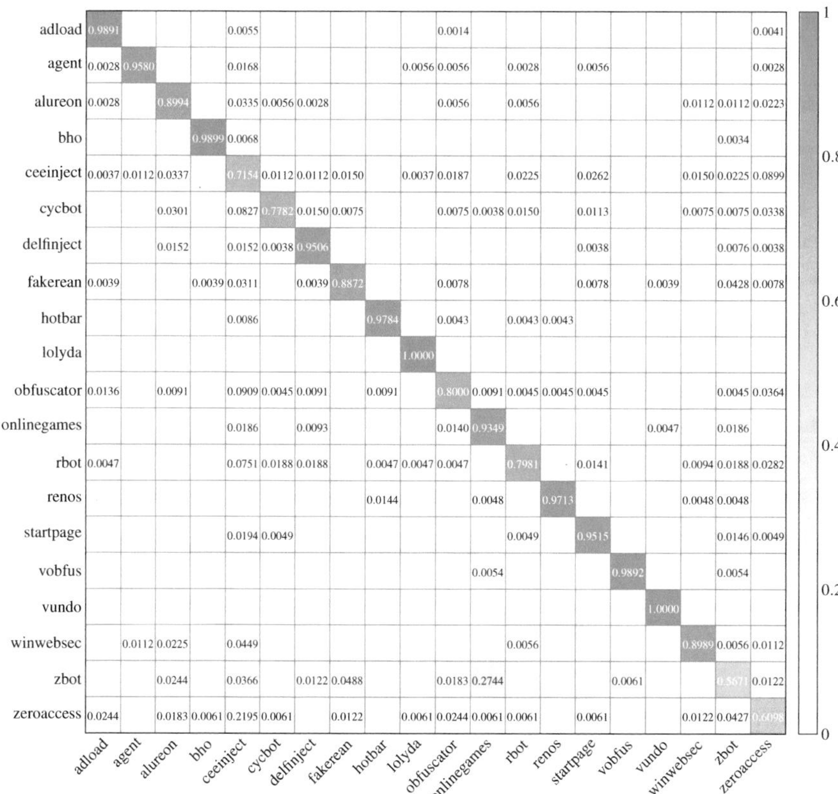

Fig. 6 Confusion matrix for GRU experiment

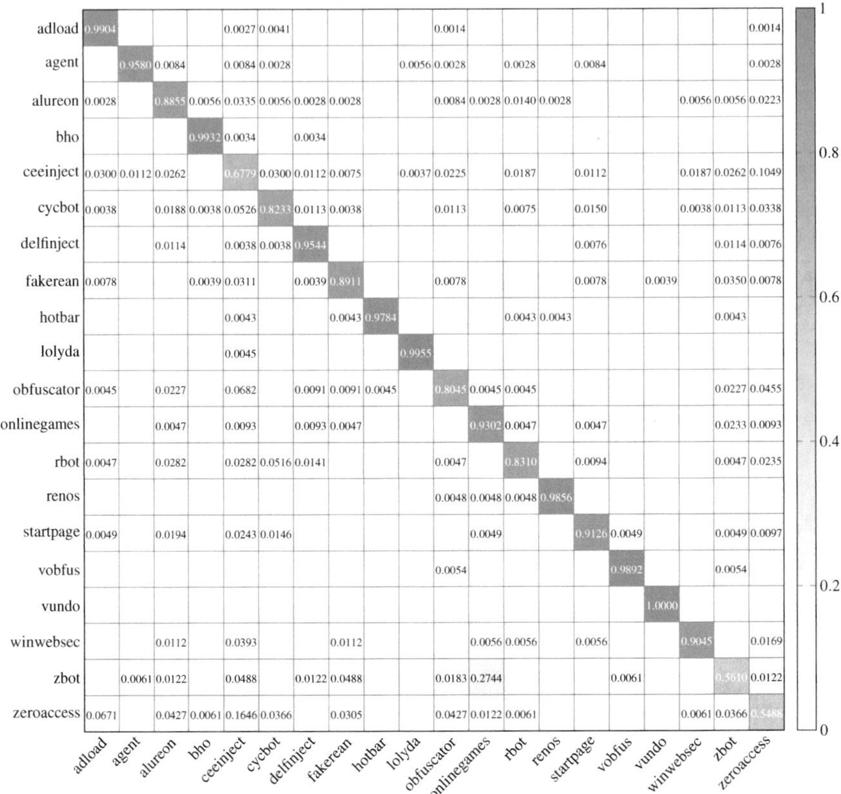

Fig. 7 Confusion matrix for stacked LSTM-GRU experiment

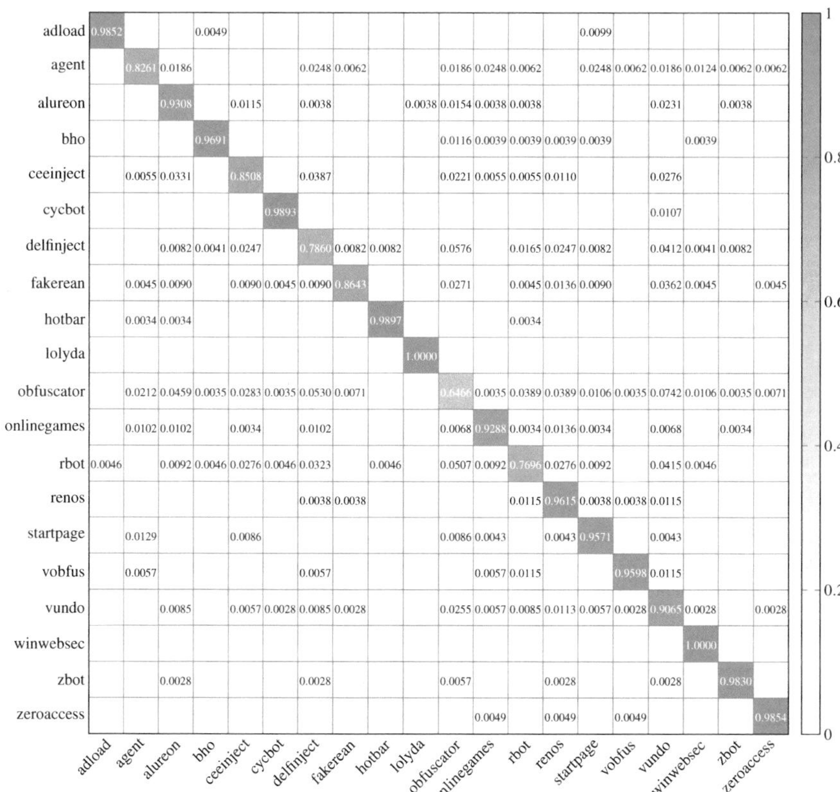

Fig. 8 Confusion matrix for VGG-19 experiment

References

1. Adware:win32/hotbar. https://www.microsoft.com/en-us/wdsi/threats/malware-encyclopedia-description?Name=Adware:Win32/Hotbar&threatId=6204.
2. Austin, Thomas H., Eric Filiol, Sébastien Josse, and Mark Stamp. 2013. Exploring hidden Markov models for virus analysis: A semantic approach. In *46th Hawaii international conference on system sciences, HICSS 2013, Wailea, HI, USA, January 7–10, 2013*, 5039–5048. IEEE Computer Society.
3. Backdoor:win32/cycbot.g. https://www.microsoft.com/en-us/wdsi/threats/malware-encyclopedia-description?Name=Backdoor:Win32/Cycbot.G.
4. Backdoor:win32/zegost.ad. https://www.microsoft.com/en-us/wdsi/threats/malware-encyclopedia-description?Name=Backdoor%3AWin32%2FZegost.AD.
5. Baysa, Donabelle, Richard M. Low, and Mark Stamp. 2013. Structural entropy and metamorphic malware. *Journal of Computer Virology and Hacking Techniques* 9 (4): 179–192.
6. Bhodia, Niket, Pratikkumar Prajapati, Fabio Di Troia, and Mark Stamp. 2019. Transfer learning for image-based malware classification. https://arxiv.org/abs/1903.11551.
7. Carrera, Ero. 2019. pefile 2019.4.18. https://pypi.org/project/pefile/.
8. Damodaran, Anusha, Fabio Di Troia, Corrado Aaron Visaggio, Thomas H. Austin, and Mark Stamp. 2017. A comparison of static, dynamic, and hybrid analysis for malware detection. *Journal of Computer Virology and Hacking Techniques* 13 (1): 1–12.
9. Gupta, Arpit. 2018. Alexa blogs: How Alexa is learning to converse more naturally. https://developer.amazon.com/blogs/alexa/post/15bf7d2a-5e5c-4d43-90ae-c2596c9cc3a6/how-alexa-is-learning-to-converse-more-naturally.
10. Jain, Mugdha, William Andreopoulos, and Mark Stamp. 2020. Convolutional neural networks and extreme learning machines for malware classification. *Journal of Computer Virology and Hacking Techniques*. To appear.
11. Khaitan, Pranav. 2016. Google AI blog: Chat smarter with Allo. https://ai.googleblog.com/2016/05/chat-smarter-with-allo.html.
12. Kim, Samuel. 2018. PE header analysis for malware detection. Master's thesis, San Jose State University. https://scholarworks.sjsu.edu/etd_projects/624/.
13. Levy, Steven. 2016. The iBrain is here—and it's already inside your phone. *Wired*. https://www.wired.com/2016/08/an-exclusive-look-at-how-ai-and-machine-learning-work-at-apple/.
14. McKinney, Wes. 2020. Pandas 1.0.5: Powerful data structures for data analysis, time series, and statistics. https://pypi.org/project/pandas/.
15. Nappa, Antonio, M. Zubair Rafique, and Juan Caballero. 2015. The malicia dataset: identification and analysis of drive-by download operations. *International Journal of Information Security* 14 (1): 15–33.
16. Nataraj, L., S. Karthikeyan, G. Jacob, and B.S. Manjunath. 2011. Malware images: Visualization and automatic classification. In *Proceedings of the 8th International Symposium on Visualization for Cyber Security*, VizSec '11.
17. Travis Oliphant. 2006. NumPy: A guide to NumPy. http://www.numpy.org/.
18. Paszke, Adam, Sam Gross, Soumith Chintala, and Gregory Chanan. 2016. PyTorch: From research to production. https://pytorch.org/.
19. Prajapati, Pratikkumar. 2020. Github repository. https://github.com/pratikpv/malware_detect2.
20. Pws:win32/delfinject. https://www.microsoft.com/en-us/wdsi/threats/malware-encyclopedia-description?Name=PWS:Win32/DelfInject&threatId=-2147241365.
21. Pws:win32/lolyda.bf. https://www.microsoft.com/en-us/wdsi/threats/malware-encyclopedia-description?Name=PWS%3AWin32%2FLolyda.BF.
22. Pws:win32/onlinegames. https://www.microsoft.com/en-us/wdsi/threats/malware-encyclopedia-description?Name=PWS%3AWin32%2FOnLineGames.
23. Pws:win32/zbot. https://www.microsoft.com/en-us/wdsi/threats/malware-encyclopedia-description?Name=PWS:Win32/Zbot&threatId=-2147368817.
24. Simonyan, Karen, and Andrew Zisserman. 2014. Very deep convolutional networks for large-scale image recognition. https://arxiv.org/abs/1409.1556.

25. Singh, Tanuvir, Fabio Di Troia, Corrado Aaron Visaggio, Thomas H. Austin, and Mark Stamp. 2016. Support vector machines and malware detection. *Journal of Computer Virology and Hacking Techniques* 12 (4): 203–212.
26. Spruston, Nelson. 2019. Pyramidal neurons: Dendritic structure and synaptic integration. *Nature Reviews Neuroscience* 9: 206–221. https://www.nature.com/articles/nrn2286.
27. Stamp, Mark. 2020. A selective survey of deep learning techniques and their application to malware analysis. In *Malware Analysis using Artificial Intelligence and Deep Learning*, chapter 1, Stamp, Mark, Mamoun Alazab, and Andrii Shalaginov, ed. 1–48. Springer.
28. Toderici, Annie H., and Mark Stamp. 2013. Chi-squared distance and metamorphic virus detection. *Journal of Computer Virology and Hacking Techniques* 9 (1): 1–14.
29. Trojandownloader:win32/adload. https://www.microsoft.com/en-us/wdsi/threats/malware-encyclopedia-description?Name=TrojanDownloader%3AWin32%2FAdload.
30. Trojandownloader:win32/agent. https://www.microsoft.com/en-us/wdsi/threats/malware-encyclopedia-description?Name=TrojanDownloader:Win32/Agent&ThreatID=14992.
31. Trojandownloader:win32/renos. https://www.microsoft.com/en-us/wdsi/threats/malware-encyclopedia-description?Name=TrojanDownloader:Win32/Renos&threatId=16054.
32. Trojan:win32/bho. https://www.microsoft.com/en-us/wdsi/threats/malware-encyclopedia-description?Name=Trojan:Win32/BHO&threatId=-2147364778.
33. Trojan:win32/startpage. https://www.microsoft.com/en-us/wdsi/threats/malware-encyclopedia-description?Name=Trojan:Win32/Startpage&threatId=15435.
34. Virtool:win32/ceeinject. https://www.microsoft.com/en-us/wdsi/threats/malware-encyclopedia-description?Name=VirTool%3AWin32%2FCeeInject.
35. Win32/alureon. https://www.microsoft.com/en-us/wdsi/threats/malware-encyclopedia-description?Name=Win32/Alureon.
36. Win32/fakerean. https://www.microsoft.com/en-us/wdsi/threats/malware-encyclopedia-description?Name=Win32/FakeRean.
37. Win32/obfuscator. https://www.microsoft.com/en-us/wdsi/threats/malware-encyclopedia-description?Name=Win32/Obfuscator&threatId=.
38. Win32/rbot. https://www.microsoft.com/en-us/wdsi/threats/malware-encyclopedia-description?Name=Win32/Rbot&threatId=.
39. Win32/vobfus. https://www.microsoft.com/en-us/wdsi/threats/malware-encyclopedia-description?Name=Win32/Vobfus&threatId=.
40. Win32/vundo. https://www.microsoft.com/en-us/wdsi/threats/malware-encyclopedia-description?Name=Win32/Vundo&threatId=.
41. Win32/winwebsec. https://www.microsoft.com/en-us/wdsi/threats/malware-encyclopedia-description?Name=Win32/Winwebsec.
42. Wong, Wing, and Mark Stamp. 2006. Hunting for metamorphic engines. *Journal in Computer Virology* 2 (3): 211–229.
43. Wu, Yonghui, Mike Schuster, Zhifeng Chen, Quoc V. Le, Mohammad Norouzi, Wolfgang Macherey, Maxim Krikun, Yuan Cao, Qin Gao, Klaus Macherey, Jeff Klingner, Apurva Shah, Melvin Johnson, Xiaobing Liu, Łukasz Kaiser, Stephan Gouws, Yoshikiyo Kato, Taku Kudo, Hideto Kazawa, Keith Stevens, George Kurian, Nishant Patil, Wei Wang, Cliff Young, Jason Smith, Jason Riesa, Alex Rudnick, Oriol Vinyals, Greg Corrado, Macduff Hughes, and Jeffrey Dean. 2016. Google's neural machine translation system: Bridging the gap between human and machine translation. https://arxiv.org/abs/1609.08144.
44. Yajamanam, S., V. R. S. Selvin, F. Di Troia, and Mark Stamp. 2018. Deep learning versus gist descriptors for image-based malware classification. In *Proceedings of the 4th International Conference on Information Systems Security and Privacy*, ICISSP 2018, 553–561.

A Novel Study on Multinomial Classification of x86/x64 Linux ELF Malware Types and Families Through Deep Neural Networks

Andrii Shalaginov and Lasse Øverlier

Abstract Through the history of desktop and server-oriented malware, Microsoft Windows was notoriously known as one of the heavily attacked Operating Systems (OS). Several factors caused this, including unobstructed installation of third-party software. Unix-like OS is considerably less susceptible to malware infections. However, there are still a few examples of successful malicious software. The challenge is that there are not that many software tools available to analyze Linux malware, including well-known automated intelligent machine learning-aided classification. Our contribution in this paper is twofolded. First, we look at the most popular approaches to analyze Linux malware into families and types. Simple binary classification is no longer efficient and it is more important to know the exact class of malware to speed up incident response. Second, we suggested methodology for multinomial Linux malware classification using deep neural network. This approach overcomes the limitation of shallow neural networks used before for multinomial Windows PE32 malware classification. Such classification has been explored successfully for MS Windows, yet, not on the Linux malware. Our focus also is specifically on desktop and server Intel-compatible Linux malware rather than affiliated ARM binaries that require designed IoT environment to run successfully. This work will serve as a stepping stone for efficient intelligent Linux malware classification using deep learning-based methods. We have created a novel dataset with 10,574 malware files labeled into 19 malware types and 442 malware families

1 Introduction

Malicious software, computer viruses, or simpler *malware* have been there for decades, targeting individual users, organizations, and critical national infrastructure. Historically, the family of Microsoft Windows Operating Systems (OS) is considered

A. Shalaginov (✉) · L. Øverlier
Norwegian University of Science and Technology, Trondheim, Norway
e-mail: andrii.shalaginov@ntnu.no

L. Øverlier
e-mail: lasse.overlier@ntnu.no

M. Stamp et al. (eds.), *Malware Analysis Using Artificial Intelligence and Deep Learning*, https://doi.org/10.1007/978-3-030-62582-5_17

to be more vulnerable and more susceptible to cyberattacks than Linux or MacOS [17, 29]. Multiple factors influenced and lead to such state of the art. Although Linux is considered to be more secure and less affected my computer viruses, there are still notorious examples of how systems are exploited [20, 40]. As a result, a very few works focus on either intelligent- or signature-based Linux malware detection [16]. This work bridges the gap between detection of Linux malware families and types and application of renewed deep learning models for similarity-based static malware detection.

Most of the researchers in the information security community work on the techniques used to identify and detect Windows malware samples among others, particularly, known and widely seen classes. They use a lot of methods, both static and dynamic analyses, to identify that malware among others, besides whether the software is malicious or benign. Very few works focus on Linux and there is a clear reason for such bias. From the global perspective of Desktop OS market share [35], we can see that during May 2019–May 2020, the overall share of MS Windows OS is 77.04%, OS X 0 18.38%, and Linux—1.68%. Market share of Linux in cloud services is much bigger—more than 90% of public cloud services run Linux [13]. However, large-scale enterprise cloud solutions have better security measures than private end users. Therefore, Linux malware analysis requires a better understanding of specific features and approaches that can be used to not only detect malware but also understand what type of malware it is. Another big challenge is malware naming. CARO (Computer Antivirus Research Organization) created a naming scheme back in 1990 [3], which was supposed to be a stepping stone for malware naming standardization. However, only Microsoft mostly uses CARO approach in their products [2] with Trend Micro moving that direction since 2018. In the literature, there can be found several sources mentioning these challenges, yet offering no comprehensive overview or even solution. As a result, one needs to use additional processing methods like majority voting [34] or AVCLass [30] tool to find our correct and realistic malware class name.

The scope of this research is (i) to reveal existing challenges that complicate Linux malware identification and cyberthreat intelligence services when it comes to feature engineering and extraction using open-source tools. To our knowledge, the topic has been approached by researchers before, however, there is no comprehensive evaluation of features that can be used for multinomial malware detection. Moreover, we aim at (ii) providing a high-level overview of possible tools and data sources for static feature contractions. Finally, (iii) we focus on multinomial Linux malware classification using deep learning with multiple abstraction layers. To our awareness, such evaluation has not been done before nor other works focused purely on Linux malware compiled specifically for Intel desktop platforms rather than ARM or MIPS. We have identified 15,101 Intel-compiled ELF files, which is the largest known Linux malware collection used in malware research. Out of those, 10,574 were labeled into malware types and families.

This paper is organized as follows. Section 2 represents the history of Linux malware and relevant background state of the art, including previous studies with applications of Machine Learning (ML). Following the relevant literature study, Sect. 3

presents the ELF file format and possible raw data that can be extracted from the file, and subsequently used to identify malware category. Section 4 provides a suggested methodology, used data and deep learning technique. Section 5 describes used experimental setup, software tools, and developed programs, while analysis of achieved results in intelligent Linux malware classification is given in Sect. 6. Conclusions and Discussions are outlined in Sect. 7.

2 State of the Art: Machine Learning for Linux Malware Detection

Linux malware is not the most mainstream attack vectors. From before, those binaries were tailored for specific attacks when an attacker seeks access to the system, e.g., outdated vulnerable web service hosted on old Linux machine [24]. Another reason for this is that Linux malware is developed for server-based production cloud systems, while desktop Linux usage pattern is far from being mainstream [35]. We can see a huge difference in available samples for Linux and Windows malware research. There were only 10,548 collected ELF Linux malware samples, including PC and ARM devices in the study by Cozzi et al. [16], out of which only 3,738 binaries were Intel-compatible. Shalaginov et al. [34] managed to collect 326,000 malicious PE32 files. Linux malware detection is not yet a Big Data problem and does not necessarily target the consumer market.

In general, the history of Linux malware is not as extensive as it can be seen in the case of MS-DOS or Microsoft Windows. One of the main reasons is that hiding malware code in open-source projects with community review is a very difficult task [8]. In a matter of fact, there are very few surveys or articles published on the evolution of Linux malware. However, we can say that early 1990s can be considered as a starting point in a growing number of Linux malware. One of the surveys [1] mentioned a few known famous Linux malware samples such as Staog (1996), Slapper (2002), and Trass Spyware (2014). Scott Granneman in 2003 article [21] defined following approximation of total number of known viruses: MS Windows—60,000, Macintosh—40, commercial Unix versions—5 and Linux—40. It means that Linux malware roughly shares 0.07% of known malware samples. In June 2020, according to VirusTotal Intelligence Platform (search command "*magic:pe32 positives:1+*"), there are PE32 malware—37.55 million, ELF—0.14 million, and Mach-O—0.025 million of samples with at least one anti-virus vendors defining file as malicious. In our particular case, it means that the number of malicious ELF files identified by VirusTotal is 0.37% out of globally known malware. It includes both Intel-compiled binaries and ARM and MIPS platforms. IoT botnets became popular in recent years, also hitting ARM portable and embedded devices running Linux [25]. Using Shodan.io service we found only 12,320 Linux devices openly connected to the Internet [9] with possible vulnerabilities and privacy concerns. Therefore, it is yet another proof that Linux malware is not a mainstream infection, however, still representing a threat to corporate and enterprise clients, especially if the software is outdated.

Cozzi et al. [16] highlighted a general lack of studies related to Linux malware, how it functions, limited availability of data, and what indicators to look for complicates the whole research. There are very few works that dissect Linux ELF malware and perform a thorough study. Cozzi et al. looked at several examples of how Linux malware implement malicious activities and what can be used for malware detection. Another distinct work in the area is ELF-miner developed by Shahzad et al. [31].

3 Linux Malware: Automated Features Extraction and Classification

Malware analysis research has gained extensive popularity and attention in the last two decades, especially focusing on automated detection using ML. However, the key factor to success and reproducible of study—data, i.e., malware binary samples. When it comes to Linux malware, the main challenge is a qualitative and descriptive ELF dataset, in contrary to millions of available Windows PE32 malware files. We investigated several sources of possible Linux malware that can be utilized in this research. First, *VxHeaven* [7] was one of the websites providing access to well-categorized binary files. However, the website was closed in 2012 by Ukrainian police forces, then worked since 2013 until 2018. Finally, it went offline, and currently several mirrors can be found online to maintain the initiative. The challenge is that most of the viruses there are already outdated. Another resource is *VirusTotal*, which was established in 2004 and now can be considered as a *de facto* standard in the information security community [5]. It provides reporting of the malware detection from 70 anti-virus vendors, which in addition to extensive threats intelligence and community reporting, giving the most extensive publicly available malware awareness. As per 03.07.2020, there have been reported 1,606,461 distinct files submitted to VirusTotal, while 836,160 files were labeled as malicious by one or more anti-virus vendors [6]. Finally, *VirusShare* [4] initiative started back in 2011 and now has collected over 400 versatile archives with different examples of malware. As per 03.07.2020, VirusShare collection offers access to 34,866,212 different malware samples available through the website. In this research, we focus on VirusShare malware files that are represented in four dedicated archives: one Linux and three ELF-specific.

When it comes to Microsoft Windows Portable Executable 32 file format, we can see that there has been done a wide range of studies [18, 28, 34, 36, 39, 41] dissecting particular features and attributes that can identify malicious behavior. In a matter of fact, one can find a well-known Windows API function calls that are attributed to the malicious behavior [10, 19, 27, 37]. Furthermore, the attack vectors and exploitation methods used famous software packages available for Windows platform [42]. Therefore, state of the art in Windows malware analysis is extensive

```
ELF Header:
  Magic:   7f 45 4c 46 01 01 01 00 00 00 00 00 00 00 00 00
  Class:                             ELF32
  Data:                              2's complement, little endian
  Version:                           1 (current)
  OS/ABI:                            UNIX - System V
  ABI Version:                       0
  Type:                              EXEC (Executable file)
  Machine:                           Renesas / SuperH SH
  Version:                           0x1
  Entry point address:               0x4001a0
  Start of program headers:          52 (bytes into file)
  Start of section headers:          99284 (bytes into file)
  Flags:                             0x9, sh4
  Size of this header:               52 (bytes)
  Size of program headers:           32 (bytes)
  Number of program headers:         3
  Size of section headers:           40 (bytes)
  Number of section headers:         10
  Section header string table index: 9
```

Fig. 1 Decoded ELF header field values extracted using "readelf -h"

and amount of available literature about possible distinct features and a corresponding set of supplementary analytics tools is overwhelming [32].

We mentioned above that Linux malware is not a mainstream research field, therefore making it challenging to have robust and resilient characteristics extraction. However, we still can find a few works focusing on the peculiarities of Linux malware samples. Generally speaking of Executable and Linkable Format (ELF) files, there is a commonly used reference structure of the file. The ELF file format was selected as the main format in x86 Unix and Linux systems in 1999 [22]. Moreover, according to ELF version 1.2 specification [14], the files are organized in the following components: (i) *ELF header* with (ii) *subsequent program header table* and (iii) *section header table*. ELF header represents 32-byte-long data structure that identifies the following parameters of the files as shown in Fig. 1.

Program headers are represented in Fig. 2. Each header has a size of 56 bytes and current example has 3 headers. Section headers and corresponding segment mapping are shown in the end.

Having in mind a general overview of the ELF file format, our main goal would extract as many features as possible to be able to perform automated analysis [12]. Header and file data can be crucial to finding dissimilarities in malicious files. Our particular goal is to investigate whether static analysis can be utilized for multi-class malware classification using ML. A study by Cozzi et al. [16] attempted to understand what kind of characteristics can be used to identify malicious behavior in Linux. The authors mostly focused on dynamic behavioral exaction and touched static analysis briefly. Only 3,738 Intel-complied ELF files were used. Bai et al. [11] applied ELF parser for classification of 8 malware classes with 763 malware samples based on the system calls. There were even attempts to provide platform-independent malware analysis for both Windows and Linux [23]. So, the main difference from previous

```
Elf file type is EXEC (Executable file)
Entry point 0x4001a0
There are 3 program headers, starting at offset 52

Program Headers:
  Type           Offset   VirtAddr   PhysAddr   FileSiz MemSiz  Flg Align
  LOAD           0x000000 0x00400000 0x00400000 0x178b8 0x178b8 R E 0x10000
  LOAD           0x018000 0x00428000 0x00428000 0x00394 0x00918 RW  0x10000
  GNU_STACK      0x000000 0x00000000 0x00000000 0x00000 0x00000 RWE 0x4

 Section to Segment mapping:
  Segment Sections...
   00     .init .text .fini .rodata
   01     .ctors .dtors .data .bss
```

Fig. 2 Overview of program header sections extracted using "readelf -l"

works is the utilization of a variety of relevant static features for malware detection for Intel-complied binaries rather than the runtime and dynamic analysis of ELF, which heavily depends on platform and OS.

Multi-class intelligent Linux malware classification requires building ML methods capable of high level of abstraction. Deep Learning and Deep Neural Networks (DNN) have been successfully used in the area of malware analysis before [15, 38, 43]. The reason for this is the ability of the DNN to model highly nonlinear relationships, especially when it comes to multiple classes rather than standard binary classification [33]. By using such an approach, the plan is to overcome the limitations of static features in the multinomial Linux classification problem due to the presence of obfuscation.

4 Methodology: Malware Analysis and Detection

In ML, any application based on the data analytics should include so-called *testing* and *training* phase according to Kononenko [26]. The fundamentals of ML-based intelligent malware detection depend on the quality of *features* or *attributes* that the classification engine is relying upon. Shalaginov et al. [32] made an extensive survey of possible way of feature extraction from malware samples, such as byte sequences, opcodes, high-level header features, etc. In our view, using high-level static features in this research can yield reliable identification of multiple classes of malware. At the same time, the dynamic behavioral analysis may not yield a sufficient amount of relevant attributes and require much more efforts to establish a testing environment. Unlike Cozzi et al. [16], we focus specifically on Linux PC malware (Intel 3086 and x86-64) and not on ARM/MIPS platforms, while having many more samples for our experimentation. So, to yield the best possible results, we developed the following methodology using static analysis based on the characteristics from Linux native tools and threat intelligence from VirusTotal platform.

Phase 0. Acquiring samples from *VirusShare*—the most comprehensive and known information security community ELF Linux malware samples that are also publicly available.

Phase 1. Filtering all files that are not completed for ELF platform, performing extraction of raw information for every malware binary file such as "md5," JSON "peframe" report, "readelf," "file," "strings," file size, and entropy. All information is being stored in MySQL database for easier subsequent access.

Phase 2. Filtering ELF Linux malware samples that have been compiled for either Intel 3086 or Intel x64-86 platform based on extracted metadata. We specifically exclude any other binaries such as ARM/MIPS to facilitate a better "ground truth" in experiments and unbiased results. Then, an extraction of the reports using VirusTotal Private API was performed.

Phase 3. Feature extraction is being performed on all types of raw data extracted at the previous phase. As a basis, the following categories of metadata and characteristics were used: "virustotal_file_report," "peframe," "readelf," "strings," "file_size," and "file_entropy." The description of 30 extracted features is shown in Table 1.

Finally, we are looking into malware types and families based on the standard developed by the Computer Anti-Virus Research Organization (CARO). The idea is to extract two classes: "type" and "family." Tools like AVCLass [30] cannot give both classes, and therefore we will be looking into VirusTotal reports generated by Microsoft, one of the very few companies following CARO naming standard.

Phase 4. In this step, we extract two labeled datasets: with class label "type" and "family" separately to investigate how those differ. This phase has to goals: (i) to investigate features and possibly select the best set and (ii) evaluate the performance of state-of-the-art ML methods and deep DNN with a large number of hidden layers.

5 Experimental Design

To investigate Linux ELF malware and understand how deep learning can be used to perform multinomial classification, there was established an experimental setup to process raw malware, extract relevant features, and build an intelligent detection model.

5.1 Dataset

As we have mentioned before, the amount of available Linux ELF malware cannot be compared to MS Windows PE32 malware binaries. Therefore, we looked for community-published datasets that can serve the purpose of multinomial malware detection and reproducible experiments. One of the famous collection has been published by VxHeaven [7]. Even though VxHeaven website offers well-classified malware samples, the resource was offline and the data was not updated for sev-

Table 1 Description of features extracted for each ELF file

Feature name	Explanation
vt_submission_names	Number of submission names
vt_times_submitted	Times the binary was submitted
vt_exif	Number of entries in "exiftool"
vt_embedded_ips	Number of embedded IPs in the binary
vt_contacted_ips	Number of IPs the binary contacted
vt_exports	Number of export functions
vt_imports	Number of import functions
vt_shared_libraries	Number of shared libraries included
vt_segments	Number of segments
vt_sections	Number of sections
vt_packers	Number of packers
vt_tags	Number of tags
vt_positives	Number of AV identified as malicious
peframe_ip	Number of identified IP addresses
peframe_url	Number of identified URLs
readelf_entry_address	Entry point address
readelf_start_prog_headers	Start of program headers
readelf_start_sec_headers	Start of section headers
readelf_number_flags	Number of flags
readelf_header_size	Size of this header
readelf_size_prog_headers	Size of program headers
readelf_number_prog_headers	Number of program headers
readelf_size_sec_headers	Size of section headers
readelf_number_section_headers	Number of section headers
readelf_sec_header_string_table_index	Section header string table index
strings_number	Number of distinct strings
strings_size	Size of all strings extracted from the file
strings_avg	Average size of each string
file_size'	Size of the file
file_entropy'	Entropy of the whole file content

eral years. Therefore, we found another alternative—popular malware distribution platform VirusShare [4]. Three archives with Linux ELF malware were acquired as described in Table 2.

Collected archives have been uploaded to VirusShare in 2014, 2016, 2019, and 2020. Following *Phase 0*, we managed to collect 56,805 unique (MD5) ELF files for further processing from the archives. After filtering all non-Intel binaries, we have shortlisted 15,101 ELF files following *Phase 1*. The top 10 types of Intel-related malware platforms are presented in Table 3. One can see that there is quite a different

Table 2 Archives extracted from VirusShare that contain Linux and ELF malware

Filename	Archive size (GB)	Number of files
VirusShare_ELF_20140617.zip	0.13	2,778
VirusShare_ELF_20190212.zip	1.24	10,426
VirusShare_ELF_20200405.zip	2.40	43,553
VirusShare_Linux_20160715.zip	10.78	9,469

Table 3 Top 10 file types using Linux "file" command

Count	Linux "file" command output
3,517	ELF 32-bit LSB executable, Intel 80386, version 1 (SYSV), statically linked, not stripped
3,194	ELF 32-bit LSB executable, Intel 80386, version 1 (SYSV), statically linked, stripped
1,785	ELF 64-bit LSB executable, x86-64, version 1 (SYSV), statically linked, not stripped
1,271	ELF 32-bit LSB executable, Intel 80386, version 1 (GNU/Linux), statically linked, stripped
610	ELF 32-bit LSB executable, Intel 80386, version 1 (SYSV), dynamically linked, interpreter /lib/ld-linux.so.2, for GNU/Linux 2.0.0, not stripped
385	ELF 32-bit LSB executable, Intel 80386, version 1 (SYSV), statically linked, missing section headers
370	ELF 64-bit LSB executable, x86-64, version 1 (SYSV), statically linked, stripped
364	ELF 32-bit LSB executable, Intel 80386, version 1 (SYSV), dynamically linked, interpreter /lib/ld-linux.so.2, for GNU/Linux 2.2.5, not stripped
315	ELF 32-bit LSB executable, Intel 80386, version 1 (SYSV), statically linked, for GNU/Linux 2.2.5, not stripped
264	ELF 32-bit LSB executable, Intel 80386, version 1 (SYSV), dynamically linked, interpreter /lib/ld-linux.so.2, for GNU/Linux 2.2.5, with debug_info, not stripped

type of files. In overall, we retrieved 700 different types of ELF files compiled for Intel platform out of 1,128 all types. During *Phase 2*, the extraction of the reports using VirusTotal Private API and corresponding characteristics using open-source and inline Linux command line software tools was performed. Finally, after *Phase 3*, we ended up having 10,574 selected ELF files that also have been labeled by Microsoft following the CARO naming convention. Reason for this selection is the ability to use trustworthy classification for both malware "types" and "families."

Moreover, all relevant files have been checked against VirusTotal to retrieve relevant static and dynamic analysis information and, most important, labels assigned by Anti-Virus vendors. However, did not find any files that have not been submitted to VT before—all were known malware samples from before. The overall raw data set extracted from selected ELF files occupied 6.8 GB in MySQL storage for the following fields in the database that are shown in the Fig. 3.

#	Name	Type	Collation	Attributes	Null	Default
1	**id**	int(11)			No	*None*
2	**md5**	char(32)	latin1_swedish_ci		No	*None*
3	**virustotal_file_report**	mediumtext	utf8_general_ci		Yes	*NULL*
4	**readelf**	mediumtext	utf8_general_ci		Yes	*NULL*
5	**peframe**	mediumtext	utf8_general_ci		Yes	*NULL*
6	**file**	text	utf8_general_ci		Yes	*NULL*
7	**strings**	mediumtext	utf8_general_ci		Yes	*NULL*
8	**size**	int(11)			Yes	*NULL*
9	**file_entropy**	double		UNSIGNED	Yes	*NULL*
10	**do_not_process**	tinyint(1)			Yes	*NULL*

Fig. 3 Raw data extracted from ELF files that are places in MySQL dataset

5.2 *Experimental Setup*

Experimental setup and all data processing have been done on Ubuntu 18.04 using Python 3.6.9 and storing data in MySql 5.7. For feature engineering, we used following tools: "ent"—entropy of the file, "strings" v 2.30—all ASCII strings in the file, "file" v 5.32— information about particular file type, "readelf" v 2.30— information about ELF formal object file, "peframe" v 6.0.3—a tool for getting JSON report on PE32 files, also giving a lot of structural information on ELF, VirusTotal Private API 2.0—all anti-virus reports. For ML part, it was used Weka 3.8.4 and RapidMiner 9.7.

6 Results and Analysis

One of the most important findings of this study is the actual distribution of the malware samples in the collected dataset. In overall, we discovered 19 distinct malware types and 442 families following ELF malware labeled by Microsoft following the CARO naming scheme. To our awareness, such descriptive statistics were not available and the summary is represented in the Table 4. It is peculiar that "DDoS" type and "Mirai" family have a considerable share in the detected x86/x64 Linux malware. Based on this, we can speculate that Linux machines are often infected for distributed attacks and botnet creation.

Table 4 Top 10 ELF x86/x64 ELF malware types and families extracted from the dataset

Malware type	Count	Malware family	Count
DDoS	4,230	Lightaidra	2,709
Trojan	2,602	Gafgyt	1,368
Backdoor	2,153	Mirai	1,346
Virus	577	Occamy	1,050
Exploit	394	RST	281
DoS	204	CoinMiner	238
TrojanDownloader	160	Setag	236
Worm	120	Berbew	220
VirTool	98	Wacatac	195
HackTool	62	Tsunami	164

6.1 Feature Selection

Even though it was extracted 30 relevant features for multinomial classification of Linux ELF malware, not all of them have the same contribution toward dissimilarities in each class. To measure the differences in such contribution, we performed feature evaluation using Information Gain method [26]. Results for 19 malware types and 442 families are shown in the Table 5.

A few peculiarities that we can see is the consistency between the most influential features when it comes to both types and families. Even though there is 10 magnitude difference in the number of classes, the following features have one of the biggest merits when compared to others: "readelf_start_sec_headers"—starting position of the section headers into ELF file measures in bytes, "file_size" in bytes, "string_size"—the size of all ASCII strings discovered in ELF file and "strings_number"—number of all distinct strings found in ELF file. The distribution of the two fist aforementioned features is represented in Figs. 4 and 5. The distribution is given for DDoS and Trojan malware types, which were found to be the most frequent in the dataset according to Table 4. We can see how distributions differ from one type to a different type, influenced by internal malicious functionality of binary files.

6.2 Classification Accuracy: State-of-the-Art Methods

To establish a "ground truth" compared to other ML methods accepted in the community and considered as state of the art, we used the following implementations in Weka with tenfold cross-validation: Naive Bayes, Support Vector Machines (SVM), multilayer perceptron, k-NN, and C4.5. The overall classification results for all malware families and types are shown in Table 6.

Table 5 Top 10 features according to Information Gain measure

Average merit	Average rank	Feature name
Malware type		
1.022 +- 0.013	1 +- 0	readelf_start_sec_headers
0.898 +- 0.037	2 +- 0	file_size
0.774 +- 0.007	3.6 +- 0.66	readelf_entry_address
0.766 +- 0.016	3.9 +- 0.83	strings_size
0.757 +- 0.019	4.5 +- 0.67	strings_number
0.675 +- 0.003	6.4 +- 0.8	readelf_sec_header_string_table_index
0.665 +- 0.009	7.3 +- 0.64	readelf_number_section_headers
0.646 +- 0.034	7.7 +- 1.27	strings_avg
0.609 +- 0.019	8.9 +- 0.54	file_entropy
0.597 +- 0.005	9.7 +- 0.46	vt_sections
Malware Family		
2.174 +- 0.026	1 +- 0	readelf_start_sec_headers
1.894 +- 0.036	2 +- 0	file_size
1.708 +- 0.03	3.1 +- 0.3	strings_size
1.671 +- 0.033	4 +- 0.45	strings_number
1.613 +- 0.025	4.9 +- 0.3	readelf_entry_address
1.478 +- 0.05	6.5 +- 0.5	readelf_number_section_headers
1.394 +- 0.178	7.5 +- 1.91	readelf_sec_header_string_table_index
1.401 +- 0.019	7.7 +- 0.46	strings_avg
1.285 +- 0.026	9.1 +- 0.7	file_entropy
1.274 +- 0.089	9.6 +- 1.02	vt_sections

One of the most accurate models is C4.5, which builds a decision tree model. However, the complexity is quite outstanding, considering that for the "type" dataset, the tree includes 693 leaves and having the size of 1,385; for the "family" dataset, the tree contains 1,148 leaves and tree size is 2,295. It makes those methods impractical.

6.3 Deep Learning

It was mentioned before that the major advantage of deep learning is the ability to model highly nonlinear data, such as multinomial classification. State-of-the-art ML methods does not perform too well on the ELF multinomial malware classification, especially hundreds of malware families cannot be classified properly. The exception is k-NN and C4.5; however, training such models on millions of malware samples will result in unmanageable and large models that are impractical in real life. To investigate the influence on number of hidden layers—and, as a result, abstraction

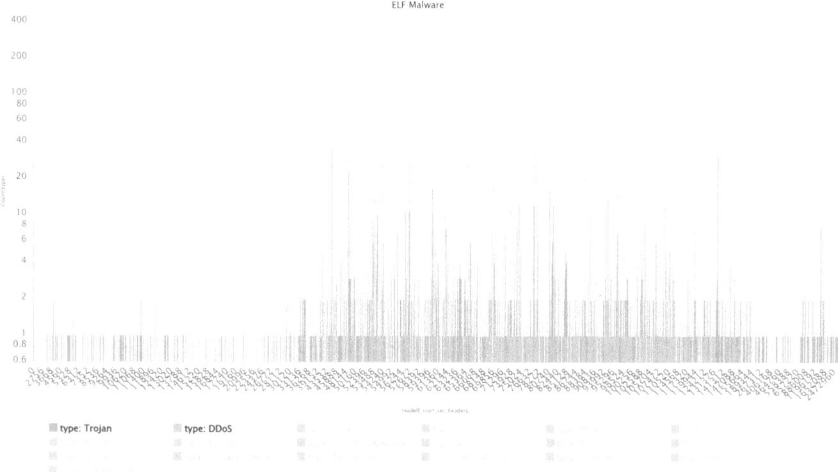

Fig. 4 Distribution of "readelf_start_sec_headers" for two most frequent ELF malware types—DDoS and Trojan (plotted in RapidMiner)

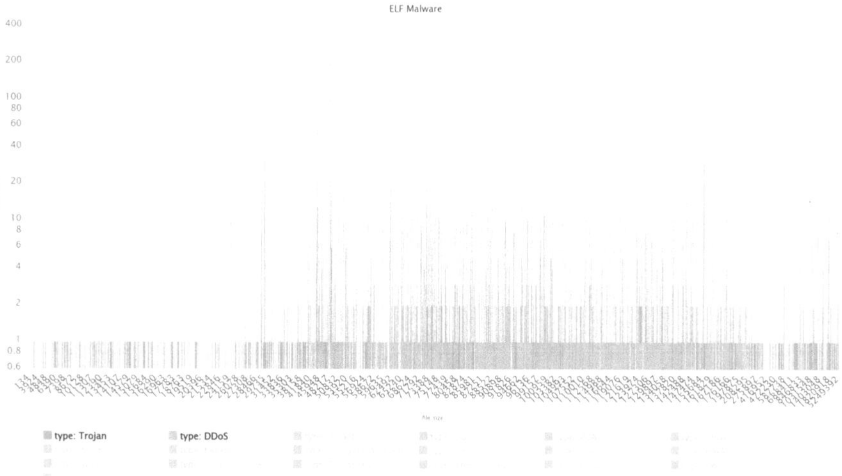

Fig. 5 Distribution of "file_size" for two most frequent ELF malware types—DDoS and Trojan (plotted in RapidMiner)

level—we performed a comparison of the DNN classification. Results for the ELF malware types are shown in Fig. 6 and—for the malware families—in Fig. 7. There is no improvement in accuracy for the family dataset for DNN with more than 10 layers while the training time and complexity grow dramatically. The conclusion that we can draw from the plots is a necessity for higher abstraction level models that can properly model features of each particular class when it comes to tens and hundreds

Table 6 Classification accuracy of ELM malware types and families

Method	Malware type, %	Malware family, %
Naive Bayes multinomial	6.6295	6.5916
Support vector machine	59.6747	47.3804
Multilayer perceptron (3 L)	63.6751	34.7172
IBk (k-NN, k = 3)	73.8415	63.8075
J48 (C4.5)	80.8587	71.5150

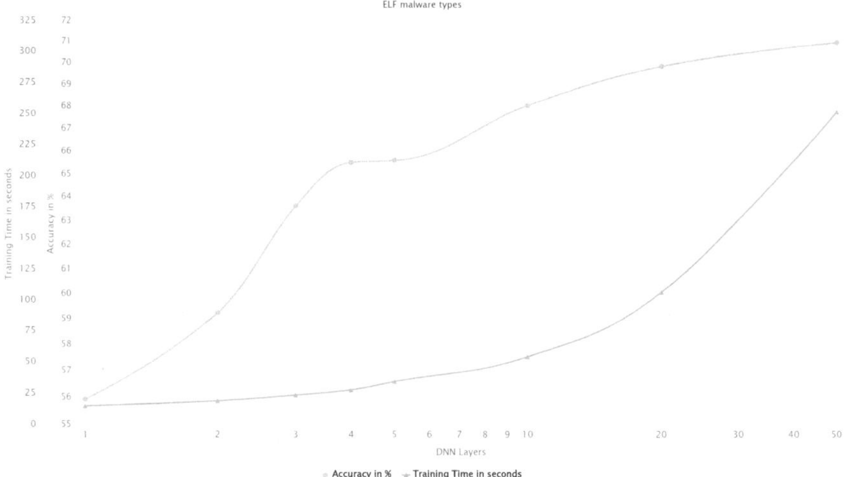

Fig. 6 Dependency of DNN accuracy and training time for 19 malware types on number of layers (plotted in RapidMiner)

of malware classes, rather than the classical problem of binary malware/goodware detection. Classification accuracy with the lower number of classes can be easily boosted by an increasing number of hidden layers; however, it is not the case if there is a considerable number of classes.

7 Discussions and Conclusions

This work describes ongoing research related to rarely explored Linux ELF malware detection and analysis designed specifically for Intel x86/x64. The general lack of tools and malware detection approaches can be explained by a much lower share of desktop Linux OS in comparison to MS Windows. Moreover, the latter has a different security mechanism when it comes to the installation of third-party software, making Linux malware targeting servers with outdated software. Moreover, there was no comprehensive properly labeled dataset available for x86/x64. One of the outcomes

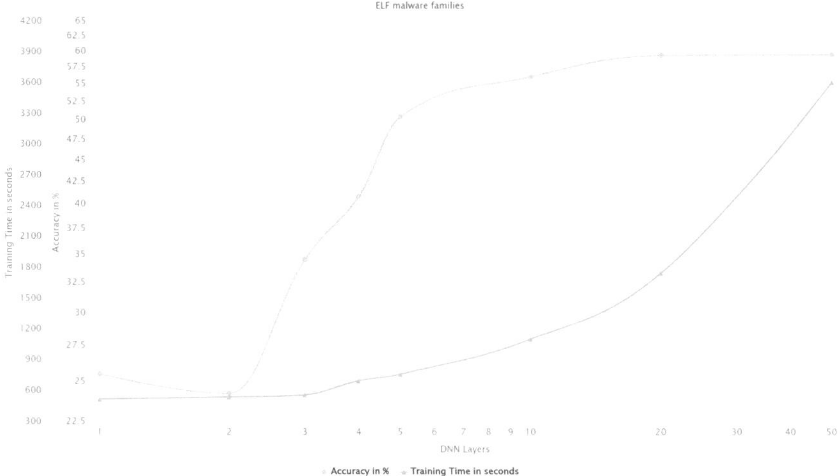

Fig. 7 Dependency of DNN accuracy and training time for 442 malware families on number of layers (plotted in RapidMiner)

of this work is corresponding ELF malware dataset containing 10,574 malware files labeled into 19 malware types and 442 malware families. Following this, we extracted 30 static features that can be used for malware classification using state-of-the-art machine learning methods. We can see that to achieve a reasonable classification accuracy and concise model size on such non-trivial dataset one needs to use deep neural networks with a large number of layers. However, additional study of static and possibly dynamic behavioral features is required to understand the better multinomial classification of Linux ELF malware binaries.

Acknowledgements Authors would like to acknowledge NTNU Malware Lab for the support and VirusTotal Intelligence Premium Services for access to data. Moreover, we are grateful to VirusShare for priceless data collection and contribution toward malware analysis community. Finally, the utilization of PEframe for ELF files processing gave valuable results making this research possible.

References

1. A brief history of linux malware. Accessed 24 June 2020.
2. Malware names. https://docs.microsoft.com/nb-no/windows/security/threat-protection/intelligence/malware-naming. Accessed 06 Feb 2020.
3. Naming scheme - caro - computer antivirus research organization. www.caro.org/naming/scheme.html. Accessed 07 Feb 2020.
4. Virusshare. https://www.VirusShare.com/. Accessed 17 Feb 2020.
5. Virustotal. https://www.virustotal.com/. Accessed 17 Feb 2020.
6. Virustotal statistics. https://www.virustotal.com/en/statistics/. Accessed 04 Feb 2020.
7. Vxheaven. Accessed 22 June 2020.
8. The short life and hard times of a linux virus. 2000. Accessed 24 June 2020.

9. Shodan.io. 2020. https://www.shodan.io. Accessed 24 June 2020.
10. Amer, Eslam, and Ivan Zelinka. 2020. A dynamic windows malware detection and prediction method based on contextual understanding of api call sequence. *Computers & Security* 92: 101760.
11. Bai, Jinrong, Yanrong Yang, Mu Shiguang, and Yu Ma. 2013. Malware detection through mining symbol table of linux executables. *Information Technology Journal* 12 (2): 380.
12. Boelen, Michael. 2019. The 101 of elf files on linux: Understanding and analysis. https://linux-audit.com/elf-binaries-on-linux-understanding-and-analysis/.
13. cbtnuggets. Why linux runs 90 percent of the public cloud workload. Accessed 12 June 2020.
14. TIS Committee et al. 1995. Tool interface standard (tis) executable and linking format (elf) specification version 1.2.
15. Coull, Scott, and Christopher Gardner. 2018. What are deep neural networks learning about malware? https://www.fireeye.com/blog/threat-research/2018/12/what-are-deep-neural-networks-learning-about-malware.html.
16. Cozzi, Emanuele, Mariano Graziano, Yanick Fratantonio, and Davide Balzarotti. 2018. Understanding linux malware. In *2018 IEEE symposium on security and privacy (SP)*, 161–175. IEEE.
17. Das, Ankush. 2018. Reasons why linux is better than windows. 11 (3): 11. https://itsfoss.com/linux-better-than-windows/.
18. Denzer, Thilo, Andrii Shalaginov, and Geir Olav Dyrkolbotn. 2019. Intelligent windows malware type detection based on multiple sources of dynamic characteristics. *NISK Journal*, 12.
19. Duncan, Rory, and Z. Cliffe Schreuders. 2019. Security implications of running windows software on a linux system using wine: A malware analysis study. *Journal of Computer Virology and Hacking Techniques* 15 (1): 39–60.
20. Eset. Linux and malware: Should you worry? Accessed 12 June 2020.
21. Granneman, Scott. 2020. Linux vs. windows viruses, 2003. Accessed 24 June 2020.
22. Hofmann, Frank. 2019. Understanding the elf file format. https://linuxhint.com/understanding_elf_file_format/.
23. Hwang, Chanwoong, Junho Hwang, Jin Kwak, and Taejin Lee. 2020. Platform-independent malware analysis applicable to windows and linux environments. *Electronics* 9 (5): 793.
24. Jayasinghe, Keshani, and Guhanathan Poravi. 2020. A survey of attack instances of cryptojacking targeting cloud infrastructure. In *Proceedings of the 2020 2nd Asia pacific information technology conference*, 100–107.
25. Kolias, Constantinos, Georgios Kambourakis, Angelos Stavrou, and Jeffrey Voas. 2017. Ddos in the iot: Mirai and other botnets. *Computer* 50 (7): 80–84.
26. Kononenko, Igor, and Matjaz Kukar. 2007. *Machine learning and data mining*. Sawston: Horwood Publishing.
27. Liu, Yingying, and Yiwei Wang. 2019. A robust malware detection system using deep learning on api calls. In *2019 IEEE 3rd information technology, networking, electronic and automation control conference (ITNEC)*, 1456–1460. IEEE.
28. Markel, Zane, and Michael Bilzor. 2014. Building a machine learning classifier for malware detection. In *2014 second workshop on anti-malware testing research (WATeR)*, 1–4. IEEE.
29. Noyes, Katherine. 2010. Why linux is more secure than windows. *Luettavissa:* http://www.pcworld.com/article/202452/why_linux_is_more_secure_than_windows.html.*Luettu*, vol 14, 2014.
30. Sebastián, Marcos, Richard Rivera, Platon Kotzias, and Juan Caballero. 2016. Avclass: A tool for massive malware labeling. In *International symposium on research in attacks, intrusions, and defenses*, 230–253. Springer.
31. Shahzad, Farrukh, and Muddassar Farooq. 2012. Elf-miner: Using structural knowledge and data mining methods to detect new (linux) malicious executables. *Knowledge and Information Systems* 30 (3): 589–612.
32. Shalaginov, Andrii, Sergii Banin, Ali Dehghantanha, and Katrin Franke. Machine learning aided static malware analysis: A survey and tutorial. In *Cyber threat intelligence*, 7–45. Cham: Springer.

33. Shalaginov, Andrii, and Katrin Franke. A deep neuro-fuzzy method for multi-label malware classification and fuzzy rules extraction. In *2017 IEEE symposium series on computational intelligence (SSCI)*, 1–8. IEEE.
34. Shalaginov, Andrii, Lars Strande Grini, and Katrin Franke. 2016. Understanding neuro-fuzzy on a class of multinomial malware detection problems. In *International joint conference on neural networks (IJCNN) 2016*, 684–691. Research Publishing Services.
35. Statcounter. Operating system market share worldwide: May 2019–May 2020. Accessed 11 June 2020.
36. Sun, Zhi, Zhihong Rao, Jianfeng Chen, Rui Xu, Da He, Hui Yang, and Jie Liu. 2019. An opcode sequences analysis method for unknown malware detection. In *Proceedings of the 2019 2nd international conference on geoinformatics and data analysis*, 15–19.
37. Tarek, Radah, Saadi Chaimae, and Chaoui Habiba. 2020. Runtime api signature for fileless malware detection. In *Future of information and communication conference*, 645–654. Springer.
38. Tobiyama, Shun, Yukiko Yamaguchi, Hajime Shimada, Tomonori Ikuse, and Takeshi Yagi. 2016. Malware detection with deep neural network using process behavior. In *2016 IEEE 40th annual computer software and applications conference (COMPSAC)*, vol 2, 577–582. IEEE.
39. Webster, George D, Bojan Kolosnjaji, Christian von Pentz, Julian Kirsch, Zachary D Hanif, Apostolis Zarras, and Claudia Eckert. 2017. Finding the needle: A study of the pe32 rich header and respective malware triage. In *International conference on detection of intrusions and malware, and vulnerability assessment*, 119–138. Springer.
40. Malware Wiki. Linux. Accessed 12 June 2020.
41. Yang, June Ho, and Yeonseung Ryu. 2015. Toward an efficient pe-malware detection tool. *Advanced Science and Technology Letters* 109: 14–17.
42. Zhoghov, Victor. 2017. The ransomware "Petya" as a challenge to the cybersecurity of Ukraine, main factors of spreading this virus in the focus of Ukraine, the steps taken by the authorities to combat this phenomenon and suggest ways to improve such activities using experience of other countries. PhD thesis, Victor Zhoghov The ransomware "Petya" as a challenge to the cybersecurity of ….
43. Zhou, Huan. 2018. Malware detection with neural network using combined features. In *China cyber security annual conference*, 96–106. Springer.

Fast and Straightforward Feature Selection Method

A Case of High-Dimensional Low Sample Size Dataset in Malware Analysis

Sergii Banin

Abstract Malware analysis and detection is currently one of the major topics in the information security landscape. Two main approaches to analyze and detect malware are *static* and *dynamic* analyses. In order to detect a running malware, one needs to perform dynamic analysis. Different methods of dynamic malware analysis produce different amounts of data. The methods that rely on low-level features produce very high amounts of data. Thus, machine learning methods are used to speed up and automate the analysis. The data that is fed into machine learning algorithms often requires preprocessing. Feature selection is one of the important steps of data preprocessing and often takes significant amount of time. In this paper, we analyze the Intersection Subtraction (IS) feature selection method that was first proposed and used on a high-dimensional dataset derived from the behavioral malware analysis. In our work, we assess its computational complexity and analyze potential strengths and weaknesses. In the end, we compare Intersection Subtraction and Information Gain (IG) feature selection methods in terms of potential classification performance and time complexity. We apply them to the dataset of memory access patterns produced by malicious and benign executables. As a result, we found that the features selected by IS and IG are very different. Nevertheless, machine learning models trained with IS-selected features performed almost as good as those trained with IG-selected features. IS allowed to achieve the classification accuracy of more than 99%. We also show, the IS feature selection method is faster than IG what makes it attractive to those who need to analyze high-dimensional datasets.

The research leading to these results has received funding from the Center for Cyber and Information Security, under budget allocation from the Ministry of Justice and Public Security of Norway.

S. Banin (✉)
Department of Information Security and Communication Technology, NTNU, Gjøvik, Norway
e-mail: sergii.banin@ntnu.no

M. Stamp et al. (eds.), *Malware Analysis Using Artificial Intelligence and Deep Learning*, https://doi.org/10.1007/978-3-030-62582-5_18

1 Introduction

Today many researchers from different research areas have to deal with big amounts of data. Various statistical methods are used to process and understand data that is too big or complex for human analysis. Part of these methods are called *machine learning*: "the automatic modeling of underlying processes that have generated the collected data" [22]. Currently, machine learning is one of the most used approaches when there is a need to predict certain qualities of objects or events. Machine learning algorithms can be divided into supervised (classification and regression) and unsupervised (clustering). In this paper, we focus on the classification: prediction of a *class* (type) of a sample based on its *features* (properties). Machine learning is widely used in different fields such as medicine, biology, manufacturing [24], or information security [4, 31]. In information security, machine learning is extensively used in production and research, as the amounts of data that need to be processed are enormous. Especially, machine learning is actively used for malware analysis and detection. According to AV-TEST Institute, there are more than 350,000 new malware samples detected every day [3]. The developers of the anti-virus solutions and researchers work on finding a way to detect malware without having to search through the entire database of already known malware. Moreover, they try to find methods that allow detecting previously unknown malware. The common practice is to find certain characteristics that are common to many malware samples. As the number of malware is very big and growing [3], the machine learning methods are used to deal with the emerging amount of data. Machine learning methods rely on *features*: properties of objects that are being studied. There are two main types of features that can be extracted from malware: static and dynamic. Static features are extracted directly from the malicious file without a need to launch it. Static features are relatively easy to extract, but at the same time it is easier to change them with a use of obfuscation or encryption [1]. However, malware becomes malicious only after it has been launched. The features that occur after the launch of malware are called *dynamic* or *behavioral* features. We can divide dynamic features into high- and low-level features [6]. File and network activity, API [2], and system calls are some of the high-level features, while opcodes, memory access operations [38], or hardware performance counters are considered to be low-level ones. We name dynamic features that emerge from the system's hardware as the low-level features [5, 21, 25]. To represent a certain behavioral event with low-level features, we need to record and process a significantly bigger amount of data. For example, to describe an API call on the high level, we only need its name and arguments passed to it on the call. However, if we decide to record a sequence of opcodes or memory access operations invoked by the API call we'll end up with hundreds if not thousands of events. In this paper, we address a problem that arises from the number of low-level features one needs to record and process while doing dynamic malware analysis.

While machine learning provides good opportunities for automation and analysis, the data that is used by machine learning algorithms has to be preprocessed. Various methods of data preprocessing are described in the literature: discretization of continuous features, attribute binarization, the transformation of discrete features into continuous, dimensionality reduction, and so on [22]. The first three of the aforementioned methods are mostly used when the chosen machine learning algorithm works only with a certain type of data. For example, the Naive Bayes classifier needs discrete data to provide a useful outcome. On its turn, dimensionality reduction is often needed, when the amount of features in the dataset is too big. Having too many features can result in increased model training times and model overfitting. There are several ways to reduce dimensionality: feature subset selection, feature extraction, and Principal Components Analysis (PCA) [22]. Feature extraction is aimed at finding a set of new features that are constructed as a function of original features. On its turn, PCA finds a new coordinate system with a focus on making the axes aligned with the highest variance of the data. These methods, however, make it harder to analyze the results achieved by the machine learning model: it is sometimes important to understand which features contribute the most toward the classification performance of a model. In such cases, in order to reduce the dimensionality, one may apply feature (subset) selection. With feature selection it is possible to select a certain amount of *best* features based on a certain feature quality measure while keeping the original features intact.

Feature selection is aimed at the dimensionality reduction. Ironically, when the amount of features becomes too big (for example, millions as in [8] or [5]) the feature selection becomes a very computationally intense task. The datasets where the number of features is much bigger than the number of learning samples are called high-dimensional low (small) sample size (HDLSS/HDSSS) datasets. Sometimes there are so many features [8] that commonly used machine learning packages simply cannot handle such datasets. Storing such a dataset in the single file or database table becomes a problem as well. Thus, the use of the common machine learning packages becomes impossible since they require data to be stored in one piece. On its turn, developing and implementation of a custom machine learning package can take more time than actual data collection and be a hard task for the researchers that don't have enough expertise in software development.

In this paper, we focus on the feature selection method that was developed and used in [8] to detect malware based on the memory access patterns. In [8], the dataset contained almost six millions of binary features and 1204 samples divided into two classes. The features represented sequences of memory access operations generated by malicious and benign software. The feature took value *1* if it was generated by a sample, and *0* if not. Utilized feature selection method was aimed at removing those features that are *present* (take value 1) in the samples of both classes. Thus, it is named Intersection Subtraction (IS) feature selection method. This method helped authors of [8] to reduce feature space from 6 M of features to 800. With the use of selected features, it became possible to train a classification model that achieved 98% classification accuracy for the two-class dataset. In this paper, we provide an additional analysis of the IS feature selection method and discuss its advantages and disadvantages. We also compare its performance with an Information Gain [22] feature selection method in a similar malware detection problem. We run our tests on the newer and larger dataset of malicious and benign executables. We show how machine learning models trained with features selected by IS feature selection perform compared to those selected by IG.

The remainder of the paper is arranged as follows. In Sect. 2, we describe the problem and provide an overview of related articles. In Sect. 3, we describe the IS feature selection method, theoretically assess its strengths and weaknesses, and explain the context in which IS might be used. In Sect. 4, we describe our experimental setup, compare feature sets selected by IS and IG, and train machine learning algorithms with the use of selected features. In Sect. 5, we discuss our findings and outline the future work. In Sect. 6, we summarize our findings and provide conclusions.

2 Background

In this section, we describe the problem area and provide an overview of the papers related to HDLSS datasets and feature selection.

2.1 Problem Description

While talking about the optimal size of the dataset to be used in machine learning model training, different authors consider different dataset sizes to be optimal. The size of the dataset consists of a number of samples and features. In various sources [15, 26], one can find suggestions that a minimal amount of samples for training should be between 50 and 80, while 200 and more samples are expected to bring increased accuracy and significantly smaller error rates. Other authors have shown that it is important to have at least 20–30 samples per class [11]. When talking about the number of features it is generally considered that the fewer features there are in the dataset, the better it is for machine learning algorithm [5, 7, 8, 22]. Some authors advise utilizing *the rule of 10*: in order to train a model with a good performance, one needs to have ten times more samples than the number of features [23]. However, in some cases, the number of features can be significantly higher than the number of learning samples. This may happen due to the context of the research and the nature of data. For example, in [8], the authors describe a novel malware detection approach. They record memory access operations performed by malicious and benign executables, split them into n-grams of various sizes, and use those n-grams as features for training the machine learning models. Each feature could take value *1* or *0* if the n-gram represented by the feature was or was not generated by the sample, respectively. The sequence of memory access operations is a sequence of *Reads* (R) and *Writes* (W). In their work, authors record a first million of memory access operations performed by each executable after it was launched. Afterward, the sequence of memory access operations is being split into the set of overlapping n-grams of a size 96. Since memory access operations take only two possible values (R and W), the potential feature space of the abovementioned approach is 2^{96} if a sequence of memory access operations would be completely random. However, as the same authors mention in their next paper [5], the memory access operations are not random. Thus, in [8], their initial feature space is "only" about 6 M of features. They had 1204 samples divided into two classes. This can be considered a good sample size based on what was suggested in [11, 15]. However, the amount of features generated under such experimental design makes it impossible to follow "the rule of 10." A straightforward approach in such conditions could be to simply use all the data for training the machine learning model. However, just the storage of a complete dataset from [8] would take more than 6 GB of space. Popular machine learning frameworks such as Weka [19] or Scikit-learn [26] are not suited to load and handle so much data. This shows a need for dimensionality reduction. In the works similar to [8] or [5], it is important to keep the original features in order to be able to interpret results. For example, having the results from [8] it might be possible to understand which memory access patterns make malicious behavior distinctive from the benign behavior. Thereby, dimensionality reduction methods such as feature extraction or PCA are not applicable in such cases. On its turn, feature selection can help to select a subset feature without hindering their original state.

Feature selection methods can be divided into several categories: filter, wrapper, and embedded methods. Filter methods choose features based on a certain quality measure such as Pearson correlation, chi-square, mutual information, and so on. Wrapper methods choose features based on the classification performance of the target machine learning model trained with the use of those features [33]. Wrapper methods are very computationally intense since for every possible feature subset there is a need to train and test the machine learning model. Embedded methods, as the name states, are embedded in the machine learning algorithms. Algorithms such as decision trees [22] perform feature selection simultaneously with model training. However, the computational overhead is higher than one of the filter methods and such algorithms are susceptible to overfitting [9] and are not suitable for high-dimensional data [33]. So for the research similar to [8], the most suitable approach for dimensionality reduction will be a filter-based method. In the case of (very) high-dimensional data, it is crucial to have a feature selection method with the lowest possible computational overhead. The perfect feature selection method will have a computational complexity of $O(n)$ that is linear to a number of features n. But such a method does not exist, since filter methods are aimed to select features that *represent* classes (and consequently samples) in the best possible way [22]. Thereby, while choosing the feature selection method to work on the high-dimensional dataset it is desirable to choose a method with the computational complexity of $O(mn)$ where m is the number of samples in the dataset.

The use of different filter-based feature selection methods is described in various papers. Information Gain [7, 24], correlation-based feature selection [5, 17], and ReliefF [17] are some of the common feature selection methods. *Information Gain* (IG) ranks features based on entropy in respect to the classes and can be described as "the amount of information, obtained from the attribute A, for determining the class C" [22]. Basically, in order to perform a feature selection based on IG, one have to calculate probabilities of an attribute to take certain values and relevant class-conditional probabilities. This results in a computational complexity around $O(mn)$, where n is the amount of features and m is the amount of samples. *Correlation-based Feature Selection method* (CFS) was proposed in [18] and is aimed at selecting the subset of features that have a high correlation to the class but low correlation between each other. By doing so it is possible to find a subset of features with minimal redundancy. The problem with this method is that it requires to calculate a pairwise correlation matrix between all of the n features and m classes which requires $m((n^2 - n)/2)$ operations. The feature selection search could require an additional $(n^2 - n)/2$ operation in a worst-case scenario. With a potential computational complexity of $O(m((n^2 - n)/2) + (n^2 - n)/2)$ the use of CFS for high-dimensional data becomes very problematic. For example, just storing of correlation matrix needed for 6 M of features in [8] would require at least 18 TB of space. Thus, in order to apply CFS on high-dimensional datasets, it might be useful to first reduce a feature space with another, less computationally intense, feature selection method, and only after apply the CFS [5]. *ReliefF* ranks features based on their ability to separate close samples from the different classes [22]. In order to perform feature selection with ReliefF, it is first important to calculate a distance matrix between all samples. The resulting computational complexity of the method can be roughly estimated as $O(n((m^2 - m)/2))$ that is almost *m/2* times more than the one of the IG. Having a large n makes the use of ReliefF less favorable than IG.

Based on the assumptions about the computational complexity of the abovementioned feature selection methods one can make a conclusion that IG might be one of the best choices when it comes to the high-dimensional datasets. The problem is that even the feature selection methods with $O(mn)$ complexity become slow with the large numbers of n. And as we mentioned above, common machine learning packages are not suitable to work with big datasets. Thus, a researcher who needs to perform feature selection on such datasets is forced to develop a custom implementation of feature selection algorithm with regard to the data in interest. In this case, inefficient implementation of the common feature selection algorithm may result in significant use of time and even inability to obtain results (e.g., due to the lack of virtual memory). For example, the Information Gain of a feature is calculated with the following formula:

$$Gain(A) = -\sum_k p_k \log p_k + \sum_j p_j \sum_k p_{k|j} \log p_{k|j}$$

where p_k is the probability of the class k, p_j is the probability of an attribute to take j_{th} value, and $p_{k|j}$ is the conditional probability of class k given j_{th} value of an attribute [22]. This shows that it is necessary to "count" how many times each attribute takes a certain value in total and when a certain class is given. Let's rewrite previously mentioned computational complexity of IG as $O(nT_{qmeaureIG})$ where $T_{qmeaureIG} = f(m)$ is the computational time needed to calculate the quality measure (Information Gain in this case) of a feature. We will need $T_{qmeaureIG}$ later to show that the IS feature selection method works faster than IG, which is important when working with high-dimensional datasets. Thus, it is easy to see that the inefficient implementation of

IG can significantly increase the time needed to obtain the results. As we will later show, it is possible to overcome this problem with a Intersection Subtraction feature selection method.

2.2 Literature Overview

In this subsection, we refer to papers where authors addressed the problems related to HDLSS datasets and feature selection on them. In [12], authors outline both *curses* and *blessings* of high dimensionality. By blessings of dimensionality, they mention the phenomenon of measure concentration and the success of asymptotic methods. While talking about curses of dimensionality authors outline several areas where they can occur: optimization, function approximation, and numerical integration. They also stress attention to the fact that many "classical" statistical methods are based on the assumption that the amount of features n is less than the amount of samples m, while $m \to \infty$. However, these methods may fail if $n > m$, especially when $n \to \infty$. Other authors in [14] outline the following challenges of high dimensionality: "(i) high dimensionality brings noise accumulation, spurious correlations, and incidental homogeneity; (ii) high dimensionality combined with large sample size creates issues such as heavy computational cost and algorithmic instability" [14]. As well as authors of [12] outline that traditional statistical methods may fail when used on high-dimensional data. The authors of [40] review the performance and limitations of several common classifiers such as Naive Bayes, linear discriminant analysis, logistic regression, support vector machines, and distance weighted discrimination in the case of two-class classification problem on HDLSS datasets. They also say that if the number of features $n \to \infty$ and both classes are from the same distribution "the probability that these two groups are 'perfectly' separable converges to 1" [40]. In simple words, it means that with a large enough amount of features it should be possible to construct a set of rules (build a classifier) that will perfectly fit (overfit) the training data. This fact outlines the importance of thorough feature selection. It will improve the capability of machine learning algorithms to create models with good *generality* and *interpretability*. The model with good generality is the model that is capable of generalizing over the dataset; such model would not be significantly changed if the number of samples in the dataset is slightly increased/decreased [40]. A model with good interpretability makes the analysis of the model itself easier. The fewer features are involved during the training the easier it is to analyze the obtained model. For example, authors of [5] underline the importance of the fact that having 29 features instead of 6 M or 15 M helps in the understanding of the underlying processes. They performed multinomial (10 class) malware classification with the use of features constructed from memory access patterns. Similar to [8], they used memory access 96-grams as features. Such feature, if found to be important in the classification, cannot be directly understood by a human analyst. Thus, in [6], they made an attempt to interpret memory access sequences with more high-level system events (API calls). Such analysis would be much harder if they had millions of features instead of 29.

Various authors addressed the problem of feature selection on HDLSS datasets more specifically. For example, same authors in [36, 37] present possible improvements to the PCA in HDLSS cases. In [36], they propose a way to estimate singular value decomposition of the cross data matrix. Later, in [37], authors explore the impact of the geometric representation of HDLSS data on a possibility to converge the dataset to an n-dimensional surface. The authors of [13] propose a nonlinear transformation of HDLSS data. They showed, how transformation based on inter-point distances helps to increase final classification accuracy. In [39], the authors propose a hybrid feature selection method that is based on ant lion optimization and gray wolf optimization methods (ALO-GWO). They evaluate the performance of the proposed method on several HDLSS datasets. The authors show that the ALO-GWO feature selection method provides a good balance between the performance of models and the ability to reduce a feature space. The abovementioned papers addressed the problem of feature selection on HDLSS. However, the number of features in the dataset used in those papers rarely exceeded several tens of thousands (e.g., in [39]). On their turn, authors of [16] during the test of their feature selection method used a dataset with more than 3 M of features. In their work, they proposed a feature selection method based on bijective soft sets (BSSReduce). They claim that the computational complexity of the method is $O(m)$ where m is the number of samples. This might have been a perfect feature selection method for the HDLSS datasets. However, after reviewing the provided algorithms, it looks like their approach relies on the precomputed bijective soft sets that have to contribute to the computational complexity as well. Nevertheless, the results of testing the BSSReduce on the several HDLSS datasets showed that it is capable of significant dimensionality reduction while keeping a competitive level of the trained model performance. It could be useful to compare BSSReduce with our method, and unfortunately authors of BSSReduce did not provide the source code of their tool. An approach different from the previously mentioned papers is present in the [5]. The authors of the paper did not focus on feature selection. However, they needed to reduce feature space in two HDLSS datasets from 6 M to 15 M of features. Authors said that "models should be simple enough" [5] to make their analysis easier. In order to reduce a large feature space, they performed feature selection in two steps. On the first step, they used custom implementation of Information Gain feature selection to reduce feature space to 50 K and fewer features. On the second step, they took the best 5 K feature selected by IG and used them in CFS implementation from Weka. This resulted in 29 features selected by CFS. The models trained with just 29 features performed almost as good as a model trained on 5 K and more features. For Naive Bayes and support vector machine algorithms, smaller feature set even allowed to increase the performance of trained models. Such approach has its own limitations. CFS is aimed at selecting features that are not correlated with each other. However, since the first feature selection step utilizes IG, there is no guarantee that features passed to the CFS does not have a strong mutual correlation. But as we mentioned above, running CFS on the HDLSS dataset with millions of features requires enormous computational resources and sometimes impossible.

3 Intersection Subtraction Selection Method

In this section, we describe the IS feature selection method and evaluate its strengths and weaknesses.

3.1 The Context

Before describing the Intersection Subtraction feature selection method we need to describe a context under which its use becomes meaningful. This method was developed during the research described in [8]. The task was to detect malware based on the memory access traces. To do this, malicious and benign executables were launched together with custom-built Intel Pin [20] tool. The raw data consisted of the first 1 M of memory access operations performed by each executable. The sequences contained *W* for each write operation and *R* for each read operation performed by an executable. These sequences were later divided into a set of overlapping n-grams of various sizes. For example, a sequence *[WWRWRR]* of a length 6 can be divided into the set of 4-grams in the following way: *[WWRW, WRWR, RWRR]*. The n-grams were directly used as features for machine learning model training. Each feature got value *1* if the corresponding n-gram was generated by the sample regardless of the number of times it was encountered in the trace of a certain sample. In other cases, the feature got value *0*. As the goal of the [8] was to be able to detect malware, it is possible to state that features that obtain *1* (are present within a certain class) pose greater interest. Such approach allows to state that *presence* of certain memory access n-grams is the sign of malicious behavior. The dataset from [8] was nearly balanced and samples were divided into two classes. So the context of the use of the proposed feature selection method is the following: two-class classification problem on a balanced dataset with binary features.

3.2 Feature Selection Algorithm

The feature space in [8] was around 6 M of unique memory access n-grams of a size 96. By the time of writing, authors were not able to implement any common feature selection method (for example, IG) to operate on such dataset. Thus, they implemented the following feature selection method. It includes the following steps:

1. Construct two vectors of features for each class. The feature is included in the vector of the class if the corresponding memory access n-gram was generated by a sample from this class.
2. Having two vectors constructed, remove from them features that are present in both vectors. Having this done we obtain two vectors of class-unique features. In other words, we *subtracted an intersection* of two feature sets from both of them.
3. Decide on the size of the final feature set *k*.
4. From each of the class-unique features vectors, select *k/2* features with the highest class-wise frequency. A class-wise frequency is the proportion of samples within the class that generate a corresponding memory access n-gram.
5. Use the *k* selected features to construct the final dataset with reduced dimensionality.

The resulting dataset is later used to build machine learning models. The operation performed in Step 2 is quite similar to the symmetric difference of two sets. However, we prefer to say that we subtract intersection from both sets, as we need those sets to be separated until the last step. It is also worth mentioning that having an intersection of two feature sets allows to explore features that fell into it. It might be useful for additional analysis of the results [8].

3.3 Computational Complexity

Let's discuss the potential computational complexity of Intersection Subtraction (IS) feature selection. As data is already labeled (samples divided into two classes), the feature vectors from the *Step 1* are ready from the beginning. *Step 2* requires finding an intersection of two sets. Imagine we have two sets A and B with cardinality of a and b, respectively. In order to find the intersection of A and B, we need to compare all elements of set A with all elements of set B. Such operation will have a computational complexity of $O(ab)$. Let's denote the intersection of A and B as $C = A \cap B$ with cardinality c. Subtracting the elements of C from A and B, similarly to the previous operation, will have the computational complexity of $O(ac + bc)$. The resulting computational complexity of $O(ab + ac + bc)$ may look quite high already, since both a and b are large in case of HDLSS datasets. However, the real implementation of IS feature selection with the use of Python programming language shows that execution of the *Step 2* does not take significant time (see Sect. 4). First of all, according to [28], subtraction *A-C* (set difference) will have computational complexity of $O(a)$. So we can already rewrite previously mentioned computational complexity of Step 2 with $O(ab + a + b)$. Moreover, if we are not interested in the intersection C itself, we can utilize two operations *A-B* and *B-A* in order to obtain sets of class-unique features. Complexity of such approach will be $O(a + b)$. *Step 4* requires the calculation of class-wise frequencies of the features. In our particular case, when features are binary, we only need to count how many samples from each class has value *1* of a certain feature. Step 4 will then have $O((a-c)m+(b-c)m)$ computational complexity. Here, m is the number of samples in the dataset, *a-c* is the amount of class-unique features from set A and *b-c*—from set B. It is also worth mentioning that Step 4 can be optimized. Let's assume that the dataset is perfectly balanced, so we have two classes with *m/2* samples. Since our IS feature selection is aimed on finding class-unique features, we can only search for *1s* among *a-c* and *b-c* features of *m/2* samples of each class. So Step 4 can be optimized to have a complexity of $O((a-c)m/2+(b-c)m/2)$. Let's now try to assess the overall computational complexity of the IS feature selection. Let us have the initial amount of features $a+b = n$ and m samples. The amount of features from intersection c is normally smaller than both a and b (here we assume that $A \not\subset B$ and $B \not\subset A$). Having this we can conclude that the complexity of *Step 2* $O(ab+a+b)$ after substitution will be smaller than $O(n^2)$ for all $a > 1$. On its turn, the complexity of *Step 4* $O((a-c)m/2+(b-c)m/2)$ should be smaller than $O(mn)$. The resulting complexity of $O(ab+a+b+(a-c)m/2+(b-c)m/2)$ should be smaller than $O(n^2+mn)$. The feature selection method where the upper boundary of computational complexity is described with n^2 is not what we outlined in Sect. 2 as a good feature selection method for HDLSS dataset. Let's now make a substitution similar to the one we made in Sect. 2. First, let's substitute m with $T_{qmeaureIS} = g(m)$ which is the time needed to calculate class-wise frequency of a feature. Second, the time T_{in} needed to find whether a certain feature from one set is present *in* another set (to find an intersection or to subtract these features from the set) is relatively small. Thus, the updated computational complexity of IS feature selection will be smaller than $O((nT_{in})^2+nT_{qmeaureIS})$ which can be smaller than $O(nT_{qmeaureIG})$ of IG. We will prove this in Sect. 4.

3.4 *Theoretical Assessment*

In this subsection, we discuss potential outcomes of the IS feature selection. As we already mentioned, IS feature selection is potentially faster than a more common IG feature selection. This makes IS attractive for the high-dimensional datasets. However, speed comes with a price. Let's look at the potential disadvantages of IS feature selection. As we described at the beginning of this section, the use of this method makes more sense when we are interested in finding features the *presence* of which poses particular interest. However, it might happen that in the dataset there will be no class-unique features. In other words, it will be impossible to say that if a certain feature of a sample takes value 1, then this sample belongs to a certain class. In such case, it will be impossible to find an intersection of two feature sets. The other problem is potential information loss due to intersection removal. Imagine we have a dataset that is represented in Table 1. It has four features and four samples labeled into two classes C1 and C2. IS feature selection will remove features f1 and f3 since they obtain value 1 (are present) in both classes. The remaining features f2 and f4 will not allow us to generate a rule that will be able to distinguish between samples s2 and s4. This example is quite small, but on the larger dataset removing a feature

Table 1 Sample dataset 1

	f1	f2	f3	f4	
s1	1	1	1	0	C1
s2	1	0	0	0	C1
s3	1	0	1	1	C2
s4	0	0	1	0	C2

that takes value *1* in all samples of one class and only in one sample of another class can lead to the inability of building a model with good performance. Such feature would be most likely selected by IG feature selection. The last disadvantage of the IS feature selection is potentially poor performance on the multinomial datasets. If we increase the number of classes we will end up in the situation of growing intersection size. In such case, the IS will remove more features from the feature space resulting in increased information loss. We begin with the description of our dataset and experimental environment. Later, we explain the basics of memory access operations and explain the way we record and process the data.

4 Experimental Evaluation

In this section, we describe experimental evaluation of the IS feature selection method. We show how IS feature selection can be applied for malware detection. During experimental evaluation we compare performance of features selected by IS and IG. In Fig. 1, we depict general data flow of our experiments. We start by recording memory access operations produced by benign and malicious executables. After, we split sequences of memory access operations into n-grams. Then we apply feature selection methods to select best features (n-grams). In the end, we use these features to train machine learning models and compare performance of the models trained with a use of features selected by different feature selection methods. Before presenting the results achieved by machine learning models, we show the experimental time complexity of the IS and IG feature selection methods. We also check how similar the feature sets selected by different methods are.

We now proceed with the description of our dataset, experimental environment, and the way we collect and process the data.

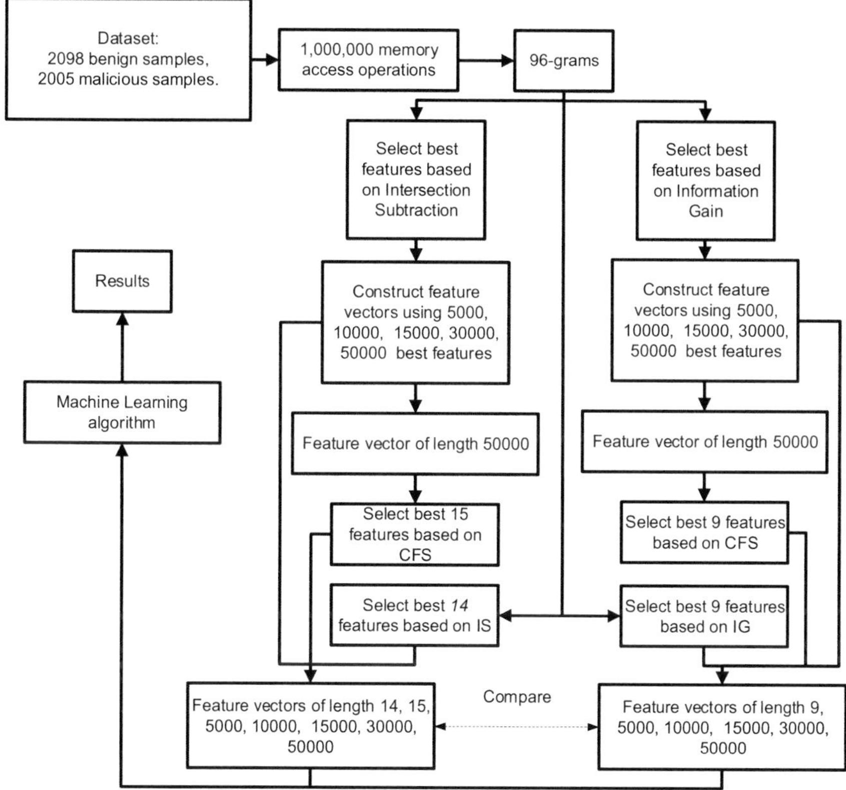

Fig. 1 The flow of data collection and feature selection

4.1 Dataset

In this work, we use dataset similar to the one used in [7]. It consists of 2098 benign and 2005 malicious Windows executables. Malicious executables were downloaded as part of *VirusShare_00360* pack available at VirusShare [35]. Malicious samples belong to the following malware families: Fareit, Occamy, Emotet, VBInject, Ursnif, Prepscram, CeeInject, Tiggre, Skeeyah, and GandCrab. According to the VirusTotal [32] reports, our samples were first seen (first submission date) between March 2018 and March 2019. Benign executables are the real software downloaded from Portable Apps [27] in September 2019.

4.2 Experimental Environment

In order to perform dynamic malware analysis, we need to avoid the influence of any environmental changes, so that all executables are launched in similar conditions. To achieve this we used an isolated Virtual Box virtual machine (VM) with Windows 10 guest operating system. VMs were launched on the Virtual Dedicated Server (VDS) with 4-core Intel Xeon CPU E5-2630 CPU running at 2.4 GHz and 32 GB of RAM with Ubuntu 18.04 as a main operating system.

4.3 Memory Access Operations

The executables used on Windows operating systems are compiled into the files in PE32 format. Files of PE32 format contain *header* and *sections*. The header contains the metadata that is used by operating system in order to properly load an executable into memory and prepare all the necessary resources. The sections contain information about imported and exported functions, resources, data, and the executable code. The executable code is stored in the binary form which can be represented as *opcodes*. Opcodes (or assembly commands) are the basic instructions that are executed by the CPU. Execution of some instructions will not require memory access. For example, execution of *MOV EAX,EBX* opcode will not result in memory access, since data is being moved between registers in the CPU. At the same time, *MOV EDI, DWORD PTR [ebp-0x20]* will generate a *Read (R)* memory access, since the data has to be read from the memory. On its turn, the *ADD DWORD PTR [EAX],ECX* will require *Reading (R)* the value from the memory location addressed by *[EAX]* and then *Writing (W)* the result of the addition to the memory. The sequences of opcodes were previously proven to be a source of effective features for malware detection [10, 30, 34]. When the sequence of opcodes is executed it generates a sequence of memory access operations. Two previous statements allow for memory access sequences to be a potential source of features for malware detection [8]. Under our experimental design we use only the type of memory access operation: *R* for read and *W* for write. We do not use the value that is transferred to or from the memory as well as the address of the memory region in use.

4.4 Data Collection

Each malware sample was launched on the clean snapshot of VM. During the execution of each sample, we recorded the first million of memory access operations produced after the launch. This was done with the help of a custom-built Intel Pin [20] tool that was launched together with the sample inside the VM. The VM had all built-in anti-virus features disabled to make malware run properly and also because they kept interrupting the work of Intel Pin. The automation of VM and data collection were performed with the help of Python 3.7 scripts.

The memory access traces were first stored in the separate files. After, they were split into the sequence of overlapping n-grams of the size 96 (96-grams). We choose n-gram size (as well as the amount of recorded memory accesses) based on the conclusions of their effectiveness drawn in [8]. The n-grams of memory access operations for each sample are then stored in the MySQL table. This table took 28.5 GB of storage.

4.5 Feature Selection and Machine Learning Algorithms

We implemented IS feature selection algorithm with Python. The custom implementation of IG feature selection algorithm was similar to one in [7]. That implementation allows to run feature selection in multiple threads, which significantly speeds up the process. We found that samples produced more than 5.5 M of unique n-grams (features). Benign samples generated more than 4.5 M of features, while malicious—more than 1 M of features. When performing IS feature selection we found that benign and malicious samples shared almost 600 K common features. Subtraction of those features resulted in almost 4 M and 430 K of class-unique benign and malicious features, respectively. According to the algorithm from Sect. 3, we selected 50, 30, 15, 10, and 5 thousands of features. We selected a similar amount of features with the IG feature selection algorithm as well. Similar to [5–7] we wanted to reduce feature space even more, so that our models are simple enough for future human analysis. Thus, we used CFS feature selection method from Weka [19] to select the most relevant and least redundant features from 50 K features selected by IS and IG. As a result, we obtained 15 features from IS-based 50 K feature set and 9 features from IG-based 50 K feature set. As CFS appends features to the feature set until the increase of its merit is no longer possible, it is impossible to control the final amount of selected features unless the GreedyStepwise search is applied. However, such search never finishes its work when applied to the larger feature sets in our experimental environment. We wanted to compare the performance of IS and IG with the CFS as well. So we tried to select the same number of features with IG and IS. However, CFS selected 15 features. And as the IS have to select equal amount of features from each class (Sect. 3) we decided to select 14 features with IS (7 from each class).

The selected features were later used to build machine learning models. The data that is actually fed into mechine learning algorithms is basically a *bitmap of presence* [8]: if a certain sample (row) generates a certain feature (column), then this feature takes value *1* for this sample. In the opposite case, the feature takes value *0*. We used the following machine learning algorithms from Weka: k-Nearest Neighbors (kNN), RandomForest (RF), Decision Trees (J48), Support Vector Machines (SVM), and Naive Bayes (NB) with the default Weka [19] parameters. We assessed the quality of the models with fivefold cross validation [22]. Accuracy (ACC) as the amount of correctly classified samples and F1-measure (F1M) that takes into account precision and recall were chosen as evaluation metrics. Further, in this section, we present the classification performance of the machine learning models.

4.6 Time Complexity

One of the reasons to use IS feature selection is that it is relatively faster than the other common methods. In this subsection, we provide time taken by IS and IG methods to select 50 K of features from the initial 5.5 M distinct features. It took 302 s (~5 min) for IS to select 50 K features. In contrast, the IG used 18,560 s (~5.15 h) to select 50 K features when running in one thread. While being launched in 16 threads, IG used 1168 s (~20 min) to select 50 K features. Further increase in the number of threads does not make sense, since this is the maximum amount of threads available at our VDS. As we can see, single-threaded IS works 3.8 times faster than IG ran with 16 threads and 61.5 times faster than IG ran with 1 thread. To find an intersection of benign and malicious feature sets, the IS used 1.18 s as the average of 1000 runs. It has used an additional 0.7 s to subtract intersection from both feature vectors. The actual implementation of our feature selection algorithms did not load the entire dataset at the same time. Thus, it is impossible to directly measure the time needed to calculate the quality measure of a single feature, since it is calculated in iterations. But indirect assessment (we divide overall time by the total amount of features to go through) showed that IS needed around $5.5 \cdot 10^{-5}$ s to assess a single feature, and IG needed $2.12 \cdot 10^{-4}$ s and $3.4 \cdot 10^{-3}$ s to assess a single feature with 16 and 1 thread, respectively. It is important to mention that the times provided are relevant to our data structure and the way we store our data. For instance, the fact that we stored memory access n-grams for each sample in a separate cell of the database table could affect the time needed to perform feature selection.

4.7 Analysis of Selected Feature Sets

Here we analyze how different are the feature sets selected by IS and IG. In Table 2, the feature amount column shows the size of the feature set for IS and IG methods; the common feature column shows the number of similar features selected by IG and IS for the corresponding feature set size; and the difference ratio column shows the ratio of the distinct features and is calculated as *(Feature amount—Common features)/Feature amount*. As we can see, most of the features selected by the IS method are different from those selected by IG. It complies with the theoretical assessment of IS (see Sect. 3), where we explained that IS may discard features with potentially high information gain only because they get value *1* in both classes. As we mentioned before, we used CFS feature selection on the feature sets of the size 50 K. It is worth mentioning that CFS selected completely different features when working

Table 2 Difference between feature sets selected by IS and IG

Feature amount	Common features	Difference ratio
50 K	994	0.98
30 K	994	0.97
15 K	979	0.93
10 K	955	0.9
5 K	812	0.84
IG/IS 9/14	0	1

with 50 K feature sets selected by IS or IG. When using IG and IS to select the same amount of features as selected by CFS we also obtained completely different feature sets.

4.8 Classification Performance

In this subsection, we present the classification performance achieved by the machine learning algorithms. Tables 3 and 4 contain evaluation metrics of machine learning models trained with the feature sets of a different lengths selected by different feature selection algorithms. Therefore, *FSL* stands for feature set length, *ACC* stands for accuracy, and F1M stands for F1-measure. As we can see, both feature vectors allowed to achieve a quite high classification accuracy. The best performing RF model that used 10 K features selected by IG managed to classify 99.9% of the samples correctly. On its turn, features selected by IS allowed to build kNN and RF models with an accuracy of 99.8%. As we can see, in most cases, models built with the use of features selected by IS have slightly lower classification performance. However, the difference in accuracy or F1-measure between IS and IG features is most of the time less than 1%. Thus, it is hard to conclude whether the features selected by IG is significantly better than those selected by IS. There is one exception for NB models built with the use of 50 K features. As it is possible to see, the NB model trained with 50 K features selected by IG has significantly lower accuracy and F1-measure than the one trained with 50 K features selected by IS. This difference might be explained by the nature of features selected by IS and the limitations of the NB method. While building the model, Naive Bayes assumes that features are independent. However, Information Gain feature selection potentially selects a lot of mutually correlated features. The IS does not take into account the mutual correlation between features as well. However, there should be less correlated features selected by IS, since one half of the features will not have *1s* in one of the classes and vice versa. These properties of Naive Bayes were studied in [29]. Even though CFS selected completely different features in IS and IG cases, the models built with those features showed a quite similar classification performance. We will discuss this in

Table 3 Classification performance with a use of features selected by IG

Method	FSL	kNN		RF		J48		SVM		NB	
		ACC	F1M	ACC	F1M	ACC	F1M	ACC	F1M	ACC	F1M
InfoGain	50 K	0.996	0.996	0.996	0.996	0.997	0.997	0.983	0.983	0.693	0.671
	30 K	0.996	0.996	0.997	0.997	0.998	0.998	0.986	0.986	0.983	0.983
	15 K	0.996	0.996	0.998	0.998	0.998	0.998	0.991	0.990	0.983	0.983
	10 K	0.998	0.998	**0.999**	**0.999**	0.998	0.998	0.992	0.991	0.983	0.983
	5 K	0.995	0.995	0.997	0.997	0.997	0.997	0.988	0.988	0.983	0.983
	9	0.988	0.988	0.988	0.988	0.988	0.988	0.988	0.988	0.988	0.988
CFS	9	0.997	0.997	0.997	0.997	0.996	0.996	0.997	0.997	0.988	0.988

Table 4 Classification performance with a use of features selected by IS

Method	FSL	kNN		RF		J48		SVM		NB	
		ACC	F1M	ACC	F1M	ACC	F1M	ACC	F1M	ACC	F1M
IS	50 K	0.991	0.991	0.997	0.997	0.997	0.997	0.983	0.983	0.982	0.982
	30 K	0.996	0.996	0.997	0.997	0.997	0.997	0.983	0.983	0.985	0.985
	15 K	**0.998**	**0.998**	**0.998**	**0.998**	0.997	0.997	0.984	0.984	0.983	0.983
	10 K	0.998	0.998	0.997	0.997	0.997	0.997	0.985	0.985	0.983	0.983
	5 K	**0.998**	**0.998**	**0.998**	**0.998**	0.997	0.997	0.985	0.985	0.983	0.983
	14(7+7)	0.983	0.983	0.983	0.983	0.983	0.983	0.983	0.983	0.983	0.983
CFS	15	0.997	0.997	0.997	0.997	0.997	0.997	0.997	0.997	0.983	0.983

Sect. 5. When we used IS and IG to select the number of features similar to CFS we found that models built with these features perform slightly worse if compared to the models built with features selected by CFS. This finding can be explained by the natures of CFS and IS algorithms. The IS will select features with higher class-wise frequency. However, such features might correlate with each other. Thus, these features might have a strong correlation with each other bringing redundant information to the model. In contrast, CFS will try to select a feature set that has as little redundant information as possible. Looking once again in Tables 3 and 4, we can conclude that both feature selection methods performed quite good under our experimental setup while selecting feature sets that are very different from each other.

Important notice. The results from Table 3 is similar to part of the results provided in [7]. This happened because our papers share the same dataset. Also the data collection processes have only minor differences: in this paper, we recorded the first million of memory access operations, while methodology of [7] is to record the first million of memory access operations unless a certain stopping criteria is met.

5 Discussion and Future Work

In this section, we discuss our findings and limitations that should be applied to the possible conclusions made based on the presented results. As we were able to see, IS feature selection works faster than IG. The main reason to this is the fact that the selection of features based on its class-wise frequency requires less computations. However, it is important to understand that all measurements of time complexity presented in this paper are specific to our conditions (available computational resource, structure of the data, implementation of feature selection algorithms) and might differ in other conditions. The theoretical assessment of the IS feature selection method predicted that features selected by IS might bring less information about samples and classes than those selected by IG. But the experimental evaluation showed only marginal difference in classification performance. Under our experimental setup, only the amount of features selected by CFS could be considered as a proof of our theoretical assessment. The CFS selected more features from IS-selected feature set to gain similar merit (what resulted in similar classification performance). As we mentioned before, CFS adds features to the feature set until its merit stops growing. These facts show that features selected by IS possess less information. Thus, on the small feature sets, we need more features selected by IS than those selected by IG. As we compared classification performance of machine learning methods, we found that under certain conditions NB might perform better when using IS-selected features. This fact can be explored more thoroughly in the future work. The method was tested on a nearly balanced dataset, and we selected the equal amount of features to represent both classes. The use of other approach in the selection of the desired amount of features or applicability on the imbalanced datasets is left for the future work.

The IS feature selection method is quite simple in implementation. However, as we discussed in Sect. 3, its applicability limited to the cases where we are interested in the fact of presence of a certain feature in the class. Thus, when features are not binary or discreet, the applicability of IS feature selection is questionable. It is possible, however, to binarize continuous variables [22], but this a separate topic and it is out of scope of this paper. There is also a number of possible improvements and modifications that can be applied to the IS feature selection method in the future. For example, we can decrease the time complexity of IS in the following way. When we calculate class-wise frequencies of features, we might limit the search space by the samples that *produce* this feature. Rough estimation suggest that it may halve the time needed to perform IS feature selection. Another modification that can be implemented in IS feature selection is introduction of the degree of membership to the intersection. For example, a certain feature f might occur in both classes *C1* and *C2*. These classes have m_{C1} and m_{C2} samples, respectively. The feature f is present in m_{C1}^{f} samples of a class *C1* and m_{C2}^{f} samples of class *C2*. For example, we may exclude feature from the intersection if

$$\frac{max(\frac{m_{C1}^f}{m_{C1}}, \frac{m_{C2}^f}{m_{C2}})}{min(\frac{m_{C1}^f}{m_{C1}}, \frac{m_{C2}^f}{m_{C2}})} > \epsilon$$

. Basically, we keep a feature if it represents ϵ times bigger fraction of samples of one class than fraction of samples of the other class. Such approach may decrease an information loss, but will contribute to the increase of computational complexity of IS feature selection method, and thus will make IS less attractive feature selection method.

It is also important to outline the following observation. IS and IG selected quite different feature sets. Moreover, CFS selected completely different features from those preselected by IS and IG. Nevertheless, classification performance of the machine learning models appeared to be very similar when using different feature sets. This raises the following question: do the mentioned feature selection methods always select the best feature set or do they find *one* of the several similarly good feature sets? This question is left open for the future studies.

6 Conclusions

In this paper, we studied the performance of Intersection Subtraction feature selection on malware detection problem. We showed that with the use of IS feature selection on HDLSS dataset, it is possible to correctly classify more than 99% of the benign and malicious samples. The main contribution of this paper is the direct comparison of IS and IG feature selection methods under the same conditions. We found that most of the features selected by IS and IG are different. The classification performance of the machine learning models trained with the use of quite different feature sets appeared to be very similar. Even though the models trained with IG-selected features showed marginally better performance, the single-thread implementation of the IS feature selection method worked 3.8 times faster than the 16-thread implementation of IG. This makes Intersection Subtraction feature selection attractive when it comes to the analysis of HDLSS datasets. The IS feature selection may help when it is not known yet whether the data is useful for the classification task at all. The number of features might so big that it is pointless to spend time running more common (also slower) feature selection methods. Thus, with certain abovementioned limitations, the IS feature selection may be successfully applied to HDLSS datasets.

References

1. Alazab, Mamoun, Sitalakshmi Venkatraman, Paul Watters, and Moutaz Alazab. 2013. Information security governance: the art of detecting hidden malware. In *IT security governance innovations: theory and research*, 293–315. IGI Global.

2. Alazab, Manoun, Robert Layton, Sitalakshmi Venkataraman, and Paul Watters. 2010. Malware detection based on structural and behavioural features of api calls.
3. AVTEST. 2020. The independent IT-Security Institute. Malware. https://www.av-test.org/en/statistics/malware/.
4. Azab, Ahmad, Mamoun Alazab, and Mahdi Aiash. 2016. Machine learning based botnet identification traffic. In *2016 IEEE Trustcom/BigDataSE/ISPA*, 1788–1794. IEEE.
5. Banin, Sergii, and Geir Olav Dyrkolbotn. 2018. Multinomial malware classification via low-level features. *Digital Investigation* 26: S107–S117.
6. Banin, Sergii, and Geir Olav Dyrkolbotn. 2019. Correlating high-and low-level features. In *International workshop on security*, 149–167. Berlin: Springer.
7. Banin, Sergii, and Geir Olav Dyrkolbotn. 2020. Detection of running malware before it becomes malicious. page To be published.
8. Banin, Sergii, Andrii Shalaginov, and Katrin Franke. 2016. Memory access patterns for malware detection. *Norsk informasjonssikkerhetskonferanse (NISK)*, 96–107.
9. Bramer, Max. 2007. *Principles of data mining*, vol. 180. Berlin: Springer.
10. Carlin, Domhnall, Philip O'Kane, and Sakir Sezer. 2017. Dynamic analysis of malware using run-time opcodes. In *Data analytics and decision support for cybersecurity*, 99–125. Berlin: Springer.
11. Dobbin, Kevin K., and Richard M. Simon. 2007. Sample size planning for developing classifiers using high-dimensional dna microarray data. *Biostatistics* 8 (1): 101–117.
12. Donoho, David L., et al. 2000. High-dimensional data analysis: The curses and blessings of dimensionality. *AMS Math Challenges Lecture* 1 (2000): 32.
13. Dutta, Subhajit, and Anil K. Ghosh. 2016. On some transformations of high dimension, low sample size data for nearest neighbor classification. *Machine Learning* 102 (1): 57–83.
14. Fan, Jianqing, Fang Han, and Han Liu. 2014. Challenges of big data analysis. *National Science Review* 1 (2): 293–314.
15. Figueroa, Rosa L., Qing Zeng-Treitler, Sasikiran Kandula, and Long H. Ngo. 2012. Predicting sample size required for classification performance. *BMC Medical Informatics and Decision Making* 12 (1): 8.
16. Gong, Ke, Xu Yong Wang, and Maozeng, and Zhi Xiao. 2018. Bssreduce an o (u) incremental feature selection approach for large-scale and high-dimensional data. *IEEE Transactions on Fuzzy Systems* 26 (6): 3356–3367.
17. Grini, Lars Strande, Andrii Shalaginov, and Katrin Franke. 2018. Study of soft computing methods for large-scale multinomial malware types and families detection. In *Recent developments and the new direction in soft-computing foundations and applications*, 337–350. Berlin: Springer.
18. Hall, M.A. 1998. *Correlation-based feature subset selection for machine learning*. PhD thesis, University of Waikato, Hamilton, New Zealand.
19. Hall, Mark, Eibe Frank, Geoffrey Holmes, Bernhard Pfahringer, Peter Reutemann, and Ian H. Witten. 2009. The WEKA data mining software: An update. *SIGKDD Explorations* 11 (1): 10–18.
20. IntelPin. 2020. A dynamic binary instrumentation tool.
21. Khasawneh, Khaled N., Meltem Ozsoy, Caleb Donovick, Nael Abu-Ghazaleh, and Dmitry Ponomarev. 2015. Ensemble learning for low-level hardware-supported malware detection. In *Research in attacks, intrusions, and defenses*, 3–25. Berlin: Springer.
22. Kononenko, Igor, and Matjaž Kukar. 2007. *Machine learning and data mining: introduction to principles and algorithms*. Sawston: Horwood Publishing.
23. Haldar, Malay. 2015. How much training data do you need? https://medium.com/@malay.haldar/how-much-training-data-do-you-need-da8ec091e956.
24. Ogorodnyk, Olga, Ole Vidar Lyngstad, Mats Larsen, Kesheng Wang, and Kristian Martinsen. 2018. Application of machine learning methods for prediction of parts quality in thermoplastics injection molding. In *International workshop of advanced manufacturing and automation*, 237–244. Berlin: Springer.

25. Ozsoy, Meltem, Khaled N. Khasawneh, Caleb Donovick, Iakov Gorelik, Nael Abu-Ghazaleh, and Dmitry Ponomarev. Hardware-based malware detection using low-level architectural features. *IEEE Transactions on Computers* 65 (11): 3332–3344.
26. Pedregosa, F., G. Varoquaux, A. Gramfort, V. Michel, B. Thirion, O. Grisel, M. Blondel, P. Prettenhofer, R. Weiss, V. Dubourg, J. Vanderplas, A. Passos, D. Cournapeau, M. Brucher, M. Perrot, and E. Duchesnay. 2011. Scikit-learn: Machine learning in Python. *Journal of Machine Learning Research* 12: 2825–2830.
27. PortableApps.com. 2020. Portableapps.com. https://portableapps.com/apps.
28. Python.org. 2020. Time complexity. https://wiki.python.org/moin/TimeComplexity.
29. Rennie, Jason D., Lawrence Shih, Jaime Teevan, and David R. Karger. 2003. Tackling the poor assumptions of naive bayes text classifiers. In *Proceedings of the 20th international conference on machine learning (ICML-03)*, 616–623.
30. Santos, Igor, Felix Brezo, Xabier Ugarte-Pedrero, and Pablo G. Bringas. 2013. Opcode sequences as representation of executables for data-mining-based unknown malware detection. *Information Sciences* 231: 64–82.
31. Shalaginov, Andrii, Sergii Banin, Ali Dehghantanha, and Katrin Franke. 2018. Machine learning aided static malware analysis: A survey and tutorial. In *Cyber Threat Intelligence*, 7–45. Berlin: Springer.
32. Virus Total. 2012. Virustotal-free online virus, malware and url scanner. https://www.virustotal.com/en.
33. Venkatesh, B., and J. Anuradha. 2019. A review of feature selection and its methods. *Cybernetics and Information Technologies* 19 (1): 3–26.
34. Vinod, P., Vijay Laxmi, and Manoj Singh Gaur. 2012. Reform: Relevant features for malware analysis. In *2012 26th international conference on advanced information networking and applications workshops*, 738–744. IEEE.
35. VirusShare. Virusshare.com. http://virusshare.com/. Accessed 09 March 2020.
36. Yata, Kazuyoshi, and Makoto Aoshima. 2010. Effective pca for high-dimension, low-sample-size data with singular value decomposition of cross data matrix. *Journal of Multivariate Analysis* 101 (9): 2060–2077.
37. Yata, Kazuyoshi, and Makoto Aoshima. 2012. Effective pca for high-dimension, low-sample-size data with noise reduction via geometric representations. *Journal of Multivariate Analysis* 105 (1): 193–215.
38. Yücel, Çağatay, and Ahmet Koltuksuz. 2020. Imaging and evaluating the memory access for malware. *Forensic Science International: Digital Investigation* 32: 200903.
39. Zawbaa, Hossam M., Eid Emary, Crina Grosan, and Vaclav Snasel. 2018. Large-dimensionality small-instance set feature selection: a hybrid bio-inspired heuristic approach. *Swarm and Evolutionary Computation* 42: 29–42.
40. Zhang, Lingsong, and Xihong Lin. 2013. Some considerations of classification for high dimension low-sample size data. *Statistical Methods in Medical Research* 22 (5): 537–550.

A Comparative Study of Adversarial Attacks to Malware Detectors Based on Deep Learning

Corrado Aaron Visaggio, Fiammetta Marulli, Sonia Laudanna, Benedetta La Zazzera, and Antonio Pirozzi

Abstract Machine learning is widely used for detecting and classifying malware. Unfortunately, machine learning is vulnerable to adversarial attacks. In this chapter, we investigate how generative adversarial approaches could affect the performance of a detection system based on machine learning. In our evaluation, we trained several neural networks for malware detection on the EMBER [3] dataset and then we built ten parallel GANs based on convolutional layer architecture (CNNs) for the generation of adversarial examples with a gradient-based method. We then evaluated the performance of our GANs, in a *gray-box* scenario, by computing the evasion rate reached by the adversarial generated samples. Our findings suggest that machine- and deep-learning-based malware detectors could be fooled by adversarial malicious samples with an evasion rate of around 99% providing further attack opportunities.

1 Introduction

Several studies have investigated the effectiveness [1, 7, 8, 17, 19, 48] and drawbacks [4, 40] of machine (and recently also deep) learning in detecting and classifying malware. Independently from the inherent limitations of malware detectors based on machine learning, the generative adversarial networks (GANs, in the remainder of the chapter) become a menace to the effectiveness of these tools.

A GAN is a tool that produces adversarial samples by using the adversarial machine learning [26]: this is a technique that leverages machine learning for fooling classifiers trained with a machine learning algorithm, leading them to wrongly classify some samples.

C. A. Visaggio · S. Laudanna · B. La Zazzera · A. Pirozzi
University of Sannio, Benevento, Italy
e-mail: visaggio@unisannio.it

F. Marulli (✉) · S. Laudanna · B. La Zazzera · A. Pirozzi
Department of Maths and Physics, University of Campania "L.Vanvitelli", Caserta, Italy
e-mail: fiammetta.marulli@unicampania.it

M. Stamp et al. (eds.), *Malware Analysis Using Artificial Intelligence and Deep Learning*, https://doi.org/10.1007/978-3-030-62582-5_19

Adversarial machine learning has been applied with a certain success especially to the field of image recognition with some surprising results [21], but also to speech recognition [2] and biometric recognition [11].

For understanding how powerful maybe this technique, we could mention the case of image recognition. The adversarial sample is an image that has been tampered within a way that cannot be distinguished by a bare eye, but that misleads the classifier. The result is that the image is not recognized at all or, even worst, is classified as a completely different image.

An exemplar case is the automatic recognition of street signs: a street sign is decoded as another street sign. Alike the fields where GANs have been experimented, they could be successfully used for generating samples of malware that are recognized by malware detectors based on machine learning as goodware.

The research community is now investigating the application of GANs to malware analysis, and so far the main result consists of some models of GANs for producing adversarial malware samples.

Our purpose is to investigate how and how much GANs are able to degrade the performance of malware detectors based on machine learning. We trained a set of classifiers using different combinations of features, obtaining a wide spectrum of performances. Thus, we built different models of GANs, observing the degradation of each detector.

This work does not help to identify how to make stronger a detector against an adversarial attack but provides data for quantifying the potential effects of a GAN on a malware detector based on machine learning.

In this chapter, we provide a brief overview of the current state of the art and some open issues related to the vulnerabilities of deep learning models adopted in designing malware recognition systems. More precisely, we focused on the weak points of these approaches when attacked by adversarial examples that are proving to be increasingly sophisticated and effective in misleading defense systems.

We provide further evidence by discussing a case study that shows how adversarial examples and generative adversarial approaches, by the means of generative adversarial neural networks (*GANs*), can degrade the detection performance of a deep learning feature-based malware detector, finally highlighting that certain features may prove to be more sensitive than others.

The chapter is organized as follows: the next section provides the background of adversarial machine learning and the most significant applications, while Sect. 3 compares the related literature. Section 4 shows the research questions we posed and the design of the case study. Section 5 discusses the obtained results, and, finally, conclusions are drawn in the last section.

2 The Deep Learning Models Adopted in Malware Detection

Machine learning (ML) and deep learning (DL) have been successfully employed for detecting malicious objects, e.g., executable files.

New malware programs appear each year in increasing amounts and hence malware detection based on signature matching is increasingly becoming an impractical approach. Machine and deep learning promise to provide valid countermeasures against modern malware because of their capability to potentially detect malware applications without specific signatures of their behavior or data.

Generally, ML-based malware detectors work on the extraction of the malware (and benign programs) features and static and/or dynamic analysis can be performed. Such systems learn from examples for creating models by which they will be able to discriminate whether a given program is a malware or not. These models are then used to estimate the likelihood that a given program is malware.

One of the bottlenecks exhibited by ML-based malware detection is represented by the high time required to learn when the number or size of features is wide or the number of sample programs is large. Although reducing the number of features could shorten the learning time, the accuracy in the detection task likely decreases. So, finding an acceptable trade-off among the detection accuracy, short learning times, and limiting the size of data, obtainable by selecting a convenient combination of sensitive features, is far from being a trivial problem.

A very accurate review of recent findings of adversarial examples in deep neural networks and a deep investigation of existing methods for generating adversarial examples is provided in [50].

2.1 The Deep Learning Models in a Nutshell

The essential background about techniques and enabling architectures of deep learning is provided in the following.

Deep learning is a kind of machine learning that makes computers to learn from experience and knowledge without explicit programming and extracts useful patterns from raw data.

Conventional machine learning algorithms exhibit some limitations since it is difficult to extract well-represented features because of the curse of dimensionality, computational bottleneck, and strong requirements of the domain and expert knowledge. Deep neural networks represent a particular kind of machine learning algorithm, leveraging several "deep" layers of networks. Furthermore, deep learning solves the problem of the representation by building multiple simple features to model a complex concept. The more the number of available training data grows, the more powerful the deep learning classifier becomes. Deep learning models solve complicated problems by complex and large models, with the help of hardware acceleration in computational time.

Traditionally, researchers build a single deep learning model using the entire dataset. However, the single deep learning model may not handle the increasing complex malware data distributions effectively since different sample subspaces representing a group of similar malware may have unique data distribution [52].

Since the performance of deep learning models keeps improving with the increasing number of samples [49], researchers build a single deep learning model using an entire data to understand the relationship between data features extracted from malware and the target [9, 27, 39, 45].

These deep learning models mainly use three types of neural network architectures:

- Convolutional neural network (CNN);
- Recurrent neural network (RNN); and
- Fully connected feedforward neural network (FC).

There are two major disadvantages in building a single deep learning model that uses a blended dataset:

- Complex data distribution;
- Scalability.

Each type of malware has unique and different characteristics, proliferation methods, and data distributions [35, 49].

Consequently, merging different types of malware into one dataset results in a very complex overall data distribution. Furthermore, the diversity and sophistication of the merged dataset continue to grow rapidly due to the large number of new malware variants that are created each year [49]. As a result, it is very challenging for a single deep learning model to understand this complex data distribution.

Additionally, the single CNN model treats malware as the image while the single RNN model considers the behavior as the sequence of events. Both models only analyze the data distribution from only one perspective. In the case of malware, the analysis of data distribution in different sample subspaces from multiple angles is preferred in order to combine the knowledge and strength of these single models effectively.

Second, building a single deep learning model for malware detection lacks scalability to train on increasingly large malware datasets. Training deep learning models on very large datasets is a computationally expensive process [20]. Since the number of new malware samples has exponentially increased through time [9, 31], building a single deep learning model requires longer computation time. This slow training process makes difficult to search and rebuild the learning model rapidly in order to adapt to the fast changing malware landscape and respond to the new techniques adopted by the malware writers. An alternative to using a single deep learning model to build malware detection systems (MDSs) is the development of ensemble-based deep learning models. Multiple deep learning models in the ensemble can work together to enhance the performance of MDSs. Researchers have developed the ensemble-based deep learning models, where each model is constructed on the whole blended dataset.

2.1.1 The Most Popular Deep Neural Network Architectures

A neural network layer includes a set of perceptrons (artificial neurons), and each one is able to map a set of inputs to output values by evaluating a simple activation function. The function of a neural network is formed in a chain $f(x) = f^{(k)}(\ldots f^{(2)}(f^{(1)}(x)))$, where $f^{(i)}$ is the function of the ith layer of the network, with $i = [1; 2; \ldots; k]$.

Convolutional neural networks (CNNs) and **recurrent neural networks** (RNNs) are the two most popular and adopted neural network architectures in recent times.

CNNs deploy convolution operations on hidden layers for weight sharing and parameter reduction. CNNs can extract local information from grid-like input data. CNNs have shown incredible successes in computer vision tasks, such as image classification [24], object detection [44] and semantic segmentation [14].

RNNs are neural networks adopted for processing sequential input data with variable length. RNNs produce an output at each step. The hidden neuron at each step is calculated based on input data and hidden neurons at a previous step. To avoid vanishing/exploding gradients of RNNs in long-term dependency, long short-term memory (LSTM) and gated recurrent unit (GRU) with controllable gates are widely used in practical applications.

Generative adversarial networks (GANs) are a type of generative model introduced by [22], where adversarial examples can be exploited to improve the representation of deep learning and perform unsupervised learning. A generative network (generator) creates artificial samples while a discriminative network (discriminator) acts as an adversary to determine if the generated samples are genuine or fake. This kind of network architectures are typically referred as generative adversarial network (GAN) and solve an optimization function described by

$$\min_{G} \max_{D} V(D, G) = \mathbb{E}_{x \sim P_r}[\log D(x)] + \mathbb{E}_{z \sim P_z}[\log D(G(z))],$$

where D and G denote the discriminator and generator, and P_r and P_z are, respectively, the distribution of input data and noise. In this competition, GAN is able to generate raw data samples that look close to the real data.

Due to the wide use and breakthrough successes, ML- and DL-based detection systems have become a major target for attacks, where adversaries are usually applied to evaluate the attack methods. Unfortunately, both ML and DL approaches to malware detection can be fooled by adversarial examples that consist of small changes to the input data causing misclassification at testing time.

3 Adversarial Attacks Against Deep Learning-Based Malware Detection System

In this section, we explore the adversarial attack techniques on ML models that have been applied to intrusion and malware attack scenarios.

Several techniques have been proposed to create adversarial examples. Most approaches suggest minimizing the distance between the adversarial example and the instance to be manipulated in order to cause the ML classifier to misclassify the testing dataset with high confidence.

Some methods require access to the gradients of the model, which typically introduce perturbations optimized for certain distance metrics between the original and perturbed samples: this kind of attack is called *white-box attack*. Other methods only require access to the prediction function, which makes these methods model-agnostic: this kind of attacks are called *black-box attack*.

A simple indiscriminate approach is gradient ascent during the training of ML model. Szegedy et al. [47] proposed a first gradient method to generate adversarial examples applied to the imaging field, using *box-constrained limited-memory Broyden-Fletc.her-Goldfarb-Shanno* (L-BFGS) optimization, an optimization algorithm that works with gradients. The adversarial examples were generated by minimizing the following function:

$$\begin{aligned} &Minimize \, \|r\|_2 \, subject \; to: \\ &1. \; f(x+r) = l \\ &2. \; x + r \in [0,1]^m , \end{aligned}$$

where x is an image represented as a vector of pixels, r represents the perturbations to be made on the pixels to create an adversarial image, l is the target label (the desired outcome class), and the parameter c is used to balance the distance between images and the distance between predictions.

Goodfellow et al. [22] proposed a simple and fast gradient-based method called *fast gradient sign method* (FGSM), using the gradient of the underlying model to find adversarial examples and the original image x is manipulated by adding or subtracting a small error ϵ to each pixel:

$$\eta = \epsilon * \mathrm{sign}\left(\nabla_x \; J\left(x, y\right)\right).$$

Here, η is the perturbed sample, ϵ is a hyperparameter controlling the amount of perturbation added to each feature (pixel), $\nabla_x \; J$ is the gradient of the models loss function with respect to the original input pixel vector x and y the target label (the true label vector for x). The sign of the gradient is positive if an increase in pixel intensity increases the error the model makes and negative if a decrease in pixel intensity increases the error. This approach requires many pixels to be changed, for this reason, Su et al. [46] demonstrated that it is actually possible to deceive image classifiers by changing a single pixel (the RGB value). The one-pixel attack uses differential evolution to find out which pixel is to be changed and how.

Brown at. al [12] proposed how to create image patches that can be added to a scene, and force a classifier into reporting a class of the attacker's choosing. This method differs from the methods aforementioned since the adversarial image isn't close to the original image but it is removed and a part of the image is replaced with a patch that can take on any shape.

Carlini and Wagner [13] modified the objective function and used a different optimizer compared with the L-BFGS attack described in [47]. Instead of using the same loss function as in L-BFGS, they solved the following box-constraint optimization problem to find an adversarial perturbation δ, making the problem more efficient to solve. CW finds the adversarial instance by finding the smallest noise $\delta \in R^{nxn}$ added to an image x that will change the classification to a class t and uses the L_2 norm (i.e., Euclidean distance) to quantify the difference between the adversarial and the original examples. Formally:

$$\text{minimize } \|\delta\|_p \text{ subject to } C\,(x+\delta)=t,\; x+\delta\in[0,1]^n,$$

where $C(x)$ is the class label returned with an image x.

While successful, gradient-based methods work only under "white-box" settings. Papernot et al. [38] showed a type of zero-knowledge attack (black-box attack) to create adversarial examples without internal model information and without access to the training data. This technique, called *Jacobian-based saliency map attack* (JSMA), unlike the previous method, proposed to use the gradient of loss with each class label with respect to every component of the input, i.e., Jacobian matrix to extract the sensitivity direction. Then a saliency map is used to select the dimension which produces the maximum error using the following equation:

$$s_t = \frac{\partial t}{\partial x_i};\; s_o = \sum_{j\neq t}\frac{\partial j}{\partial x_i};\; s(x_i) = s_t\,|s_o|\cdot(s_t<0)\cdot(s_o>0).$$

In the previous formula , s_t represents the Jacobian of target class t and s_o represents the sum of Jacobian values of all non-target class. Changing the selected pixel will significantly increase the probability of the model labeling the image as the target class. The purpose of JSMA attack is to optimize the L_o distance metric (the amount of perturbed features).

Moosavi-Dezfooli et al. [33] proposed an algorithm, *DeepFool*, to compute adversarial examples using an iterative linearization of the classifier to generate minimal perturbations that are sufficient to change the classification labels. Starting with a binary classification problem, this method creates an adversarial example computing the Euclidean distance between perturbed samples and original samples in an iterative manner until $sign(f(x)) \neq sign(f(x+r))$ where r is the minimum perturbation required.

Zeroth-order optimization attack (ZOO) was proposed by Chen et al. [15] and consists of approximating the full gradient via a random gradient estimate using the difference between the predicted probability of the target model and the desired class label. Precisely, the method uses zeroth-order stochastic coordinate descent to optimize the malicious sample by adding perturbations to each feature and querying the classifier to estimate the gradient and Hessian of the different features. In this scenario, solving the optimization problem is computationally expensive and the authors proposed a ZOO-Adam algorithm to find the optimal perturbations for the target sample.

Most of the attacks presented have been initially tested on image domains by introducing perturbations to existing images but they can equally be applied to other types of data, such as datasets with a limited number of features since these attacks are not data-type dependent. In a cybersecurity scenario, a malicious user could access any type of data used by a classifier and produce adversarial examples.

4 Generative Adversarial Attacks Against Malware Detection Systems

In this section, we examine existing generative adversarial algorithms used to attack malware detectors.

Generative adversarial algorithms have been mainly applied to image recognition, where generative adversarial networks (GANs) were used to generate images that were indistinguishable from real ones. In the process of image generation, for example, the GAN network modifies some features like pixels, while a human eye does not perceive the difference from an original one. Using GAN to create a binary file poses more difficulties than an image, because changing a bit in a binary may corrupt the file. For this reason, generating executable files with a GAN could be challenging.

The main difference between image and malware is that images are continuous while malware features are binary. Changing byte arbitrarily could break semantics and syntax of portable executable (PE) so we are limited in the types of modification that can be done without breaking the malware functionality. For this reason, different approaches have been proposed in the literature such as adding padding bytes (adversarial noise) at the end of a file beyond PE boundaries [30]. Another approach consists of injecting the adversarial noise in an unused PE region that is not mapped in memory [32]. Most works in literature simply ignore this problem. In order to overcome this limitation, attackers must have a white-box model in which the type of the ML algorithm used and the features to be used are known. One of the first demonstrations of an adversarial creation of a PE is the work [25]; in this paper, authors adopt a gray-box model in which they only know the set of used features based on API calls but do not know the ML model used by the classifier. In MalGAN, authors generate adversarial examples by adding some irrelevant features to the binary files because removing features may crack the executable or its intended behavior. The adversarial generated example is expressed by the following formula:

$$m' = m|o',$$

where m is the initial feature vector, each element of m corresponds to the presence/absence of a particular feature in a malware, and then this input vector is fed into a multi-layer feedforward neural network with weights

$$\theta_g.$$

The output layer of this network has M neurons and has a sigmoid as an activation function which is continuous in the range (0,1) as the last layer. The output of this network is denoted as o.

Since malware features are binary, the output in the continuous space must be transformed into the binary space with a transformation called *binarization*. This procedure generates a new binary vector *o'*. Then the resulting *m'* is a binary vector obtained by the initial m vector through an OR bit-wise operation with the o' binary vector.

The non-zero element of the binary vector *o'* acts as an irrelevant feature to be added to the original malware. While MalGAN and the detector use the same API as features quantity and this could affect the performance of avoidance, in [28] authors add some noise to malware, extracting features (API list) from clean malware and input them to the generator.

In another work [30], the authors present a gradient-based attack to generate adversarial malware binaries but their limit is the manipulation to the padding bytes appended at the end of the PE to guarantee that the malware integrity is preserved. With this approach, they reach an evasion rate of 60% against *raff2017malware* used as a classifier. GANs are also used to generate a malicious document. In [51], the authors propose a method based on *Wasserstein generative adversarial network* (WGAN) to generate a malicious PDF with an evasion rate of 100% as stated. A malicious PDF is a document that embeds and executes malicious code. In this work, the authors generate adversarial examples by modifying 68 features extracted from various attributes: size, metadata, and structural attributes.

5 Case Study

We carried out a case study to examine the effectiveness of adversarial models against malware detectors based on deep learning. To this aim, we considered a cooperative system of generative adversarial networks, where multiple GANs (couples of generators and discriminators) run in parallel for supporting a multistage black-box attack.

Under the realistic hypothesis that an attacker knows very little about the system he wants to attack (the case of a black-box attack), the attacker could set up some sort of brute-force attack by deploying a pool of specific generators built for interacting with a corresponding pool of specific targets (discriminators).

The attack strategy envisages multiple stages (steps). As first, the only knowledge owned by the attacker consists of knowing that the target victim could behave according to a ML or DL model and the kind of inputs it could accept, so the attacker trains several generators working over different groups of features (possibly, he could try over all the sensible combinations) for refining the generation of artificial adversarial samples.

This training stage is performed without effectively interacting yet with the real target. Discriminators play the role of the potential victims, as substitutes of real victim systems. In the middle of attacking time, the attacker will start a smooth interaction with its victim, this time represented by a black box. By carefully analyzing responses from the black box, it is able to figure out what features are used by the malware detector black box. Adversarial test cases are produced by exploiting all the trained generators in the attacker's wallet.

Most of these samples will be harmless since they will not act on the right set of features but we can suppose that almost one of these generators will be able to generate samples that will produce some effects. By this way, the attacker will gain knowledge about its adversary and can implement a gray-box attack, basing on the features set its victim works over.

At the real attack time, the attacker will exploit only the right generator and proceed to attack and refine its generation model until its target is reached out. The case study we propose should not be regarded as exhaustive but it can be regarded as proof of the concept that adversary attacks pointing ML and DL systems can be implemented in many alternative and successful ways, for tampering with real existing defense systems.

5.1 Case Study Design

As first, we trained ten parallel GANs simultaneously, where both the detectors and the generators were realized by adopting deep neural network models. In particular, the models used for the discriminators implement a fully connected feedforward network architecture (FFNNs) while, for models of the generators, we adopted a convolutional layer architecture (CNNs).

For implementing each GAN comprised in our cooperative system, we took our cue from the general system architecture and the generators neural network architecture, implemented as a CNN, suggested in MalGAN [25]. Unlike MalGAN, we don't use a black-box detector and a substitute detector, since we designed our case study from the perspective of a "patient attacker" deploying a multistage attack. Our approach differs from MalGAN also in the kind of features considered both for training detectors and generators and in the generation strategy of the adversarial samples, as it will be detailed in the following sections.

Then, in the first stage, we assume that the attacker knows at least the features adopted by its victim for distinguishing between goodware and malware and has access also to its gradients. In this way, we could directly exploit the gradient's information provided by the detectors for training the generators and refining the capability to artificially generate samples that look like genuine ones.

With regard to the type of samples we analyzed and the kind of analysis performed by the victim detectors, specifically, we considered the surface features extracted from binary files applications, so the detectors answering to adversarial attacks are trained to perform a *static analysis* over the inputs they are fed in.

Correspondingly, adversarial examples will be generated by crafting these surface features. Since the surface feature space is discrete, we will apply a transformation to continuous space, in order to apply a gradient-based method for improving the probability that the generated adversarial examples will go undisturbed through the detection system. The approach suggested in [23] allows us to work in discrete and binary input domains, differently from most of the other proposed approaches [30] that operate only in continuous and differentiable domains.

Furthermore, static analysis has the advantage that does not require the execution of samples in a sandbox or safe environment for studying their behaviors, and the features for training the detector and/or classifiers can be extracted over specific subsets of features. Conversely, the dynamic analysis could reveal more information about malicious behaviors by the applications (e.g.., actions relationships and patterns) but the operative conditions are more difficult to achieve. Challenging results obtained by adopting static analysis in training machine and deep learning algorithms for malware detection are described in [3], where the authors provided, as first, an open-source dataset, namely, "EMBER," consisting in a collection of surface features extracted from a little under a million of malicious applications targeting Windows O.S. environment; furthermore, they provide experiments that compare a baseline gradient boosted decision tree model trained using LightGBM [29] with default settings to MalConv [43], an end-to-end but featureless deep learning model for malware detection, which recently became a very popular benchmark in this kind of experiments.

In the case of malware detection, unlike other application domains, like image and speech recognition, manipulating bytes can severely compromise application functionalities and validity; therefore, generating adversarial examples is not straightforward. An unavoidable requirement that should lay down every manipulation strategy consists in adopting generation techniques that are able to guarantee the preservation of malware functionality in the adversarially manipulated samples.

In our evaluation we trained, validated, and tested discriminator models for malware detection, by adopting for all the same samples, randomly extracted from the EMBER [3] dataset. Finally, we selected the first ten ones that obtained the best accuracy in the detection task. Then, we build ten parallel GANs, and we trained ten generators for the corresponding trained detectors (discriminators). For training the generators, we adopted a descendent gradient-based strategy and we adopted the maximum mean discrepancy (MMD) [5] as distance function for evaluating sample distributions similarity during the training process.

We examined the results obtained in the different stages of our experiment for measuring the effectiveness of the adversarial strategy and the robustness of the malware detectors to these kinds of attacks.

We adopted, for evaluating the reached performances, the following metrics: accuracy, sensitivity, specificity, and evasion rate. The evasion rate represents a measure of the success rate obtained by generator networks in fooling their opponent discriminators; it can be computed as the ratio between the number of adversarial examples that were misclassified as benign samples (also referred in the following as "*goodware*") by each detector, over the total amount of adversarial samples submitted to the discriminators.

By adopting the EMBER dataset, we were able to fit the detection performance obtained from the state of the art. Then, by using the adversarial crafting algorithm, we were able to mislead, on average, the ten detectors by decreasing the average accuracy over all the models ranging from a minimum of 20.63% (best case) to a maximum of 40.8% (worst case), by mixing genuine samples with adversarial one's samples and acting over the surface features.

Our preliminary experiments revealed, at a first sight analysis, that the byte distribution (byte histogram) is among the most sensitive features. This finding could suggest that machine- and deep-learning-based malware detectors, which work on static and surface features, could be fooled by adversarial malicious samples that are able to reach a bytes distribution with a high level of likelihood with the goodware bytes distribution.

5.2 General Architecture

The overall architecture of the system we propose corresponds to the general schema of a GAN (Fig. 1), where each couple made of a generator (G) and a discriminator (D) acts independently from each other.

5.2.1 The Discriminator Network Model

Following the approaches suggested in [36, 42, 43], we adopted a fully connected multilayered feedforward neural network as the base architecture for our discriminator's models. All the models we trained for obtaining the detection systems, as detailed in the following sections, share the same number of dense hidden layers, their size and the size of the output layer, set to 1, since the detection acts as a binary classification task (e.g., malware or goodware). All the trained models differ in the input layer size, since we performed several experiments by changing the size and the values of the input vectors, according to the combinations of features that we

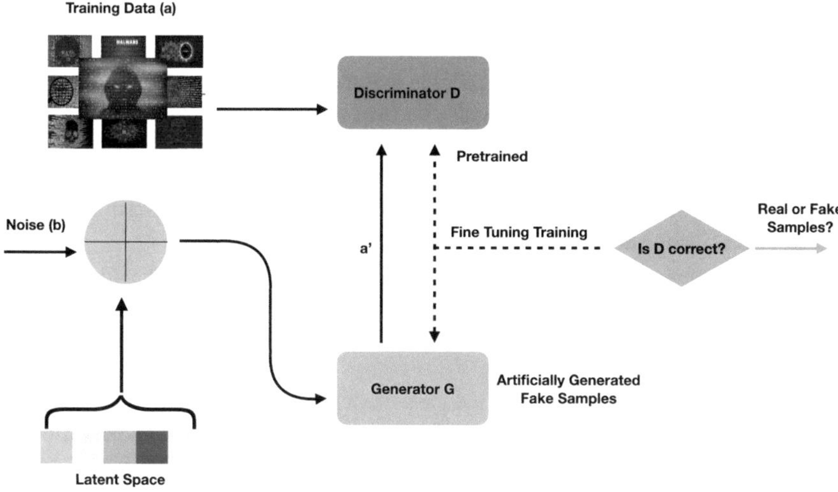

Fig. 1 GANs logic and building blocks of the proposed GANs-based architecture

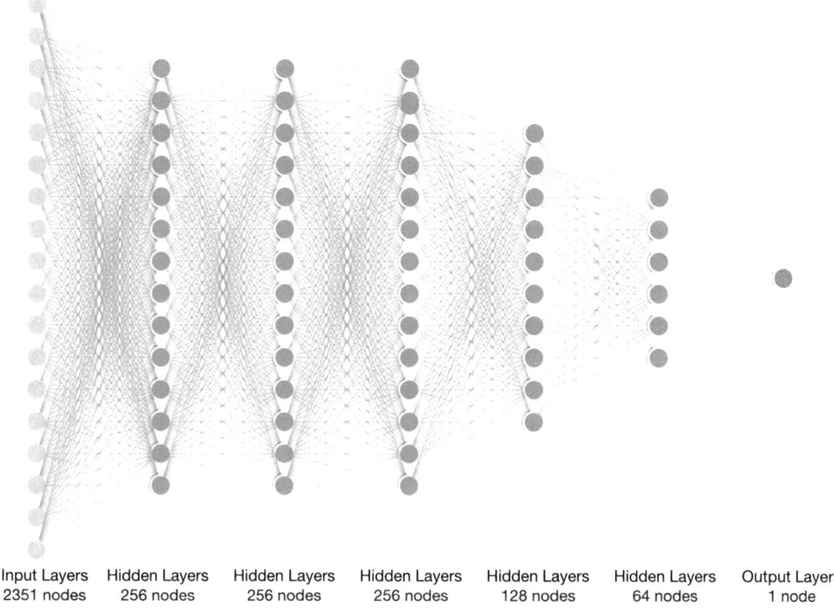

Fig. 2 The discriminator network: malware detector architectural schema

aimed to test. Figure 2 shows the general architecture of the discriminator network we adopted in our study.

The basic model adopted for each discriminator of our pool includes five hidden dense and fully connected layers characterized by decreasing size *(256-256-256-128-*

64); for each discriminator, the input size was variable, according to the subset of features we considered for each model. The maximum size of the input layer was set to *2,351*, when we consider all the available features provided in the EMBER dataset for implementing a static malware analysis over the input samples. In addition, we adopted the *Adam* algorithm as optimization function and the *binary cross entropy* as loss function. Finally, as an activation function, we adopted the *ReLU* that allowed to alleviate the vanishing gradient issues and is faster when compared with other non-linear activation functions.

We performed all the training cycles for *250 epochs* with *batch size* set to *64*. We adopted different learning rates l_r varying in the range [0.01;0.5]; finally, all the discriminators models were able to converge to an accuracy rate $a \geq 80\%$ and a *false positive rate* $(FPR) \leq 1\%$ (where *FPR* is computed as the number of benign samples misclassified as malicious over the total number of malicious samples detected), by adopting a $l_r = 0.05$ and a number of iterations = *250 epochs*. The performance metrics values were cross validated over the validation and the test sets. All the experiments were repeated five times and the average values obtained in these experiments were considered as the values of the final hyperparameters for tuning the networks.

The reason underlying the strategy of training several models was dictated also by the need to apply a reduction to the whole set of the features provided by PEs files; even though the best accuracy is performed when a detection model is trained over the whole features set, we need also to limit the performance decay, in terms of data size and training time, in order to make this approach feasible for real-world scenarios. So, we applied a strategy for reducing features and we were able to obtain a trade-off among accuracy, data size, and learning time. Anyway, we didn't investigate more space and time complexity on this occasion, but it will be the object of further and necessary investigations.

5.2.2 The Generator Network Model

For the generation network model, we followed the general setting adopted in [25].

The model we adopted for the generators is represented by a convolutional neural network (CNN) trained on a sample fraction extracted by the EMBER dataset. We split the Ember dataset in order to save a fraction of samples, made of benign and malicious samples that were not included in the training set of our detectors. We trained the generators until all of them reached at least an accuracy rate $a \geq 98\%$, when artificially reproducing the original samples, as it will be detailed in the following of this section. For the generation of adversarial examples (AEs), we set two main constraints:

- ***Functionality preserving***: Adding noise for generating adversarial examples should not break the sample's behavior.
- ***Features probability distributions invariance***: Since we worked only on surface features, we don't manipulate the content of binaries but we try to change the

surface information as their probability distribution looks like more close to the distribution of good samples.

The CNNs we adopted for generators are characterized by layer sizes set to X-256-X, where X represents the variable size for both the input and the output, according to the input dimension that has to be transformed and the adversarial sample size that has to be produced. Noise vectors adopted for manipulating genuine inputs have the same dimensions of the input, according to the number of features that are considered in each couple of generator-discriminator. The Adam optimizer was selected as an optimization function. Each generator was trained for 500 epochs with a learning rate set to 0.05. These training parameters were obtained after several experiments until the best tuning that guarantees convergence for all the generators with an accuracy rate over the original ground truth stabilized to 98%. This accuracy was cross validated also over the validation and test set. All the experiments were repeated five times and the average values obtained in these experiments were considered as the values of the final hyperparameters for tuning the networks.

The generation of AEs is usually done by adding small perturbations to the original input in the direction of the gradient. The gradient-based methods work for continuous input sets but they fail in the case of discrete input sets. If we denote the set of the features as $X \subseteq [0, N-1]$, where $N = 2351$, the features comprised in the PEs files can be arbitrarily represented as scalars in a set X [0, N - 1], where $N = 2,351$. So, AEs can be generated in a continuous embedding space E and reconstructed them to original X.

5.2.3 Adversarial Example Generation Problem

Given a trained deep learning model f, an original input data sample x, generating an adversarial example x', can generally be described as a box-constrained optimization problem:

$$\begin{aligned} \mathbf{min\ x'}\ \left\|\mathbf{x'}-\mathbf{x}\right\| \\ \mathbf{s.t.\ m\ (x)} = \mathbf{l} \\ \mathbf{m}\left(\mathbf{x'}\right) = \mathbf{l'} \\ \mathbf{l} \neq \mathbf{l'} \\ \mathbf{x'} \in [\mathbf{0},\ \mathbf{1}], \end{aligned}$$

where

- x is the genuine input sample;
- x' is the artificial input sample;
- m (·) represents the trained deep learning model;
- l and l' represent, respectively, the output labels produced by the model m (·) when processing x and x'; and
- ||·|| denotes the distance between two samples.

δ is the difference between x' and x, and represents the perturbation (noise) added to x. This optimization problem minimizes the perturbation while misclassifying the prediction with a constraint of input data. Other variants of this optimization problem can be considered in different scenarios and assumptions. For instance, in the image recognition domain, some adversaries consider that if $\delta < \epsilon$, the perturbation is small enough to be unnoticeable to humans and it is viewed as a constraint. The optimization objective function becomes the distance of the targeted prediction score from the original prediction score.

5.3 *Adversary Logic*

As described in [50], the adversarial examples can be categorized in a taxonomy along seven axes. In our study, we followed the axes of the *adversarial falsification* and the *iterative attack*. For the first dimension, we were interested in training generators able to lead a decay in the detection accuracy of each detector, as it will be shown in the results subsection. For the second dimension, by exploiting the transferability of adversarial examples [37], we divided our attack into multiple stages until we reach a fine-tuned generator for addressing the victim's vulnerabilities. We considered as:

- *False positive*: The negative examples artificially generated that are misclassified as positive samples.
- *False negative*: The positive examples artificially generated that are misclassified as negative samples.

In the case of the malware detection task, a benign software being classified as malware is a false positive. Conversely, a false negative is a malware (usually considered as positive) that cannot be identified by the trained model. This is also known as *machine/deep learning evasion*.

5.3.1 Threat Model

We define the threat model as follows:

- The adversaries can attack only at the testing/deploying stage. They can tamper with only the input data in the testing stage after the victim deep learning model is trained. Further, we assume that neither the trained model nor the training dataset can be modified. The adversaries may have knowledge of the trained model (architectures and parameters) but not allowed to modify the model, which is a common assumption for many online machine learning services. We are not considering attacks at the training stage (e.g., training data poisoning [16, 34]), even if they are another interesting topic to explore.
- Since we considered adversarial attacks for deep neural networks, the adversaries target only the integrity of their inputs. In general, integrity is essential to a deep

learning model, although other security issues related to confidentiality and privacy have drawn attention in deep learning. Anyway, in the case we considered PE files, the integrity of the input is crucial. So, we focused on the attacks that degrade the performance of deep learning models for malware detection: attacks cause the increase of false positives and false negatives.

5.3.2 Adversarial Examples Generation

AEs are artificial inputs that are generated by modifying legitimate inputs so as to fool the classification models. In the fields of image and speech recognition, modified inputs are considered adversarial when they are indistinguishable by humans from the legitimate inputs, and yet they fool the model. Conversely, discrete sequences are inherently different than speech and images, as changing one element in the sequence may completely alter its meaning. For example, changing one word in a sentence may hinder its gradient in a binary file, where the input is a discrete sequence of bytes, changing one byte may result in invalid bytecode or different runtime functionality. In malware detection, an AE is a binary file that is generated by modifying an existing malicious binary. While the original file is correctly classified as malicious, its modified version is misclassified as benign. Recent works as [23] have shown that AEs cause catastrophic failures in malware detection systems, trained on a set of handcrafted features such as file headers and API calls. Our experiment (contribution) is focused on changing surface features by keeping the same original distribution of benign samples.

5.4 Dataset

We chose EMBER released by Endgame [3] as the dataset for our case study. EMBER is a collection of features extracted from a large corpus of Windows portable executables.

The first version of the dataset is a collection of 1.1 million PEs that were all scanned by VirusTotal in 2017. The second EMBER dataset release consisted of features extracted from samples collected in or before 2018.

The set of binary files is divided as follows:

- 900,000 training samples grouped in:
  - 300,000 malicious;
  - 300,000 benign; and
  - 300,000 unlabeled.
- 200,000 test samples grouped in:
  - 100,000 malicious;
  - 100,000 benign.

```
{
  "sha256": "04637...", "appeared": "2017-01",
  "label": 1,
  "histogram": [ 3818, 155, 135, ... ],
  "byteentropy": [ 0, 0, 0, ... ],
  "strings": { "numstrings": 170,
               "avlength": 8.170588235294117,
               ... },
  "general": { "size": 33334, "vsize": 45056,
               "has_debug": 0, ... },
  ...
}
```

Fig. 3 Code snippet from Ember dataset JSON files describing PEs features

The dataset is made up of JSON files. Each sample includes

- the sha256 hash of the original file as a unique identifier;
- the month the files was first seen;
- a label, which may be 0 for benign, 1 for malicious, or -1 for unlabeled; and
- eight groups of raw features that include both parsed values and format-agnostic histograms.

A code snippet from the JSON file is shown in Fig. 3.

5.4.1 Raw Features

The raw features include both parsed features and format-agnostic histograms and counts of strings. Parsed features, extracted from the PE file, are

- *General file*: Information including the file size and basic information obtained from the PE header.
- *Header information*: Reporting the timestamp, the target machine (string), and a list of image characteristics (list of strings). From the optional header, the target subsystem (string); DLL characteristics (a list of strings); the file magic as a string (e.g., "PE32"); major and minor image versions; linker versions; system versions and subsystem versions; and the code, headers, and commit sizes are provided.
- *Imported functions*: After having parsed the import address table, the imported functions by the library are reported.
- *Exported functions*: The raw features include a list of the exported functions.
- *Section information*: Properties of each section are provided, including the name, the size, the entropy, the virtual size, and a list of strings representing section characteristics.

The EMBER dataset also includes three groups of features that are format-agnostic, as they do not require parsing the PE file:

- Byte histogram contains 256 integer values, representing the count of each byte value within the file. The byte histogram is normalized to a distribution, since the file size is represented as a feature in the general file information.
- Byte-entropy histogram approximates the joint distribution p(H,X) of entropy H and byte value X.
- String information reported is the number of strings, their average length, a histogram of the printable characters within those strings, and the entropy of characters across all the printable strings.

5.5 Performance Metrics

Four metrics were used to evaluate the detectors (*accuracy, sensitivity, specificity, and evasion rate*) obtained under different testing conditions. We provide the general standard definitions for these metrics while reserving us to improve the explanation about how they were specifically computed in the specific sections. As first we provide definitions for true positives, true negatives, false positives, and false negatives.

True positive: (TP) = the number of malicious samples correctly identified as malicious;

False positive: (FP) = the number of benign (goodware) samples incorrectly identified as malicious;

True negative: (TN) = the number of benign samples correctly identified as benign; and

False negative: (FN) = the number of malicious samples incorrectly identified as benign.

By combining these observations it is possible to compute further indicators, whose general meaning is provided as follows:

Accuracy: The accuracy of a test is defined as its ability to differentiate the benign and malicious samples correctly. To estimate the accuracy of a test, we compute the proportion of true positive (TP) and true negative (TN) in all the evaluated cases. Mathematically, this can be stated as follows:

$$\text{Accuracy} = \frac{TP + TN}{(TP + FP + TN + FN)}.$$

Sensitivity: The sensitivity of a test is its ability to determine the malicious cases correctly. To estimate it, we should calculate the proportion of true positive (TP) in malicious cases. Mathematically, this can be stated as follows:

$$\text{Sensitivity} = \frac{TP}{(TP + FN)}.$$

Specificity: The specificity of a test is its ability to determine the good cases correctly. To estimate it, we compute the proportion of true negative among good cases. Mathematically, this can be stated as follows:

$$\text{Specificity} = \frac{TN}{(TN + FP)}.$$

Finally, we consider another indicator of sensitiveness known as the ***evasion rate***. When a dataset for testing the ability exhibited by a system in detection and/or classification task is poisoned with carefully designed adversarial examples, there are two adversary perspectives: the victim (detector) and the attacker (generator) ones. So, if we are interested to estimate the robustness of a detection system by computing its performance decay under an adversarial attack (e.g., an accuracy decay), we are interested in the estimation of the ability of the generator to produce adversarial examples that are misclassified. In this perspective, the evasion rate can be adopted as an indicator for measuring the generation ability and is defined, according to the definition provided in [10], as follows:

$$\text{Evasion Rate (EV)} = \frac{FN_{\text{AEs}}}{N_{\text{AEs}}},$$

where N_{AEs} represents the number of artificially generated adversarial samples of malware submitted to the detector and FN_{AEs} is the fraction of the overall counted false negatives (malware incorrectly classified as goodware) represented by adversarial samples set (that is to say, artificially generated malware incorrectly classified as goodware).

5.6 *Case Study Treatments*

The case study was conducted on an Ubuntu 18.04 platform, running on a cluster composed of five machines, with the same hardware configuration, equipped with an Intel Xeon E5-2620 processor and 128 GB RAM. We exploited the GPU functionalities of 5 NVIDIA GeForce RTX 2080 boards, by using the CUDA toolkit 9.0 and cuDNN with a TensorFlow-GPU v.1.13.1 version, running with Python 3.7. We further adopted the Ember script tools version 0.1.0, LightGBM 2.1.0, scikitlearn 0.19.1, NumPy 1.14.2, and SciPy 1.0.0, Matplotlib 3.2.2 for plotting results.

Table 1 PEs surface feature groups in the ember dataset

Feature group ID	Description and original name	Number of features
FG00	All	2351
FG01	General file info (General)	10
FG02	Header info (Header)	62
FG03	Imported functions (Imports)	1280
FG04	Exported functions (Exports)	128
FG05	Section info (Section)	255
FG06	Byte histogram (Histogram)	256
FG07	Byte-entropy histogram (Byte entropy)	256
FG08	String info (Strings)	104

5.6.1 Malware Detector Training: The Method

The total number of features comprised in each PE is equal to 2,351, grouped in eight families, according to the PE specifications[41]. Families' names and their quantity are provided in Table 1. We added, for convenience of comparison, the 9th group (FG00) representing the group including all the feature families, that is to say, 2,351 features.

For our case study, we started from considering all the features belonging to a group as a unit, so we always selected all the features in a feature group or we selected none.

Each combination was evaluated according to the following information:

- selected feature groups;
- accuracy and false positive rates (FPR) computed by varying the threshold of malware-likelihood scores by 0.01.

As described in the performance metric subsection, we define the accuracy as the ratio of the number of correct answers to the number of all answers, and FPR as the ratio of the number of malware-determination answers to the number of good samples.

For each feature combination, we associated a set of feature vectors with a ground-truth label and trained a different model; finally, we performed testing (malware-likelihood computation) operations. After performing training, validation, and tests, we selected the ten best detectors, according to the best accuracy values in the detection and a FP rate less than the limit threshold of 0.01.

5.6.2 Adversarial Examples Generator: The Method

As for the generation strategy for adversarial examples, we worked on the surface features of binary files and we focused on producing small changes on the most sensitive features groups, in order to reproduce, for the artificially generated samples, the distributions of the same features exhibited by goodware samples.

We want to remark that, for the purposes of the case study, we only considered applications metadata, extracted by the original binary files and conveniently provided in the EMBER dataset. We were interested to provide further evidence that feature-based models for malware detection, even if realized by the means of deep neural networks, may be broken by adversarial samples properly designed. We haven't considered the whole binary files, because manipulating the content of a binary file, even also changing a small number of byte, can severely compromise the behavior and the functionalities of the application. This aspect, also investigated in the works of [30, 32], will be a matter of further investigations, possibly joining both surface features and payload of binary files.

With previous works, we share the common approach of generating AEs by adding small perturbations to the original malicious inputs, in order to follow the probability distributions of the selected features groups, in the direction of the descent gradient, for reducing the distance between probabilities distribution.

Since we considered surface features, we observed that some features are more sensitive than others. So, our generation strategy consisted in following the probability distribution trend of these sensitive features in genuine goodware samples, thus producing a noise able to make closer the surface features probability distributions of the followed model (the genuine benign sample) with the probability distributions of the following model (the malware sample that has to be manipulated).

5.6.3 Training, Validation, and Test Sets Composition

In this section, details about the size of the training, validation, and test sets employed for performing the case study are provided. The samples composing these sets have been randomly extracted as a subset of the EMBER files collection, only excluding the adversarial samples that were artificially generated.

- **Training set for discriminators**: 300,000 genuine samples, divided into 150,000 goodware and 150,000 malware (X_{trainD}).
- **Training set for generators**: 300,000 genuine samples, divided into 150,000 goodware and 150,000 malware (X_{trainG}); this training set is intersectionless with the set adopted for training discriminators:

$$X_{trainD} \cap X_{trainG} = \emptyset.$$

- **Validation set for discriminators**: 50,000 genuine samples (GEs), divided into 25,000 goodware and 25,000 malware ($X_{validationD}$).

- **Test set for discriminators (excluding the adversarial samples)**: 45,000 genuine samples, divided into 15,000 goodware and 30,000 malware ($X_{testGEs}$).
- **Test set for GANs (discriminators including the adversarial scenario):** 45,000 samples, divided into 15,000 genuine benign samples, 15,000 genuine malicious samples (the same of discriminators without attack), 15,000 artificially generated adversarial examples of malware ($X_{testAEs}$).

5.7 *Case Study Results and Performance Evaluation*

In order to provide a clear and convenient explanation of our case study and its results, we decide to present the results splitting them into two scenarios, in order to compare how the malware detector performances degrade when attacked with adversarial samples. Results of our tests will be summarized in terms of *accuracy*, *sensitivity*, and *specificity* metrics.

Regarding the first scenario, sensitivity corresponds to the *true positive rate (TPR)*, where we considered as true positives all the malicious samples that were correctly identified as malware in the detection task. Finally, we considered the *false positive rate (FPR)* obtained in the malware detection task, computed according to the following equation:

$$FPR = \frac{FP}{FP + TP}.$$

Regarding the second scenario, instead, accuracy is computed as the success rate, i.e., the *evasion rate* (ER) obtained from the generator against his opponent (the detector), and measures the number of adversarial examples that pass undisturbed. In this scenario, the ER (coinciding with the TPR) is computed as the number of adversarial malicious examples that are misclassified as "good guys" (goodware); it corresponds to the ratio between the number of adversarial examples that successfully pass as "good guys" and the total number of adversarial examples submitted to the detector (discriminator).

5.7.1 Scenario 1: Discriminator Performance Excluding the Adversarial Attack

Results are shown only for the best ten trained models, according to the described criteria for the accuracy and FPR. In addition to these criteria, we performed two different tests for obtaining a further indication of the sensitiveness of the considered feature groups. Defining as $n_{C_{K_i}}$, with $K \in [A, B]$, the maximum number of feature groups considered in the ith combination C_i, Tables 2 and 3 show the accuracy scores of the best ten models (plus the 11th case of selecting all the available different features), respectively, in the case in which we set the additional conditions in the training models to

Table 2 Accuracy scores for the best ten combinations of features by considering the combinations of 4 different feature groups at most(C_A)

Combination ID	Selected feature group combination	Total number of feature	Accuracy rate (%)
C_{A_1}	General, Header, Histogram, Section	583	**92.27**
C_{A_2}	General, Header, Histogram, Strings	432	91.84
C_{A_3}	General, Header, Section, Strings	431	90.66
C_{A_4}	General, Header, Histograms	328	89.45
C_{A_5}	Header, Section, Strings	421	88.23
C_{A_6}	General, Header, Byte entropy	328	87.12
C_{A_7}	General, Section, Strings	369	86.35
C_{A_8}	General, Header, Strings	176	85.73
C_{A_9}	Section, Strings	359	83.07
$C_{A_{10}}$	General, Section	265	80.24
C_{A_0}	**All**	2351	**98.32**

$$C_S : n_{C_i} = 1$$
$$C_A : 1 \leq n_{C_i} \leq 4$$
$$C_B : 5 \leq n_{C_i} \leq 8.$$

The combination named *All* corresponds to all the eight groups *(Header, Imports, Section, Histogram, General, Exports, Byte entropy, Strings)*, including all the 2,351 features.

The accuracy metric was computed by adopting the test set denoted as ($X_{testGEs}$), comprising 45,000 genuine samples divided into 15,000 benign and 30,000 malicious samples.

By analyzing the results shown in Table 3, we can observe that the highest value for the accuracy is scored by the combination C_{B_0}, including all the features, while the closest score to this combination is obtained with a reduced set of features (combination C_{B_1}), with a difference in accuracy that is at minimum 1.43% (C_{B_0} versus C_{B_1}) and at maximum 2.58% (C_{B_0} versus C_{B_9}).

The feature groups *Header, Imports, Section*, and *Histogram* revealed to be particularly sensitive in biasing the accuracy score.

Table 3 Accuracy scores for the best ten combinations of features by considering at least five and at most seven different feature groups (C_B)

Combination ID	Selected feature group combination	Total number of feature	Accuracy rate (%)
C_{B_1}	Header, Imports, Section, Histogram, General, Strings	1967	**96.89**
C_{B_2}	Header, Imports, Section, Histogram, General, Byte entropy, Strings	2223	96.55
C_{B_3}	Header, Imports, Section, Histogram, Byte entropy, String	2213	96.39
C_{B_4}	Header, Imports, Section, Histogram, General, Exports, Byte entropy	2247	96.28
C_{B_5}	Header, Imports, Section, Histogram, String	1957	96.12
C_{B_6}	Header, Imports, Section, Histogram, Exports, Byte entropy	2237	96.07
C_{B_7}	Header, Imports, Section, Histogram, General, Exports, String	2095	95.96
C_{B_8}	Header, Imports, Section, Histogram, Byte entropy	2109	95.89
C_{B_9}	Header, Imports, Section, Histogram, General, Exports	1991	95.74
C_{B_0}	All	2351	**98.32**

Table 4 Accuracy scores for singleton feature group combinations (C_S)

Combination ID	Selected feature group combination	Total number of feature	Accuracy rate (%)
C_{S_1}	Imports	1280	82.79
C_{S_2}	Section	255	73.46
C_{S_3}	Histogram	256	73.14
C_{S_4}	Byte entropy	256	65.81
C_{S_5}	Strings	104	64.73
C_{S_6}	General	10	**61.59**
C_{S_7}	Header	62	54.13
C_{S_8}	Exports	128	20.45

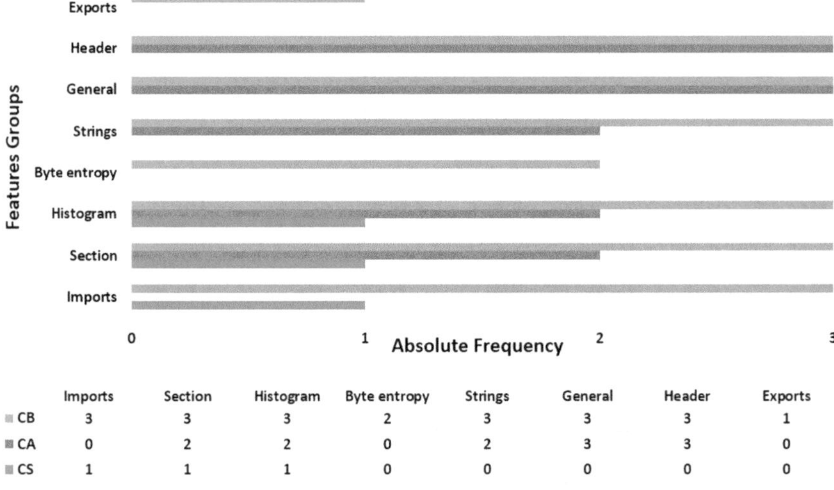

	Imports	Section	Histogram	Byte entropy	Strings	General	Header	Exports
CB	3	3	3	2	3	3	3	1
CA	0	2	2	0	2	3	3	0
CS	1	1	1	0	0	0	0	0

CB CA CS

Fig. 4 Feature groups' absolute frequency in the three most accurate models

Particularly, we observed that the feature group *Histogram* appears in the best scores both in the reduced (C_A) and in the extended (C_B) feature group combinations.

Finally, since the information about feature sensitiveness to the accuracy is crucial for designing the generation strategies for adversarial samples that will be effective, we performed the last test considering only single group combinations, as shown in Table 4.

We can observe that the feature groups that scored the best accuracy values were *imports, section*, and *histogram*. These three groups were also included in all the ten best ranking models considered in Table 3, where the best results were generally obtained. Figure 4 summarizes the absolute frequency scored for all the eight features groups over the three best ranked models for each of the three training we performed.

5.7.2 Generator and Discriminator Performance Including Adversarial Attack

For generating adversarial examples and testing the pool comprising the ten most accurate discriminator models described in the previous scenario, we trained correspondingly ten generators.

Given the initial working hypothesis of having knowledge, at this stage, of discriminators gradients generated during the training over the genuine dataset, we had the opportunity to exploit them in combination with the inner gradients of generators, in order to apply a semi-direct training process for the generators. To be clearer, we could adopt both a direct method for training the generators and an indirect one.

The direct method does not require to involve the discriminators during the training of the generators that are trained by simply comparing the difference elapsing between the probability distributions of genuine samples and the artificially generated (adversarial) samples. This method is practicable in this case because we have the true genuine data (a kind of white-box attack at the first stage) available.

In the second and most realistic stage of the attack scenario, we imagined (a black-box attack) genuine data aren't available and the direct method for training generators can't be applied yet. In this situation, the generators can be trained by submitting, at each training iteration, the generated outputs to the victim and collecting the response, for computing a step for making descendant the gradient function.

For reasons of simplicity, we adopted the direct method, since our aim was to provide evidence that feature-based deep learning models for malware detection could work with high accuracy even if the static analysis is performed; anyway, as other kinds of deep learning models, also performing a dynamic analysis of samples, they are affected by adversarial examples carefully designed.

So, the generators were trained by performing the comparison between the probability distribution of its generated samples with a "genuine" training set and back-propagating the difference (the error) through the network, at each iteration of the training process. To compute the distance (or similarity measure), we adopted the maximum mean discrepancy (MMD) [5, 6, 18], able to compare effectively two distributions.

Then, the training process of the generative networks develops as follows. Given a random variable with uniform probability distribution as input, we want the probability distribution of the generated output to be the "genuine data set probability distribution"; we considered two subcases:

- the first one in which the genuine dataset is the same adopted for training the discriminators;
- the second one, in which the genuine dataset is represented by a different and intersection fewer dataset from the one adopted for training discriminators.

The training process for each of the generators follows the basic idea to optimize its inner network by repeating the following steps:

- to generate some random inputs of the same size as the corresponding discriminator;
- to make these inputs go through the generator and the discriminator and collect both generated outputs;
- to compare the "genuine probability distribution" and the artificially generated one, by computing the MMD distance between the true samples and the generated ones; and
- to adopt backpropagation to make one step of gradient descent to lower the MMD distance between the truly genuine and artificially generated distributions.

We discuss here how to manipulate a source malware sample x into an adversarial malware binary x^* by slightly changing the surface feature values. Generators aim to minimize the confidence associated with the malicious class (i.e., it maximizes the probability of the adversarial malware sample being classified as benign), under the constraint that q_{max} is the maximum amount of noise (changes) that can be added to the original sample for being effective. The deep network implementing each generator produces the probability of the generic sample x being malware, denoted in the following with *f(x)*. If $f(x) \geq 0.5$, the input file is thus classified as malware (and as benign, otherwise).

This can be characterized as the following constrained optimization problem:

$$min_x f(x)$$
$$s.t. d(x, x^*) \leq q_{max},$$

where

- x denotes the genuine sample distribution;
- x^* denotes the generated sample; and
- d(x, x^*) is the distance function computed as the MMD distance.

We solve this problem with a gradient-descent algorithm over the generator networks by adopting as loss function the distance between the true and the generated distributions at the current iteration.

We trained each generator, for both the genuine datasets, for 500 epochs, and we also adopted a learning rate set to 0.05. These hyperparameters for the training process were obtained after all the ten generators were able to converge and we stopped when the error reached the threshold value of 0.02 (2%) (corresponding to an accuracy rate in validation and testing of 98%). We were not able to reach lower error rates, because we trained generators for being able to produce just over 15.000 adversarial samples, in order to ensure the same numerosity of the genuine test examples when testing the discriminators. The overall time for training the ten generators until all of them converge to a similarity rate of 98%, estimated between the truly genuine and the adversarial generated samples distributions, lasted about 1 day and a half (about 37 h). We repeated the training process five times and we considered as assessed the generator models after a time of about 10 d.

In this way, we obtained 30,000 adversarial examples, divided into two sets G_{EQ} and G_{NEQ} comprising, respectively, 15,000 adversarial malicious samples generated (AEs$_{EQ}$)from comparison with the genuine training set adopted for discriminator models and 15,000 artificially generated samples (AEs$_{NEQ}$) computed by evaluating the difference from a different dataset from the one adopted for training discriminator models.

Finally, we addressed the adversarial attack to the ten discriminators with both the two sets of generated adversarial samples and we provide a brief discussion over the results we observed.

Like in Scenario 1, for computing performance metrics, we tested the discriminators with two variants of the test set denoted as $(X_{test AEs})$; each variant comprises 45,000 samples, divided into 15,000 genuine benign samples and 30,000 malicious samples, in turn divided into 15,000 genuine malware and 15,000 adversarial malware samples. The two classes of test sets differ only for the kind of adversarial malware samples included. In the first class, we included adversarial malware samples generated by comparison with the same training set adopted for training discriminator models; in the second class, we included adversarial malware samples generated by adopting a different training set from the one adopted for training discriminators.

So, we discussed these two cases of AE attack and we compared them with the original accuracy scored by each discriminator when excluding AEs from its test set.

To verify the efficacy of the attack, for each test we measured beyond the accuracy and the sensitivity, also the evasion rate [10], computed as the percentage of malicious samples that managed to evade the network [30].

For each of the three cases shown in Table 5, accuracy was computed considering a test set comprising 45,000 samples; anyway, these tests were differently composed for allowing, respectively, the case excluding the adversarial examples and the two cases including AEs generated by comparing or not comparing to the genuine training set adopted for discriminators. These cases include

- Excluding AEs: No AEs attack is performed against the discriminators; the test set is made of 45,000 genuine samples only, divided into 15,000 goodware samples and 30,000 malware samples.
- Including AEs trained over the training set adopted by the discriminator: AEs attack is performed against the discriminators; the test set is made of 45,000 samples, among which 30,000 genuine samples are divided into 15,000 goodware and 15,000 malware; the remaining 15,000 represent adversarial examples; this test set will be called as follows: ***AEs$_{EQ}$***.
- Including AEs trained over a different training set from the one adopted by the discriminator: AEs attack is performed against the discriminators; the test set is made of 45,000 samples, among which 30,000 genuine samples are divided into 15,000 goodware and 15,000 malware; the remaining 15,000 represent adversarial examples; this test set will be called in the following as ***AEs$_{NEQ}$***.

The results that we obtained in terms of evasion rate and accuracy decay for each of the ten discriminators are summarized in Tables 5 and 6.

Table 5 Accuracy rate reached by attacking discriminators with adversarial examples from sets AEs_{EQ} and AEs_{NEQ}

	TEST 1	TEST 2	TEST 3
Discriminator ID	Accuracy excluding AEs (%)	Accuracy including AEs_{EQ} (%)	Accuracy including AEs_{NEQ} (%)
C_{B_1}	**96.89**	58.77	73.32
C_{B_2}	96.55	56.54	78.73
C_{B_3}	96.39	57.87	77.17
C_{B_4}	96.28	60.59	74.43
C_{B_5}	96.12	59.25	74.31
C_{B_6}	96.07	62.58	76.87
C_{B_7}	95.96	56.55	75.77
C_{B_8}	95.89	59.75	76.38
C_{B_9}	95.74	56.74	76.17
C_{B_0}	**98.32**	57.52	74.68

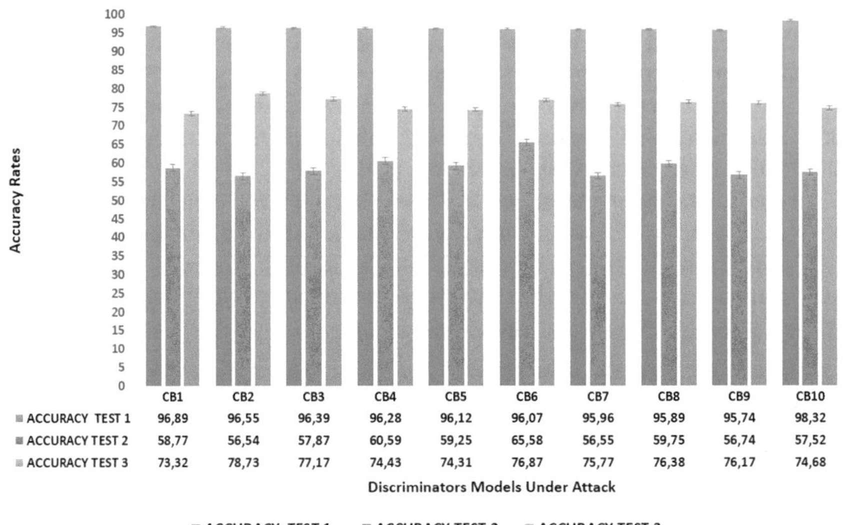

Fig. 5 Accuracy rate distributions for discriminators under AEs attack

In Figs. 5, 6, and 7 are reported, respectively, the accuracy rate distributions and the trend line of the accuracy decay, computed over the ten discriminators and the two types of AEs considered, when discriminators are under AEs attack.

We can observe that AEs perform worse (test set AEs_{NEQ}) than the other AEs adversarial set, producing, over the ten tested models for discriminators, an average decay of accuracy valued to Delta $(a_{AEs_{EQ}}) = 20.63$ points. Minimum loss $min_{loss} = 19.20$ points and maximum loss $max_{loss} = 23.57$ points.

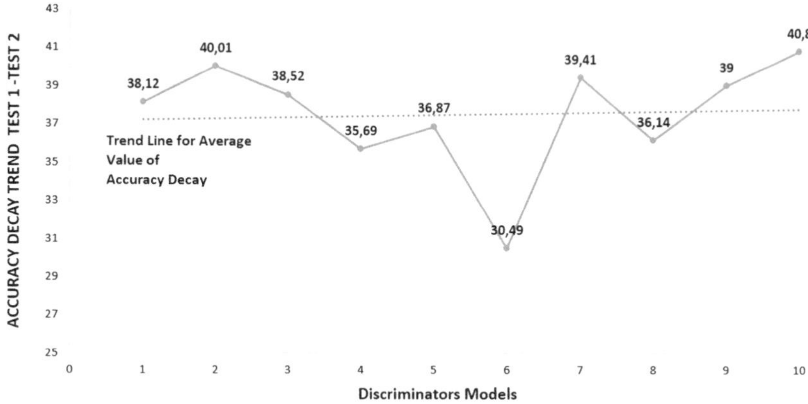

Fig. 6 Trend of accuracy decay of discriminators under AEs attack (TEST 1–TEST 2)

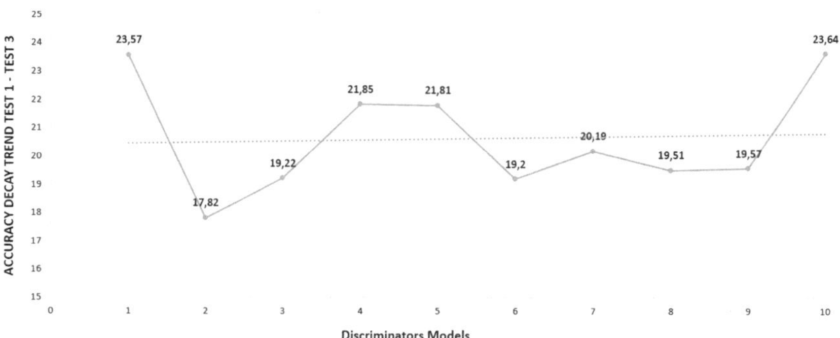

Fig. 7 Trend of accuracy decay of discriminators under AEs attack (TEST 1–TEST 3)

Instead, when adversarial attack is performed by adopting adversarial examples produced by generators trained over the same dataset adopted for training the discriminators, AEs perform better (test set AEs_{EQ}) than the other AEs adversarial set, affecting over the ten tested models for discriminators an average decay of accuracy valued to Delta $(a_{AEs_{EQ}}) = 37.50$ points. Minimum loss $min_{loss} = 30.49$ points and maximum loss $max_{loss} = 40.80$ points.

6 Conclusions

Attacks and defenses on adversarial examples draw great attention. The vulnerability to adversarial examples becomes one of the major risks for applying DNNs in safety-critical environments.

Table 6 Evasion rate computed attacking discriminators with adversarial examples (AEs) from sets A_{EQ} and A_{NEQ}

Discriminator ID	Evasion rate with AEs_{EQ} (%)	Evasion rate with AEs_{NEQ} (%)
C_{B_1}	**35.84**	**26.69**
C_{B_2}	33.72	27.38
C_{B_3}	34.65	25.44
C_{B_4}	32.61	26.14
C_{B_5}	34.89	25.98
C_{B_6}	33.77	26.29
$C)_{B_7}$	31.52	27.66
C_{B_8}	29.05	26.39
C_{B_9}	27.39	24.61
C_{B_0}	**39.12**	**32.45**

Adversarial perturbations can easily fool deep neural networks (DNNs) in the testing/deploying stage exploiting blind spots in the ML engine. The effectiveness of an adversarial system is measured in terms of *evasion rate* and it depends upon a specific group of features considered for the input set. Applied to the creation of malware, GANs are able to generate a new instance of a malware family without knowing an explicit model of the initial distribution of the data.

So an attacker could use GANs to fool detection systems, just by sampling the provided data. On the other hand, GANs are also useful to build more robust machine learning models helping in the development of a better training set. Real defense technologies such as AV or EDR must take into account an acceptable trade-off among the detection accuracy, short learning times, and limit the size of data obtainable by selecting a convenient combination of the sensitive feature. The effectiveness of an attack on the ML model also depends on the knowledge of the system by the attacker. In this case study, we conducted a gray-box attack in which the features of the training set are known: this permits us to reach a very high evasion rate (about 98%).

References

1. Alazab, Mamoun, Sitalakshmi Venkatraman, Paul Watters, and Moutaz Alazab. 2013. Information security governance: the art of detecting hidden malware. In *IT security governance innovations: theory and research*, 293–315. IGI Global.
2. Alzantot, Moustafa, Bharathan Balaji, and Mani Srivastava. 2018. Did you hear that? adversarial examples against automatic speech recognition. arXiv:1801.00554.
3. Anderson, Hyrum S., and Phil Roth. 2018. Ember: an open dataset for training static pe malware machine learning models. arXiv:1804.04637.

4. Apruzzese, Giovanni, Michele Colajanni, Luca Ferretti, Alessandro Guido, and Mirco Marchetti. 2018. On the effectiveness of machine and deep learning for cyber security. In *2018 10th international conference on cyber Conflict (CyCon)*, pages 371–390. IEEE, 2018.
5. Arbel, Michael, Dougal Sutherland, Mikołaj Bińkowski, and Arthur Gretton. 2018. On gradient regularizers for mmd gans. *Advances in neural information processing systems* 6700–6710.
6. Arjovsky, Martin, and Léon Bottou. 2017. Towards principled methods for training generative adversarial networks. arXiv:1701.04862.
7. Azab, Ahmad, Mamoun Alazab, and Mahdi Aiash. 2016. Machine learning based botnet identification traffic. In *2016 IEEE Trustcom/BigDataSE/ISPA*, 1788–1794. IEEE.
8. Azab, Ahmad, Robert Layton, Mamoun Alazab, and Jonathan Oliver. 2014. Mining malware to detect variants. In *2014 fifth cybercrime and trustworthy computing conference*, 44–53. IEEE.
9. Benchea, Răzvan, and Dragoş Teodor Gavriluţ. 2014. Combining restricted boltzmann machine and one side perceptron for malware detection. In *International conference on conceptual structures*, 93–103. Springer.
10. Biggio, Battista, Igino Corona, Davide Maiorca, Blaine Nelson, Nedim Šrndić, Pavel Laskov, Giorgio Giacinto, and Fabio Roli. 2013. Evasion attacks against machine learning at test time. In *Joint European conference on machine learning and knowledge discovery in databases*, 387–402. Springer.
11. Biggio, Battista, Paolo Russu, Luca Didaci, Fabio Roli, et al. 2015. Adversarial biometric recognition: A review on biometric system security from the adversarial machine-learning perspective. *IEEE Signal Processing Magazine* 32 (5): 31–41.
12. Brown, Tom B., Dandelion Mané, Aurko Roy, Martín Abadi, and Justin Gilmer. 2017. Adversarial patch. arXiv:1712.09665.
13. Carlini, Nicholas, and David Wagner. 2017. Towards evaluating the robustness of neural networks. In *2017 IEEE symposium on security and privacy (sp)*, 39–57. IEEE.
14. Chen, Liang-Chieh, George Papandreou, Florian Schroff, and Hartwig Adam. 2017. Rethinking atrous convolution for semantic image segmentation. arXiv:1706.05587.
15. Chen, Pin-Yu, Huan Zhang, Yash Sharma, Jinfeng Yi, and Cho-Jui Hsieh. 2017. Zoo: Zeroth order optimization based black-box attacks to deep neural networks without training substitute models. In *Proceedings of the 10th ACM workshop on artificial intelligence and security*, 15–26.
16. Chen, Xinyun, Chang Liu, Bo Li, Kimberly Lu, and Dawn Song. 2017. Targeted backdoor attacks on deep learning systems using data poisoning. arXiv:1712.05526.
17. Damodaran, Anusha, Fabio Di Troia, Corrado Aaron Visaggio, Thomas H. Austin, and Mark Stamp. 2017. A comparison of static, dynamic, and hybrid analysis for malware detection. *Journal of Computer Virology and Hacking Techniques* 13 (1): 1–12.
18. Dziugaite, Gintare Karolina, Daniel M Roy, and Zoubin Ghahramani. Training generative neural networks via maximum mean discrepancy optimization. arXiv:1505.03906.
19. Firdausi, Ivan, Alva Erwin, Anto Satriyo Nugroho, et al. 2010. Analysis of machine learning techniques used in behavior-based malware detection. In *2010 second international conference on advances in computing, control, and telecommunication technologies*, 201–203. IEEE.
20. Gibert, Daniel. 2016. *Convolutional neural networks for malware classification*. Tarragona, Spain: University Rovira i Virgili.
21. Goodfellow, Ian, Patrick McDaniel, and Nicolas Papernot. 2018. Making machine learning robust against adversarial inputs. *Communications of the ACM* 61 (7): 56–66.
22. Goodfellow, Ian J., Jonathon Shlens, and Christian Szegedy. 2014. Explaining and harnessing adversarial examples. arXiv:1412.6572.
23. Grosse, Kathrin, Nicolas Papernot, Praveen Manoharan, Michael Backes, and Patrick McDaniel. 2017. Adversarial examples for malware detection. In *European symposium on research in computer security*, 62–79. Springer.
24. He, Kaiming, Xiangyu Zhang, Shaoqing Ren, and Jian Sun. 2015. Delving deep into rectifiers: Surpassing human-level performance on imagenet classification. In*Proceedings of the IEEE international conference on computer vision* 1026–1034.

25. Hu, Weiwei, and Ying Tan. 2017. Generating adversarial malware examples for black-box attacks based on gan. arXiv:1702.05983.
26. Huang, Ling, Anthony D Joseph, Blaine Nelson, Benjamin IP Rubinstein, and J Doug Tygar. 2011. Adversarial machine learning. In *Proceedings of the 4th ACM workshop on Security and artificial intelligence*, 43–58.
27. Jung, Wookhyun, Sangwon Kim, and Sangyong Choi. 2015. Poster: deep learning for zero-day flash malware detection. In *36th IEEE symposium on security and privacy*, vol. 10, 2809695–2817880.
28. Kawai, Masataka, Kaoru Ota, and Mianxing Dong. 2019. Improved malgan: Avoiding malware detector by leaning cleanware features. In *2019 international conference on artificial intelligence in information and communication (ICAIIC)*, 040–045. IEEE.
29. Ke, Guolin, Qi Meng, Thomas Finley, Taifeng Wang, Wei Chen, Weidong Ma, Qiwei Ye, and Tie-Yan Liu. 2017. Lightgbm: A highly efficient gradient boosting decision tree. In *Advances in neural information processing systems* 3146–3154.
30. Kolosnjaji, Bojan, Ambra Demontis, Battista Biggio, Davide Maiorca, Giorgio Giacinto, Claudia Eckert, and Fabio Roli. 2018. Adversarial malware binaries: Evading deep learning for malware detection in executables. In *2018 26th European signal processing conference (EUSIPCO)*, 533–537. IEEE.
31. Kolosnjaji, Bojan, Apostolis Zarras, George Webster, and Claudia Eckert. 2016. Deep learning for classification of malware system call sequences. In *Australasian joint conference on artificial intelligence*, 137–149. Springer.
32. Kreuk, Felix, Assi Barak, Shir Aviv-Reuven, Moran Baruch, Benny Pinkas, and Joseph Keshet. 2018. Deceiving end-to-end deep learning malware detectors using adversarial examples. arXiv:1802.04528.
33. Moosavi-Dezfooli, Seyed-Mohsen, Alhussein Fawzi, and Pascal Frossard. 2016. Deepfool: a simple and accurate method to fool deep neural networks. In*Proceedings of the IEEE conference on computer vision and pattern recognition* 2574–2582.
34. Muñoz-González, Luis, Battista Biggio, Ambra Demontis, Andrea Paudice, Vasin Wongrassamee, Emil C Lupu, and Fabio Roli. 2017. Towards poisoning of deep learning algorithms with back-gradient optimization. In *Proceedings of the 10th ACM workshop on artificial intelligence and security*, 27–38.
35. Obeis, Turki, and Wesam Bhaya Nawfal. 2016. Review of data mining techniques for malicious detetion. *Research Journal of Applied Sciences* 11 (10): 942–947.
36. Oyama, Yoshihiro, Takumi Miyashita, and Hirotaka Kokubo. 2019. Identifying useful features for malware detection in the ember dataset. In *2019 seventh international symposium on computing and networking workshops (CANDARW)*, 360–366. IEEE.
37. Papernot, Nicolas, Patrick McDaniel, and Ian Goodfellow. 2016. Transferability in machine learning: from phenomena to black-box attacks using adversarial samples. arXiv:1605.07277.
38. Papernot, Nicolas, Patrick McDaniel, Ian Goodfellow, Somesh Jha, Z Berkay Celik, and Ananthram Swami. 2017. Practical black-box attacks against machine learning. In *Proceedings of the 2017 ACM on Asia conference on computer and communications security*, 506–519.
39. Pascanu, Razvan, Jack W Stokes, Hermineh Sanossian, Mady Marinescu, and Anil Thomas. 2015. Malware classification with recurrent networks. In *2015 IEEE international conference on acoustics, speech and signal processing (ICASSP)*, 1916–1920. IEEE.
40. Pendlebury, Feargus, Fabio Pierazzi, Roberto Jordaney, Johannes Kinder, and Lorenzo Cavallaro. 2019. {TESSERACT}: Eliminating experimental bias in malware classification across space and time. In *28th {USENIX} Security Symposium ({USENIX} Security 19)*, 729–746.
41. Pietrek, Matt. 2002. Inside windows-an in-depth look into the win32 portable executable file format. *MSDN Magazine* 17 (2): 80–90.
42. Puranik, Piyush Aniruddha. 2019. Static malware detection using deep neural networks on portable executables.
43. Raff, Edward, Jon Barker, Jared Sylvester, Robert Brandon, Bryan Catanzaro, and Charles Nicholas. 2017. Malware detection by eating a whole exe. arXiv:1710.09435.

44. Redmon, Joseph, Santosh Divvala, Ross Girshick, and Ali Farhadi. 2016. You only look once: Unified, real-time object detection. In *Proceedings of the IEEE conference on computer vision and pattern recognition* 779–788.
45. Saxe, Joshua, and Konstantin Berlin. 2015. Deep neural network based malware detection using two dimensional binary program features. In *2015 10th international conference on malicious and unwanted software (MALWARE)*, 11–20. IEEE.
46. Su, Jiawei, Danilo Vasconcellos Vargas, and Kouichi Sakurai. 2019. One pixel attack for fooling deep neural networks. *IEEE Transactions on Evolutionary Computation* 23 (5): 828–841.
47. Szegedy, Christian, Wojciech Zaremba, Ilya Sutskever, Joan Bruna, Dumitru Erhan, Ian Goodfellow, and Rob Fergus. 2013. Intriguing properties of neural networks. arXiv:1312.6199.
48. Ucci, Daniele, Leonardo Aniello, and Roberto Baldoni. 2019. Survey of machine learning techniques for malware analysis. *Computers & Security* 81: 123–147.
49. Ye, Yanfang, Tao Li, S. Donald Adjeroh, and Sitharama, and Iyengar. 2017. A survey on malware detection using data mining techniques. *ACM Computing Surveys (CSUR)* 50 (3): 1–40.
50. Yuan, Xiaoyong, Pan He, Qile Zhu, and Xiaolin Li. 2019. Adversarial examples: Attacks and defenses for deep learning. *IEEE Transactions on Neural Networks and Learning Systems* 30 (9): 2805–2824.
51. Zhang, Jinlan, Qiao Yan, and Mingde Wang. 2019. Evasion attacks based on wasserstein generative adversarial network. In *2019 Computing, communications and IoT applications (ComComAp)*, 454–459. IEEE.
52. Zhong, Wei, and Gu Feng. 2019. A multi-level deep learning system for malware detection. *Expert Systems with Applications* 133: 151–162.

Related Topics

Detecting Abusive Comments Using Ensemble Deep Learning Algorithms

Ravinder Ahuja, Alisha Banga, and S C Sharma

Abstract Today, there is an avalanche of data on social networking sites. Technology has facilitated our way of Internet usage and provided us with great liberty to do what, when, and how we like. In just one click, we can share, like, comment any post on social media, but this liberty has caused a severe threat to humans; unfortunately, the online interaction among users with such ease involves harassment, abuse, and bullying actions. The concern over this problem has triggered to build up better models for classifying the abusive comments. In this chapter, we have applied four classification algorithms: Naive Bayes, Random Forest, Decision Tree, and Support Vector Machine, with Bag of Words features. Deep learning algorithms: Convolutional Neural Network (CNN), Long Short-Term Memory (LSTM), and an ensemble of LSTM and CNN are applied using GloVe and fastText word embedding to classify the comments into six categories: toxic, severe toxic, obscene, threat, insult, and identity hate. We have taken data set from Kaggle competition. We conducted experiments by using Keras library and TensorFlow at the back end and taken accuracy as performance parameter. We found that CNN, LSTM blend with fastText word embedding performs better out of all the algorithms applied with an accuracy of 98.46%.

1 Introduction

Social media provides an environment where people are given complete freedom to post, comment, engage in discussions, and share their opinions. Such online platforms have given us an entirely new dimension to communicate and express our thoughts.

R. Ahuja (✉) · A. Banga · S. C. Sharma
Indian Institute of Technology Roorkee Saharanpur Campus, Saharanpur, India
e-mail: ahujaravinder022@gmail.com

A. Banga
e-mail: alishabanga47@gmail.com

S. C. Sharma
e-mail: scs60fpt@gmail.com

M. Stamp et al. (eds.), *Malware Analysis Using Artificial Intelligence and Deep Learning*, https://doi.org/10.1007/978-3-030-62582-5_20

With time, online networking sites like Facebook and Twitter have proved to be an integral part of our social lives. They are handy for expressing opinions, thoughts, and views of people present all around the globe. In 2017 only, the daily Facebook post volume is 4.3 billion, and 656 million tweets are posted daily on Twitter [28]. There are many applications of analyzing contents on social media like knowing person's dietary preferences [38]. With so much importance of social media, it is a matter of concern for the authorities to make these platforms safe for the masses. With such a wide variety of users, social media platforms are getting miserable because of a constant increase in the toxicity of comments, posts, and thoughts. Functionalities to report comments as abusive or toxic, block people, report discussions, and remove such comments have been provided. Still, there is a lack of a moderating functionality that flags such comments. There are users all around the globe with different languages and writing styles. Not only language, but they are becoming more innovative with the use of URLs and unique uses of emoticons. The high usage of these social networking sites demands a well-modeled automated approach to handle such vast data and detect toxicity in the content present with reasonable accuracy. There have been criticisms of online social networking sites for the negligence of cyber bullying and their incapability to classify and remove toxic comments [10]. As these sites have a wider audience, such comments and posts spread like wildfire, which is a matter of concern for Social Networking Sites. Although efforts have been made to increase the online environment's safety based on the crowd sourcing techniques in most cases, these techniques have failed in detecting toxicity. As per cyber Bullying Research Center data of 2016 [5], 26% of the sample reported cyberbullying related to different types of bullying, two or more times within a time interval of just 30 d. 73% of adult users on the Internet have seen online harassment as per 2014 Pew Report [7]. It is becoming a toxic place buried under tonnes of rubbish and obsolete comments and posts. It is reported that the person who faced online harassment has decreased participation, which has occurred in the next project [32]. There are so many cases that children were so bullied on social media that they are depressed and lead to suicide. Therefore, the work we are proposing is centered around this very thought to make social media a safer place. Automatic detection and classification of the comments, posts, or messages at the correct time are of paramount importance. Considering the humongous amount of comments being produced everyday classifying comments manually would not be a feasible approach. Therefore, an optimal model is required for text classification. We describe a method to efficiently perform this task using deep learning models and detecting toxicity in the comments through our work. Our job is intended to classify comments into six classes as obscene, identity hate, toxic, severe toxic, insult, and threat so that social media can now select the type of toxicity they are trying to resolve. The contribution of this chapter is as follows: (i) Machine learning algorithms—Support Vector Classifier, Decision Tree, Naive Bayes, and Random Forest classifier with a bag of words is applied to classify the text into six categories (ii) Deep learning algorithms—CNN, LSTM, and their blends is used with GloVe and fastText word embedding. The rest of the chapter is organized as follows: Sect. 2 contains literature survey, Sect. 3 contains material and methods, Sect. 4 contains approach used, and Sect. 5 contains experimental results and analysis, followed by the conclusion section.

2 Literature Survey

Abusive text classification research was first represented in the paper [35] in which supervised machine learning technique SVM was applied on TF-IDF features. In paper [20], toxic comment classification is performed using SVM. The data was collected from Youtube.com, and 2665 English comments were collected. Out of which 1451 comments were classified as neutral or positive. Other 1214 were stated to be abusive or spam comments and achieved an accuracy of 86.95% using tenfold cross-validations. In the paper [10], attacks on perspective API created by Google and the Jigsaw team for a toxic detection system is proposed (2017). The Perspective website provides some sample phrases, and an attack is applied to these phrases. This paper shows that toxicity scores can be reduced to the level of non-toxic phrases. Further, this perspective API tool is only for the English language. Recently, researchers have been applying deep learning models like Convolution Neural nets and LSTM to optimize text classification. In the paper [8], CNN with word2vec word embedding was compared against a bag of words in which SVM and Naive Bayes were applied on designed DTMs (Document Term Matrices). LSTM for tweets was introduced for twitter sentiment production in the paper [31], which gave us the idea that LSTM might work pretty well for our dataset and solve a vanishing gradient problem. In the paper [6], word, character, and sentence level representations were used to perform sentiment analysis. They used two datasets and found that character level embedding has performed better in the case of the first dataset, but in case of other data, all of the embeddings perform much better. In the paper [33], authors have applied traditional machine learning algorithms on 4029 messages to detect profanity related texts on Twitter. They reported that Logistic regression is performing well among all algorithms applied. In the paper [16], authors have applied the SVM and Naïve Bayes algorithm to identify abusive comments from text or images from social media platforms. In the paper [21], authors have applied fifteen transformations on the toxic data to determine whether using these transformations can improve the models' performance. But they found that applying these transformations do not increase the performance of the model. Their suggestion was to select the best model instead of wasting time on the transformation of data. In the paper [27], the authors have applied logistic regression, convolutional neural network, long short-term memory, and CNN-LSTM to classify toxicity in the comments. They reported that CNN-LSTM is giving the highest accuracy of 98.20%. In the paper[25], authors have considered the TF-IDF features and applied three classification algorithms, namely Logistic Regression, Random Forest, and Gradient Boosting. Out of the three algorithms applied, logistic regression has given the highest accuracy of 97.20%. In the report [30], authors have applied LSTM and CNN on the dataset taken from Kaggle competition and used various word embedding like word2vec and GloVe. In the Thesis [19], Siyuan Li applied word2vec, Glove, and skip-gram embedding vectors in Gated Recurrent Unit, and Bi-LSTM (bi-directional long short-term memory). They applied sampling techniques and penalizing loss to handle the imbalance issue of the dataset. He concluded that using pre-trained word embedding is not necessarily will improve the performance. Sampling technique and penalizing loss increase the performance of the model. In the paper [1], authors have applied deep learning techniques on three datasets from three different social media platforms. Firstly, they have applied traditional classification algorithms with char n-gram and word unigram and computed the performance of these algorithms. They have also applied CNN, LSTM, and Bi-LSTM with random, GloVe, and SSWE word embedding. They have applied transfer learning also to know whether the model trained on one dataset can be applied to another dataset or not. In the paper [24], authors have applied J48, Jrip, SVM with different kernels, KNN, Naïve Bayes, Random Forest, and CNN with one hidden layer and two hidden layers for cyberbullying detection. They reported that CNN, with two hidden layers, outperforms all the algorithms applied. In the paper [15], authors have applied support vector machine, multinomial naive bayes, GausseanNB, back propagation multilabel neural networks algorithms on Bangla language text for toxic comments classification. They evaluated the models on the basis log loss and hamming log and found that back propagation multilabel neural network is performing better among all the algorithms. In the paper [22], authors have presented a review of various machine learning techniques used in toxic comment classification from 2012 to 2015. In addition to this, they have also presented two tools for detecting abusive comments and their advantages and limitations. In paper [30], authors have applied CNN, LSTM, Bi-directional LSTM, Bi-directional GRU, Bi-directional GRU with attention, and using Glove and fastText word embedding. They have also applied logistic regression using char n-gram and word n-gram. Further, authors ensemble all these algorithms on Wikipedia and Twitter datasets and reported the highest performance parameters AUC of 98.3%, F1 score of 79.1%, the precision of 74%, and recall of 88%. In the paper [2], authors have applied decision tree, random forest, support vector machines, gradient-based decision tree, and deep neural network with Glove word embedding to detect hate speech. In the paper [11], Mai Ibrahim et al. applied data augmentation to remove the data's imbalance effect. Further, an ensemble of three algorithms, i.e., convolutional neural network, bi-directional long short-term memory, and bi-directional gated recurrent unit. They have applied two classifiers: one is used to identify whether the comment is toxic or not. If the comment is toxic, then another multi-classifier is applied to classify different types of toxicity. They reported the highest f-score of 82.82% in the case of toxic/non-toxic classifiers and an f-score of 87.24% in the case of toxicity-type classifiers.

3 Materials and Methods

3.1 Dataset

We have taken a dataset from www.Kaggle.com, which consists of 159571 Wikipedia comments labeled by humans as a training set and 153164 comments for testing. These ratings have classified toxicity in 6 classes as toxic, severe toxic, obscene, threat, insult, and identity hate. Figure 1 shows the count of different tags in the dataset. Word cloud for comments under these six categories is shown in Fig. 2.

3.2 Data Pre-processing

Text pre-processing is an important step before applying any classification algorithms. Authors in the paper [13] compared various text pre-processing techniques. The following pre-processing techniques have been applied. Technique 1: Unnecessary characters like!, ", ()*+,-./: etc. and stop words are removed. Technique 2: All the letters are converted to lowercase. Technique 3: Tokenization. Technique 4: Stemming. We have pre-processed the data using Python regular expression, stemming from Porter Stemmer, removing stop words with NLTK libraries.

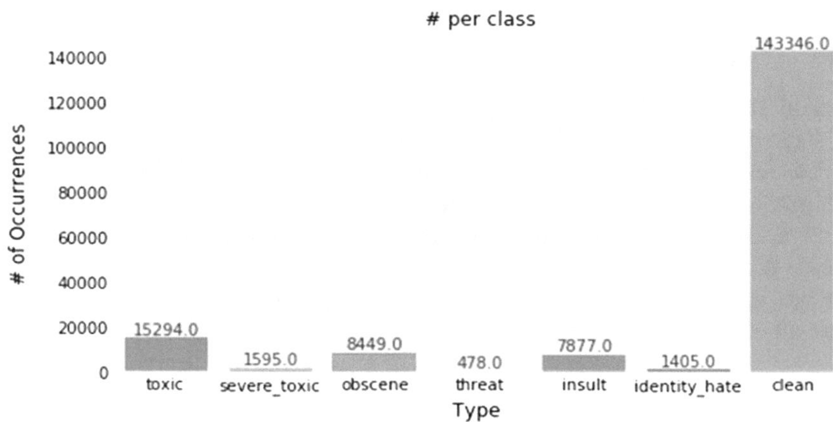

Fig. 1 Count of different tags in a dataset

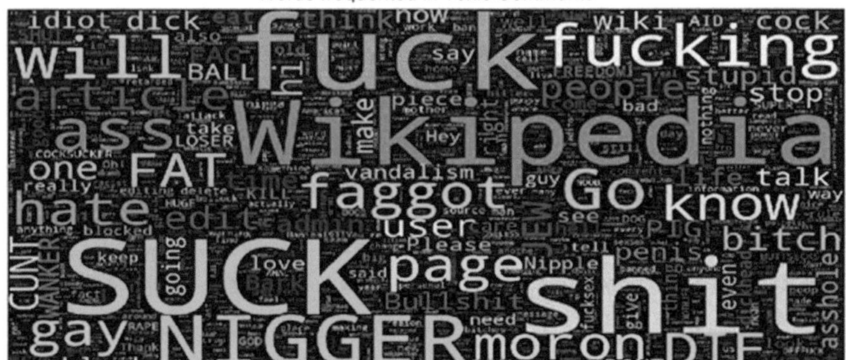

Fig. 2 Word cloud for toxic comments

3.3 Text Representation Techniques

Text can be represented by various techniques such as a bag of words (BoW), Term Frequency-Inverse Document Frequency (TF-IDF), and different pre-trained word embeddings such as word2vec, GloVe, and fastText. We have used following text representation schemes in our study:

3.3.1 Bag of Words (BoW)

Bag of Words is a Natural Language Processing model where only the vocabulary of documents is considered instead of the structure of the material. It is, therefore, called a bag of words because the structure of the document is wholly disregarded. It is extensively being used for document classification, where each word frequency is taken into consideration. It is also known as the Vector Space Model. This model is only concerned about the word's occurrence rather than wherein the document, the term occurs. This model's limitation is that two documents will be similar if they have the same bag of words.

3.3.2 Term Frequency-Inverse Document Frequency (TF-IDF)

In the TF-IDF [14] term is taken as the weighing factor for the word importance in a document or corpus. We have found the top 30 words related to each type of toxicity in the comments using TF-IDF approach as shown in Fig. 3. TF (Term Frequency)—which measures how frequently a word is present in a document and dividing it by document length, results in normalization as given in Eq. (1).

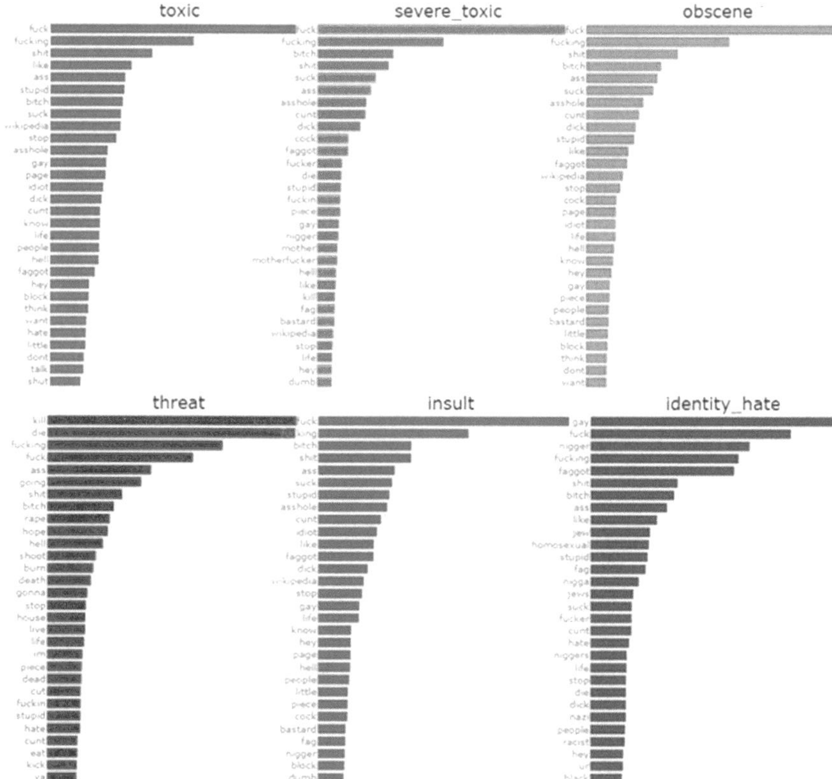

Fig. 3 Top thirty words in different six categories of comments

$$\text{TF}(t) = \frac{\text{Number of occurrences of the token in the document}}{\text{Total token count in the document}} \quad (1)$$

IDF (Inverse Document Frequency) measures the term importance in the given document. In this methodology, rare terms are given more weightage as compared to frequent terms. The intuition behind this is that the words like is, an, and the occurs very frequently but have very little importance for inference something whereas the abusive terms like idiot and moron would be less in frequency but of prime importance in the comment to make strong predictions. In a set of documents, U-IDF weight for a token t is computed as given in Eq. (2):

$$\text{IDF}(t) = \frac{\log(\text{Total document count})}{\text{Count of documents in } U \text{that contain } t} \quad (2)$$

In post j, weight for term i is given in Eq. (3):

$$\text{TF-IDF}_{ij} = \text{TF}_{ij} \times \text{IDF} \quad (3)$$

Therefore, each comment can be represented using vector, which further can be represented by the TF-IDF value.

3.3.3 Word Embedding

Word embedding are numerical representations of the text, and they map words to vectors using a dictionary such that words with similar meaning have similar description. There are various pre-trained embedding like word2vec, GloVe, ElMo, fastText. Word embeddings are used for a variety of tasks in deep learning, such as semantic analysis, syntactic parsing, named entity recognition, etc. Embedding provides a more sophisticated way to represent words in digital space. They provide a measure of similarity between words and phrases. The two word embedding used in our study is as follows:

Global Vectors for Words Representation (GloVe): The GloVe [23] is a count-based model and can be easily used for larger data compared to the bag of words model. The main idea of this model is that the word's ratio to word co-occurrence probabilities can give us useful information as some words will have a higher likelihood of co-occurring more with some particular words. It consists of the following steps:

Step 1 The co-occurrence of a word concerning the other words is collected in the form of a matrix X. Each element Xij in this matrix represents how often word I appears in the context of word j. A window size is used before the term and after the term. Less weight is given for more distant words, using the formula given in Eq. (4).

$$\text{Decay} = \frac{1}{\text{offset}} \tag{4}$$

Step 2 Soft constraints are defined for each word pair as given in Eq. (5):

$$w_i^T w_j + b_i + b_j = \log(X_{ij}) \tag{5}$$

W_i is the vector for the main word, w_j vector for the context word, b_i, b_j are scalar biases for the main and context words.

Step 3 The cost function is defined as given in Eq. (6)

$$J = \sum_{i=1}^{V} \sum_{j=1}^{V} f(X_{ij})(w_i^T w_j + b_i + b_j - \log(X_{ij})^2 \tag{6}$$

Here, f is a weighting function and takes care that prevalent words are not used for the learning process, and V is the size of the vocabulary. The Euclidean distance (cosine similarity) between two vectors provides information regarding the similarity between the words. The similarity metrics used for this nearest neighbor evaluation produce a single scalar, but this could be a problem because a unique number should not represent the relationship between two words. GloVe uses vector differences of the two words such that they are able to capture as much meaning as possible. The GloVe embedding is trained for working on non-zero entries of the word–word co-occurrence matrix. Although for large corpora, this computation can be expensive, it is a one-time-up-front cost. Training iterations are much faster as the number of non-zero entries in the matrix is much smaller as compared to the vocabulary of the corpus.

fastText Embedding: Pre-trained word vectors for 294 languages were released by Facebook trained on Wikipedia in 2017 called fastText. fastText [3] library is very efficient in terms of word representation learning and classification of the sentence. This model is a continuation of the skip-gram Word2vec model. fastText is helpful in finding the vector representation for rare words, and it assumes word to be formed by n-grams of character, thus taking into account sub-word information. N-grams within a range of 3 to 6 characters were chosen. GloVe fails where words are not in the dictionary. fastText can give vector representations for words not found in the dictionary, i.e., that are not in the training set called OOV (Out of Vocabulary) words since these can also be broken down into character n-grams. For this, they averaged the vector representation of its n-grams. Words were broken into chunks and using the vectors for these chunks to create a final vector for the word. We have used 300-dimension fastText word embedding.

3.4 *Traditional Machine Learning Methods*

Classification algorithms are now being used in various applications and producing good results. In the paper [12], authors have applied classification algorithms on text classification. Following machine learning algorithms are implemented considering the BoW features.

3.4.1 Support Vector Machine

Support Vector Machine (SVM) [29] is a supervised learning algorithm. It is used for classification as well as regression problems by finding a hyperplane that differentiates the classes most optimally. Each data point is marked in n-dimensional space. Finding the right hyperplane is the challenge in SVM. Hyperplanes can be linear, quadratic, etc. depending on the data points. SVM works for text classification tasks as well, as they can handle high-dimensional input space. SVM can also handle nonlinear classifications by using kernel tricks where inputs are mapped into high-dimensional space. New data is then mapped into one of the categories as per the hyperplane. The regularization and gamma are used for tuning the model.

3.4.2 Decision Trees

Decision trees [26] are supervised learning algorithms wherein decisions are made using some criteria, and the data are classified in the fashion of a tree. The data is split into subsets until the point each leaf node is assigned a class variable. The nodes are the points where the data is split, and decisions are made. Decision trees are simple to implement and work well with both numerical and categorical variables. The feature that divides the data most broadly is kept at the root node, and further decisions are made accordingly as which feature best divides the data. This algorithm is simple to implement and do not require much pre-processing. Generally, the criteria for decision-making can be a Gini index, information gain. In this chapter, we have used the Gini index for splitting.

3.4.3 Random Forest

Random forest [4] is an ensemble of decision tree algorithm. An ensemble algorithm uses the results of multiple models, whether same or different. Therefore, for this model, multiple decision trees are used for classification, and therefore called a forest. The larger the number of trees that are used for making a decision, the more likely it is to get an accurate decision. In the case of a random forest, the splitting is done randomly, i.e., the features chosen for decision-making are chosen randomly as opposed to the concept of a decision tree. They also overcome the problem of over fitting by introducing randomness. The mode of classes of individual trees is taken as the criteria for deciding the class label of the whole forest. The parameters that can be tweaked for this model are the number of decision trees.

3.4.4 Naive Bayes Algorithm

Naïve Bayes [36] is a supervised learning algorithm that uses the concept of Bayes probability theorem based on conditional and class probabilities. Thus, it has also been used for text classification tasks. Bayes theorem gives us the probability of an event given that an event has already occurred. It assumes that the features are independent of each other. It is extremely useful with large data with a lesser number of features. It is simple to perform and a fast algorithm. The drawback of this algorithm is that the assumption of independence among features is rarely true in practical cases due to which it is called a naive algorithm. It is a good algorithm for categorical data compared to numerical data and highly efficient in terms of accuracy, speed, and simplicity. It is good for linear classification.

3.5 Deep Learning Methods

Deep Learning algorithms are nowadays widely used in different applications like image processing, healthcare, sentiment analysis, etc. We have applied CNN, LSTM, and their blend in our approach.

3.5.1 Convolutional Neural Network (CNN)

CNN's were recently used in NLP systems and achieved remarkable results [17, 34, 37]. In the case of artificial neural nets, each input neuron was connected to hidden layers, and when they reach out to output neurons, but in our task, we are not considering them as they miss out on the spatial features of our text. ANN would consider the relationship between vectors. However, in reality, each vector has a different relationship with other vectors. CNN performs well on data with the high locality. We expect high locality for our dataset because comments would be based on a specific idea or content. As shown in Fig. 4, we have applied CNN in the comments. Convolutional neural nets have a stack of various input layers and multiple hidden layers that are passed on to the output layer. Hidden layers itself contain a convolutional layer, pooling layers, connected layers full of connections, and normalization layers, with the use of nonlinear activation function like ReLU or tanh. A convolutional neural net has three layers: (1) Convolution Layers—performs convolutions on the complete width and depth of input by sliding a window, the window (filters) performs different operations on the data. Each cell of the matrix is a tokenized character. Unlike 2D orientations in computer vision, texts have one-dimensional structure only, and here, the word sequence matters. So, we fix up this one dimension of the filter, and thus, it matches the word vector and can vary in the number of rows. Rows are representing the word present in the sentence matrix that would be filtered. Each filter computes the dot product of its entries and corresponding input. After performing the computation, it produces a 2D activation map of the filter. All the convolutions and computations result in training the network about filters.

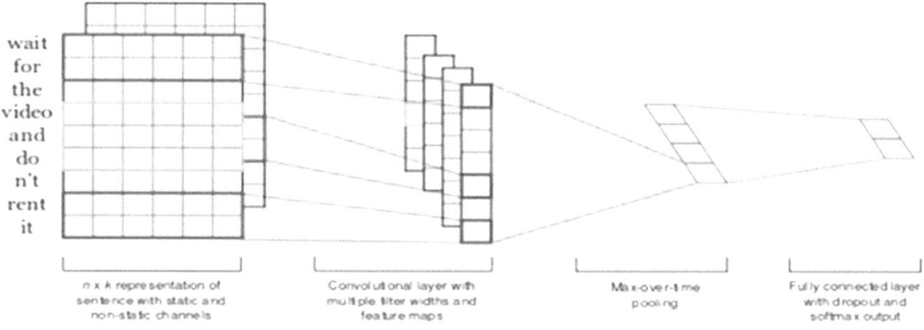

Fig. 4 CNN with word embedding architecture [17]

These filters activate whenever they detect any special kind of features at a particular spatial input position. Thus, stacking all the activation maps of filters, along with input dimensions, resulting in a full output of the convolutional layer. One can now interpret every entry of output as a neuron output, which only looks at a small input region and sharing parameters with other connected neurons. The extent of connectivity to the smaller regions of input is known as the neuron's receptive field. (2) Pooling Layers—Pooling layers are used for nonlinear dimensionality reduction of the input feature map keeping the most salient information. It partitions the input volume into non-overlapping regions. For all such regions, based on the pooling function we are using, it outputs a value. Pooling functions can be max-pooling, min polling, and average pooling. The most common and better is the max-pooling function, which considers the local neighborhood's maximum. The intuition behind using the pooling layer is the importance of a feature's rough location concerning its neighboring feature. Here, we are not concerned about the exact position of the input. Thus, the pooling layer reduces the input volume dimensions, reduces computations, and controls over fitting. In our context, based on the value, some characters from each filter would be selected by the max-pooling technique. (3) Fully connected Layer—Finally, after the convolutional layer and pooling layer, we reach out to a fully connected layer that performs a high level of reasoning. Every neuron of a fully connected layer is connected with every other neuron in the next layer. All connections have activation function involved with some random weights and biases, which they will learn with training. Its output to softmax function with cross-entropy loss and results in providing probabilities for each class.

3.5.2 Long Short-Term Memory Network (LSTM)

Recurrent Neural Network (RNN) is being used in text classification [18], but it is not able to handle long-term dependencies. LSTM is a special variant of RNN introduced by Hochreiter and Schmidhuber [9]. LSTM is powerful to handle long-term dependencies. LSTM has memory blocks connected as a set of recurrent subnets. In each block, LSTM has memory cells along with some special units—put, output, and forget gate that control, protect, and let information flow through the cell. Cell state is analogous to a conveyor belt that runs an entire chain of information flow with some liner interactions. The gates regulate the information to the cell state. Sigmoid Layer (with either 0 or 1 as output) and pointwise multiplication operations compose these gates. As shown in Fig. 5, the memory block has one cell with three gates as nonlinear units collecting information and controlling other cells' activation with multiplication. Input, output, and the previous state of the cell are multiplied by input, output, and forget the gate.

The cell usually has sigmoid activation, which simply looks for the gate closed (when 0) and the gate open (when 1) as given in Eq. (7).

$$i = \sigma(x_t U^i + h_{(t-1)} W^i) \tag{7}$$

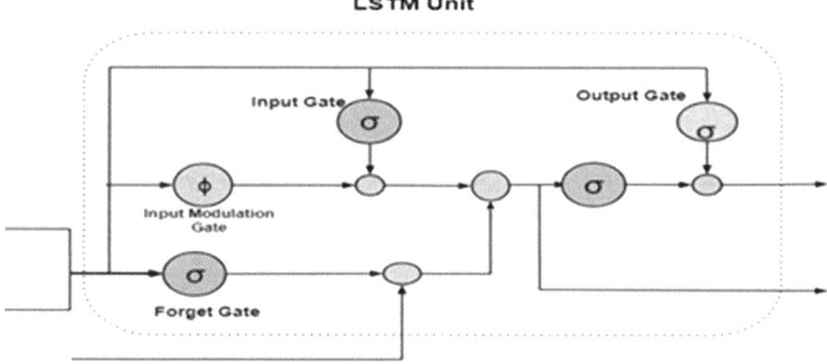

Fig. 5 LSTM block diagram [9]

In the above equation, σ is the sigmoid activation, xt is the input to the cell, $h-(t-1)$ is the previous hidden state, and W is the weight matrix. The initial decision in LSTM is regarding the information LSTM is throwing out of the cell state and forget gate help us doing that using the sigmoid layer. This layer looks for the $ht-1$ and xt output either 0 or 1. 0 means to completely ignore, and one means completely retain information from a cell state, $Ct-1$, as given in Eqs. (8) and (9).

$$i = \sigma(x_t U^i + h_{(t-1)} W^i) \tag{8}$$

$$c_t = c_{(t-1)} * f + g * i \tag{9}$$

σ is the sigmoid activation, xtis the input to the cell, $h(t-1)$ is the previously hidden state, ctis the current cell state, $c(t-1)$ is the previous cell state, and g is the external input gate. If the gates let information in then, LSTM decides what new information will be stored in the cell state. This decision needs two tasks to perform -Input (sigmoid) gate layer deciding which information will update and a tan hyperbolic function creating a new value Ct added to the state. Now, decisions are made, and it's time for LSTM to update the old state, $Ct-1$, into Ct, forgetting the things initially decided. After all the steps are performed, LSTM will get its updated value for each state, as given in Eq. (10).

$$h_t = \tanh(c_t) \times o \tag{10}$$

$$o = \sigma(x_t U^o + h_{t-1} W^o) \tag{11}$$

Finally, after performing the entire above task comes the decision of output from the cell state. This somewhat works like input operations, the sigmoid layer and tan hyperbolic will do their work. One decides whether to output the part of the information or not through 0 and 1, and others will decide which part of the information is important to output. The output formula is given in Eq. (11).

As shown in Fig. 6, data to be fed into the LSTM is converted to numeric form. Here we use LSTM with pre-trained word embedding GloVe and fastText. The embedding of each word is passed sequentially into an LSTM. LSTM takes words in a sequence

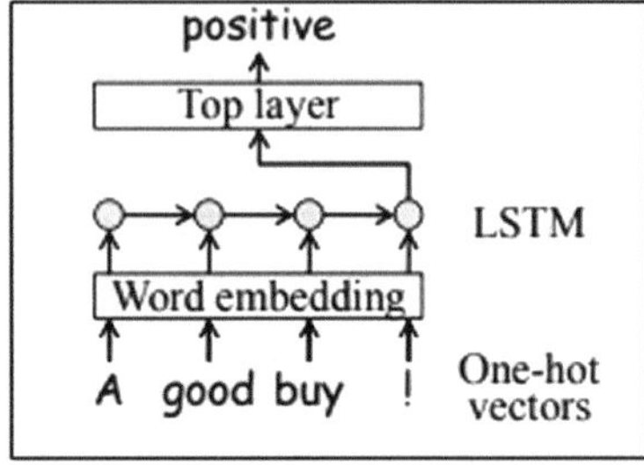

Fig. 6 Working of LSTM with word embedding

one by one, i.e., at time t, it takes input the ith word and the output from $t-1$. Therefore, it can learn long-term dependencies as opposed to traditional RNNs. That is, LSTM can be used to embed text regions of variable and large sizes.

3.5.3 CNN-LSTM Blend

A more optimized approach can be to use the features of both the models (CNN and LSTM). Deep Neural nets such as densely connected neural networks (DNNs), convolutional neural networks (CNNs), and recurrent neural nets (RNNs) are good performers in their respective fields. CNN's are good for image recognition and computer vision problems, but they can also be used for text classification tasks; on the other hand, RNNs are good for language modeling and speech recognition issues. Therefore, it is interesting to know whether one kind of deep learning model can learn from others to improve performance. Model blending is combining the models that

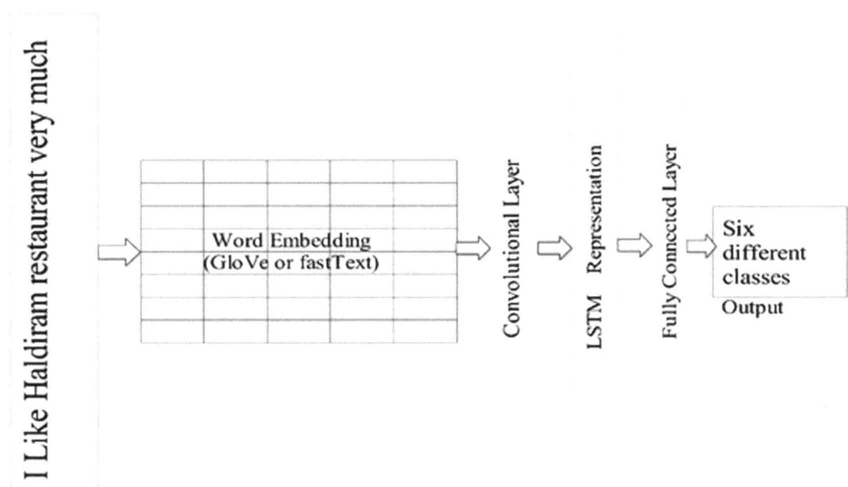

Fig. 7 Depicts CNN+LSTM Blend Architecture

have the same dependent variables and or the same independent variables. Both LSTM and CNN are quite powerful models, but their functionalities and learning processes are quite different.

CNN models are powerful for detecting patterns in the data, and LSTMs are good for capturing language context. As shown in Fig. 7, the output from the CNN layer is passed to the LSTM network. CNN performs very well in feature extraction, and LSTM is used for sequence modeling. We have used CNN-LSTM with fastText and GloVe embeddings. The difference between CNN and RNN is that both use different structures for knowing the contextual information. CNN uses a fixed window of words to learn the text context, whereas RNNs use a recurrent structure. CNN can select more discriminative features through the max-pooling layer, and then the CNN output is passed on to LSTM, where more discriminative features are learned.

4 Methodology Used

Figure 8 shows the overall methodology used in our study. We have used a dataset from the Kaggle competition. The dataset is pre-processed to get useful information out of it and to reduce the noise. The processed dataset is converted into numeric values using a bag of words and word embeddings (GloVe and fastText). We have applied four classification algorithms (Random Forest, Naïve Bayes, Support Vector Machine, and Decision Tree) on the bag of word representation and evaluated the performance based on accuracy. We have applied two deep learning algorithms (CNN and LSTM) and their blend CNN-LSTM on the text. Text is represented by GloVe and fastText word embedding and evaluated the performance based on accuracy.

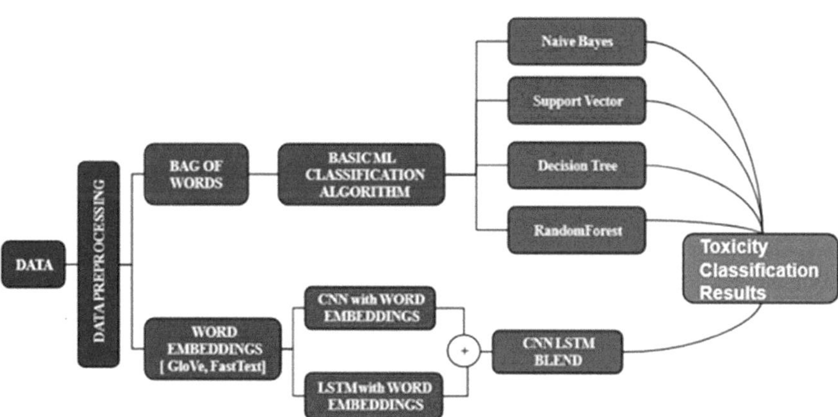

Fig. 8 Flowchart of the Methodology Used

CNN is good for feature extraction, whereas LSTM is good for sequence modeling. Therefore, to use the advantages of both these models, we blended LSTM and CNN to increase the performance.

5 Experimental Results and Analysis

We have applied four classification algorithms (SVM, Naïve Bayes, Random Forest, and Decision Tree) with BoW, and two deep learning algorithms (CNN, LSTM, and their blend CNN and LSTM) with the two-word embedding GloVe and fastText. The performance parameter considered is accuracy. With this approach, detecting toxicity and then categorizing the toxic comment into six different categories—toxic, severe toxic, obscene, threat, insult, and identity hate are given in Table 1.

As shown in Table 1, we have achieved an accuracy of 89.92% with SVM, 90.24% with a decision tree, 93.45% with naive Bayes classifier, and 91.78% with random forest classifier. Naive Bayes gives the best results out of all these classification models because it is a linear model. The other reason for naive Baye's performance is that our data does not have a binary classification. This model can handle this type of task very well as it is based on Gaussian probability, so giving better probabilities of each class. The performance of deep learning algorithms (CNN, LSTM, and CNN-LSTM) with word embeddings (GloVe, and fastText) is also presented in Table 1. With GloVe embedding, we have achieved an accuracy of 95.42% with CNN, 97.06% with LSTM, and 98.07% with CNN-LSTM blend.

The fastText embedding proves to be a better option in terms of word embedding because we have achieved an accuracy of 95.98% with CNN, 97.64% with LSTM,

Table 1 Results of various approaches applied

Machine Learning Models with BoW		
Sr. No.	Models used	Accuracy (%)
1	Support vector classifier	89.92
2	Decision tree	90.24
3	Naïve Bayes	93.45
4	Random forest	91.78
GloVe word embedding		
5	Convolutional neural network	95.42
6	Long short-term memory	97.06
7	CNN+LSTM blend	98.07
fastText word embedding		
5	Convolutional neural network	95.98
6	Long short-term memory	97.64
7	CNN+LSTM blend	98.46

Table 2 Comparison of our results with the existing results

Sr. No.	Research Paper	Accuracy (%)
1	Saif, M. A., Medvedev, A. N., Medvedev, M. A., Atanasova, T. [16]	96.45
2	Pallam Ravi, Hari Narayana Batta, Greeshma S, and Shaik Yaseen [17]	97.92
3	Georgakopoulos, S. V., Tasoulis, S. K., Vrahatis, A. G., Plagianakos, V. P. [10]	91.20
4	In'es BEN AMAR, Antoine COPPIN, and Emerick LECOMTE (2017) [18]	97.72
5	Our Approach	98.46

and 98.46% with CNN-LSTM Blend. The fastText is better for this problem due to their function of considering the substructure of the word. The reason for this embedding going well with our models is its ability to work well on new words being used by users on social media. There is no rule of using strict standard rules of English for expression on social media. This is a very important aspect when we are working on data generated by highly evolved users. They are more in the habit of using abbreviations, hashtags, and code words, which are sometimes entirely different from the standard language structure fastText embedding takes care of. One more positive point of this embedding is that it is multilingual and not necessarily restricted to English; so, this model can be used for other languages. The maximum accuracy was achieved for CNN-LSTM blend with fastText. Thus, it can be seen that the CNN -LSTM blend has significantly improved the accuracy both for fastText as well as GloVe word embedding. The LSTM model performs better than CNN due to its functionality of good sequential modeling and recurrent memory blocks. CNN is through powerful, as its results show, in comparison to our other approaches, it is least accurate because of its weakness to work only close to the data and unable to handle sequential modeling. But when CNN is combined with LSTM, it performed extremely well because CNN models are very efficient in detecting patterns in the data, and for capturing language context, we have LSTM. As shown in Table 2, our approach is better than the state-of-the-art approaches existing in the literature.

6 Conclusion and Future Work

Nowadays, scientists and researchers are concerned about making social media platforms a better place for users to interact freely in ways they think to be appropriate. Toxicity and abusiveness in comments is a big hurdle for online social media networking sites. Identifying toxic, abusive, offensive comments in social media is a huge problem that needs to be tackled and explored urgently. An efficient approach or solution to this problem is still in its infancy. Hence, our work is devoted to curbing social media abuse, toxicity, and cyberbullying. In this study, we have implemented four classification models—SVM, Naive bayes, random forest, decision tree with BoW features and deep learning models—CNN and LSTM along with their blends using GloVe and fastText word embedding to classify a comment as toxic. Naive Bayes proves to a better approach among all the classification techniques applied in our study and giving an accuracy of 93.45%. Our optimal blending of CNN and LSTM on both fastText and GloVe enhanced our results. We have achieved promising results of 98.46% accuracy with CNN-LSTM blend with fastText word embedding. CNN and the LSTM blend proved to achieve better results by using each model's advantages. The use of word embedding has increased the performances of these models with fastText giving remarkable results. In future more deep learning models like Gated Recurrent Unit (GRU), Bi-LSTM, Bi-GRU, and BERT. An ensemble of different word embedding can be done to improve performance. Working on better data with more features and considering parameters like user's past posts, profile, and general use of language.

List of Abbreviations

API: Application Programming Interface
ANN: Artificial Neural Network
BoW: Bag of Words
Bi-GRU: Bi-directional Gated Recurrent Unit
Bi-LSTM: Bi-directional Long Short-Term Memory
CNN: Convolutional Neural Network Modified
DTM: Document Term Matrices
GloVe: Global Vector for Word Representation
GRU: Gated Recurrent Unit
IR: Information Retrieval
LSTM: Long Short-Term Memory
LDA: Linear Discriminant Analysis
ML: Machine Learning
NLTK: Natural Language Toolkit
OOV: Out of Vocabulary
PLSA: Probabilistic Latent Semantic Analysis
RNN: Recurrent Neural Network
ReLU: Rectifier Linear Unit
NLTK: Natural Language Toolkit
SVM: Support Vector Machine
SSWE: Sentiment Specific Word Embedding
TF-IDF: Term Frequency-Inverse Document Frequency
URL: Uniform Resource Locator

References

1. Agrawal, Sweta, and Amit Awekar. 2018. Deep learning for detecting cyberbullying across multiple social media platforms. In *European conference on information retrieval*, 141–153. Springer.
2. Badjatiya, Pinkesh, Shashank Gupta, Manish Gupta, and Vasudeva Varma. 2017. Deep learning for hate speech detection in tweets. In *Proceedings of the 26th international conference on world wide web companion*, 759–760, 2017.
3. Bojanowski, Piotr, Edouard Grave, Armand Joulin, and Tomas Mikolov. 2017. Enriching word vectors with subword information. *Transactions of the Association for Computational Linguistics* 5: 135–146.
4. Breiman, Leo. 2001. Random Forests. *Machine Learning* 45 (1): 5–32.
5. Brooks, Bryony. 2018. *Cyberbullying and cyberbullicide: The role of linguistic features in suicides by text*. Hofstra University.
6. Santos, Cicero Dos, and Maira Gatti. 2014. Deep convolutional neural networks for sentiment analysis of short texts. In *Proceedings of COLING 2014, the 25th international conference on computational linguistics: technical papers*, 69–78.
7. Duggan, Maeve, L. Rainie, A. Smith, C. Funk, A. Lenhart, and M. Madden. 2014. Online harassment. Washington, DC: Pew research center.
8. Georgakopoulos, Spiros V., Sotiris K. Tasoulis, Aristidis G. Vrahatis, and Vassilis P. Plagianakos. 2018. Convolutional neural networks for toxic comment classification. In *Proceedings of the 10th hellenic conference on artificial intelligence*, 1–6.
9. Hochreiter, Sepp, and Jürgen Schmidhuber. 1997. Long short-term memory. *Neural Computation* 9 (8): 1735–1780.
10. Hosseini, Hossein, Sreeram Kannan, Baosen Zhang, and Radha Poovendran. 2017. Deceiving google's perspective api built for detecting toxic comments. arXiv:1702.08138.
11. Ibrahim, Mai, Marwan Torki, and Nagwa El-Makky. 2018. Imbalanced toxic comments classification using data augmentation and deep learning. In *2018 17th IEEE international conference on machine learning and applications (ICMLA)*, 875–878. IEEE.
12. Ikonomakis, M., Sotiris Kotsiantis, and V. Tampakas. 2005. Text classification using machine learning techniques. *WSEAS Transactions on Computers*, 4(8):966–974.
13. Jianqiang, Zhao, and Gui Xiaolin. 2017. Comparison research on text pre-processing methods on twitter sentiment analysis. *IEEE Access* 5: 2870–2879.
14. Jing, Li-Ping, Hou-Kuan Huang, and Hong-Bo Shi. 2002. Improved feature selection approach tfidf in text mining. In *Proceedings international conference on machine learning and cybernetics*, vol. 2, 944–946. IEEE.
15. Jubaer, A.N.M., Abu Sayem, and Md Ashikur Rahman. 2019. Bangla toxic comment classification (machine learning and deep learning approach). In *2019 8th international conference system modeling and advancement in research trends (SMART)*, 62–66. IEEE.
16. Kansara, Krishna B., and Narendra M. Shekokar. 2015. A framework for cyberbullying detection in social network. *International Journal of Current Engineering and Technology*, 5(1):494–498.
17. Kim, Yoon. 2014. Convolutional neural networks for sentence classification. arXiv:1408.5882.
18. Lai, Siwei, Liheng Xu, Kang Liu, and Jun Zhao. 2015. Recurrent convolutional neural networks for text classification. In *Twenty-ninth AAAI conference on artificial intelligence*.
19. Li, Siyuan. 2018. *Application of recurrent neural networks in toxic comment classification*. Ph.D. thesis, UCLA.
20. Maus, Adam. 2009. Svm approach to forum and comment moderation. *Class Projects for CS*.
21. Mohammad, Fahim. 2018. Is preprocessing of text really worth your time for online comment classification? arXiv:1806.02908.
22. Parekh, Pooja, and Hetal Patel. 2017. Toxic comment tools: A case study. *International Journal of Advanced Research in Computer Science* 8 (5)

23. Pennington, Jeffrey, Richard Socher, and Christopher D Manning. 2014. Glove: Global vectors for word representation. In *Proceedings of the 2014 conference on empirical methods in natural language processing (EMNLP)*, 1532–1543.
24. Ptaszynski, Michal, Juuso Kalevi Kristian Eronen, and Fumito Masui. 2017. Learning deep on cyberbullying is always better than brute force. In *LaCATODA@ IJCAI*, 3–10.
25. Ravi, Pallam, Greeshma S Hari Narayana Batta, and Shaik Yaseen. 2019. Toxic comment classification. *International Journal of Trend in Scientific Research and Development (IJTSRD)*.
26. Rasoul Safavian, S., and David Landgrebe. 1991. A survey of decision tree classifier methodology. *IEEE Transactions on Systems, Man, and Cybernetics* 21 (3): 660–674.
27. Saif, Mujahed A., Alexander N. Medvedev, Maxim A. Medvedev, and Todorka Atanasova. 2018. Classification of online toxic comments using the logistic regression and neural networks models. In *AIP conference proceedings*, vol. 2048, 060011. AIP Publishing LLC.
28. Schultz, Jeff. 2017. How much data is created on the internet each day. *Micro Focus Blog*, 10.
29. Suykens, Johan A.K., and Joos Vandewalle. 1999. Least squares support vector machine classifiers. *Neural processing letters* 9 (3): 293–300.
30. van Aken, Betty , Julian Risch, Ralf Krestel, and Alexander L.öser. 2018. Challenges for toxic comment classification: An in-depth error analysis. arXiv:1809.07572.
31. Wang, Xin, Yuanchao Liu, Cheng-Jie Sun, Baoxun Wang, and Xiaolong Wang. 2015. Predicting polarities of tweets by composing word embeddings with long short-term memory. In *Proceedings of the 53rd annual meeting of the association for computational linguistics and the 7th international joint conference on natural language processing (Volume 1: long papers)*, 1343–1353.
32. Wulczyn, Ellery, Nithum Thain, and Lucas Dixon. 2017. Ex machina: Personal attacks seen at scale. In *Proceedings of the 26th international conference on world wide web*, 1391–1399.
33. Xiang, Guang, Bin Fan, Ling Wang, Jason Hong, and Carolyn Rose. 2012. Detecting offensive tweets via topical feature discovery over a large scale twitter corpus. In *Proceedings of the 21st ACM international conference on Information and knowledge management*, 1980–1984.
34. Xiao, Yijun, and Kyunghyun Cho. 2016. Efficient character-level document classification by combining convolution and recurrent layers. arXiv:1602.00367.
35. Yin, Dawei, Zhenzhen Xue, Liangjie Hong, Brian D. Davison, April Kontostathis, and Lynne Edwards. 2009. Detection of harassment on web 2.0. *Proceedings of the content analysis in the WEB* 2: 1–7.
36. Zhang, Harry. 2004. The optimality of naive bayes, 2004. *American Association for Artificial Intelligence (www. aaai. org)*.
37. Zhang, Xiang, Junbo Zhao, and Yann LeCun. 2015. Character-level convolutional networks for text classification. In*Advances in neural information processing systems*, 649–657.
38. Zhou, Qingqing, and Chengzhi Zhang. 2018. Detecting users' dietary preferences and their evolutions via Chinese social media. *Journal of Database Management (JDM)* 29 (3): 89–110.

DURLD: Malicious URL Detection Using Deep Learning-Based Character Level Representations

Sriram Srinivasan, R. Vinayakumar, Ajay Arunachalam, Mamoun Alazab, and KP Soman

Abstract Cybercriminals widely use Malicious URL, a.k.a. malicious website as a primary mechanism to host unsolicited content, such as spam, malicious advertisements, phishing, and drive-by exploits, to name a few. Previous studies used blacklisting, regular expression, and signature matching approaches to detect malicious URLs. However, these approaches are limited to detect variants of existing or newly generated malicious URLs. Over the last decade, classic machine learning techniques have been used to detect malicious URLs. In this work, we evaluate various state-of-the-art deep learning-based character level embedding methods for malicious URL detection. To leverage and transform the performance improvement, we propose DeepURLDetect (DURLD) in which raw URLs are encoded using character level embedding. To capture several types of information in URL, we used the hidden layers in deep learning architectures to extract features from character level embedding and then employ a non-linear activation function to estimate the probability of the URL as malicious or not. Experimental evaluation demonstrates that DURLD can detect variants of malicious URLs, and it is computationally inexpensive when compared to various relevant deep learning-based character level embedding methods.

S. Srinivasan (✉) · K. Soman
Center for Computational Engineering and Networking, Amrita School of Engineering, Amrita Vishwa Vidyapeetham, Coimbatore, India
e-mail: sri27395ram@gmail.com

K. Soman
e-mail: kpsoman2000@gmail.com

R. Vinayakumar
Division of Biomedical Informatics, Cincinnati Children's Hospital Medical Center, Cincinnati, OH, USA
e-mail: Vinayakumar.Ravi@cchmc.org

A. Arunachalam
Centre for Applied Autonomous Sensors Systems (AASS), Örebro University, Örebro, Sweden
e-mail: ajay.arunachalam@oru.se

M. Alazab
Charles Darwin University, Darwin, Australia
e-mail: mamoun.alazab@cdu.edu.au

M. Stamp et al. (eds.), *Malware Analysis Using Artificial Intelligence and Deep Learning*, https://doi.org/10.1007/978-3-030-62582-5_21

1 Introduction

Malicious Uniform Resource Locator (URL) host unsolicited information and attackers use malicious URLs as one of a primary tool to carry out cyber attacks. E-mail and social media resources such as Facebook, Twitter, and WhatsApp are the most commonly used applications to spread malicious URLs [3, 4, 35]. They host unsolicited information on the web page. Whenever an unsuspecting user visits that website unknowingly through the URL, the host may get compromised, making them victims of various types of frauds including malware installation, data, and identity theft. Every year, malicious URLs have been causing billions of dollars worth of losses [12]. These factors force the development of efficient techniques to detect malicious URLs promptly and give an alert to the network administrator.

Most of the commercial products existing in markets are based on the blacklisting method [5]. This method relies on a database that contains a list of malicious URLs. The blacklists are continually updated by the anti-virus group through scanning and crowdsourcing solutions. The blacklisting method can be used to detect the malicious URLs which are already present in the database. But, they completely fail to detect the variants of the existing malicious URLs or entirely new malicious URLs. In recent days, cyberattackers follow mutation techniques to generate several variants of existing malware. To cope with this, machine learning techniques are employed.

In recent days, the most commonly used approach is applying domain knowledge to extract lexical features of URL, followed by applying machine learning models. The most commonly used feature engineering technique is Bag-of-words (BoW) and the most commonly used machine learning model is the support vector machine (SVM) [29]. Though machine learning-based solution can be used instead of blacklisting methodology, it suffers from many issues:

1. The conventional URL representation methods fail to capture the sequential patterns and relationships among the characters.
2. Conventional machine learning models rely on manual feature engineering. This requires extensive domain knowledge in the cybersecurity domain and it is considered a daunting task.
3. Fails to hold unrevealed features and it doesn't generalize on the test data. Additionally, the number of unique words is immensely large and as a result, the machine learning model faces memory constraints while training.

To alleviate the aforementioned issues, this work proposes a model named DeepURLDetect (DURLD) which uses a modern machine learning technique, typically called "deep learning" with character embedding. Deep learning uses multiple hidden layers in which each layer does non-linear projection to learn representations of multiple levels of abstraction and they are applied to many cybersecurity applications [2, 6, 8, 9, 26–28, 34, 37–39]. The main contributions of the proposed work are

1. Detailed investigation and analysis of various benchmark deep learning architectures are performed for malicious URL detection.

2. Various types of data sets are used in the experimental analysis to find out how generalizable the models are. The difference between the time-split and random-split of data splitting methods is shown in the experimental analysis.
3. Experiments are shown for character level embedding and n-gram representation with various deep learning architectures

The rest of the sections of this chapter are organized as follows. Section 2 discusses the related works of malicious URL detection. Section 3 provides information about URLs. Section 4 discusses the background details of benchmark text classification models and the proposed model. Section 5 provides the major shortcomings in malicious URL detection. Section 6 includes a description of the data set. The working flow of malicious URL detection is discussed in Sect. 7. Section 8 contains information on proposed architecture. Details of performance measures are discussed in Sect. 9. The results are discussed in Sect. 10. At last, the conclusion and future works are placed in Sect. 11.

2 Related Works

For the detailed literature survey of machine learning-based malicious URL detection, see [29]. This section discusses the most important works in malicious URL detection.

At the beginning stages, blacklisting, regular expression, and signature matching approaches are most commonly used for malicious URL detection. These methods completely fail to detect new or variant of existing URLs. Moreover, the signature database has to be updated frequently to handle new patterns of malicious URLs. Later, machine learning algorithms were used to effectively detect new types of malicious URLs. Conventional machine learning algorithms depend on feature engineering to extract a list of features from URLs. This feature engineering requires extensive domain knowledge of URL in cybersecurity and a list of good features has to be carefully chosen through feature selection. There are various types of features which are used in the published works for malicious URL detection. This includes blacklist features [18, 22], lexical features [20, 22, 23], host-based features [14, 24], content features, context- and popularity-based features [13, 15, 21]. Blacklist features are estimated through checking its presence of a URL in a blacklist. This could serve as a strong feature in identifying malicious URLs. Lexical features are estimated through the string properties of the URL, e.g., the number of special characters, length of URL, etc. Host-based features are obtained from the hostname properties of the URL. This includes information related to WHOIS information, IP Address, Geographic location, etc. Content features are derived from the HTML and JavaScript when an unsuspecting user visits a webpage through the malicious URL. Content features include information related to their ranking, popularity scores, and source of sharing. Many existing studies have used separate feature category and as well as a combination of these features which were continually determined through domain experts.

Feature engineering is a daunting task with considering the security threats. For example, obtaining context-based features consumes more time and it is highly risky too. Moreover, feature selection requires extensive domain knowledge. The information which is obtained directly from the raw URL is a well-known approach [22, 23]. From the published results, obtaining the lexical feature is easier in comparison to other features and it gave good performances [11]. Statistical properties of the URL string such as length of the URL and number of special characters [20] have been most commonly used and other most popular features were BoW, and term frequency methods such as term document matrix (TDM) and term frequency and inverse document frequency (TF-IDF) and n-gram features [11, 20, 22]. All these features are not effective in extracting sequential order and semantics of URL. This completely disregards the information from unseen characters. Moreover, malicious URL detection solution based on the feature engineering with conventional machine learning can be easily broken by an adversary.

In recent days, the application of deep learning with character level embedding has been used for malicious URL detection. In [41], we compared a detailed analysis of deep learning with character level embedding and conventional machine learning with feature engineering methods for malicious and phishing URL detection. Deep learning architectures performed well in comparison to the conventional algorithms. The application of the recurrent neural network (RNN) and long short-term memory (LSTM) is applied to phishing URL detection [10]. For comparative analysis, lexical features and statistical URL analysis were used with random forest classifier. Both models performed well, the performance of LSTM was good in comparison to the conventional machine learning method. In [31], we used convolutional neural network (CNN) with character level Keras embedding for detecting malicious URLs, file paths, and registry keys. This study showed how a unique deep learning architecture could be used on different cybersecurity problems. Like this, there are so many benchmark deep learning architectures that exist. In this work, we evaluate the performance of various deep learning architectures for malicious URL detection.

3 An Overview of Uniform Resource Locator (URL)

A uniform resource locator (URL) is a part of the uniform resource identifier (URI) which is used to identify and retrieve a resource from the Internet service. A URL is composed of three parts as shown in Fig. 1. The first part defines the type of protocol, for example, http or https, the second part defines the domain name or IP address and the third part defines the path and its parameters to a specific resource on the web. The protocol is separated by a double slash from the other parts of the URL and it is followed by the domain name. The path and its parameters are separated by a single slash. A sample URL is given in Fig. 1. An adversary may use the URL as the main source to host malicious activities. Most commonly, the malicious URLs are spread

Table 1 Character level deep learning architectures

Name	Architecture	Task
Endgame [7]	LSTM	Detecting and categorizing domain names that are generated by DGAs
Invincea [31]	CNN	To detect malicious URLs, file paths and registry keys
CMU [17]	Bidirectional recurrent structures	Social media text classification, Twitter
MIT [44]	Hybrid of CNN and LSTM	Social media text classification, Twitter
NYU [43]	Stacked CNN layers	Text classification

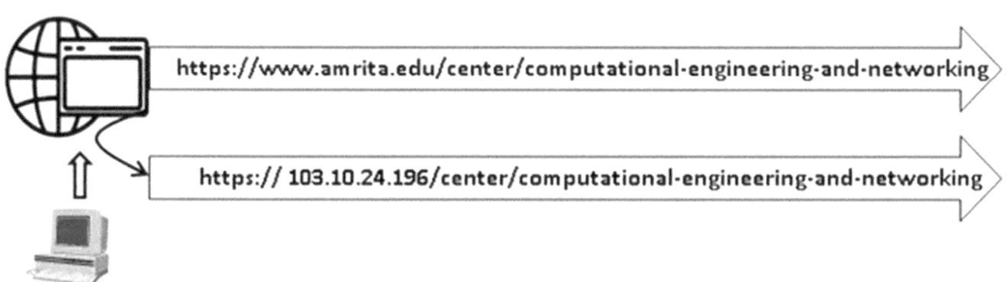

Fig. 1 Uniform resource locator (URL) components

via email and other social media apps. Once an unsuspecting user visits a malicious URL, the host system may get compromised. Thus, detecting the nature and type of URL is considered as a significant task.

4 Background Details of Deep Learning Models

4.1 Hybrid Architecture—Convolutional Neural Network and Long Short-Term Memory (CNN-LSTM) with Character Level Keras Embedding

Convolutional neural network (CNN) is very similar to Deep Neural Network (DNN) and it uses convolution operation to extract features from the input. The example of DNN and CNN network is shown in Figs. 2 and 3, respectively. CNN based on character (CNN-C) level is a minimal variant of the deep CNN based on character level [44]. CNN-C primarily uses 1D convolution and pooling operations also called temporal convolution and temporal pooling, respectively. CNN-C extracts optimal features from the character level representation of URLs. For character level rep-

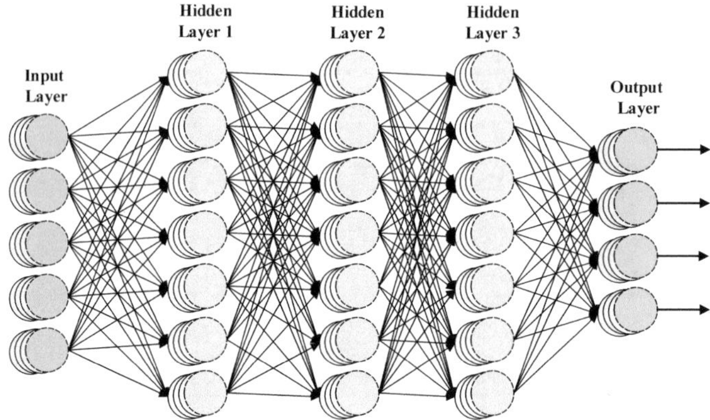

Fig. 2 Deep Neural Network architecture

resentation, the character level Keras embedding representation is used. This takes three parameters such as dictionary size, the maximum length of the character vector, and embedding vector length. Initially, the character level Keras embedding weights can be initialized as a hyperparameter. The weights are optimized during backpropagation. The CNN features are passed into LSTM which facilitates to learn character level sequence representation.

4.2 Character-Based Models

Character-based model takes an input text as a string of characters and automatically extracts features. These features can be used for performing different tasks (e.g., text classification). There are different character-based models that exist in the field of natural language processing (NLP), and in this work, the efficacy of them are evaluated for cybersecurity application, namely, malicious URL detection. All models use embedding as the first layer to transform the URLs into numeric vectors. The details of the various character-based models are given in Table 1. The details of the various character-based models are given below

1. **Character level models based on RNN**
   Endgame Architecture: Reference [7] It uses LSTM with character level Keras embedding for modeling domain generation algorithms (DGAs) to detect and categorize the domain names that are generated by DGAs. The character level Keras embedding facilitates to learn the sequence of characters of domain names and it helps to preserve the order of character in domain names. Moreover, it completely avoids the feature engineering method that is an important step for classical machine learning methods. The method has better performance when compared to the other methods such as hidden Markov model, feature engineering,

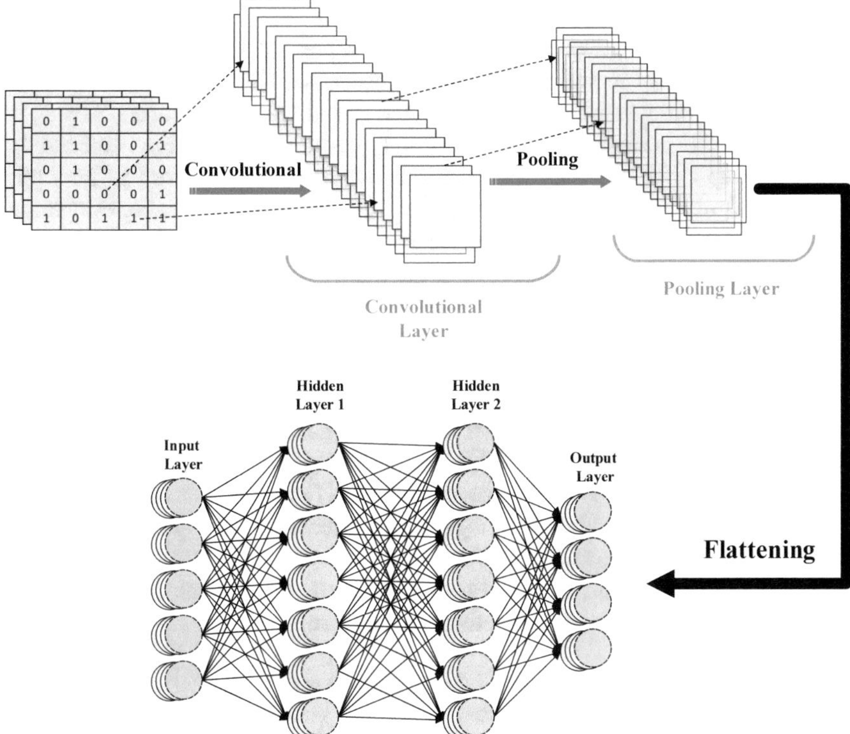

Fig. 3 Convolutional Neural Network architecture

and bigrams with classical machine learning classifiers. The proposed LSTM network is composed of an embedding layer for URL representation, LSTM layer for optimal feature extraction, and logistic regression for classification. The embedding layer maps each character to shape 128 and passes into LSTM for feature learning and logistic regression for assigning a probability score for each domain name.

CMU Architecture: Reference [17] CMU Architecture is named as Tweet2vec for tweet representation and classification for social media data. It uses a bidirectional gated recurrent unit (BGRU) to learn feature representation of Twitter data. The tweets are tokenized into a stream of characters and each character is represented by using one-hot character encoding. These one-hot representations are mapped into a character space and fed into the BGRU model. The model contains forward and backward GRU which facilitates to learn the sequence of characters in the domain name. Both the forward and backward GRU layers are combined using a fully connected layer and a $softmax$ non-linear activation function was used for tweet classification, particularly to predict hashtags of tweets. For comparative study, the tweet2vec is evaluated on the word level tweet representation.

2. **Character level models based on CNN**
   NYU Architecture: Convolutional neural network (CNN) is most commonly used in the field of image processing. In recent days, 1D CNN has been mapped into text classification [43]. They have used word based CNN and LSTM. The CNN of NYU is stacked CNN. With CNN, pre-trained embedding, embedding, and lookup tables are used as text representation methods. With LSTM, pre-trained word embedding is used as a text representation method. For evaluation, they used different types of large-scale data sets. They claimed that the character level CNN model performed well in comparison to the classical and deep learning models. The efficacy of deep learning models is evaluated on the classical text representation methods such as BoW, n-gram with TF-IDF.
   Invincea Architecture: To model short character strings such as URLs, file paths, or registry keys of cybersecurity data, [31] proposed CNN network. This CNN network is composed of character level Keras embedding layer, parallel CNN layer, followed by three fully connected layers. All three fully connected layers contain 1,024 units and $ReLU$ as an activation function. The architecture uses batch normalization and dropout regularization techniques to speed up the model training and prevent overfitting. To classify the short character strings as either legitimate or malicious, the CNN network contains a fully connected layer with unit 1 and $sigmoid$ non-linear activation function.

3. **Character level models based on hybrid CNN and RNN**
   MIT Architecture: Reference [44] This is an extension of the NYU model for tweet classification. It is composed of stacked CNN layers followed by an LSTM layer. The stacked CNN layer results in overfitting. To alleviate this, a minimum number of parameters are used.

4.3 Problem Formulation

The objective of this work is to classify a given URL as either legitimate or malicious and classification problem is binary. Let us consider a set of URLs $U = \{(u_1, y_1), (u_2, y_2), \cdots (u_n, y_n)\}$ where u represents URL and y represents '0' for legitimate and '1' for malicious.

There are two steps involved in the classification procedure, firstly, the optimal feature representation and secondly, the prediction function. Feature representation forms n-dimensional vector representation x_n which can be passed into prediction function as input $y_n = sign(f(x_n))$. The main aim is to minimize the total number of misclassification. This can be achieved by minimizing the loss function. This type of loss function can also include a regularization term. In this work, f is represented as deep learning architectures.

5 Shortcomings in Malicious URL Detection

There are no publically available benchmark data sets for research in malicious URL detection. Most of the published results on malicious URL detection have used their own private data sets in evaluating the efficacy of various conventional machine learning algorithms and deep learning architectures. These private data sets are collected from various sources such as Alexa, DMOZ, Phishtank, OpenPhish, MalwareDomains, MalwareDomainList, and many others. Though, these approaches cannot be regarded as generic methods due to the uncommon data sets. Most of the published results haven't given any importance to the time-split methodology to divide the data into train and test. Recently, [30] discussed the importance of time-split in dividing the data into train and test sets. The data splitting methodology based on time-split is very important to meet the zero day malware detection. Recently, the background reason for not deploying machine learning-based solutions for security is discussed by [33]. The detailed test cases that should be considered in test experiments are discussed in detail by [36]. These different test cases help to evaluate the robustness of machine learning-based solutions. Moreover, they have discussed the difficulty behind applying data science techniques for cybersecurity.

6 Description of Data Set

It is necessary to test different forms of URL to assess the performance of various conventional machine learning classifiers and deep learning architectures. There are two types of data sets that are used. They are Data set 1 and Data set 2. The Data set 1 is collected from publically available sources such as Alexa.com, DMOZ directory, MalwareDomainlist.com and MalwareDomains.com, CEN Amrita Vishwa Vidyapeetham research internal network backbone. Data set 2 is from Sophos research [32]. The most commonly used methodology for dividing data into train and test is random-split [25]. Data set 1 follows random-split. The classifier which is modeled using a random-split approach is not an efficient splitting methodology to meet zero day malware detection [30]. Data set 2 follows both the random-split and time-split [30]. In the domain of cybersecurity, it is good to follow the time-split [19, 42]. This facilitates to enhance zero day malware detection. The detailed statistics of both Data set 1 and Data set 2 are reported in Table 2.

7 Model Configuration of Malicious URL Detection Engine

The pseudo-code of the malicious URL detection engine is given as Algorithm 1. It is composed of three different sections. They are (1) preprocessing, (2) optimal features extraction, and (3) classification.

Table 2 Detailed statistics URL data set

Data set	Category	Legitimate	Malicious
Data set 1	Train	212,751	175,121
Data set 1	Test	122,406	101,616
Data set 2 random-split	Train	43,771	43,430
Data set 2 random-split	Test	18,516	18,857
Data set 2 time-split	Train	39,271	47,044
Data set 2 time-split	Test	23,016	15,243

In preprocessing, the URLs are transformed into feature vector using text representation methods and the optimal features from the numeric vectors are extracted using various benchmark models such as Invincea, NYU, MIT, CMU, Endgame, and proposed model, DeepURLDetect (DURLD), and finally, classification is done using fully connected layer with non-linear activation function.

Algorithm 1: Malicious URL Detection Engine

Input: A set of URLs $U_1, U_2, ..., U_n$.
Output: Deep URL Detect Model Labels $y_1, y_2, .., y_n$ (0: legitimate or 1: Malicious) and BenchMarkModels prediction p.

VectorizedURLs = Dataprocessing(Ui) `// URLs into numerical vectors using text representation method`
Predicions p = BenchMarkModels(VectorizedURL) `// Invincea, NYU, MIT, CMU, Endgame`
for *each URLs* U_i **do**
  $lcURL$= lowerCase(U_i)
  ZeroPaddedURL Z_i = Padding($lcURL$)
  E = Character level Keras Embedding (Z_i)
  C = CNN(E)
  L = LSTM(C)
  Compute y_i = Sigmoid(L)
end for

The characters in URLs are converted into lower case. This is due to the reason domain names are case insensitive, and differentiating between capital and small letters may cause regularization issue. Otherwise, the models have to be run for more number of epochs to learn the patterns of all possibilities of characters that exist in the URLs. The Data set 1 corpus contains 150 unique characters, dictionary size and the maximum length of the URL is 2,307. The Data set 2 random-split and time-split corpus contains 42 unique characters and the maximum length of the URL is 246. The URL which is lesser than the maximum length is padded with 0. The detailed architecture and configuration details of DURLD is shown in Tables 3 and 4 for Data

Table 3 Detailed configuration parameter information of DURLD for Data set 1

Layer (type)	Output shape	Param #
embedding_1 (Embedding)	(None, 2307, 128)	19200
conv1d_1 (Conv1D)	(None, 2306, 128)	32896
max_pooling1d_1 (MaxPooling1)	(None, 1153, 128)	0
lstm_1 (LSTM)	(None, 70)	55720
dense_1 (Dense)	(None, 1)	71
activation_1 (Activation)	(None, 1)	0
Total params: 107,887		
Trainable params: 107,887		
Non-trainable params: 0		

set 1 and Data set 2 random-split and time-split, respectively. In DURLD, character level Keras embedding contains 128 as embedding size, as each character is mapped into 128 dimensions. This helps to learn the similarity among characters by mapping the semantics of similar characters to similar vectors. All models contain character level Keras embedding as a URL representation method and the dimensionality of the embedding size is set to the same size to conduct a fair comparative evaluation strategy. To know the effectiveness of character level Keras embedding, a 3-gram text representation method is mapped into domain names. The features of 3-grams are hash it into a vector of length 1,000 using feature hashing. These 1,000-dimensional vectors are passed into DNN for optimal feature extraction and classification; the detailed configuration parameter details of DNN is available in Table 5. To avoid overfitting and speed up the training process, dropout and batch normalization are used, respectively, in between the DNN layers. Followed by the embedding layer, DURLD contains the convolution layer. This layer contains 64 filters with filter length 5, activation function $ReLU$. The convolutional layer follows maxpooling with pool length 4, LSTM with 70 memory blocks. Finally, the optimal features are passed into a fully connected layer which contains $sigmoid$ non-linear activation function which results in 0 for legitimate and 1 for spam. The loss function is binary cross-entropy and is defined mathematically as given below.

$$loss(pd, ed) = -\frac{1}{N}\sum_{i=1}^{N}[ed_i \log pd_i + (1 - ed_i)\log(1 - pd_i)], \quad (1)$$

where pd is a vector of predicted probability for all samples in testing data set, ed is a vector of the expected class label, values are either 0 or 1.

Table 4 Detailed configuration parameter information of DURLD for Data set 2 random-split and time-split

Layer (type)	Output shape	Param #
embedding_1 (Embedding)	(None, 246, 128)	5376
conv1d_1 (Conv1D)	(None, 245, 128)	32896
max_pooling1d_1 (MaxPooling1	(None, 122, 128)	0
lstm_1 (LSTM)	(None, 70)	55720
dense_1 (Dense)	(None, 1)	71
activation_1 (Activation)	(None, 1)	0
Total params: 94,063		
Trainable params: 94,063		
Non-trainable params: 0		

Table 5 Detailed configuration details of DNN

Layer (type)	Output shape	Param #
dense_1 (Dense)	(None, 128)	128128
batch_normalization_1 (Batch	(None, 128)	512
activation_1 (Activation)	(None, 128)	0
dropout_1 (Dropout)	(None, 128)	0
dense_2 (Dense)	(None, 96)	12384
batch_normalization_2 (Batch	(None, 96)	384
activation_2 (Activation)	(None, 96)	0
dropout_2 (Dropout)	(None, 96)	0
dense_3 (Dense)	(None, 64)	6208
batch_normalization_3 (Batch	(None, 64)	256
activation_3 (Activation)	(None, 64)	0
dropout_3 (Dropout)	(None, 64)	0
dense_4 (Dense)	(None, 32)	2080
batch_normalization_4 (Batch	(None, 32)	128
activation_4 (Activation)	(None, 32)	0
dropout_4 (Dropout)	(None, 32)	0
dense_5 (Dense)	(None, 16)	528
batch_normalization_5 (Batch	(None, 16)	64
activation_5 (Activation)	(None, 16)	0
dropout_5 (Dropout)	(None, 16)	0
dense_6 (Dense)	(None, 1)	17
Total params: 150,689		
Trainable params: 150,017		
Non-trainable params: 672		

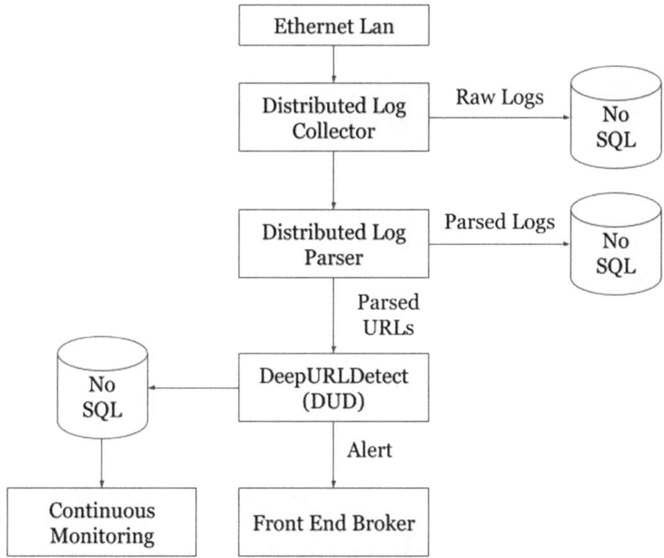

Fig. 4 Proposed architecture—DeepURLDetect (DURLD)

8 Proposed Architecture—DeepURLDetect (DURLD)

The proposed architecture for malicious URL detection in the Ethernet level is shown in Fig. 4. It is called as DeepURLDetect (DURLD). DURLD is a hybrid of convolution and long short-term memory in-house model. This module can be added to the existing scalable framework for cyber threat situational awareness to enhance the malicious detection rate [40]. The architecture consists of three main modules (1) Data collection, (2) Identifying malicious URL, and (3) Continuous monitoring

A distributed log collector collects URL logs from different sources inside an Ethernet LAN in a passive way and pass it into a distributed database. Following, the URLs are parsed using distributed log parser and fed into the deep learning module. This classifies the URLs into either malicious or legitimate. A copy of the preprocessed URLs is stored in a distributed database for further use. The deep learning module has a Front End Broker to display detailed information about the URL analysis. The framework contains a continuous monitoring module that monitors detected malicious URLs. This monitors the targeted URLs once every 30 s. This helps to detect the malicious URL which is generated using Digitally Generated Algorithms (DGA).

Table 6 Test results

Model	Accuracy (%)	Precision (%)	Recall (%)	F1-score (%)
Data set 1 (both train and test from public sources)				
Invincea [31]	99.0	99.5	98.4	98.9
NYU [43]	97.7	98.0	96.8	97.4
MIT [44]	97.9	98.8	96.6	97.7
CMU [17]	99.1	99.2	98.7	99.0
Endgame [7]	99.1	99.3	98.7	99.0
DeepURLDetect (proposed)	97.2	97.4	96.4	96.9
3-gram with DNN 5 layer (proposed)	95.4	96.8	93.0	94.9
Data set 2 random-split				
Invincea [31]	96.4	97.9	93.1	95.4
NYU [43]	96.1	97.9	92.2	95.0
MIT [44]	96.0	96.1	93.6	94.9
CMU [17]	95.4	95.1	93.3	94.2
Endgame [7]	96.6	97.2	94.1	95.6
DeepURLDetect (proposed)	95.4	97.4	90.8	94.0
3-gram with DNN 5 layer (proposed)	95.0	96.2	90.9	93.5
Data set 2 time-split				
Invincea [31]	96.1	95.9	94.4	95.1
NYU [43]	95.0	95.8	91.3	93.5
MIT [44]	93.3	97.7	85.2	91.0
CMU [17]	94.1	95.9	89.0	92.3
Endgame [7]	97.1	97.6	95.0	96.3
DeepURLDetect (proposed)	93.1	94.5	87.8	91.1
3-gram with DNN 5 layer (proposed)	93.0	96.3	85.6	90.7

Table 7 Test results. TPR and FPR are w.r.t. a threshold 0.5

Model	TPR (%)	FPR	AUC
Data set 1 (both train and test from public sources)			
Invincea [31]	88.9	0.087	0.9995
NYU [43]	89.5	0.10	0.9974
MIT [44]	85.9	0.124	0.9980
CMU [17]	87.7	0.105	0.9995
Endgame [7]	87.1	0.081	0.9996
DeepURLDetect (proposed)	87.6	0.116	0.9964
3-gram with DNN 5 layer (proposed)	87.9	0.121	0.9909
Data set 2 random-split			
Invincea [31]	93.9	0.142	0.9914
NYU [43]	94.2	0.143	0.9918
MIT [44]	95.0	0.146	0.9918
CMU [17]	94.4	0.145	0.9915
Endgame [7]	93.6	0.139	0.9913
DeepURLDetect (proposed)	95.3	0.146	0.9922
3-gram with DNN 5 layer (proposed)	95.3	0.143	0.9922
Data set 2 time-split			
Invincea [31]	85.9	0.023	0.9938
NYU [43]	83.3	0.031	0.9896
MIT [44]	82.7	0.027	0.9815
CMU [17]	89.5	0.027	0.9865
Endgame [7]	92.1	0.034	0.9962
DeepURLDetect (proposed)	79.9	0.032	0.9823
3-gram with DNN 5 layer (proposed)	71.4	0.042	0.9812

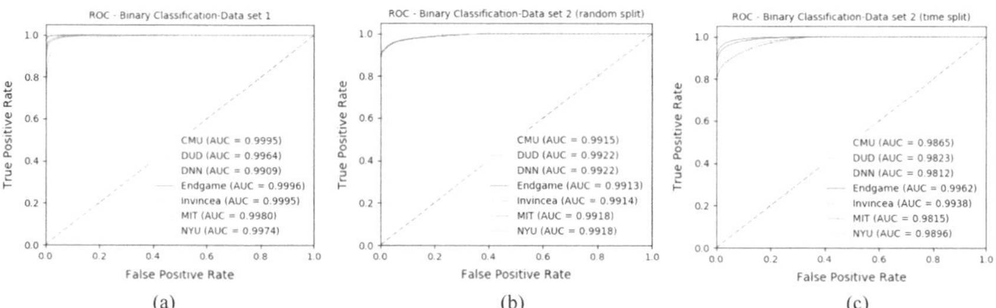

Fig. 5 ROC curve for **a** Data set 1, **b** Data set 2 (random-split) **c** Data set 2 (time-split)

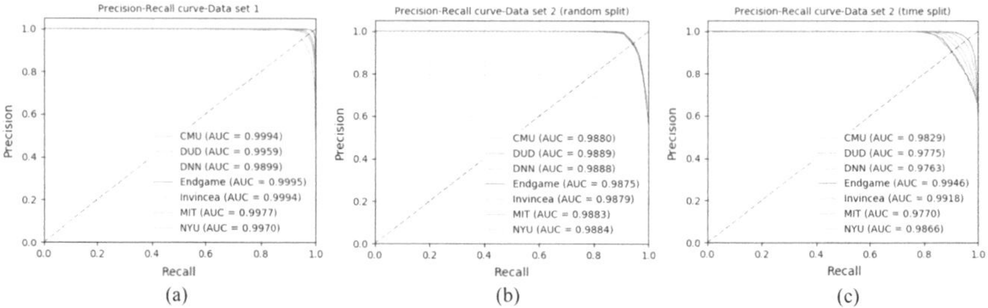

Fig. 6 ROC curve for **a** Data set 1, **b** Data set 2 (random-split) **c** Data set 2 (time-split)

9 Performance Measures

The main objective of this work is to classify whether the URL is either benign or malicious. To identify the performance of the deep learning architectures, we have used the following statistical metrics.

- True positive (TP): malicious URL that is correctly classified as malicious URL.
- True negative (TN): benign URL that is correctly classified as benign URL.
- False positive FP:benign URL that is incorrectly classified as malicious URL.
- False negative (FN): malicious URL that is incorrectly classified as benign URL.

The above metrics are obtained from the confusion matrix. In the confusion matrix, each row indicates the URL samples in a predicted class and each column indicates the URL samples in an actual class. We estimate the statistical measures such as Accuracy, Precision, Recall, F1-score, true positive rate (TPR), and false positive rate (FPR) from confusion matrix, and they are defined mathematically as follows:

$$Accuracy = \frac{TP + TN}{TP + TN + FP + FN} \tag{2}$$

$$\text{Re}call = \frac{TP}{TP + FN} \tag{3}$$

$$\text{Pr}ecision = \frac{TP}{TP + FP} \tag{4}$$

$$F1 - score = \frac{2 * \text{Re}call * \text{Pr}ecision}{\text{Re}call + \text{Pr}ecision} \tag{5}$$

$$TPR = \frac{TP}{TP + FN} \tag{6}$$

$$FPR = \frac{FP}{FP + TN} \tag{7}$$

The accuracy estimates the ratio of the total number of correct classifications. The precision estimates the number of correct classifications penalized by the number of incorrect classifications. The recall estimates the number of correct classifications penalized by the number of missed entries. The recall is also called sensitivity or true positive rate. The F1-score estimates the harmonic mean of precision and recall, which serves as a derived effectiveness measurement. The receiver operating characteristic (ROC) curve denotes the performance of the classifier which is plotted using TPR on Y-axis and FPR on X-axis. Generally, the ROC curve is used when the classes are balanced and for imbalanced classes precision-recall curve is used. To generate the precision-recall curve, we estimated the trade-off between the precision and recall across varying thresholds in the range of [0, 1].

10 Evaluation Results and Observations

All deep learning architectures are implemented using TensorFlow [1] with Keras [16] and conventional machine learning algorithms are implemented using Scikit-learn [25]. The graphical processing unit (GPU)-enabled machine is used for experimental purposes. Initially, all the models are trained with Data set 1. To evaluate the performance, the trained model is tested with the test data set of Data set 1. Likewise, same approach is followed for Data set 2 random-split and Data set 2 time-split. Most of the models performed well on Data set 1 when compared to the models trained on Data set 2 random-split and time-split. Moreover, the performance of various models on Data set 2 random-split is good when compared to the models trained on Data set 2 time-split. This is because the samples of test data of Data set 2 time-split is unseen during training. The detailed results are reported in Tables 6 and 7. The receiver operating characteristic (ROC) curve for various models on different test data sets are shown in Fig. 5a for Data set 1, Fig. 5b for Data set 2 random-split, and Fig. 5c for Data set 2 time-split with comparing two operating characteristics such as true positive rate and false positive rate across varying threshold in the range [0.0–1.0]. Likewise, the precision-recall curve for various models on different test data sets are shown in Fig. 6a for Data set 1, Fig. 6b for Data set 2 random-split, and Fig. 6c for Data set 2 time-split with comparing two operating characteristics such as precision and recall across varying thresholds in the range [0.0–1.0]. Obtaining better AUC in the precision-recall curve indicates that the models predict more accurately. The performance of all models has a marginal difference in terms of accuracy and AUC, and thus voting methodology can be employed to distinguish whether the URL is legitimate or malicious. This can further enhance the malicious URL detection rate. This remains as one of the significant direction toward future work. The models that used character level Keras embedding as the text representation method performed well when compared to DNN with a 3-gram text representation method. Deep learning with embedding-based malicious URL detection can be a robust solution over handcrafted features with conventional machine learning-based solutions. This is because the attacker can utilize domain knowledge to learn the handcrafted features to evade detection. They can make use of generative adversarial networks in deep learning; the details are discussed in [7].

11 Conclusion

In this work, a comparative analysis of various deep learning-based character level embedding models for malicious URL detection is done. All deep learning architectures have a marginal difference in terms of accuracy. Among five models, two are based on RNN, two are based on CNN, and one is based on a hybrid of CNN and LSTM. All the models performed well and achieved around 93–98% malicious URL detection rate with a false positive rate of 0.001. For comparative analysis, DNN with n-gram is used. In all test cases, deep learning-based character level models performed well when compared to the other models. All deep learning-based character level embedding-based models have the potential to handle variants of malicious URLs. Though deep learning has performed well, it is good to have conventional methods such as blacklisting using regular expression, signature matching method, and conventional machine learning-based solutions as an initial gateway followed by deep learning-based character level embedding models. The DeepURLDetect (DURLD) model can be made more robust by adding auxiliary modules such as registration services, website content, network reputation, file paths, and registry keys. This can be considered as one of the significant directions for future work.

Acknowledgements This work was supported by the Department of Corporate and Information Services, Northern Territory Government of Australia and in part by Paramount Computer Systems and Lakhshya Cyber Security Labs. We are grateful to NVIDIA India, for the GPU hardware support to the research grant. We are also grateful to Centre for Computational Engineering and Networking (CEN), Amrita School of Engineering, Coimbatore, for encouraging this research.

References

1. Abadi, Martín, Paul Barham, Jianmin Chen, Zhifeng Chen, Andy Davis, Jeffrey Dean, Matthieu Devin, Sanjay Ghemawat, Geoffrey Irving, and Michael Isard. 2016. Tensorflow: A system for large-scale machine learning. In *12th* {USENIX} *symposium on operating systems design and implementation (*{OSDI} *16)*, 265–283.
2. Alazab, M., R. Layton, R. Broadhurst, and B. Bouhours. 2013. Malicious spam emails developments and authorship attribution. In *2013 fourth cybercrime and trustworthy computing workshop*, 58–68.
3. Alazab, Mamoun, and Roderic Broadhurst. 2016. Spam and criminal activity. *Trends and Issues in Crime and Criminal Justice (Australian Institute of Criminology)* (526). https://www.aic.gov.au/publications/tandi/tandi526.
4. Alazab, Mamoun, Robert Layton, Roderic Broadhurst, and Brigitte Bouhours. 2013. Malicious spam emails developments and authorship attribution. In *2013 fourth cybercrime and trustworthy computing workshop*, 58–68. IEEE, 2013.

5. Alazab, Mamoun, Sitalakshmi Venkatraman, Paul Watters, and Moutaz Alazab. 2010. Zero-day malware detection based on supervised learning algorithms of api call signatures.
6. Alazab, Mamoun, Sitalakshmi Venkatraman, Paul Watters, and Moutaz Alazab. 2013. Information security governance: the art of detecting hidden malware. In *IT security governance innovations: theory and research*, 293–315. IGI Global.
7. Anderson, Hyrum S., Jonathan Woodbridge, and Bobby Filar. 2016. Deepdga: Adversarially-tuned domain generation and detection. In *Proceedings of the 2016 ACM workshop on artificial intelligence and security*, 13–21.
8. Azab, A., M. Alazab, and M. Aiash. 2016. Machine learning based botnet identification traffic. In *2016 IEEE Trustcom/BigDataSE/ISPA*, 1788–1794.
9. Azab, A., R. Layton, M. Alazab, and J. Oliver. 2014. Mining malware to detect variants. In *2014 fifth cybercrime and trustworthy computing conference*, 44–53.
10. Bahnsen, A.C., E.C. Bohorquez, S. Villegas, J. Vargas, and F.A. González. 2017. Classifying phishing urls using recurrent neural networks. In *2017 APWG symposium on electronic crime research (eCrime)*, 1–8.
11. Blum, Aaron, Brad Wardman, Thamar Solorio, and Gary Warner. 2010. Lexical feature based phishing url detection using online learning. In *Proceedings of the 3rd ACM Workshop on Artificial Intelligence and Security*, 54–60.
12. Broadhurst, Roderic, Peter Grabosky, Mamoun Alazab, Brigitte Bouhours, and Steve Chon. 2014. An analysis of the nature of groups engaged in cyber crime. *An Analysis of the Nature of Groups engaged in Cyber Crime, International Journal of Cyber Criminology* 8 (1): 1–20.
13. Cao, Jian, Qiang Li, Yuede Ji, Yukun He, and Dong Guo. 2016. Detection of forwarding-based malicious urls in online social networks. *International Journal of Parallel Programming* 44 (1): 163–180.
14. Chiba, Daiki, Kazuhiro Tobe, Tatsuya Mori, and Shigeki Goto. 2012. Detecting malicious websites by learning ip address features. In *2012 IEEE/IPSJ 12th international symposium on applications and the internet*, 29–39. IEEE.
15. Choi, Hyunsang, Bin B. Zhu, and Heejo Lee. 2011. Detecting malicious web links and identifying their attack types. *WebApps* 11 (11): 218.
16. Chollet, François. 2015. keras.
17. Dhingra, Bhuwan, Zhong Zhou, Dylan Fitzpatrick, Michael Muehl, and William W Cohen. 2016. Tweet2vec: Character-based distributed representations for social media. arXiv:1605.03481.
18. Felegyhazi, Mark, Christian Kreibich, and Vern Paxson. 2010. On the potential of proactive domain blacklisting. *LEET* 10: 6.
19. Harikrishnan, N.B., R. Vinayakumar, K.P. Soman, and Prabaharan Poornachandran. 2019. Time split based pre-processing with a data-driven approach for malicious url detection. In *Cybersecurity and secure information systems*, 43–65. Springer.
20. Kolari, Pranam, Tim Finin, and Anupam Joshi. 2006. Svms for the blogosphere: Blog identification and splog detection. In *AAAI spring symposium on computational approaches to analysing weblogs*.
21. Lee, S., and J. Kim. 2013. Warningbird: A near real-time detection system for suspicious urls in twitter stream. *IEEE Transactions on Dependable and Secure Computing* 10 (3): 183–195.
22. Ma, Justin, Lawrence K. Saul, Stefan Savage, and Geoffrey M. Voelker. 2009. Beyond blacklists: learning to detect malicious web sites from suspicious urls. In *Proceedings of the 15th ACM SIGKDD international conference on Knowledge discovery and data mining*, 1245–1254.
23. Ma, Justin, Lawrence K. Saul, Stefan Savage, and Geoffrey M. Voelker. 2009. Identifying suspicious urls: an application of large-scale online learning. In *Proceedings of the 26th annual international conference on machine learning*, 681–688.
24. Kevin McGrath, D., and Minaxi Gupta. 2008. Behind phishing: An examination of phisher modi operandi. *LEET* 8: 4.
25. Pedregosa, Fabian, Gaël Varoquaux, Alexandre Gramfort, Vincent Michel, Bertrand Thirion, Olivier Grisel, Mathieu Blondel, Peter Prettenhofer, Ron Weiss, and Vincent Dubourg. 2011. Scikit-learn: Machine learning in python. *the Journal of Machine Learning Research*, 12: 2825–2830.

26. R., V., M. Alazab, A. Jolfaei, S. K.P., and P. Poornachandran. 2019. Ransomware triage using deep learning: Twitter as a case study. In *2019 cybersecurity and cyberforensics conference (CCC)*, 67–73
27. S, S., V. R, M. Alazab, and S. KP. 2020. Network flow based iot botnet attack detection using deep learning. In *IEEE INFOCOM 2020 - IEEE conference on computer communications workshops (INFOCOM WKSHPS)*, 189–194.
28. S, S., V. R, S. V, M. Alazab, and S. KP. 2020. Multi-scale learning based malware variant detection using spatial pyramid pooling network. In *IEEE INFOCOM 2020 - IEEE conference on computer communications workshops (INFOCOM WKSHPS)*, 740–745.
29. Sahoo, Doyen, Chenghao Liu, and Steven CH Hoi. 2017. Malicious url detection using machine learning: A survey. arXiv:1701.07179.
30. Sanders, Hillary, and Joshua Saxe. 2017. Garbage in, garbage out: How purport-edly great ml models can be screwed up by bad data. *Technical report.*
31. Saxe, Joshua, and Konstantin Berlin. 2017. expose: A character-level convolutional neural network with embeddings for detecting malicious urls, file paths and registry keys. arXiv:1702.08568.
32. Schiappa, Madeline. 2009. *Machine learning: How to build a better threat detection model.* Accessed July 3, 2020.
33. Sommer, R., and V. Paxson. 2010. Outside the closed world: On using machine learning for network intrusion detection. In *2010 IEEE symposium on security and privacy*, 305–316.
34. Srinivasan, S., V. Ravi, S. V., M. Krichen, D. Ben Noureddine, S. Anivilla, and S. K. P. 2020. Deep convolutional neural network based image spam classification. In *2020 6th conference on data science and machine learning applications (CDMA)*, 112–117.
35. Tran, Khoi-Nguyen, Mamoun Alazab, and Roderic Broadhurst. 2014. Towards a feature rich model for predicting spam emails containing malicious attachments and URLs.
36. Verma, Rakesh. 2018. Security analytics: Adapting data science for security challenges. In *Proceedings of the fourth ACM international workshop on security and privacy analytics*, 40–41.
37. Vinayakumar, R., M. Alazab, K.P. Soman, P. Poornachandran, A. Al-Nemrat, and S. Venkatraman. 2019. Deep learning approach for intelligent intrusion detection system. *IEEE Access* 7: 41525–41550.
38. Vinayakumar, R., M. Alazab, K.P. Soman, P. Poornachandran, and S. Venkatraman. 2019. Robust intelligent malware detection using deep learning. *IEEE Access* 7: 46717–46738.
39. Vinayakumar, R., M. Alazab, S. Srinivasan, Q. Pham, S.K. Padannayil, and K. Simran. 2020. A visualized botnet detection system based deep learning for the internet of things networks of smart cities. *IEEE Transactions on Industry Applications* 56 (4): 4436–4456.
40. Vinayakumar, R., Prabaharan Poornachandran, and K.P. Soman. 2018. Scalable framework for cyber threat situational awareness based on domain name systems data analysis. In *Big data in engineering applications*, 113–142. Springer.
41. Vinayakumar, R., K.P. Soman, and Prabaharan Poornachandran. 2018. Evaluating deep learning approaches to characterize and classify malicious url's. *Journal of Intelligent & Fuzzy Systems*, 34(3):1333–1343.
42. Vinayakumar, R., K.P. Soman, Prabaharan Poornachandran, Mamoun Alazab, and Sabu Thampi 2019. Amritadga: a comprehensive data set for domain generation algorithms (dgas) based domain name detection systems and application of deep learning. In *Big data recommender systems-Volume 2: application paradigms*, 455–485. Institution of Engineering and Technology (IET).
43. Vosoughi, Soroush, Prashanth Vijayaraghavan, and Deb Roy. 2016. Tweet2vec: Learning tweet embeddings using character-level cnn-lstm encoder-decoder. In *Proceedings of the 39th international ACM SIGIR conference on research and development in information retrieval*, 1041–1044.
44. Zhang, Xiang, Junbo Zhao, and Yann LeCun. Character-level convolutional networks for text classification. In *Advances in neural information processing systems*, 649–657.

Sentiment Analysis for Troll Detection on Weibo

Zidong Jiang, Fabio Di Troia, and Mark Stamp

Abstract The impact of social media on the modern world is difficult to overstate. Virtually all companies and public figures have social media accounts on popular platforms such as Twitter and Facebook. In China, the micro-blogging service provider, Sina Weibo, is the most popular such service. To influence public opinion, Weibo trolls—the so-called Water Army—can be hired to post deceptive comments. In this chapter, we focus on troll detection via sentiment analysis and other user activity data on the Sina Weibo platform. We implement techniques for Chinese sentence segmentation, word embedding, and sentiment score calculation. In recent years, troll detection and sentiment analysis have been studied, but we are not aware of previous research that considers troll detection based on sentiment analysis. We employ the resulting techniques to develop and test a sentiment analysis approach for troll detection, based on a variety of machine learning strategies. Experimental results are generated and analyzed. A Chrome extension is presented that implements our proposed technique, which enables real-time troll detection when a user browses Sina Weibo.

1 Introduction

Social media plays a significant role in the ongoing development of the Internet, as people tend to acquire more information from social media than other platforms. Deceptive comments created by trolls present a challenging problem in social media applications. Trolls can be hired to publish misleading comments in an effort to affect public opinion related to events or people, or even to negatively influence the economy of a country.

Z. Jiang · F. Di Troia · M. Stamp (✉)
San Jose State University, San Jose, CA, USA
e-mail: mark.stamp@sjsu.edu

F. Di Troia
e-mail: fabio.ditroia@sjsu.edu

M. Stamp et al. (eds.), *Malware Analysis Using Artificial Intelligence and Deep Learning*, https://doi.org/10.1007/978-3-030-62582-5_22

Sina Weibo is a widely used micro-blogging social media platform in China. A majority of Weibo posts are written in Chinese and, like Twitter, most posts published on Weibo are short—until recently, there was a 140 character limit. With the number of daily active users in excess of 200 million (as of 2019), Weibo is one of the largest social media platforms in China.

Weibo is based on weak relationships, in the sense that a user can share content that is visible to all of the user base. Therefore, many celebrities, businesses, and Internet influencers all over the world register as Weibo users in an effort to expand their exposure to the Chinese public. Weibo has become a platform where government and businesses can communicate more efficiently with the general public.

The Chinese Water Army refers to a group of people who can be hired to post deceptive comments on Weibo. Such troll activity is difficult to detect, due in part to the unsegmented characteristic of the Chinese sentences. In some cases, Chinese sentences can be segmented in different ways to yield different meanings.

Recent research has shown that hidden Markov models (HMMs) are effective for sentiment analysis of English text [37]. Chinese word segmentation can also be accomplished using HMMs [4, 21, 32]. In this research, we use HMMs, Word2Vec, and other learning techniques to perform word segmentation and sentiment analysis on Sina Weibo "tweets" for the purpose of detecting potential troll activity. We use Word2Vec and HMMs for Chinese text segmentation, we employ HMMs and naïve Bayes for sentiment analysis, and we use XGBoost and support vector machines (SVMs) for troll detection.

We have generated a large training dataset by crawling the Sina Weibo and Tencent Weibo platforms. Using an HMM-based Chinese sentence segmentation model comparable to that in [32], we pre-process each post into a list of words. Then, following the approach in [10], we construct a Word2Vec similarity scoring matrix based on the word list that we have generated. A baseline of sentiment is determined from the corpus that we have collected.

For sentiment analysis, we use a Word2Vec based technique to calculate sentiment scores. We use extracted features from Weibo comments as observations to train HMM models for each emotion, and we use the trained models to determine the emotions of each comment. We use an XGBoost model to aggregate sentiment analysis results with user activity data to build the troll detection model. As a point of comparison, we experiment with an approach based on support vector machines.

Finally, we present a Chrome extension that we have developed. This Chrome extension implements our troll detection model, and it enables us to detect potential troll activity on Weibo in real-time.

The remainder of this paper is organized as follows. In Sect. 2, we discuss relevant background topics. Section 3 contains an overview of selected previous work. In Sect. 4, we consider data sources and data collection methods. In Sect. 5, we provide implementation details and includes experimental results. Lastly, in Sect. 6, we give a summary of our work, including a brief discussion of possible directions for future development.

2 Background

In this section, we discuss several relevant background topics. First, to motivate this research, we discuss trolls in the context of social media. Then, we introduce machine learning models that are used in this research. We conclude this section with a brief discussion of the evaluation metric that we employ.

2.1 *Trolls*

Troll users publish misleading, offensive, or trivial following-up content in online communities. The content of a troll posting generally falls into one of several categories. It may consist of an apparently foolish contradiction of common knowledge, a deliberately offensive insult to the readers of a newsgroup or mailing list, or a broad request for trivial follow-up postings. The result of such posting is frequently a flood of angry responses. In some cases, the follow-up messages posted in response to a troll can constitute a large fraction of the contents of a newsgroup or mailing list on a particular topic over an extended period of time. These messages may be transmitted around the world to vast numbers of computers, wasting network resources, and costing resources. Troll threads frustrate people who are trying to carry on substantive discussions [12].

Organized troll activity on the Sina Weibo platform was first detected in 2013. This initial group of troll users consisted of about 20,000 individuals in 50 ICQ chat groups associated with a person nicknamed “Daxia.” Subsequently, troll activity became an online business on the Weibo platform. Trolls can be hired by businesses to publish negative comments against their competitors or to generate anonymous good reviews or positive comments. Prior to 2015, much of the troll activity on Weibo was designed to adversely affect the reputation of businesses. After 2015, stricter controls were set on speech on the Internet in China, and Sina Weibo developed a more sophisticated infrastructure to filter such troll comments. Currently, most of the troll activity on Weibo turned is designed to promote celebrities and companies.

Troll users on the Weibo platform can be categorized by their source of content. Traditionally, trolls use automated fake accounts to post repeated messages in an effort to dominate the comments. An example of such activity is shown in Fig. 1. However, the Weibo platform has recently improved their infrastructure to block these repeated messages from users, based on proxy detection, combined with message filters for repeated comments.

Recently, troll users have become more sophisticated. Some organized Weibo trolls are supervised by a management group who controls what, when, and where they reply on the Weibo platform. Specific details of the comments that each troll account publishes are made by individual troll users rather than being copied from the management group. The management group only gives out the overall emotional trend that the comments should convey. Thus, content made by troll users is repetitive

自从上次杨紫事件以后我就再也不敢随便评论明星了 有些粉丝真的是恶心 没三观没脑子就算了还要强迫改变你的看法 妈个逼你管老子啊真是

爷爷_ex
自从上次杨紫事件以后我就再也不敢随便评论明星了 有些粉丝真的是恶心 没三观没脑子就算了还要强迫改变你的看法 妈个逼你管老子啊真是

龍哥yx1
自从上次杨紫事件以后我就再也不敢随便评论明星了 有些粉丝真的是恶心 没三观没脑子就算了还要强迫改变你的看法 妈个逼你管老子啊真是

Fig. 1 Weibo comments dominated by troll activity [26]

but not monotonously so. This fact makes troll detection on Weibo challenging, since troll comments are composed and published by real human users. Furthermore, in recent years, trolls are mostly hired by companies and celebrities to make positive comments toward themselves.

2.2 *Machine Learning Techniques*

Content-based troll detection usually utilizes natural language processing (NLP) using machine learning to analyze and categorize text. This is accomplished by constructing language processing models on comments and posts so as to label comments with a high polarity of emotion or repetitiveness. By applying sentiment analysis methods, we can filter comments with either high or low sentiment scores representing extreme positive or negative sentiment. This is accomplished by calculating word relevance, and by analyzing correlations using word embedding techniques, such as Word2Vec. We can then mark potential troll comments or pass along user information behind such comments to the next stage of a troll detection model. A key point of this research is to use sentiment analysis in troll detection.

Classifying specific comments as troll activity is challenging. Therefore, utilizing user behavioral information to discern deceptive activity is a popular trend in troll detection. Like most social media platforms, Weibo has numerous user relationship data, such as the number of followers, number that a user is following, user rank, and number of original Weibo tweets. Also, trolls commonly make attacks in a small time window following a specific tweet [11]. We can utilize this fact in combination with other user relationship information in a troll detection system. The goal of including such data in our troll detection approach is to reduce the false negatives that affect strictly content-based detection methods.

Next, we introduce the various machine learning techniques used in this research. Specifically, we discuss hidden Markov models, Word2Vec, XGBoost, and support vector machines.

2.2.1 Hidden Markov Models

Hidden Markov models (HMMs) are well known for their use in pattern prediction and for deriving hidden states from observations. An HMM (of order one) is a stochastic model representing states where each future state depends only on the current state, and not on states further in the past. By training an HMM on an observation sequence, we can obtain the probability of the transitions between each hidden state and probability distributions for the observations, based on those hidden states. A generic HMM is illustrated in Fig. 2.

In Fig. 2, the matrix A drives the Markov process for the hidden states, while the matrix B probabilistically relates the hidden states to the observations $\mathcal{O}_i$. The HMM notation is summarized in Table 1.

Applications of HMMs are extremely diverse, but for our purposes, two relevant uses are English text analysis and speech recognition [28]. Other applications of HMMs range from classic cryptanalysis [30] to malware detection [7, 29]. In this research, we use HMMs for both Chinese word segmentation and for emotion classification.

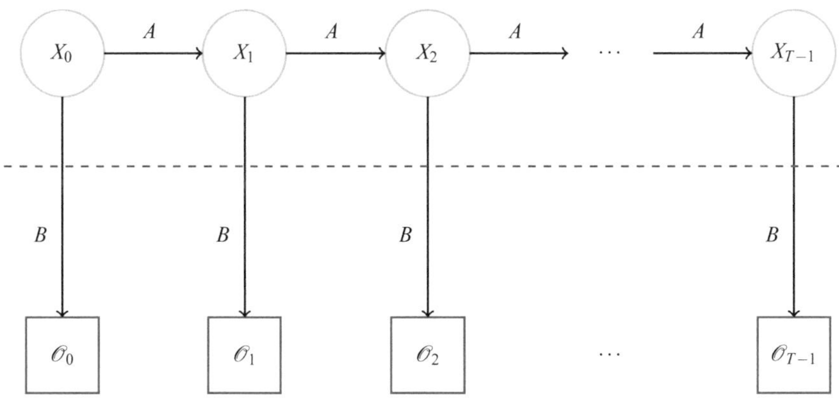

Fig. 2 Hidden Markov model

Table 1 HMM notation

Notation	Description
T	Length of the observation sequence
N	Number of states in the model
M	Number of observation symbols
Q	Distinct states of the Markov process, $q_0, q_1, \ldots, q_{N-1}$
V	Possible observations, assumed to be $0, 1, \ldots, M-1$
A	State transition probabilities
B	Observation probability matrix
π	Initial state distribution
$\mathcal{O}$	Observation sequence, $\mathcal{O}_0, \mathcal{O}_1, \ldots, \mathcal{O}_{T-1}$

2.2.2 Word2Vec

Word2Vec has recently gained considerable popularity in natural language processing (NLP) [18]. This word embedding technique is based on a shallow neural network, with the weights of the trained model serving as embedding vectors—the trained model itself serves no other purpose. These embedding vectors capture significant relationships between words in the training set. Word2Vec can also be used beyond the NLP context to model relationships between more general features or observations.

When training a Word2Vec model, we must specify the desired vector length, which we denote as N. Another key parameter is the window length W, which represents the width of a sliding window that is used to extract training samples from the data.

Certain algebraic properties hold for Word2Vec embeddings. For example, suppose that we train a state-of-the-art Word2Vec model on English text. Further, suppose that we let

$$w_0 = \text{“king”},\ w_1 = \text{“man”},\ w_2 = \text{“woman”},\ w_3 = \text{“queen”},$$

and we define $V(w_i)$ to be the Word2Vec embedding of word w_i. Then according to [18], the vector $V(w_3)$ is closest to

$$V(w_0) - V(w_1) + V(w_2)$$

where “closeness” is in terms of cosine similarity. Results such as this indicate that in the NLP context, Word2Vec embeddings capture meaningful aspects of the semantics of the language.

In this research, we train Word2Vec models Chinese text. These models are then used for sentiment analysis of Weibo tweets.

2.2.3 XGBoost

Boosting is a general technique for constructing a stronger classifier from a large collection of relatively weak classifiers [6]. XGBoost typically uses decision trees as the base classifiers. To generate our models, we use the XGBoost package in Python. With this implementation, it is easy to analyze the significance of each individual feature relative to the overall model, and thus eliminate ineffective features.

2.2.4 SVM

In [2], support vector machine (SVM) is described as "a rare example of a methodology where geometric intuition, elegant mathematics, theoretical guarantees, and practical algorithms meet." The essential ideas behind SVMs are the following.

Separating hyperplane —We seek a hyperplane that will separate two labeled classes.
Maximize the margin —We want to find hyperplane that maximizes the "margin" between two classes, where margin is defined as the minimum distance.
Work in higher dimensional space —By shifting the problem to a higher dimensional space, there is a better chance that we can find a separation hyperplane, or separating hyperplane with a larger margin.
Kernel trick —Perhaps surprisingly, we are able to work in a higher dimensional space without paying any significant penalty with respect to computational complexity. This is a powerful "trick" and is the key reason why SVM is one of the most popular machine learning techniques available.

In this research, SVM serves as a comparison to XGBoost for troll detection. We find that XGBoost performs better on one of our datasets, while SVM is superior on another dataset.

2.3 *Evaluation Metric*

We use accuracy as the primary measure of success for all of our classification experiments. The accuracy is computed as

$$\text{accuracy} = \frac{\text{TP} + \text{TN}}{\text{TP} + \text{TN} + \text{FP} + \text{FN}},$$

where TP is number of true positive cases, TN is true negatives, FP is false positives, and FN is the number of false negatives. Accuracy can be seen to simply be the ratio of correct classifications to the total number of classifications.

Finally, we note that cross-validation is used in all of our experiments. Cross-validation is a popular technique that serves to smooth any bias in the data, while also maximizing the number of datapoints. Specifically, we employ fivefold cross-validation.

3 Related Work

Related work in sentiment analysis includes [36], where a combination of emotional orientation and logistical regression is used to analyze Amazon.com reviews. By filtering the training dataset by text length, vocabulary complexity, correlation with the product, sentiment similarity, and transition words, the proposed model achieved 91.2% accuracy. For the problem of fake Weibo tweet detection—as opposed to the troll detection we consider in this research—an XGBoost model based on user activity achieved 93% accuracy in [17].

From our review of the literature, it appears that only [24] applies sentiment analysis to the troll detection problem. The work in [24] applies domain adaptation techniques to a recursive neural tensor network (RNTN) sentiment analysis model to detect trolls that post repetitive, destructive, or deceptive comments. This previous work achieves 78% accuracy. The results in [24] serve as a baseline for our research.

Sentiment analysis is widely used for mining subjective information in online posts. In [14], Kim, et al., use hidden Markov models with syntactic and sentiment information for sentiment analysis of Twitter data. This differs from classic approaches that use n-grams and polarity lexicons, as they group words based on similar syntactic and sentiment groups (SIG), then build HMMs, where the SIGs define the hidden states. Zhao and Ohsawa [37] propose a two-dimensional HMM to analyze Amazon reviews in Japanese. For our purposes, this work illustrates an important method for converting segmented Japanese text into word vectors using Word2Vec. Feng and Durdyev [10] implemented three types of classification models (SVM, XGBoost, LSTM) for the aspect-level sentiment analysis of restaurant customer reviews in Chinese. According to the research in [10], LSTM yields better F-1 scores and accuracy, as compared to SVM and XGBoost. Further related research can be found in Liu et al. [15], which uses a self-adaptive HMM.

Troll detection based on user characteristic data is considered in Zhang et al. [11]. In this paper, Weibo troll detection is based on a Bayesian model and genetic algorithm. The proposed technique includes novel features (as compared to previous work) such as the ratio of followers, average posts, and Weibo credibility, and achieves an accuracy of about 90%.

Liu et al. [17] use XGBoost to detect fake Weibo posts based on features such as a user's number of posts, description, gender, followers, and reposts. The authors attain an accuracy of more than 95%. Both [11] and [17] use data beyond Weibo post text itself, and achieve good results. This previous work serves as inspiration for some of the features considered in this paper.

The special interest group for Chinese language processing (SIGHAN) of the Association for Computational Linguistics organizes competitions for Chinese word segmentation. In the first SIGHAN bake-off event in 2003, Zhang et al. [35] proposed a word segmentation approach using hierarchical HMMs to form a Chinese lexical analyzer, ICTCLAS. In 2005, Masayuki et al. [1] presented three word segmentation models, including a character tagging classifier based on support vector machines (SVMs) that also used maximum expropriation Markov models and conditional random fields. These models were based on previously proposed methods, with a different combination of out-of-vocabulary (OOV) extraction techniques being used.

In general, for Chinese word segmentation, character-based models perform better than word-based models. Wang et al. [33] highlighted that OOV techniques for word extraction perform poor for in-vocabulary (IV) words. They proposed a generative model that performs well on both OOV and IV words, and achieved good results on the popular SIGHAN datasets. Chen et al. [5] report the use of Gibbs sampling in combination with both word-based hierarchical process models and character-based HMMs. Their solution achieved better performance (in terms of F_1 score) than the state-of-the-art models at that time.

4 Datasets

We acquired data and generated additional data for the various parts of this research. We have Chinese segmentation data, sentiment analysis data, data consisting of Weibo comments, and user data corresponding to the Weibo comments data. This data is split into three datasets, namely, a Chinese segmentation dataset, a sentiment analysis dataset, and a troll detection dataset. Next, we discuss each of these three datasets.

4.1 Chinese Segmentation Dataset

For Chinese sentence segmentation, we acquired the dataset used in the SIGHAN 2005 Competition for Chinese sentence processing [9]. This dataset includes training, testing, and validation data. The training data consists of approximately 860,000 segmented Chinese sentences. Most of these sentences are from newspapers and published books. The test set includes about 22,000 unsegmented sentences from similar sources, while the validation set contains the segmentation of all of the sentences in the test set. Table 2 gives additional statistics for this dataset.

We consider the character-based generative model proposed in [32]. In this model, the features from the training data consist of the positions of each character in each segmented word. The beginning character in each segment is marked as B, any

Table 2 Chinese segmentation dataset (SIGHAN 2nd Bakeoff 2005)

Source	Training	Testing
Academia Sincia	708,953	14,432
Peking University	19,056	1,944
City University of Hong Kong	53,019	1,492
Microsoft Research Asia	86,924	3,985

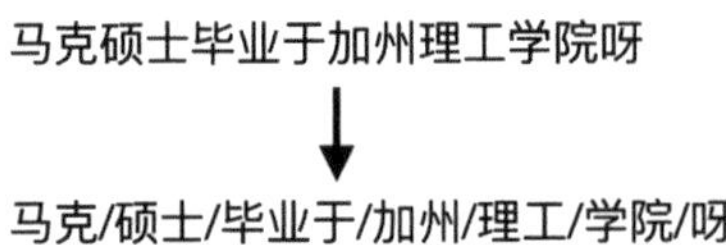

Fig. 3 Sample Chinese sentence segmentation

Table 3 State sequence for Chinese sentence

马	克	硕	士	毕	业	于	加	州	理	工	学	院	呀
B	E	B	E	B	M	E	B	E	B	E	B	E	S

middle character or characters are marked as M, and the ending character is marked as E. On the other hand, all one-character words are marked as S.

For example, consider the sample Chinese sentence in Fig. 3, which includes the correct segmentation for this sentence. Table 3 gives the states corresponding to the sentence in Fig. 3.

4.2 Sentiment Analysis Dataset

For sentiment analysis, we use the sentiment training dataset from the Python SnowNLP package [34]. This particular dataset includes 16,548 sentences with positive sentiment and 18,574 with negative sentiment. The source for this dataset is Chinese online shopping, movie, and book reviews. However, this data might not accurately represents tweets and comments appearing in Weibo.

Since there is no public datasets for Weibo, we crawled about 5 million Sina Weibo posts to obtain additional data for our sentiment model. This data includes terms and slang that are commonly seen on Weibo. From this data, we created a collection of 2,325,644 sentences with positive sentiment and 960,899 sentences with negative sentiment.

From all of our Weibo crawled data, we manually extracted 500 tweets for each of the six emotions of interest, namely, happiness, surprise, fear, anger, disgust, and sadness [3]. This data will be used to train an HMM for each emotion.

We process each comment using the Pandas package in Python to remove stop words, numbers, nonsense emoji, and single-word comments. A language detection method was implemented to detect non-Chinese comments and translate English comments into Chinese using the Translator package in Google Translate. We eliminate all comments that are in languages other than Chinese and English, which results in a negligible loss of data. In addition, we removed pure reposts and tagging that is not relevant to our analysis.

For positive and negative sentiment analysis, we used the Word2Vec embedding method. The resulting embedding vectors enable us to calculate a word sentiment score, after segmenting a Weibo comment into a list of word. For sentiment analysis (based on six basic emotions), we use the features introduced in [15], which are then used to train HMMs for sentiment classification. These features are mutual information, χ^2 distance, and term frequency inverse document frequency, which are defined as follows.

MI —Mutual information (MI) is based on correlations between two terms. In this research, MI is used to determine the relevance between words and emotions. The formula for MI representing the correlation between emotion e and text t is given by

$$\mathrm{MI}(t, e) = \log \frac{P(t \mid e)}{P(e)}.$$

CHI — We use a χ^2 distance measure (CHI) to quantify the dependence between emotion e and text t. The higher the CHI value, the more dependent the text t is on the emotion e. We calculate CHI as

$$\mathrm{CHI}(t, e) = \frac{N(AD - BC)^2}{(A + B)(C + D)(A + C)(B + D)},$$

where A is the prevalence of word t in comments with emotion e, B is the prevalence of word t in comments with emotions other than e, C is the absence of word t in comments with emotion e, D is the absence of word t in comments with emotions other than e, and N is the total number of comments.

TF-IDF —Term frequency inverse document frequency (TF-IDF) was originally developed to extract keywords from text, for purposes such as indexing. We use TF-IDF to determine keywords with respect to the various emotions under consideration. We compute the TF-IDF as

$$\text{TF-IDF}(t, e) = \frac{N_{e,t}}{\sum_k N_{k,t}} \log\left(\frac{N}{n_e} + 0.01\right),$$

where $N_{e,t}$ is the number of times word t appears in a comment with emotion e, N is the total number of comments, and n_e is the number of comments in which the emotion e appears.

4.3 Troll Detection Dataset

After some initial experiments, we realized that there are limitations to the features specified in [15, 24]. Therefore, we introduced more user information related features that we obtained by mining Weibo comment data. When crawling the Weibo data, we use the JavaScript object notation (JSON) packet returned from representational state transfer (REST) calls to the Weibo mobile site [27], which includes user-related information. Typical operations under the REST API include GET, POST, UPDATE, and DELETE. We extract the user information listed in Table 4. We also include a small dataset of 673 normal users and 75 trolls from a Kaggle data source [16].

All user information and corresponding comments are grouped by original tweet ID and stored in CSV format. One CSV file contains all of the comments regarding one tweet, and each entry represents all of the information listed in Table 4. We selected eight tweets with a total of 31,980 comments from Sina Weibo accounts belonging primarily to business owners and celebrities. The detailed tweet information and statistics for this dataset are listed in Table 5.

We manually labeled the data for rows 1, 2, 3, and 6 in Table 5 as troll or non-troll by examining the content of each comment. Combining these results with fake account data from [16], we have about 3500 comment entries for our initial training and testing data for the troll detection model. This manual labeling is extremely tedious. To accelerate this process, we created a bot based on Selenium [25] to help open each user's Weibo page based on the UID that we provide from the dataset.

Table 4 List of user-related information from comment data

Field Name	Dataset	Description
`uid`	UID	Unique User ID for User Account in Weibo
`screen_name`	Username	Displayed User Nickname
`followers_count`	Follower	User's follower count
`follow_count`	Following	User's following count
`status_count`	Original_post	User's original composed tweet count
`urank`	User_rank	User's rated rank by user activity in Weibo
`verified`	Verified	Whether user is verified celebrity or business
`description`	Description	User's own description in headline
`like_count`	Like_count	Like count of this comment
`floor_number`	Floor_number	Location where the comment is at
`text`	Comment	Comment content

Table 5 Statistics of troll detection tweets and comments crawled from Weibo

	Tweet ID	User details	Number
1	44275283	LeEco CEO YT Jia declared bankrupt	812
2	44317480	Actress Yiyan Jiang volunteered teaching in rural	829
3	44564209	Yong actress Zi Yang suspected done plastic surgery	335
4	44718878	Reporting fraud in singer Hong Han Foundation receiving donation	1210
5	44651702	Singer Hong Han Foundation donation to Wu Han Coronavirus battle	3379
6	44650056	Criticism of multiple celebrities' donation to Coronavirus battle	814
7	43961306	Suspected breakup of Han Lu and Xiaotong Guan (Actor/ress)	8371
8	43961306	Han Lu and Xiaotong Guan (Actor/ress) showoff their same sweatshirt	16,230

To extract features from the data listed in Table 4, we use Python Pandas [20] and Numpy [19]. Some of the available features proved to be of little use, and these were dropped, as discussed below. In order to have better features for our models, we also perform some feature engineering. For example, we note whether users provide a self-description or not.

Features such as follower count, following count, and the number of original composed tweets clearly have a high significance in our analysis. However, we found that building models with quantitative numbers from these categories biases the model, due to the large differences across users. Weibo users typically only follow a fairly small number of accounts, while troll users typically follow a large number of accounts. Therefore, we dropped the follower and following count in the raw dataset and instead compute the ratio of following to follower and use this as a feature. Similarly, we introduce a feature consisting of the ratio of original posts to followers to help identify troll users, who often make a large number of posts without a commensurate increase in their follower count—we use this engineered feature in place of the composed post feature from the raw data.

When crawling Weibo, we noticed that some users frequently comment on the same tweet rather than replying to other comments under a tweet. Therefore, we select users who have more than one comment under a tweet, and we count the comments made for each such user. Then, we computed the median of these comment counts. Following this approach, a "frequent comment" feature is generated based on users who made more comments than the median number.

We have a total of 19 features that we use in our XGBoost model. One of the engineered features related to the sentiment score is denoted as `diffOriginalSenti`. This feature is the score for a comment minus the sentiment score of the original tweet. Table 6 lists the complete set of features that we obtain by combining sentiment analysis result and user information data in Table 4.

Table 6 Features considered for troll detection model

	Feature	Description	Source
F0	`follower`	Follower count	Crawled Weibo dataset
F1	`following`	Following count	Crawled Weibo dataset
F2	`original_post`	Number of original tweets	Crawled Weibo dataset
F3	`urank`	Rank by user activity in Weibo	Crawled Weibo dataset
F4	`verified`	User certified or not	Crawled Weibo dataset
F5	`like_count`	Like count for a comment	Crawled Weibo dataset
F6	`floor_number`	Comment location	Crawled Weibo dataset
F7	`description`	Self-description (1 or 0)	Engineered feature
F8	`freqComment`	Frequent comments	Engineered feature
F9	`ffRatio`	`following` divided by `follower`	Engineered feature
F10	`foRatio`	`original_post` divided by `follower`	Engineered feature
F11	`sentiment`	Comment sentiment score (0 to 1)	Engineered feature
F12	`diffOriginalSenti`	`sentiment` minus sentiment of original	Engineered feature
F13	`happy`	Happiness score (0 to 1)	Engineered feature
F14	`sad`	Sadness score (0 to 1)	Engineered feature
F15	`anger`	Anger score (0 to 1)	Engineered feature
F16	`disgust`	Disgust score (0 to 1)	Engineered feature
F17	`fear`	Fear score (0 to 1)	Engineered feature
F18	`surprise`	Surprise score (0 to 1)	Engineered feature

We would like to maximize our troll detection accuracy while minimizing the number of features needed. To achieve this, we perform feature analysis in order to rank the significance of features, so that we can drop features. This feature reduction process is discussed in Sect. 5.5.

5 Implementation and Results

In this section, we give our results. First, we discuss the Weibo crawler that we have implemented. Then, we consider our Chinese word segmentation results, followed by our emotion classification technique, both of which are based on hidden Markov models. Then, we consider our Word2Vec-based sentiment score and our XGBoost and SVM-based troll detection results. We conclude this section with a discussion of a Chrome extension that implements this troll detection system.

5.1 Weibo Crawler

As mentioned above, in order to have sufficient training and testing data, we developed a crawler to obtain such data directly from the Weibo platform. Our crawler extracts posts, comments, and user information.

To extract posts, the crawler certain considers a number of tweets under specific Weibo accounts. Note that Weibo tweets are similar to Twitter tweets, in that users can retweet other users' posts to their own Weibo account. The crawler disregards retweets and only keeps original posts.

Comment crawling is used to obtain additional information related to posts. Most comments contain repetitive messages and include username and hash-tags. We remove this extraneous data with the Python Panda Dataframe function before pipelining the comments into the word segmentation stage. Also, it is very common to see bilingual comments in Weibo, where most of the text is Chinese, but some English is included. Therefore, we incorporate a language detection module extended from the Google language library, which uses naïve Bayes to filter and translate English to Chinese.

Our comment crawler works on Weibo mobile [27] data, where the tweets and comments page are slightly simplified. The crawler makes HTTP requests such as

```
https://m.weibo.cn/comments/hotflow?id=TWEETID&mid=TWEETID&max_id=
```

which yields JSON data from the Weibo platform containing comments related to the specified tweet.[1] We parse the resulting comment data contained in the JSON packet to extract all of the raw data, including user information and comment content using the BeautifulSoup package in Python [22]. After all of the entries have been collected, they are saved into a CSV file, organized by tweet. Feature extraction and model training are based on these CSV files.

5.2 HMM for Chinese Segmentation

As illustrated in Table 3, when segmenting a Chinese sentence, we consider four states, namely, B for begin, M for middle, E for end, and S for single. Thus, we train an HMM with four hidden states. The observation sequence consists of Chinese characters in the training dataset, and the hidden states correspond to B, M, E, and S. It follows that the hidden state transition matrix of the HMM is 4 $\times$ 4 and of the form

$$\begin{bmatrix} \mathrm{B} \rightarrow \mathrm{B} & \mathrm{B} \rightarrow \mathrm{E} & \mathrm{B} \rightarrow \mathrm{M} & \mathrm{B} \rightarrow \mathrm{S} \\ \mathrm{E} \rightarrow \mathrm{B} & \mathrm{E} \rightarrow \mathrm{E} & \mathrm{E} \rightarrow \mathrm{M} & \mathrm{E} \rightarrow \mathrm{S} \\ \mathrm{M} \rightarrow \mathrm{B} & \mathrm{M} \rightarrow \mathrm{E} & \mathrm{M} \rightarrow \mathrm{M} & \mathrm{M} \rightarrow \mathrm{S} \\ \mathrm{S} \rightarrow \mathrm{B} & \mathrm{S} \rightarrow \mathrm{E} & \mathrm{S} \rightarrow \mathrm{M} & \mathrm{S} \rightarrow \mathrm{S} \end{bmatrix}. \tag{1}$$

We implemented this HMM-based Chinese text segmentation, which is similar to that in [32]. When training, the first character of each segmented word is marked as a beginning state (B). Then characters are marked as middle states (M), until the last character is read, which is marked as an end state (E), with any single-character words marked as such (S). The emission probability, the state transition probability, and the initial state probability are then used to update the state transition probability matrix in (1). Subsequently, we use the trained HMM to segment Weibo posts and comments line by line.

5.3 HMM for Emotion Classification

For each word in a tweet or comment, we can calculate a three-dimensional vector based on its MI, CHI, and TF-IDF scores, as discussed in Sect. 4.2. After calculating the feature vectors for each emotion, we obtain a mean value of each feature over all tweets labeled by each specific emotion. This mean value is used as an observation. The transition feature between states S_{k-1} and S_k is computed as

[1] One obstacle we encountered was a change in the Weibo mobile site at the beginning of 2020. To avoid being blocked when crawling a large number of comments, we were forced to modify the crawler to use the "max_ID" property for the current comment page.

Table 7 Features for words w_0, w_1, and w_2 with respect to each emotion

Word	Emotion	MI	CHI	TD-IDF
w_0	Happiness	0.0012	0.0247	0.0009
	Anger	0.0012	0.0247	0.0070
	Sadness	0.0015	0.0100	0.0450
	Surprise	0.0080	−0.0050	0.0220
	Disgust	0.0020	0.0470	0.0117
	Fear	0.2200	0.0700	0.0009
w_1	Happiness	0.0167	0.0064	0.1045
	Anger	−0.0012	0.0247	0.0009
	Sadness	0.0200	−0.1416	0.0009
	Surprise	0.0012	−0.0247	−0.0009
	Disgust	−0.0012	0.0247	0.0009
	Fear	0.0012	0.0247	−0.0009
w_2	Happiness	0.0012	0.0247	0.0009
	Anger	0.0012	0.0247	0.0070
	Sadness	0.0015	0.0100	0.0450
	Surprise	0.3693	0.0820	−0.0119
	Disgust	0.0526	0.0247	0.0008
	Fear	0.0012	0.0247	0.0007

$$P(S_k = s_p \mid S_{k-1} = s_q) = \begin{cases} 1 \text{ if } p = q + 1 \\ 0 \text{ otherwise} \end{cases},$$

where the HMM states S_k correspond to the features MI, CHI, and TF-IDF. This determines how close a feature vector in the test tweet is to those in the training set, with respect to the various emotions. The emission probability

$$P(y_k \mid S_k^{e_i}) = J(y_k \mid S_k^{e_i}) = \frac{M_{11}}{M_{11} + M_{10} + M_{01}}$$

can be calculated by Jaccard similarity [13], which measures the correlation between the feature vector y_k and the state S_k, where M_{11} is the total number of tweets containing feature vector y_k and state S_k with respect to emotion e_i, M_{10} is the number of tweets containing only state S_k with respect to emotion e_i, and M_{01} is the number of tweets containing only feature vector y_k with respect to emotion e_i.

Table 7 gives an example of the relationship between three consecutive words w_0, w_1, and w_2 in a particular test case.

We train an HMM for each of the six emotions. Then, we score a sample against each model, and assign an emotion to the tweet based on the largest probability.

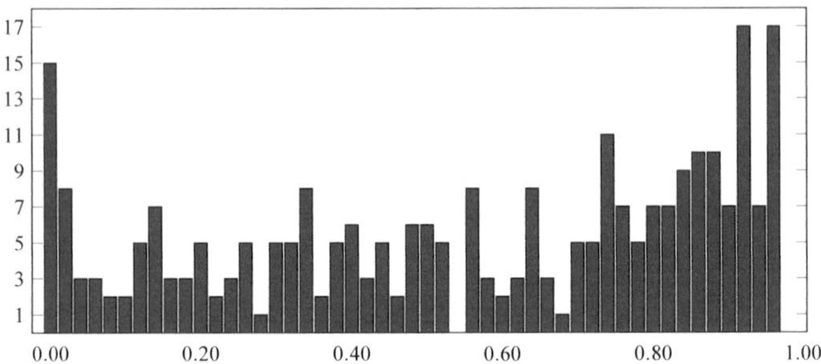

Fig. 4 Sentiment score distribution for comments in Table 6, Row 1

5.4 Sentiment Score Calculation

We first construct a Word2Vec model based on the 35,124 online shopping reviews contained in the sentiment analysis dataset discussed in Sect. 4.2. Note that this Word2Vec model is based on segmented Chinese text. We use the GenSim package in Python [31] to train this Word2Vec model.

Next, the resulting Word2Vec embeddings are used to assign a sentiment score to segmented Chinese words, based on 3,286,543 tweets that we crawled from Weibo. These sentiment scores are determined using naïve Bayes. Specifically, we use naïve Bayes to compute

$$P(c_1 \mid w_1, \ldots, w_n) = \frac{P(w_1, \ldots, w_n \mid c_1)P(c_1)}{P(w_1, \ldots, w_n \mid c_1)P(c_1) + P(w_1, \ldots, w_n \mid c_2)P(c_2)},$$

where c_1 represents the event that a word is positive and c_2 represents the event that a word is negative. Our sentiment score is computed as the product of $P(c_1 \mid w_1, \ldots, w_n)$ and the word similarity score computed using the Word2Vec model. The resulting score can be viewed as the probability of a word in a tweet being positive, where 0 represents an extremely negative word, while 1 represents an extremely positive word. Figure 4 illustrates a sample sentiment score distribution for one of our training datasets. Note that this particular bar graph is based on the 812 comments corresponding to Weibo tweet ID 44275293, as listed in Table 5. Figure 4 shows the sentiment score frequency count distribution for all 812 comments with brackets of width 0.02 over the range of 0 to 1.

5.5 Troll Detection with XGBoost and SVM

Our troll detection model is based on XGBoost. We train our XGBoost models using Python under the Jupyter Notebook environment. For these XGBoosting troll

Table 8 Troll detection statistics crawled from Weibo

	Feature	Description
F0	`follower`	Follower count
F1	`following`	Following count
F2	`original_post`	Number of original tweets
F3	`ffRatio`	`following` divided by `follower`
F4	`foRatio`	`original_post` divided by `follower`
F5	`urank`	User activity rank in Weibo
F6	`verified`	User certified or not
F7	`description`	User's self-description (1 or 0)
F8	`freqComment`	User comments frequently or not
F9	`like_count`	Like count for comment
F10	`floor_number`	Location of comment
F11	`sentiment`	Sentiment score of the comment (0 to 1)
F12	`diffOriginalSenti`	`sentiment` minus sentiment of original

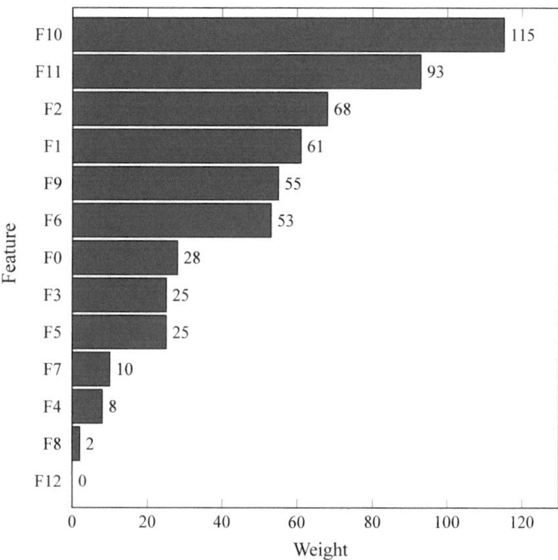

Fig. 5 Initial XGBoost features ranking

detection models, we drop non-quantitative features, which leaves us with the features listed in Table 8. Training an XGBoost model on all of these features, we achieve about 80% accuracy.

Ranking the features in this full-feature XGBoost model, we obtain the results in Fig. 5. Note that F12 in Fig. 5 has a weight of 0, which implies that F12 (the `diffOriginalSenti` feature) contributed nothing to the classification.

Next, we consider recursive feature elimination (RFE), where we drop the lowest ranked feature, then retrain the model. Our RFE results are given in Fig. 6. We observe

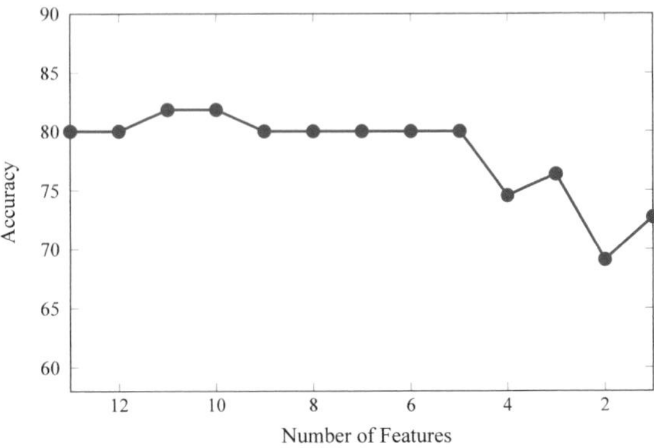

Fig. 6 XGBoost model accuracy versus numbers of features

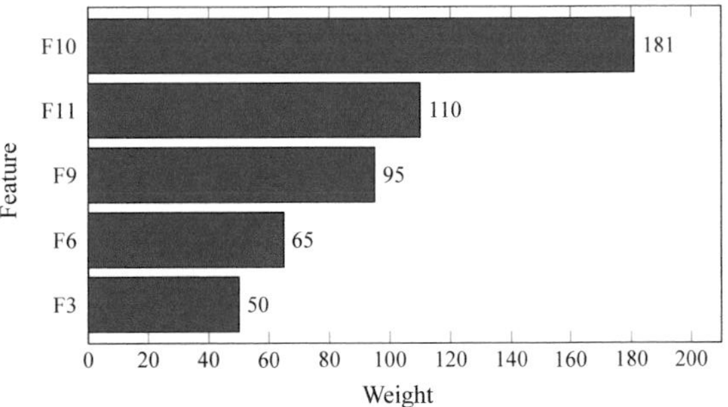

Fig. 7 XGBoost features and rankings

that the model improves when we drop the two lowest ranked features, namely, F12 and F8, but beyond that, the model will lose accuracy if we drop additional features. Hence, our optimal XGBoost model uses all of the features in Table 8, except F8 and F12. With this model, we achieve an accuracy of about 82%.

The features used and their relative importance rank are given in Fig. 7. Note that these are the features in the XGBoost model.

For comparison with our XGBoost classifier, we also experiment with an SVM classifier. We utilize the Python scikit-learn package [23] to train our SVM classifier. We compare the results of these SVM classification experiments to the XGBoost experiments, based over two different datasets and various sets of features. Next, we summarize these results.

As discussed above, using XGBoost as our classification method and RFE, we achieve an accuracy of about 82%. With some additional feature engineering, we were able to increase this troll detection accuracy to 83.64% on our Weibo crawled dataset, using only the three features labeled as F9, F10, F11 in Fig. 7. In addition, using our SVM classifier, we achieve 87.27% accuracy on the same dataset, based on the same three features.

As another experiment, we compare our XGBoost and SVM models using the SnowNLP sentiment dataset for the sentiment score calculation. By using this training dataset for sentiment analysis, and with the addition of features F3 and F6, the

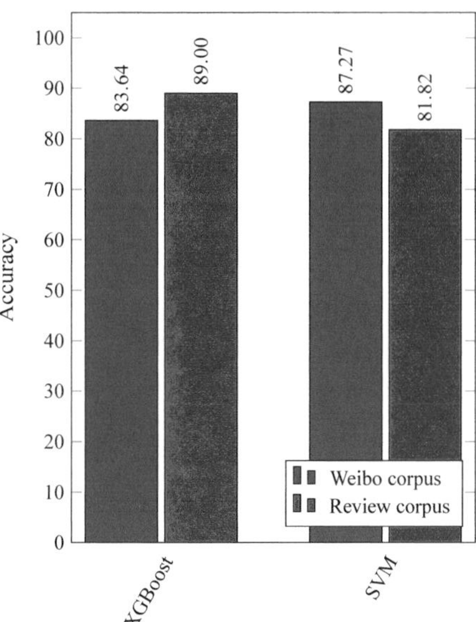

Fig. 8 Comparison of XGBoost and SVM

accuracy for XGBoost is 89%. However, with this same dataset and feature set, the SVM model achieves an accuracy of only 81.82%. These accuracy are summarized in the form of a bar graph in Fig. 8.

Our experimental results show that we can achieve an accuracy as high as 89%, which far exceeds the 78% accuracy obtained in the comparable previous work in [24], and matches the accuracy in the previous work [11]. A significant advantage of our approach is that it only requires a small number of easily obtained features, as compared to any previous research. This makes our troll detection technique highly efficient, and thus suitable for real-time troll detection, as validated by the Chrome extension discussed in Sect. 5.6.

5.6 Chrome Extension for Troll Detection

Since our troll detection mechanism is written in Python, for real-time troll detection on the Sina Weibo mobile website, we created a Chrome extension using HTML and JavaScript. In this extension, we pass the JSON packet with Weibo comment information to the back-end, which is built on the Django framework [8]. This back-end implements our troll detection model, as discussed above. The overall workflow for the plug-in is summarized as follows.

1. Run the crawler script against all the comments currently displayed in the browser under one tweet and send the packet to a server-side portal (currently running as localhost).

Fig. 9 Chrome extension employing troll detection model

2. On the server-side, sort the essential user information and comment text from the returned JSON packet, as generated by the crawler.
3. On the server-side, run our sentiment analysis classifier against the comment text and acquire the text sentiment scores.
4. On the server-side, aggregate the sentiment scores and other user information, feed these into our troll detection model, and return the troll detection result to the client-side plug-in.
5. Modify the CSS style sheet for any detected troll comments by adding an orange background behind the text.

A screenshot showing this plug-in in action is given in Fig. 9. Note that in this implementation, tweets and comments that have been flagged as potential troll activity are blurred.

6 Conclusion and Future Work

The widespread use of social media enables information transfer to occur much faster than ever before. However, troll activities detract from the utility of social media. Trolls have a variety of motivations, ranging from deception to profits, and it is not likely that these motivating factors will diminish in the future. Therefore, intelligent defenses against trolls are essential.

In this research, we utilized a variety of machine learning techniques to analyze comment content and user information on the Sina Weibo platform. By conducting sentiment analysis and by including user data aggregation, we were able to efficiently identify troll comments on Sina Weibo with higher accuracy, as compared to previous work. We developed a Chrome extension that served to highlight the practicality of our approach.

For future work, more user data and other features can be considered. In addition, deep learning techniques that utilize sequential information, such as long short-term memory (LSTM) networks, could prove useful. The Chrome extension that we have developed could be extended to support the Weibo desktop platform.

References

1. Asahara, Masayuki, Kenta Fukuoka, Ai Azuma, Chooi-Ling Goh, Yotaro Watanabe, Yuji Matsumoto, and Takashi Tsuzuki. 2005. Combination of machine learning methods for optimum Chinese word segmentation. In *Proceedings of the fourth SIGHAN workshop on Chinese language processing*.
2. Bennett, Kristin P., and Colin Campbell. 2000. Support vector machines: Hype or hallelujah? *SIGKDD Explorations* 2: 1–13.
3. Calvo, Rafael, and Sunghwan Kim. 2012. Emotions in text: Dimensional and categorical models. *Computational Intelligence*, early view.

4. Chen, Miaohong, Baobao Chang, and Wenzhe Pei. 2014. A joint model for unsupervised chinese word segmentation. *Proceedings of the 2014 conference on empirical methods in natural language processing*, 854–863.
5. Chen, Miaohong, Baobao Chang, and Wenzhe Pei. 2014. A joint model for unsupervised Chinese word segmentation. In *Proceedings of the 2014 conference on empirical methods in natural language processing*, EMNLP 2014, 854–863, Association for Computational Linguistics, Stroudsburg.
6. Chen, Tianqi, and Carlos Guestrin. XGBoost: A scalable tree boosting system. http://arxiv.org/abs/1603.02754.
7. Dhanasekar, Dhiviya, Fabio Di Troia, Katerina Potika, and Mark Stamp. 2018. Detecting encrypted and polymorphic malware using hidden Markov models. In *Guide to vulnerability analysis for computer networks and systems: an artificial intelligence approach*, ed. Simon Parkinson, Andrew Crampton, and Richard Hill, 281–299, Springer, Berlin.
8. Django Python web framework. 2020. https://www.djangoproject.com/.
9. Emerson, Thomas. 2005. The second international Chinese word segmentation bakeoff. https://pdfs.semanticscholar.org/65e9/0d9f6754d32db464f635e7fdec672fad9ccf.pdf.
10. Feng, Suofei, and Eziz Durdyev. 2018. Fine-grained sentiment analysis of restaurant customer reviews in Chinese language. http://cs229.stanford.edu/proj2018/report/195.pdf.
11. Huang, Yingying, Mengyi Zhang, Yuqing Yang, Shijie Gan, and Yanmei Zhang. 2016. The Weibo spammers' identification and detection based on Bayesian-algorithm. In *2016 2nd workshop on advanced research and technology in industry applications*, WARTIA-16.
12. Indiana University. 2018. What is a troll? https://kb.iu.edu/d/afhc.
13. Jaccard, Paul. 1908. Nouvelles recherches sur la distribution florale. *Bulletin de la Societe Vaudoise des Sciences Naturelles* 44: 223–270.
14. Kim, N.-R., Kyoungmin Kim, and Jong-Heon Lee. 2017. Sentiment analysis in microblogs using hmms with syntactic and sentimental information. *International Journal of Fuzzy Logic and Intelligent Systems* 17: 329–336.
15. Liu, Li, Dashi Luo, Ming Liu Liu, Jun Zhong, Ye Wei, and Letian Sun. 2015. A self-adaptive hidden Markov model for emotion classification in Chinese microblogs. *Mathematical Problems in Engineering* 2015: 987189.
16. Liu, Linqing, Yao Lu, Ye Luo, Renxian Zhang, Laurent Itti, and Jianwei Lu. 2016. Detecting "smart" spammers on social network: A topic model approach. In *Proceedings of the NAACL Student Research Workshop*, 45–50, Association for Computational Linguistics, San Diego, CA.
17. Liu, Yixin, Xinhua Wang, and Wen Long. 2019. Detection of false Weibo repost based on XGBoost. In *IEEE/WIC/ACM International Conference on Web Intelligence*, WI '19, 97–105, ACM, New York.
18. Mikolov, Tomas, Kai Chen, Greg S. Corrado, and Jeffrey Dean. 2013. Efficient estimation of word representations in vector space. http://arxiv.org/abs/1301.3781.
19. Numpy. 2020. https://numpy.org/.
20. pandas Python data analysis library. 2020. https://pandas.pydata.org/.
21. Haiyun, Peng, Erik Cambria, and Amir Hussain. 2017. A review of sentiment analysis research in chinese language. *Cognitive Computation* 8: 423–435.
22. Richardson, Leonard. 2020. BeautifulSoup Python package. https://www.crummy.com/software/BeautifulSoup/bs4/doc/.
23. scikit-learn Python package. 2019. https://scikit-learn.org/stable/.
24. Seah, Chun Wei, Hai Leong Chieu, Kian Ming A. Chai, Loo-Nin Teow, and Lee Wei Yeong. 2015. Troll detection by domain-adapting sentiment analysis. In *18th International Conference on Information Fusion*, Fusion 2015, 792–799.
25. Selenium Projects. 2020. https://www.selenium.dev/projects/.
26. Sina-Entertainment. 2016. Weibo of Zi Yang dominated by troll. http://ent.sina.com.cn/s/m/2016-11-03/doc-ifxxneua4008428.shtml.
27. Sina-Weibo. 2009. Weibo mobile site. https://m.weibo.cn/.

28. Stamp, Mark. 2004. A revealing introduction to hidden Markov model. https://www.cs.sjsu.edu/~stamp/RUA/HMM.pdf.
29. Stamp, Mark and S. Venkatachalam. 2011. Detecting undetectable metamorphic viruses. In *Proceedings of 2011 International Conference on Security & Management*, 340–345.
30. Vobbilisetty, Rohit, Fabio Di Troia, Richard M. Low, Corrado Aaron Visaggio, and Mark Stamp. 2017. Classic cryptanalysis using hidden Markov models. *Cryptologia* 41(1): 1–28.
31. Řehůřek, Radim, and Petr Sojka. 2010. Software framework for topic modelling with large corpora. In *Proceedings of the LREC 2010 Workshop on New Challenges for NLP Frameworks*, 45–50.
32. Wang, Kun, Chengqing Zong, and Keh-Yih Su. 2009. Which is more suitable for chinese word segmentation, the generative model or the discriminative one? *Proceedings of the 23rd Pacific Asia Conference on Language, Information and Computation*, 827–834.
33. Wang, Kun, Chengqing Zong, and Keh-Yih Su. 2009. Which is more suitable for Chinese word segmentation, the generative model or the discriminative one? In *Proceedings of the 23rd Pacific Asia Conference on Language, Information and Computation*, 827–834, City University of Hong Kong.
34. Wang, Rui. 2018. SnowNLP Python package. https://github.com/isnowfy/snownlp.
35. Zhang, Hua-Ping, Hong-Kui Yu, De-Yi Xiong, and Qun Liu. 2003. HHMM-based Chinese lexical analyzer ICTCLAS. In *Proceedings of the Second SIGHAN Workshop on Chinese Language Processing*, 184–187, Association for Computational Linguistics, Stroudsburg.
36. Zhao, Jun, and Hong Wang. 2016. Detection of fake reviews based on emotional orientation and logistic regression. *Journal of CAAI Transactions on Intelligent Systems* 13: 336–342.
37. Zhao, Xiaoyi, and Yukio Ohsawa. 2018. Sentiment analysis on the online reviews based on hidden Markov model. *Journal of Advances in Information Technology* 9: 33–38.

Log-Based Malicious Activity Detection Using Machine and Deep Learning

Katarzyna A. Tarnowska and Araav Patel

Abstract This chapter describes the application of intelligent computational techniques to the problem of malicious activity detection. It is proposed to embed machine and deep learning models for malicious activity detection into the framework of a log-based decision support system (DSS) for information security administrators. It is expected that such a solution will enable organizational-wide protection of informational assets, by providing accurate and comprehensive real-time insights into violations of information security policies. In this work, we present experiments and results on database systems' log analysis using traditional machine learning (ML) methods and deep learning (DL) on the synthetic log dataset simulating user activity in a hypothetical company.

1 Introduction

The organization of the chapter is as follows. First, the background information is presented: (1) the problem area of malicious activity, such as unauthorized access to data or configuration changes; (2) insider threats, such as masquerader-based attacks; (3) information security audit, including audit logging for security and compliance; (4) intrusion detection systems (IDSs), with the focus on host-based IDS and anomaly-based detection. The second section provides a review of current IDS solutions and research specifically for relational database management systems (RDBMSs). Section three describes the proposed methods for anomaly detection in system event logs. Within the experimental design, traditional ML approaches such as distance-based outlier detection and support vector machines (SVMs) are compared with the

K. A. Tarnowska (✉)
The Department of Computer Science, San Jose State University, One Washington Square, San Jose, CA, USA
e-mail: katarzyna.tarnowska@sjsu.edu

A. Patel
The Department of Computer Science, University of California at Berkeley, Berkeley, CA, USA
e-mail: araavp@berkeley.edu

M. Stamp et al. (eds.), *Malware Analysis Using Artificial Intelligence and Deep Learning*, https://doi.org/10.1007/978-3-030-62582-5_23

DL method based on the sequential autoencoder model. The fourth section presents and discusses the results. Examples of detected incidents' scenarios are presented. The chapter concludes with a combined ML-DL approach, limitations, and future work to integrate the models into the framework of DSS for the information security audit.

Data security Data has become a valuable asset for most companies. As such it has become important for companies to protect and manage this asset. Sensitive and confidential data include personal records, medical records, company's contracts, clients, financial transactions, etc. On the other hand, we have seen a surge in data-targeted attacks. The consequences of data breaches are serious: lawsuits, fines, loss of trust, reputation, and customer base. According to the 2013 IOUG Enterprise Security Survey, the three greatest threats to data security are (1) human error (77%), (2) internal hackers or unauthorized users (63%), and (3) malicious code/viruses (49%) [31]. In the Kroll 2012 Global Fraud Survey, it is reported that 60% of frauds are committed by insiders, increasing from 55% in the previous year [25]. Likewise, the 2012 Cybercrime report by PwC states that the most serious fraud cases were committed by insiders [34]. Therefore, the problem to be solved within this work is human malicious/insider activity.

Secondly, the database is considered the most vulnerable layer (54%) surpassing the network (51%) and the server layer (51%) in terms of the severity of potential damage [31]. On the other hand, most IT resources are allocated to protect the network layer (65%), the server layer (59%), and the database layer as third (57%) in terms of budget, staff time, etc. [31]. In this work, the focus is on the underresourced, but most vulnerable—the database layer.

Thirdly, data security is mandated by regulations, such as Sarbanes-Oxley-Act (SOX)—58%, local state/government data protection laws (49%), Health Insurance Portability and Accountability Act HIPAA/HITECH for healthcare electronic records (37%), PCI DSS for payment card industry (30%), SAS 70 (18%), or Federal Information Security Management Act FISMA (11%) [31]. The General Data Protection Regulation (GDPR) adopted by the EU in 2016 mandates any organization handling personal identifiable information (PII) of EU citizens to leverage security, auditing, and intrusion detection mechanisms under heavy fine for compliance failure.

Malicious activity detection *Malicious activity* is the harmful act initiated by trusted *insiders*, that is, users authorized to access an organization's network, system, or data. A malicious insider is defined as a current or former employee, contractor, or business partner who intentionally exceeded or misused that access in a manner that negatively affected the *confidentiality*, *integrity*, or *availability* of the organization's information or information systems [5]. The motivations and behaviors of insider threats vary widely; however, the damage of insiders can inflict is significant. Detecting compromised user accounts and insiders within the company who may have malicious intent has become a key problem in enterprise security. In *masquerader attacks*, users hide their identity by impersonating other people [46]. Such attacks became one of the most frequent forms of security attacks, including the database domain.

Log-based audit control A *log file* is a file that records either events that occur in an operating system or other applications. Logs are composed of log entries; each entry contains information related to a specific event that has occurred within a system or network [24]. Almost every software generates logs, mostly for debugging purposes. The logs generated daily by software and hardware are in a massive volume. It is challenging to keep track and analyze large heterogeneous logs. Detecting and preventing cyber-attacks, which become even more sophisticated, require to generate a holistic view rather than analyzing logs individually. The process of an *information security audit* is a formal check if the system is meeting security requirements as well as system and organizational policies [33]. According to the National Institute of Standards and Technology, *audit logs* are records of events based on applications, users, and systems, while the *audit trails* involve audit logs of applications, users, and systems. In many critical applications, such as in hospitals or banking, collecting audit trail is required by law [47]. The human inspection of audit information is generally tedious as well as time-consuming. While the traditional audit process is highly manual, performed by specialized auditors, there exist automated tools that support the information security audit. So-called security information and event management (SIEM) systems aggregate and analyze data. However, they use proprietary (non-standard) data formats and the analytics is limited to statistical methods. Moreover, they are complicated to deploy and expensive.

Intrusion detection systems Log analysis for *intrusion detection* is understood as the process to detect attacks on a specific environment using system logs as the primary source of information. A *host-based intrusion detection system* (HIDS) is a system that monitors a computer on which it is installed to detect an intrusion and/or misuse and responds by logging the activity and notifying the designated authority. HIDS monitors the system audit and event logs and notifies the system administrator accordingly with alert messages. The analysis of data captured by IDS should be preferably analyzed in real time (rather than in batch-mode later). Intrusion detection approaches include statistical modeling, data mining-based methods, signature analysis, rule-based systems, genetic algorithms, etc. The notification mechanism based on *anomaly (outlier) detection* techniques has the capability of detecting new or unknown attacks by using the classification techniques, as opposed to more traditional signature-based or rule-based detection.

Anomaly-based detection An *outlier* is defined as an observation in a dataset that appears to be inconsistent with the remainder of that set of data [22]. *Anomalies* are defined as events that deviate from the standard and happen rarely. These are patterns in data that do not conform to expected behavior or do not "follow the rest of the pattern" [6]. In anomaly-based approaches, malicious activity is detected as a deviation from the normal behavior of the users in the systems. Its benefit is in detecting unknown intrusions. The models analyze current sessions and log entries by comparing them against the profile representing a normal behavior. If a deviation is found during the comparison, the notification is sent to the information security authority. The "normal" behavior is modeled using either supervised or unsupervised techniques [7]. Supervised learning builds a predictive model based on the instances labeled as normal and as an anomaly, but the problem is that anomalous instances are usually few in the real-world datasets. Machine learning algorithms designed specifically for anomaly detection tasks are isolation forests, DBScan, one-class SVMs, elliptic envelopes, the local outlier factor, and others. These methods are rooted in traditional machine learning. A common problem is a high rate of false positives and over-fitting the "normal" profiles. There is a trade-off between relaxing these margins and a higher rate of missed attacks. Therefore, fine-tuning the anomaly detection models to find the optimum threshold is problematic.

Database systems security *Database systems* are systems designed to store and manage data effectively. Database intrusion is understood as any activity that violates data *integrity*, data *confidentiality*, or data *availability*. Traditional database security mechanisms offer basic security features such as authentication, authorization, access control, data encryption, data masking, and auditing. These methods are not sufficient to guarantee data security, especially guard against malicious data access, as they were mostly designed to prevent intrusions, not to detect them. These database security mechanisms need to be complemented by suitable ID mechanisms to address especially the problem of insider threats. The goal of the intrusion detection, as a "second line of defense", is to minimize the harm done by malicious activity by early detection and notification. Most log-based IDS have been designed for network-based intrusion detection [4, 37, 50, 56].

Although a database management system (DBMSs) is a vulnerable IT system layer that contains sensitive information, to date, there have been few ID mechanisms proposed that are specifically tailored to function within a generic DBMS. IDS designed for operating systems or network are not suitable for protecting DBMS against insider threats. These are more difficult to defend against, as they come from subjects that are legitimate users of the system.

The malicious user threat can come from the high-privileged database administrator (DBA) accounts. Within this research, it is proposed to gather cross-organizational and cross-system user events data and apply intelligent computational techniques to detect and notify about security incidents. IDS notifications should be provided to the Information Security Administrator, who oversees IT systems/database administrators (see Fig. 1).

2 Related Work

Machine Learning methods for Intrusion Detection Systems Prior work on IDS systems recommends a machine learning approach rather than a rule-based due to the increased efficiency, scalability, and the generalization in contrary to over-fitting.

A sample rule-based approach to host-based IDS for log analytics was described in [35]. A framework for time-series based pattern mining for anomaly detection was proposed in [12].

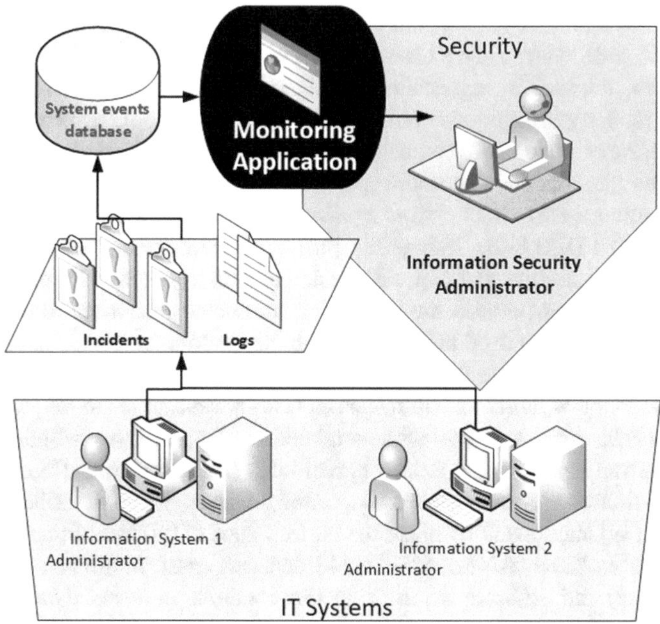

Fig. 1 The design of the decision support system for information security administrator, with the log-based application monitoring the activity of the system users, including high-privileged system users, such as administrators

Research on IDS for DBMS proposes employing both unsupervised and supervised ML methods. The major drawback of the supervised methods is that they require to work with labeled data. Traditional classification architectures are not sufficient for effective anomaly detection. As they are not meant to be used in an unsupervised manner, they do not handle well severe class imbalance and, therefore, struggle to correctly recall the outliers. Work in [23] uses clustering to represent normal user behavior and outlier detection techniques to identify behavior that deviates from these profiles. The COBWEB clustering method with SQL query mining was used in [44] to determine deviations from normal profiles. A multi-stage log-analysis approach based on Kibana for pattern-matching and Bayes Net for ML was proposed to detect SQL-injection attacks in [32]. The research in [21] proposed protecting databases from internal and external attacks using a hidden Markov model (HMM); however, having only information on known attacks, the system left the database vulnerable to novel intrusions. The problem of masquerade detection was tackled with profile hidden Markov models (PHMMs) based on user-issued UNIX commands in [20]. One-class SVM (OCSVM) was proposed for anomaly detection of user behavior for the database security audit in [27]. The same method was applied for detecting anomalous windows registry accesses in [17]. Windows event logs were pre-processed and analyzed using statistical methods based on standard deviation in [11]. A singular value decomposition (SVD)-based algorithm for user and entity behavior analytics within an intelligent platform for malicious activity is described in [48]. Naive Bayes classifier was used in [8] to evaluate the legitimacy of a database transaction represented as a hexplet data structure constructed from SQL queries. They also proposed binding all the queries of the same transaction in one behavior to reduce the false positive, similarly as in [19]. Work in [49] utilized Naive Bayes classifier and maximum aposteriori probability to detect intrusions. The relevant features were extracted from a parse-tree representation of SQL commands. A system called DBSAFE, designed in [43], uses database audit log files to train a hybrid binary classifier and Naive-Bayesian classifier model for profiling normal users. Predetermined policies are added for automated responses to detect intrusions. Work in [40] proposed an IDS based on the random forests (RF) method with weighted voting to balance the impact of each tree. RF is graph-based and can be used for modeling SQL queries. K-means algorithm was applied in [38] to recognize anomalous behavior in a dataset with email notification; however, without the capability to analyze incoming log entries in real time. Big data-based IDS categorizing logs in real time as either high, medium, or low in severity was described in [36].

Work in [39] attempts to compare the performance of different data mining techniques (such as KNN, ANNs, SVMs, J48 decision trees, multilayer perceptions, random forest) and different feature selection methods to detect database anomalous behavior. The anomaly detection solution presented in [51] is DBMS-specific (MS SQL Server). A general ID solution is needed to accommodate different types of DBMSs that exist within the enterprise infrastructure. Work in [54] attempted a generic and customizable approach for database exploitation detection using reinforcement learning in conjunction with neural network and association rule mining.

Deep Learning for Intrusion Detection Although deep learning (DL) is a subset of machine learning, it is a newer and more complex way of learning than the norm [2]. DL allows quick detection of attacks without having to retrain the entire model for incoming log entries [13]. The application of an artificial neural network (ANN) for the user's behavioral analysis system was proposed in [1]. A general-purpose framework for online log anomaly detection and diagnosis in [10] utilized long short-term memory (LSTM) network and self-adaptation. The authors used natural language processing by handling log entries as sequential elements that followed patterns and rules. However, it proved somehow difficult to analyze the entire log messages from different systems. A stacked LSTM network for anomaly detection in times series was described in [30]. A hybrid LSTM autoencoder model was used in [15] for anomaly detection in application log data. The authors also provided a comparison of its accuracy and generalization with unsupervised ML methods, such as KNN and K-means. Autoencoders with nonlinear dimensionality reduction for anomaly detection tasks were proposed in [42], which demonstrated that autoencoders are able to detect subtle anomalies.

Current insider attacks are becoming more sophisticated, and anomaly detection that includes logs from different layers of IT system (OS, database, application) can provide the most accurate insights and event correlation analysis. For example, an attempt to analyze and correlate events from different system layers (OS, kernel, application, and web application data) within host-based IDS was proposed in [53]. A multifaceted and comprehensive approach to a cyber defense system that utilizes the biological immune system was described in [9]. Visual techniques have the potential to enhance machine learning for decision-making about potential threats [26, 52].

3 Methods

This research attempts to develop a cross-platform approach to intrusion detection based on machine and deep learning methods. The goal is to identify unexpected access patterns by authorized users, including masqueraders. This work proposes mining database audit logs with the goal of detecting violations of access control. We consider a DBMS layer that implements role-based access control (RBAC) model [45]. Under this model system *permissions* are associated with *roles*, grouping several *users*, rather than with single users. Since RBAC is standardized and adopted in various DBMS products, the proposed IDS solution is generic. Both traditional machine learning and deep learning methods are employed to model normal database access behavior using audit-log data with the goal of recognizing intruders.

3.1 Solution Design

The architecture for the proposed log-based IDS system supporting insider threat detection consists of the following components (see Fig. 2):

- Log aggregation—log files from different IT systems and different IT layers are aggregated into one central repository.
- Log data pre-processing and feature extraction—data is pre-processed accordingly for anomaly detection models.
- Modeling user behavior/profile—using deep learning models, which will be updated periodically.
- Anomaly detection and alerting—comparing the current profile with the base profile to detect statistical deviations. Corresponding notifications are generated and sent as an alert to the security administrator.

3.2 User Behavior Modeling

Users' behavior within systems can be tracked over time using the log trail. These datasets can be used to learn a *baseline* profile of the user's behavior. Any deviations from this behavior can be flagged as potential *anomalies* that warrant further investigation. Within this research, we propose to compare traditional ML with novel deep

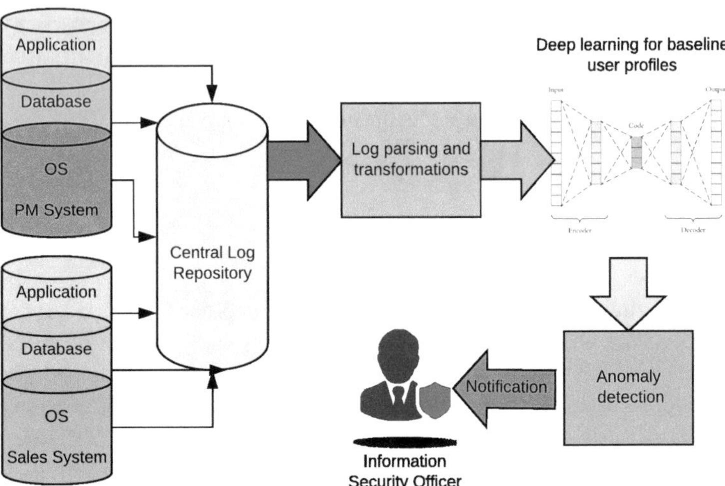

Fig. 2 The proposed log-based anomaly detection system for malicious user activity within the IT enterprise infrastructure (here, Project Management, and Sales systems). Deep learning methods are proposed to learn "baseline" user behavior and detect deviations, which will result in notifying Information Security Officer

learning (DL) techniques. They will be applied to system log datasets including audit trail and operating system event logs to identify anomalous behavior of users. There are two approaches to anomaly detection:

- Historical baseline—the user's behavior is compared to its own behavior in the past.
- Peer baseline—the user's behavior is compared to users with similar roles/ privileges in the systems ("peers").

User behavior is defined as follows. First, the time-granularity for behavior analytics has to be defined (hourly, daily, weekly, etc.). Each log entry is described by timestamp with precision up to seconds, and this timestamp feature will be used to derive granularity. Secondly, content features of the logs, such as the command itself, or data objects interacted with, are used for analysis.

3.3 Anomaly Detection

We propose two methods for anomaly detection:

1. Traditional machine learning—distance-based—ML method based on the distance metric, and support vector machines (SVMs) algorithm.

2. Deep learning-based—learning normal user behavior with artificial neural network (ANN) based on the autoencoder model and using it to detect anomalous behavior.

Support vector machines (SVMs) are being applied to anomaly detection in the one-class setting, that is, using one class to learn a region that contains the training data instances (a boundary). Kernels, such as radial basis function (RBF) kernel, can be used to learn complex regions. If a test instance falls within the learned region, it is declared as normal, else it is declared as anomalous.

The DL methods can be particularly useful in user modeling because of its ability to self-learn and to adapt. The input to ANN will be encoded data about the user actions. Autoencoders will be used as models of an artificial neural network model for the task of anomaly detection. Autoencoders are a type of unsupervised neural network that accepts an input set of data, compresses the data into a latent-space representation (an encoder), and reconstructs the input data from the latent representation (a decoder). These types of ANN are versatile because they learn compressed data encoding in an unsupervised manner [2]. They can be implemented efficiently by training one layer at a time. Autoencoders are well-suited for the anomaly detection problem. Namely, mean squared error (MSE), which measures the reconstruction loss, can be also indicative of an anomaly. Autoencoder is expected to minimize MSE on the trained data, but for unexpected or anomalous data, which the model has never has seen, it will yield higher MSE. If the MSE of the reconstruction is high, then likely the data instance is an outlier and in our case, an anomaly.

3.4 Scenario

The dataset of the generated system logs simulates system user activity in a medium-sized hypothetical company, which specializes in weapon production. The case study is described in more detail in [41, 52]. The classified information is related to products, clients, and orders. Sensitive information includes sales data, and HR data, such as employee data. Two sample database systems were designed and deployed to simulate enterprise infrastructure in the considered company. It includes systems for project management (PM) and Sales system, both running on Oracle 11g DBMS. The company employs a role-based access model to restrict access to data *objects* according to the job *role*. Each *user* is assigned role corresponding to their job function, for example, Project Manager, Developer, Human Resources, System Administrator. Each role is assigned privileges to read or modify corresponding data objects, such as Employees, Payments, Products, and Projects tables. There are 18 users, 8 roles, and 14 objects (tables) in the PM database system, and 16 users, 10 roles, and 5 tables in the Sales system.

Table 1 Sample log entries generated by simulating malicious user activity in the source systems. Audit trail includes the following information: the system's component, the command's content, type of operation, object, and subject (user) involved in the operation, its result, and timestamp

Attribute	Sample log 1	Sample log 2	Sample log 3
System	PM-Oracle	Sales-Oracle	Sales-Linux
Command	exec dbms_session.set_ sql_trace(true);	update Admin.Employees set "Salary" = 2500 where "Employee" = 'Skala Peter'	Echo ORACLE_SID
Operation	execute procedure	UPDATE	N/A
Object	dbms_ session	PMO.EMPLOYEES	N/A
User	SYSTEM	PSKALA	Root
Result	Successful	Unsuccessful	Successful
Date	2/13/20	4/12/2020	11/30/2020
Time	20:45:00	17:37:01	2:12:45

3.4.1 Logging Elements

Log entries generated for the described scenario contain the following elements:

- *System's* name—host name, system component, or resource;
- The *command's* content;
- The type of *operation* (i.e., login, read, write, delete);
- *Object*—the resource on which operation performed (i.e., identity or name of affected data, component, such as DB table);
- The *user* performing the operation (i.e., user ID);
- *Result status*—if the operation was successful or not;
- *Date/Time*—stamp consisting of day-month-year and hour-minutes-second.

Sample log entries, related to insider threats, are presented in Table 1. In sample log 1, the masquerader is executing a malicious procedure in the database from the privileged System account. In Sample log 2, the malicious employee is attempting to perform an illegal operation of increasing its own salary. In Sample log 3, the attacker is trying to obtain the database SID in the OS-level command. All these operations were taking place outside of normal business hours (9–17).

4 Experiments

The preliminary experiments on the dataset described in Sects. 3.4 and 4.1.1 were conducted on distance-based outlier detection in RapidMiner. The second series of experiments involved supervised machine learning based on SVM (in Python scikit-learn) and unsupervised deep learning models (in Python Tensorflow). These involved generating a synthetic dataset of train and test data generated in Python and described in more detail in Sect. 4.2.1. The results of each experiment are presented and discussed in the next sections.

4.1 Distance-Based Outlier Detection

The experiments on distance-based outlier detection were conducted using Rapid-Miner. The effectiveness of the approach was measured with common metrics for an IDS, including the rate of false positives, false negatives, and accuracy [3].

4.1.1 Data Pre-Processing

One challenge in academic research on log-based intrusion detection is the limited availability of large datasets for event analysis [16], as opposed to commonly available malware datasets [2]. The dataset used in this research was generated by simulating normal and malicious activity in the monitored systems (PM and Sales). One of the problems in anomaly detection is that anomalies only typically occur 0.001–1% of the time, which causes a massive imbalance in class labels. The dataset reflects that imbalance with 17 records labeled as part of security incidents out of 1510 records in total (around 0.011 or 1.1%). The anomalous data was generated by simulating common database security threats, such as excessive privilege abuse, legitimate privilege abuse, privilege elevation, SQL injection, weak authentication/password attacks, and database communication protocol vulnerabilities. The operations were logged using operating systems' logging facilities (Event Log in Windows XP and syslog in Linux-Red Hat 5) and switching on Oracle's AUDIT option on the specified operations, objects, and users. The logs from both monitored systems were automatically extracted, transformed (standardized), and loaded into Microsoft SQL Server 2008 repository using ETL processes implemented in SQL Server Integration Services 2008. The following attributes were used as input to the distance-based outlier detection algorithm: operation command, date, time, object, user, operation type, and system. The date attribute was transformed into the day of the week attribute (1–7) and the quarter (1–3).

4.1.2 Results

The detected outliers were compared against the ground truth, that is, whether the log was part of the injected incident event in the hypothetical company setting. Table 2 shows a list of correctly identified incidents (true positives-TP), missed incidents (false negatives—FN), as well as log entries incorrectly flagged as incidents

Table 2 Distance-based outlier detection results in RapidMiner. Examples of events' entries detected by the algorithm as outliers (in a 1 and 2% models) versus the ground truth (incident or not)

#	Log entry (command @ user/system @ time)	Outlier (1%)	Outlier (2%)	Actual
1	select ∗ from DBA_USERS @ SYSTEM/Oracle_PMO @ 22:35	Yes	Yes	Yes
2	login @ jczolg/Oracle_PMO @ 4:23	Yes	Yes	Yes
3	exec dbms_session.set_sql_trace(true); @ SYSTEM/Oracle_PMO @ 20:45	Yes	Yes	Yes
4	ECHO ORACLE_SID @ root/Linux @ 2:12	Yes	Yes	Yes
5	lsnrctl status @ root/Linux @ 3:12	Yes	Yes	Yes
6	revoke create any procedure from rwidawsi @ ADMIN/Oracle_PMO @ 23:34	Yes	Yes	No
7	LOGOFF BY CLEANUP @ jczolg/ADMIN @ 19:58	Yes	Yes	No
8	alter table CLIENTS modify Client_name NVARCHAR2(50) not null @ KTARNOWS/Oracle_PMO @ 18:05	Yes	Yes	No
9	insert into PROJECT_DETAILS VALUES(1,1,'Atomic bomb construction') @ ZFENICKI/Oracle_PMO @ 17:30	Yes	Yes	No
10	insert into PRODUCTS values (1,1,Bomb,1000, null) @ SBASZEL/Oracle_Sales @ 17:30	Yes	Yes	No
11	LOGON @ WMSYS/Oracle_Sales @ 22:50	Yes	Yes	Yes
12	select username,password from dba_users where password=EXTERNAL; @ SYS/Oracle_Sales @ 15:45	Yes	Yes	Yes
13	select ∗ from projects @ Admin/Oracle_PMO @ 0:39	No	No	Yes
14	select ∗ from project_details @ Admin/Oracle_PMO @ 0:40	No	Yes	Yes
15	update Employees set Salary=2500 where Name='Skala Peter' @ PSKALA/Oracle_PMO @ 17:37	No	Yes	Yes
16	update Employees set salary=2000 where employee_id=1 @ MWARZYCHA/Oracle_PMO @ 17:35	No	Yes	No
17	select ∗ from Clients @ KTARNOWS/Oracle_PMO @ 18:06	No	Yes	Yes
18	insert into Salesperson values (1,1,Karolina Uszka) @ MWARZYCHA/Oracle_Sales @ 16:38	No	Yes	No
19	insert into Salesperson values (1,1,Karolina Uszka) @ GOSTROW/Oracle_Sales @ 16:28	No	Yes	No

(false positives—FP). The two tested models of distance-based outlier detection included the outliers percentage of 1% (default in RapidMiner) and 2%.

4.1.3 Discussion

Table 2 shows the log entries from the OS and at the DBMS level, were either detected or missed as incidents. "True negatives", which comprise the largest group of events that represent normal user behavior and not flagged as outliers by the tested model, are not included in the table. **Detected incidents** The first group of events listed (#1-#5;#11-#12) represents correctly detected security incidents ("true positives"). The first correctly detected anomaly is the operation of reading from the system table DBA_USERS, which is a part of the attack, where the malicious user has cracked the SYSTEM user password. In the presented event, the masquerader looks for user accounts in the system with the goal of abusing object privileges and read classified or sensitive data. Log entry #2 in Table 2 denotes unsuccessful attempts to login into the *jczolg* account, which is an account of the Information Security Officer. One reason it has been detected as an anomaly is atypical hours of login (late in the evening), another might be the atypical result of the login operation for that user. Log entry #3, also correctly detected as an outlier, is related to a SQL-injection attack, where a malicious user exploits the dbms_session package to inject malicious code. Log entries #4-#5 are elements of an attack where a masquerader is trying to obtain Oracle instance SID and check a DBMS listener status from the compromised root account.

False alarms The second group of presented events (#6-#10) consists of so-called "false positives" (commonly known as "false alarms")— marked by the algorithm as outliers, but not actual security incidents. For example, in log entry #8 *KTARNOWS* with the *Developer* role is legally modifying the *Clients* table. The same, in log entry #9 *ZFENICKI* with the *Project_manager* role is adding a new project, which is a legal operation for his role. Similarly, in log entry #10, *SBASZEL* is authorized to add new products to the *SALES* database, as he is the R&D Director.

Undetected incidents Finally, the third group of the presented log observations (#13-#15;#17) contains so-called "false negatives", which are the actual security incidents but not detected by the system. For example, in log entry #15, user *PSKALA* made an illegal, but unsuccessful attempt to change his own salary. On the other hand, log entry # 16 represents the same operation but with an authorized user—*MWARZYCHA*, who has an *HR* role.

One can see, that few events were detected in the model with the percentage of outliers set as 2%, but not detected in the model with 1% outliers, for example, in log #14 (masquerader attack) and log #17 (attempt of illegal read). While the 2%-model improved incident detection (TP), it also incorrectly labeled events that were not an attack (FP). For example, logs #18 and #19 (data update) were not the actual incidents, but the 2%-model has flagged them as outliers. As a result, detection capability improved but generated more "false alarms" as well. Table 3 summarizes the results with the numbers for TP, FP, FN, TN, and the metrics for accuracy, precision, recall, and false alarm rate (FAR). The table also compares two models in three metrics: accuracy, recall, precision, and false alarm rate (FAR). Accuracy (1) was computed as the ratio of correctly classified examples to all events. Recall (2) can be also interpreted as the *detection rate* or *probability of detection* (the ratio of

Table 3 The summaries of distance-based outlier detection experiments in RapidMiner. The table presents the number of true positives (TP)—detected as outliers and actual incidents, false positives (FP)—detected as outliers, but not part of the incident, false negatives (FN)- not outliers, but actual incidents, true negatives (TN), and the derived metrics of accuracy, recall, precision, and false alarm rate

Model	#TP	#FP	#FN	#TN	Accuracy (%)	Recall (%)	Precision (%)	FAR (%)
Model with 1% outliers	8	7	5	1489	99.2	61.5	53.3	0.5
Model with 2% outliers	11	17	2	1479	98.7	84.6	39.3	1.3

correctly classified "normal" events to all actual normal events). Precision (3) is the ratio of items correctly classified incidents to all events classified as incidents. FAR (4) represents items incorrectly labeled as incidents.

$$Accuracy = \frac{TP + TN}{TP + TN + FP + FN}, \tag{1}$$

$$Recall = \frac{TP}{TP + FN}, \tag{2}$$

$$Precision = \frac{TP}{TP + FP}, \tag{3}$$

$$FalseAlarmRate(FAR) = \frac{FP}{TN + FP}. \tag{4}$$

The goal is highly accurate detection of attacks while minimizing false positives and false negatives. Accuracy and precision are higher in 1%-model, while 2% model correctly detects more incidents (higher recall), but also results in more "false alarms", deteriorating precision and accuracy. Improving the rate of correct detection of an intrusion at the cost of increased false alarms is a known problem in intrusion detection systems.

4.2 *Machine and Deep Learning for Anomaly Detection*

The experiments on supervised ML using SVM were conducted with Python scikit-learn library was used. For deep learning Python Tensorflow Keras library was used.

4.2.1 Synthetic Dataset

Training dataset The training dataset simulates the intrusion-free database trail within PMO and Sales systems as described in the scenario in Sect. 3.4. This dataset, which simulates normal user behavior, was generated within the algorithm implemented in Python. Log audit records represent the recorded sequence of events in database systems within the Jan–Aug 2020 timeframe and 9 am–5 pm working hours. The attributes include source system, timestamp (date: year, month, and day, time: hours, minute, and second), operation type, user, and object (if applicable). Types of operations include login, select (read), update, insert, delete, and logoff. Randomized 50,000 log records were used for training the ML/DL models.
Test dataset Numerous, fresh, never-before-seen attacks were generated to check the generalization aptitude of the ID system. The intrusion/anomalous log entries were generated with a Python script, which implements an algorithm that represents violations against RBAC policy in the considered company, malicious attacks, such as *brute force* attack, denial of service, and anomalous behavior, such as unapproved operations and operations outside normal business hours. The subset of randomized incident logs was used as a test dataset for ML/DL models.

4.2.2 Feature Pre-Processing

Since the attributes are categorical or text, there is a need for transformation before making it suitable for deep learning models. It is critical to frame the problem correctly for DL models to be useful in anomaly detection. Autoencoders have proven to work well with dependent data; however, the logs consist of both independent and dependent variables. The system/user/operation/object relationship is dependent, whereas the date and times are independent. Including dependent and independent attributes to train autoencoders yielded poor results in terms of accuracy (32–65%). Tuning the parameters (such as changing the number of epochs and batch size) did not improve these results significantly. Therefore, the log data was split between independent and dependent attributes, using the former as input for SVM model, and the latter as input to the DL model. A system, user, operation, and object attributes were transformed into numeric (integer) representations. Similarly, date (year-month-day) and time (hour-minute-second) were transformed into separate integer attributes.

4.2.3 Results

Supervised machine learning The SVM model was used as a supervised ML method with log entries labeled as either anomalous or not, with train size 10,000 or 50,000. Table 4 presents the process of parameter tuning for the 1:1 anomalous data ratio. The model was tested with different types of the kernel (linear/polynomial/RBF), and different values of c-parameter for the kernel.

Table 4 Parameter tuning for SVM-based machine learning model using different kernels and c-parameter values to optimize accuracy. The table presents changes leading to improvement only

Kernel	C-Param	Train size	Accuracy (%)	Precision (%)	Recall (%)
Linear	2	10,000	70.09	72.15	64.74
Polynomial	20	10,000	90.68	95.10	85.87
Polynomial	2	10,000	91.24	96.44	85.92
Polynomial	6	10,000	91.24	95.97	86.01
RBF	8	50,000	99.72	99.99	99.45
RBF	**20**	**50,000**	**99.97**	**100.00**	**99.94**

Table 5 Comparing accuracy, precision, and recall of an optimal model (RBF, c = 20) for different ratios of an anomaly in train data and different sets of attributes (timestamp vs. the whole log)

Log	Anomaly ratio	Accuracy (%)	Precision (%)	Recall (%)
Timestamp	1:1	99.97	100.00	99.94
Timestamp	1:1000	99.96	100.00	64.00
Whole log	1:1000	99.92	80.00	24.00

The first attempted SVM model with a linear kernel yielded an accuracy of 70% and has not changed significantly after trying different parameter values. After changing the kernel from linear to polynomial, the accuracy improved to around 90% and the optimal parameters yielded an accuracy of 92.1%. The best results were obtained for the SVM with the RBF kernel with an accuracy of 99.995% after parameter tuning. The experiments with different parameters were repeated for the different ratio of anomaly data (1:1000), and for the entire log entry as input data (WL). Table 5 presents the accuracy, precision, and recall of the optimal model (SVM with RBF kernel on 50,000 training data points) for different ratios of normal to anomaly logs (1:1 and 1000:1) and different feature selection.

As one can see, decreasing the ratio of anomaly data (from 1:1 to 1:1000) to meet the realistic scenarios deteriorates the recall metric. Further, when the whole log was used as train input data, as opposed to timestamp data only, the accuracy, precision, and recall metrics were lower, with 99.92, 80, and 24% accordingly.

Unsupervised deep learning The deep sequential autoencoder model was used as an unsupervised DL method that was trained on normal log entries only. The log entries consist of the *system*, *user*, *operation*, and *object* portion of the log. The model was tested with different values for the following parameters:

- Layers—number of layers for input and the first hidden layer;
- Activation—activation method (relu/sigmoid);
- Epochs—number of training epochs;
- Batch size—number of training examples in one forward/backward pass;
- Validation—whether test data will be used as validation;
- Shuffle—whether the log data will be shuffled prior to model implementation.

Table 6 Parameter tuning for deep learning model based on sequential autoencoder. The Layers column includes the encoding layers and for each row, there is a symmetrical decoding layer after the 2nd 3-node layer. The Normal and Anomaly Avg columns represent the average scores of the normal and anomaly logs with the Metric ($RMSE$ / MAE^2)

Layers	Activation	Batch	Valida	Shuff	Metric	Normal Avg	Anomaly Avg	Detect (%)
[3, 25, 3]	sigmoid: 2nd 25	32	N/A	N/A	RMSE	$8.9 * 10^{-5}$	$1.2 * 10^{-1}$	55.6
[3, 48, 24, 12, 6, 3]	tanh: 2nd 48	32	N/A	N/A	RMSE	$8.1 * 10^{-1}$	$1.0 * 10^{1}$	70.6
[3, 30, 15, 7, 3]	sigmoid: 2nd 30	32	N/A	N/A	RMSE	1.2	$1.1 * 10^{1}$	73.9
[3, 48, 24, 12, 6, 3]	sigmoid: 2nd 48	32	N/A	N/A	RMSE	3.5	$1.8 * 10^{1}$	75.5
[3, 48, 24, 12, 6, 3]	sigmoid: 2nd 48	256	N/A	N/A	RMSE	2.1	8.6	75.5
[3, 48, 24, 12, 6, 3]	sigmoid: 2nd 48	128	N/A	N/A	RMSE	1.1	$1.1 * 10^{1}$	75.7
[3, 30, 15, 7, 3]	sigmoid: 2nd 30	128	N/A	N/A	RMSE	$8.7 * 10^{-2}$	$5.4 * 10^{-1}$	77.7
[3, 30, 15, 7, 3]	sigmoid: 2nd 30	64	N/A	N/A	RMSE	$1.1 * 10^{-1}$	1.9	79.5
[3, 30, 15, 7, 3]	sigmoid: 2nd 30	80	N/A	N/A	RMSE	1.4	6.2	86.1
[3, 30, 15, 7, 3]	linear: middle 3, sigmoid: 2nd 30	80	TRUE	TRUE	RMSE	$2.0 * 10^{-1}$	1.3	86.8
[3, 30, 15, 7, 3]	linear: middle 3, sigmoid: 2nd 30	80	TRUE	TRUE	MAE^2	$1.5 * 10^{-4}$	$9.4 * 10^{-4}$	89.2
[3, 48, 24, 12, 6, 3]	**linear: middle 3, sigmoid:** 2nd 48	**80**	**TRUE**	**TRUE**	MAE^2	$6.8 * 10^{-3}$	$1.5 * 10^{-1}$	**92.5**

The decoding accuracy was measured with the root mean standard/squared absolute error ($RMSE$, MAE^2):

$$RMSE = \sqrt{(\frac{1}{n})\sum_{i=1}^{n}(y_i - x_i)^2}, \quad (5)$$

$$MAE^2 = [(\frac{1}{n})\sum_{i=1}^{n}|y_i - x_i|]^2. \quad (6)$$

Table 6 presents the results of the error averages for the normal and anomalous data. The detection accuracy was measured with the percent of anomalies that had a metric score greater than the average normal metric score. It clearly shows that the reconstruction error becomes far bigger for anomalous data.

All models in the table were performed with 200 epochs and *relu* as the default activation for each layer if specified otherwise. The first attempted autoencoder model had the following architecture, in terms of layers: [3, 25, 10, 3, 10, 25, 3] and "relu" activation functions for each layer, except "sigmoid" activation for the 2nd 25-node layer. The batch size was the default (32), and there was no validation nor shuffling of log data. Though the root mean squared error (RMSE) averages for anomaly data were $1.3 * 10^3$ times smaller than the normal data (training and testing data), only 55.6% of anomalies had an RMSE score larger than the normal testing data RMSE average. Some of the anomalies had very high RMSE values, skewing the average RMSE value for all the anomalies. This made it very important to not only consider the difference in RMSE scores between normal and anomaly logs, but also to consider detection capability. After increasing the number of hidden layers and changing their respective nodes to [3, 48, 24, 12, 6, 3, 6, 12, 24, 48, 3] with "sigmoid" activation for the 2nd 48-node layer, the anomaly detection accuracy increased to 74.5%. With these sets of layers and activations, tuning the batch size to 128 made the accuracy rise to 77.7%. The set of hidden layers: [3, 30, 15, 7, 3, 7, 15, 30, 3] with sigmoid activation on the 2nd 30-node layer gave very similar results to the hidden layers: [3, 48, 24, 12, 6, 3, 6, 12, 24, 48, 3]. Both these sets of layers were tested with each tune of parameters and the higher value was recorded in the results table. Validating the model with the test data, shuffling the log entries prior to encoding and decoding the data and further tuning of the batch size to an optimal 80 provided an accuracy of 78.1%. However, the results improved drastically to 86.8% after implementing a linear activation function for the middle 3-node layer. The best result of 92.5% detection capability was obtained with hidden layers: [3, 48, 24, 12, 6, 3, 6, 12, 24, 48, 3] with a "linear" activation at the middle 3-node layer and "sigmoid" activation at the 2nd 48-node layer, batch size of 80, validation of the model with test data, and shuffling of log entries prior to encoding and decoding.

4.2.4 Discussion

In our proposed hybrid solution, we utilize the strengths of the ML and DL models. The incoming log entry is first pre-processed by converting the data to integer representations. The SVM model performs timestamp-based classification on the log data. The role-based features (system, user, operation, object) are analyzed with the deep sequential autoencoder model. The results of both models determine the nature of the incoming log entry. The SVM model will label the timestamp portion of the log entry as either normal or anomalous. The deep sequential autoencoder will provide a value that is represented by the square of the mean absolute error and, based on a Gaussian distribution, will label the role-based features as either normal or anomalous. Combining the results of both models will determine the anomalous nature of the log: whether it is normal, anomalous in the role-based features, anomalous within the timestamp, or anomalous in both the role-based features and the timestamp.

The last two columns in Table 7 mark whether each part of the hybrid model determined the log entry to be anomalous. The first eight log entries are labeled

Table 7 Sample log entries fed into the hybrid (machine and deep learning) model. The table presents different types of incoming log entries (prior to pre-processing) and whether each model detected it as a normal or anomalous log

#	System	User	Operation	Object	Date	Time	Anom (SVM)	Anom (ANN)
#1	PM	apuzon	LOGON	none	2020-08-05	23:05:03	Yes	No
#2	SALES	kuszka	INSERT	ORDERS	2020-01-20	21:45:55	Yes	No
#3	SALES	gnyski	INSERT	ORDERS	2020-05-22	8:26:15	Yes	No
#4	PM	gmjarcinska	INSERT	PROJECT_ DETAILS	2020-08-17	1:30:55	Yes	No
#5	PM	ktarnows	LOGON	none	2020-07-25	16:16:31	Yes	No
#6	PM	adynka	SELECT	TASKS	2020-06-21	14:45:23	Yes	No
#7	PM	kuszka	DELETE	CLIENTS	2020-01-12	11:26:53	Yes	No
#8	SALES	amaly	SELECT	PRODUCTS	2020-07-04	12:48:02	Yes	No
#9	SALES	mwarzycha	SELECT	ORDERS	2020-01-03	9:01:33	No	Yes
#10	SALES	apuzon	SELECT	DISTRI-BUTORS	2020-04-10	14:23:04	No	Yes
#11	PM	ktarnows	INSERT	TASKS	2020-03-19	14:34:40	No	Yes
#12	SALES	kespiel	UPDATE	ORDERS	2020-07-23	14:47:51	No	Yes
#13	PM	adynka	LOGON	none	2020-06-24	10:05:45	No	No
#14	PM	adynka	LOGON	none	2020-06-24	11:07:23	No	Yes
#15	PM	ktarnows	ALTER	PROJECT_ TYPE	2020-01-01	9:28:47	No	No
#16	SALES	sbaszel	SELECT	WARE-HOUSES	2020-05-05	15:39:51	No	No
#17	PM	mwarzycha	SELECT	DEPART-MENTS	2020-05-29	10:49:53	No	No
#18	PM	aosinska	LOGON	none	2020-08-07	10:56:23	No	No

anomalous based on timeframes outside the normal working hours/days. Though these first eight rows may not be malicious, they are marked as anomalous due to the timestamp at which they were logged. The hybrid model aims to reduce false negatives, but at the cost of increasing the number of false positives, as in the case of marking logs outside of work hours as anomalous. Rows #9-#12 are also anomalous as they are unapproved operations, meaning that the users who performed the operations and objects were not privileged to do so. This can mean that they were privileged without the approval, either by themselves or another user, or there was an intrusion into the RBAC database system. For example, user *mwarzycha* falls under the HR role; however, she performed the operation SELECT on the ORDERS object. This operation and object pair is a common log for the WAREHOUSEMAN role and is not permitted for the HR role. This means that the user *mwarzycha* was privileged without the approval of the WAREHOUSEMAN role or was given access to just the SELECT and ORDERS operation and object pair. The rows #9-#12 were marked as anomalous by the deep sequential autoencoder part of the hybrid detection model. Rows #13-#14 show an example of an intrusion that was caught the ANN part of the model. The user *adynka* logged in in row #13, and then the user logged in again, without logging off, in row #14. This shows that *adynka* had logged in from multiple accounts, suggesting that her account has been breached. The last four rows (#15-#18) are marked as normal logs, as they are within the RBAC model and the timestamp of the log is within working hours. For example, user *mwarzycha*, who is under the HR role, is performing the SELECT operation on the DEPARTMENTS object on a Friday at 10:49 am. This is an authorized operation and object pair under the HR role that is logged during normal work hours. For these last four rows, the deep sequential autoencoder marked the event as normal, and the SVM model marked the timestamp as normal.

5 Conclusions

This chapter proposed a log-based anomaly detection system in application, database, and operating system layers. Our research has demonstrated that deep learning methods, specifically, autoencoders can be successfully applied to detect deviations from normal user behavior in data-based systems. The proposed IDS determines role intruders that deviate from normal behavior. If the error between the original input data and the model's decoded representation is within a high-level of confidence of the Gaussian distribution, the incoming log entry is considered normal, and otherwise, abnormal. Our experiments included manual tuning of the parameters of autoencoders, such as layers, activation, batch size, validation, and shuffling. Further, we have proposed a hybrid method combining ML models and DL models to further optimize anomaly detection. Although the performance of autoencoders was already good enough without temporal information, we augment the anomaly detection with the SVM method trained on timestamp data. It performs a combined anomaly detection at per log entry-level, rather than at per session-level as many previous methods are limited to. The proposed generic solution is tailored for any role-based access control (RBAC) database system. The limitations of this work were also identified. The technique was tested on an artificial dataset, but ideally, the method should be tested on a real world, highly unbalanced log dataset from the organization. It is necessary to test the approach on a greater variety of possible insider threat scenarios, such as weak audit trail or backup data exposure. The experiments within this work focused on database system logs. The goal is to extend and evaluate this methodology for real-world log data accumulated from different platforms, DBMS systems, and applications. The aggregated log analytics will involve handling known challenges for log management, such as many log sources, inconsistent log content, inconsistent timestamps, and inconsistent log formats [24].

5.1 Future Work

Future works include testing and incorporating other deep learning techniques (i.e., different types of recurrent neural networks, restricted Boltzmann machines) and integrating log data from different database systems (i.e., SQL Server) and applications to perform more comprehensive detection. Parameters for neural networks need further detailed investigation. The future work also includes text mining techniques on parsing and analyzing the commands' contents (such as proposed in [28, 55]). Evaluating the approach for efficiency will involve testing on very large log event datasets, such as described in [29]/[14].

As there is a gap in both research and in academia, but most notably in the industry offerings of comprehensive decision support systems (DSSs) for information security (IS) administrators, the future work includes incorporating the tested and verified models into the framework of intelligent DSS for IS. Such as system is expected to enable better enforcement of organizational policies and compliance with the mandated regulations. The prototype DSS was implemented for a hypothetical medium-sized company. The target organizations of this system are small-medium companies who cannot afford expensive to deploy and more complicated SIEM systems, but that are mandated by various regulations to meet the audit compliance because of sensitive/classified data stored within their systems. The benefits of the proposed systems are customization to the organizational needs, but supporting a variety of platforms existing in the enterprise IT infrastructure. It is proposed to collect and analyze the comprehensive log data including (1) the IT infrastructure across the entire organization, meaning heterogeneous IT systems; and (2) all layers of each IT system: from the operating system level, through database system-level, up to the application level. In the DSS customization process, the ETL process extracting logs from different systems/platforms has to be implemented accordingly to the organization's needs, as well as the central repository of transformed and standardized log records has to be set up locally. Such a DSS tool will help organizations in undergoing security audits and in compliance with security standards defined by organizations, such as ISO international information security standards. A DSS framework incorporating a combination of ML algorithms with neural network-based DL approaches offers rich opportunity to detect new variants of malicious and zero-day attacks. The new types of attacks and vulnerabilities exploitations will help organizations craft and implement better security policies and defense measures.

While many organizations concentrate the majority of their resources toward securing the perimeter of their networks, they often neglect the most critical company asset, databases [18]. Database security logging and monitoring are more difficult because the data is often sensitive, there are legitimate privileged users or varying degrees. Databases contain the most valuable data companies own—customer, employee, financial, and intellectual property to name a few categories. Protecting, logging, and monitoring database data should be a core activity of every business, unfortunately, many businesses fail to provide adequate security logging and monitoring for their databases. This research attempts to help close this gap.

References

1. Anashkin, E., and M Zhukova. 2020. An implementation of artificial neural networks into behavioral analysis system. *IOP Conference Series: Materials Science and Engineering* 734: 121–161.
2. Berman, Daniel, Anna Buczak, Jeffrey Chavis, and Cherita Corbett. 2019. A survey of deep learning methods for cyber security. *Information* 10 (4).
3. Britel, Merieme. 2018. Big data analytic for intrusion detection system. In *2018 International Conference on Electronics, Control, Optimization and Computer Science, ICECOCS*, 1–5.
4. Camacho, José, José Manuel García-Giménez, Noemí Marta Fuentes García, and Gabriel Maciá-Fernández. 2019. Multivariate big data analysis for intrusion detection: 5 steps from the haystack to the needle. *CoRR*, arXiv:abs/1906.11976.
5. Cappelli, Dawn M, Andrew P Moore, and Randall F Trzeciak. 2012. *The CERT guide to insider threats: how to prevent, detect, and respond to information technology crimes (Theft, Sabotage, Fraud)*. Addison-Wesley.
6. Chandola, Varun, Arindam Banerjee, and Vipin Kumar. 2009. Anomaly detection: A survey. *ACM Computing Surveys (CSUR)* 41 (3): 1–58.
7. Ashok Kumar D, and Venugopalan Srinivasagopalan Rajan. 2017. Intrusion detection systems: A review. *International Journal of Advanced Research in Computer Science* 8, 10.
8. Darwish, Saad M. 2016. Machine learning approach to detect intruders in database based on hexplet data structure. *Journal of Electrical Systems and Information Technology* 3: 261–269.
9. Dasgupta, Dipankar. 2007. Immuno-inspired autonomic system for cyber defense. *information Security Technical Report* 12 (4): 235–241.
10. Du, Min, Feifei Li, Guineng Zheng, and Vivek Srikumar. 2017. Deeplog: Anomaly detection and diagnosis from system logs through deep learning. In *Proceedings of the 2017 ACM SIGSAC Conference on Computer and Communications Security, CCS '17*, 1285–1298. Association for Computing Machinery.
11. Dwyer, John, and Traian Marius Truta. 2013. Finding anomalies in windows event logs using standard deviation. https://www.nku.edu/~trutat1/papers/CollaborateCom13_dwyer.pdf.
12. Feremans, Len, Vincent Vercruyssen, Wannes Meert, Boris Cule, and Bart Goethals. 2019. A framework for pattern mining and anomaly detection in multi-dimensional time series and event logs. In *International Workshop on New Frontiers in Mining Complex Patterns*, 3–20. Springer.
13. Fontaine, Jaron, Chris Kappler, Adnan Shahid, and Eli De Poorter. 2019. Log-based intrusion detection for cloud web applications using machine learning. In *Advances on P2P, Parallel, Grid, Cloud and Internet Computing, 3PGCIC 2019*, ed. L Barolli, P Hellinckx, and J Natwichai, vol. 96, 197–210. Springer.
14. Glasser, Joshua, and Brian Lindauer. 2013. Bridging the gap: A pragmatic approach to generating insider threat data. In *2013 IEEE Security and Privacy Workshops*, 98–104. IEEE.
15. Grover, Aarish. 2018. Anomaly detection for application log data. Master's thesis, San Jose State University.
16. He, Shilin, Jieming Zhu, Pinjia He, and Michael R Lyu. 2016. Experience report: System log analysis for anomaly detection. In *2016 IEEE 27th International Symposium on Software Reliability Engineering (ISSRE)*, 207–218. IEEE.
17. Heller, Katherine, Krysta Svore, Angelos D Keromytis, and Salvatore Stolfo. 2003. One class support vector machines for detecting anomalous windows registry accesses. https://academiccommons.columbia.edu/doi/10.7916/D85M6CFF.
18. Horwath, Jim. 2012. Setting up a database security logging and monitoring program.
19. Hu, Yi, and Brajendra Panda. 2003. Identification of malicious transactions in database systems. In *Seventh International Database Engineering and Applications Symposium, 2003. Proceedings*, 329–335. IEEE.
20. Huang, Lin, and Mark Stamp. 2011. Masquerade detection using profile hidden markov models. *Computers and Security* 30 (8): 732–747.

21. Islam, Mohammad Saiful, Mehmet Kuzu, and Murat Kantarcioglu. 2015. A dynamic approach to detect anomalous queries on relational databases. In *Proceedings of the 5th ACM Conference on Data and Application Security and Privacy*, 245–252.
22. Johnson, Richard Arnold, and Dean W. Wichern. 2002. *Applied Multivariate Statistical Analysis*, 5th ed. Prentice Hall.
23. Kamra, Ashish, Evimaria Terzi, and Elisa Bertino. 2008. Detecting anomalous access patterns in relational databases. *The VLDB Journal* 17 (5): 1063–1077.
24. Kent, Karen, and Murugiah Souppaya. 2006. Guide to computer security log management. *NIST Special Publication* 92: 1–72.
25. Kroll global fraud report 2011/12. https://www.slideshare.net/abaytelman/kroll-global-fraud-report-2011-2012.
26. Legg, Philip A. 2017. Human-machine decision support systems for insider threat detection. In *Data Analytics and Decision Support for Cybersecurity*, 33–53. Springer.
27. Li, Yong, Tao Zhang, Yuan Yuan Ma, and Cheng Zhou. 2016. Anomaly detection of user behavior for database security audit based on ocsvm. In *2016 3rd International Conference on Information Science and Control Engineering (ICISCE)*, 214–219. IEEE.
28. Lin, Qingwei, Hongyu Zhang, Jian-Guang Lou, Yu Zhang, and Xuewei Chen. 2016. Log clustering based problem identification for online service systems. In *2016 IEEE/ACM 38th International Conference on Software Engineering Companion (ICSE-C)*, 102–111. IEEE
29. Lindauer, Brian, Joshua Glasser, Mitch Rosen, Kurt C Wallnau, and L ExactData. 2014. Generating test data for insider threat detectors. *Journal of Wireless Mobile Networks, Ubiquitous Computing Dependable Application* 5 (2): 80–94.
30. Malhotra, Pankaj, Lovekesh Vig, Gautam Shroff, and Puneet Agarwal. 2015. Long short term memory networks for anomaly detection in time series. In *Proceedings*, vol. 89, 89–94. Presses universitaires de Louvain.
31. Joseph McKendrick. Data security: Leaders vs. laggards - 2013 IOUG enterprise data security survey.
32. Moh, Melody, Santhosh Pininti, Sindhusha Doddapaneni, and Teng-Sheng Moh. 2016. Detecting web attacks using multi-stage log analysis. In *2016 IEEE 6th International Conference on Advanced Computing (IACC)*, 733–738. IEEE.
33. Nieles, Michael, Kelley Dempsey, and Victoria Pillitteri. 2017. *An introduction to information security*. Technical report. National Institute of Standards and Technology.
34. PricewaterhouseCoopers LLP. 2011. Protecting against the growing threat — Events and trends. https://www.pwc.com.cy/en/events/assets/economic-crime-survey.pdf.
35. Raut, Umesh K. 2018. Log based intrusion detection system. *IOSR Journal of Computer Engineering*, 20 (5): 15–22.
36. Reghunath, K. 2017. Real-time intrusion detections system for big data. *International Journal of Peer to Peer Networks (IJP2P)* 8 (1).
37. Ring, Markus, Sarah Wunderlich, Dominik Gruedl, Dieter Landes, and Andreas Hotho. 2017. A toolset for intrusion and insider threat detection. In *Data Analytics and Decision Support for Cybersecurity: Trends, Methodologies and Applications*, ed. Ivn Palomares Carrascosa, Harsha Kumara Kalutarage, and Yan Huang, 1st ed., 3–31. Springer Publishing Company, Incorporated.
38. Rodrigues, A.J. 2013. Automated log analysis using ai: intelligent intrusion detection system. *Computer Journal* 132.
39. Ronao, Charissa Ann, and Sung-Bae Cho. 2014. A comparison of data mining techniques for anomaly detection in relational databases. In *International Conference on Digital Society*.
40. Ronao, Charissa Ann, and Sung-Bae Cho. 2015. Random forests with weighted voting for anomalous query access detection in relational databases. In *Artificial Intelligence and Soft Computing*, ed. Leszek Rutkowski, Marcin Korytkowski, Rafal Scherer, Ryszard Tadeusiewicz, Lotfi A. Zadeh, and Jacek M. Zurada, 36–48. Cham: Springer International Publishing.
41. Rudowski, Michal, and Katarzyna Tarnowska. 2016. Decision support system for information systems security audit (WABSI) as a component of IT infrastructure management. *Information Systems in Management* 5 (3): 389–400.

42. Sakurada, Mayu, and Takehisa Yairi. 2014. Anomaly detection using autoencoders with non-linear dimensionality reduction. In *Proceedings of the MLSDA 2014 2nd Workshop on Machine Learning for Sensory Data Analysis, MLSDA'14*, 4–11. Association for Computing Machinery.
43. Sallam, Asmaa, Elisa Bertino, Syed Rafiul Hussain, David Landers, Robert Michael Lefler, and Donald Steiner. 2017. DBSAFE - an anomaly detection system to protect databases from exfiltration attempts. *IEEE Systems Journal* 11 (2): 483–493.
44. Sallam, Asmaa, Daren Fadolalkarim, Elisa Bertino, and Qian Xiao. 2016. Data and syntax centric anomaly detection for relational databases. *Wiley International Review of Data Mining and Knowledge Discovery* 6 (6): 231–239.
45. Sandhu, Ravi, David Ferraiolo, and Richard Kuhn. 2000. The NIST model for role-based access control: Towards a unified standard. In *Proceedings of the Fifth ACM Workshop on Role-Based Access Control, RBAC '00*, 47–63. Association for Computing Machinery.
46. Schonlau, Matthias, William DuMouchel, Wen-Hua Ju, Alan F. Karr, Martin Theusan, and Yehuda Vardi. 2001. Computer intrusion: Detecting masquerades. *Statistical Science* 16 (1): 58–74.
47. HHS Office of the Secretary and Office for Civil Rights (OCR). Security rule guidance material, Aug 2017.
48. Shashanka, M., M. Shen, and J. Wang. 2016. User and entity behavior analytics for enterprise security. In *2016 IEEE International Conference on Big Data (Big Data)*, 1867–1874.
49. Shebaro, Bilal, Asmaa Sallam, Ashish Kamra, and Elisa Bertino. 2013. Postgresql anomalous query detector. In *Proceedings of the 16th International Conference on Extending Database Technology, EDBT '13*, 741–744. Association for Computing Machinery.
50. Shenfield, Alex, David Day, and Aladdin Ayesh. 2018. Intelligent intrusion detection systems using artificial neural networks. *ICT Express* 4 (2): 95–99.
51. Spalka, Adrian, and Jan Lehnhardt. 2005. A comprehensive approach to anomaly detection in relational databases. In *Data and Applications Security XIX*, ed. Sushil Jajodia and Duminda Wijesekera, 207–221. Berlin: Springer.
52. Tarnowska, Katarzyna. 2013. System security audit.
53. Torkaman, Atefeh, Marjan Bahrololum, and Mohammad Hesam Tadayon. 2014. A threat-aware host intrusion detection system architecture model. *7th International Symposium on Telecommunications (IST'2014)*, 929–933.
54. Wee, Chee Keong, and Richi Nayak. 2019. A novel machine learning approach for database exploitation detection and privilege control. *Journal of Information and Telecommunication* 3 (3): 308–325.
55. Xu, Wei, Ling Huang, Armando Fox, David Patterson, and Michael I. Jordan. 2009. Detecting large-scale system problems by mining console logs. In *Proceedings of the ACM SIGOPS 22nd Symposium on Operating Systems Principles, SOSP '09*, 117–132. Association for Computing Machinery.
56. Yen, Ting-Fang, Alina Oprea, Kaan Onarlioglu, Todd Leetham, William Robertson, Ari Juels, and Engin Kirda. 2013. Beehive: Large-scale log analysis for detecting suspicious activity in enterprise networks. In *Proceedings of the 29th Annual Computer Security Applications Conference, ACSAC '13*, 199–208. Association for Computing Machinery.

Image Spam Classification with Deep Neural Networks

Ajay Pal Singh and Katerina Potika

Abstract Image classification is a fundamental problem of computer vision and pattern recognition. We focus on images that contain spam. Spam is unwanted bulk content, and image spam is unwanted content embedded inside the images. Image spam potentially creates a threat to the credibility of any email-based communication system. While a lot of machine learning techniques are successful in detecting textual based spam, this is not the case for image spams, which can easily evade these textual-spam detection systems. In our work, we explore and evaluate four deep learning techniques that detect image spams. First, we train deep neural networks using various image features. We explore their robustness on an improved dataset, which was especially build in order to outsmart current image spam detection techniques. Finally, we design two convolution neural network architectures and provide experimental results for these, alongside the existing VGG19 transfer learning model, for detecting image spams. Our work offers a new tool for detecting image spams, usage of a bigger dataset, and is compared against recent related tools.

1 Introduction

Over the last decade, email and Internet is flooded with spam content. A spam can be defined as unwanted content, distributed mostly via emails. Due to the effluence of spam emails over the Internet a lot of techniques have surfaced which classify the spam from the valid content. Reports from Symantec [18] indicated that 90.4% of the emails include spam content. These spam emails can include phishing links, malware, advertisements, adult content, and others, which may impose a significant threat to the security of the user's privacy.

Spam initial was only in the form of texts. With the advent of machine learning, many classifiers were developed to filter such spam based on email content. Lai

A. P. Singh (✉) · K. Potika
San Jose State University, One Washington Square, San Jose, CA 95192, USA
e-mail: ajay.id.4.ms@gmail.com

K. Potika
e-mail: katerina.potika@sjsu.edu

M. Stamp et al. (eds.), *Malware Analysis Using Artificial Intelligence and Deep Learning*, https://doi.org/10.1007/978-3-030-62582-5_24

and Tsai [11] used four different machine learning techniques, including K-nearest neighbors (KNN), support vector machines (SVM), and Naïve Bayes, that used email messages to filter spam emails. These classifiers were able to classify text-based spam with approximately 95% accuracy. Hence, over the years detecting content-based spam emails became very easy. Google, Microsoft, and Yahoo use techniques that perform very accurately to classify spam emails from the authentic emails.

However, over time, spammers came up with novel ideas to fool these content-based classification techniques. Thus, image spam was developed, where unwanted textual information was delivered in the form of images. To detect these types of image text, optical character recognition (OCR) techniques [4] were developed which were able to extract the text from the images. It involves segmentation of the textual region within the images and using techniques to extract text from these regions. However, these text-based classifiers were not always successful in detecting image spam. One reason was that segmentation of textual area within these images in itself is a difficult task [17]. Also, spammers started using obfuscation techniques, which made the OCR techniques less effective.

A more direct approach was used by Annadatha et al. [1] and Aneri Chavda et al. [2], where they consider properties of the image itself to classify spam images. They used image processing techniques in conjunction with various machine learning models. We use different deep learning techniques on various image properties as compared to these previous work [1] and [2]. We train neural networks and deep neural networks on these image properties, instead of using the machine learning techniques that were used previously in [2]. We then divulge deeper and experiment with other deep learning techniques, such as convolutional neural networks (CNN) based on raw images. Finally, we discuss the use of transfer learning and train a VGG19 model on our dataset. The main focus of this work is to quantify the robustness of these techniques on an improved spam dataset created by Aneri et al. [2].

The remainder of the report is organized as follows. In Sect. 2, we discussed the problem statement and the motivation behind it. Section 3 focuses on the related work done so far in this domain. Section 4 describes the essential background, topics, and terminologies needed to understand this project. Section 5 discusses the various datasets used in this work, the steps involved in pre-processing these datasets and the architecture used to train the deep learning models. Section 6 presents the experimental results. Finally, Sect. 7 concludes and provides scope for future work.

2 Problem Statement and Motivation

This section defines the problem statement and scope for this work. It also focuses on the motivation and purpose of solving this problem.

We focus on binary image classification. Anything that contains the marketing, sexual, or other unwanted content embedded within the images is called a SPAM image, whereas anything other than that is considered a HAM image. HAM is a keyword specified in the and used in previous papers [1, 2], so in order to maintain the consistency we will also use the same terminology.

The goal is to use specifically Neural Networks and Deep Neural Networks on the problem of spam image classification in order to obtain better results as obtained in the previous papers. There are two main parts towards that direction:

- Classification based on image features: extraction of 38 features from the images as described in [2] and then use the different Neural Networks and Deep Neural Network architectures with the motive to improve the results.
- Use of Raw Images: usage of deep learning techniques, such as Convolution Neural Networks, with different architectures and then with pre-built models based on transfer learning on the spam image classification problem.

Using different approaches as mentioned above the results are presented in the form of tables, graphs, and other metrics to give a quantification of this work.

The Internet is flooded with spam content, whether it is in the form of text or unwanted text in the images. Previous techniques are good to detect textual spam but the spammers are coming up with new ways to fool such techniques. We try to solve this hard problem of image spams. We discuss the results obtained to classify spam images by leveraging the power of neural networks and deep learning. Since the advent of deep learning, there is not much research done on this domain. By using our approach potentially administrators of email systems or other systems can minimize the spam content that is even embedded in images.

3 Background

Let us present the essential background and terminology that we need.

3.1 Spam Categories

In general spam detection techniques are partitioned into the following two categories:

1. Content-based spam: spam in emails that are in textual form; classifiers in this case deal with the actual content of the email extracted from email headers, keywords, body of email, etc. Wide variety of machine learning techniques are available for such spam classification [3].
2. Non-content-based spam: spam that use advanced forms; such one is an image spam. For image spam classification we can look at the properties of an image, or with the advent of deep learning we can use images in their raw byte form. There are different generation of image spams ranging from first generation to third generation. Images in the first generation contain plain spam images, but in

the second and third generation images are obfuscated using noise, by overlaying background of images to make them more resistant to OCR techniques. OCR techniques are capable of segmenting the part of the images that contain specific object for further purposes, for example, extraction of text or object detection.

3.2 Classification Techniques

Here we mention the techniques that we use and are implemented for the experiments described in Sect. 6.

3.2.1 Neural Networks

Artificial neural networks (ANNs) are algorithms that are modeled after the neuronal structure of cerebral cortex of the human brain but on much smaller scales. A neural network (NN) structure is divided into different layers: the input layer, one or many hidden layers, and the output layer. Each layer comprises nodes or neurons. A basic unit of computation in a neural network is a neuron, which receives an input from previous layer nodes along with their specific weights, and performs a function on them. This function is also called the activation function. These neurons are activated by using different activation functions. Some examples of activation functions: the sigmoid, the tanh, and the Relu function. A bias is usually added with each layer to provide regularization and move the function graph by some constant from the center. The goal of the ANN's is to decrease the loss function, which is derived from the dataset.

As compared to support vector machines (SVM), which use only one function, neural networks provide non-linearity due to the structure of its layers. There are different types of neural networks, for example, feedforward NN, single layer perceptron, multi-layer perceptron, and so on. The output layer in case of a classification problem usually consists of a sigmoidal activation function to provide probabilities for different classes, or labels for the dataset.

3.2.2 Deep Neural Networks

They are differentiated from basic NN by their depth; that is, the number of node layers through which data passes in a multi-step process of pattern recognition. In this each layer of a network is supposed to learn specific features as received from the previous layer. The further the layer is, nodes are able to understand more complex features, since they aggregate and recombine those features from previous layers. Deep learning networks perform automatic feature extraction without human intervention, unlike most traditional machine learning algorithms.

3.2.3 Convolution Neural Networks

The idea behind Convolution Neural Networks (CNN) was derived from NNs with neurons that learn from weights and biases. Also, each layer calculates a non-linear function using teh dot product and some activation functions. CNN's work on the images itself in the input layer. The normal NN's does not scale very well on the raw images, because they are not able to learn enough features from them. The architecture of CNN's introduces different types of layers, each of which learn specific features from the previous layer. Thus, the general idea is that the starting layers are able to learn more generic features, such as the curves and edges from the images, then as the architecture grows these layers becomes more specific, for example, detecting the ears of an animal. Unlike a NN a CNN have neurons arranged in three dimensions, namely, width, height, and depth. The depth here refers to the channels in an image, for a color image the depth is 3, whereas for a grayscale image it is 1. The CNN architecture is built from different types of layers, which are repeated as necessary to build the deep CNN. These layers are

1. Convolutional Layer: Tries to learn the features from the images by preserving the spatial relationship among pixels. It uses the concept of filters. The images are divided into small squares on which these filters are projected. These filters contain different values of pixels, for example, a filter can be used to find edges of text in a spam image. So, if there exist edges in the speculated square, those pixels are activated. The pixels on the filter are multiplied on the input image area under consideration and a sum is performed over those activated pixels within the filter to check the intensity of the filter. These filters work in sync with the depth of the images, so if the image is RGB then there are 3 filters used for each depth of given sizes. There are three parameters that control the size of the Convolution layers: stride, depth, and padding. The depth is mostly determined by the depth of the raw images or based on the previous layer input. The stride defines the number of pixels the filters are slided from left to right. Sometimes the input layer features are padded with zeros to maintain proportion with the size of the filter and to control the size of the output layer. After sliding the filter over all locations of the input array we get an activation map or feature map.
2. RELU (Rectified Linear Units) Layer: To provide non-linearity after each Convolution layer it is suggested to apply a RELU layer. A RELU layer works far better in terms of performance as compared to the sigmoid or tanh function without compromising the accuracy. This layer also overcomes the problem of vanishing gradient. In vanishing gradient the lower layers train much slowly, because the gradient due to back-propagation decreases exponentially through the layers. The RELU function is given as

$$F(x) = \max(0,x)$$

This function changes all negative values to 0 and increases the non-linearity of the model without affecting the output of the Convolution layer.

Input -> Conv -> ReLU -> Conv -> ReLU -> Pool -> ReLU -> Conv -> ReLU -> Pool ->Fully Connected

Fig. 1 A simple CNN architecture composed of different layers

3. Pooling (or down-sampling) Layer. It uses a filter of a given size, moves across the input from previous layer, and applies a given function. For example, in a max pooling layer, a max of all the filter values is given as output. There are other types of pooling layers as well such as average pooling and L_2-norm pooling. This layer serves two purposes: it decreases the amount of computation by decreasing the amount of parameters of weights, and it overcomes overfitting in the model.
4. Dropout Layer: This layer helps in overcoming the problem of overfitting. Overfitting means the weights and parameters are so tuned to the training examples after training the network, that they perform very bad on the new examples. So this layer drops out a random set of activation in a given layer by setting their values to 0 [16].
5. Fully Connected Layer: it takes as input from any of the Convolution layer, Pooling or RELU layer and outputs a N dimensional vector. This N depends on the number of classes you want to classify. In case of the image spam classification problem the value of $N = 2$, i.e., whether the image is a SPAM or a HAM.

A classic CNN architecture is composed of the above layers repeated in some fashion as necessary. A simple example of such architecture is given in Fig. 1.

3.2.4 Transfer Learning

Data is an essential part of deep learning community. As you train your networks on large amount of data the network becomes more redundant and efficient in generalizing the results to new datasets. Thus in case you have a small amount of dataset to actually work on, transfer learning overcomes this caveat. It is a process of using pre-trained models, which are trained on millions or billions of samples of generalized datasets and then fine-tune these models on your own datasets. Rather than training the whole network we use a pre-trained model weights and freeze them and focus on training only specific lower level of layers which are more specific to our dataset. If your dataset is different from the pre-trained model dataset then in that case training more layers of the model is preferred. We focus on using two such pre-trained models VGG16 and VGG19.

A technique called data augmentation is widely used to overcome the problem of having less samples of dataset. It performs different image transformation on the images to produce new images and hence augment the dataset. Some transformations may include scaling the image by some ratio, rotating them, skewing, flipping, and cropping the images.

3.3 Quality Metrics

The following terms will be used to quantify the results:

- True Positive (TP): an image is a SPAM image and classifier marks it is as a SPAM
- True Negative (TN): an image is a HAM image and classifier marks it is as a HAM
- False Positive (FP): an image is a HAM image and classifier marks it is as a SPAM
- False Negative (FN): an image is a SPAM image and classifier marks it is as a HAM
- Confusion Matrix: This is a matrix between TP, TN, FP, and FN. A Fig. 2 taken from [2] is shown below.
- Accuracy: It is a metric used to determine how well a classifier works. It is defined as

$$Accuracy = \frac{TP + TN}{P + N} \tag{1}$$

  where P (Positive) $= TP + FN$ and N (Negative) $= TN + FP$.
- ROC and AUC: Receiver Operating characteristics (ROC) and Area under the Curve (AUC) can be obtained from a trained classifier. A ROC curve is plotted against True positive rate (TPR) and False Positive rate (FPR) for varied threshold values. An area under this ROC curve is known as AUC value. TPR and FPR are determined as

$$TPR = \frac{TP}{P} = \frac{TP}{TP + FN} \tag{2}$$

$$FPR = \frac{FP}{N} = \frac{FP}{FP + TN} \tag{3}$$

  An AUC close to 1 is considered as a good classifier.
- K-fold cross validation: In this technique the classifier is trained K times. The dataset is divided into K subsets. Each time a classifier is trained, one of the K subsets is used as the validation or test and the remaining $K - 1$ subsets are used as the training set. The accuracy over all these K classifiers is averaged to provide the average accuracy. Cross validation techniques are generally used to overcome overfitting within the dataset.
- Stratified K-fold cross validation: It is a slight variation in the K-fold cross validation technique. In this each fold is created in such a fashion that each subset contains approximately the same percentage of each target class. This is used in cases where the dataset classes are skewed, i.e., one class predominates the other.

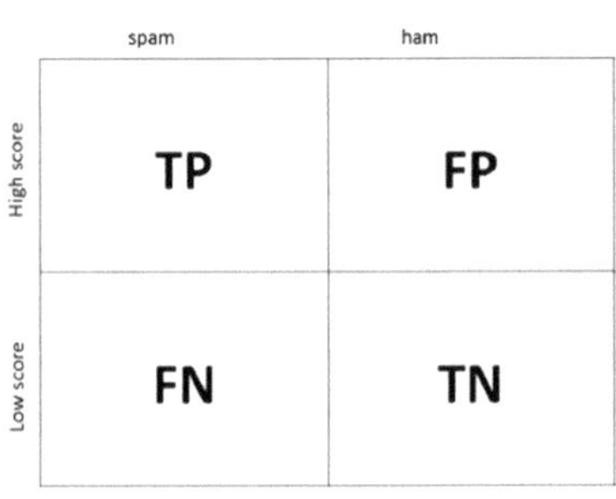

Fig. 2 A confusion matrix [2]

4 Related Work

Image spam detection can be done using various techniques. One such approach is to use content-based detection, i.e., to segment the content using OCR techniques and then classify it. Other approaches include to detect image spams based on the properties or features extracted from these images. Different machine learning algorithms are also used in conjunction with image processing to generate strong classifiers. Additionally, deep learning techniques, such as CNN, can also be used on the raw images to detect image spams.

A content-based image spam detection technique is discussed in [1], which uses OCR techniques on spam images. These techniques extract the text from the segmented image and perform textual analysis on the extracted text to determine whether the image is a SPAM or a HAM.

Another paper by Gevaryahu, Elias-Bachrach et al. [4] use the image metadata properties, such as file size and other properties to detect SPAM images. The work presented in [5] use probabilistic boosting trees for the classification by working on the color and gradient histogram features and achieve an accuracy of 94%.

Sanjushree, Suhasini et al. [10] use SVMs and particle swarm optimizations (PSO) on ten metadata features and three textual features for SPAM image classification. Using the combination of these techniques they were able to achieve a 90% accuracy on the Dredze dataset (discussed below) using 380 test and 300 training images. The authors of [5] use cluster-based filtering techniques with client and server-based models and claim to achieve a 99% accuracy.

A fuzzy inference system was used to analyze multiple features in [8], claiming to have achieved an accuracy of 85%. Annadatha et al. [1] used a linear SVM classifier on 21 image properties and each property was associate with a weight for the classification. The author performed feature reduction and selection based on these weights. Two datasets [5] and Dredze [4] are used by them and they achieved an accuracy of 97% and 99%, respectively. Moreover, they also developed a new challenging dataset for their classifier.

Aneri et al. [2] used 38 features extracted for the Dredze and the Image Spam hunter (ISH) datasets (discussed below). In their approach they use SVM kernels and achieved an accuracy of 97% with linear, 96% with Radial Basis Function (RBF), and 95% with a polynomial kernel on the ISH dataset. Respectively, the results are 98% using linear, 98% using RBF, and 95% using a polynomial kernel on the Dredze dataset. Additionally, feature reduction techniques are used, like univariate feature selection (UFS) and recursive feature elimination (RFE), to reduce the initial 38 features and provide better results. Expectation maximization (EM) clustering techniques are some of their other approaches, but these did not performed very well and achieved an accuracy of 87% on the ISH and 70% on the Dredze dataset. One of the main contributions of this work is the creation of a new challenging dataset, that we also use in this work. The accuracy was not very good for it being 52%. The number of samples that we use is relatively large from these datasets. A big part of our work also discusses various data processing techniques used to extract images from the spam archive, which was an unprocessed dataset as provided by Dredze. Hence, the amount of data the experiments and results are based on are large as compared to all previous papers.

In a recent work, Sharma et al. [14] use CNNs along with other machine learning techniques to solve the SPAM image classification problem. They consider it similar to our real-world image spam and challenging datasets. The best results they get are for their CNNs with an accuracy of 99.02% for the ISH dataset, 83.13% for the improved dataset of [2], and 71.83% for the challenging dataset of [1] based on the Dredze dataset.

Soranamageswari et al. [15] use backpropagation neural network based on only color features and was able to achieve a 92.82% accuracy. An interesting aspect used was that of splitting the image into different blocks and using them as features, but this approach only achieves a 89.32% accuracy.

Hong-Gang Zhang and Er-Xin Shang [13] use CNNs to classify into 7 categories of unwanted content embedded in images. They worked on a dataset containing around $52K$ images. The 7 categories targeted by them include commodities, spam images, political content images, adult content, recipe images, scenes, and everything else. They used five-convolution alongside max pooling layers. The output of these layers is given to the fully connected layer and a SVM classifier is used on this N sized feature vector. They resized all the images to a 256×256 size and were able to achieve an average accuracy of 75.1%. Following this approach we use a custom CNN architecture, but only on a binary classification, namely, SPAM and HAM.

5 Framework

5.1 Datasets

We experiment with the three different datasets that are used by Aneri et al. [2]. However, the number of images used is much larger compared to the images used in that approach. We focus on making use of different formats of images such as gif, jpeg, jpg, png, tif, and bmp. The gif images were processed and the first frame was extracted from the gif images and converted to png files.

5.1.1 Dredze Dataset

Dredze et al. [4] created a SPAM image archive dataset as well as a personalized SPAM image archive. The personalized SPAM image archive contained a lot of unprocessed files in different formats such as gif, txt, and jpg. We pre-processed this archive as well to augment our dataset. Then, the experiments were performed on the combination of the Dredze SPAM archive and the Dredze personalized archive. Earlier papers [1, 2] focused only on a subset of the personalized SPAM images archive. After the pre-processing step, the personalized SPAM images were 3165 with 1760 HAM images, and the SPAM images obtained were 10937. Totally, $14,103$ SPAM images and $12,565$ HAM images.

5.1.2 Image Spam Hunter (ISH)

The image spam hunter dataset contained both HAM and SPAM images [5]. We extracted and processed 922 SPAM and 810 HAM images.

5.1.3 Improved Dataset

This dataset was created by Aneri et al. [2]. This dataset was created by performing transformation on the HAM images to make them SPAM. The HAM images were resized to the size of SPAM images to align their metadata features. Noise was introduced in the SPAM images to make their edge detection difficult, since SPAM images generally have less noise as compared to HAM images. These noise-induced SPAM images were overlayed on top of the HAM images to generate the improved dataset. We experimented with the additional $1,030$ improved SPAM images.

5.1.4 Combined Dataset

In general CNN requires large amount of datasets to converge and perform better. So instead of experimenting with individual datasets mentioned above we combined together all these datasets to augment the number of SPAM image samples. In order to account for the HAM images we downloaded images belonging to different categories, that are not SPAM, to make our dataset balanced.

5.2 *Data Pre-processing*

The Dredze dataset archive contained a lot of unwanted files and corrupted images from which features could not be extracted. There were a lot of images in different formats and almost 60–70% of the images were in gif formats. These gif images were processed to extract the first frame and then saved in a png format which helped in augmenting the dataset. The steps in order to achieve this objective are described next:

1. All unwanted formats, such as .txt, were removed.
2. All .gif images were converted to .png files. All frames of the .gif images were extracted and the first frame was saved as a SPAM image. The rest of the frames did not contain a SPAM image and were discarded.
3. All corrupted files were removed.

In order to achieve the above objectives, different proofs of concepts were tried. We created bash scripts to perform all the above steps and keep track of the results after each step to get a clean augmented SPAM images archive.

5.3 *Image Features*

The first part of the experiments used NNs and DNNs on features extracted from the images from the different datasets. We use the 38 features as mentioned in [2]. The features are classified into five big categories: metadata, color, texture, shape, and noise features. Figure 3 below describes the different features belonging to each category. The different categories of features are discussed below.

5.3.1 Metadata Properties

These properties contain the image height, width, aspect ratio, bit depth, and compression ration of the image files. Compression of an image is defined as

$$\text{Compression Ratio} = \frac{\text{height} * \text{width} * \text{channels}}{\text{file size}} \tag{4}$$

5.3.2 Color Properties

These properties include mean, skew, variance, and entropy values of different properties of an image such as RGB colors, kurtosis, hue, brightness, and saturation. Mean can be a basic color feature that represents the average pixel value of the image. That is, it is useful for determining an image background. A SPAM compared to a HAM

Feature Domain	Feature	Description
Metadata	height	Height of the image
	width	Width of image
	aspect ratio	Ratio of height and width
	compression ratio	How compressed is image
	file size	Size on disk
	image area	Area of image
Color	entr-color	Entropy of color histogram
	r-mean	Mean of red histogram
	g-mean	Mean of green histogram
	b-mean	Mean of blue histogram
	r-skew	Skew of red histogram
	g-skew	Skew of green histogram
	b-skew	Skew of blue histogram
	r-var	Variance of red histogram
	g-var	Variance of green histogram
	b-var	Variance of blue histogram
	r-kurt	Kurtosis of red histogram
	g-kurt	Kurtosis of green histogram
	b-kurt	Kurtosis of blue histogram
	entr-hsv	Entropy of hsv histogram
	h-mean	Mean hue of hsv histogram
	s-mean	Mean saturation of hsv histogram
	v-mean	Mean brightness of hsv histogram
	h-var	Variance of hue hsv histogram
	s-var	Variance of saturation hsv histogram
	v-var	Variance of brightness hsv histogram
	h-skew	Skew of hue hsv histogram
	s-skew	Skew of saturation hsv histogram
	v-skew	Skew of brightness hsv histogram
	h-kurt	Kurtosis of hue hsv histogram
	s-kurt	Kurtosis of saturation hsv histogram
	v-kurt	Kurtosis of brightness hsv histogram
Texture	lbp	Entropy of LBP histogram
Shape	entr-hog	Entropy of HOG
	edges	Total number of edges in an image
	avg-edge-length	Average edge length
Noise	snr	Signal to noise ratio
	entr-noise	Entropy of noise

Fig. 3 Different image features [2]

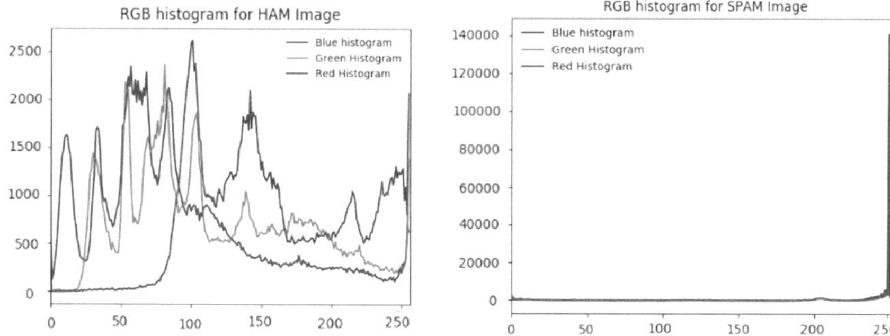

Fig. 4 SPAM verses HAM RGB's histogram

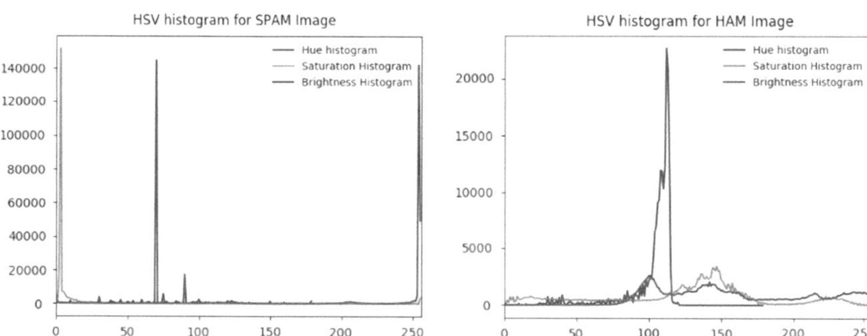

Fig. 5 SPAM verses HAM HSV's histogram

image has different histogram properties for these features as depicted in the examples of Fig. 4. These histograms show the reasoning behind selecting these properties of color for the classification task. Similarly, in the examples of Fig. 5 one can see the histogram for HSV's values of a SPAM and a HAM image. In images, a glossier surface has more positive skew values as compared to a lighter and matte surface. Hence, we can use skewness in making judgments about image surfaces. Kurtosis values are interpreted in combination with noise and resolution measurement. High kurtosis values go hand in hand with low noise and low resolution. SPAM images usually have high kurtosis values.

5.3.3 Texture Properties

The local binary pattern (LBP) is used to determine similarity and information of adjacent pixels. The LBP would appear to be a strong feature for detecting an SPAM image that is simply text set on a white background. In the case of SPAM images these histograms will have high intensity for specific values rather than being scattered.

5.3.4 Shape Properties

Histogram of oriented gradients (HOG) determines how intensity gradient changes in an image. HOG descriptors are mainly used to describe the structural shape and appearance of an object in an image, making them excellent descriptors for object classification. Edges are one of the most important features to detect SPAM images. They serve to highlight the boundaries in an image. Canny edge filters are used to find the edges. Fig. 8 shows the contrast in canny edges for SPAM and HAM images. Figures 6 and 7 show the hog features for a HAM and a SPAM image.

5.3.5 Noise Properties

These features include signal to noise ratio (SNR) and entropy of noise. SPAM images tend to have lesser noise as compared to HAM images. SNR is defined as the ratio of mean to standard deviation of an image.

5.4 Techniques Used

Let us briefly describe the various techniques and architectures that we are going to use.

5.4.1 Neural Networks

A backpropagation neural network with 1 hidden layer with 20 neurons was used. The input layer consisted of the 38 features. The hidden layer used the RELU activation function and the output layer consists of the sigmoid activation function with one neuron. A K-fold stratified cross validation, with $K = 10$, was used with this. An architecture of the model is shown in Fig. 8.

5.5 Deep Neural Networks

The previous neural network was extended to introduce another hidden layer with 10 neurons and with the RELU activation function. Binary cross entropy was used as the loss function and again with K-fold stratified cross validation. We make use of two CNN architectures. We name them as CNN1 and CNN2. CNN1 was trained for 30 iterations, whereas CNN2 was run with 25 iterations. Both of them were trained with a batch size of 64 images. The training set contained 19924 images, whereas the validation set contained 2681 images.

Fig. 6 **a** HAM original Image **b** HAM Grayscale Image **c** HAM HOG **d** HAM Canny edges

5.5.1 CNN1 Architecture

The CNN1 architecture as defined below was used as the third model to draw results. The images were first rescaled to 128x128x3 and then fed to the network.

1. 96 filters of size $3 \times 3 \times 3$ were used to the input layer with a stride of 1 followed by the RELU function.

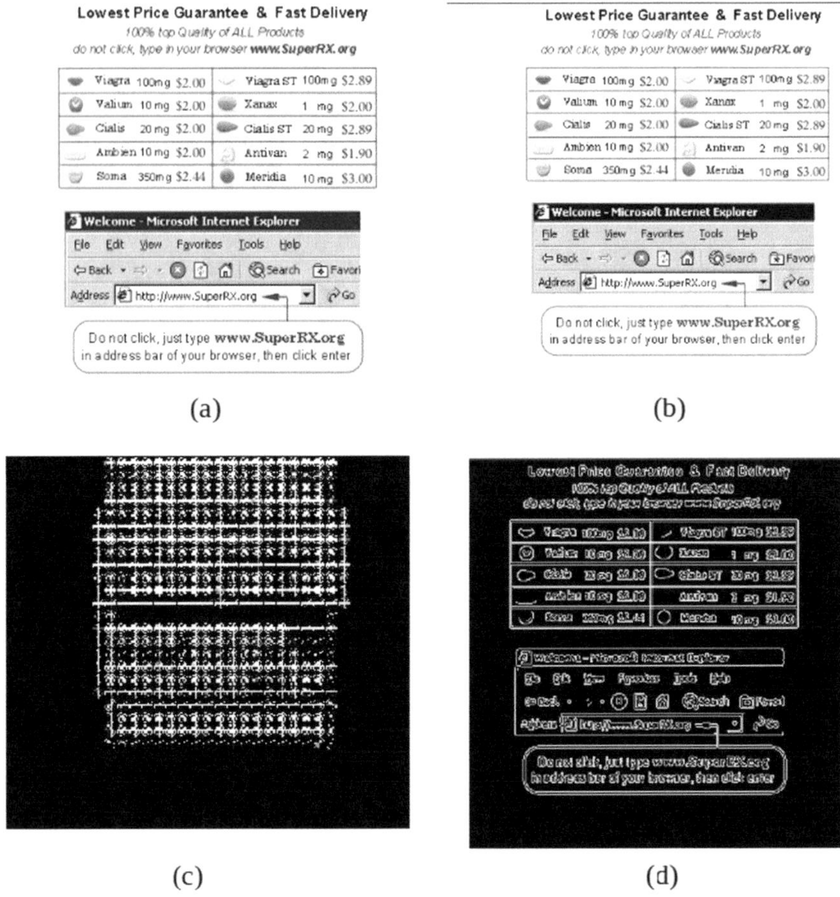

Fig. 7 **a** SPAM original Image **b** SPAM Grayscale **c** SPAM HOG **d** SPAM Canny edges

2. On the output of $96 \times 126 \times 126$ from the previous layer a max pool layer is used taking the maximum value from a 3×3 area with stride of 2×2.
3. On the input of $96 \times 62 \times 62$ from previous layer another convolution layer with 128 filters, with 3×3 filter size and stride of 1, and no padding is used followed by a RELU activation function.
4. On $128 \times 60 \times 60$ input from previous layer another max pooling layer with a 3×3 area and stride of 2 is used on the input of previous layers to produce an output of size $128 \times 29 \times 29$.
5. The input of the previous layer is flattened and given to a fully connected layer. The N vector obtained from the input layer is of size 107648. On this N vector a dense layer of size 256 is used with RELU as activation function and a dropout layer with probability of 0.1. Another dense layer of size 1, which acts as the output layer is added to the end of this with sigmoid activation function.

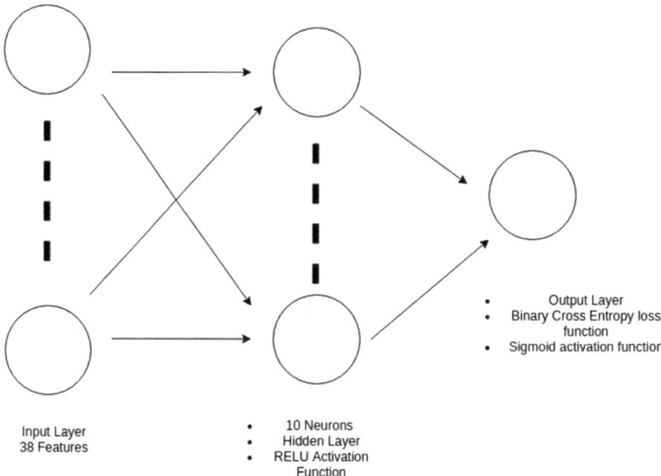

Fig. 8 Neural network architecture

5.5.2 CNN2 Architecture

The CNN2 architecture is defined below and was used as the fourth model to draw results. The images were again first rescaled to $128 \times 128 \times 3$ and then fed to the network.

1. 128 filters of size $3 \times 6 \times 6$ were used to the input layer with a stride of 2 followed by the rectilinear linear operator (RELU) function.
2. On the output of $128 \times 62 \times 62$ from previous layer a max pool layer is used taking the maximum value from a 4×4 area with stride of 1.
3. On the input of $128 \times 59 \times 59$ from previous layer another convolution layer with 128 filters with 4×4 filter size and stride of 1 and no padding is used followed by a RELU activation function.
4. On $128 \times 56 \times 56$ input from previous layer another max pooling layer with a 3×3 area and stride of 1 is used on the input of previous layers to produce an output of size $128 \times 54 \times 54$.
5. On the previous layer input another convolution layer with 256 filters with 3×3 filter size and stride of 2 is used followed by a RELU activation function.
6. On the $256 \times 26 \times 2$ input from the previous layer another max pooling layer with a 5×5 area and stride of 2 is used on the input of previous layers to produce an output of size $256 \times 12 \times 12$.
7. The input of the previous layer is flattened and given to a fully connected layer. The N vector obtained from the input layer is of size 36864. On this N vector a dense layer of size 1024 is used with RELU as activation function and a dropout layer with probability of 0.2. Another dense layer of size 128 is added after that with a dropout layer with probability of 0.1 and RELU activation function. The

final layer is of 1 neuron, which acts as the output layer with sigmoid activation function.

5.6 *Transfer Learning*

There are pre-trained models that are open sourced. These models are trained on billions of images such as the ImageNet database [9]. Transfer learning is used to decrease the computation time to train your own network and to make you make use of these pre-trained networks. We can freeze some layers based on our requirements and only train a subset of those layers on our own dataset. There are two such pre-trained models available as open source, VGG16 and VGG19 [7, 12]. However, here we only discuss the VGG19 model and the assumption is that VGG16 would have performed similar to VGG19 with some minor difference in accuracy.

5.6.1 VGG19

In the VGG19 architecture we added 3 fully connected layers at the bottom with 1024, 512, and 1 layer, respectively and added a dropout layer with probability of 0.3 with 1024 neurons layer. We freeze all the layers of the network and just trained this fully connected layer added to the end in 50 iterations.

6 Experimental Results

We first discuss the NNs results, then we go deeper and show our results for DNNs. After that, we will show an alternative approach, of using raw images from our dataset, and explain the results obtained for the CNN1 and the CNN2 architecture. Finally, we conclude with the results obtained from the VGG19 model, which uses the transfer learning approach. All the results were obtained by using the datasets discussed in Sect. 5. Specifically, NNs and DNNs were trained and tested on the Dredze, the ISH dataset, and the improved dataset. Whereas CNN1, CNN2, and VGG19 were run on the combined dataset.

6.1 *Neural Network Results*

We created a neural network with the architecture discussed in Sect. 5 and ran it for the ISH, Dredze, and Improved Dataset.

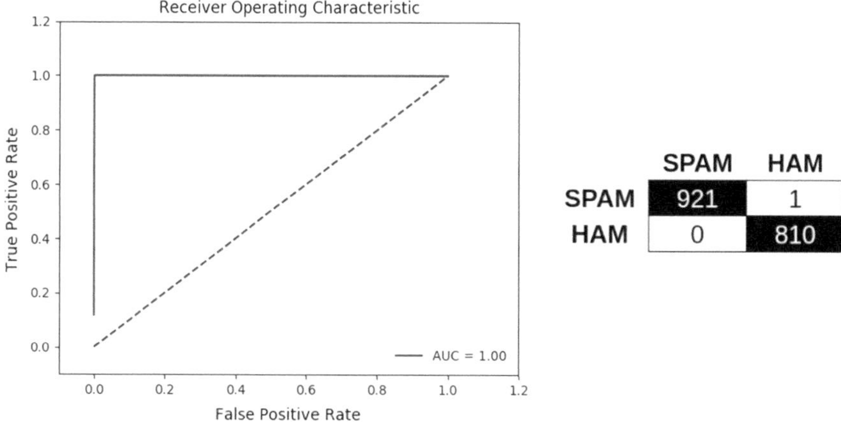

	SPAM	HAM
SPAM	921	1
HAM	0	810

Fig. 9 ROC and confusion matrix for ISH dataset trained on NN

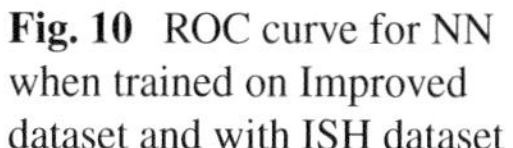

Fig. 10 ROC curve for NN when trained on Improved dataset and with ISH dataset

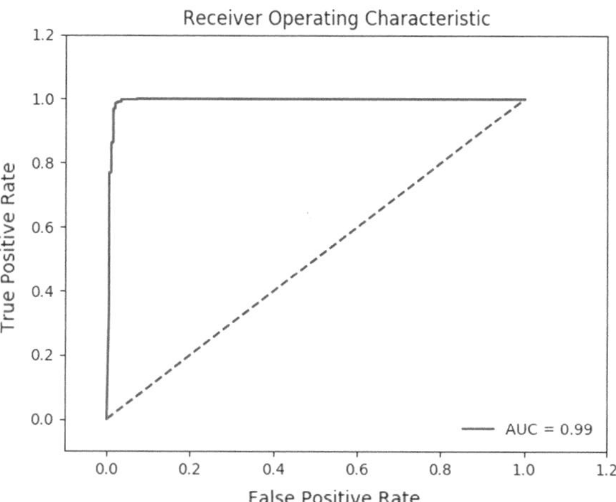

6.1.1 ISH Dataset

The NN was run with 100 mini batch size and for 500 iterations with 10-fold stratified cross validation. The mean accuracy obtained after training the model was 99.07%. Figure 9 shows the AUC achieved by the best classifier over the whole ISH dataset and the confusion matrix obtained with a 0.7 threshold value is shown in Fig. 9. The FP rate obtained was 0.12%.

When the above-trained model was tested on the improved dataset, it gave a very low accuracy of 5.5%, which was expected as the improved dataset was meant to fool such classifiers. So, in the next experiment the ISH dataset and with the improved dataset and then trained on the NN, which gave an accuracy of 98.29% and an area under curve of 0.99. The ROC curve regarding the same is given in Fig. 10.

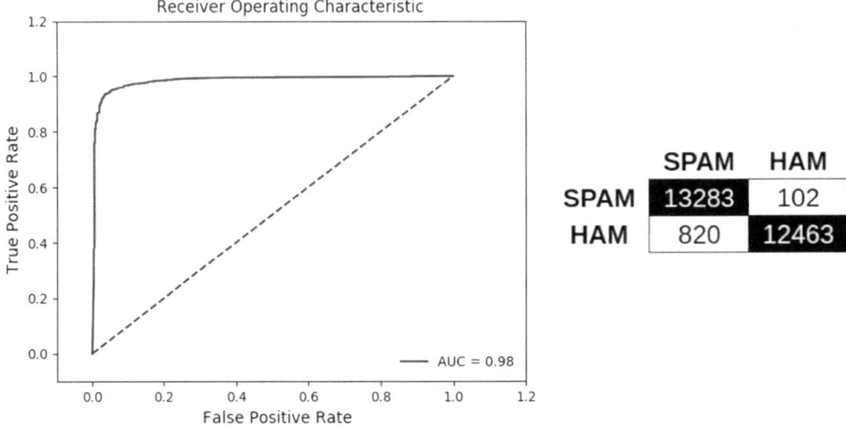

	SPAM	HAM
SPAM	13283	102
HAM	820	12463

Fig. 11 ROC and confusion Matrix for Dredze dataset trained on NN

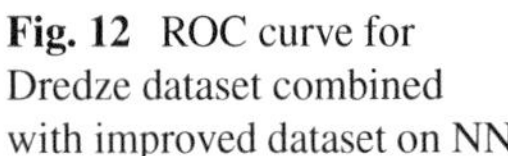
Fig. 12 ROC curve for Dredze dataset combined with improved dataset on NN

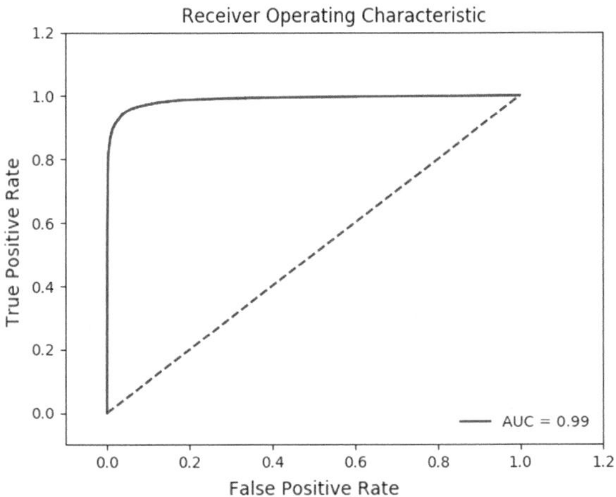

6.1.2 Dredze Dataset

The same NN was run on the Dredze personalized dataset and on the Dredze spam archive combined with 10-fold stratified cross validation. We got mean accuracy of 98.9% and 96.71%, respectively. The ROC curve when the NN was run on the Dredze spam archive and with the personalized dataset is shown in Fig. 11 alongside its confusion matrix. The FPR achieved in this case was 0.8%. When this model was tested on the improved dataset we achieved an accuracy of 4.2%.

When the whole Dredze dataset was combined with the improved dataset and then trained on the NN we achieved an accuracy of 94.42%. Figure 12 below shows the ROC curve obtained for the same experiment.

Datasets	Accuracy
ISH	99.07
ISH + Improved	98.29
Dredze Personalized	98.9
Dredze Personalized + Spam archive	96.71
Dredze Personalized + Spam archive + Improved	94.42

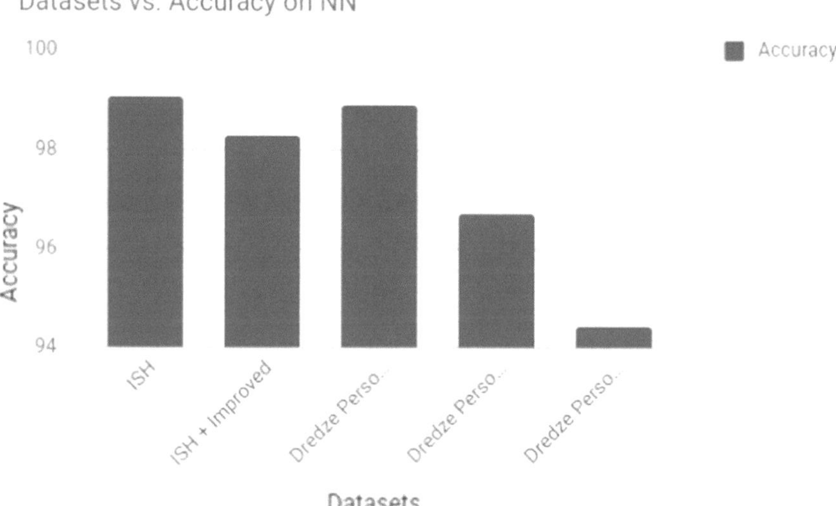

Fig. 13 Summarized Result of NN trained on different datasets

The summary of all the results when trained with different combination of datasets on the NN is given in Fig. 13.

As shown above, NN gave best results for the ISH dataset. It performed worse when trained on the ISH or the Dredze dataset, and then tested on the improved dataset. However, when the two datasets were combined with the improved dataset, the NN was still able to perform better, however, decreased the overall accuracy of the other datasets as it acted as noise for them.

6.2 *Deep Neural Network Results*

The purpose of using a DNN was to compare the results obtained from the NN and see if the introduction of extra hidden layers actually affects the results or not.

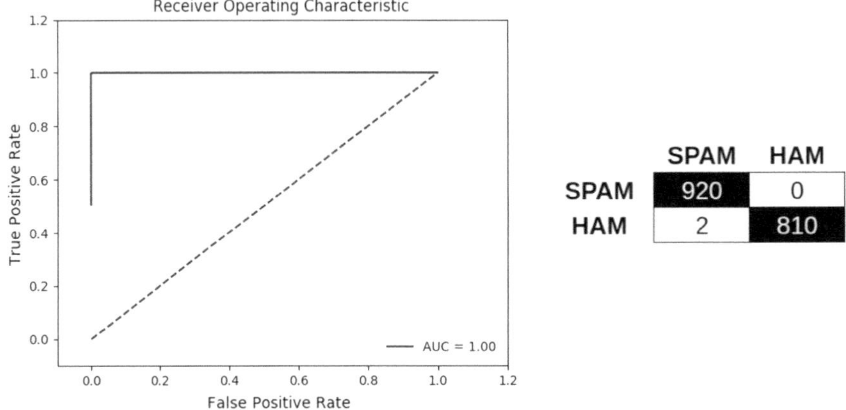

Fig. 14 ROC and confusion matrix for ISH dataset trained on DNN

The experiments are performed on the same datasets and their combination with the improved dataset as done in the NN approach.

6.3 *Image Spam Hunter*

Two experiments were performed on the ISH with the same configuration as discussed in Sect. 5. When the DNN was trained on the ISH dataset alone we achieved a mean accuracy of 98.78%, the ROC curve and the confusion matrix are shown in Fig. 14. The FPR in this case was 0% and when this model was tested on the improved dataset we achieved an accuracy of 5.24%. When the DNN was trained on the improved dataset and with the ISH dataset we achieve a mean accuracy of 98.13% and an ROC curve as shown in Fig. 15.

6.4 *Dredze Dataset*

After training it on the personalized dataset we achieved an accuracy of 98.95%. When the same model was trained on the personalized combined with the SPAM archive we obtained an accuracy of 96.82% and a FPR of 1%. When this model was tested on the improved dataset we achieved an accuracy of 5.9%. The ROC curve and confusion matrix obtained for the latter case is shown in Fig. 16.

When the Dredze dataset was combined with the improved dataset we achieved the following ROC curve of Figure 17 and a mean accuracy of 95.63%.

The summary of all the results when trained with different combination of dataset on the DNN is shown in Fig. 18.

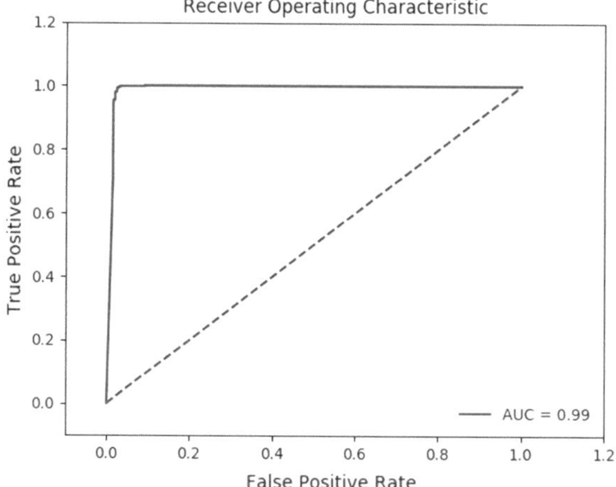

Fig. 15 ROC curve for ISH dataset and with improved dataset and trained on DNN

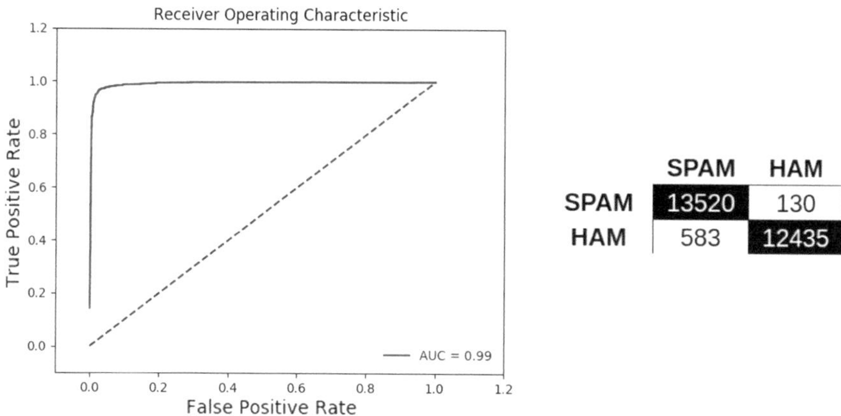

	SPAM	HAM
SPAM	13520	130
HAM	583	12435

Fig. 16 ROC and confusion matrix for Dredze dataset trained on DNN

After comparing the results we obtain from NNs and DNNs, we can conclude that the introduction of an extra layer indeed increases the accuracies for Dredze dataset with more samples but decreases the accuracy of the ISH dataset with comparable lesser samples. It also became more robust with the improved dataset.

6.5 Convolution Neural Networks and Transfer Learning Results

We trained the CNN1 and the CNN2 architectures on 19924 images of SPAM and HAM, and test on 2681 images. The CNN1, CNN2, and VGG19 are trained on a GPU machine with GeForce GTX 960M, Cuda Version 8.0 and compute capability—5.0

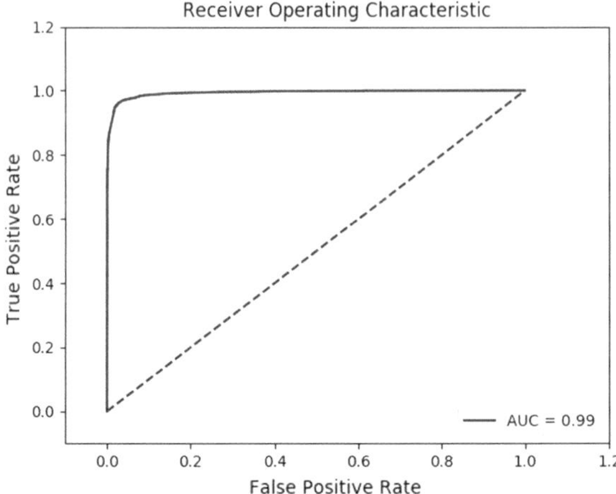

Fig. 17 ROC curve for Dredze dataset combined with improved dataset on DNN

Datasets	Accuracy
ISH	98.78
ISH + Improved	98.13
Dredze Personalized	98.95
Dredze Personalized + Spam archive	96.82
Dredze Personalized + Spam archive + Improved	95.63

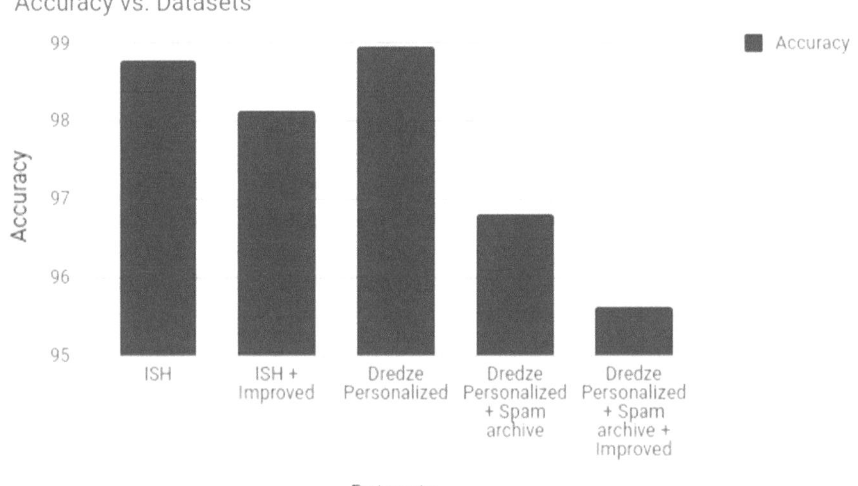

Fig. 18 Summarized Result of DNN trained on different datasets

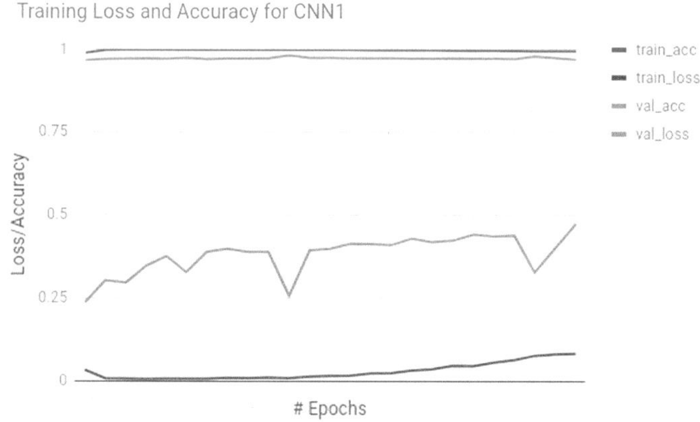

Fig. 19 Accuracy verses Loss for the CNN1 model

Models	Accuracy	#Improved Images detected (out of 1029)
CNN1	97.5	102 (9.91%)
CNN2	97.9	161 (15.64%)
VGG19	98.2	390 (37.9%)

Fig. 20 CNNs and transfer learning accuracies

configuration, and each model took an average of 4 days to train. CNN1 was trained for 30 iterations, CNN2 for 25 iterations, and VGG19 for 50 iterations. For all of these models Adam optimizer was used along with binary cross entropy. Figure 19 shows the training accuracy, training loss, validation accuracy, and validation loss obtained over the 40 epochs the CNN1 model was trained on.

Then, Fig. 20 shows the accuracy results for the three models when trained on the combined dataset, and when tested on the improved dataset.

From the above table it can be concluded that as the CNN2 performed little bit better as compared to CNN1 as it had more layers. Also VGG19 performed better than the other two because it was a pre-trained model on a much larger dataset. Transfer learning hence is preferable in such scenarios when there are lesser time and resources available to train your own model.

7 Conclusion and Future Work

In this work we make use of different real-world image spam datasets and provide strong classifiers based on neural networks, deep neural networks, and convolution neural networks. We compare our results to the ones presented by Aneri et al. [2]. These techniques were able to learn even from the improved dataset provided by them. We performed different experiments with different combinations of datasets which were derived from Dredze (image archive and personalized), the image spam hunter, and the improved dataset. In the CNN experiments, we matched and kept the datasets for SPAM and HAM balanced by randomly sampling HAM files from different categories over the Internet. All the experiments, especially the ones with the convolution neural networks, showed really promising results because the size of the used dataset was comparatively much larger to the previous experiments performed in the past and the diversity of the HAM files.

With the advent of deep learning which make use of big data available across the Internet, even more strong classifiers are feasible. Techniques like generative adversarial networks (GAN's), introduced in year 2012 by Ian Goodfellow [6] can be used for this purpose. Using GAN's which is based on Nash equilibrium, more stronger and robust classifiers can be built. Also, object segmentation using CNN and RNN (Recurrent Neural Networks) [19] can be used to detect the segmented region of SPAMS and remove them from the images by extrapolating a background from ham images. Using such techniques, SPAM images can be converted to ham dynamically. Also, different experiments with different architectures within the CNN can be used to quantify different results. We can also use recursive feature elimination (RFE) and univariate feature selection (UFS), as done in [2] on the image features, when trained to neural networks and deep neural network to decrease the number of features under consideration.

References

1. Annadatha, A., and M. Stamp. 2016. Image spam analysis and detection. *Journal of Computer Virology and Hacking Techniques*.
2. Chavda, Aneri, Katerina Potika, Fabio Di Troia, and Mark Stamp. Support vector machines for image spam analysis. In *Proceedings of the 15th international joint conference on e-Business and telecommunications, ICETE 2018 - Volume 1: DCNET, ICE-B, OPTICS, SIGMAP and WINSYS, Porto, Portugal, July 26-28, 2018*, 597–607.
3. Dhanaraj, S., and V. Karthikeyani. 2013. A study on e-mail image spam filtering techniques. In *2013 international conference on pattern recognition, informatics and mobile engineering (PRIME)*, 49–55. IEEE.
4. Dredze, Mark, Reuven Gevaryahu, and Ari Elias-Bachrach. 2007. Learning fast classifiers for image spam. In *CEAS*, 2007–487.
5. Gao, Yan, Ming Yang, Xiaonan Zhao, Bryan Pardo, Ying Wu, Thrasyvoulos N Pappas, and Alok Choudhary. 2008. Image spam hunter. In *ICASSP 2008 IEEE international conference on acoustics, speech and signal processing*, 1765–1768. IEEE.
6. Goodfellow, Ian, Jean Pouget-Abadie, Mehdi Mirza, Bing Xu, David Warde-Farley, Sherjil Ozair, Aaron Courville, and Yoshua Bengio. 2014. Generative adversarial nets. In *Advances in neural information processing systems*, 2672–2680.
7. He, Kaiming, Xiangyu Zhang, Shaoqing Ren, and Jian Sun. 2015. Delving deep into rectifiers: Surpassing human-level performance on imagenet classification. In *Proceedings of the IEEE international conference on computer vision*, 1026–1034.
8. Jain, U., and S. Dhavale. 2015. *Image spam detection technique based on fuzzy inference system*, Master's Report. Department of Computer Engineering, Defense Institute of Advanced Technology.

9. Krizhevsky, Alex, Ilya Sutskever, and Geoffrey E. Hinton. 2012. Imagenet classification with deep convolutional neural networks. In *Advances in neural information processing systems*, 1097–1105.
10. Kumaresan, T., S. Sanjushree, K. Suhasini, and C. Palanisamy. 2015. Image spam filtering using support vector machine and particle swarm optimization. *International Journal of Computer Applications in Technology* 1: 17–21.
11. Lai, Chih-Chin, and Ming-Chi Tsai. 2004. An empirical performance comparison of machine learning methods for spam e-mail categorization. In *2004 HIS'04 fourth international conference on hybrid intelligent systems*, 44–48. IEEE.
12. Long, Jonathan, Evan Shelhamer, and Trevor Darrell. 2015. Fully convolutional networks for semantic segmentation. In *Proceedings of the IEEE conference on computer vision and pattern recognition*, 3431–3440.
13. Shang, Er-Xin, and Hong-Gang Zhang. 2016. Image spam classification based on convolutional neural network. In *2016 international conference on machine learning and cybernetics (ICMLC)*, vol. 1, 398–403. IEEE.
14. Sharmin, Tazmina, Fabio Di Troia, Katerina Potika, and Mark Stamp. 2020. Convolutional neural networks for image spam detection. *Information Security Journal: A Global Perspective* 29 (3): 103–117.
15. Soranamageswari, M., C. Meena. 2010. Statistical feature extraction for classification of image spam using artificial neural networks. In *Second international conference on machine learning and computing (ICMLC)*, 101–105. IEEE.
16. Srivastava, Nitish, Geoffrey Hinton, Alex Krizhevsky, Ilya Sutskever, and Ruslan Salakhutdinov. 2014. Dropout: A simple way to prevent neural networks from overfitting. *The Journal of Machine Learning Research* 15 (1): 1929–1958.
17. Stamp, Mark. 2011. *Information security: principles and practice*. Wiley.
18. Whitney, Lance. 2009. Report: Spam now 90 percent of all e-mail. *CNET News*, 26.
19. Zheng, Shuai, Sadeep Jayasumana, Bernardino Romera-Paredes, Vibhav Vineet, Zhizhong Su, Dalong Du, Chang Huang, and Philip HS Torr. Conditional random fields as recurrent neural networks. In *Proceedings of the IEEE international conference on computer vision*, 1529–1537.

Universal Adversarial Perturbations and Image Spam Classifiers

Andy Phung and Mark Stamp

Abstract As the name suggests, image spam is spam email that has been embedded in an image. Image spam was developed in an effort to evade text-based filters. Modern deep learning-based classifiers perform well in detecting typical image spam that is seen in the wild. In this chapter, we evaluate numerous adversarial techniques for the purpose of attacking deep learning-based image spam classifiers. Of the techniques tested, we find that universal perturbation performs best. Using universal adversarial perturbations, we propose and analyze a new transformation-based adversarial attack that enables us to create tailored "natural perturbations" in image spam. The resulting spam images benefit from both the presence of concentrated natural features and a universal adversarial perturbation. We show that the proposed technique outperforms existing adversarial attacks in terms of accuracy reduction, computation time per example, and perturbation distance. We apply our technique to create a dataset of adversarial spam images, which can serve as a challenge dataset for future research in image spam detection.

1 Introduction

E-mail, or electronic mail, is one of the most popular forms of communication in the world, with over 3.9 billion active email users [4]. As a side effect of this rapid growth, the number of unwanted bulk email messages—i.e., spam messages—sent with commercial or malicious intent has also grown. According to [4], 60 billion spam emails will be sent each day for the next 3 years.

While text-based spam filtering systems are in use by most, if not all, e-mail clients [8], spammers can embed messages in attached images to evade such

A. Phung
Independence High School, San Jose, CA, USA
e-mail: phungandy0080@students.esuhsd.org

M. Stamp (✉)
San Jose State University, San Jose, CA, USA
e-mail: mark.stamp@sjsu.edu

M. Stamp et al. (eds.), *Malware Analysis Using Artificial Intelligence and Deep Learning*, https://doi.org/10.1007/978-3-030-62582-5_25

systems—such messages are known as image spam. Image spam detectors based on optical character recognition (OCR) have been deployed to combat such e-mail. As a countermeasure, spammers can modify images so as to disrupt OCR-based techniques [9].

In recent years, deep learning models, such as multi-layer perceptrons and convolutional neural networks, have been successfully applied to the image spam problem [1, 2, 6, 9, 12, 24, 25]. Note that these techniques do not rely on OCR, but instead detect image spam directly, based on characteristics of the images.

With the recent development of perturbation methods, the possibility exists for spammers to utilize adversarial techniques to defeat image-based machine learning detectors [26]. To date, we are not aware of perturbation techniques having been used by image spammers, but it is highly likely that this will occur in the near future.

The main contributions of our research are the following.

- We show that the universal perturbation adversarial attack is best suited for the task of bypassing deep learning-based image spam filters.
- We propose a new image transformation-based attack that utilizes the maximization of layer activations to produce spam images containing universal perturbations. This technique focuses perturbations in the most salient regions, as well as concentrating natural features in the remaining regions.
- We compare our proposed adversarial technique to existing attacks and find that our approach outperforms all others in terms of accuracy reduction, computation time per example, and perturbation magnitude.
- We generate a large dataset containing both non-spam and adversarial spam images using our proposed attack. The authors will make this dataset available to researchers.

The remainder of this chapter is organized as follows. In Sect. 2, we provide an overview of relevant research and related work. In Sect. 3, we evaluate adversarial attacks in the context of image spam, and in Sect. 4, we present our proposed attack. Finally, Sect. 5 concludes this chapter, where we have included suggestions for future work.

2 Background

2.1 Image Spam Filtering

The initial defenses against image spam relied on optical character recognition (OCR). In such OCR-based systems, text is extracted from an image, at which point a traditional text-based spam filter can be used [3]. As a reaction to OCR-based techniques, spammers introduced images with slight modifications, such as overlaying a light background of random artifacts on images, which are sufficient to render OCR ineffective. The rise of learning algorithms, however, has enabled the creation of image spam filtering systems based directly on image features.

In 2008, a filtering system using a global image feature-based probabilistic boosting tree was proposed and achieved 89.44% detection rate with a false positive rate of 0.86% [9]. Two years later, an artificial neural network for image classification was proposed [25]. These latter authors used were able to classify image spam with 92.82% accuracy based on color histograms, and 89.39% accuracy based on image composition extraction.

The two image spam detection methods presented in [2] rely on the principal component analysis (PCA) and support vector machines (SVM). In addition, the authors of [2] introduce a new dataset that their methods cannot reliably detect. Two years later, the authors of [6] improved on the results in [2] by training a linear SVM on 38 image features, achieving 98%, accuracy in the best case. The authors also introduce a challenge dataset that is even more challenging than the analogous dataset presented in [2].

The recent rise of deep learning, a subfield of machine learning, coupled with advances in computational speed has enabled the creation of filtering systems capable of considering not only image features but also entire images at once. In particular, convolutional neural networks (CNNs) are well suited to computer vision tasks due to their powerful feature extraction capabilities.

In recent years, CNNs have been applied to the task of image spam detection. For example, in [1], a CNN is trained on an augmented dataset of spam images, achieving 6% improvement in accuracy, as compared to previous work. Similarly, the authors of [12] consider a CNN, which achieved 91.7% accuracy. In [24], a CNN-based system is proposed, which achieves an accuracy of 99% on a real-world image spam dataset, 83% accuracy on the challenge dataset in [2] (an improvement over previous works), and 68% on the challenge dataset in [6].

From the challenge datasets introduced in [2, 6], we see that the accuracy of machine learning-based filtering systems can be reduced significantly with appropriate modifications to spam images. In this research, we show that the accuracy of such systems can be reduced far more by using the adversarial learning the approach that we present below.

2.2 Adversarial Learning

The authors of [26] found that by applying an imperceptible filter to an image, a given neural network's prediction can be arbitrarily changed. This filter can be generated from the optimization problem

$$\begin{aligned}&\textbf{minimize } \|r\|_2\\&\textbf{subject to } f(x+r)=l \text{ and } x+r\in[0,1]^m,\end{aligned}$$

where f is the classifier, r is the minimizer, l is the target label, and m is the dimension of the image. The resulting modified images are said to be *adversarial examples*, and the attack presented in [26] is known as the *L-BFGS Attack*. These adversarial examples generalize well to different network architectures and networks.

More recently, many advances have been made in both adversarial example generation and detection. For example, in [28] a taxonomy is proposed for generation and detection methods, as well as a threat model. Based on this threat model, the task of attacking neural network-based image spam detectors requires an attack that is false negative (i.e., generative of positive samples misclassified as negative) and black box (i.e., the attacker does not have access to the trained model). Attacks on image spam classifiers must satisfy these two criteria.

After the introduction of the L-BFGS Attack, the authors of [10] built on their work in [26] by introducing the *Fast Gradient Sign Method* (FGSM). This method uses the gradient of the loss function with respect to a given input image to efficiently create a new image that maximizes the loss, via backpropagation. This can be summarized with the expression

$$\text{adv}_x = x + \varepsilon \,\text{sign}\big(\nabla_x J(\theta, x, y)\big),$$

where θ is the parameters of the model, x is the input image, y is the target label, and J is the cost function used to train the model. These authors also introduce the notion that adversarial examples result from linear behavior in high-dimensional spaces.

The authors of [5] introduce *C&W's Attack*, a method designed to combat *defensive distillation*, which consists of training a pair of models such that there is a low probability of successively attacking both models. C&W's Attack is a non-box constrained variant of the L-BFGS Attack that is more easily optimized and effective against both distilled and undistilled networks. They formulate adversarial example generation as the optimization problem

$$\begin{aligned}&\textbf{minimize } D(x, x+\delta) + c \cdot f(x+\delta)\\ &\textbf{such that } x+\delta \in [0,1]^n,\end{aligned}$$

where x is the image, D is one of the three distance metrics described below, and c is a suitably chosen constraint (the authors choose c with binary search). The authors also utilize three distance metrics for measuring perturbation: L_0 (the number of altered pixels), L_2 (the Euclidean distance), and L_∞ (the maximum change to any of the coordinates), and introduced three subvariants of their attack that aim to minimize each of these distance metrics.

It is important to note that the previously mentioned attacks require knowledge of the classifier's gradient and, as such, cannot be directly deployed in a black-box attack. In [19], the authors propose using a surrogate model for adversarial example generation to enable the transferability of adversarial examples to attack black-box models. Differing from gradient-based methods, the authors of [7] introduced a method, *Zeroth-Order Optimization* (ZOO), which is inspired by the work in [5].

The ZOO technique employs gradient estimation, with the most significant downside being that it is computationally expensive.

The paper [14] introduces the *DeepFool* attack, which aims to find the minimum distance from the original input images to the decision boundary for adversarial examples. They found that the minimal perturbation needed for an affine classifier is the distance to the separating affine hyperplane, which is expressed (for differentiable binary classifiers) as

$$\begin{aligned} &\mathbf{argmin}_{\eta_i} \|\eta_i\|_2 \\ &\textbf{such that } f(x_i) + \nabla f(x_i)^T \eta_i = 0, \end{aligned}$$

where i denotes the iteration, η is the perturbation, and f is the classifier. In comparison to FGSM, DeepFool minimizes the magnitude of the perturbation, instead of the number of selected features. This would appear to be ideal for spammers, since it would tend to minimize the effect on an image.

The *universal perturbation* attack presented in [13] is also suited to the task at hand. We believe that universal adversarial examples are most likely to be deployed by spammers against black-box models due to their simplicity and their transferability across architectures. Generating universal perturbations is an iterative process, as the goal is to find a vector v that satisfies

$$\|v\|_p \le \xi \text{ and } \mathbb{P}_{x\sim\mu}(\hat{k}(x+v) \ne \hat{k}(x)) \ge 1-\delta,$$

where μ is a distribution of images, $\hat{k}$ is a classification function that outputs for each image x and a label $\hat{k}(x)$. The results in [13] show that universal perturbations are misclassified with high probability, suggesting that the existence of such perturbations are correlated to certain regions of the decision boundary of a deep neural network.

Finally, the authors of [11] propose input restoration with a preprocessing network to defend against adversarial attacks. The authors' defense improved the classification precision of a CNN from 10.2% to 81.8%, on average. These results outperform existing input transformation-based defenses.

3 Evaluating Adversarial Attacks

3.1 Experimental Design

The two multi-layer perceptron and convolutional neural network architectures presented in [24] are each trained on both of the datasets presented in [9], which henceforth will be referred to as the *ISH Dataset*, and the dataset presented in [6], which henceforth will be referred to as the *MD Dataset* (modified Dredze). We use TensorFlow [27] to train our models—both architectures have been trained as they

Table 1 Image spam datasets

Name	Spam images	Non-spam images
ISH dataset	928	830
MD dataset	810	784
Total	1738	1613

were presented in their respective articles on each of the datasets. NumPy [17] and OpenCV [18] are used for numerical operations and image processing tasks, respectively. All computations are performed on a laptop with 8GB ram, using Google Colaboratory's Pro GPU.

The ISH Dataset contains 928 spam images and 830 non-spam images, while the MD Dataset contains 810 spam images and 784 non-spam images; all images in both datasets are in *jpg* format. These datasets are summarized in Table 1.

Dataset preprocessing for the networks presented in [24] consists of downsizing each of the images such that their dimensions are 32×3, applying zero-parameter Canny edge detection [20] to a copy of the downsized image, and concatenating the downsized image with the copy that had Canny edge detection applied. This process results in 64 images, which are used to train the two neural networks, one for the ISH dataset, and one for the MD dataset. The four resulting models achieved accuracies within roughly 7% of the accuracies reported in [24].

To enable the generation of adversarial examples, four larger models with an input size of 400x400 are also trained on the original datasets. The first few layers of each of these models are simply used to downscale input images such that the original architectures can be used after downscaling. These four alternative models achieve accuracy roughly equivalent to the original models. The four adversarial attacks (FGSM, C&W's Attack, DeepFool, and universal perturbation) utilize these four alternative models to generate adversarial examples that can then be formatted as the original datasets to attack the original four models. This procedure attempts to exploit the transferability of adversarial examples to similar architectures.

The IBM Adversarial Robustness Toolbox (ART) [16] is used to implement C&W's Attack, DeepFool, and universal perturbations, while FGSM was implemented independently from scratch. An attempt was made to optimize the parameters of each technique—the resulting parameters are summarized in Table 2. Note that for the universal perturbation attack, FGSM was used as the base attack, as the IBM ART allows any adversarial attack to be used for computing universal perturbations.

The metrics used to evaluate each of the four attacks are the average accuracy, area under the curve (AUC) of the receiver operating characteristic (ROC) curve, average L_2 perturbation measurement (Euclidean distance), and average computation time per example for each of the four models. Scikit-learn [21] was used to generate the ROC curves for each attack.

We use 251 data points for accuracy and L_2 distances collected for the FGSM, DeepFool, and universal perturbation experiments, in accordance with the full size

Table 2 Attack parameters

Attack	Description	Value
FGSM	Perturbation magnitude	0.1
C&W's attack	Target confidence	0
	Learning rate	0.001
	Binary search steps	20
	Maximum iterations	250
	Initial trade-off	100
	Batch size	1
DeepFool	Max iterations	500
	Overshoot parameter	10^{-6}
	Class gradients	10
	Batch size	1
Universal perturbation	target accuracy	0%
	Max iterations	250
	Step size	64
	Norm	∞

of the test dataset, which contains 251 examples for generating adversarial examples. However, only 28 data points were collected from the C&W's Attack experiment due to a large amount of time required to generate each data point (roughly five minutes per data point). The technique that will be used as the basis of our proposed attack will be selected based on the performance of each attack, as presented in the next section.

3.2 *Analysis*

The mean accuracy, computation time per example, and L_2 distance were recorded for each of the four models attacked by each of the attack methods. This data was compiled into the tables discussed in this section.

From Table 3, we see that for FGSM, the accuracy of the attacked models is shown to vary inconsistently while Fig. 1 shows that the distribution of the L_2 distances of the generated adversarial examples skew right. Based on these results and corresponding density plots of the accuracy and L_2 distance distributions, the FGSM attack can be ruled out as a candidate due to poor accuracy.

The mean L_2 (Euclidean) distances of the adversarial examples are given in Table 4. The distribution of distances appears to be roughly equivalent across all attacks.

DeepFool can also be ruled as a candidate, as the attack has been seen to be only marginally better than the FGSM attack in terms of performance while also having a

Table 3 Mean accuracy per adversarial example

Model	FGSM (%)	C&W's attack (%)	DeepFool (%)	Universal perturbation (%)
MLP (ISH)	95.2	89.2	98.8	98.7
CNN (ISH)	36.2	49.6	61.5	49.9
MLP (MD)	69.7	75.6	93.5	94.3
CNN (MD)	82.8	77.2	14.5	8.4

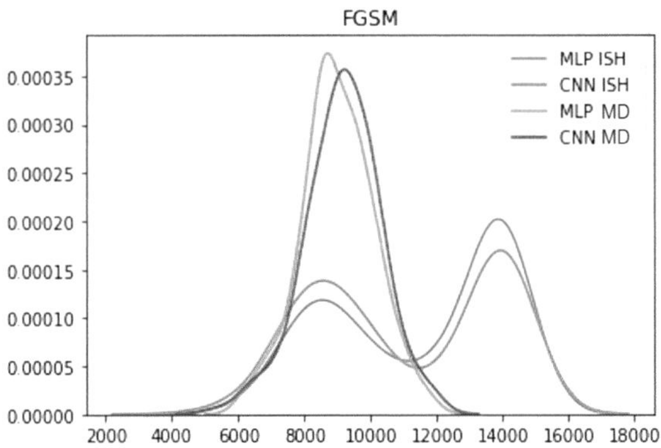

Fig. 1 Density plot of L_2 (Euclidean) distances (Fast Gradient Sign Method)

Table 4 Mean L_2 (Euclidean) distance of adversarial examples from original images

Model	FGSM attack	C&W's	DeepFool perturbation	Universal
MLP (ISH)	11537.55	10321.77	11513.26	11483.72
CNN (ISH)	11108.44	10924.14	11216.19	11416.58
MLP (MD)	8998.71	9185.04	9566.02	9490.56
CNN (MD)	9144.49	9009.91	9128.99	9381.15

Table 5 Mean computation time per adversarial example

Model attack	FGSM	C&W's	DeepFool perturbation	Universal
MLP (ISH)	0.180	269.65	19.90	4.37
CNN (ISH)	0.038	251.01	4.75	2.87
MLP (MD)	0.164	270.58	36.30	3.71
CNN (MD)	0.165	244.47	1.48	5.23

significantly higher average computation time per adversarial example. This can be observed in Table 5, where the computation time per example varies greatly.

In contrast, C&W's Attack shows consistent performance in all three metrics at the cost of high computation time (roughly five minutes per adversarial example). The

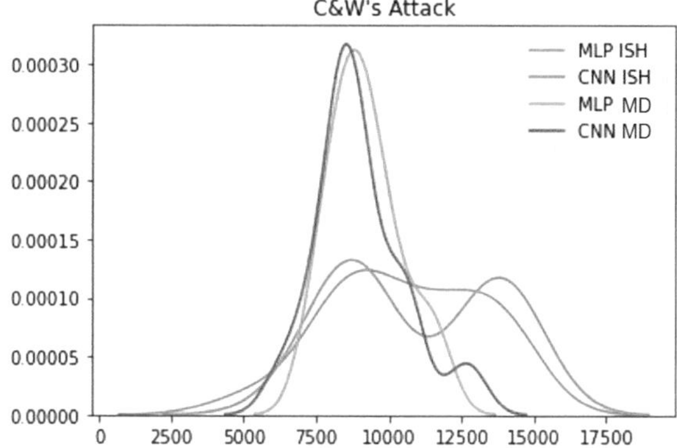

Fig. 2 Density plot of L_2 (Euclidean) distances (C&W's Attack)

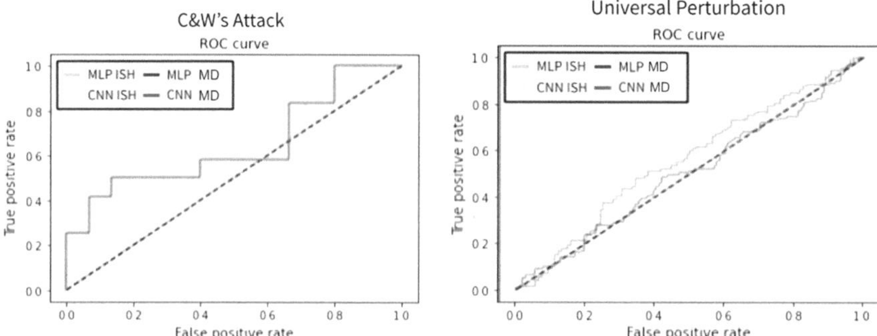

Fig. 3 ROC curves of C&W's Attack and the universal perturbation when used to attack the four classifiers

consistency of this attack is ideal from a spammer's perspective, though the trade-off is a relatively high computation time. In addition, the left skew of this attack with respect to L_2 distance, as presented in Fig. 2, indicates that the perturbation made to spam images is much lower in comparison to the other attacks.

The universal perturbation attack is inconsistent in terms of accuracy, as shown in Table 3, where the mean accuracy across the four models is clearly shown to fluctuate wildly, but this is simply due to the fact that only one perturbation (albeit with varying success across architectures) is applied to all spam images, which is highly advantageous for spammers. The generation and application of this perturbation to an image takes roughly four seconds, which would result in greater performance in a real-world spam setting in comparison to C&W's Attack.

To further compare C&W's Attack and the universal perturbation attack, the ROC curves of the two are presented in Fig. 3. These ROC curves can be used to quantify the diagnostic ability of the models attacked by each method.

Table 6 Mann–Whitney U test results comparing C&W's attack and the universal perturbation attack

Model	Accuracy p-value	L_2 distance p-value
MLP ISH	0.000 (H0 is rejected)	0.034 (H0 is rejected)
CNN ISH	0.384 (H0 is not rejected)	0.098 (H0 is not rejected)
MLP MD	0.000 (H0 is rejected)	0.057 (H0 is not rejected)
CNN MD	0.000 (H0 is rejected)	0.016 (H0 is rejected)

The ROC curve for C&W's attack is much noisier due to being generated from only 28 data points. Taking this into consideration, it can be inferred that both C&W's Attack and the universal perturbation attack are able to reduce the areas under the ROC curve (AUC) of the attacked models to values close to 0.5. This suggests that both attacks are able to reduce the class separation capacity of attacked image spam classifiers to essentially random.

To analyze the differences in distribution of the accuracy and L_2 distance data collected from the trials conducted on C&W's Attack and the universal perturbation attack, the Mann–Whitney U test was utilized via its implementation in SciPy [22]. The Mann–Whitney U test compares two populations—in this case, the accuracy and L_2 distance data from both attacks for each attacked model. The null hypothesis (H0) for the test is that the probability is 50% that a randomly drawn value from the first population will exceed a value from the second population. The result of each test is a Mann–Whitney U Statistic (not relevant in our case) and a p-value. We use the p-value to determine whether the difference between the data is statistically significant, where the standard threshold is $p = 0.05$. The results of these tests are given in Table 6.

The results in Table 6 imply that the performance of these two attacks (C&W's Attack and the universal perturbation attack) are nearly identical when attacking a CNN trained on the ISH dataset, as evidenced in the second row, where the null hypothesis is not rejected. However, the L_2 distance measurement for spam images that have had the universal perturbation applied should remain constant relative to the original spam image. Therefore, the results of these tests suggest that the universal perturbation attack is able to achieve similar performance to C&W's Attack, in terms of perturbation magnitude, with a much lower computation time per example in comparison to C&W's Attack.

Given the above evidence, the universal perturbation attack is the best choice for image spam, as it is unrivaled in terms of potential performance in a real-world setting. The key advantages of the universal perturbation attack include that it generates a single perturbation to be applied to all spam images and its relatively fast computation time per adversarial example. Therefore, universal perturbation will be used as a basis for our image transformation technique, as discussed and analyzed in the remainder of this paper. A sample adversarial spam image generated with the universal perturbation attack is presented in Fig. 4.

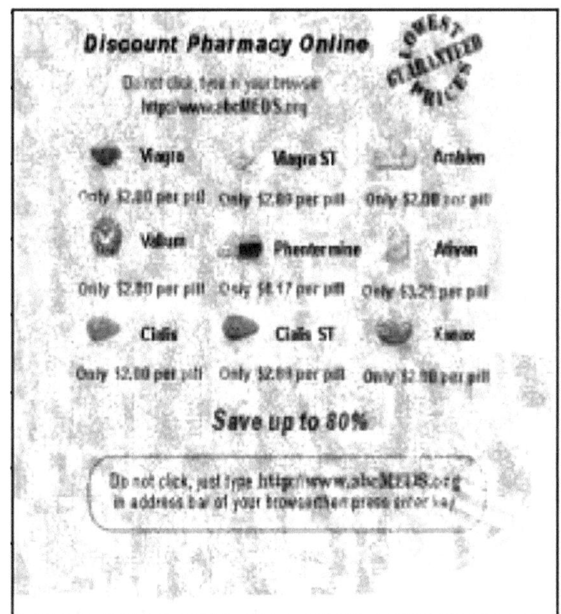

Fig. 4 Adversarial spam image generated with the universal perturbation attack

4 Inceptionism-Augmented Universal Perturbations

4.1 Procedure

Based on the results and discussion above, a transformation that is applied to spam images prior to generating adversarial examples, since perturbations cannot be transformed after application, should meet the following conditions.

- Lower the misclassification rate
- Preserve adversarial effects after image resizing
- Make non-spam features more prominent while retaining legibility

Given the above criteria, a reasonable approach would be to maximize the presence of "natural features" in a given spam image. That is, the features characteristic of non-spam images learned by classifiers should be maximized while retaining legibility. To accomplish this, the procedure for maximizing the activation of a given output neuron (in this case, the non-spam output neuron), as introduced in [15], dubbed "DeepDream," can be used to increase the number of natural features in all images from the non-spam subsets of the ISH and MD datasets. This is accomplished by maximizing the activations of the convolutional layer with the greatest number of parameters and output layer in the corresponding CNNs. The resulting two sets of images that have had DeepDream applied ("dreamified" images) are then grouped into batches of four images. The weighted average of the four images in each batch can then be taken to produce two processed non-spam datasets of images with high concentrations of natural features, as batches of greater than four images, may result in high noise. Each of the images in the resulting two non-spam datasets are henceforth referred to as *natural perturbations*.

To preserve the adversarial effect that the universal perturbation introduces, the Gradient-weighted Class Activation Mapping (Grad-CAM) technique introduced in [23] is used to generate a class activation map for each spam image in each dataset. The inverse of each such map is used with a natural perturbation generated from the same dataset to remove the regions of the natural perturbation where the class activation map is highest. By superimposing the resulting natural perturbations onto the corresponding spam images, the regions where the universal perturbation is most effective are left intact while the regions of the spam images affected by the natural perturbations benefit by being more non-spam like. The presence of natural features in the resulting spam images should also result in robustness against resizing prior to inference by a deep learning-based image spam detection model, as the natural features should be still be somewhat preserved even after being shrunken.

The universal perturbation is then applied to each of the resulting spam images. The result is that we potentially reduce a deep learning-based image spam detector's accuracy due to the presence of a natural perturbation and a universal adversarial perturbation and retain some sort of adversarial effect in the case of resizing. This procedure also allows for the retention of legible text within spam images.

4.2 Implementation

To generate our two sets of "dreamified" images, the CNN architecture presented in [24] is trained on both the ISH and MD datasets, with inverted labels to allow for the maximization of the activations of the neurons corresponding to non-spam images, as the activations for spam images would be maximized if the labels weren't inverted. These two models are trained with the TensorFlow Keras API, with the hyperparameters given in [24]. For each of the models, the convolutional layer with the highest number of parameters and the output layer were chosen as the layers in which the activation should be maximized via gradient ascent, as the aforementioned convolutional layer is responsible for recognizing the most complex natural features. Each of the images from the non-spam subsets of the ISH and MD datasets were used for inference on the two CNN models. The CNN models use the losses of the chosen layers to iteratively update the non-spam images with gradient ascent so that the number of non-spam features is maximized. Each non-spam image is updated for 64 iterations with an update size of 0.001. The resulting "dreamified" images are then grouped into batches of 4 and blended via evenly distributed weighted addition to produce a total of 392 grayscale images, each of size $400 \times 400 \times 1$. These 392 grayscale images are evenly split between the ISH dataset and MD datasets.

To utilize GradCAM, the CNN architecture presented in [24] is trained on both the ISH and MD datasets with normal labels. For each image from the spam subsets

Table 7 Mean accuracy of each model with spam images created by the proposed method

Images	MLP (ISH) (%)	CNN (ISH) (%)	MLP (MD) (%)	CNN (MD) (%)
Modified spam images	80.1	98.8	98.4	75.3
Modified spam images with universal perturbations	72.2	50.4	78.7	23.7

of the ISH and MD datasets, GradCAM is used to generate a corresponding class activation map based on the activations of the last convolutional layer in each of the two models. This is accomplished by computing the gradient of the top predicted class with respect to the output feature map of the last convolutional layer, using the mean intensity of the gradient over specific feature map channels. OpenCV [18] is then used to upscale each of the class activation maps to 400×400, convert them to binary format and invert the result to allow the class activation maps to be applied to the natural perturbations such that only the areas with the highest activation will contain the natural perturbations. The bitwise AND of each processed class activation map and a randomly selected natural perturbation can then be used to generate two sets of processed natural perturbations, which are superimposed on the corresponding spam images from each of the two spam subsets. This procedure results in two subsets of spam images with natural perturbations.

Lastly, the universal perturbation is generated and applied to all images within the two spam image subsets that have had natural perturbations applied. For this operation, we use the IBM Adversarial Robustness Toolbox [16]. The hyperparameters for the Universal Perturbation attack remain the same as those given in Table 2, above.

4.3 Performance Evaluation

The mean accuracy, computation time per example, and L_2 distance were recorded for each of the four models attacked using spam images with modified universal perturbations. This is analogous to what was done during the attack selection process. This data has been compiled into the tables discussed in this section.

As can be seen from the results in Table 7, the proposed method for generating adversarial spam images is capable of lowering a learning-based model's accuracy to 23.7%. In addition, on average, our proposed technique is much more effective while being evenly distributed in terms of accuracy on similar learning-based models.

From Table 8, we see that in contrast to C&W's Attack, which on average takes 258.93 s per example, the time necessary to generate adversarial spam images with natural perturbations is significantly lower and comparable to that of the original Universal Perturbation attack. This is another advantage of our proposed attack.

Table 8 Mean computation time per adversarial spam image (in seconds)

MLP (ISH)	CNN (ISH)	MLP (MD)	CNN (MD)
5.46	5.15	5.87	4.80

Table 9 Mean L_2 (Euclidean) distance of modified adversarial spam images from original images

MLP (ISH)	CNN (ISH)	MLP (MD)	CNN (MD)
11392.02	11309.40	9440.69	9628.61

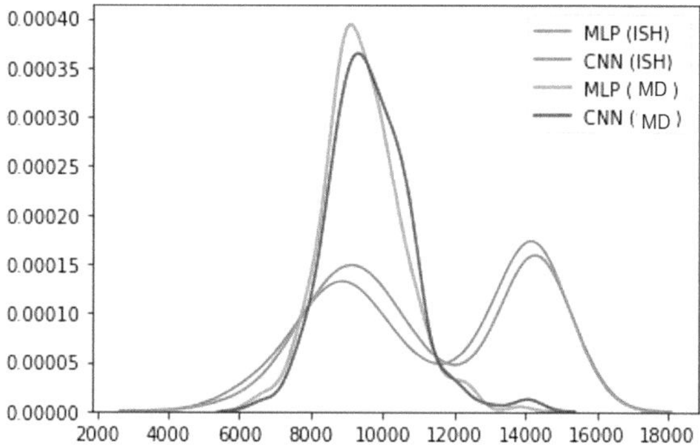

Fig. 5 Density plot of L_2 (Euclidean) distances of the modified adversarial spam images from the original images

The mean L_2 distances and the distribution of the L_2 distances of the modified adversarial spam images are given in Table 9. From Fig. 5, we see that the distributions of these distances are, on average, not skewed, indicating that the natural perturbations have had a slightly negative effect on the spam image L_2 distances, as the distributions for the original Universal Perturbation attack were skewed to the left.

The ROC curves of the models attacked by the proposed method, which appear in Fig. 6, are slightly worse in comparison to that of the original Universal Perturbation attack, suggesting once more that the attack is capable of reducing the class separation capacity of attacked image spam classifiers to essentially random.

4.4 Proposed Dataset Analysis

Figure 7 contains an example of modified adversarial spam images. From this image, we observe that the proposed method was able to effectively utilize class activation maps generated with GradCAM to selectively apply a random natural perturbation to the spam image. As discussed in the previous section, this decreases classification accuracy even prior to the application of a universal perturbation.

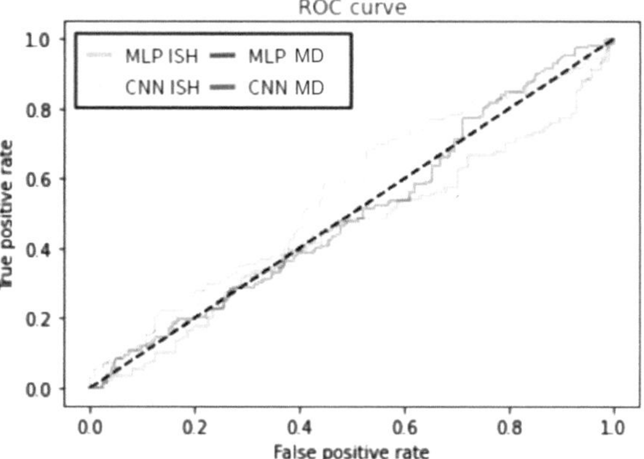

Fig. 6 ROC curves of each of the four models attacked by the modified spam images generated with the proposed method

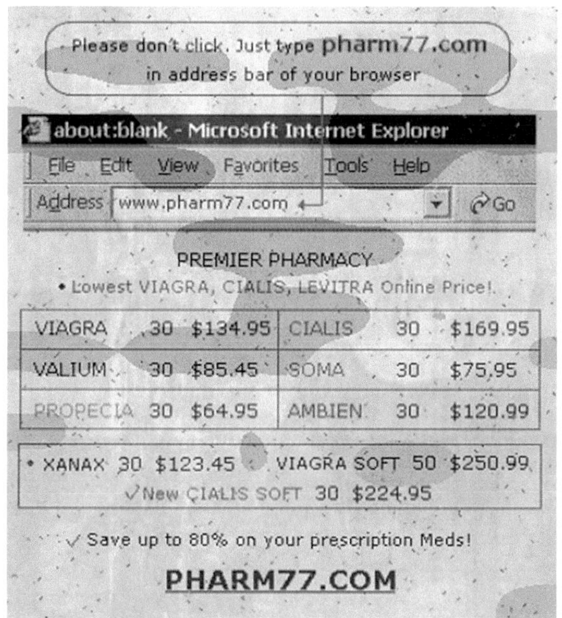

Fig. 7 Example of modified adversarial spam image generated with the proposed method

To fully evaluate the effect of the modified adversarial spam images from the two modified datasets, two sets of class activation maps are generated from the spam subsets of the two datasets using GradCAM and the corresponding CNN models. These activation maps are then averaged to obtain two heatmaps from the class activation maps, as shown in Figs. 10 and 11. For comparison, the same process was applied to the original datasets to obtain Figs. 8 and 9.

As can be seen in Figs. 8 and 9, the activation regions for spam images from the original ISH and MD datasets are skewed towards the top and bottom. The narrow shape of these regions represent the regions in spam images that generate the highest

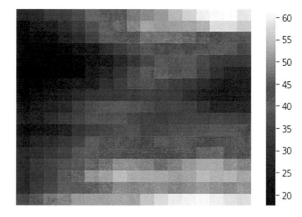

Fig. 8 ISH spam data

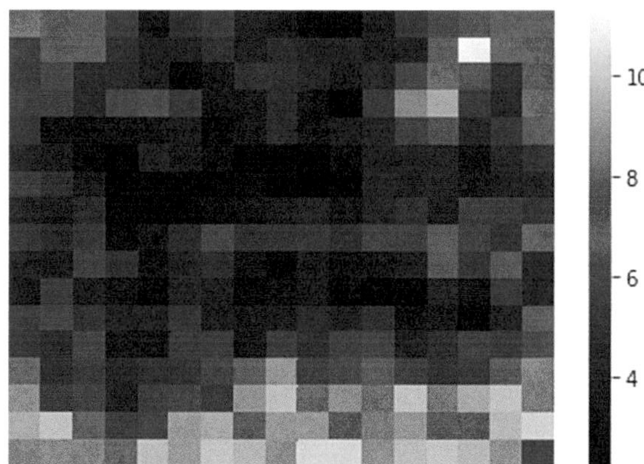

Fig. 9 MD spam data

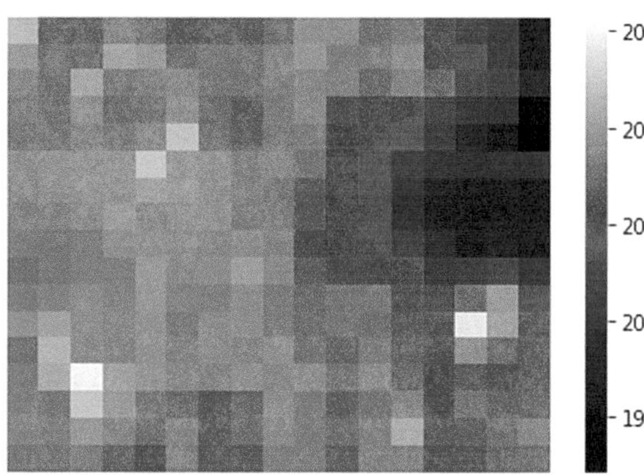

Fig. 10 Modified ISH spam data

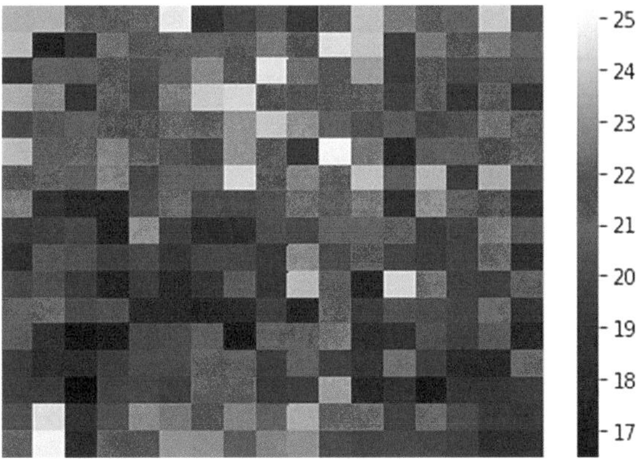

Fig. 11 Modified MD spam data

activations in the neurons of the deep learning-based classifier. The central region of the average class activation map for spam images from the MD dataset is much darker in comparison to that of spam images from the ISH dataset due to the superimposition of natural images directly onto spam features, as described in [6].

In contrast, Figs. 10 and 11 indicate that the introduction of natural and universal adversarial perturbations are able to more evenly distribute the activation regions. This result shows that the spam images from the modified datasets are much closer—in terms of natural features—to non-spam images. This also suggests that the proposed method outperforms the procedure used to generate the original MD dataset as outlined in [6].

5 Conclusion and Future Work

Modern deep learning-based image spam classifiers can accurately classify image spam that has appeared to date in the wild. However, spammers are constantly creating new countermeasures to defeat anti-spam technology. Consequently, the eventual use of adversarial examples to combat deep learning-based image spam filters is inevitable.

In this chapter, four adversarial attacks were selected based on specific restrictions and constraints of the image spam problem. These adversarial attacks were evaluated on the CNN and MLP architectures introduced in [24]. For training data, we used the dataset presented in [9] and [6]. The Fast Gradient Sign Method (FGSM) attack, C&W's Attack, DeepFool, and the Universal Perturbation attack were all evaluated based on mean accuracy reduction, mean computation time per adversarial spam image, mean L_2 distance from the original spam images, and ROC curves of the attacked classifiers. Through further statistical analysis, the Universal Perturbation was chosen as a base for our proposed image transformation attack, due to its versatility and overall high performance in terms of accuracy reduction and computation time.

To maximize the number and intensity of natural features in an attack, the approach introduced in [15] for maximizing activations of certain layers in a deep neural network was used. This technique serves to generate sets of "natural perturbations" from the non-spam subsets of the image spam datasets. These natural perturbations were then modified via the class activation maps of all spam images in both datasets. The class activations were generated using GradCAM from the two convolutional neural networks trained on the ISH and MD datasets. These activation maps allow the regions in spam images recognized to contribute most to the spam classification to benefit from a universal adversarial perturbation.

Our technique resulted in comparable—if not greater—accuracy reduction as compared to C&W's Attack. In addition, our approach is computation much more efficient than C&W's Attack. Furthermore, the nature of our attack implies that the only potential computational bottleneck is generating the modified natural perturbations. This aspect of the attack would not be an issue in practice, unless a spammer generates vast numbers (i.e., in the millions) of modified adversarial spam images.

A dataset of modified adversarial spam images has been generated by the authors by applying the proposed attack to the spam subsets of the ISH and MD datasets. This dataset will be made freely available to researchers.

Future work will include evaluating the ability of adversarial attack defense methods. We will consider defensive distillation against adversarial spam images generated with our proposed attack. The goal of this research will be to develop defenses specifically designed for natural perturbation-augmented adversarial spam images. For example, the subtraction of predicted adversarial perturbations is one path that we intend to pursue.

References

1. Aiwan, Fan, and Yang Zhaofeng. 2018. Image spam filtering using convolutional neural networks. *Personal and Ubiquitous Computing* 22 (5–6): 1029–1037.
2. Annadatha, Annapurna, and Mark Stamp. 2016. Image spam analysis and detection. *Journal of Computer Virology and Hacking Techniques* 14 (1): 39–52.
3. Apache. SpamAssassin. https://spamassassin.apache.org/.
4. Campaign Monitor. 2019. Email Usage Statistics in 2019. https://www.campaignmonitor.com/blog/email-marketing/2019/07/email-usage-statistics-in-2019/.
5. Carlini, Nicholas, and David Wagner. 2017. Towards evaluating the robustness of neural networks. In *2017 IEEE symposium on security and privacy (sp)*, 39–57, IEEE.
6. Chavda, Aneri, Katerina Potika, Fabio Di Troia, and Mark Stamp. 2018. Support vector machines for image spam analysis. In *Proceedings of the 15th international joint conference on e-business and telecommunications*.
7. Chen, Pin-Yu, Huan Zhang, Yash Sharma, Jinfeng Yi, and Cho-Jui Hsieh. 2017. Zoo: Zeroth order optimization based black-box attacks to deep neural networks without training substitute models. In *Proceedings of the 10th ACM workshop on artificial intelligence and security*, 15–26.

8. Dada, Emmanuel Gbenga, Joseph Stephen Bassi, Haruna Chiroma, Shafii Muhammad Abdulhamid, Adebayo Olusola Adetunmbi, and Opeyemi Emmanuel Ajibuwa. 2019. Machine learning for email spam filtering: review, approaches and open research problems. *Heliyon* 5 (6).
9. Gao, Yan, Ming Yang, Xiaonan Zhao, Bryan Pardo, Ying Wu, Thrasyvoulos N. Pappas, and Alok Choudhary. 2008. Image spam hunter. In *2008 IEEE international conference on acoustics, speech and signal processing*.
10. Goodfellow, Ian J, Jonathon Shlens, and Christian Szegedy. 2014. Explaining and harnessing adversarial examples. arXiv:1412.6572.
11. Jiang, Jianguo, Boquan Li, Min Yu, Chao Liu, Weiqing Huang, Lejun Fan, and Jianfeng Xia. 2019. Restoration as a defense against adversarial perturbations for spam image detection. In *International conference on artificial neural networks*, 711–723. Springer.
12. Kumar, Amara Dinesh, Soman KP, et al. 2018. Deepimagespam: Deep learning based image spam detection. arXiv:1810.03977.
13. Moosavi-Dezfooli, Seyed-Mohsen, Alhussein Fawzi, Omar Fawzi, and Pascal Frossard. 2017. Universal adversarial perturbations. In *Proceedings of the IEEE conference on computer vision and pattern recognition*, 1765–1773.
14. Moosavi-Dezfooli, Seyed-Mohsen, Alhussein Fawzi, and Pascal Frossard. 2016. Deepfool: A simple and accurate method to fool deep neural networks. In *Proceedings of the IEEE conference on computer vision and pattern recognition*, 2574–2582.
15. Mordvintsev, Alexander, Christopher Olah, and Mike Tyka. Inceptionism: Going deeper into neural networks. 2015. https://research.googleblog.com/2015/06/inceptionism-going-deeper-into-neural.html.
16. Nicolae, Maria-Irina, Mathieu Sinn, Minh Ngoc Tran, Beat Buesser, Ambrish Rawat, Martin Wistuba, Valentina Zantedeschi, Nathalie Baracaldo, Bryant Chen, Heiko Ludwig, Ian Molloy, and Ben Edwards. 2018. Adversarial robustness toolbox v1.2.0. arXiv:1807.01069.
17. NumPy. 2020. https://numpy.org/.
18. OpenCV. 2020. https://opencv.org/.
19. Papernot, Nicolas, Patrick McDaniel, Ian Goodfellow, Somesh Jha, Z Berkay Celik, and Ananthram Swami. 2017. Practical black-box attacks against machine learning. In *Proceedings of the 2017 ACM on Asia conference on computer and communications security*, 506–519.
20. Rosebrock, Adrian.2020. Zero parameter automatic canny edge detection. https://pyimagesearch.com/2015/04/06/zero-parameter-automatic-canny-edge-detection-with-python-and-opencv/.
21. Scikit-learn. 2020. Machine learning in python. https://scikit-learn.org/.
22. SciPy. 2020. https://scipy.org/.
23. Selvaraju, Ramprasaath R, Michael Cogswell, Abhishek Das, Ramakrishna Vedantam, Devi Parikh, and Dhruv Batra. 2017. Grad-cam: Visual explanations from deep networks via gradient-based localization. In *Proceedings of the IEEE international conference on computer vision*, 618–626.
24. Sharmin, Tazmina, Fabio Di Troia, Katerina Potika, and Mark Stamp. 2020. Convolutional neural networks for image spam detection. *Information Security Journal: A Global Perspective* 29 (3): 103–117.
25. Soranamageswari, M, and C. Meena. 2010. Statistical feature extraction for classification of image spam using artificial neural networks. In *2010 second international conference on machine learning and computing*.
26. Szegedy, Christian, Wojciech Zaremba, Ilya Sutskever, Joan Bruna, Dumitru Erhan, Ian Goodfellow, and Rob Fergus. 2013. Intriguing properties of neural networks. arXiv:1312.6199.
27. TensorFlow. 2020. https://tensorflow.org/.
28. Yuan, Xiaoyong, Pan He, Qile Zhu, and Xiaolin Li. 2019. Adversarial examples: Attacks and defenses for deep learning. *IEEE Transactions on Neural Networks and Learning Systems* 30 (9): 2805–2824.

Printed in the United States
by Baker & Taylor Publisher Services